Plant–Microbe Interactions for Sustainable Growth and Resilience

This book reviews dynamic relationships between plants and microbes, highlighting their roles in enhancing crop resilience, growth, and productivity. Integrating research and practical insights, the book provides solutions to pressing agricultural challenges, including soil degradation, pest resistance, and nutrient inefficiency. It discusses plant–microbe interactions; regulation of the host–microbe interaction by small RNA molecules and their mechanisms; and applications of sustainable crop production and climate-resilient agriculture. The book covers a broad spectrum of plant–microbe interactions—from root symbioses to shoot colonization—and provides a holistic understanding of how these interactions impact crop health and productivity, helping readers optimize their strategies for improving crop yields and sustainability.

Key Features:

- Ensures a thorough presentation for understanding plant–microbe dynamics—from basic interactions to molecular mechanisms and practical applications.
- Offers a comprehensive exploration of how plant–microbe interactions can revolutionize sustainable agriculture.
- Presents innovative strategies for harnessing natural partnerships to improve crop yields and sustainability.

NextGen Agriculture: Novel Concepts and Innovative Strategies

Series Editor
Chittaranjan Kole

Allele Mining for Genomic Designing of Fruit Crops
Chittaranjan Kole, Kenta Shirasawa, and Anil Kumar Singh

Allele Mining for Genomic Designing of Cereal Crops
Chittaranjan Kole, Sharat Kumar Pradhan, and Vijay K Tiwari

Allele Mining for Genomic Designing of Vegetable Crops
Chittaranjan Kole, Tusar Kanti Behera, and Prashant Kaushik

Allele Mining for Genomic Designing of Oilseed Crops
Chittaranjan Kole, Manish Kumar Pandey, and Naveen Puppala

Sustainable Urban Agriculture: New Frontiers
Kheir Al-Kodmany, Madhav Govind, Sharmin Khan, and Chittaranjan Kole

Allele Mining for Genomic Designing of Grain Legume Crops
Chittaranjan Kole, C. Bharadwaj, and Abhimanyu Sarkar

The Vertical Farm: Scientific Advances and Technological Developments
Kheir Al-Kodmany, Andrew Keong Ng, Abel Tablada, and Chittaranjan Kole

Next Generation Food Crops for Human Health
Edited by Sangam Dwivedi, Rodomiro Ortiz, and Chittaranjan Kole

Smart Agriculture: Concepts, Strategies, and Case Studies
Edited by Verónica Saiz-Rubio, Francisco Rovira-Más, and Chittaranjan Kole

Tailoring the Tomato Genome for NextGen Agriculture
Edited by Dilip R. Panthee and Chittaranjan Kole

Plant–Microbe Interactions for Sustainable Growth and Resilience
Edited by Vishnu Mishra, Saurabh Pandey, Vishal Varshney, and Chittaranjan Kole

For more information, please visit our website: https://www.routledge.com/Nextgen-Agriculture/book-series/NGA

Preface to the Series

Agriculture is going through a phase of transformation now. It has both challenges and opportunities. One of the major challenges it is facing is obviously food security. A population of about 9.3 billion projected by 2050 necessitates an increase of crop production by 60–70%. Another 10–15% increase would be needed to counter the reduction of crop production due to global warming and climate change. Nutrition security is another challenge with more than 800 million people suffering from malnutrition. Depletion of fossil fuel and increasing dependence on biofuel from crop residues is another challenge. Loss of biodiversity and land degradation also add to the list of challenges. The cultivable area is also decreasing fast due to housing, transport, and industrialization. Above all, climate change and global warming have a huge impact on agriculture, health, and environment.

Several novel concepts and strategies are emerging to address the above challenges. For example, practice of precision farming will lead to precise and timely use of inputs resulting in higher yield and better quality of produce on the one hand and reduction of loss of inputs on the other. Similarly, urban agriculture and vertical agriculture will add a lot of cultivable areas in future. Protected farming will facilitate growing crops under controlled conditions leading to reduction in biotic and abiotic stresses and precise utilization of inputs.

Agriculture requires maintenance of the philosophy farmers have been following since inception over thousands of years which are being sold today in new packages with sophisticatedly coined labels such as traditional agriculture, natural agriculture, sustainable agriculture, conservation agriculture, integrated agriculture, regenerative agriculture, organic agriculture, etc.

In the changing scenarios, next generation agriculture will need designed crop varieties and breeds produced through designed practices. This book series aims at deliberations on several novel concepts and innovative strategies developed for next generation agriculture and will hopefully benefit agricultural education, research, and outreach.

Chittaranjan Kole
Kolkata
January 10, 2024

Special Acknowledgement to
Phullara Kole
For her outstanding assistance in editing this book series

Plant–Microbe Interactions for Sustainable Growth and Resilience

Edited by Vishnu Mishra, Saurabh Pandey,
Vishal Varshney and Chittaranjan Kole

CRC Press
Taylor & Francis Group
Boca Raton London New York

CRC Press is an imprint of the
Taylor & Francis Group, an **informa** business

Designed cover image: Shutterstock

First edition published 2026
by CRC Press
2385 NW Executive Center Drive, Suite 320, Boca Raton FL 33431

and by CRC Press
4 Park Square, Milton Park, Abingdon, Oxon, OX14 4RN

CRC Press is an imprint of Taylor & Francis Group, LLC

ISBN: 978-1-032-98982-2 (hbk)
ISBN: 978-1-041-00081-5 (pbk)
ISBN: 978-1-003-60808-0 (ebk)

DOI: 10.1201/9781003608080

Typeset in Times
by Apex CoVantage, LLC

Dedicated to
Dr. Norman Ernest Borlaug
The Father of Green Revolution, Nobel Peace Prize Winner, and the Founder of the World Food Prize

Who inspired my research and academic works
with his generous appreciations and advice

Contents

Preface

Plants, as sessile organisms, face a myriad of challenges from biotic stresses, such as pathogens and pests, and abiotic stresses, including drought, salinity, and extreme temperatures. These stresses significantly impair plant growth, yield, and quality, posing threats to global food security and ecosystem stability. To thrive in such dynamic environments, plants have evolved intricate mechanisms to interact with both their beneficial and pathogenic microbial counterparts. These plant–microbe interactions play a pivotal role in modulating plant resilience, nutrient acquisition, and stress tolerance, offering promising avenues for sustainable agriculture.

The book *Plant–Microbe Interactions for Sustainable Growth and Resilience* aims to elucidate the complex interplay between plants and microbes, highlighting their significance in promoting sustainable growth and enhancing plant resilience. Microorganisms, including bacteria, fungi, and viruses, engage in diverse interactions with plants, ranging from mutualistic symbioses to antagonistic encounters. Beneficial microbes, such as nitrogen-fixing bacteria and mycorrhizal fungi, enhance nutrient uptake and stress tolerance, while pathogenic microbes challenge plant defenses. Understanding these interactions at the molecular, physiological, and ecological levels is critical for harnessing their potential to develop resilient, high-yielding crops that require fewer chemical inputs.

This volume explores the latest advancements in plant–microbe interactions, covering topics such as the molecular mechanisms of symbiosis, microbial contributions to nutrient cycling, biocontrol of pathogens, and the role of the plant microbiome in mitigating abiotic stresses. Cutting-edge technologies, including high-throughput sequencing, metagenomics, and synthetic biology, have revolutionized our ability to study these interactions, revealing the intricate networks that govern plant–microbe communication. The book also addresses the application of these insights in agricultural practices, such as biofertilizers, biopesticides, and microbiome engineering, to promote sustainable farming systems that align with global efforts to combat climate change and ensure food security.

Plant–Microbe Interactions for Sustainable Growth and Resilience is a comprehensive, timely, and thought-provoking compilation designed to bridge the gap between fundamental research and practical applications. It brings together contributions from leading global experts in plant microbiology, molecular biology, and agronomy to provide an up-to-date perspective on the field. The chapters cover the discovery and characterization of plant-associated microbes, their roles in stress amelioration, and innovative strategies for leveraging these interactions to enhance crop productivity and resilience.

This book is intended for researchers, students, agronomists, and policymakers interested in sustainable agriculture and plant science. It aims to provide a thorough understanding of how plant–microbe interactions can be harnessed to develop crops that are better equipped to withstand environmental challenges while reducing reliance on synthetic fertilizers and pesticides. By presenting the latest methodologies and interdisciplinary approaches, this volume seeks to inspire further research and innovation in this rapidly evolving field.

We hope that *Plant–Microbe Interactions for Sustainable Growth and Resilience* will serve as a valuable resource for readers, fostering a deeper appreciation of the critical role that microbes play in plant health and sustainability. Our goal is to equip the scientific community and agricultural practitioners with the knowledge and tools needed to advance the development of resilient, eco-friendly agricultural systems for a sustainable future.

We express our heartfelt gratitude to the contributing authors for their expertise and dedication in shaping this volume. Their rigorous research and insightful perspectives have made this book a comprehensive resource. We also thank the editorial team, reviewers, and publishers for their unwavering support in bringing this project to fruition. Finally, we acknowledge the global research community for their tireless efforts in advancing our understanding of plant–microbe interactions, paving the way for sustainable agricultural innovations.

This book is structured to provide a logical progression from fundamental concepts to applied strategies. Each chapter is designed to stand alone while contributing to the overarching theme of plant–microbe interactions. We encourage readers to explore the diverse perspectives presented here and to consider how these insights can be applied to address pressing challenges in agriculture and environmental sustainability.

Editor Bios

Dr. Vishnu Mishra is a distinguished researcher with a comprehensive plant biotechnology and molecular biology background. He completed his master's in plant biotechnology from the prestigious Banaras Hindu University. Furthermore, he demonstrated his expertise by qualifying for national-level exams such as the Council of Scientific & Industrial Research (CSIR), Graduate Aptitude Test in Engineering (GATE), and Department of Biotechnology-Junior Research Fellowship (DBT-JRF). Dr. Mishra pursued his Ph.D. at the National Institute of Plant Genome Research, affiliated with Jawaharlal Nehru University, where he developed a strong foundation in plant biology and plant–microbe interactions. From 2022 to 2025, Dr. Mishra continued his scientific journey as a postdoctoral researcher at the Delaware Biotechnology Institute, University of Delaware, USA, where he expanded his research in genomics, biotechnology, and molecular biology, emphasizing plant–pathogen communication and plasmodesmata biology.

Dr. Mishra has been awarded a postdoctoral fellowship by the Dirección General de Asuntos del Personal Académico (DGAPA), Universidad Nacional Autónoma de México (UNAM), under the UNAM Postdoctoral Program (POSDOC). Currently, Dr. Mishra serves as a DGAPA postdoctoral fellow at the Laboratorio de Genómica Funcional de Leguminosas, Departamento de Biología, Facultad de Estudios Superiores Iztacala, Universidad Nacional Autónoma de México (UNAM), Mexico. His ongoing research focuses on plant–microbe interactions for sustainable crop growth and resilience, particularly exploring molecular mechanisms that enhance legume adaptability under nutrient stress conditions.

Dr. Mishra's scholarly contributions are extensive, encompassing research articles, review papers, and book chapters in leading international journals and publishing houses. He has reviewed over 50 scientific manuscripts and serves on the editorial board of *Frontiers in Plant Science*, reflecting his commitment to academic excellence and scientific integrity.

As an ambassador and active member of the American Society of Plant Biologists (ASPB), Dr. Mishra has participated in numerous international symposia and scientific events, contributing to the global dialog on plant science innovation. His dedication to advancing fundamental and applied plant biology continues to make a significant impact on the field.

Dr. Saurabh Pandey is currently working as an assistant professor in the Department of Plant Molecular Biology and Biotechnology at Indira Gandhi Krishi Vishwavidyalaya, Raipur, Chhattisgarh, India. He has over 10 years of professional experience in diverse areas of plant biotechnology, including plant stress biology, virus–host interactions, abiotic stress tolerance, molecular breeding, plant transformation, and nutritional biofortification in tomato and rice crops. He completed his Ph.D. in plant virology from the National Institute of Plant Genome Research (NIPGR), New Delhi, with the degree awarded by Jawaharlal Nehru University, New Delhi. His doctoral research focused on the functional analysis of RNA helicases and their role in multi-stress tolerance, particularly under viral infections. Previously, he completed his M.Sc. (agriculture) in plant biotechnology from the University of Agricultural Sciences, GKVK, Bangalore, India, earning a gold medal for academic excellence. His undergraduate degree (B.Sc. agriculture) was completed at Indira Gandhi Agricultural University, Raipur, India.

Dr. Pandey's current research emphasizes biotic stress management in key food crops through molecular interventions. His professional journey has included extensive work on stress-resilient crop improvement, development of molecular markers, Single Nucleotide Polymorphism (SNP)-based association studies, and genome-wide identification of key regulatory genes. He is involved in introducing stress-responsive genes, developing functional markers, and elucidating regulatory networks that enhance stress tolerance. He has published over 40 research and review articles in high-impact journals, including *Plant Physiology and Biochemistry*, *Frontiers in Plant Science*,

Environmental and Experimental Botany, *Plant Science*, and the *Journal of Plant Growth Regulation*. Dr. Pandey serves as an associate editor for the *Journal of the Science of Food and Agriculture*. He is also an editorial board member in *Frontiers in Plant Science* (Abiotic Stress section), *BMC Plant Biology*, *Discover Plants*, and *Discover Agriculture*, among others. He also served as a guest editor for *Frontiers in Ecology and the Environment*, compiling a special issue. He is also an active reviewer for reputed journals, including *Physiologia Plantarum*, *Plant Physiology and Biochemistry*, and the *South African Journal of Botany*. His academic output includes five edited books published by internationally recognized publishers, such as CRC Press and Springer. He has authored more than 15 popular science articles aimed at disseminating research insights to farming and extension communities. His Google Scholar profile reports an H-index of 21 and around 1,800 citations, reflecting the high impact of his work. He has authored and contributed to several book chapters on molecular markers, genomics, plant breeding, and plant virology, addressing both academic and applied agricultural research. Dr. Pandey has been an active member of professional societies, including the Indian Virological Society, the Indian Academy of Horticultural Sciences, and the Society of Plant Biochemistry and Biotechnology.

In terms of technical expertise, he is proficient in gene expression analysis (qPCR and RACE-PCR), molecular cloning, Southern blotting, protein expression and purification, virus-induced gene silencing (VIGS), yeast-two-hybrid and bimolecular fluorescence complementation (BiFC) assays, confocal microscopy, and microbial culture techniques. He is also well-versed in bioinformatics tools for genome-wide analysis, promoter mining, and primer designing. His achievements include a university gold medal, the Best PG Thesis Award, and several national fellowships, such as National Talent Scholarship (NTS), Junior Research Fellowship (JRF), and Senior Research Fellowship (SRF), from the Indian Council for Agricultural Research (ICAR) and Department of Biotechnology (DBT), Government of India. His doctoral work has been featured in national science communication platforms such as Science Monitor (Rajya Sabha TV), *The Hindu* newspaper, and DBT/DST's official media. Dr. Pandey has also contributed to sequence submissions to National Centre for Biotechnology Information (NCBI), playing a critical role in resource development for crop improvement.

He has actively participated in several international and national training programs, symposia, and conferences. His collaborative and interdisciplinary research approach continues to make significant contributions to the advancement of plant molecular science and agricultural biotechnology.

Dr. Vishal Varshney is a distinguished researcher in the plant molecular biology field. He is currently a postdoctoral fellow at the Institut de Biologie Moléculaire des Plantes du CNRS (IBMP-CNRS) in Strasbourg, France, where he began in April 2024. Previously, he served as an assistant professor of botany at the Govt. Shaheed Gendsingh College in Charama, Chhattisgarh, from 2022 to 2024. Dr. Varshney earned his Ph.D. from NIPGR in New Delhi, where he conducted significant research on the molecular and functional characterization of an F-box protein (SKIP31) in *Arabidopsis thaliana* under the supervision of Dr. Manoj Majee. He completed his M.Sc. and B.Sc. in botany from Aligarh Muslim University, both with gold medal honors. His research primarily focuses on the ubiquitin–proteasome pathway and its role in enhancing crop yield by optimizing seed agronomic traits. Dr. Varshney has published extensively in high-impact journals, contributing to the understanding of seed maturation, vigor, and stress responses in plants. Notable publications include research on the *Arabidopsis* F-box protein SKP1-INTERACTING PARTNER 31, which modulates seed maturation and vigor, published in *The Plant Cell*. In addition to his research, Dr. Varshney has received several prestigious awards and fellowships, including the Inspiring Science Award (2024) from TNQ Technologies, Senior and Junior Research Fellowships from the CSIR and University Grants Commission (UGC) for his Ph.D. studies, and multiple gold medal awards for his academic excellence. In addition to his research, Dr. Varshney has reviewed over 20 articles for various reputed journals and serves as guest topic coordinator for a special issue "The Dynamics of Environmental Stresses and Seed Physiology: A Complex Interaction in Plant Systems" in *Frontiers* *in Plant Science Journal*, further underscoring his expertise and influence in the field. Dr. Varshney is skilled in various technical competencies such as molecular biology, cell biology, protein biochemistry, microscopy, metabolomics, and bioinformatics. His ultimate goal is to contribute to crop improvement by enhancing the understanding of the molecular mechanisms underlying plant and seed biology, aspiring to become an independent researcher.

Prof. Chittaranjan Kole is an internationally reputed scientist with an illustrious professional career spanning over 40 years and original contributions in the fields of plant genomics, biotechnology, and molecular breeding, leading to the publication of more than 160 quality research articles. He has edited over 197 books for the leading publishers of the world, including Springer Nature, Wiley-Blackwell, and Taylor & Francis Group. His scientific contributions and editing acumen have been appreciated by seven Nobel Laureates, including Profs. Norman Borlaug, Arthur Kornberg, Werner Arber, Phillip Sharp, Günter Blobel, Lee Hartwell, and Roger Kornberg. He has been honored with a number of fellowships, honorary fellowships, and national and international awards, including the Outstanding Crop Scientist award conferred by the International Crop Science Society. He was awarded the Raja Ramanna Fellowship by the Department of Energy, Government of India. He has served in all prestigious positions in academia, including as vice-chancellor of Bidhan Chandra Krishi Viswavidyalaya, project coordinator of Indo-Russian Center for Biotechnology in India, and director of research of the Institute of Nutraceutical Research of Clemson University in the USA. He has also worked at Pennsylvania State University and Clemson University as a visiting professor. He has been heading the International Climate-Resilient Crop Genomics Consortium, the International Phytomedomics and Nutriomics Consortium, and Genome India International as their founder president. He is also the founder chairman of Prof. Chittaranjan Kole Foundation for Science and Society.

Contributors

Pinky Agarwal
National Institute of Plant Genome Research
New Delhi, India

Shashi Bhushan Agrawal
Laboratory of Air Pollution and Global Climate Change
Department of Botany
Banaras Hindu University
Varanasi, Uttar Pradesh, India

Zishan Ahmad
State Key Laboratory of Tree Genetics and Breeding, Co-Innovation Centre for Sustainable Forestry in Southern China, Bamboo Research Institute, Key Laboratory of National Forestry and Grassland Administration on Subtropical Forest Biodiversity Conservation, School of Life Sciences
Nanjing Forestry University
Nanjing, Jiangsu, China

Monika Asthana
Department of Biotechnology, School of Life Sciences, Swami Vivekanand Campus Khandari
Dr. Bhimrao Ambedkar University
Agra, India

Siddhartha Barun
School of Forensic Science
National Forensic Sciences University
Gandhinagar, Gujarat, India

Sana Basri
Department of Botany
Aligarh Muslim University
Aligarh, Uttar Pradesh, India

Sabhyata Bhatia
National Institute of Plant Genome Research Aruna Asaf Ali Marg
New Delhi, India

Ritu Chaudhary
Department of Botany
Aligarh Muslim University
Aligarh, Uttar Pradesh, India

Neha Chaurasiya
Department of Botany
Aligarh Muslim University
Aligarh, Uttar Pradesh, India

Jaishri Rubina Das
Université Paris-Saclay, INRAE, AgroParisTech, Institute Jean-Pierre Bourgin for Plant Sciences (IJPB)
Versailles, France

Joyati Das
Institute of Advanced Studies in Science and Technology (IASST), Vigyan Path, Paschim Boragaon, Garchuk, Guwahati
Assam, India

Vindhesh Dixit
Department of Biotechnology, School of Life Sciences, Swami Vivekanand Campus Khandari
Dr. Bhimrao Ambedkar University
Agra Uttar Pradesh, India

Shahla Faizan
Environmental Physiology Laboratory
Department of Botany, Aligarh Muslim University
Aligarh, Uttar Pradesh, India

Poonam Gupta
School of Forensic Science
National Forensic Sciences University
Gandhinagar, Gujarat, India

Priya Jaiswal
National Institute of Plant Genome Research
New Delhi, India

G. Jeeva
Department of Oilseeds
Tamil Nadu Agricultural University
Coimbatore, Tamil Nadu, India

M. Jincy
School of Agriculture
Lovely Professional University
Phagwara, Punjab, India

Akshara Kampurath
Department of Horticulture
Central University of Tamil Nadu
Thiruvarur, India

Preedhi Kapoor
Plant Physiology and Biochemistry Section
Department of Botany
Aligarh Muslim University
Aligarh, India

Charanpreet Kaur
Department of Plant and Soil Sciences
Delaware Biotechnology Institute
University of Delaware
Newark, Delaware, USA

Simran Kauts
Department of Zoology, School of Bioengineering and Biosciences
Lovely Professional University
Phagwara, Punjab, India

Pramod Kumar
Department of Biotechnology, School of Life Sciences, Swami Vivekanand Campus Khandari
Dr. Bhimrao Ambedkar University
Agra, India

Ranjna Kumari
Department of Botany, School of Bioengineering and Biosciences
Lovely Professional University
Phagwara, Punjab, India

Rinkee Kumari
Environmental Physiology Laboratory
Department of Botany, Aligarh Muslim University
Aligarh, Uttar Pradesh, India

Ritu Kumari
Department of Botany, School of Bioengineering and Biosciences
Lovely Professional University
Phagwara, Punjab, India

Maridul Kundan
Plant Sciences and Agro-Technology Division CSIR–Indian Institute of Integrative Medicine, Canal Road
Jammu, India

Subramanian Manivannan
Department of Horticulture
Central University of Tamil Nadu
Thiruvarur, India

Muhammad Aamir Manzoor
Department of Plant Science, School of Agriculture and Biology
Shanghai Jiao Tong University
Shanghai, China

Bipin Maurya
Department of Botany
Maharaja Suhel Dev University
Azamgarh, Uttar Pradesh, India

Bilal Ahmad Mir
Department of Botany, School of Bioengineering and Biosciences
Lovely Professional University
Phagwara, Punjab, India

Vishnu Mishra
Laboratorio de Genómica Funcional de Leguminosas
Departamento de Biología, Facultad de Estudios Superiores Iztacala
Universidad Nacional Autónoma de México
Tlalnepantla, Mexico

Nishu Mittal
Faculty of Biosciences, Institute of Biosciences and Technology
Shri Ramswaroop Memorial University
Barabanki, Uttar Pradesh, India

Jay Hind Nishad
BRIC–National Agri-Food and Biomanufacturing Institute (Formerly NABI & CIAB)
Mohali, India

Avantika Pandey
Department of Botany
University of Lucknow
Lucknow, Uttar Pradesh, India

Ekta Pandey
Environmental Physiology Laboratory
Department of Botany
Aligarh Muslim University
Aligarh, Uttar Pradesh, India

Niharika Pandey
Govind Ballabh Pant University of Technology and Agriculture (GBPUAT)
Pantnagar, Uttarakhand, India

Saurabh Pandey
Department of molecular Biology and Biotechnology
Indira Gandhi Agricultural University
Raipur, Chhattisgarh, India

Yogesh Kumar Pandey
Department of Chemistry
Bareilly College, Bareilly
Uttar Pradesh, India

Zubair Ahmad Parrey
Department of Biochemistry, School of Bioengineering and Biosciences
Lovely Professional University
Phagwara, Punjab, India

Lydia Pramitha
School of Agricultural Sciences
Karunya Institute of Technology and Sciences Coimbatore
Tamil Nadu, India

Gurmeen Rakhra
Department of Biochemistry, School of Bioengineering and Biosciences
Lovely Professional University
Phagwara, Punjab, India

Afiya Rajan Selin
Department of Horticulture
Central University of Tamil Nadu
Thiruvarur, India

Senthilkumar Shanmugasundram
Department of Horticulture
Central University of Tamil Nadu
Thiruvarur, India

Diksha Sharma
GBPUAT, Pantnagar
Uttarakhand, India

Akanksha Shukla
School of Bioengineering and Biosciences
Lovely Professional University
Phagwara, Punjab, India

Dhananjay Singh
DSS Takara Bio India Pvt. Ltd., A-5 Mohan Cooperative Industrial Estate, Mathura Road
New Delhi, India

Dharmendra Singh
Department of Zoology
Shiv Harsh Kisan P.G. College
Basti, Uttar Pradesh, India

Simarjot Singh
Department of Agronomy, School of Agriculture
Lovely Professional University
Phagwara, Punjab, India

Perinba C. Suresh
Department of Horticulture
North-Eastern Hill University
Shillong, India

Virevol Thakro
Department of Life Sciences
Pachhunga University College
Mizoram Central University
Aizawl, Mizoram, India

Ankita Thakur
Department of Botany, School of Bioengineering and Biosciences
Lovely Professional University
Phagwara, Punjab, India

Ayushi Verma
Faculty of Biotechnology, Institute of Biosciences and Technology
Shri Ramswaroop Memorial University
Barabanki, Uttar Pradesh, India

Manisha Yadav
National Institute of Plant Genome Research Aruna Asaf Ali Marg
New Delhi, India

Introduction

Plants in nature coexist with a vast diversity of microorganisms that colonize their rhizosphere, phyllosphere, and endosphere. These microbes ranging from bacteria, fungi, actinomycetes, and archaea to viruses play a vital role in influencing plant growth, development, and resilience under varying environmental conditions. The multifaceted interactions between plants and their associated microbiota form the foundation of plant health and productivity, particularly in the face of climate change, soil degradation, and the growing demand for sustainable agricultural practices. In recent decades, the concept of the plant holobiont has reshaped our understanding of plant biology. Plants and their associated microbial communities function as an integrated system that collectively contributes to nutrient acquisition, stress tolerance, disease resistance, and overall ecosystem stability. Beneficial microorganisms such as plant growth-promoting rhizobacteria (PGPR), arbuscular mycorrhizal fungi (AMF), endophytes, and biocontrol agents have demonstrated remarkable potential to replace or complement chemical fertilizers and pesticides, aligning with the global vision of environmentally sustainable agriculture.

The present volume, *Plant–Microbe Interactions for Sustainable Growth and Resilience*, aims to bring together recent advances, emerging insights, and applied strategies that harness beneficial plant–microbe associations to enhance crop productivity and stress tolerance. The book highlights molecular, physiological, and ecological perspectives of plant–microbe interactions, including mechanisms of signaling, nutrient exchange, induced systemic resistance, and genetic or biotechnological approaches for improving symbiotic efficiency.

Each chapter is contributed by distinguished experts from diverse institutions and backgrounds, offering a comprehensive view of the subject from fundamental mechanisms of root colonization and signaling to the application of microbial consortia in precision agriculture. The volume also emphasizes microbial genomics, metagenomics, and bioinformatics tools, which are revolutionizing our capacity to explore and manipulate plant-associated microbial communities for sustainable crop improvement. Furthermore, this book underscores the importance of climate-resilient agriculture through microbial interventions. Beneficial microbes can mitigate the impacts of abiotic stresses, such as drought, salinity, heat, and heavy metal toxicity, while also enhancing carbon sequestration and soil fertility. Such microbial-based innovations pave the way for eco-friendly, low-input farming systems capable of supporting global food security under changing climatic conditions.

This edited volume will serve as a valuable resource for researchers, academicians, students, and professionals working in plant science, microbiology, biotechnology, and environmental sustainability. By integrating current knowledge with emerging trends, this book seeks to inspire new research directions and collaborative efforts aimed at developing resilient, sustainable, and productive agroecosystems through the intelligent use of plant-associated microbes.

Abbreviations

(ACC)	1-Aminocyclopropane-1-carboxylic acid
(TTC)	2,3,5-Triphenyltetrazolium chloride
(2,4-DAPG)	2,4-Diacetyl phloroglucinol
(ABA)	Abscisic acid
(ASR)	Acquired systemic resistance
(ACC deaminase)	Aminocyclopropane-1-carboxylate deaminase
(NH_3)	Ammonia
(AMT)	Ammonium transporters
(APCNF)	Andhra Pradesh Community Managed Natural Farming
(AP2)	APETALA2
(AMF)	Arbuscular mycorrhizal fungi
(AGO)	Argonaute
(AGO1)	ARGONAUTE1
(amiRNA)	Artificial miRNA
(AO)	Ascorbate oxidase
(Asn)	Asparagine
(AON)	Autoregulation of nodulation
(ARF)	Auxin response factor
(APOLO)	AUXIN-REGULATED PROMOTER LOOP
(Bt)	*Bacillus thuringiensis*
(BKK1)	BAK1-LIKE1
(BTH)	Benzothiadiazole
(BCAs)	Biological control agents
(BNF)	Biological nitrogen fixation
(BGA)	Blue-Green algae
(BIK1)	*Botrytis*-induced kinase 1
(BAK1)	BRI1-associated kinase
(CdS)	Cadmium sulfide
(CCaMK)	Calcium/calmodulin-dependent serine/threonine kinase
(CDPKs)	Calcium-dependent protein kinases
(CML)	Calmodulin-like
(CaMs)	Calmodulins
(CNTs)	Carbon nanotubes
(Cas9)	CRISPR-associated protein 9
(CAT)	Catalase
(CERK1)	Chitin elicitor receptor kinase 1
(CTV)	Citrus tristeza virus
(ClYVV)	Clover yellow vein virus
(CRISPR)	Clustered regularly interspaced short palindromic repeats
(COLDAIR)	Cold-assisted intronic noncoding RNA
(COOLAIR)	Cold-induced long antisense intragenic RNA
(COLDWRAP)	Cold of winter-induced noncoding RNA from the promoter
(CS)	Common scab
(CSSP)	Common symbiosis signaling pathway
(CT)	Computed tomography
(CLSM)	Confocal laser scanning microscopy

(Cu)	Copper
(CRN)	CORYNE
(CEPs)	C-Terminally encoded peptides
(CEPD)	C-Terminally encoded peptide downstream
(CEPR)	C-Terminally encoded peptide receptor
(CBS1)	Cystathionine-β-synthase-like 1
(CPs)	Cysteine proteases
(DAMPs)	Damage-associated molecular patterns
(DNF1)	Defective in nitrogen fixation 1
(DCL1)	Dicer-like protein 1
(DMI3)	Does not make infections 3
(DSB)	Double-strand break
(dsRNA)	Double-stranded RNA
(ENOD)	Early nodulin
(ETI)	Effector-triggered immunity
(EBEs)	Effector-binding elements
(ETS)	Effector-triggered susceptibility
(ELENA1)	ELF18-INDUCED LONG-NONCODING RNA1
(eTMs)	Endogenous target mimics
(ENMs)	Engineered nanomaterials
(ET)	Ethylene
(ERFs)	Ethylene response factors
(EPS)	Exopolysaccharide
(EPS)	Extracellular polysaccharides
(Ecp6)	Extracellular protein 6
(FMT)	Fecal microbiota transplantation
(FIB2)	FIBRILLARIN 2
(flg22)	Flagellin
(FLS2)	FLAGELLIN-SENSING 2
(F3H)	Flavanone 3-hydroxylase
(FLC)	FLOWERING LOCUS C
(FISH)	Fluorescence *in situ* hybridization
(FISH-TAMB)	Fluorescent *in situ* hybridization of transcript-annealing molecular beacons
(1-FFT)	Fructan 1-fructosyltransferase
(GC-MS)	Gas chromatography–mass spectrometry
(GM)	Genetic modification
(GMOs)	Genetically modified organisms
(GE)	Genome editing
(GA)	Gibberellic acid
(Gln)	Glutamine
(GSH)	Glutathione
(Gly)	Glycine
(GV)	Granulovirus
(GRF)	Growth-regulating factor
(HSPs)	Heat shock proteins
(HPLC)	High-performance liquid chromatography
(HDR)	Homology-directed repair
(H_2O_2)	Hydrogen peroxide
(HCA)	Hydroxycinnamic acid
(OH–)	Hydroxyl radical
(HR)	Hypersensitive response

(IAA)	Indole acetic acid
(IPS1)	Induced by phosphate starvation 1
(ISR)	Induced systemic resistance
(ITs)	Infection threads
(C4/C5)	Inner cortex
(IPM)	Integrated pest management
(IPD3)	Interacting protein of DMI3
(i-PGPR)	Intracellular PGPR
(lncRNAs)	Intronic lncRNAs
(IRLC)	Inverted repeat-lacking clade
(Fe)	Iron
(FeNPs)	Iron nanoparticles
(FeMo-co)	Iron–molybdenum cofactors
(iTRAQ)	Isobaric tags for relative and absolute quantitation
(ICS1)	Isochorismate synthase 1
(INA)	Isonicotinic acid derivatives
(IPTG)	Isopropyl β-D-1-thiogalactopyranoside
(ICAT)	Isotope-coded affinity tags
(JAZ)	Jasmonate ZIM-domain
(JA)	Jasmonic acid
(LBD)	Lateral organ boundaries domain
(PbS)	Lead sulfide
(LecRLKs)	Lectin receptor-like kinases
(LRR)	Leucine-rich repeat
(LM)	Light microscope
(LHP1)	LIKE HETEROCHROMATIN PROTEIN 1
(LCOs)	Lipo-chitooligosaccharides
(LPS)	Lipopolysaccharides
(LOX1)	Lipoxygenase 1
(LOX2)	Lipoxygenase 2
(LC-MS/MS)	Liquid chromatography–tandem mass spectrometry
(lncRNAs)	Long non-coding RNAs
(LMT1)	Low-molecular-weight tristeza 1
(LIN)	Lumpy infections
(LysM)	Lysine motif
(LYK3)	LysM-receptor-like kinase 3
(MRI)	Magnetic resonance imaging
(MLD)	Malectin-like domain
(MDA)	Malondialdehyde
(Mn)	Manganese
(MS)	Manuscript
(MAS)	Marker-assisted selection
(MS)	Mass spectrometry
(MoA)	Mechanisms of action
(MT)	Metallothioneins
(MAMPs)	Microbe-associated molecular patterns
(MISR)	Microbe-induced systemic resistance
(miRNAs)	microRNAs
(C3)	Middle cortex
(MLO)	Mildew resistance locus O
(MYA)	Million years ago

(MAPK)	Mitogen-activated protein kinase
(MPs)	Molecular patterns
(Mo)	Molybdenum
(MoFe)	Molybdenum–Iron
(MudPIT)	Multidimensional protein identification technology
(NP)	Nanoparticle
(NATs)	Natural antisense transcripts
(NiCK4)	NFR5-interacting cytoplasmic kinase 4
(NHPG)	NHP-O-glucoside
(NHP)	N-Hydroxypipecolic acid
(NO)	Nitric oxide
(N_2)	Nitrogen
(nif)	Nitrogen fixation
(NFC)	Nitrogen-fixing clade
(NFP)	Nod factor perception
(NFR1)	Nod factor receptor 1
(NFR5)	Nod factor receptor 5
(NFs)	Nod factors
(NFs)	Nodulation factors
(NIN)	Nodulation inception
(Nops)	Nodulation outer proteins
(NPL)	Nodulation pectate lyase
(NodD)	Nodulation protein D
(NSP1)	Nodulation signaling pathway proteins 1
(NSP2)	Nodulation signaling pathway proteins 2
(NAM)	Nodule apical meristem
(NVB)	Nodule vascular bundle
(NCR)	Nodule-specific cysteine-rich
(ncRNAs)	Noncoding RNAs
(NPR1)	NON-EXPRESSOR OF PR GENES 1
(NHEJ)	Nonhomologous end-joining
(NRPS)	Non-ribosomal peptide synthetase
(NDRE)	Normalized Difference Red Edge Index
(NDVI)	Normalized Difference Vegetation Index
(NF-YA1)	Nuclear factor-YA1
(NMR)	Nuclear magnetic resonance
(NPV)	Nucleopolyhedrovirus
(NB)	Nucleotide binding
(NLRs)	Nucleotide-binding leucine-rich repeat receptors
(NBS-LRR)	Nucleotide-binding site-leucine-rich repeat
(OGs)	Oligogalacturonides
(OEC)	Oxygen-evolving complex
(PAMPs)	Pathogen-associated molecular patterns
(PR)	Pathogenesis-related
(PR1)	*Pathogenesis-related gene 1*
(PRRs)	Pattern recognition receptors
(PTI)	Pattern-triggered immunity
(POX)	Peroxidase
(PAL)	Phenylalanine ammonia lyase
(PSMs)	Phosphate-solubilizing microbes
(P)	Phosphorus

(PALM)	Photoactivated localization microscopy
(PS II)	Photosystem II
(piRNAs)	Piwi-interacting RNAs
(PBs)	Plant bio-stimulants
(PGPB)	Plant growth-promoting bacteria
(PGPF)	Plant growth-promoting fungi
(PGPMs)	Plant growth-promoting microorganisms
(PGPR)	Plant growth-promoting rhizobacteria
(PLT)	PLETHORA
(PRC2)	Polycomb repressive complex 2
(K)	Potassium
(KSEs)	Potassium-solubilizing enzymes
(PVY)	Potato virus Y
(pri-miRNA)	Primary miRNA
(PBZ)	Probenazole
(Pro)	Proline
(PPIs)	Protein–protein interactions
(QTLs)	Quantitative trait loci
(QS)	Quorum sensing
(RGS)	Raw garlic stalk
(ROS)	Reactive oxygen species
(RLKs)	Receptor-like kinases
(RLPs)	Receptor-like proteins
(ERN)	Required for nodulation
(RPG)	Required for polar growth
(R)	Resistance
(RBOH)	Respiratory burst oxidase
(RR)	Response regulator
(rRNA)	Ribosomal RNA
(RNAi)	RNA interference
(RNA Pol II)	RNA polymerase II
(RISC)	RNA-induced silencing complex
(RHD6)	ROOT HAIR DEFECTIVE 6
(RNS)	Root nodule symbiosis
(ROLB)	Root onset lateral branching
(SA)	Salicylic acid
(SABC1)	Salicylic acid biogenesis controller 1
(SOS)	Salt overly sensitive
(SAG)	SA-O-β-glucoside
(SEM)	Scanning electron microscopy
(FISH-SIMS)	Secondary ion mass spectrometry
(siRNAs)	Short interfering RNAs
(SiNPs)	Silica nanoparticles
(AgNPs)	Silver nanoparticles
(sgRNA)	Single guide RNA
(scRNA-seq)	Single-cell RNA sequencing
(ssRNAs)	Single-stranded RNA
(snRNAs)	Small nuclear RNAs
(snoRNAs)	Small nucleolar RNAs
(SUNA1)	snoRNA-related lincRNA
(STEDM)	Stimulated emission depletion microscopy

(STORM)	Stochastic optical reconstruction microscopy
(SIM)	Structured illumination microscopy
(O2–)	Superoxide
(SOD)	Superoxide dismutase
(S)	Susceptibility
(SYM)	Symbiosome membrane
(SNF)	Symbiotic nitrogen fixation
(SynComs)	Synthetic microbial communities
(SAR)	Systemic acquired resistance
(TCP4)	Teosinte branched 1/Cycloidea/proliferating cell factor 4
(ASCO)	The ALTERNATIVE SPLICING COMPETITOR
(LTAR)	The Long-Term Agroecosystem Research
(TEV)	Tobacco etch virus
(TMV)	Tobacco mosaic virus
(ToLCNDV)	Tomato leaf curl New Delhi virus
(TYLCV)	Tomato yellow leaf curl virus
(TML1/2)	Too Much Love 1/2
(tracrRNA)	Trans-activating crRNA
(TALENs)	Transcription activator-like effector nucleases
(TFs)	Transcription factors
(tRNAs)	Transfer RNAs
(TRAP-Seq)	Translating ribosome affinity purification sequencing
(TEM)	Transmission electron microscopy
(TEs)	Transposable elements
(TuMV)	Turnip mosaic virus
(2DDIGE)	Two-dimensional difference gel electrophoresis
(2DE)	Two-dimensional gel electrophoresis
(T3Es)	Type III effectors
(T3SS)	Type III secretion systems
(T4SS)	Type IV secretion systems
(UHPLC-DAD-QTOF)	Ultra-high-performance liquid chromatography-diode array detector-quadrupole time-of-flight
(VPE)	Vacuolar processing enzyme
(V)	Vanadium
(VPY)	Vapyrin
(VIM1)	Variant in methylation 1
(VBNC)	Viable but Nonculturable
(VSRs)	Viral suppressors of RNAi
(VOCs)	Volatile organic compounds
(WIMPs)	Wound-induced molecular patterns
(Y2H)	Yeast two-hybrid
(Zn)	Zinc
(ZFNs)	Zinc-finger nucleases
(ZnO NPs)	Zinc oxide nanoparticles
(ZYMV)	Zucchini yellow mosaic virus
(BGAL1)	β-Galactosidase 1

1 Unraveling the Mysteries of Root Nodule Symbiosis
Insights and Applications

Rinkee Kumari, Ekta Pandey,
Shahla Faizan, and Saurabh Pandey

1.1 INTRODUCTION

Root nodule symbiosis (RNS) is a mutually beneficial relationship between a host plant, typically a legume, and nitrogen-fixing bacterial symbionts, often called rhizobia. This intricate symbiotic relationship culminates in the development of specialized legume-associated nodular structures on the host plant's roots. Within these nodules, the rhizobia bacteria reside intracellularly, converting atmospheric nitrogen (N_2) into a form that the plant can efficiently utilize, namely, ammonia (NH_3). In exchange, the host plant provides the rhizobia with carbohydrates and other essential nutrients derived from photosynthesis, creating a symbiotic exchange of resources (Battenberg and Hayashi, 2022). This symbiotic relationship needs proximity between the rhizobia and the legume, which is essential to facilitate the reciprocal exchange of nutrients. Across all crop species, nitrogen is a fundamental nutrient; however, their symbiotic interaction with rhizobia enables the remarkable adaptation that allows legumes to obtain nitrogen from the air directly. Studies have explored a hypothetical method that controls the symbiotic association in non-leguminous plants, in addition to a mechanism that may enhance the nitrogen-fixing capacity of legumes, with both processes relying on the signaling interplay between rhizobia and their leguminous partners (Nelson and Sadowsky, 2015). The host plant secretes specific chemical signals, such as flavonoids, detected by compatible rhizobia species. Nod factors (NFs), which are lipo-chitooligosaccharide (LCO) signaling molecules, are produced by the bacteria in response to this recognition. Once these NFs access plant receptors, a series of signaling events begin, which ultimately leads infection threads (ITs) and root nodules to emerge (Mbengue et al., 2020).

For numerous significant reasons, it is crucial to comprehend the nitrogen-fixing rhizobacteria–legume–seaweed interface. First, synthetic nitrogen-based fertilizers are less needed since rhizobacteria can fix atmospheric nitrogen into a plant-available form via biological nitrogen fixation (BNF) (Monib et al., 2024). Sustainable agriculture has significant ramifications because of the potential for reduced pollution, reduced consumption of nonrenewable resources, and fewer greenhouse gas emissions. Harnessing this symbiosis can contribute to more environmentally friendly and economically viable farming practices (Qiao et al., 2019; Harman and Uphoff, 2019). Moreover, the nitrogen-rich nodules and the decomposition of legume residues can increase the water-holding capacity and overall soil health and improve soil structure. Hence, it can increase crop yields and resilience, particularly in resource-constrained or degraded agricultural systems (Jena et al., 2022). Understanding the mechanisms that govern this symbiosis can help optimize the benefits for crop production, making it a valuable asset in the quest for sustainable intensification of agriculture (Porter and Sachs, 2020). Beyond the agronomic benefits, legumes and their associated rhizobia play a crucial role in the global nitrogen cycle, influencing the availability of this essential nutrient in terrestrial ecosystems (Goyal et al., 2021a). Establishing legume–rhizobia symbiosis can be a valuable factor in habitat restoration and land reclamation efforts, as it can help kickstart soil

DOI: 10.1201/9781003608080-1

regeneration and establish a diverse plant community. This understanding is crucial in addressing the global challenges of environmental sustainability and preserving natural habitats (Abd-Alla et al., 2023). Deciphering the molecular mechanisms and signaling pathways that govern RNS can also inform genetic engineering approaches to enhance the efficiency of nitrogen fixation, expand the host range of rhizobia–legume interactions, and develop more effective inoculants for agricultural applications (Goyal et al., 2021b).

This book chapter explores the RNS between legumes and nitrogen-fixing rhizobial bacteria. It begins by examining the mechanisms underlying this relationship, including the recognition and signaling processes, nodule formation and development, the nitrogen fixation process, and the nutrient exchange and regulation in symbiosis. The chapter also explores the latest research advancements and applications, such as genetic engineering and the integration of RNS into sustainable agriculture. Finally, it addresses this field's challenges and future directions, improving symbiotic efficiency and expanding the host range of rhizobia–legume interactions.

1.2 MOLECULAR MECHANISM OF SYMBIOSIS

Leguminous plants and nitrogen-fixing rhizobial bacteria form complex and tightly orchestrated molecular interactions at the root nodule level. It begins with an intricate signaling exchange between the two organisms, mediated by the secretion of specific chemical compounds. The following sections will delve deeper into the step-by-step molecular mechanism governing these remarkable symbiotic interactions.

1.2.1 Signal Exchange and Recognition

In the soil, rhizobia are attracted to the rhizosphere by the complex mixture of chemicals released continually by host plant roots. This mixture includes flavonoids, which encourage the bacteria to create LOCs, known as NFs. These flavonoids serve as crucial signaling molecules in the symbiotic relationship between plants and microbes (Olanrewaju et al., 2019). Flavonoids are prevalent secondary metabolites synthesized via the phenylpropanoid biosynthetic pathway and display various structural variations (Liu et al., 2021). When soil nitrogen availability is insufficient, flavonoid production and accumulation are significantly enhanced (Figure 1.1). Rhizobial NF biosynthesis is triggered in the rhizosphere by flavonoids produced by the host plant, primarily via NodD (Shumilina et al., 2023). This, in turn, increases the expression of other *nod* genes (*nodA*, *nodB*, and *nodC*) involved in LCO production. Microsymbiont compatibility is impacted by flavonoid diversity (Khokhani et al., 2021). The activation of the *nod* gene is, however, inhibited by several flavonoids. Sustaining *nod* gene induction could be facilitated by a balance of flavonoids' inhibitory and stimulatory activities (Shah and Smith, 2020). Two critical receptors, the Nod factor perception (NFP) (MtNFP) and the LysM domain receptor-like kinase 3 (MtLYK3), regulate the perception of NFs in the legume species *Medicago truncatula. Lotus japonicus*, another legume, has LjNFR1 and LjNFR5 as Nod factor receptors (NFRs) (Yang and Wang, 2023) (Figure 1.2).

The legume NFRs are distinctive. They have three LysM domains: An intracellular kinase domain, a transmembrane helix, and a juxtamembrane domain (Wong et al., 2019). Interestingly, the pseudokinase MtNFP/LjNFR5's intracellular kinase domain lacks the conserved amino acids required for activation. Symbiotic reactions need MtLYK3/LjNFR1's intracellular domain kinase activity, which may phosphorylate LjNFR5 *in vitro* (Radutoiu et al., 2007). Symbiotic signal transduction occurs at the plasma membrane, where MtNFP and MtLYK3 form homomeric and heteromeric complexes. LjNFR1 and LjNFR5 expression in *Medicago truncatula* and *Lotus filicaulis* allowed the host to recognize the natural symbionts of *L. japonicus*, *Mesorhizobium loti*, or DZL strains, showing that host recognition depends on NF structure (Yang et al., 2022).

Arbuscular mycorrhizal fungi (AMF) signal rice LysM receptor heterodimer OsMYR1 and OsCERK1. *Medicago* mutants with nodular morphologies were supplemented by replacing these

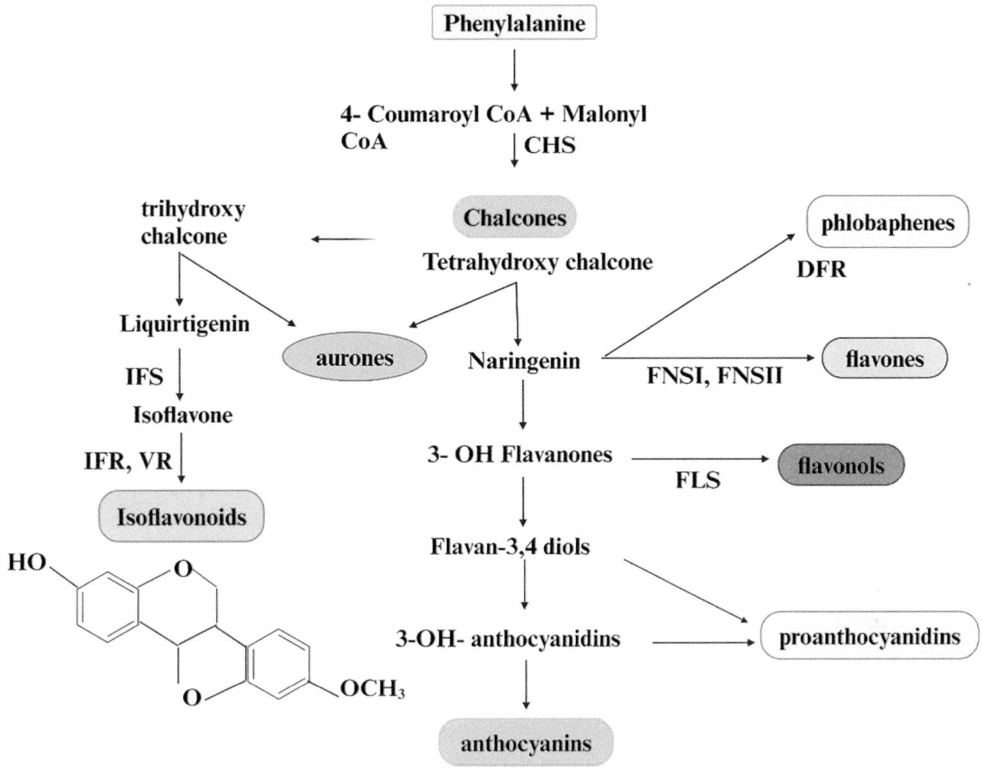

FIGURE 1.1 Biosynthesis pathway of flavonoids.

receptors' extracellular domains with those of MtNFP and MtLYK3, revealing the role of extracellular LysM domains in NF signal perception (Wang et al., 2022a). Further research showed that MtLYK3/LjNFR1's LysM1 domain produces an NF ligand-binding pocket. By transferring these areas to LjCERK6, NF binding was improved (Yang et al., 2022).

Commonly, plant signaling involves receptor-like kinase (RLK)–receptor-like cytoplasmic kinase (RLCK) complexes that transduce signals from RLKs. RLCK NiCK4 binds and phosphorylates the cytoplasmic domains of the NF receptors LjNFR5 and LjNFR1, coupling NF sensing to nodule organogenesis (Hailemariam et al., 2024). *L. japonicus* nick4 mutants still developed nodules, indicating additional RLCKs are involved in LjNFR1–LjNFR5 signaling (Yang and Wang, 2023). Heterotrimeric G-protein complexes and regulator of G-protein signaling (RGS) proteins also regulate lysM receptors. GmNFR1 NF receptor phosphorylates RGS proteins, preserving Gα proteins in an inactive trimeric form and triggering calcium spiking. The process is currently under study (Aittaleb et al., 2010). Notably, specific legume–rhizobia symbioses occur without conventional NF signaling. Cytokinin-like chemicals could trigger nodule organogenesis in *Aeschynomene* species in *Bradyrhizobium* strains lacking the *nodABC* genes for NF synthesis (Sharma et al., 2020). The USDA61 strain's type III effectors can directly activate soybeans, providing a symbiotic signaling pathway without NFs. Alternative rhizobia signals may activate the cysteine-rich receptor-like kinase AeCRK, vital to NF-independent symbiosis in *Aeschynomene evenia* (Yang et al., 2022).

1.2.2 Plant Perception and Response

Nodulation in *L. japonicus* requires LysMRLKs LjNFR1 (LYK) and LjNFR5 (LYR). The NFR1 and NFR5 mutants of *L. japonicus* exhibited no root hair deformation, calcium spiking, or nodulin

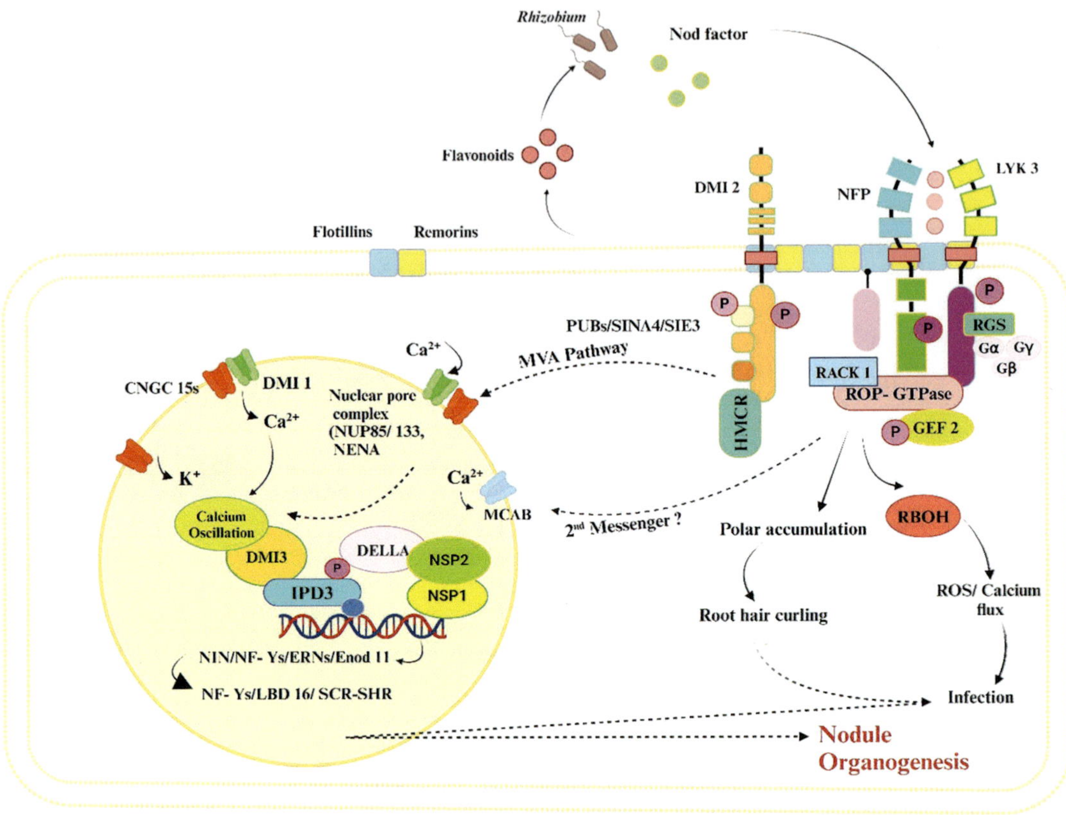

FIGURE 1.2 *Medicago truncatula* common symbiosis signaling pathway (CSSP) model: Flavonoids from legume roots trigger NF secretion by the rhizobium. NFP, LYK3, and DMI2 create a receptor complex (organized by flotillins and remorins) to modulate the perception of NFs. NFP binds to NFs and phosphorylates by LYK3, activating Rho of plants guanosine triphosphatases (ROP-GTPases) to induce IT formation through reactive oxygen species (ROS) bursts via respiratory burst oxidases (RBOHs) and calcium flux in the cytoplasm. Root hair curling is caused by ROP-GTPases interacting with LYK3 and NFP to distribute receptor complexes in root hair tips due to phosphorylated Rho of Plants Guanine Nucleotide Exchange Factors. RACK1, a scaffold protein that interacts with active ROP9, also has a role in signal transduction. G-proteins, Gα, Gβ, and Gγ, control nodule growth by interacting with LYK3 via RGS. PUBs, SINA4, and SIE3 are E3 ubiquitin ligases interacting with DMI2 to regulate turnover. DMI2 and 3–3-hydroxy-3-methylglutaryl-CoA reductase combine to produce mevalonate, which is used as a secondary messenger to transmit signals from the cell membrane to the nuclear envelope-localized CNGC15a/b/c, causing calcium flow in the nucleus. From the karyoplasm to the nuclear envelope, MCA8 carries calcium ions. The potential interaction between DMI1 and CNGC15s involves transporting potassium or calcium ions to modulate nodule organogenesis. NiCK4 or other RLCKs activate the symbiosis-specific nucleoporins NUP85, NUP133, and NENA, which contribute to calcium oscillations in the nucleus. DMI3 activates downstream gene expression by influencing calcium oscillations and phosphorylating IPD3, forming a transcriptional complex with transcription factors (TFs) DELLA, NSP2, and NSP1. Important components downstream of the CSSP include NIN, NF-Ys, ERNs, LBD16, and the SHR-SCR module—important early response factor Enod11.

gene expression when exposed to *Mesorhizobium loti*. LjNFR1 and LjNFR5 form a receptor complex with nanomolar concentrations of *Mesorhizobium loti* NFs and autophosphorylate and transphosphorylate each other, triggering the signaling pathway vital to rhizobial symbiosis (Patra and Mandal, 2022). LjNFR1 and LjNFR5 overexpressed under the ubiquitin promoter stimulate spontaneous nodule formation even without rhizobia, revealing their significance in the plant signaling mechanism during symbiosis (Tsyganova and Tsyganov, 2018). *Medicago truncatula*'s symbiosis involves MtNFP (LYR) as the signaling receptor and MtLYK3 as the way receptor. MtNFP starts early infection, whereas MtLYK3 aids entry (Smit et al., 2007). Primarily, *lyk3* mutants undergo root hair deformation and calcium spiking while interacting with *Sinorhizobium meliloti* but cannot coil root hair tightly or produce ITs (Liu, 2019). Conversely, *nfp* mutants show no response to *S. meliloti* nodulation factors. The expression of NFP, detected through pNFP: GUS, occurs in various cells during different stages of rhizobial inoculation, suggesting a role in the infection process. Rare Nod+ RNAi lines with reduced NFP expression exhibit abortive ITs, indicating that optimal NFP levels are crucial for proper IT development (Bensmihen et al., 2011). Interestingly, expressing NFP specifically in the epidermis of *Mtnfp* mutants restores cortical symbiotic events, proving that epidermal expression alone is enough to trigger nodule organogenesis signaling (Gough, 2015). Studies on the nodulation process in various legumes reveal essential differences and similarities in their NF receptor functions (Garg and Geetanjali, 2009). In *Pisum sativum*, the NF receptor PsSYM37 aligns with a two-receptor model, similar to *Medicago truncatula*. PsSym37 inactivation affects IT formation rather than early symbiotic responses, leading to reduced nodule formation (Igolkina et al., 2018). While MtNFP in *Medicago truncatula* and PsSYM10 in *P. sativum* are inactive kinases requiring active kinase partners for signaling, these partners are not MtLYK3 or PsSYM37, as mutants lacking these do not affect early symbiotic signaling. MtLYK3 facilitates the development and growth of ITs, and nodule type does not affect the two parallel routes for cell division and rhizobial invasion involved in nodulation (Rudaya et al., 2021). As seen in *Medicago truncatula* and *P. sativum*, NF structure specificity is crucial, with specific alleles affecting rhizobia recognition and nodulation efficacy, such as *Sym2A* and its interaction with *nodX* (Singh et al., 2021a). Structural specificity also varies in *Sesbania rostrata* depending on growth conditions, highlighting NF roles in rhizobial entry modes. For *Glycine max*, a palaeoalloploid, two pairs of NFRs are involved. Studies indicate that *GmNFR1α*, *GmNFR5α*, and *GmNFR5β* can rescue nodulation deficiencies, with *GmNFR1α* abundance influencing nodule numbers (Ghantasala and Roy, 2022).

1.2.3 Infection and Nodule Formation

Forming ITs is a crucial downstream process that facilitates the accommodation of rhizobia within the host plant. Through symbiotic signaling events and subsequent transcriptional cascades, the formation of these ITs is mediated by the activation of structural genes (Cao et al., 2017). The primary vectors for rhizobial invasion are plant-derived ITs produced by the plant cell wall material and the plasma membrane's inward folding. These ITs pave the way for rhizobia to penetrate plant tissues (Van Zeijl, 2017). A comprehensive understanding of this crucial mechanism in the legume–rhizobia symbiosis has been provided by Tsyganova et al. (2021), who revealed a systematic analysis of the molecular, physiological, and morphological regulation of IT development and elongation. Extensive alteration of cell walls and membrane expansion are involved in forming the infection chamber and ITs. Research into the *Medicago truncatula* "infection" has shown that regulatory genes for hormone and flavonoid production and those involved in the cell cycle are activated when the rhizobial infection begins (Breakspear et al., 2014). The host plant's exocytosis machinery delivers essential components for developing the infection chamber and ITs, including vesicle-associated membrane proteins (Harrison and Ivanov, 2017). Although their function is unreliable, critical regulators of IT development, including the putative E3 ligase LUMPY INFECTIONS (LIN)/CERBERUS and the protein VAPYRIN (VPY), were recently found via genetic studies (Liu et al., 2019). The LIN–VPY–Exo70H4 protein complex at the tips of elongated pre-ITs indicates that exocytosis promotes the polarized development of *Medicago truncatula* ITs (Liu et al., 2020a). During rhizobial symbiosis, CERBERUS/LIN stabilizes LjVPY1 and LjVPY2 in the trans-Golgi network and early endosome, modulating endocytic trafficking and signaling complex recycling (Yang et al., 2022). Additionally, the ROP regulates signaling. GTPase is necessary to begin the IT and its extension from the tip of the root hair to the nodule primordium. The interaction between ROP-GTPases and LysM-RLKs positively affects rhizobial infection and nodulation (Dalla Via et al., 2017). Recent research has shown that LysM-RLK receptor kinases phosphorylate RopGEF2, activating ROP-GTPase in response to the symbiotic signal. Activated ROPs and LysM-RLKs generate a polar distribution, a complex that controls the deformation of root hairs and the formation of ITs during rhizobial symbiosis (Feiguelman et al., 2018). Furthermore, normal IT elongation requires FLOT4 and SYMREM1, which bind to membrane nanodomains at the ends of elongating ITs. FLOT4 and SYMREM1 may create a particular nanodomain for optimum rhizobial infection, with SYMREM1 perhaps providing to recruit and maintain critical receptors inside the nanodomain (Zhou et al., 2024). Exploring the potential connection between the ROP-GTPases and nanodomain components such as SYMREM1 and flotillin FLOT4 to control rhizobial infection in greater detail would be a fascinating effort (Jhu and Oldroyd, 2023). It has been shown that the NF-induced infection-specific leucine-rich repeat (LRR) receptor-like kinase, LjRINRK1, and the membrane-spanning receptor-like protein, LjEPR3, improve nodulation-induced rhizobium infection and legume–rhizobium recognition (Fabre et al., 2015). In the final stage of infection, rhizobia is released into cortical cells of plants, forming a symbiosome where nitrogen fixation occurs. This involves the collapse of the cellulose sheath at the IT apex, allowing direct interaction with the host plasma membrane, potentially mediated by plant glycoproteins and glycolipids (Venado et al., 2020). Mutations affecting rhizobial release, such as those involving the NFRs NFP and LYK3, have been identified (Singh and Verma, 2023). In indeterminate nodules, proper rhizobial release and differentiation, essential for bacterial survival, involve rhizobial factors like bacA and lpsB and plant factors such as nodule-specific cysteine-rich (NCR) peptides. NCR peptides, found only in inverted repeat-lacking clade (IRLC) legumes like *Medicago*, facilitate bacterial differentiation and are processed by the DNF1 signal peptidase complex (Figure 1.3) (Saeki, 2011).

1.2.4 Nodule Development and Function

The legume root cortex dedifferentiates and forms nodules in response to rhizobial NFs. Indeterminate and determinate nodules have persisting meristems. Middle cortical cells form a persistent meristem at the apex of indeterminate nodules in *Medicago* and peas (Xiao, 2015). When this meristem separates, the indeterminate nodule grows elongated. In contrast, determinate nodules such as soybean and lotus lack a meristem. The outer cortical cells form these spherical nodules by cell expansion rather than proliferating (Brar, 2020). A map of the cellular fate of uncertain nodules in *Medicago truncatula* and genes implicated in distinct phases of nodule organogenesis divided epidermal nodule formation into five stages in response to an infection (Kang et al., 2019). Anticlinal cell divisions in pericycle cells opposite the root's protoxylem pole begin nodule primordium priming. To initiate a nodule, the inner cortex (C4/C5) and the middle cortex (C3) undergo anticlinal and periclinal divisions, which lead to the nodule apical meristem (NAM) (Magne, 2017). Nodule development, the third stage, is characterized by a decrease of the epidermis, inhibition of mitosis in the inner cortex, and significant proliferation of the middle cortex to generate a primordium on the roots with an active NAM and IT penetration (Yang, 2020). Fourth, a bacteriophilic or infected zone and a peripheral nodule vascular bundle (NVB) system are formed during nodule development. Nodule senescence is the final stage, characterized by decreased N_2 fixation activity and the simultaneous death of bacteroids and plant cells (Kazmierczak et al., 2020). Nodules that have reached maturity and are not yet infected consist of the following: The nodule parenchyma, endodermis, cortex, peripheral vascular bundles, and the inner symbiotic area (Mergaert, 2020). Zone I is the meristematic apical region; Zone II is the zone of symbiotic cell differentiation and IT penetration; Zone III is the zone of nitrogen fixation; and Zone IV is the zone of senescence, according to a differentiation gradient (Yang, 2020). However, definite nodules produce primordia in the outer or middle root cortex adjacent to the epidermis. Mature determinate nodules have parenchyma, endodermis, and cortical outer tissues. These nodules have a nitrogen-fixing zone around a senescent zone in their inner symbiotic region. Many lenticels develop on determinate nodules as they mature (Figure 1.4) (Liu et al., 2020b).

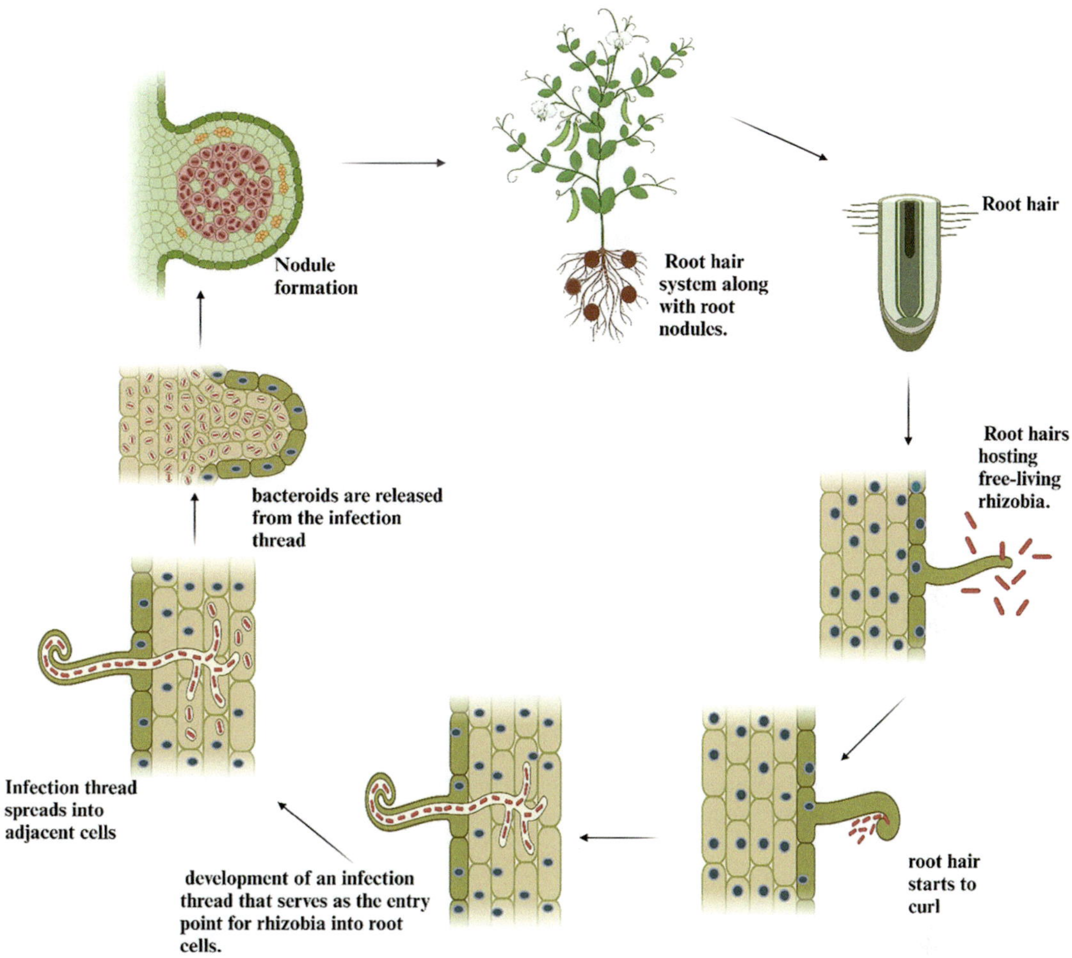

FIGURE 1.3 Symbiotic journey: The process of root nodule formation in plants. This diagram illustrates the step-by-step development of root nodules, highlighting the interaction between root hairs and rhizobia for nitrogen fixation.

Recent reviews by Liu et al. (2018a) and Lin et al. (2020) indicate that phytohormones regulate nodule growth. Nodule organogenesis requires hormone regulation and maintenance.

Cytokinins have a positive role in nodule organogenesis by regulating the division of cortical cells. Leguminous plants are the only ones that can be stimulated to produce nodule-like structures and cortical cell divisions by introducing exogenous cytokinin (Dolgikh et al., 2016). Cortical cell divisions and spontaneous nodule growth were enhanced by overexpressing the *LOG4* gene, which controlled endogenous cytokinin levels (Reid et al., 2017). *Rhizobium* or NF management quickly induces several cytokinin production and activation genes, consistent with the discovery that cytokinins accumulate during nodulation: Transcriptomic and quantitative real-time PCR analysis (Lin et al., 2021). Having unusually low or high cytokinin levels reduced nodule numbers, showing that *de novo* cytokinin homeostasis is crucial for nodule organogenesis (Hnatuszko-Konka et al., 2021). Also, gain-of-function cytokinin receptor mutants spontaneously form nodules. Epidermis and cortical nodulation may be affected differently by cytokinin. β-Glucuronidase (GUS) expression mediated by the cytokinin receptor gene promoter exhibited insignificant, intermittent signals in the epidermis but significant signals in dividing cortical cells (Witchel, 2021). Aborted nodule

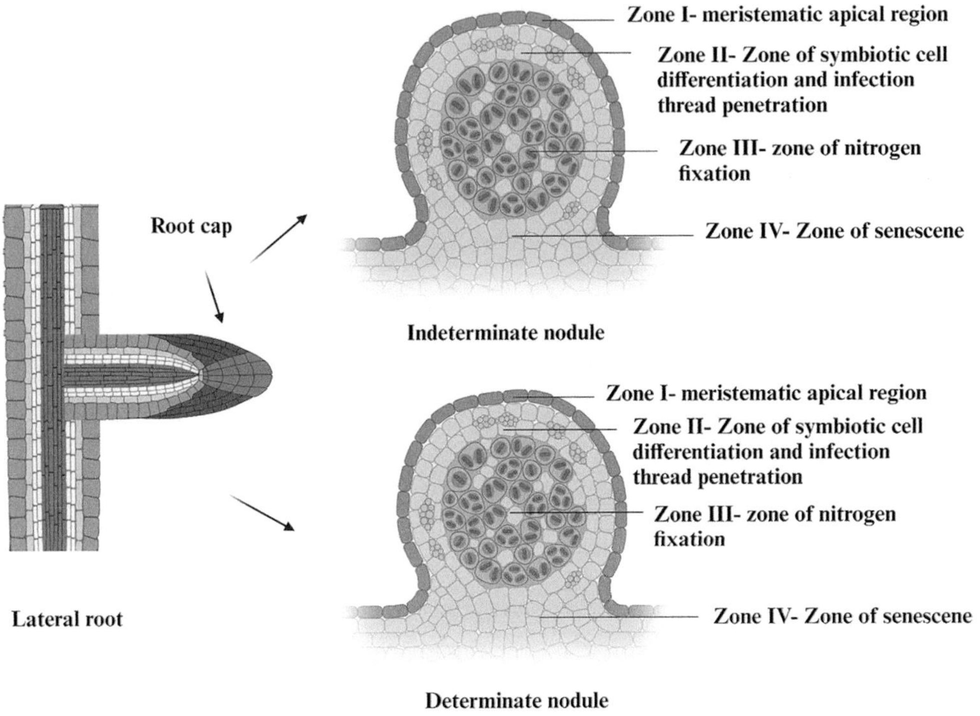

FIGURE 1.4 Exploring root nodules: Structure of lateral roots alongside two types of root nodules—indeterminate and determinate. Each nodule type is organized into distinct zones: The meristematic apical region, zones of differentiation and infection, nitrogen fixation, and senescence.

primordia and root hair hyper-infection led to *LHK1* gene deficiency. Cortical cell-specific AtCKX3 expression decreased nodulation, but epidermis-specific expression enhanced nodule development. NF treatment increased numerous cytokinin response regulator (RR) genes, and reducing multiple RRs significantly reduced nodule number (Buhian and Bensmihen, 2018). Since cytokinin may operate as both a local and a long-distance signal, it would be fascinating to comprehend whether it translocates across cell layers to coordinate cortical cell division and formation of ITs on symbiosis signaling (Roy et al., 2020).

WOX5 and *PLETHORA* (*PLT*) genes within the start sites are expressed, forming nodules and lateral roots in response to local auxin accumulation. Founder cells stimulated by auxin accumulation provide growth to lateral roots; however, nodules have not been observed (Wardhani, 2020). Auxin biosynthesis, auxin-responsive genes, and signaling-related genes were stimulated by rhizobium or NFs. Auxin and auxin transport inhibitors may promote cell division and pseudonodule development without rhizobial inoculation, the most conclusive growth binding auxin to nodule organogenesis (Aryal, 2020). Depletion of auxin receptors (GmTIR1/AFB3) reduced nodule number and IT development in *G. max*, but overexpression of GmTIR1 increased nodule formation (Chen et al., 2022). Activated auxin production below the rhizobial infection site may modulate rootward polar auxin transport, causing localized cell division, proliferation, and nodule organogenesis (Gao et al., 2021). CRE1-mediated cytokinin sensing enhances localized auxin accumulation essential for nodule organogenesis by inhibiting polar auxin transport below the site of rhizobial infection (Molla et al., 2023). The expression patterns of the auxin-responsive genes *DR5* and *GH3* revealed that, during nodule organogenesis, auxin accumulates in the proliferating cortex cells, underlining this process's significance (Xiao, 2015). Nodule primordium formation after bacterial infection is hindered by flavonoids because they prevent polar auxin transfer. Flavonoids

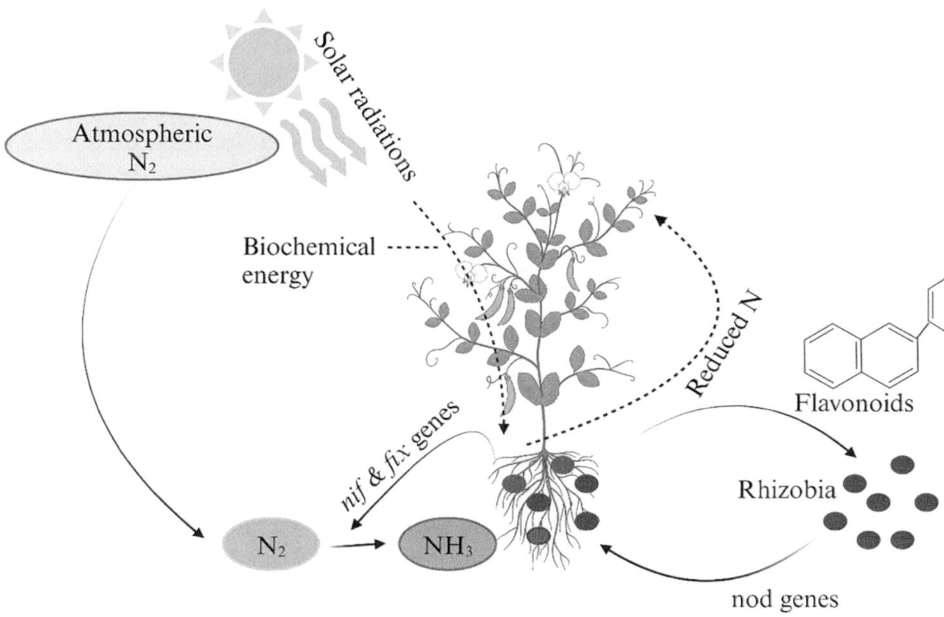

FIGURE 1.5 Concise overview of the symbiotic nitrogen fixation process in legumes facilitated by rhizobia.

promote nodulation in certain legumes by increasing cytokinin synthesis during nodule organogenesis (Dong and Song, 2020).

Nodulation in legumes is negatively affected by ethylene. The early phases of nodulation include ethylene, as *L. japonicus* showed increased ethylene synthesis 6 hours after *Rhizobium* bacterium inoculation (Guinel, 2015). Aminoethoxyvinyl glycine, an ethylene biosynthesis inhibitor, also increased ITs and nodules in legume species. Nodule organogenesis and NF-induced calcium spikes were dramatically suppressed by the ethylene precursor 1-aminocyclopropane-1-carboxylic acid (ACC) (Shaharoona, 2011). *Mtsickle* (*ein2*) in *Medicago truncatula* and *Ljein2a ein2b* in *L. japonicus* are ethylene-insensitive mutants with hyperinflation and hypernodulation. These results imply that legume nodule formation and development depend on perfectly controlled ethylene levels during nodulation (Liu et al., 2018b).

1.3 NUTRIENT EXCHANGE IN SYMBIOSIS

1.3.1 Nitrogen Fixation

Legumes and nitrogen-fixing soil bacteria, rhizobia, provide the most physiologically accessible nitrogen in natural and cultivated environments. Rhizobia–legume interaction creates root nodules, where the bacteroid rhizobia may convert atmospheric N_2 to ammonia (Wang et al., 2019). Root nodule bacteria can convert gaseous nitrogen (N_2) from the atmosphere into forms the plant can utilize, such as ammonia or amino acids (Figure 1.5). In this mutualistic relationship, plants supply the bacteria with dicarboxylic acids for carbon (Kumar et al., 2020). This symbiosis allows the plant to obtain increased nitrogen and amino acids, while rhizobia receives carbon compounds and other nutrients derived from photosynthesis. Inside the nitrogen-fixing nodules, plant cells and their bacterial partners create the necessary environment for nitrogen reduction, assimilation, and the exchange of metabolites (Bhatla and Lal, 2023). This partnership is sustained by coordinated changes in gene expression within both plants and rhizobia, adapting their biochemical and molecular processes to support each other. Nitrogen fixation is catalyzed by a large metalloprotein complex that includes the enzyme nitrogenase (Gresshoff, 2012). Depending on the metal co-factor, three distinct nitrogenases are found in bacteria that fix nitrogen. *Azotobacter* species consist of nitrogenases, which can utilize either iron (Fe) or vanadium (V) (Bellenger et al., 2020). In contrast, the nitrogenases in rhizobia associated with legumes typically contain molybdenum (Mo) (Bonilla and Bolanos, 2010). The catalytic component of the MoFe protein, called dinitrogenase, is responsible for binding nitrogen gas (N_2). Meanwhile, a minor dinitrogenase reductase donates electrons to the MoFe protein. This two-component system is typical of nitrogenases. The complex is oxygen-sensitive and operates in a practically anoxic environment inside the nodule (Chen, 2019). The opposite is true for reducing N_2 to NH_4^+, which requires an abundant oxygen supply via aerobic respiration, producing ATP. The oxygen-binding and oxygen-transporting protein leghemoglobin (Lb) production helps resolve this paradox by controlling oxygen levels inside the nodule (Torres and Stacey, 2003).

1.3.2 Carbon Transfer

Rhizobial BNF requires much energy. A minimum of 16 ATP molecules and 8 electrons with a minimal energy level are essential for transforming nitrogen gas into ammonia (Soumare et al., 2020). For BNF to remain active, plants must continuously provide rhizobia with energy through reduced carbon molecules. Although they form an insignificant portion of a legume plant's overall biomass, root nodules may absorb almost a quarter of 0.25 of its photosynthetically derived nutrients (Raza et al., 2020). Sucrose is one of the main carbohydrates transported by the plant's phloem to its sink organs—the root nodules (Hennion et al., 2019). Much remains unclear, however, about the regulation of this process and the mechanism by which carbon gets transported to the root nodules (Liu et al., 2018b). Root hair cells infected with the infection, the meristem, the invasion zone, and the nodule vascular system all express the nodule-specific sucrose transporter MtSWEET11 of *Medicago truncatula*. There may be a role for MtSWEET11 in transporting sucrose to root nodules, but it does not exist for BNF (Kryvoruchko et al., 2016). Soybean nodules have elevated sucrose uptake transporter GmSUT1 levels, mainly in the vascular bundles and the peripheral fixation zone. Some have hypothesized that GmSUT1 could assist with the transport of sucrose to the root nodules, as overexpressing the gene raised both the nodule count and the plant's nitrogen content (Deng et al., 2022).

The symbiosome membrane (SYM) is driven by H+-ATPases, which bring protons into the symbiosome space to stimulate bacteroids and plants to take nutrients up. SYM contains 197 proteins, some implicated in solute transport, membrane trafficking, protein folding and degradation, and metabolism (including hydrogen ions, sulfate, calcium, peptide/dicarboxylate, and nitrate transporters) (Clarke et al., 2014). While C4-dicarboxylates may be the primary carbon source for active root nodule bacteroids, N_2 fixation is reduced by mutations that affect SYM dicarboxylate transport (Schulte et al., 2021). On C4-dicarboxylate transport protein A (*dctA*), *dctB*, and *dctD*, the SYM's well-established dicarboxylate transport system is found. Two-component regulatory mechanisms are encoded by *dctB* and *dctD*, whereas *dctA* encodes a structural protein for C4-dicarboxylate transport. In rhizobial Dct systems, aspartate, succinate, oxaloacetate, malate, and fumarate are the main substrates (Crang, 2019). Plants absorb NH_3 from bacteroids; however, they also emit aspartate and alanine. Transferring BNF-produced NH_3 involves passive diffusion across the bacteroid membrane and the nodulin 26 channel. While an aquaporin superfamily member, nodulin 26 prefers ammonia permease. NH_3 is protonated to NH_4^+ in the acidic symbiosome and taken in by the host via a cation channel (Ramesh et al., 2023).

1.4 REGULATION OF NUTRIENT EXCHANGE

1.4.1 Oxygen Regulation

During symbiosis, rhizobia tackle zones within the nodule with low oxygen content. Symbiosis depends on regulating oxygen levels as oxygen reduces nitrogenase activity (Abd-Alla et al., 2023).

The transport barrier regulates oxygen levels to produce nitrogen fixation optimal concentrations. Bacteroid-induced signaling and physiological changes in bacteria enhance symbiosis and nitrogen fixation (Lamouche et al., 2019). It is well-established that low oxygen levels activate the two-component system FixJL, which enhances the transcription of most nitrogen fixation-related genes (Hawkins and Oresnik, 2022). Recent studies have identified three proteins in *Rhizobium leguminosarum*—hFixL, FnrN, and NifA—that function as oxygen sensors. Each of these proteins is closely regulated by time; for instance, in indeterminate nodule zone I (the meristem zone) and zone II (the invasion zone), hFixL triggers the production of FnrN. In response to air oxygen levels approaching anaerobic conditions, FnrN triggers the expression of fixNOQP in zone III, also known as the nitrogen-fixing zone (Rutten et al., 2021). The activation of the genes required for nitrogenase synthesis in near-anaerobic conditions shows both the activity of the protein and the signaling of symbiosis by microaerobic stress. Studies have also shown that oxygen concentration controls lipopolysaccharide (LPS) production and modification (Hawkins and Oresnik, 2022). This may aid adaptation to environments with low oxygen levels. Certain rhizobia determine their host range for symbiosis via strain-specific LPS production and modification (Ledermann et al., 2021). The inability to produce or appropriately regulate LPS has been associated with problems in symbiotic establishment. Symbiosis is an environmental context in which LPS modification is likely to occur because of the environment-dependent nature of LPS content and decoration (Mote, 2020). It has recently been shown that flavonoids can induce changes in the decoration of LPS and that these changes are important for symbiosis and survival to occur under such conditions (Wang et al., 2022b).

1.4.2 Signaling and Feedback Mechanism

In all host–microbe symbioses, the nitrogen-fixing *Rhizobium*–legume interaction is most characterized. Bacteria and plants interact during development. Genetics, cellular, and biochemical studies unravel these interactions (Granada Agudelo et al., 2023). Specific examples are the symbioses between *Sinorhizobium meliloti* and *Medicago* (*Medicago Sativa* and *Medicago truncatula*), and *Mesorhizobium* and *L. japonicus* (Shimamura et al., 2022; Maillet et al., 2020). First, legumes export signals: in *Medicago sativa*, luteolin interacts with NodD, a *Sinorhizobium meliloti* transcription factor, to activate rhizobial nodulation genes. These encode NF-synthesizing enzymes (Stambulska and Bayliak, 2020). Common nod genes synthesize NF chito-oligomeric networks and attach N-acyl groups, whereas host-specific *nod* genes add side groups. Plant proteins respond to the NF signal via access and signaling receptors with three extracellular LysM domains (Krönauer and Radutoiu, 2021). SymRK, a third protein, contains extracellular LRRs. This heterotrimeric complex seems to be kinase-active. NF side groups affect binding/signaling specificity, whereas receptors specify NF structure (Oelmüller et al., 2023). NFR contact leads to ionic and chemical processes in the plant nucleus, including calcium spikes decoded by CCamK, Cyclops, and other proteins, including transcription factors (Singh et al., 2021b). The presence of NF molecules affects the location of NFRs, yet their expression is constitutively expressed. Proteins associated with the plant membrane, such as flotillins and remorins, are expected to form complexes with NFRs on the cytoplasmic surface. These proteins may mediate interactions among the NFRs, arrays of cytoskeleton components, and the machinery responsible for cell wall building (Mbengue et al., 2020). Infection requires living bacteria to generate extracellular polysaccharides (EPSs), although NF is enough to trigger morphogenesis and early gene expression. A two-component regulatory cascade co-regulates EPS biosynthesis enzyme genes in *Sinorhizobium meliloti* (Acosta-Jurado et al., 2021). The sensor histidine kinase ExoS phosphorylates ChvI, which is subsequently inhibited by the periplasmic regulatory protein ExoR. In addition to regulating EPS, the ExoR–ExoS–ChvI system controls sessile vs motile habit and other functions. Bacterial EPS may have structurally specific functions (Chaves-Olarte et al., 2023). LysM-domain receptor kinases differentiate normal and altered-structure EPS in *Lotus*. ESR3 is generated via early NF signaling. Infection occurs when bacteria promote the plant to build a thread of expanding plant cell walls and membranes (Awwanah, 2020). A developing IT must interact dynamically with the cytoskeleton, cell membrane, and wall synthesis machinery but specifics are few (Bozsoki et al., 2020). ITs start in root hairs and transverse the cell length, emerging into the apoplastic (intercellular) region and infecting further cell layers. ITs end in plant cells further in the growing nodule (Gao et al., 2024). Bacteria are endocytosed into plant plasma membrane vesicles, a previously unknown mechanism. The symbiosome forms in the plant membrane around the bacterium. The newly produced plant proteins Lb and NCR peptides regulate bacterial activity (Schulte, 2021). In *Medicago* but not *Lotus*, NCRs regulate bacterial cell division and differentiation (Zhang et al., 2023). As a result, *Sinorhizobium meliloti* inhibits cell division, endoduplicates its genome, and differentiates (Xing et al., 2022). Molecular oxygen is bound by Lb, resulting in high oxygen concentration but low free O_2. Reduction triggers FixL, a hemoprotein sensor kinase, which phosphorylates transcription factor FixJ, activating nitrogen-fixation genes (Amiri, 2021). A sophisticated bacterial cytochrome system and metabolic pathways that export net nitrogen for plant sugars modify the bacterium to feed the host cell. N-rich molecules are exported via symbiosis-specific N-assimilation mechanisms (Baslam et al., 2020).

Feedback systems in the *Rhizobium*–legume symbiosis finely regulate interactions between plants and bacteria. The NF starts nodulation and may control receptor availability by supplying to plant receptors, producing a feedback loop that alters plant responses (Singh et al., 2021a). Crucially for infection, bacterial EPS synthesis is under control by the ExoR–ExoS–ChvI system, which modulates bacterial activity depending on plant signals and generates a responsive feedback loop (Geiger et al., 2021). NF signaling also affects plant hormone pathways and immunological responses, thereby balancing defense and symbiont acceptance through hormonal cross-action. Lb controls oxygen levels inside nodules to facilitate nitrogen fixation; feedback changes production to meet the metabolic demand (Rutten and Poole, 2019). In *Medicago*, NCR peptides control bacterial differentiation and might react to plant nitrogen conditions, thus guiding bacteroid growth (Haag and Mergaert, 2019). Acting as a feedback loop, the FixL–FixJ mechanism ensures that nitrogen fixation occurs only under ideal oxygen levels (Rutten, 2021). These feedback mechanisms guarantee a symbiotic interaction wherein bacteria locate an appropriate habitat and plants get fixed nitrogen, maximizing reciprocal advantages.

1.4.3 Symbiotic Efficiency

The symbiosis between specific plant species and nitrogen-fixing bacteria, most commonly *Rhizobium* and *Bradyrhizobium* genera, is known as root–nodule symbiosis (Mahmud et al., 2020). The efficient conversion of atmospheric nitrogen into a form that plants can use as a nutrient is rendered possible by this symbiosis, which is vital for the plants and bacteria involved (Kumar et al., 2020). Root–nodule symbiosis is more likely to succeed when the plant and the bacteria interact. Plants are filled with specialized structures known as nodules, providing bacteria a habitat on their roots (Mathesius, 2022). The plant provides the bacteria with carbs and other nutrients within these nodules, which they use to fuel their nitrogen-fixing abilities (Fahde et al., 2023). In exchange, nitrogenase enzymes produced by nitrogen-fixing bacteria convert atmospheric nitrogen into ammonia, which plants may absorb for essential nutritional development and growth (Santhosh et al., 2019). Because of this mutualistic relationship, the plant can develop, especially in soil low in nitrogen, since it obtains a vital nutrient that might otherwise be unavailable (Prasad et al., 2021). The specificity of the plant–bacteria relationship further enhances the efficiency of root–nodule symbiosis. The plant and bacteria exchange specific signaling molecules to activate nitrogen fixation and guarantee nodule formation. This specificity assists in producing the most efficient use of the invested resources of plants and bacteria (Nakei et al., 2022). With all factors assessed, root–nodule symbiosis is an efficient instance of a symbiotic relationship that reveals how important it is for two organisms to work together to benefit both (Table 1.1). This symbiosis is vital to nitrogen cycling and ecosystem productivity because it efficiently converts atmospheric nitrogen into a form

TABLE 1.1
Factors Affecting Root Nodule Symbiosis Efficiency.

Determinants of Efficiency	Factors	Description/Impact on Symbiosis	References
Biological factors	Plant species	Different legume species vary in their ability to form efficient nodules and fix nitrogen	Mathesius (2022)
	Rhizobial strains	The compatibility of the specific rhizobial strain with the host plant can significantly influence efficiency	Allito et al. (2015)
	Plant health	The overall health of the plant (disease resistance and stress response) impacts nodule formation and function	Shinwari et al. (2019)
	Symbiotic partner diversity	A diverse community of rhizobia may enhance symbiotic performance through various interactions	Fields et al. (2021)
	Nodule development	Factors like nodule size, number, and morphology impact the efficiency of nitrogen fixation	Li et al. (2018)
Environmental factors	Soil pH	Soil acidity or alkalinity affects nutrient availability and microbial activity, influencing nodulation	Ferreira et al. (2016)
	Gene expression and signal transduction	The expression of specific genes involved in signaling pathways is critical for initiating and maintaining nodulation. Hormonal signaling (e.g., auxins and cytokinins) can influence nodule development and function, affecting overall symbiotic efficiency	
	Moisture availability	Water stress can limit plant growth and rhizobial activity, reducing symbiotic efficiency	Mouradi et al. (2016)
	Light intensity	Adequate sunlight is necessary for photosynthesis, which supports the energy needs of both plants and rhizobia	Friel and Friesen (2019)
	Soil aeration	Good aeration supports root health and rhizobial activity; compacted soil can hinder both	Bodner et al. (2021)
Agricultural factors	Crop rotation	Rotating legumes with nonlegumes can enhance soil health and rhizobial populations over time	Sauvadet et al. (2021)
	Fertilization practices	Excessive use of nitrogen fertilizers can reduce plants' reliance on symbiotic nitrogen fixation	Khan et al. (2019)
	Seeding rate	High-density planting may lead to competition for resources and affect nodule formation	Liao et al. (2022)
	Cover cropping	Cover crops can improve soil structure and nutrient availability, enhancing nodulation in subsequent crops	Zhou et al. (2017)
Soil-related factors	Soil texture	Soil type (clay, sandy, or loamy) affects water retention, aeration, and root penetration	Cassel (2019)
	Soil nutrient availability	The availability of micronutrients (like molybdenum) is vital for adequate nitrogen fixation	Kafeel et al. (2023)
	Soil microbial community	The presence of beneficial or pathogenic microbes can influence rhizobial survival and efficiency	Chernov and Semenov (2021)
	Soil organic matter	High organic matter content supports microbial diversity and nutrient availability for plants and rhizobia	Biswas and Kole (2017)
	Compaction and soil structure	Compacted soils may restrict root growth and impair nodule function	Wu et al. (2024)

organisms can use (Tsiknia et al., 2021). Rhizobial strains from relict legumes show significant genetic diversity compared to commercial strains, influencing symbiosis efficiency. Genomic analysis can help select complementary strains and improve agricultural legume–rhizobia efficiency (Safronova et al., 2021). Research indicates these rhizobia provide host resistance to various hazardous metals, including cadmium, nickel, mercury, chromium, arsenic, aluminum, copper, and lead (Ayub et al., 2022). Various leguminous crops' growth, development, and yield under heavy metal regimes improved when inoculated with suitable rhizobial species. In such conditions, legumes may establish roots with naturally existing metal-tolerating rhizobia, promoting their diversity (Goyal and Habtewold, 2023).

1.5 APPLICATIONS AND IMPLICATIONS

RNS, the complex relationship between legumes and nitrogen-fixing bacteria, fascinates scientists. These symbiotic interactions enhance leguminous crop growth and production, with enormous potential for addressing global agricultural and environmental sustainability issues. Research on RNS is utilized to produce biofertilizers. Researchers may develop more effective nitrogen-fixing bacteria, crop seeds, or soil inoculants through studies on root nodule production and function. These biofertilizers could substitute for energy-intensive, hazardous, synthetic nitrogen fertilizers. Sustainable agriculture using biofertilizers improves soil conditions and reduces crop production's carbon impact. RNS knowledge about more legumes could be used to develop productive symbiotic interactions. Researchers may develop nonleguminous crops with comparable symbiotic connections with nitrogen-fixing bacteria by identifying genetic and molecular factors. This might enhance food production and reduce synthetic fertilizer use by expanding the crops that can benefit from biological nitrogen fixation. RNS may mitigate climate change. Beans may absorb atmospheric nitrogen and integrate it into the soil via their symbiotic interaction with nitrogen-fixing bacteria, reducing the need for energy-intensive nitrogen fertilizers. This method reduces greenhouse gas emissions from synthetic fertilizer production and use, rendering agriculture more environmentally friendly. Research on RNS may also produce new biotechnological applications (Table 1.2), such as nitrogen-fixing bacteria for soil bioremediation or high-value chemical production. By studying the molecular concepts that regulate the symbiotic interaction, researchers might discover alternative methods to utilize these bacteria.

1.6 FUTURE DIRECTION AND CHALLENGES

Terrestrial plants interact with microbes in mutualistic, commensal, or parasitic ways. As we study this complicated biological process, various prospective possibilities and challenges might open new horizons in sustainable agriculture, environmental remediation, and more. The critical future goal is to uncover the genetic and molecular pathways that develop and sustain symbiosis. Understanding signaling pathways, gene expression patterns, and regulatory networks might assist academics in more efficient and reliable symbiotic systems. This might require developing new inoculants or genetically modifying legumes to build solid and beneficial interactions with nitrogen-fixing bacteria. This approach might be applied to many agricultural systems by studying non-legume species' symbiotic interactions. Studying root nodule microbiome dynamics is another fascinating path. The complicated interaction among the host plant, nitrogen-fixing bacteria, and the microbial community might reveal symbiosis stability and production variables. This knowledge may help develop comprehensive soil health and nutrient cycle management strategies for more sustainable and ecologically friendly agriculture. High-throughput sequencing, metabolomics, and computational modeling could accelerate RNS research. These techniques may identify new symbiotic partners, reveal metabolic pathways, and anticipate ideal symbiosis circumstances. Combining these technological developments with field-based studies and real-world applications may connect vital research and practical implementation. Subsequently, RNS applies beyond agriculture. Legumes fix atmospheric nitrogen and enhance soil fertility, which may be used to heal destroyed land and reduce greenhouse gas emissions.

TABLE 1.2
Genetic Innovations in Enhancing Symbiosis.

Genetic Innovation	Description	Applications	Potential Benefits	References
CRISPR/Cas9	Gene-editing technology that allows precise modifications to specific DNA sequences	Enhancing nodule formation in legumes	Improved nitrogen-fixing efficiency	Rasheed et al. (2022)
Transgenic approaches	Involves the introduction of foreign genes into plants	Development of legumes with enhanced symbiotic traits	Expanding compatibility with different rhizobial strains	Chandrashekharaiah et al. (2021)
Mutagenesis	Random mutations are induced to create genetic variation	Identifying beneficial traits in various legumes	Faster breeding cycles and increased gene diversity	Begna (2021)
Synthetic biology	Engineering synthetic pathways for novel trait development	Creation of new symbiotic abilities, potentially in nonlegumes or different environments	Potential to enhance resilience and nutrient uptake	Toth (2022)
Genomic selection	Selecting plants based on genomic data for desirable traits	Speeding up breeding programs targeting nitrogen fixation	More efficient breeding for higher yields and better nutrient use	Singh et al. (2020)
RNA silencing	Technique to downregulate specific gene expression	Reducing negative feedback in nodule development	Increased nodule number and function	Reyero-Saavedra et al. (2017)
Marker-assisted selection	Using molecular markers linked to desired traits for breeding	Selecting for varieties with superior nodulation characteristics	More targeted breeding efforts and increased plant performance	Chandrashekharaiah et al. (2021)

Studying RNS may also help develop biotechnological solutions like biofuels and sustainable materials. To understand RNS, we must collaborate across disciplines, integrate varied data sources, and use scientific knowledge. We can maximize these unique biological phenomena and contribute to a more sustainable and resilient future by embracing these paths and dealing with these issues.

REFERENCES

Abd-Alla, M.H., Al-Amri, S.M. and El-Enany, A.W.E. 2023. Enhancing rhizobium–legume symbiosis and reducing nitrogen fertilizer use are potential options for mitigating climate change. *Agriculture*, 13(11), p.2092. https://doi.org/10.3390/agriculture13112092

Acosta-Jurado, S., Fuentes-Romero, F., Ruiz-Sainz, J.E., Janczarek, M. and Vinardell, J.M. 2021. Rhizobial exopolysaccharides: Genetic regulation of their synthesis and relevance in symbiosis with legumes. *International Journal of Molecular Sciences*, 22(12), p.6233. https://doi.org/10.3390/ijms22126233

Aittaleb, M., Boguth, C.A. and Tesmer, J.J. 2010. Structure and function of heterotrimeric G protein-regulated Rho guanine nucleotide exchange factors. *Molecular Pharmacology*, 77(2), pp.111–125. https://doi.org/10.1124/mol.109.061234

Allito, B.B., Nana, E.M. and Alemneh, A.A. 2015. Rhizobia strain and legume genome interaction effects on nitrogen fixation and yield of grain legume: A review. *Molecular Soil Biology*, 6. https://doi.org/10.5376/msb.2015.06.0002

Amiri, A.A. 2021. *Regulation of nitrogen fixation in rhizobia–legume symbioses* (Doctoral dissertation, Murdoch University).

Aryal, S. 2020. *Nodule zone-specific gene expression in soybean*. South Dakota State University.

Awwanah, M. 2020. *Characterization of Populus x canescens LysM-receptor like kinases LYK4/LYK5 and LysM-receptor like protein LYM2 and their roles in chitin signaling* (Doctoral dissertation, Göttingen, Georg-August Universität).

Ayub, M.A., ur Rehman, M.Z., Umar, W., Zuhra, N. and Shabaan, M. 2022. Role of legumes in phytoremediation of heavy metals. In *Advances in legumes for sustainable intensification*, pp.345–360. Academic Press. https://doi.org/10.1016/B978-0-323-85797-0.00011-2

Baslam, M., Mitsui, T., Sueyoshi, K. and Ohyama, T. 2020. Recent advances in carbon and nitrogen metabolism in C_3 plants. *International Journal of Molecular Sciences*, 22(1), p.318. https://doi.org/10.3390/ijms22010318

Battenberg, K. and Hayashi, M. 2022. Evolution of root nodule symbiosis: Focusing on the transcriptional regulation from the genomic point of view. *Plant Biotechnology*, 39(1), pp.79–83. https://doi.org/10.5511/plantbiotechnology.22.0127a

Begna, T. 2021. Application of mutation in crop improvement. *International Journal of Research in Agronomy*, 4(2), pp.01–08. https://doi.org/10.33545/2618060X

Bellenger, J.P., Darnajoux, R., Zhang, X. and Kraepiel, A.M.L. 2020. Biological nitrogen fixation by alternative nitrogenases in terrestrial ecosystems: A review. *Biogeochemistry*, 149, pp.53–73. https://doi.org/10.1007/s10533-020-00666-7

Bensmihen, S., De Billy, F. and Gough, C. 2011. Contribution of NFP LysM domains to the recognition of Nod factors during the *Medicago truncatula/Sinorhizobium meliloti* symbiosis. *PloS One*, 6(11), p.e26114. https://doi.org/10.1371/journal.pone.0026114

Bhatla, S.C. and Lal, M.A. 2023. Nitrogen metabolism. In *Plant physiology, development and metabolism*, pp.295–334. Singapore: Springer Nature Singapore.

Biswas, T. and Kole, S.C. 2017. Soil organic matter and microbial role in plant productivity and soil fertility. In *Advances in soil microbiology: Recent trends and future prospects: Volume 2: Soil-microbe-plant interaction*, pp.219–238. http://www.springer.com/series/14379

Bodner, G., Mentler, A. and Keiblinger, K. 2021. Plant roots for sustainable soil structure management in cropping systems. In *The root systems in sustainable agricultural intensification*, pp.45–90. https://doi.org/10.1002/9781119525417.ch3

Bonilla, I. and Bolanos, L. 2010. Mineral nutrition for legume-rhizobia symbiosis: B, Ca, N, P, S, K, Fe, Mo, Co, and Ni: A review. In *Organic farming, pest control and remediation of soil pollutants: Organic farming, pest control and remediation of soil pollutants*, pp.253–274. https://doi.org/10.1007/978-1-4020-9654-9

Bozsoki, Z., Gysel, K., Hansen, S.B., Lironi, D., Krönauer, C., Feng, F., de Jong, N., Vinther, M., Kamble, M., Thygesen, M.B. and Engholm, E. 2020. Ligand-recognizing motifs in plant LysM receptors are major determinants of specificity. *Science*, 369(6504), pp.663–670.

Brar, A.K. 2020. *The fate of nitrogen-fixing and non-nitrogen-fixing strains of the symbiotic bacterium Sinorhizobium meliloti in root nodules of host Medicago species*. The Florida State University.

Breakspear, A., Liu, C., Roy, S., Stacey, N., Rogers, C., Trick, M., Morieri, G., Mysore, K.S., Wen, J., Oldroyd, G.E. and Downie, J.A. 2014. The root hair "infectome" of *Medicago truncatula* uncovers changes in cell cycle genes and reveals a requirement for auxin signaling in rhizobial infection. *The Plant Cell*, 26(12), pp.4680–4701. www.plantcell.org/cgi/doi/10.1105/tpc.114.133496

Buhian, W.P. and Bensmihen, S. 2018. Mini-review: Nod factor regulation of phytohormone signaling and homeostasis during rhizobia-legume symbiosis. *Frontiers in Plant Science*, 9, p.1247.k, p.342. https://doi.org/10.3389/fpls.2018.01247

Cao, Y., Halane, M.K., Gassmann, W. and Stacey, G. 2017. The role of plant innate immunity in the legume–rhizobium symbiosis. *Annual Review of Plant Biology*, 68(1), pp.535–561. https://doi.org/10.1146/annurev-arplant-042916-041030

Cassel, D.K. 2019. Effects of soil characteristics and tillage practices on water storage and its availability to plant roots. In *Crop reactions to water and temperature stresses in humid, temperate climates*, pp.167–186. CRC Press.

Chandrashekharaiah, P.S., Paul, V., Kushwaha, S., Sanyal, D. and Dasgupta, S. 2021. Biotechnological approaches for enhancing stress tolerance in legumes. In *Sustainable agriculture reviews 51: Legume agriculture and biotechnology*, Vol. 2, pp.247–293. https://doi.org/10.1007/978-3-030-68828-8_9

Chaves-Olarte, E., Meza-Torres, J., Herrera-Rodríguez, F., Lizano-González, E., Suárez-Esquivel, M., Baker, K.S., Rivas-Solano, O., Ruiz-Villalobos, N., Villalta-Romero, F., Cheng, H.P. and Walker, G.C. 2023. A sensor histidine kinase from a plant-endosymbiont bacterium restores the virulence of a mammalian intracellular pathogen. *Microbial Pathogenesis*, 185, p.106442. https://doi.org/10.1016/j.micpath.2023.106442

Chen, T. 2019. *Understanding substrate and inhibitor interactions with Mo-nitrogenase* (Doctoral dissertation, University of Oxford).

Chen, Y., Wang, Z., Ye, H., Liu, S., Nguyen, H.T., Lam, H.M. and Siddique, K.H. 2022. Root physiology and morphology of soybean in relation to stress tolerance. In *Advances in botanical research*, Vol. 102, pp.77–103. Academic Press.

Chernov, T.I. and Semenov, M.V. 2021. Management of soil microbial communities: Opportunities and prospects (a review). *Eurasian Soil Science*, 54, pp.1888–1902. https://doi.org/10.1134/S1064229321120024

Clarke, V.C., Loughlin, P.C., Day, D.A. and Smith, P.M. 2014. Transport processes of the legume symbiosome membrane. *Frontiers in Plant Science*, 5, p.699. https://doi.org/10.3389/fpls.2014.00699

Crang, N. 2019. *Investigating the links between N_2 fixation and metabolism in rhizobia* (Doctoral dissertation, University of Oxford).

Dalla Via, V., Traubenik, S., Rivero, C., Aguilar, O.M., Zanetti, M.E. and Blanco, F.A. 2017. The monomeric GTPase RabA2 is required for progression and maintenance of membrane integrity of infection threads during root nodule symbiosis. *Plant Molecular Biology*, 93, pp.549–562. https://doi.org/10.1007/s11103-016-0581-5

Deng, L., Zhao, S., Yang, G., Zhu, S., Tian, J. and Wang, X. 2022. Soybean GmSUT1 transporter participates in sucrose transport to nodules during rhizobial symbiosis. *Plant Growth Regulation*, pp.1–11. https://doi.org/10.1007/s10725-021-00764-y

Dolgikh, E.A., Kirienko, A.N., Leppyanen, I.V. and Dolgikh, A. 2016. Role of phytohormones in the control of symbiotic nodule development in legume plants. I. Cytokinins. *Sel'skokhozyaistvennaya biologiya*, 3, pp.285–298. Russian. https://doi.org/10.15389/agrobiology.2016.3.285rus

Dong, W. and Song, Y. 2020. The significance of flavonoids in the process of biological nitrogen fixation. *International Journal of Molecular Sciences*, 21(16), p.5926. https://doi.org/10.3390/ijms21165926

Fabre, S., Gully, D., Poitout, A., Patrel, D., Arrighi, J.F., Giraud, E., Czernic, P. and Cartieaux, F. 2015. Nod factor-independent nodulation in *Aeschynomene evenia* required the common plant-microbe symbiotic toolkit. *Plant Physiology*, 169(4), pp.2654–2664. https://doi.org/10.1104/pp.15.01134

Fahde, S., Boughribil, S., Sijilmassi, B. and Amri, A. 2023. Rhizobia: A promising source of plant growth-promoting molecules and their non-legume interactions: Examining applications and mechanisms. *Agriculture*, 13(7), p.1279. https://doi.org/10.3390/agriculture13071279

Feiguelman, G., Fu, Y. and Yalovsky, S. 2018. ROP GTPases structure-function and signaling pathways. *Plant Physiology*, 176(1), pp.57–79. https://doi.org/10.1104/pp.17.01415

Ferreira, T.C., Aguilar, J.V., Souza, L.A., Justino, G.C., Aguiar, L.F. and Camargos, L.S. 2016. pH effects on nodulation and biological nitrogen fixation in *Calopogonium mucunoides*. *Brazilian Journal of Botany*, 39, pp.1015–1020. https://doi.org/10.1007/s40415-016-0300-0

Fields, B., Moffat, E.K., Friman, V.P. and Harrison, E. 2021. The impact of intra-specific diversity in the rhizobia–legume symbiosis. *Microbiology*, 167(4), p.001051. https://doi.org/10.1099/mic.0.001051

Friel, C.A. and Friesen, M.L. 2019. Legumes modulate allocation to rhizobial nitrogen fixation in response to factorial light and nitrogen manipulation. *Frontiers in Plant Science*, 10, p.1316. https://doi.org/10.3389/fpls.2019.01316

Gao, J.P., Liang, W., Liu, C.W., Xie, F. and Murray, J.D. 2024. Unraveling the rhizobial infection thread. *Journal of Experimental Botany*, 75(8), pp.2235–2245. https://doi.org/10.1093/jxb/erae017

Gao, Z., Chen, Z., Cui, Y., Ke, M., Xu, H., Xu, Q., Chen, J., Li, Y., Huang, L., Zhao, H. and Huang, D. 2021. GmPIN-dependent polar auxin transport is involved in soybean nodule development. *The Plant Cell*, 33(9), pp.2981–3003. https://doi.org/10.1093/plcell/koab183

Garg, N. and Geetanjali. 2009. Symbiotic nitrogen fixation in legume nodules: Process and signaling: A review. *Sustainable Agriculture*, pp.519–531.

Geiger, O., Sohlenkamp, C., Vera-Cruz, D., Medeot, D.B., Martínez-Aguilar, L., Sahonero-Canavesi, D.X., Weidner, S., Pühler, A. and López-Lara, I.M. 2021. ExoS/ChvI two-component signal-transduction system activated in the absence of bacterial phosphatidylcholine. *Frontiers in Plant Science*, 12, p.678976. https://doi.org/10.3389/fpls.2021.678976

Ghantasala, S. and Roy Choudhury, S. 2022. Nod factor perception: An integrative view of molecular communication during legume symbiosis. *Plant Molecular Biology*, 110(6), pp.485–509. https://doi.org/10.1007/s11103-022-01307-3

Gough, C. 2015. Dissecting the roles in outer and inner root cell layers of plant genes that control rhizobial infection and nodule organogenesis. In *Biological nitrogen fixation*, pp.583–592. https://doi.org/10.1002/9781119053095.ch59

Goyal, R.K. and Habtewold, J.Z. 2023. Evaluation of legume–rhizobial symbiotic interactions beyond nitrogen fixation that help the host survival and diversification in hostile environments. *Microorganisms*, 11(6), p.1454. https://doi.org/10.3390/microorganisms11061454

Goyal, R.K., Mattoo, A.K. and Schmidt, M.A. 2021a. Rhizobial–host interactions and symbiotic nitrogen fixation in legume crops toward agriculture sustainability. *Frontiers in Microbiology*, 12, p.669404. https://doi.org/10.3389/fmicb.2021.669404

Goyal, R.K., Schmidt, M.A. and Hynes, M.F. 2021b. Molecular biology in the improvement of biological nitrogen fixation by rhizobia and extending the scope to cereals. *Microorganisms*, 9(1), p.125. https://doi.org/10.3390/microorganisms9010125

Granada Agudelo, M., Ruiz, B., Capela, D. and Remigi, P. 2023. The role of microbial interactions on rhizobial fitness. *Frontiers in Plant Science*, 14, p.1277262. https://doi.org/10.3389/fpls.2023.1277262

Gresshoff, P.M., ed. 2012. *Nitrogen fixation: Achievements and objectives*. Springer Science & Business Media.

Guinel, F.C. 2015. Ethylene, a hormone at the center-stage of nodulation. *Frontiers in Plant Science*, 6, p.1121. https://doi.org/10.3389/fpls.2015.01121

Haag, A.F. and Mergaert, P. 2019. Terminal bacteroid differentiation in the *Medicago*–rhizobium interaction–a tug of war between plant and bacteria. In *The model legume Medicago truncatula*, pp.600–616. https://doi.org/10.1002/9781119409144.ch75

Hailemariam, S., Liao, C.J. and Mengiste, T. 2024. Receptor-like cytoplasmic kinases: Orchestrating plant cellular communication. *Trends in Plant Science*, 29(10), pp.1113–1130.

Harman, G.E. and Uphoff, N. 2019. Symbiotic root-endophytic soil microbes improve crop productivity and provide environmental benefits. *Scientifica*, 2019(1), p.9106395. https://doi.org/10.1155/2019/9106395

Harrison, M.J. and Ivanov, S. 2017. Exocytosis for endosymbiosis: Membrane trafficking pathways for development of symbiotic membrane compartments. *Current Opinion in Plant Biology*, 38, pp.101–108. https://doi.org/10.1016/j.pbi.2017.04.019

Hawkins, J.P. and Oresnik, I.J. 2022. The rhizobium–legume symbiosis: Co-opting successful stress management. *Frontiers in Plant Science*, 12, p.796045. https://doi.org/10.3389/fpls.2021.796045

Hennion, N., Durand, M., Vriet, C., Doidy, J., Maurousset, L., Lemoine, R. and Pourtau, N. 2019. Sugars en route to the roots. Transport, metabolism and storage within plant roots and towards microorganisms of the rhizosphere. *Physiologia Plantarum*, 165(1), pp.44–57.

Hnatuszko-Konka, K., Gerszberg, A., Weremczuk-Jeżyna, I. and Grzegorczyk-Karolak, I. 2021. Cytokinin signaling and de novo shoot organogenesis. *Genes*, 12(2), p.265. https://doi.org/10.3390/genes12020265

Igolkina, A.A., Porozov, Y.B., Chizhevskaya, E.P. and Andronov, E.E. 2018. Structural insight into the role of mutual polymorphism and conservatism in the contact zone of the NFR5-K1 heterodimer with the nod factor. *Frontiers in Plant Science*, 9, p.344. https://doi.org/10.3389/fpls.2018.00344

Jena, J., Maitra, S., Hossain, A., Pramanick, B., Gitari, H.I., Praharaj, S., Shankar, T., Palai, J.B., Rathore, A., Mandal, T.K. and Jatav, H.S. 2022. Role of legumes in cropping system for soil ecosystem improvement. In *Ecosystem services: Types, management and benefits*, p.415. Nova Science Publishers, Inc.

Jhu, M.Y. and Oldroyd, G.E. 2023. Dancing to a different tune, can we switch from chemical to biological nitrogen fixation for sustainable food security? *PLoS Biology*, 21(3), p.e3001982. https://doi.org/10.1371/journal.pbio.3001982

Kafeel, U., Jahan, U. and Khan, F.A. 2023. Role of mineral nutrients in biological nitrogen fixation. In *Sustainable plant nutrition*, pp.87–106. Academic Press.

Kang, Y., Li, M., Sinharoy, S. and Verdier, J. 2019. A snapshot of functional genetic studies in *Medicago truncatula*. In *The model legume Medicago truncatula*, pp.7–30. https://doi.org/10.1002/9781119409144.ch02

Kazmierczak, T., Yang, L., Boncompagni, E., Meilhoc, E., Frugier, F., Frendo, P., Bruand, C., Gruber, V. and Brouquisse, R. 2020. Legume nodule senescence: A coordinated death mechanism between bacteria and plant cells. *Advances in Botanical Research*, 94, pp.181–212. https://doi.org/10.1016/bs.abr.2019.09.013

Khan, M.N., Ijaz, M., Ali, Q., Ul-Allah, S., Sattar, A. and Ahmad, S. 2019. Biological nitrogen fixation in nutrient management. In *Agronomic crops: Volume 2: Management practices*, pp.127–147. Springer Singapore.

Khokhani, D., Carrera Carriel, C., Vayla, S., Irving, T.B., Stonoha-Arther, C., Keller, N.P. and Ané, J.M. 2021. Deciphering the chitin code in plant symbiosis, defense, and microbial networks. *Annual Review of Microbiology*, 75(1), pp.583–607. https://doi.org/10.1146/annurev-micro-051921-114809

Krönauer, C. and Radutoiu, S. 2021. Understanding Nod factor signalling paves the way for targeted engineering in legumes and non-legumes. *Current Opinion in Plant Biology*, 62, p.102026. https://doi.org/10.1016/j.pbi.2021.102026

Kryvoruchko, I.S., Sinharoy, S., Torres-Jerez, I., Sosso, D., Pislariu, C.I., Guan, D., Murray, J., Benedito, V.A., Frommer, W.B. and Udvardi, M.K. 2016. MtSWEET11, a nodule-specific sucrose transporter of *Medicago truncatula*. *Plant Physiology*, 171(1), pp.554–565. https://doi.org/10.1104/pp.15.01910

Kumar, N., Srivastava, P., Vishwakarma, K., Kumar, R., Kuppala, H., Maheshwari, S.K. and Vats, S. 2020. The rhizobium–plant symbiosis: State of the art. In *Plant microbe symbiosis*, pp.1–20. https://doi.org/10.1007/978-3-030-36248-5_1

Lamouche, F., Chaumeret, A., Guefrachi, I., Barrière, Q., Pierre, O., Guérard, F., Gilard, F., Giraud, E., Dessaux, Y., Gakière, B. and Timchenko, T. 2019. From intracellular bacteria to differentiated bacteroids:

Transcriptome and metabolome analysis in *Aeschynomene* nodules using the *Bradyrhizobium* sp. strain ORS285 bclA mutant. *Journal of Bacteriology*, 201(17), pp.10–1128. https://doi.org/10.1128/jb.00191-19

Ledermann, R., Schulte, C.C. and Poole, P.S. 2021. How rhizobia adapt to the nodule environment. *Journal of Bacteriology*, 203(12), pp.10–1128. https://doi.org/10.1128/jb.00539-2

Li, X., Zheng, J., Yang, Y. and Liao, H. 2018. INCREASING NODULE SIZE1 expression is required for normal rhizobial symbiosis and nodule development. *Plant Physiology*, 178(3), pp.1233–1248. https://doi.org/10.1104/pp.18.01018

Liao, Z., Zeng, H., Fan, J., Lai, Z., Zhang, C., Zhang, F., Wang, H., Cheng, M., Guo, J., Li, Z. and Wu, P. 2022. Effects of plant density, nitrogen rate and supplemental irrigation on photosynthesis, root growth, seed yield and water-nitrogen use efficiency of soybean under ridge-furrow plastic mulching. *Agricultural Water Management*, 268, p.107688. https://doi.org/10.1016/j.agwat.2022.107688

Lin, J., Frank, M. and Reid, D. 2020. No home without hormones: How plant hormones control legume nodule organogenesis. *Plant Communications*, 1(5). https://doi.org/10.1016/j.xplc.2020.100104

Lin, J., Roswanjaya, Y.P., Kohlen, W., Stougaard, J. and Reid, D. 2021. Nitrate restricts nodule organogenesis through inhibition of cytokinin biosynthesis in *Lotus japonicus*. *Nature Communications*, 12(1), p.6544. https://doi.org/10.1038/s41467-021-26820-9

Liu, A., Contador, C.A., Fan, K. and Lam, H.M. 2018a. Interaction and regulation of carbon, nitrogen, and phosphorus metabolisms in root nodules of legumes. *Frontiers in Plant Science*, 9, p.1860. https://doi.org/10.3389/fpls.2018.01860

Liu, C.W., Breakspear, A., Stacey, N., Findlay, K., Nakashima, J., Ramakrishnan, K., Liu, M., Xie, F., Endre, G., de Carvalho-Niebel, F. and Oldroyd, G.E. 2019. A protein complex required for polar growth of rhizobial infection threads. *Nature Communications*, 10(1), p.2848. https://doi.org/10.1038/s41467-019-10029-y

Liu, H., Zhang, C., Yang, J., Yu, N. and Wang, E. 2018b. Hormone modulation of legume–rhizobial symbiosis. *Journal of Integrative Plant Biology*, 60(8), pp.632–648. https://doi.org/10.1111/jipb.12653

Liu, J., Liu, M.X., Qiu, L.P. and Xie, F. 2020a. SPIKE1 activates the GTPase ROP6 to guide the polarized growth of infection threads in *Lotus japonicus*. *The Plant Cell*, 32(12), pp.3774–3791. https://doi.org/10.1105/tpc.20.00109

Liu, M. 2019. *Investigation of the role of the AP2/ERF transcription factor ERN1 in the root nodule symbiosis signaling network of* Lotus japonicus (Doctoral dessertation, SOKENDAI – The Graduate University for Advanced Studies).

Liu, S., Ratet, P. and Magne, K. 2020b. Nodule diversity, evolution, organogenesis and identity. In *Advances in Botanical Research*, Vol. 94, pp.119–148. Academic Press. https://doi.org/10.1016/bs.abr.2019.09.009

Liu, W., Feng, Y., Yu, S., Fan, Z., Li, X., Li, J. and Yin, H. 2021. The flavonoid biosynthesis network in plants. *International Journal of Molecular Sciences*, *22*(23), p.12824. https://doi.org/10.3390/ijms222312824

Magne, K. 2017. *The roles of the NOOT-BOP-COCH-LIKE genes in the symbiotic organ identity and in plant development* (Doctoral dissertation, Université Paris Saclay (COmUE)).

Mahmud, K., Makaju, S., Ibrahim, R. and Missaoui, A. 2020. Current progress in nitrogen fixing plants and microbiome research. *Plants*, 9(1), p.97. https://doi.org/10.3390/plants9010097

Maillet, F., Fournier, J., Mendis, H.C., Tadege, M., Wen, J., Ratet, P., Mysore, K.S., Gough, C. and Jones, K.M. 2020. *Sinorhizobium meliloti* succinylated high-molecular-weight succinoglycan and the *Medicago truncatula* LysM receptor-like kinase MtLYK10 participate independently in symbiotic infection. *The Plant Journal*, 102(2), pp.311–326. https://doi.org/10.1111/tpj.14625

Mathesius, U. 2022. Are legumes different? Origins and consequences of evolving nitrogen fixing symbioses. *Journal of Plant Physiology*, 276, p.153765. https://doi.org/10.1016/j.jplph.2022.153765

Mbengue, M.D., Hervé, C. and Debellé, F. 2020. Nod factor signaling in symbiotic nodulation. In *Advances in botanical research*, Vol. 94, pp.1–39. Academic Press. https://www.sciencedirect.com/science/article/pii/S0065229619300953

Mergaert, P. 2020. Differentiation of symbiotic nodule cells and their rhizobium endosymbionts. In *Advances in botanical research*, Vol. 94, pp.149–180. Academic Press. https://doi.org/10.1016/bs.abr.2019.09.008

Molla, F., Kundu, A. and DasGupta, M. 2023. Sucrose-induced auxin conjugate hydrolase restores symbiosis in a Medicago cytokinin perception mutant. *Plant Physiology*, 191(4), pp.2447–2460. https://doi.org/10.1093/plphys/kiad045

Monib, A.W., Niazi, P., Barai, S.M., Sawicka, B., Baseer, A.Q., Nikpay, A., Fahmawi, S.M.S., Singh, D., Alikhail, M. and Thea, B. 2024. Nitrogen cycling dynamics: Investigating volatilization and its interplay with N_2 fixation. *Journal for Research in Applied Sciences and Biotechnology*, 3(1), pp.17–31.

Mote, R. 2020. *From metabolites and microbes to grazing beef wellbeing and productivity: A systems biology approach to bovine fescue toxicosis* (Doctoral dissertation, University of Georgia).

Mouradi, M., Farissi, M., Bouizgaren, A., Makoudi, B., Kabbadj, A., Very, A.A., Sentenac, H., Qaddourya, A. and Ghoulam, C. 2016. Effects of water deficit on growth, nodulation and physiological and biochemical processes in *Medicago sativa*-rhizobia symbiotic association. *Arid Land Research and Management*, 30(2), pp.193–208. https://doi.org/10.1080/15324982.2015.1073194

Nakei, M.D., Venkataramana, P.B. and Ndakidemi, P.A. 2022. Soybean-nodulating rhizobia: Ecology, characterization, diversity, and growth promoting functions. *Frontiers in Sustainable Food Systems*, 6, p.824444. https://doi.org/10.3389/fsufs.2022.824444

Nelson, M.S. and Sadowsky, M.J. 2015. Secretion systems and signal exchange between nitrogen-fixing rhizobia and legumes. *Frontiers in Plant Science*, 6, p.491. https://doi.org/10.3389/fpls.2015.00491

Oelmüller, R., Tseng, Y.H. and Gandhi, A. 2023. Signals and their perception for remodelling, adjustment and repair of the plant cell wall. *International Journal of Molecular Sciences*, 24(8), p.7417. https://doi.org/10.3390/ijms24087417

Olanrewaju, O.S., Ayangbenro, A.S., Glick, B.R. and Babalola, O.O. 2019. Plant health: Feedback effect of root exudates–rhizobiome interactions. *Applied Microbiology and Biotechnology*, 103, pp.1155–1166. https://doi.org/10.1007/s00253-018-9556-6

Patra, D. and Mandal, S. 2022. Nod factors are dispensable for nodulation: A twist in *Bradyrhizobia*–legume symbiosis. *Symbiosis*, 86(1), pp.1–15. https://doi.org/10.1007/s13199-021-00826-9

Porter, S.S. and Sachs, J.L. 2020. Agriculture and the disruption of plant–microbial symbiosis. *Trends in Ecology & Evolution*, 35(5), pp.426–439.

Prasad, S., Malav, L.C., Choudhary, J., Kannojiya, S., Kundu, M., Kumar, S. and Yadav, A.N. 2021. Soil microbiomes for healthy nutrient recycling. In *Current trends in microbial biotechnology for sustainable agriculture*, pp.1–21. Springer Singapore.

Qiao, H., Zheng, F., Jiang, H. and Dong, K. 2019. The greenhouse effect of the agriculture-economic growth-renewable energy nexus: Evidence from G20 countries. *Science of the Total Environment*, 671, pp.722–731. https://doi.org/10.1016/j.scitotenv.2019.03.336

Radutoiu, S., Madsen, L.H., Madsen, E.B., Jurkiewicz, A., Fukai, E., Quistgaard, E.M., Albrektsen, A.S., James, E.K., Thirup, S. and Stougaard, J. 2007. LysM domains mediate lipochitin–oligosaccharide recognition and *Nfr* genes extend the symbiotic host range. *The EMBO Journal*, 26(17), pp.3923–3935. https://doi.org/10.1038/sj.emboj.7601826

Ramesh, R.D., Selvaraj, K., Muthusamy, K., Lakshmanan, L., Francis, S.P., Tharumasivam, S.V. and Selvakumar, V. 2023. Endurance of microbes against nitrogen starvation by altering the biochemical and physiological activities of plants. In *Plant–microbe interaction: Recent advances in molecular and biochemical approaches*, pp.33–63. Academic Press. https://doi.org/10.1016/B978-0-323-91876-3.00014-2

Rasheed, A., Barqawi, A.A., Mahmood, A., Nawaz, M., Shah, A.N., Bay, D.H., Alahdal, M.A., Hassan, M.U. and Qari, S.H. 2022. CRISPR/Cas9 is a powerful tool for precise genome editing of legume crops: A review. *Molecular Biology Reports*, 49(6), pp.5595–5609. https://doi.org/10.1007/s11033-022-07529-4

Raza, A., Zahra, N., Hafeez, M.B., Ahmad, M., Iqbal, S., Shaukat, K. and Ahmad, G. 2020. Nitrogen fixation of legumes: Biology and physiology. In *The plant family Fabaceae: Biology and physiological responses to environmental stresses*, pp.43–74. Springer Singapore.

Reid, D., Nadzieja, M., Novák, O., Heckmann, A.B., Sandal, N. and Stougaard, J. 2017. Cytokinin biosynthesis promotes cortical cell responses during nodule development. *Plant Physiology*, 175(1), pp.361–375. https://doi.org/10.1104/pp.17.00832

Reyero-Saavedra, M.D.R., Qiao, Z., Sánchez-Correa, M.D.S., Díaz-Pineda, M.E., Reyes, J.L., Covarrubias, A.A., Libault, M. and Valdés-López, O. 2017. Gene silencing of Argonaute5 negatively affects the establishment of the legume–rhizobia symbiosis. *Genes*, 8(12), p.352. https://doi.org/10.3390/genes8120352

Roy, S., Liu, W., Nandety, R.S., Crook, A., Mysore, K.S., Pislariu, C.I., Frugoli, J., Dickstein, R. and Udvardi, M.K. 2020. Celebrating 20 years of genetic discoveries in legume nodulation and symbiotic nitrogen fixation. *The Plant Cell*, 32(1), pp.15–41. www.plantcell.org/cgi/doi/10.1105/tpc.19.00279

Rudaya, E.S., Kozyulina, P.Y., Pavlova, O.A., Dolgikh, A.V., Ivanova, A.N. and Dolgikh, E.A. 2021. Regulation of the later stages of nodulation stimulated by IPD3/CYCLOPS transcription factor and cytokinin in pea *Pisum sativum* L. *Plants*, 11(1), p.56. https://doi.org/10.3390/plants11010056

Rutten, P.J. 2021. *Characterisation of oxygen regulation mechanisms in Rhizobium leguminosarum for repurposing as tools in the engineering of nitrogen fixation* (Doctoral dissertation, University of Oxford).

Rutten, P.J. and Poole, P.S. 2019. Oxygen regulatory mechanisms of nitrogen fixation in rhizobia. *Advances in Microbial Physiology*, 75, pp.325–389. https://doi.org/10.1016/bs.ampbs.2019.08.001

Rutten, P.J., Steel, H., Hood, G.A., Ramachandran, V.K., McMurtry, L., Geddes, B., Papachristodoulou, A. and Poole, P.S. 2021. Multiple sensors provide spatiotemporal oxygen regulation of gene expression in a *Rhizobium*-legume symbiosis. *PLoS Genetics*, 17(2), p.e1009099.

Saeki, K. 2011. Rhizobial measures to evade host defense strategies and endogenous threats to persistent symbiotic nitrogen fixation: A focus on two legume–*Rhizobium* model systems. *Cellular and Molecular Life Sciences*, 68, pp.1327–1339. https://doi.org/10.1007/s00018-011-0650-5

Safronova, V., Sazanova, A., Kuznetsova, I., Belimov, A., Guro, P., Karlov, D., Yuzikhin, O., Chirak, E., Verkhozina, A., Afonin, A. and Andronov, E. 2021. Increasing the legume–rhizobia symbiotic efficiency due to the synergy between commercial strains and strains isolated from relict symbiotic systems. *Agronomy*, 11(7), p.1398. https://doi.org/10.3390/agronomy11071398

Santhosh, K.S., Akhila, D.S., Dechamma, M.M., Rajeshwari, V. and Reddy, A.D. 2019. An integrative approach to understand the role of the nitrogen fixing microbial consortia in the environment. *Journal of Pharmacognosy and Phytochemistry*, 8(2S), pp.909–915.

Sauvadet, M., Trap, J., Damour, G., Plassard, C., Van den Meersche, K., Achard, R., Allinne, C., Autfray, P., Bertrand, I., Blanchart, E. and Deberdt, P. 2021. Agroecosystem diversification with legumes or non-legumes improves differently soil fertility according to soil type. *Science of the Total Environment*, 795, p.148934. https://doi.org/10.1016/j.scitotenv.2021.148934

Schulte, C. 2021. *Modelling rhizobial metabolism in Rhizobium–legume symbioses* (Doctoral dissertation, University of Oxford).

Schulte, C.C., Borah, K., Wheatley, R.M., Terpolilli, J.J., Saalbach, G., Crang, N., de Groot, D.H., Ratcliffe, R.G., Kruger, N.J., Papachristodoulou, A. and Poole, P.S. 2021. Metabolic constraints on nitrogen fixation by rhizobia in legume nodules. *bioRxiv*, pp.2021-02. https://doi.org/10.1101/2021.02.16.431433

Shah, A. and Smith, D.L. 2020. Flavonoids in agriculture: Chemistry and roles in, biotic and abiotic stress responses, and microbial associations. *Agronomy*, *10*(8), p.1209. https://doi.org/10.3390/agronomy10081209

Shaharoona, B., Imran, M., Arshad, M. and Khalid, A. 2011. Manipulation of ethylene synthesis in roots through bacterial ACC deaminase for improving nodulation in legumes. *Critical Reviews in Plant Sciences*, 30(3), pp.279–291. https://doi.org/10.1080/07352689.2011.572058

Sharma, V., Bhattacharyya, S., Kumar, R., Kumar, A., Ibañez, F., Wang, J., Guo, B., Sudini, H.K., Gopalakrishnan, S., DasGupta, M. and Varshney, R.K. 2020. Molecular basis of root nodule symbiosis between *Bradyrhizobium* and 'crack-entry' legume groundnut (*Arachis hypogaea* L.). *Plants*, 9(2), p.276. https://doi.org/10.3390/plants9020276

Shimamura, M., Kumaki, T., Hashimoto, S., Saeki, K., Ayabe, S.I., Higashitani, A., Akashi, T., Sato, S. and Aoki, T. 2022. Phenolic acids induce nod factor production in *Lotus japonicus*–Mesorhizobium symbiosis. *Microbes and Environments*, 37(1), p.ME21094. https://doi.org/10.1264/jsme2.ME21094

Shinwari, Z.K., Tanveer, F. and Iqrar, I. 2019. Role of microbes in plant health, disease management, and abiotic stress management. In *Microbiome in plant health and disease: Challenges and opportunities*, pp.231–250. Springer Singapore.

Shumilina, J., Soboleva, A., Abakumov, E., Shtark, O.Y., Zhukov, V.A. and Frolov, A. 2023. Signaling in legume–rhizobia symbiosis. *International Journal of Molecular Sciences*, 24(24), p.17397. https://doi.org/10.3390/ijms242417397

Singh, A., Singh, N.B., Yadav, V., Bano, C., Khare, S. and Yadav, R.K. 2021a. Nod factor signaling in legume–*Rhizobium* symbiosis: Specificity and molecular genetics of nod factor signaling. In *Abiotic stress and legumes*, pp.33–67. Academic Press. https://doi.org/10.1016/B978-0-12-815355-0.00003-5

Singh, J. and Verma, P.K. 2023. Role of Nod factor receptors and its allies involved in nitrogen fixation. *Planta*, 257(3), p.54. https://doi.org/10.1007/s00425-023-04090-7

Singh, R.K., Prasad, A., Muthamilarasan, M., Parida, S.K. and Prasad, M. 2020. Breeding and biotechnological interventions for trait improvement: Status and prospects. *Planta*, 252, pp.1–18. https://doi.org/10.1007/s00425-020-03465-4

Singh, V.P., Singh, S., Tripathi, D.K., Prasad, S.M., Bhardwaj, R. and Chauhan, D.K., eds. 2021b. *Abiotic stress and legumes: Tolerance and management*. Academic Press.

Smit, P., Limpens, E., Geurts, R., Fedorova, E., Dolgikh, E., Gough, C. and Bisseling, T. 2007. Medicago LYK3, an entry receptor in rhizobial nodulation factor signaling. *Plant Physiology*, 145(1), pp.183–191. https://doi.org/10.1104/pp.107.100495

Soumare, A., Diedhiou, A.G., Thuita, M., Hafidi, M., Ouhdouch, Y., Gopalakrishnan, S. and Kouisni, L. 2020. Exploiting biological nitrogen fixation: A route towards a sustainable agriculture. *Plants*, 9(8), p.1011. https://doi.org/10.3390/plants9081011

Stambulska, U.Y. and Bayliak, M.M. 2020. Legume-rhizobium symbiosis: Secondary metabolites, free radical processes, and effects of heavy metals. In *Co-evolution of secondary metabolites*, pp.291–322. Springer International Publishing.

Torres, M.A. and Stacey, G. 2003. Physiology and genetics nitrogen fixation. In *Advances in plant physiology*, Vol. 6, p.413. Scientific Publisher (Jodhpur).

Toth, T.D. 2022. *Engineering plant–microbe communication for synthetic symbioses* (Doctoral dissertation, Massachusetts Institute of Technology).

Tsiknia, M., Tsikou, D., Papadopoulou, K.K. and Ehaliotis, C. 2021. Multi-species relationships in legume roots: From pairwise legume–symbiont interactions to the plant–microbiome–soil continuum. *FEMS Microbiology Ecology*, 97(2), p.fiaa222. https://doi.org/10.1093/femsec/fiaa222

Tsyganova, A.V., Brewin, N.J. and Tsyganov, V.E. 2021. Structure and development of the legume-rhizobial symbiotic interface in infection threads. *Cells*, 10(5), p.1050.

Tsyganova, A.V. and Tsyganov, V.E.E. 2018. Plant genetic control over infection thread development during legume–*Rhizobium* symbiosis. In *Symbiosis*; Rigobelo, E.C., Ed., pp.23–52; London, UK: IntechOpen.

Van Zeijl, A. 2017. *Dissecting hormonal pathways in nitrogen-fixing rhizobium symbioses* (Doctoral dissertation, Wageningen University and Research).

Venado, R.E., Liang, J. and Marín, M. 2020. Rhizobia infection, a journey to the inside of plant cells. In *Advances in botanical research*, Vol. 94, pp.97–118. Academic Press. https://doi.org/10.1016/bs.abr.2019.09.007

Wang, D., Dong, W., Murray, J. and Wang, E. 2022a. Innovation and appropriation in mycorrhizal and rhizobial symbioses. *The Plant Cell*, 34(5), pp.1573–1599. https://doi.org/10.1093/plcell/koac039

Wang, E.T., Tian, C.F., Chen, W.F., Young, J.P.W., Chen, W.X. and Wang, E.T. 2019. Symbiosis between rhizobia and legumes. In *Ecology and evolution of rhizobia: Principles and applications*, pp.3–19. Springer Singapore.

Wang, L., Chen, M., Lam, P.Y., Dini-Andreote, F., Dai, L. and Wei, Z. 2022b. Multifaceted roles of flavonoids mediating plant–microbe interactions. *Microbiome*, 10(1), p.233. https://doi.org/10.1186/s40168-022-01420-x

Wardhani, T.A.K. 2020. *Divergence of rhizobium-induced cytokinin signalling in nodulating species* (Doctoral dissertation, Wageningen University and Research).

Witchel, S.F. 2021. *Oscar Escobar| Nursen Gurtunca| Pushpa Viswanathan|. Zitelli and Davis' atlas of pediatric physical diagnosis: Zitelli and Davis' atlas of pediatric physical diagnosis*, E-Book. Elsevier Health Sciences.

Wong, J.E., Nadzieja, M., Madsen, L.H., Bücherl, C.A., Dam, S., Sandal, N.N., Couto, D., Derbyshire, P., Uldum-Berentsen, M., Schroeder, S. and Schwämmle, V. 2019. A *Lotus japonicus* cytoplasmic kinase connects Nod factor perception by the NFR5 LysM receptor to nodulation. *Proceedings of the National Academy of Sciences*, 116(28), pp.14339–14348.

Wu, Q., Yang, L., Liang, H., Liu, M., Chen, Y., Chen, D. and Shen, P. 2024. Impacts of soil compaction and phosphorus levels on the dynamics of phosphate-solubilizing and nitrogen-fixing bacteria in the peanut rhizosphere. *Agronomy*, 14(9), p.1971. https://doi.org/10.3390/agronomy14091971

Xiao, T.T. 2015. *Root and nodule: Lateral organ development in* N_2 *-fixing plants*. Wageningen University and Research. https://edepot.wur.nl/340643

Xing, S., Zheng, W., An, F., Huang, L., Yang, X., Zeng, S., Li, N., Ouenzar, K., Yu, L. and Luo, L. 2022. Transcription regulation of cell cycle regulatory genes mediated by NtrX to affect *Sinorhizobium meliloti* cell division. *Genes*, 13(6), p.1066. https://doi.org/10.3390/genes13061066

Yang, H. and Wang, E. 2023. Dynamic regulation of symbiotic signal perception in legumes. *Science Bulletin*, 68(7), pp.670–673. https://doi.org/10.1016/j.scib.2023.03.028 2095-9273

Yang, J., Lan, L., Jin, Y., Yu, N., Wang, D. and Wang, E. 2022. Mechanisms underlying legume–rhizobium symbioses. *Journal of Integrative Plant Biology*, 64(2), pp.244–267. https://doi.org/10.1111/jipb.13207

Yang, L. 2020. *Regulation of nodule senescence in Medicago truncatula* (Doctoral dissertation, Université Côte d'Azur).

Zhang, S., Wang, T., Lima, R.M., Pettkó-Szandtner, A., Kereszt, A., Downie, J.A. and Kondorosi, E. 2023. Widely conserved AHL transcription factors are essential for NCR gene expression and nodule development in Medicago. *Nature Plants*, 9(2), pp.280–288.

Zhou, N., Li, X., Zheng, Z., Liu, J., Downie, A. and Xie, F. 2024. RinRK1 promotes accumulation of Nod factor receptors in nanodomain-like structures at the tip of infected root hairs. *bioRxiv*, pp.2024-02. https://doi.org/10.1101/2024.02.23.581682

Zhou, Y., Zhu, H. and Yao, Q. 2017. Improving soil fertility and soil functioning in cover cropped agroecosystems with symbiotic microbes. In *Agro-environmental sustainability: Volume 1: Managing crop health*, pp.149–171. https://doi.org/10.1007/978-3-319-49724-2_8

2 Advanced Techniques in Studying Plant–Microbe Interactions

Ritu Chaudhary, Sana Basri, and Neha Chaurasiya

2.1 INTRODUCTION

Among the most critical issues facing farming today is how to increase crop productivity effectively while maintaining food availability for a global populace that is expanding quickly. Furthermore, due to the escalation of adverse environmental conditions like heat, salt, and drought, global warming poses a serious danger to agricultural output. These variables have negatively influenced agrarian production, especially in the Asian subcontinent, which emphasizes the requirement for creative ways to increase agricultural yields. Because they are so important to plant wellness and illness control, an awareness of microorganism–plant relationships is necessary (Swarupa et al., 2016; Kumar et al., 2021a; Ramlal et al., 2023). During their lifespans, plants in their native surroundings engage in countless relationships with different microbes. Comprehending the dynamics between hosts and pathogens is essential as it offers a comprehension of the physiology of the host, one or more infections, and the cause of illness. To survive, all living things form relationships with their neighbours, which are made possible by the chemical interaction between the microorganisms and their hosts. In order to interact successfully with other rhizosphere species, plant roots emit chemicals during those interactions. Figure 2.1 illustrates the idea of plant–pathogen relationships. Being the site of most plant–microbe relationships, the rhizosphere's microorganisms are very significant. These associations can generally be categorized as impartial, harmful, or advantageous. Although many microorganisms have little influence on plants, they can nonetheless be quite crucial because they can produce energy from organic molecules produced from plants, help with the cycling of nutrients, and alter the microenvironment in which plants grow. There are many different places where helpful bacteria may be found, including the plant roots, soil, surfaces, and even the insides of plant tissues, where it can be challenging to discern between good and harmful microorganisms. However, helpful bacteria significantly improve the uptake of nutrients, generate growth promoters, and create substances that impede pathogens, all of which contribute to the well-being of plants. On the other hand, pathogenic microorganisms, including viruses, bacteria, fungi, and oomycetes, are known for their detrimental impacts and frequently cause plant illnesses. Additional abiotic stressors that affect the results of these relationships between plants and microbes include salt, temperature swings, drought, and acidity of the water pooling and soil (Swarupa et al., 2016). Plant–microbe interactions represent a complex and ongoing relationship that has existed since plants first colonized Earth. Together, these interactions have produced a diversified community of host and nonhost species over millions of years, which is referred to as a "holobiont" (Dolatabadian, 2020). In the ongoing struggle that persists between viruses and plants, both parties are constantly adjusting to surpass each other in the face of an environment that is continually changing (Swarupa et al., 2016; Kumar et al., 2021a; Ramlal et al., 2023). Microorganisms are found within crops and underneath the ground and can have beneficial or detrimental impacts on plants. Plants are natively home to various microorganisms (Imam et al., 2016; Sharma et al., 2020). Whenever plant defense is either retarded or is not stimulated,

DOI: 10.1201/9781003608080-2

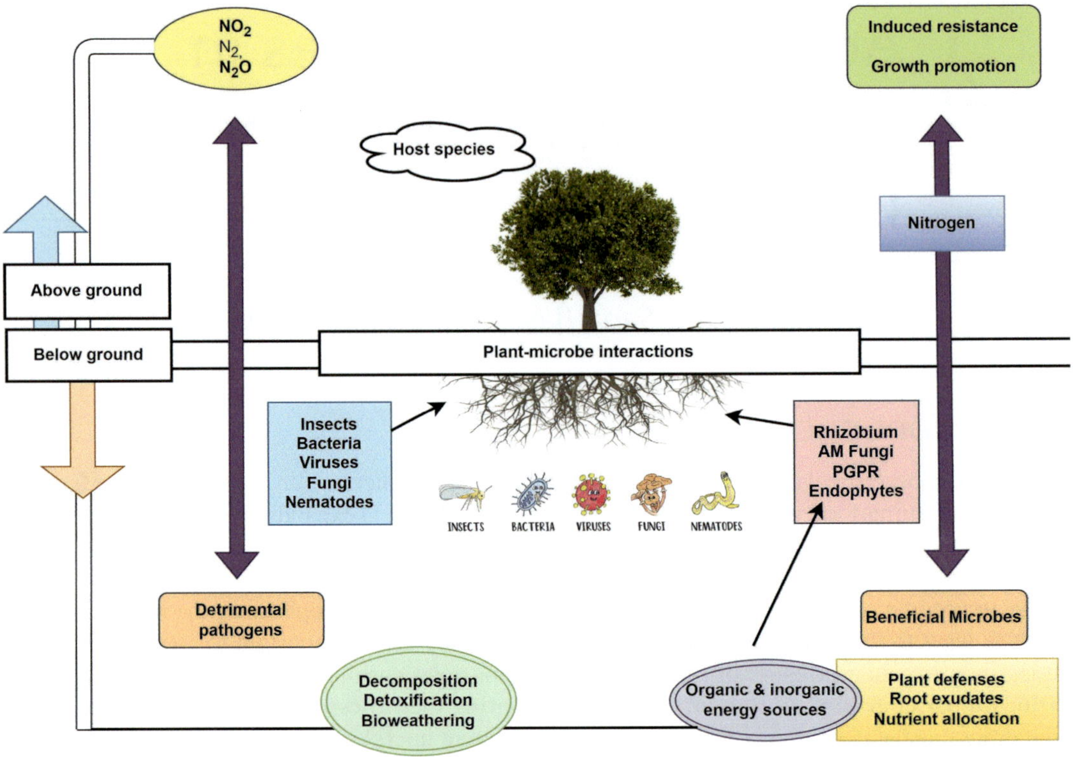

FIGURE 2.1 The complicated web of relationships among microbes and plant roots at the root–soil junction is explained by the model known as "plant–microbe interactions". Microbes that are advantageous use an array of processes, such as enhanced nutrient collection, the generation of growth regulators, and synthesizing compounds that hinder pathogens from encouraging plant development and suppressing plant illnesses.

several bacteria cause illnesses that significantly reduce the production and output of economically relevant crops, resulting in substantial financial losses (Macho and Zipfel, 2014; Ellouze et al., 2020; Gorshkov and Tsers, 2022). There is significant variation in the ways that bacteria and plants associate. Some microorganisms and plants build positive partnerships, including those that function as biological control agents (Kumar et al., 2021b) or form arbuscular mycorrhizal associations (Mishra et al., 2018; Ellouze et al., 2018). In contrast, other plants and microorganisms play destructive roles or stay inert. Both advantageous and detrimental microorganisms, including bacteria and fungi, are commonly encountered by plants in various environments, comprising agricultural and untouched surroundings. Beneficial interactions take many forms. For example, rhizobia and mycorrhizal fungi enable the exchange of nutrients, which improves plants' availability of nitrogen and other vital minerals. These interactions can also shield plants from infections, help them adapt to environmental stressors, and stimulate plant development by producing phytohormones. On the other hand, when pathogenic microorganisms infiltrate plants, detrimental interactions can result in illness and tissue damage (Dolatabadian, 2020). Plant–microbe relationships have long been a fascinating and dynamic field of research, but little is known about how complicated these relationships are and how little is yet unknown (Ayyagari et al., 2023). Knowing these relationships between microorganisms and plants is essential for learning how microbes may benefit or impede plant health. With the advancement of fast-paced sequencing of the genomes of microorganisms and plants, novel techniques have created fresh possibilities for investigating the processes underpinning the detection, signal transduction, and development of symbiosis as a plant defensive response by bacteria (Dolatabadian, 2020).

The goal of ongoing research is to elucidate how plants discriminate between harmful and helpful microorganisms and how these interactions affect the biological functions of plants. Numerous experimental techniques have been used to clarify and investigate the positive relationships between plants and microbes. While few techniques concentrate on altering the crop, others examine the biochemistry and physiology of the beneficial plant growth-promoting bacteria (PGPB). Presently, we summarize some of the most recent methods used to investigate how plants communicate with PGPB. These methods include investigating plant microbiomes and deciphering genes using DNA genome sequencing expressed by PGPB; investigating PGPB and its constituent parts using proteomics, metabolomics, transcriptomics, and gene profiling of plants; growth-promoting bacteria genome editing and cultures encapsulated before usage to cure plants; PGPB nitrogenase tests; plant and PGPB imaging; and the application of specialized chambers of development for the cultivation and observation of plants contaminated with bacteria (Gamalero et al., 2022).

2.2 HIGH-THROUGHPUT SEQUENCING TECHNOLOGIES

Microbes and plants have long been useful partners in promoting different aspects of life. The classic examples of plant–microbe-mediated function include basic processes such as bioremediation, N_2 fixation, nutrient mobilization, and different biotic and abiotic stress management (Singh et al., 2011; Vimal et al., 2017). The field of emerging techniques is required to study and understand the aforementioned basic mechanisms to develop methods for improving the techniques for the good of humanity. Conventional techniques certainly offer the initial and basic elements, but more sophisticated techniques are needed for a deeper approach. So, we are fortunate for the most recent "omics" technology, which helped us examine the genome of an organism. Following the first publication of a novel DNA sequencing technique by Sanger et al. (1977), which is based on polymerase terminating the DDNTP chain during DNA replication, numerous attempts were undertaken to increase the efficacy of techniques. Maxam and Gilbert (1977), again in the same year, presented a marginally novel technique based on nucleotide cleavage and chemical modification. Nevertheless, compared to single fragment sequencing in Sanger's methods, next-generation sequencing (NGS) provides numerous fragments of parallel sequencing every cycle run. In addition to this, "omics-driven methodologies" also facilitate the comprehension of the plant–microbe interactions in terms of disease existence and symbiotic relationship (King et al., 2019). In addition to substantially advancing traditional omics techniques and the sequencing of millions of DNA fragments, NGS also surely shortened the entire process, total duration, and cost. The field of "Omics", which originated from the collaborative analysis concept, contributes to the molecular unraveling of intricate plant–microbe interactions (Diwan et al., 2022). In order to comprehend the intricate relationship between plans and the corresponding microorganisms, the application of multi-omics approach is crucial in exploring the molecular pathways and deciphering the functionality behind them (Doni et al., 2022). Identification of important regulatory networks, metabolic pathways, and molecular markers implicated between the host and microbes is made possible by various integrated multi-omics methods (López-Mondéjar et al., 2017). Despite some knowledge gaps in this aspect of plant–microbe interactions, more research has recently started to use an omics approach to examine other microbiome traits including activity and functional capability.

2.2.1 METAGENOMICS

"Metagenomics", a popular omic approach, is the process of analyzing environmental sample genomes. Metagenomic investigation with NGS revealed that only approximately 5% of bacteria can be cultured using current techniques, indicating that many microorganisms, particularly bacteria, remain undiscovered and have unidentified activities (Braga et al., 2016). Additionally, 16S recombinant DNA sequencing from the surface of leaves revealed that environmental conditions have a greater impact on the composition of the microbial community than plant species. These metagenomic research not only assigned functional attributes to microbes but also identified a number of viruses that are related to plants yet have no harmful impacts. Our understanding of the molecular underpinnings of the interactions between plants and microbes, including plant genotype and environmental factors influencing microbiome composition, organization, and evolution, has greatly increased with the availability of large-scale plant–microbe genetic databases (Levy et al., 2018a). Targeted or shotgun sequencing-derived high-throughput screening (HTS) datasets have revealed more details about the makeup and roles of the core plant microbiomes, including the existence of beneficial and pathogenic microbes (Hu et al., 2020). Another effective method for determining gene function in plant–microbe relationships is transposon sequencing or Tn-Seq (Cole et al., 2017). Rhizosphere structure and function can be changed by applying genome-wide association (GWAS) study and metagenome investigations to the identification of particular microbiological taxa and prospective plant genes. While it is still difficult to identify the primary plant loci responsible for the recruitment and regulation of microbiota, current research has made an effort in this direction (Bergelson et al., 2019; Trivedi et al., 2020; Escudero-Martinez et al., 2022).

Since the genetic interactions between different kingdoms of plants and microbes are highly complex, multi-omics techniques are necessary, and further study is required to determine the molecular basis of symbiosis or the pathogenicity of the bacteria that associate with plants. These integrative methods provide a way to more precisely identify potential genes implicated in plant–microbe interactions. Using multi-omics techniques, several genes linked to plant adaptation have been found in various plant-associated microbiota.

2.2.2 Metatranscriptomics

Metatranscriptomics is the study of how functional genes are expressed in given circumstances to distinguish the microbiome's ability to support the host under stress. The Illumina Genome Analyzer IIx and the Roche Genome Sequencer FLX system are two popular NGS systems for meta-transcriptomics research (Schenk et al., 2012). Whole transcriptome sequencing can now be used to study and find isoforms, new transcripts, alternative splice variants, and, consequently, genomic variants because of high-throughput sequencing technology; moreover, bioinformatics offers the additional benefit of concurrent data analysis, which is utilized to carry out computational research for a more detailed knowledge of the protein–protein interaction modulators (PPI M) (Aragona et al., 2022). Moreover, studies have shown that metatranscriptomics of enriched microbial RNA enables the monitoring of crop microbiomes and transcriptional activity, hence contributing to plant health maintenance (Haveman et al., 2021). The metatranscriptome data for most plants shared a similar pattern of expression for most genes (about 80%) in response to fungal inocula at 12–96 hours post-inoculation, uncovering transcriptional plasticity of the distinctly responding genes. This is an extremely interesting finding from a thorough study (Guo et al., 2021). Additionally, their findings demonstrate how differences in plant responses to pathogens and fungal endophytes fine-tune gene regulation, a highly desired conclusion for a metatranscriptomic investigation. Metatranscriptomics has been utilized to interpret interactions between plants and microbes in order to comprehend the functions of exudates from roots in influencing the microbiome of plants and to provide insight into the transcriptional temporal dynamics involved in the modulation of the host immune system, particularly under biotic stress (Hu et al., 2018; Chai and Schachtman, 2022). However, active and passive members of the microbial community may be precisely identified and their transcribed pathways of metabolism quantified using metatranscriptomics techniques. Thus, combining metatranscriptomics and metaproteomics in the research on plant–microbe interactions can advance the understanding of gene and protein functions down to the species level. Previous research has emphasized the chemical and physical mechanisms bacteria employ to invade plant roots (Antolín-Llovera et al., 2012; Bi et al., 2018). One important finding was the identification of a transcription factor (MYC2) that, depending on the light conditions present, regulates a complicated network involving salicylic acid, jasmonic acid, and gibberellic acid to set up either microbiome-induced protection or proliferation. Radial cell layers comprise plant roots, and the functional significance of this concentric arrangement is still unclear. RNA sequencing and translating ribosome affinity purification sequencing (TRAP-Seq) are utilized to evaluate the expression of different cell-type-tagged ribosomes. This method enabled the recovery of mRNA bound to ribosomes for use in integrated omics investigations of plant microbiomes to obtain cell layer translatomes (Salas-González et al., 2021; Kawa and Brady, 2022). Fröschel et al. (2020).

2.2.3 Case Studies and Applications

According to a thorough investigation by Huang et al. (2019), *Arabidopsis* plants generate specialized triterpenes that regulate and facilitate the assembly of the microbiota living in the roots of the plants. The scientists discovered distinct metabolic networks composed of various triterpene biosynthesis gene clusters that are expressed exclusively in the roots of *Arabidopsis* through the use of metabolomics, metagenomics, and mutant lines. They observed that, in contrast to the wild type, the mutant lines with defective triterpene biosynthesis gene clusters harbored distinct microbiota. The direct functions that root exudates play in the construction of the root microbiota are demonstrated in this study, which also presents prospects to apply integrated omics in microbiota engineering to enhance plant growth and health.

Plants have developed many kinds of defense systems in response to the diverse ways in which viruses attack their hosts. One well-known substance for infection signaling is salicylic acid. Dieryckx et al. (2015) discussed the potential effects of salicylic acid and methyl salicylic acid on the growth of *Botrytis cinerea* as a model organism in the context of comparative proteomics. The results of this study indicate that salicylic acid has the potential to be a phytohormone that can fight disease since two sulfosalicylic acid derivatives, methylsalicylic and acetylsalicylic acid, were discovered to cause significant metabolic changes in cells exposed to salicylic acid. Salicylic acid, zeatin, and jasmonic acid were established as biomarker phytohormones, along with hallmark molecules such aromatic amino acids, flavonoids, and products of the phenylpropanoid pathway and shikimic acid. Due to the incapacity of plants to regulate the outnumbering transcripts of plants over those of bacteria, transcriptomics studies of plant-associated bacteria have often been conducted independently from plants.

2.3 "OMICS" APPROACH IN PLANT–MICROBE INTERACTION

Although parallel methods for profiling enable concurrent evaluation of multiple genes, proteins, transcripts, or metabolites, and the application of molecular profiling methods in plant biological development and interactions with microbes is still in its infancy, they significantly impact our knowledge of plant biology. The investigation of intricate worldwide biological networks that underpin a variety of plant functionalities has been made possible by "omics"-based methodologies. These high-tech advancements have also sped up the production of broad-scale genomic assets for burgeoning and established ideal crops, enabling applied research by bridging information between various plant–microbe relationships (Swarupa et al., 2016).

Our comprehension of plant–microbe interactions has been dramatically improved by omics technological advances, such as transcriptomics, proteomics, metabolomics, and genomics (Jain et al., 2024). Thanks to these state-of-the-art instruments, scientists can decipher the complex molecular mechanisms controlling plant symbiosis, pathogen defense, and nutrient uptake. Advanced sequencing methods have shown the genetic underpinnings of symbiotic relationships by enabling thorough mapping of microbial populations in the rhizosphere. The study of microbial genomes allows researchers to identify important genes involved in vital functions such disease resistance, phosphorus solubilization, and nitrogen fixation (Jain et al., 2024; Pang et al., 2024). Proteomic investigations have discovered important proteins mediating plant–microbe interactions, whereas metabolomic profiling has provided insight into the small molecules transferred between bacteria and plants. The creation of microbial inoculants suited to particular crop requirements is made possible by these methods, which provide a comprehensive knowledge of the biological processes underlying these interactions (Jain et al., 2024; Xie et al., 2024). These innovative techniques expand our understanding of the interactions between plants and microbes and propel precise farming forward by using microbial capabilities to improve crop sustainability and production. Table 2.1 presents a handful of current research that used "omics" techniques to analyze plant–microbe interactions.

TABLE 2.1
Some Recent Studies That Applied "Omics" Techniques to Study Plant–Microbe Relationships.

Host	Pathogen	Reference(s)
Proteomics		
Solanum lycopersicum	*Pseudomonas syringae* pv. *Tomato*	Parker et al. (2013)
Musa acuminata	*Fusarium oxysporum* f. sp. *Cubense*	Li et al. (2013)
Actinidia chinensis	*Pseudomonas syringae* pv. *actinidiae*	Petriccione et al. (2013)
Vigna mungo	Mung bean yellow mosaic virus	Kundu et al. (2013)
Vitis vinifera	*Candidatus Phytoplasma vitis*	Margaria et al. (2013)
Hordeum vulgare	*Puccinia hordei*	Bernardo et al. (2012)
Metabolomics		
Triticum	*Fusarium graminearum*	Pasquet et al. (2014)
Glycine max	*Fusarium tucumaniae*	Scandiani et al. (2015)
Nicotiana tabacum	*Phytophthora nicotianae*	Ibáñez et al. (2010)
Asparagus officinalis	*Fusarium proliferatum*	Waśkiewicz et al. (2013)
Oryza sativa	*Magnaporthe grisea*	Jones et al. (2011)
Arabidopsis thaliana	*Pseudomonas syringae* pv. *Tomato*	Allwood et al. (2010)
Glycine max	*Rhizoctonia solani*	Aliferis et al. (2014)
Hordeum vulgare	*Fusarium graminearum*	Kumaraswamy et al. (2011)
Vitis vinifera	*Botrytis cinerea*	Hong et al. (2012)
Nicotiana benthamiana	*Pseudomonas syringae* pv. *Tabaci*	Lee et al. (2013)
Nicotiana tabacum	*Phytophthora nicotianae*	Ibáñez et al. (2010)
Transcriptomics		
Pisum sativum	*Sclerotinia sclerotiorum*	Zhuang et al. (2012)
Musa acuminata	*Mycosphaerella musicola*	Passos et al. (2013)
Solanum lycopersicum	*Phytophthora capsici*	Jupe et al. (2013)
Arabidopsis thaliana	*Hyaloperonospora arabidopsidis*	Asai et al. (2014)
Brassica napus	*Leptosphaeria maculans* 'brassicae'	Lowe et al. (2014)
Lactuca sativa	*Botrytis cinerea*	De Cremer et al. (2013)

2.3.1 Proteomics

Proteomics is the study of proteins, focusing on their abundance, transformations, and interactions with other molecules. Studies in agriculture and medicine have significantly progressed as a result. Proteins are studied using a variety of approaches, such as gel-based techniques like fluorescent two-dimensional difference gel electrophoresis (2DDIGE) and two-dimensional gel electrophoresis (2DE), as well as gel-free techniques like isobaric tags for relative and absolute quantitation (iTRAQ), multidimensional protein identification technology (MudPIT), isotope-coded affinity tags (ICAT), and mass spectrometry (MS) and matrix-assisted laser desorption/ionization time-of-flight (MALDI-TOF), which are frequently used in research. The high sensitivity of MS is one of its main benefits (Swarupa et al., 2016). Transcript levels and protein sequences can be obtained by genomics, but posttranslational modifications are not captured. The quantity of protein synthesized might vary depending on how rapidly or improperly an mRNA is translated. Therefore, transcript profiles might not provide a thorough knowledge of cellular regulation on their own. Due to this constraint, research is now primarily focused on proteins, especially when studying proteins connected to plants that may be employed in agricultural improvement initiatives (Swarupa et al., 2016). When a pathogen is encountered, a number of kinases and receptors are essential in initiating signaling cascades. As soon as a signal is perceived, kinases like MAP, calcium-dependent protein kinases (CDPKs), and receptor kinases (such as flagellin receptor FLS2) start downstream reactions, which include phosphorylation events (Mehta et al., 2008; Quirino et al., 2010). In order to obtain a systems-level comprehension of the relationships between plants and microbes, it is imperative to broaden, incorporate, and create protein databases while utilizing protein-level datasets in conjunction with genetic screening along with other large-scale high-throughput methods (Jayaraman et al., 2012). Proteomics plays a critical role in investigating protein–protein interactions (PPIs). Proteomics and metaproteomics methods allow the partially quantitative investigation of bacterial protein variations among particular ecological specimens. These methods frequently depend on liquid chromatography–tandem mass spectrometry (LC-MS/MS). In soil metaproteomics, proteins are investigated to comprehend the roles played by microbes in the functioning of the soil ecology. This helps identify the microbial species and secretory enzymes that are in charge of different metabolic processes. These methods entail several crucial processes, including the extraction, isolation, and fractionation of proteins following sample collection, as well as MS evaluation and comparison with a proteome database for recognition and functionality (Levy et al., 2018b). The research by Cheng et al. (2009) provides a noteworthy illustration of how proteomics is being utilized to enhance our comprehension of plant–bacterial relationships further. According to the researchers, the existence of canola root exudates dramatically changed the expression levels of 72 PGPB proteins. Most of these proteins were linked to the uptake and transportation of nutrients, the production of the cell membrane, and the control of translation or transcription. After a more thorough analysis, it was shown that three of the four proteins displayed notable variations in expression that significantly affected the communication between plants and bacteria. The phyllosphere metaproteome of woodland trees has been investigated using metaproteomics (Lambais et al., 2017), which revealed proteins released differentially by strains of PGPB in reaction to root secretions (Kierul et al., 2015). The proteins and organisms accountable for methane oxidation and nitrogen fixation in the fields of rice have been identified (Levy et al., 2018b; Bao et al., 2014). In recent work (Bona et al., 2019), where the bacteriome had already been characterized by metabarcoding (Novello et al., 2017), a metaproteomic technique was utilized to provide a more profound knowledge of the active bacterial population in a wine estate during effective pest control. Investigators discovered that there was a discrepancy between the most engaged participants of the ecosystem and the dominating bacterial members in both the grapevine rhizosphere and heavy soil. Combining metaproteomics and metabarcoding techniques on identical samples from the environment, two fundamental issues of microbial ecology were resolved: "Who?" and "what?" This allowed for a thorough knowledge of the bacterial population and its roles in the wine estate ecosystem.

Proteomics and metabolomics are effective methods for examining the mutualistic relationships between nitrogen-fixing bacteria and root nodules. Profound studies have thoroughly studied the variety of proteins and metabolites released by both partners throughout this connection. Secondary metabolites, including phenolic acids, flavonoids, isoprenoids, and alkaloids, are among the signaling molecules that both microbes and plants release into the soil during the early stages of their interactions. Root and microbe receptors recognize these substances, which sets off a series of morphological and physiological alterations necessary for symbiosis to succeed. Researching these signal molecules and the reactions they elicit is essential to improving farming practices. Nitrogen fixation is the most well-studied symbiotic interaction, which depends on highly synchronized signaling between the two partners. As well-established signaling metabolites and major chemoattractants, flavonoids, in particular, are essential for starting and controlling this interaction (Peters and Verma, 1990). They cause the rhizobial *nod* genes to become active. The function of flavonoids in the nodulation process is the subject of several research (Schmidt et al., 1994; Peck et al., 2006; Zhang et al., 2009). Lery et al. (2011) conducted a MS-based proteome analysis of the interaction between *Gluconacetobacter diazotrophicus* and sugarcane, revealing key molecular mechanisms underlying nitrogen fixation. They discovered an overexpression of signal cascade proteins in the sugarcane genotype SP70–1143, which is well-known for its excellent nitrogen-fixing capability. Furthermore, the effectiveness of nitrogen metabolism was demonstrated by the presence of glutamate ammonia lyase in SP70–1143 plants colonized by *G. diazotrophicus*. Additionally, they discovered nine bacterial proteins that were activated in the roots by plant signals. Similarly, when *Medicago truncatula* formed nodules in response to the interaction with *Sinorhizobium meliloti*, Van Noorden et al. (2007) discovered 131 proteins in the plant. Their research also showed that auxin treatment increased the expression of a number of redox-related proteins that are essential for the development of nodules, such as isoflavone reductase and late embryogenesis-like proteins.

Proteins are essential for signaling networks, defense responses, and cellular homeostasis during plant–microbe interactions. Proteomic analysis provides valuable information about these interactions, especially in microorganisms that cannot be cultured. While Jain et al. (2021) utilized proteomics to discover secreted proteins in *Erwinia amylovora*-induced fire blight disease in apples, Khatabi et al. (2019) used proteomics to investigate pathogen effectors and host proteins in *Phytophthora infestans*-induced illness in potatoes. By using "Prey" and "Bait" proteins, functional proteomics uses the yeast two-hybrid (Y2H) system to study PPIs. Protein connections and functions may be identified and studied, thanks to mass spectrometric data analysis programs like MultiIdent, PeptIdent, MASCOT, SEQUEST, and Sherpa, which make proteomic analysis easier. The scope of proteome research is further increased by programs like ProFound, MOWSE, MS-Fit, PepSea, PepFrag, and MS-Tag, which provide specialized analysis of MALDI-TOF and MS/MS spectra (Chandok et al., 2022). Proteomic analyses using 2D-LC/MS/MS detected important plant-related proteins in *Medicago truncatula* nodules and demonstrated that rice tissues treated with *Sinorhizobium meliloti* exhibit distinct protein expression levels. The different protein secretion patterns of the phytopathogen *Pseudomonas syringae* DC3000 and the symbiotic *Sinorhizobium meliloti* in response to root exudates from different host plants have also been clarified by comparative proteomic analysis. Furthermore, studies on *Fhb1* near-isogenic line alleles have shed important light on resistance to *Fusarium graminearum*. Research on helpful microorganisms in the pea rhizosphere has also shown that infection with the necrotrophic fungus *Sclerotinia sclerotiorum* causes distinct proteome responses. These results highlight how critical proteomic methods are for deciphering the complex molecular dynamics underlying interactions between microbes and plants (Jain et al., 2021). Comparative proteomics of the *B. cinerea*-secreted proteome identified altered proteins, mainly pectinases, that are responsible for cell wall disintegration. Additionally, the secreted proteome is dominated by serine proteases, followed by threonine proteases and metalloproteases (Sarim et al., 2020). The control system against *Magnaporthe oryzae* in both susceptible and resistant rice cultivars was identified by iTRAQ proteomics.

2.3.2 Metabolomics

As the byproduct of metabolic processes in cells, metabolomics offers detailed metabolite data and is frequently used in conjunction with other "omics" techniques (Saito and Matsuda, 2010). Metabolic profile research employs several methods, such as nuclear magnetic resonance (NMR), high-performance liquid chromatography (HPLC), and chromatography coupled with MS, which facilitates the detection of a wider variety of complex substances. Gas chromatography–mass spectrometry (GC-MS) and liquid chromatography–mass spectrometry (LC-MS) are the two most often utilized methods (Swarupa et al., 2016). One can use targeted or untargeted metabolomics to evaluate how different therapies affect the levels of particular metabolites (Levy et al., 2018b). In particular, microbial metabolism or host plants can be directly impacted by bacterial components such as nodulation (*nod*) genes, which produce nod factors (Levy et al., 2018b). To guarantee compliance with MS, specific methods of extraction suited to each type of molecule are typically needed for lipidomic and metabolomic investigations based on MS (Domon and Aebersold, 2010). Due to these complications, techniques for concurrent extraction of metabolites and lipids have been developed, allowing for the utilization of smaller sample weights or volumes. This technique increases precision and expedites the process of getting ready for extensive analysis. NMR metabolomics, which may be employed to identify pre-symptomatic indicators of illness and stress and evaluate an organism's reaction to diverse stressful circumstances, is another technique for researching soil metabolites (Nakayasu et al., 2016). This method helps discover putative markers of soil quality since it can be used for both fingerprint and targeted metabolomics. For example, Rochfort et al. (2005) employed NMR, or proton NMR spectroscopy, to analyze the metametabolome of soils from both organic and farmed sources. They characterized both exogenous and internal elements of soil by using sonication-based isolation and grinding in their metabolomic technique (Rochfort, 2005). The metabolomic analysis is less affordable than DNA sequencing since it demands higher-end instruments and more excellent technical proficiency than metagenomics. Furthermore, the amount of open-source databases that include references to individual metabolites is minimal, sometimes making it difficult to attribute observed metabolites to particular species appropriately. Notwithstanding these drawbacks, metabolomics is an effective method for finding critical metabolic pathways connected to relationships between plants and microbes and for detecting and quantifying small compounds. This method offers a thorough understanding of the intricate relationships found in both the host plants and the microbiota (Levy et al., 2018b; Gupta et al., 2022). A metabolomic technique was used to study the effects of chemicals, especially aromatic carboxylic acids, produced by root secretion from grass species such as *Avena barbata* on the building and continuity of rhizosphere communities (Zhalnina et al., 2018). In recent years, *Oryza sativa* cv. Nipponbare underwent metabolome analysis using ultra-high-performance liquid chromatography-diode array detector-quadrupole time-of-flight (UHPLC-DAD-QTOF) after inoculation with 11 plant growth-promoting bacterial strains, 8 from the genus *Azospirillum*, including *Paraburkholderia phytofirmans* PsJN, *Herbaspirillum seropedicae* SmR1, and *Burkholderia glumae* AU6208. All PGPB-inoculated plants had the same metabolic profile, identified by high levels of three hydroxycinnamic acid (HCA) derivatives (N-p-coumaroylputrescine, N-feruloylputrescine, and N-feruloylquinic acid) and low levels of three alkylresorcinols. The rice pathogen *Burkholderia glumae* AU6208 was injected into the plants by the investigators in order to examine its influence on the production of HCA derivatives, as it is well known that HCA-based compounds affect plant defense. The findings showed that, in contrast to PGPB inoculation, pathogen inoculation resulted in a reduced buildup of HCA derivatives, indicating that plants may distinguish between valuable and dangerous bacteria based on changes in the buildup of these HCA molecules (Valette et al., 2020).

The numerous plant-related processes that are supported by intricate worldwide biological networks can now be addressed, thanks to "omics"-based methodologies. By combining information across plant–microbe species, these technical advancements have also facilitated translational research and expedited the establishment of genome-scale assets for current and upcoming model plant species, as illustrated in Figure 2.2.

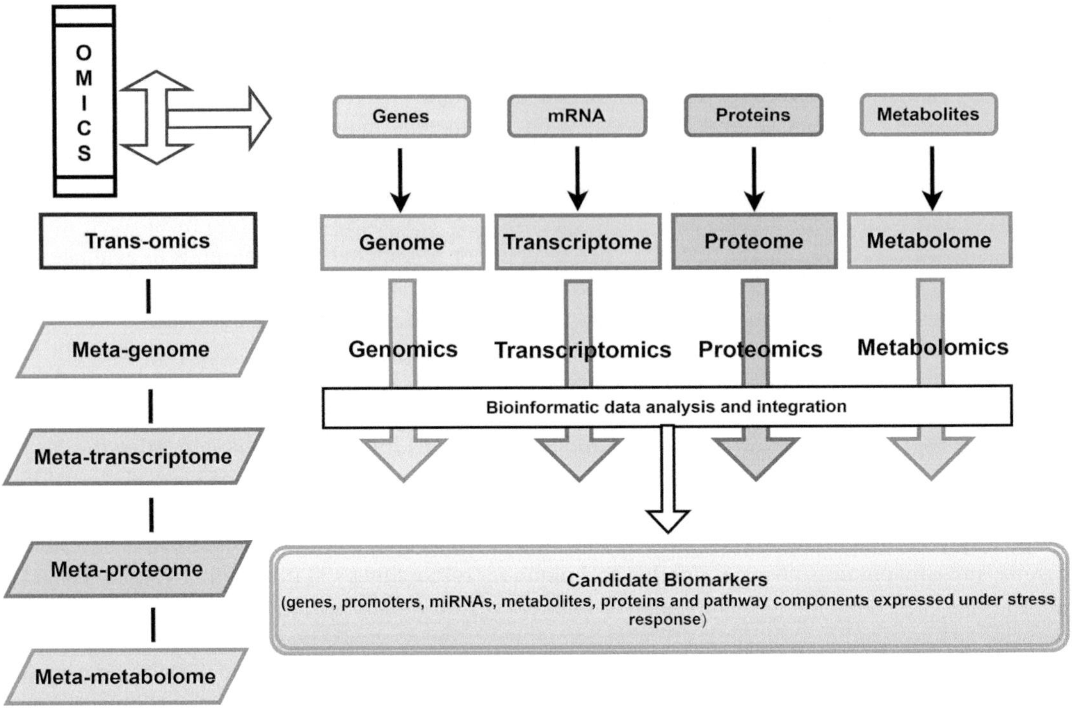

FIGURE 2.2 Plant–microbe interactions using an omics methodology. With the use of cutting-edge "omics" technology, this framework demonstrates how a coordinated strategy is used to comprehend the extent and intricacy of plant–microbe associations. These methods would yield more data regarding how certain gene products contribute to developing relationships with the host and how microbes respond to plant signals.

2.4 IMAGING TECHNIQUES

All plant–microbe interactions begin at the cellular level, and it has long been understood that these interactions can be intricate and important for determining a plant's susceptibility or resistance to microbial development. However, in circumstances where microbe–plant interactions are not simultaneous or uniform within the plant, biochemical studies of extracts from tissues fail to offer high resolution of cellular responses; in these circumstances, therefore, the ability to visually identify interactions at a single cell level becomes especially crucial. After its invention, the electron microscope was the preferred instrument for studying how microorganisms and plants interact below the cellular level. Over time, the development of numerous immunological, cytochemical, and analytical techniques to pinpoint and identify certain atoms, chemical structures, or enzyme activities within the cell has increased its usage.

The use of light microscopes (LMs) in plant study has increased significantly, despite the fact that they cannot match the electron microscope's resolution. Monitoring gene expression at the cellular level using *in situ* hybridization or applying visually detectable reporter genes like *GUS* (β-glucuronidase) is now common for the LM. More importantly, recent parallel advancements in computerized image processing, confocal laser scanning, and video microscopy, and an expanding selection of fluorescent probes that may be used on living cells have all contributed to the LM's increased versatility in plant study. When taken as a whole, these have created fresh and fascinating possibilities for viewing cellular structures and functions directly. Furthermore, the growing simplicity of creating transgenic plants and cells has made it possible to visualize fluorescently labeled cell components in cells that underwent the least amount of invasive treatment.

These novel imaging techniques provide many prospects, but they can only be used in a limited number of plant–microbe systems since they require living cells. Although they offer the best visuals, studies using protoplasts or plant cell suspension cultures might not fully capture what occurs in whole plants. Nearly all investigations involving living plant cells in intact tissues employ epidermal cells on the plant surface because most cells in intact plants are not available for direct observation. By using microparticle bombardment to generate transient gene expression systems, such cells can also be used for thorough cytological studies of the impact of the expression of putative defense genes on fungal ingress and proliferation (Bushnell et al., 1998; Schweizer et al., 1999).

With the use of imaging tools, the knowledge of the intricate relationships that exist between plants and the microbes that live within them has been transformed. The following are a few of the most popular techniques.

2.4.1 Light Microscopy

An essential tool in contemporary cell biology is light microscopy. The resolution of light microscopy is compatible to the sizes of subcellular structures; a wide variety of fluorescent probes are available to mark proteins, organelles, and additional structures for imaging; and the comparatively non-perturbing nature of light allows for long-term imaging of living cells to monitor their dynamics. These features make light microscopy an excellent tool for imaging biology in living cells (Thorn, 2016).

2.4.1.1 Confocal Laser Scanning Microscopy

Plants create intricate mosaics of various microhabitats that are inhabited and modified by microorganisms, encompassing both beneficial and pathogenic species. In a unique instance of plant–microbe interactions, pathogen virulence is determined by a confluence of ecological and genetic characteristics. Typically, infections only happen when the microbial genes that cause them are expressed, and they only affect particular plant tissues. To investigate these coordinated mechanisms and help comprehend the complexities of the infection process, confocal laser scanning microscopy (CLSM) is being employed as a tool (Haupt et al., 2006). Furthermore, without invasive experimental modification, transgenic plants and cells can be created with fluorescently labeled specified components. By using these techniques to study plant–microbe interactions, researchers have determined how microbes affect cytosolic levels of calcium, cytoskeleton rearrangements, DNA cleavage, and the comprehensive resolution of viral component intracellular and intercellular trafficking (Heath, 2000). Researchers studied the mechanism underlying the degenerative disease of *Vitis vinifera* caused by *Xylella fastidiosa*. They discovered that the bacterial cells' vessel-to-vessel migration causes this disease by using CLSM (Newman et al., 2003). Following the downstream insertion of *GFP* genes, some used confocal microscopy to investigate the modification of the expression of *vir* genes in *Agrobacterium tumefaciens* during the infection process (Li et al., 1999).

Optimizing a basic trypan blue and aniline blue dual-stain and applying it to two model pathosystems—*Phytophthora infestans*-tomato and *Pseudoperonospora cubensis*-cucumber—have resulted in a dramatic boost in confocal microscopy observation yield. The "mild" approach proved the most effective when comparing two dual-stain techniques for confocal microscopy investigations of cucumber leaves challenged with *Pseudoperonospora cubensis*. This approach enables the simultaneous identification of many pathogen structures layered with plant defense responses. Furthermore, this approach helps address host plant resistance reactions and examine the pathogenicity of the specific isolate using ImageJ's assistance in quantifying plant defense responses. Several fungal infection structures and plant defense responses were detected when this approach was used on tomato leaf samples challenged with *Phytophthora infestans*. Additionally, the dual-stain made it possible to identify a strange substance that is sensitive to aniline blue in the pathogen cell walls near the location where its hyphae emerge through the leaf stomata. The findings provided here suggest that this approach can be used for in-depth (perhaps quantitative) analyses of various plant–fungal pathosystems (Nowicki et al., 2012).

Using certain dyes to stain fungal structures is frequently necessary for the study of arbuscular mycorrhiza. Although they produce great results, fluorescent dyes like acid fuchsin and wheat germ agglutinin conjugates are either extremely costly or dangerous. The arbuscular mycorrhizal fungus (AMF) *Glomus intraradices* forms intraradical fungal structures in three plant species: carrot, *Casuarina equisetifolia*, and *Medicago truncatula*. These structures can be effectively stained using Uvitex 2B, a safer and less expensive dye. Uvitex 2B's intensity and stability enable the use of both epifluorescence microscopy in conjunction with image deconvolution and CLSM to obtain high-quality images. Additionally, in the context of promoter activation research, arbuscular mycorrhizal structures can be revealed by using Uvitex 2B and β-glucuronidase staining because they are compatible (Diagne et al., 2011).

2.4.1.2 Fluorescence Microscopy

In biology, fluorescence microscopy has emerged as a crucial tool for studying living tissues or cells. This technique is frequently used to find bacteria inside plant tissues, despite requiring more sophisticated and costly equipment than traditional transmitted-light microscopy. Fluorescence microscopy makes it possible to determine the location of fluorescent materials inserted into living cells in the past. Numerous fluorescent dyes, protein tags, and other techniques for fluorescent labeling of cells are available that can be utilized for monitoring a microbial inoculant (Romano et al., 2020). Some researchers suggested using serological techniques like immunofluorescence and the double-antibody sandwich enzyme-linked immunosorbent test to examine bioinoculant colonization behavior. Using immunofluorescence, they identified the presence of *Azotobacter chroococcum* Mac 27 L in root segments of wheat plants cultivated hydroponically (Narula et al., 2007). But one of the most popular techniques for monitoring endophytic implanted bacteria in plant tissues is the use of green fluorescence protein (GFP), which, when exposed to blue or near-ultraviolet (UV) light, produces a bright green light (Wang et al., 2015).

2.4.2 Electron Microscopy

Classical transmission electron microscopy (TEM) and scanning electron microscopy (SEM) were the principal instruments for observing microscopic creatures for many years. These tools are still essential for determining the nature of interactions between these organisms (Czymmek et al., 2023).

2.4.2.1 Scanning Electron Microscopy

SEM is frequently used to examine the surface chemistry and structure of various synthetic and biological materials at the micrometer to nanometer scale. Because of its high resolution, high depth of field, ease of use, generally simple sample preparation, and ability to perform microchemical and crystallographic analysis, SEM is one of the most effective and adaptable characterization techniques available today. SEM is, in fact, an essential instrument for the advancement of nanotechnology and the characterization of materials with nanostructures. Effective SEM imaging is nevertheless required, as seen by the widespread use of this technology by individuals with a wide range of technical backgrounds, including those in engineering, life science, forensics, materials science, mineralogy, and other areas of government, business, and academia (Ul-Hamid, 2018).

The interaction between *Gluconacetobacter diazotrophicus*, a nitrogen-fixing bacterium that does not form nodules and is found in the intercellular spaces of sugarcane, and various plants such as *Arabidopsis thaliana*, maize (*Zea mays*), rice (*Oryza sativa*), wheat (*Triticum aestivum*), oilseed rape (*Brassica napus*), tomato (*Lycopersicon esculentum*), and white clover (*Trifolium repens*) has been studied. Our research demonstrates that inoculating seedlings grown in a sugar-rich culture medium with very few *Gluconacetobacter diazotrophicus* leads to significant intracellular colonization of root meristems and gradual systemic root colonization. Observations through light microscopy of thin sections from resin-embedded root tips of *Arabidopsis* and the aforementioned crop plants, which were inoculated with GUS-labeled and NifH promoter-GUS-labeled *Gluconacetobacter diazotrophicus*, revealed blue-stained bacteria within the cytoplasm of the root cells, suggesting that the intracellular environment was conducive for the expression of nitrogenase genes. Electron microscopy showed that the blue-stained intracellular *Gluconacetobacter diazotrophicus* was located inside membrane-enclosed vesicles. Research has been conducted on the possibility of these new inoculations with *Gluconacetobacter diazotrophicus* to facilitate non-nodular endosymbiotic nitrogen fixation and on whether these inoculations might also offer a plant model to examine the endosymbiotic theory regarding the emergence of eukaryotic organelles (Cocking et al., 2006).

2.4.2.2 Transmission Electron Microscopy

TEM is a useful method for displaying the ultrastructure of tissues and cells. The resolution of the electron microscope is roughly 1,000 times higher than that of the LM, which is the primary benefit of TEM over LM. The electron microscope can now see the ultrastructure (substructure) of every cell due to its higher resolving capacity. The basic idea behind TEM is that electrons go through a material (like resin) to create an image of the specimen. Since the electron beam can only successfully penetrate a section to a depth of about 100 nm, the tissue must be sectioned to a thickness of less than 100 nm to maximize the resolution of the device and produce a high-quality image (Woods and Stirling, 2018).

2.4.3 Other Imaging Techniques

2.4.3.1 X-ray Computed Tomography

In modern times, X-ray computed tomography (CT) is frequently used as an effective method to gather intricate three-dimensional (3D) structural data (Yang et al., 2017; Earles et al., 2019). X-ray CT uses a stack of radiography pictures acquired while rotating the sample to create 3D depictions of plant organs or tissues. Benefits include minimal sample preparation, X-ray optics and photon energy, scalable field of view and resolution from less than 1 μm to 1 mm, wide penetration depth depending on photon energy, and a noninvasive nature that allows for nondestructive and live imaging. The technique of X-ray attenuation is dependent on density, and since intercellular air spaces are ubiquitous in many plant organs, this makes them particularly well-suited for plant imaging. Additionally, because many plant organs have unique shapes and microstructures, the movement of gases and liquids is fundamentally a 3D process that necessitates the collection, examination, and reconstruction of various transport paths, many of which have distinct 3D topologies (Piovesan et al., 2021).

2.4.3.2 Magnetic Resonance Imaging

Tools with a high level of temporal and geographical resolution are necessary to comprehend how plants grow, develop, and interact with their surroundings. One major advantage of magnetic resonance imaging (MRI) is that it is a noninvasive technology that can be utilized to track physiological processes occurring *in vivo*. MRI can also be used to visualize interior structures and metabolites (Borisjuk et al., 2012).

2.4.3.3 Raman Spectroscopy

In the last 15 years or more, Raman spectroscopy has become more widely recognized as a sophisticated tool that enables the quick, noninvasive characterization and identification of microorganisms. It stands out from other currently used techniques due to its low cost, convenience of use, quick analysis, and extensive information on the structure, relationships, and chemical makeup of biomolecules within microorganisms. The Raman spectra of the microorganisms show their entire molecular makeup, including a variety of vibrational modes of both species-unspecific and dependable taxonomic biomarkers (DNA/RNA, lipids, proteins, and carbohydrates) (Stöckel et al., 2016). It is an extremely accurate technique for the examination of extremely complex plant tissues because it gives spatial information on vibrations from complicated biological materials. For an extremely broad variety of chemicals, Raman spectra can be utilized as a fingerprinting method. Raman spectroscopy makes it possible to trace all of the polymers that go into making a plant's cell wall at the same time, which makes it easier to analyze the molecular structure and composition of the cell wall. This approach is utilized for comparative examinations because of its strong sensitivity to even small structural alterations (Saletnik et al., 2021).

2.4.4 Advanced Techniques

2.4.4.1 Super-Resolution Microscopy

The ability to resolve structures below the diffraction limit of light makes super-resolution microscopy techniques like photoactivated localization microscopy (PALM), stimulated emission depletion microscopy (STED), stochastic optical reconstruction microscopy (STORM), and structured illumination microscopy (SIM) promising tools for further research on membrane compartmentalization. Sadly, most super-resolution methods require completely static objects and are not suitable for live-cell imaging approaches. Since PALM, STORM, and SIM rely on repeatedly acquiring the same picture plane for later reconstruction, they are typically applied to fixed tissue samples. However, platforms for live-cell SIM, STED, and STORM have been created, and the initial effective methods show that live-cell imaging can benefit from iterative super-resolution microscopy techniques (Konrad and Ott, 2015).

Several types of imaging are involved that include components such as fluorochromes, molecular markers, and other imaging modalities. Following the release of PGPB as biocontrol or biofertilizer agents into the environment, it is important to monitor the bacteria's survival, behavior, and activity in the soil. Separating the newly introduced variant from the autochthonous microbiota is typically the first step toward monitoring bacteria in complex environments. As a result, even in vast fields, markers identifying the added strains should be stable and distinctive. Throughout the 1980s and 1990s, several molecular and serological markers were created (Gamalero et al., 2003; Ahmad et al., 2011).

Antibiotic resistance, particularly to kanamycin, rifampicin, and streptomycin, is one of the earliest and most extensively used indicators. It relies on the choosing of random chromosomal antibiotic-resistant mutants or after Tn5 mutagenesis. Once the PGPB strain has been identified as having antibiotic resistance, colonies measured on selective media containing a particular antibiotic can be used to assess the strain's density in soil or the rhizosphere. This goes without saying that this method solely monitors the strain's culturable fraction. However, it should be remembered that naturally occurring strains identified as having antibiotic resistance can present significant risks regarding the potential horizontal gene transfer that could lead to the transmission of antibiotic resistance genes in the environment. Therefore, it is not advised to intentionally release PGPB tagged with genes that resist antibiotics (Kang et al., 2017; Miller et al., 2022).

Visualizing using a fluorochrome by flow cytometry, microscopy, or bacterial cell counting is easy, affordable, and well-established. Furthermore, it is feasible to distinguish between living and dead cells by combining fluorochromes with varying degrees of membrane permeability (e.g., propidium iodide labeling cells with ruptured membranes and SYBR green staining all living cells). A method for optimizing staining techniques has been devised based on these two fluorochromes to count the number of bacterial cells within environmental samples using flow cytometry. However, fluorochromes are limited in their use because they do not enable researchers to differentiate the imported bacterial strain from the native microflora (Nescerecka et al., 2016).

Several researchers have reported studies of the root colonization dynamics of labeled PGPB using immunofluorescence, offering both qualitative and quantitative characterization of this phenomenon. Here, antibody specificity is the primary variable to take into consideration. The aforementioned limitation can be overcome by the use of strain-specific antibodies aimed contrary to the introduced bacterial strain (usually brought up against cell membrane proteins), coupled with enzymes, fluorochrome, or radioisotopes (Kragelund and Nybroe, 1996; Troxler et al., 1997; Sørensen et al., 2011; Gamalero et al., 2004a).

The usage of molecular marker genes, for instance, the one encoding GFP, obtained from the jellyfish *Aequorea victoria*, subsequently aroused a great deal of attention (Chalfie et al., 1994). Several additional fluorescent proteins, such as enhanced GFP (EGFP), enhanced cyan fluorescent protein (ECFP), and enhanced yellow fluorescent protein (EYFP), have been created since the GFP gene was isolated (Bloemberg et al., 2000). Despite the impossibility of obtaining a red variant of GFP, this problem was solved when DsRed was isolated from the coral *Discosoma striata* (Matz et al., 1999). The coupling of autofluorescent proteins and the visualization of multiple distinct metabolic processes in a single cell or in different cells within a complex system using fluorescence microscopy and CLSM were made possible by the availability of numerous proteins with varying excitation and emission wavelengths (Bloemberg et al., 2000). Complementing the data acquired by microscopic inspection is the analysis of digital images using specialized software that can yield information on the morphology, shape, number, volume, and surface area of cells. This type of examination includes the following steps: Using digital technology to capture the image; removing background noise from the matrix (which can be particularly strong in soil, roots, and leaves); and identifying and characterizing the bacterial cells that need to be examined (Daims and Wagner, 2007).

Transconjugation is a simple method for obtaining GFP-expressing bacterial cells, and commercial kits are available to facilitate this process (Tombolini et al., 1999; Tombolini et al., 1997). Constitutively expressed GFP is stable in aerobic settings, does not typically impact the host's fitness, and is not altered by the bacterial host cells' metabolic processes (Bloemberg, 2007).

Using flow cytometry, GFP-expressing cells may be visualized and quantified. For instance, the quantity of culturable, viable but nonculturable (VBNC), and whole cell count of *Pseudomonas fluorescens* 92rkG5 was found along the primary root of tomato plants cultivated under controlled conditions by combining microscopic observation, dilution plating, and flow cytometry. Nevertheless, examining bacterial biofilms, where oxygen deprivation results from bacterial metabolism, is not a good use for GFP markers (Gamalero et al., 2005). It has been suggested that the fluorescence-activating and absorption-shifting tag (FAST) tiny tag be used for this type of application. This tag is sensitive to oxygen shortage, but it becomes luminous upon the addition of an exogenous fluorescent dye (Monmeyran et al., 2018).

It is highly appealing to visualize bacterial cells at the precise moment when they are displaying a certain metabolic activity when researching rhizosphere ecology. Numerous reporter genes were developed and designed to achieve this goal (Lindow, 1995). A reporter gene is, by definition, a gene that can be inserted intentionally ahead of a gene of interest in a bacterial chromosomal structure or plasmid, giving the bacterial cells the ability to be recognized and seen with ease. The most often utilized genes in environmental ecology are *GFP* and its derivatives (which are driven by an inducible promoter), luminescence (*luxAB* and *luc*), and chromogenic (*lacZ* and *gusA*) genes (Sørensen and Nybroe, 2006) (Figure 2.3).

The luciferase system has been widely used as a reporter gene since the *lux* operon from the marine bacteria *Vibrio fisheri* was discovered and its regulation was understood. Likewise, the luciferase *luc* genes extracted from *Photinus pyralis*, the firefly (De Wet et al., 1987), and *Pyrophorus plagiophthalamus*, the luminous click beetle, may be readily expressed in bacterial cells (Wood and DeLuca, 1987). For instance, a *Pseudomonas fluorescens* strain that was isolated from the rhizosphere of a birch tree (*Betula pendula*) had a single copy of a *mer-luc* marker gene cassette inserted into its chromosome. In a microcosm system, the ability to survive off the strain in forest soil exposed to varying temperatures was evaluated by counting the total number of light-producing colonies on a medium comprising $HgCl_2$ and by amplifying the *luc* gene using PCR (Björklöf and Jørgensen, 2001).

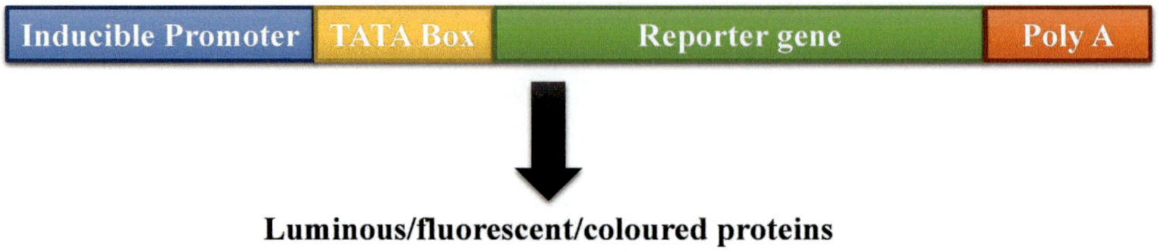

FIGURE 2.3 Typical reporter system construction. A promoter that can induce transcription typically controls the reporter gene. When the altered cells carry out a certain function, the gene expression causes them to appear fluorescent, luminous, or colored.

When grown on a medium containing the inducer isopropyl β-D-1-thiogalactopyranoside (IPTG) and substrate analog X-gal (5-bromo-4-chloro-3-indolylβ-d-galactopyranoside), bacteria expressing the *E. coli GUS* gene producing beta-glucuronidase appear blue. Comparably, the deep blue product that results from the conversion of the colorless substrate 5-bromo-4-chloro-3-indolyl-3-o-galactoside to the *lacZ* gene expressing beta-galactosidase in *E. coli* has been employed as a reporter gene in a variety of organisms (Gamalero et al., 2022).

Assessing living, active bacterial cells is a basic problem in the study of microbial ecology (Kumar and Ghosh, 2019). Using probes that target the 16S ribosomal RNA (rRNA), traditional fluorescence *in situ* hybridization (FISH) enables the visualization of live cells using fluorescent or confocal imaging (Amann et al., 1995). A standard methodology includes three steps: permeabilization (to help the fluorescent probe internalize and combine with the target), hybridization (to determine whether the cells are viable), and cell fixation (to maintain cell viability). Live bacterial cells from intricate communities can be observed using probes tailored to a particular taxonomic group. Many changes to the initial methodology have been suggested since this technique was first developed. Fluorescent *in situ* hybridization of transcript-annealing molecular beacons (FISH-TAMB) is a recent technique that uses living cells because it does not require the fixation phase. This method works well for labeling intracellular mRNA targets, distinguishing between various transcriptional stages, finding rare and active taxa, and preserving the viability of cells (Harris et al., 2022). Furthermore, the ability of the original FISH methodology to concurrently detect only a limited number of taxonomically distinct bacteria is one of its primary shortcomings. A group of researchers created a multicolor FISH method that uses probes labeled with eight fluorochromes with varying excitation and emission wavelengths. This method is effective in characterizing intricate communities of bacteria in a variety of environmental samples (Lukumbuzya et al., 2019).

FISH has been used in conjunction with other methods, including secondary ion mass spectrometry (FISH-SIMS), Raman microspectroscopy (Raman–FISH), and microautoradiography (MAR–FISH), to visualize, identify, and determine the incorporation of labeled substrates by specific microorganisms in complex microbial communities (Musat et al., 2012).

2.5 BIOINFORMATICS AND MACHINE LEARNING

2.5.1 WinRhizo System

The physical, biological, and chemical properties of soil vary greatly within a given region, making it a complicated system. Plant root development in any given area in this setting indicates how well the plant can take advantage of the inconsistent nutritional supplies. Plant roots are influenced by a variety of factors, including pH; water availability; oxygen gradient; soil temperature; the microorganisms that live inside and on the surface of roots (endophytes and epiphytes); the formation of symbioses with rhizobia or mycorrhizal fungi, or both; the arrangement of macro- and micronutrients; and the expulsion of an extensive variety of distinct compounds by the soil microflora (McBratney et al., 2014). These factors work together to shape the growth and development of a certain plant's root system. At least four fields of plant biology are dedicated to studying root system morphology (Lynch, 1995).

Root morphology examines "the exterior attributes of an individual root axis as an organ, including root hairs, root diameter, root cap, the manner of appearance of daughter roots, waves of the root axis, and cortical senescence, among other characteristics of the epidermis". "How single root axes are interconnected with each other through branching" is the subject of root topology research. The term "root distribution" describes "the existence of roots in a geographical gradient or grid", and it is associated with the biomass or length of the roots as a function of many parameters such as soil depth, distance from the stem, and spacing between adjacent plants. Finally, the "spatial configuration of the root system" is examined by root architecture (Lynch, 1995). Since root architecture is a descriptor of several axes, topology and distribution are known once we know the architecture of the root.

An essential predictor of a plant's yield and overall health, root architecture analysis is based on the significant relationship between root development and the efficiency of the absorption of xylem sap and transport, and consequently plant production. Many image analysis software tools have been developed to readily analyze root features while avoiding human error. One of the most well-known of them is WinRHIZO™ (Regent Instruments Inc., Quebec City, QC, Canada). WinRHIZO™ is made up of an image acquisition system that includes a specialized, high-quality scanner and a lighting system, together with software that can easily convert root photographs into information regarding root morphology, including overall root length, diameter, area, volume, and branching; architecture; topology; and color analyses. The same hardware and software can also be used to compute data for foliage, such as the total projected area (Arsenault et al., 1995). Although this technique is undoubtedly a significant improvement over more antiquated and imprecise manual procedures, not all laboratories will be able to afford its high cost. To prevent breaking the finer side roots in this analysis, extreme caution must be exercised when washing and handling the roots while also removing any associated substrate (soil, sand, peat, etc.). Additionally, to guarantee a high-quality study and minimize the amount of overlap between distinct root segments, the roots must be meticulously opened. Although this kind of analysis can yield extremely precise information, the process can be challenging, laborious, and time-consuming. Currently, this method is being used to examine how roots develop in various plant species, with a focus on how PGPB and AMF modify roots (Gamalero et al., 2002, 2004b, 2008, 2010; Mougel et al., 2006; Cesaro et al., 2020).

It has been suggested that in addition to monitoring root developmental features, color examination is an analytical function of WinRHIZO in conjunction with 2,3,5-triphenyltetrazolium chloride (TTC) staining used to assess the metabolically active and inactive portions of plant roots. For instance, active clover (*Trifolium repens* L.) roots turned dark red, bright red, or pale red, but inactive ones did not show any staining (Sturite et al., 2005). Recently, a micrograph of seedlings of *Chengiopanax sciadophylloides*, a woody species capable of accumulating radioactive cesium, was examined using the same function to assess the arbuscular mycorrhizal colonization rate (the ratio of the fungal body to the plant root area). The sample was taken in the Yamakiya district of Fukushima, Japan, several years after the 2011 nuclear disaster (Seitaro Deguchi et al., 2017).

2.5.2 Other Imaging Systems

ImageJ (previously NIH Image) is a software program for image processing created at the US National Institutes of Health. This method is predicated on converting the image under consideration into a binary image, which is then subjected to additional analysis through the manual establishment of a threshold level. Sadly, there is nothing user-friendly about this software (Kimura et al., 1999). To resolve this problem, Tajima and Kato compared the outcomes produced by the

WinRHIZO system with those produced by 16 alternative image processing algorithms found in the ImageJ package. The Triangle algorithm was found to be the most effective binary conversion technique among them, although it underestimated the root length when compared to the data supplied by the WinRHIZO system. Thankfully, it was discovered that increasing the data acquired using Image J by two-thirds may produce comparable numbers for the two systems (Tajima and Kato, 2011). Consequently, an open-source version of picture J, or Image J Rhizo, was created as a macro, giving researchers who are unable to afford the price of proprietary software programs a new, precise tool for root picture analysis. When Image J Rhizo's efficiency was compared to WinRHIZO's, the results showed a linear correlation between the values produced for total root length (Pierret et al., 2013).

A group of researchers (Delory et al., 2017) conducted a comparative analysis of the root length data acquired by WinRHIZO™, Image J Rhizo, and Tennant's manual line intersect approach (Tennant, 1975). Overall, the findings demonstrated the need for cautious interpretation of the data obtained using various techniques. These authors suggested applying a stain to improve the contrast between the background and the finer roots and to minimize root overlaps during the study.

Another root image analysis tool has been created and released as open-source software. It is named RhizoVision Explorer (https://www.rhizovision.com/, accessed on 22 March 2022) and is distinguished by its user-friendly interface, quick picture processing, and accurate measurements. Except for root volume, which WinRHIZO significantly overestimated, the outcomes of the two methodologies on the examined parameters showed a good overlap, indicating that RhizoVision Explorer had been properly validated (Seethepalli et al., 2021).

Given that image analysis may be used in field settings for precision agriculture, it has drawn a lot of interest. Precision agriculture refers to scientific endeavors that utilize advanced sensors and analytical tools to improve agricultural yields and support management decisions. These kinds of systems have lately been used to identify diseases that are transmitted through the soil and are caused by microorganisms that are phytopathogenic to plants. For instance, Flavescence Dorée and Esca diseases in grapevine (*Vitis vinifera* L. cv. Dolcetto) were detected using a commercial optical sensor. The disease Flavescence Dorèe, which is caused by the transportation of phytoplasma by the insect vector *Scaphoideus titanus*, is typified by the inhibition of plant growth and the reddening of leaves that roll downward from the veins. Foliar inter-venal necrosis and chlorosis are the primary symptoms of Esca disease, a fungal infection. Plant yield can be significantly reduced by each of these diseases. An experienced plant pathologist will typically identify diseased plants in a vineyard by observing their symptoms and verifying the diagnosis using with molecular analysis. Commercial optical sensors (OptRx®) placed on a transport system (often a quadbike, although drones are also utilized) and linked to a mobile PC with GIS and RTK-GNSS software were used to monitor the health condition of grapevines in three vineyards. By detecting reflectance values from plant surfaces, this optical sensor collects data on the normalized difference vegetation index (NDVI) and normalized difference red edge index (NDRE). Negative signals are typical of healthy plants, while lower NDVI and NDRE are associated with sick plants. The data collected by the optical sensor were validated later by molecular analysis, indicating the technique's enormous potential (Daglio et al., 2022).

2.6 CRISPR-CAS 9 GENE EDITING

Chromosome modification technique based on clustered regularly interspaced short palindromic repeats (CRISPR) is a naturally occurring prokaryotic mechanism that shields bacteria from invasion by foreign DNA originating from either bacteriophages or foreign plasmids. The CRISPR-Cas system, which employs Cas as a nuclease enzyme to identify complementary regions on DNA strands, may be modified to add or remove genes in the genomes of a broad range of prokaryotic and eukaryotic organisms, as well as to edit genomes, meaning to remove or change specific nucleotides (Jinek et al., 2012; Cong et al., 2023; Qi et al., 2013; Hsu et al., 2014). As a tool for genome engineering, the CRISPR-Cas system from *Streptococcus pyogenes* has been created due to its relatively low complexity in comparison to other bacterial systems. A hybrid of trans-activating crRNA (tracrRNA) and crRNA exists in the natural *Streptococcus pyogenes* system, and it is this hybrid that guides the Cas9 endonuclease to the target site. The two RNAs merge into a single-guide RNA (sgRNA) for convenience of usage in genome editing. The endonuclease breaks the target double-stranded DNA after attaching to the intended DNA sequence. The endonuclease Cas9 from the bacterium *Streptococcus pyogenes* (although Cas endonucleases from other bacteria have also been utilized) and a single RNA guide molecule (sg) that matches a portion of the target DNA sequence make up the most basic form of the CRISPR genome modification technology. The sgRNA directs the Cas9 endonuclease complex to its target DNA, whereupon the target chromosomal DNA is cleaved in both strands by the Cas9 nuclease activity. Using the CRISPR-Cas9 system, multiple genes can be altered simultaneously, potentially eliminating unnecessary pathways in a bacterium. Hence, the nucleotides in the sgRNA must be altered to confer different targeting specificity. A study showed that *Bacillus subtilis* strain HS3, which synthesizes the volatile organic substance 2,3-butanediol and the antifungal lipopeptides fengycin and surfactin, selectively colonizes the root hairs of the grass *Lolium perenne* in a hydroponic setting (Yi et al., 2018). This study evaluated whether the previously mentioned lipopeptides were the cause of the bacterium's antifungal activity by using the CRISPR-Cas system to delete the bacterial gene *sfp*, which codes for 4'-phosphopantetheinyl transferase, an enzyme required for the synthesis of the lipopeptides in question. The antifungal properties of this biocontrol strain against the fungal diseases caused by *Rhizoctonia solani* and *Fusarium culmorum* were eliminated upon *sfp* gene knockout, indicating the significance of fengycin and surfactin in this biocontrol strain's antifungal activity.

2.7 CURRENT CHALLENGES IN PLANT–PATHOGEN ASSOCIATIONS

Similar to people, crops are regularly targeted and confronted with a variety of beneficial and harmful microbes. Plant defense systems stop many pathogenic incursions that cause illness. However, the advent and progress of novel pathogens may aid in the pathogen's escape from the host's robust natural defenses (Stukenbrock and Bataillon, 2012; Misra and Chaturvedi, 2015). According to Britton and Liebhold (2013) and Misra and Chaturvedi (2015), both plants and infections develop in reaction to one another. This co-adaptive conflict between arms and farming practices leads to the spread and colonization of pathogens in the new host local ecosystems in which they have no past ancestral histories. Exotic infections can cause devastating eruptions in environments due to various circumstances, including the co-emergence of viruses and hosts, human meddling, global commerce, and population growth. The emergence of novel or developed human migration, pathogen species, plant diversification and speciation, plant vulnerability and prosperity, hybridization between extant pathogens, and, most significantly, global warming are some of the significant factors promoting the increase of emerging plant diseases and pathogens (Garrett et al., 2010; Misra and Chaturvedi, 2015). The aforementioned conditions have resulted in an increased frequency of novel disease appearance in the past few years, which has caused the decline of several wild animals, an overall decrease in biodiversity, and a decrease in agricultural yields (Cobb et al., 2012; Fisher et al., 2012). Consequently, it is essential to recognize and understand newly developing infections and create potent defense mechanisms against them. Plant species preservation, availability of food, and methods of agriculture have all been seriously threatened by the advent of diseases. Identifying novel viruses, comprehending their formation processes, and promptly implementing prevention strategies to avert their spread are more crucial than ever (Thynne et al., 2015). Intricate relationships exist between plants and diseases, and the emergence of novel infections poses an ongoing obstacle to researchers globally. The worldwide circulation and variety of plant diseases have been significantly altered by trading and human mobility of plants and plant-based products. Understanding the migratory paths can facilitate monitoring pathogen mobility, facilitating efficient disease control and quarantine protocols (Goss, 2015). Thanks to genomics-based genetic marker discoveries, the advanced study of plant pathogens allows the distinct population genetic information to be obtained. Our knowledge of fast genome adaptability and the capacity of pathogens to induce plant illnesses have improved due to the growing accessibility of pathogenic genome sequences (Benson et al., 2006, 2012; Thynne et al., 2015). The procedure by which viruses modify their course of pathogenicity in response to novel settings is known as "accelerated genome adaptation". Horizontal gene transfer (HGT) and interspecies hybridization are the two most extensively researched processes of rapid genome adaptation (Stukenbrock et al., 2007; Keeling and Palmer, 2008; Giraud et al., 2008; Raffaele and Kamoun, 2012). These techniques have been thoroughly evaluated (Misra and Chaturvedi, 2015; Thynne et al., 2015). Novel approaches to combat these dangers have been developed as a result of continuing study on the development of novel infections and their effects on pathogenicity. In order to better understand the adaptive development of plant diseases and develop disease control measures, population genomics has proven to be a helpful technique. Additional strategies that will help prevent new plant diseases include encouraging agricultural heterogeneity and limiting the exchange and transportation of plant components. Moreover, the successful management of new phytopathogens will depend on improved collaboration and interaction among ecologists, plant pathologists, epidemiologists, university researchers, and other pertinent stakeholders.

2.8 CONCLUSION

Our understanding of plant–microbe interactions has been completely transformed by cutting-edge approaches, which offer previously unheard-of insights into the intricate and dynamic linkages between these two entities. Utilizing state-of-the-art technologies such as omics technologies (genomics, transcriptomics, proteomics, and metabolomics) to provide high-resolution visualization of microbial colonization and microscopy techniques (e.g., confocal, super-resolution, and electron microscopy) to provide comprehensive examinations of the genetic, transcriptional, protein, and metabolic characteristics of both plants and microbes allows a system-level understanding of their interactions. Structural and molecular activities within plant tissues, and imaging and modeling methods (such as computational modeling and X-ray tomography) make it possible to simulate intricate biological processes and research relationships between plants and microbes in three dimensions. With the use of synthetic biology technologies, new plant–microbe interactions could be engineered for use in agriculture that enhance improved nutrient uptake or disease resistance. Researchers have discovered a new species of bacteria and their essential functions in plant disease and health, and they have been enabled to decipher the complex molecular pathways underpinning beneficial and detrimental interactions. Focused approaches to agricultural development, such as improved nutrient uptake and the use of biocontrol agents, have been developed that help deepen our understanding of the ecological importance of the interactions between plants and microbes in natural ecosystems. We should anticipate many more groundbreaking findings in this area as technology develops. These developments will not only progress basic science but also have real-world implications for environmental science, human health, and agriculture. Modern methods have completely changed the understanding of microbe–plant interactions. Through the provision of an unparalleled understanding of the molecular, cellular, and ecological dimensions of these associations, these techniques have paved the way for novel directions in agricultural investigation and advancement.

REFERENCES

Ahmad, F., Husain, F.M. and Ahmad, I., 2011. Rhizosphere and root colonization by bacterial inoculants and their monitoring methods: A critical area in PGPR research. In *Microbes and Microbial Technology*; Ahmad, I., Ahmad, F. and Pichtel, J., Eds. Springer, New York.

Aliferis, K.A., Faubert, D. and Jabaji, S., 2014. A metabolic profiling strategy for the dissection of plant defense against fungal pathogens. *PLoS One*, 9(11), e111930.

Allwood, J.W., Clarke, A., Goodacre, R. and Mur, L.A., 2010. Dual metabolomics: A novel approach to understanding plant–pathogen interactions. *Phytochemistry*, 71(5–6), 590–597.

Amann, R.I., Ludwig, W. and Schleifer, K.H., 1995. Phylogenetic identification and in situ detection of individual microbial cells without cultivation. *Microbiological Reviews*, 59(1), 143–169.

Antolín-Llovera, M., Ried, M.K., Binder, A. and Parniske, M., 2012. Receptor kinase signaling pathways in plant-microbe interactions. *Annual Review of Phytopathology*, 50(1), 451–473.

Aragona, M., Haegi, A., Valente, M.T., Riccioni, L., Orzali, L., Vitale, S., Infantino, A., 2022. New-generation sequencing technology in diagnosis of fungal plant pathogens: A dream comes true? *Journal of Fungi*, 8(7), 737.

Arsenault, J.-L., Pouleur, S., Messier, C. and Guay, R., 1995. WinRHIZO™, a root-measuring system with a unique overlap correction method. *Horticultural Science*, 30, 906.

Asai, S., Rallapalli, G., Piquerez, S.J., Caillaud, M.C., Furzer, O.J., Ishaque, N., Wirthmueller, L., Fabro, G., Shirasu, K. and Jones, J.D., 2014. Expression profiling during Arabidopsis/downy mildew interaction reveals a highly-expressed effector that attenuates responses to salicylic acid. *PLoS Pathogens*, 10(10), e1004443.

Bao, Y.Y., Qin, X., Yu, B., Chen, L.B., Wang, Z.C. and Zhang, C.X., 2014. Genomic insights into the serine protease gene family and expression profile analysis in the planthopper, *Nilaparvata lugens*. *BMC Genomics*, 15, 1–17.

Benson, D.A., Cavanaugh, M., Clark, K., Karsch-Mizrachi, I., Lipman, D.J., Ostell, J. and Sayers, E.W., 2012. GenBank. *Nucleic Acids Research*, 41(D1), D36–D42.

Benson, D.A., Karsch-Mizrachi, I., Lipman, D.J., Ostell, J. and Wheeler, D.L., 2006. GenBank. *Nucleic Acids Research*, 34(suppl_1), D16–D20.

Bergelson, J., Mittelstrass, J. and Horton, M.W., 2019. Characterizing both bacteria and fungi improves understanding of the *Arabidopsis* root microbiome. *Scientific Reports*, 9, 24.

Bernardo, L., Prinsi, B., Negri, A.S., Cattivelli, L., Espen, L. and Valè, G., 2012. Proteomic characterization of the Rph15 barley resistance gene-mediated defence responses to leaf rust. *BMC Genomics*, 13, 1–16.

Bi, G., Zhou, Z., Wang, W., Li, L., Rao, S., Wu, Y., . . . Zhou, J.M., 2018. Receptor-like cytoplasmic kinases directly link diverse pattern recognition receptors to the activation of mitogen-activated protein kinase cascades in *Arabidopsis*. *The Plant Cell*, 30(7), 1543–1561.

Björklöf, K. and Jørgensen, K.S., 2001. Applicability of non-antibiotic resistance marker genes in ecological studies of introduced bacteria in forest soil. *FEMS Microbiology Ecology*, 38(2–3), 179–188.

Bloemberg, G.V., 2007. Microscopic analysis of plant–bacterium interactions using autofluorescent proteins. *European Journal of Plant Pathology*, 119, 301–309.

Bloemberg, G.V., Wijfjes, A.H.M., Lamers, G.E.M., Stuurman, N. and Lugtenberg, B.J.J., 2000. Simultaneous imaging of *Pseudomonas fluorescens* WCS365 populations expressing three different autofluorescent proteins in the rhizosphere: New perspectives for studying microbial communities. *Molecular Plant–Microbe Interactions*, 13, 1170–1176.

Bona, E., Massa, N., Novello, G., Boatti, L., Cesaro, P., Todeschini, V., Magnelli, V., Manfredi, M., Marengo, E., Mignone, F. and Berta, G., 2019. Metaproteomic characterization of the *Vitis vinifera* rhizosphere. *FEMS Microbiology Ecology*, 95(1), fiy204.

Borisjuk, L., Rolletschek, H. and Neuberger, T., 2012. Surveying the plant's world by magnetic resonance imaging. *The Plant Journal*, 70(1), 129–146.

Braga, R.M., Dourado, M.N. and Araújo, W.L., 2016. Microbial interactions: Ecology in a molecular perspective. *Brazilian Journal of Microbiology*, 47, 86–98.

Britton, K.O. and Liebhold, A.M., 2013. One world, many pathogens!

Bushnell, W.R., Somers, D.A., Giroux, R.W., Szabo, L.J. and Zeyen, R.J., 1998. Genetic engineering of disease resistance in cereals. *Canadian Journal of Plant Pathology*, 20, 137–220.

Cesaro, P., Massa, N., Cantamessa, S., Todeschini, V., Bona, E., Berta, G., Barbato, R. and Lingua, G., 2020. Tomato responses to *Funneliformis mosseae* during the early stages of arbuscular mycorrhizal symbiosis. *Mycorrhiza*, 30, 601–610.

Chai, Y.N. and Schachtman, D.P., 2022. Root exudates impact plant performance under abiotic stress. *Trends in Plant Science*, 27, 80–91.

Chalfie, M., Tu, Y., Euskirchen, G., Ward, W.W. and Prasher, D.C., 1994. Green fluorescent protein as a marker for gene expression. *Science*, 263(5148), 802–805.

Chandok, I.K., Afreen, H., Afreen, R., Haider, S., Moharana, D.P., Hussain, T., Aamir, M., Shanmugam, V. and Malik, M.Z., 2022. Functional genomics tools for studying microbe-mediated stress tolerance in plants. In *Mitigation of Plant Abiotic Stress by Microorganisms* (pp. 175–204). Academic Press.

Cheng, Z., Duan, J., Hao, Y., McConkey, B.J. and Glick, B.R., 2009. Identification of bacterial proteins mediating the interactions between *Pseudomonas putida* UW4 and *Brassica napus* (Canola). *Molecular Plant–Microbe Interactions*, 22(6), 686–694.

Cobb, R.C., Filipe, J.A., Meentemeyer, R.K., Gilligan, C.A. and Rizzo, D.M., 2012. Ecosystem transformation by emerging infectious disease: Loss of large tanoak from California forests. *Journal of Ecology*, 100(3), 712–722.

Cocking, E.C., Stone, P.J. and Davey, M.R., 2006. Intracellular colonization of roots of *Arabidopsis* and crop plants by *Gluconacetobacter diazotrophicus*. *In Vitro Cellular & Developmental Biology-Plant*, 42(1), 74–82.

Cole, B.J., Feltcher, M.E., Waters, R.J., Wetmore, K.M., Mucyn, T.S., Ryan, E.M., . . . Visel, A., 2017. Genome-wide identification of bacterial plant colonization genes. *PLoS Biology*, 15(9), e2002860.

Cong, L., Ran, F.A., Cox, D., Lin, S., Barretto, R., Habib, N., . . . Zhang, F., 2013. Multiplex genome engineering using CRISPR/Cas systems. *Science*, 339(6121), 819–823.

Czymmek, K.J., Duncan, K.E. and Berg, H., 2023. Realizing the full potential of advanced microscopy approaches for interrogating plant–microbe interactions. *Molecular Plant–Microbe Interactions*, 36(4), 245–255.

Daglio, G., Cesaro, P., Todeschini, V., Lingua, G., Lazzari, M., Berta, G. and Massa, N., 2022. Potential field detection of *Flavescence dorée* and Esca diseases using a ground sensing optical system. *Biosystems Engineering*, 215, 203–214.

Daims, H. and Wagner, M., 2007. Quantification of uncultured microorganisms by fluorescence microscopy and digital image analysis. *Applied Microbiology and Biotechnology*, 75, 237–248.

De Cremer, K., Mathys, J., Vos, C., Froenicke, L., Michelmore, R.W., Cammue, B.P.A. and De Coninck, B., 2013. RNA seq-based transcriptome analysis of *Lactuca sativa* infected by the fungal necrotroph *Botrytis cinerea*. *Plant, Cell & Environment*, 36(11), 1992–2007.

De Wet, J.R., Wood, K.V., DeLuca, M., Helinski, D.R. and Subramani, S., 1987. Firefly luciferase gene: Structure and expression in mammalian cells. *Molecular and Cellular Biology*, 7, 725.

Deguchi, S., Matsuda, Y., Takenaka, C., Sugiura, Y., Ozawa, H. and Ogata, Y., 2017. Proposal of a new estimation method of colonization rate of arbuscular mycorrhizal fungi in the roots of *Chengiopanax sciadophylloides*. *Mycobiology*, 45(1), 15–19.

Delory, B.M., Weidlich, E.W.A., Meder, L., Lutje, A., Van Duijnen, R., Weidlich, R. and Temperton, V.M., 2017. Accuracy and bias of methods used for root length measurements in functional root research. *Methods in Ecology and Evolution*, 8, 1594–1606.

Diagne, N., Escoute, J., Lartaud, M., Verdeil, J.L., Franche, C., Kane, A., . . . Svistoonoff, S., 2011. Uvitex2B: A rapid and efficient stain for detection of arbuscular mycorrhizal fungi within plant roots. *Mycorrhiza*, 21(4), 315–321.

Dieryckx, C., Gaudin, V., Dupuy, J.W., Bonneu, M., Girard, V. and Job, D., 2015. Beyond plant defense: Insights on the potential of salicylic and methylsalicylic acid to contain growth of the phytopathogen *Botrytis cinerea*. *Frontiers in Plant Science*, 6, 859.

Diwan, D., Rashid, M.M. and Vaishnav, A., 2022. Current understanding of plant–microbe interaction through the lenses of multi-omics approaches and their benefits in sustainable agriculture. *Microbiology Research*, 265, 127180.

Dolatabadian, A., 2020. Plant–microbe interaction. *Biology*, 10(1), 15.

Domon, B. and Aebersold, R., 2010. Options and considerations when selecting a quantitative proteomics strategy. *Nature Biotechnology*, 28(7), 710–721.

Doni, F., Miranti, M., Mispan, M.S., Mohamed, Z. and Uphoff, N., 2022. Multi-omics approaches for deciphering the microbial modulation of plants' genetic potentials: What's known and what's next? *Rhizosphere*, 24, 100613.

Earles, J.M., Buckley, T.N., Brodersen, C.R., Busch, F.A., Cano, F.J., Choat, B., Evans, J.R., Farquhar, G.D., Harwood, R., Huynh, M. and John, G.P., 2019. Embracing 3D complexity in leaf carbon–water exchange. *Trends in Plant Science*, 24(1), 15–24.

Ellouze, W., Hamel, C., Singh, A.K., Mishra, V., DePauw, R.M. and Knox, R.E., 2018. Abundance of the arbuscular mycorrhizal fungal taxa associated with the roots and rhizosphere soil of different durum wheat cultivars in the Canadian prairies. *Canadian Journal of Microbiology*, 64(8), 527–536.

Ellouze, W., Mishra, V., Howard, R.J., Ling, K.S. and Zhang, W., 2020. Preliminary study on the control of cucumber green mottle mosaic virus in commercial greenhouses using agricultural disinfectants and resistant cucumber varieties. *Agronomy*, 10(12), 1879.

Escudero-Martinez, C., Coulter, M., Alegria Terrazas, R., Foito, A., Kapadia, R., Pietrangelo, L., . . . Bulgarelli, D., 2022. Identifying plant genes shaping microbiota composition in the barley rhizosphere. *Nature Communications*, 13(1), 3443.

Fisher, M.C., Henk, D.A., Briggs, C.J., Brownstein, J.S., Madoff, L.C., McCraw, S.L. and Gurr, S.J., 2012. Emerging fungal threats to animal, plant and ecosystem health. *Nature*, 484(7393), 186–194.

Fröschel, C., Komorek, J., Attard, A., Marsell, A., Lopez-Arboleda, W.A., Le Berre, J., . . . Dröge-Laser, W., 2021. Plant roots employ cell-layer-specific programs to respond to pathogenic and beneficial microbes. *Cell Host & Microbe*, 29(2), 299–310.

Gamalero, E., Berta, G., Massa, N., Glick, B.R. and Lingua, G., 2008. Synergistic interactions between the ACC deaminase-producing bacterium *Pseudomonas putida* UW4 and the AM fungus *Gigaspora rosea* positively affect cucumber plant growth. *FEMS Microbiology Ecology*, 64, 459–467.

Gamalero, E., Berta, G., Massa, N., Glick, B.R. and Lingua, G., 2010. Interactions between *Pseudomonas putida* UW4 and *Gigaspora rosea* BEG9 and their consequences on the growth of cucumber under salt stress conditions. *Journal of Applied Microbiology*, 108, 236–245.

Gamalero, E., Bona, E. and Glick, B.R., 2022. Current techniques to study beneficial plant–microbe interactions. *Microorganisms*, 10(7), 1380.

Gamalero, E., Lingua, G., Berta, G. and Lemanceau, P., 2003. Methods for studying root colonization by introducing beneficial bacteria. *Agronomie*, 23, 407–418.

Gamalero, E., Lingua, G., Giusy Caprì, F., Fusconi, A., Berta, G. and Lemanceau, P., 2004a. Colonization pattern of primary tomato roots by *Pseudomonas fluorescens* A6RI characterized by dilution plating, flow cytometry, fluorescence, confocal and scanning electron microscopy. *FEMS Microbiology Ecology*, 48, 79–87.

Gamalero, E., Lingua, G., Tombolini, R., Avidano, L., Pivato, B. and Berta, G., 2005. Colonization of tomato root seedling by *Pseudomonas fluorescens* 92 rkG5: Spatio-temporal dynamics, localization, organization, viability, and culturability. *Microbial Ecology*, 50, 289–297.

Gamalero, E., Martinotti, M.G., Trotta, A., Lemanceau, P. and Berta, G., 2002. Morphogenetic modifications induced by *Pseudomonas fluorescens* A6RI and *Glomus mosseae* BEG12 in the root system of tomato differ according to plant growth conditions. *New Phytologist*, 155, 293–300.

Gamalero, E., Trotta, A., Massa, N., Copetta, A. Martinotti, M.G. and Berta, G., 2004b. Impact of two fluorescent pseudomonads and an arbuscular mycorrhizal fungus on tomato plant growth, root architecture and P acquisition. *Mycorrhiza*, 14, 185–192.

Garrett, K.A., Jumpponen, A., Montaño, L.G., 2010. Emerging plant diseases: What are our best strategies for management? *Trends in Ecology & Evolution*, 25(8), 456–465.

Getzke, F., Hassani, M.A., Crüsemann, M., Malisic, M., Zhang, P., Ishigaki, Y., . . . Schulze-Lefert, P., 2023. Cofunctioning of bacterial exometabolites drives root microbiota establishment. *Proceedings of the National Academy of Sciences*, 120(15), e2221508120.

Giraud, T., Refrégier, G., Le Gac, M., de Vienne, D.M. and Hood, M.E., 2008. Speciation in fungi. *Fungal Genetics and Biology*, 45(6), 791–802.

Gorshkov, V. and Tsers, I., 2022. Plant susceptible responses: The underestimated side of plant–pathogen interactions. *Biological Reviews*, 97(1), 45–66.

Goss, E.M., 2015. Genome-enabled analysis of plant-pathogen migration. *Annual Review of Phytopathology*, 53(1), 121–135.

Guo, L., Yu, H., Wang, B., Vescio, K., DeIulio, G.A., Yang, H., Berg, A., Zhang, L., Edel-Hermann, V., Steinberg, C., Kistler, H.C. and Ma, L.J., 2021, September. Metatranscriptomic comparison of endophytic and pathogenic *Fusarium–Arabidopsis* interactions reveals plant transcriptional plasticity. *Molecular Plant–Microbe Interactions*, 34(9), 1071–1083. https://doi.org/10.1094/MPMI-03-21-0063-R. Epub 2021 Oct 11. PMID: 33856230; PMCID: PMC9048145.

Gupta, S., Schillaci, M. and Roessner, U., 2022. Metabolomics as an emerging tool to study plant–microbe interactions. *Emerging Topics in Life Sciences*, 6(2), 175–183.

Harris, R.L., Vetter, M.C.Y.L., van Heerden, E., Cason, E., Vermeulen, J.G., Taneja, A., Kieft, T.L., DeCoste, C.J., Laevsky, G.S. and Onstott, T.C., 2022. FISH-TAMB, a fixation-free mRNA fluorescent labeling technique to target transcriptionally active members in microbial communities. *Microbial Ecology*, 84, 182–197.

Haupt, B.J., Pelling, A.E. and Horton, M.A., 2006. Integrated confocal and scanning probe microscopy for biomedical research. *The Scientific World Journal*, 15, 1609–1618.

Haveman, N.J., Khodadad, C.L., Dixit, A.R., Louyakis, A.S., Massa, G.D., Venkateswaran, K. and Foster, J.S., 2021. Evaluating the lettuce metatranscriptome with MinION sequencing for future spaceflight food production applications. *npj Microgravity*, 7(1), 22.

Heath, M.C., 2000. Advances in imaging the cell biology of plant–microbe interactions. *Annual Review of Phytopathology*, 38, 443–459.

Hong, Y.S., Martinez, A., Liger-Belair, G., Jeandet, P., Nuzillard, J.M. and Cilindre, C., 2012. Metabolomics reveals simultaneous influences of plant defence system and fungal growth in *Botrytis cinerea*-infected *Vitis vinifera* cv. Chardonnay berries. *Journal of Experimental Botany*, 63(16), 5773–5785.

Hsu, P.D., Lander, E.S. and Zhang, F., 2014. Development and applications of CRISPR-Cas9 for genome engineering. *Cell*, 157, 1262–1278.

Hu, J., Wei, Z., Kowalchuk, G.A., Xu, Y., Shen, Q. and Jousset, A., 2020. Rhizosphere microbiome functional diversity and pathogen invasion resistance build up during plant development. *Environmental Microbiology*, 22, 5005–5018.

Hu, L., Robert, C.A., Cadot, S., Zhang, X.I., Ye, M., Li, B., . . . Erb, M., 2018. Root exudate metabolites drive plant-soil feedbacks on growth and defense by shaping the rhizosphere microbiota. *Nature Communications*, 9(1), 2738.

Huang, A.C., Jiang, T., Liu, Y.X., Bai, Y.C., Reed, J., Qu, B., . . . Osbourn, A., 2019. A specialized metabolic network selectively modulates *Arabidopsis* root microbiota. *Science*, 364(6440), eaau6389.

Ibáñez, A.J., Scharte, J., Bones, P., Pirkl, A., Meldau, S., Baldwin, I.T., Hillenkamp, F., Weis, E. and Dreisewerd, K., 2010. Rapid metabolic profiling of Nicotiana tabacum defence responses against *Phytophthora nicotianae* using direct infrared laser desorption ionization mass spectrometry and principal component analysis. *Plant Methods*, 6, 1–16.

Imam, J., Singh, P.K. and Shukla, P., 2016. Plant microbe interactions in post genomic era: Perspectives and applications. *Frontiers in Microbiology*, 7, 1488.

Jain, A., Sarsaiya, S., Singh, R., Gong, Q., Wu, Q. and Shi, J., 2024. Omics approaches in understanding the benefits of plant–microbe interactions. *Frontiers in Microbiology*, 15, 1391059.

Jain, A., Singh, H.B. and Das, S., 2021. Deciphering plant–microbe crosstalk through proteomics studies. *Microbiological Research*, 242, 126590.

Jayaraman, D., Forshey, K.L., Grimsrud, P.A. and Ané, J.M., 2012. Leveraging proteomics to understand plant–microbe interactions. *Frontiers in Plant Science*, 3, 44.

Jinek, M., Chylinski, K., Fonfara, I., Hauer, M. and Doudna, J.A., 2012. A programmable dual-RNA-guided endonuclease in adaptive bacterial immunity. *Science*, 337, 816–821.

Jones, O.A., Maguire, M.L., Griffin, J.L., Jung, Y.H., Shibato, J., Rakwal, R., Agrawal, G.K. and Jwa, N.S., 2011. Using metabolic profiling to assess plant–pathogen interactions: An example using rice (*Oryza sativa*) and the blast pathogen *Magnaporthe grisea*. *European Journal of Plant Pathology*, 129, 539–554.

Jupe, F., Witek, K., Verweij, W., Śliwka, J., Pritchard, L., Etherington, G.J., Maclean, D., Cock, P.J., Leggett, R.M., Bryan, G.J. and Cardle, L., 2013. Resistance gene enrichment sequencing (RenSeq) enables reannotation of the NB-LRR gene family from sequenced plant genomes and rapid mapping of resistance loci in segregating populations. *The Plant Journal*, 76(3), 530–544.

Kang, Y., Shen, M., Xia, D., Ye, K., Zhao, Q. and Hu, J., 2017. Caution of the intensified spread of antibiotic resistance genes by inadvertent introduction of beneficial bacteria into soil. *Acta Agriculturae Scandinavica, Section B—Soil & Plant Science*, 67, 576–582.

Kawa, D. and Brady, S.M., 2022. Root cell types as an interface for biotic interactions. *Trends in Plant Science*, 27, 1173–1186.

Keeling, P.J. and Palmer, J.D., 2008. Horizontal gene transfer in eukaryotic evolution. *Nature Reviews Genetics*, 9(8), 605–618.

Khatabi, B., Gharechahi, J., Ghaffari, M.R., Liu, D., Haynes, P.A., McKay, M.J., Mirzaei, M. and Salekdeh, G.H., 2019. Plant–microbe symbiosis: What has proteomics taught us? *Proteomics*, 19(16), 1800105.

Kierul, K., Voigt, B., Albrecht, D., Chen, X.H., Carvalhais, L.C. and Borriss, R., 2015. Influence of root exudates on the extracellular proteome of the plant growth-promoting bacterium *Bacillus amyloliquefaciens* FZB42. *Microbiology*, 161(1), 131–147.

Kimura, K., Kikuchi, S. and Yamasaki, S.I., 1999. Accurate root length measurement by image analysis. *Plant Soil*, 216, 117–127.

King, E., Wallner, A., Rimbault, I., Barrachina, C., Klonowska, A., Moulin, L. and Czernic, P., 2019. Monitoring of rice transcriptional responses to contrasted colonizing patterns of phytobeneficial *Burkholderia sl* reveals a temporal shift in JA systemic response. *Frontiers in Plant Science*, 10, 1141.

Konrad, S.S. and Ott, T., 2015. Molecular principles of membrane microdomain targeting in plants. *Trends in Plant Science*, 20(6), 351–361.

Kragelund, L. and Nybroe, O., 1996. Competition between *Pseudomonas fluorescens* Ag1 and *Alcaligenes eutrophus* JMP134 (pJP4) during colonization of barley roots. *FEMS Microbiology Ecology*, 20, 41–51.

Kumar, J., Ramlal, A., Kumar, K., Rani, A. and Mishra, V., 2021. Signaling pathways and downstream effectors of host innate immunity in plants. *International Journal of Molecular Sciences*, 22(16), 9022.

Kumar, J., Ramlal, A., Mallick, D. and Mishra, V., 2021. An overview of some biopesticides and their importance in plant protection for commercial acceptance. *Plants*, 10(6), 1185.

Kumar, S.S. and Ghosh, A.R., 2019. Assessment of bacterial viability: A comprehensive review on recent advances and challenges. *Microbiology*, 165(6), 593–610.

Kumaraswamy, K.G., Kushalappa, A.C., Choo, T.M., Dion, Y. and Rioux, S., 2011. Mass spectrometry based metabolomics to identify potential biomarkers for resistance in barley against fusarium head blight (*Fusarium graminearum*). *Journal of Chemical Ecology*, 37, 846–856.

Kundu, S., Chakraborty, D., Kundu, A. and Pal, A., 2013. Proteomics approach combined with biochemical attributes to elucidate compatible and incompatible plant–virus interactions between *Vigna mungo* and Mungbean Yellow Mosaic India Virus. *Proteome Science*, 11, 1–14.

Lambais, M.R., Barrera, S.E., Santos, E.C., Crowley, D.E. and Jumpponen, A., 2017. Phyllosphere metaproteomes of trees from the Brazilian Atlantic forest show high levels of functional redundancy. *Microbial Ecology*, 73, 123–134.

Lee, S., Yang, D.S., Uppalapati, S.R., Sumner, L.W. and Mysore, K.S., 2013. Suppression of plant defense responses by extracellular metabolites from *Pseudomonas syringae* pv. *tabaci* in *Nicotiana benthamiana*. *BMC Plant Biology*, 13, 1–13.

Lery, L.M., Hemerly, A.S., Nogueira, E.M., von Krüger, W.M. and Bisch, P.M., 2011. Quantitative proteomic analysis of the interaction between the endophytic plant-growth-promoting bacterium *Gluconacetobacter diazotrophicus* and sugarcane. *Molecular Plant–Microbe Interactions*, 24(5), 562–576.

Levy, A., Conway, J.M., Dangl, J.L. and Woyke, T., 2018a. Elucidating bacterial gene functions in the plant microbiome. *Cell Host & Microbe*, 24, 475–485.

Levy, A., Salas Gonzalez, I., Mittelviefhaus, M., Clingenpeel, S., Herrera Paredes, S., Miao, J., Wang, K., Devescovi, G., Stillman, K., Monteiro, F. and Rangel Alvarez, B., 2018b. Genomic features of bacterial adaptation to plants. *Nature Genetics*, 50(1), 138–150.

Li, L., Li, Y., Lim, T.M. and Pan, S.Q., 1999. GFP-aided confocal laser scanning microscopy can monitor *Agrobacterium tumefaciens* cell morphology and gene expression associated with infection. *FEMS Microbiology Letters*, 179, 141–146.

Li, X., Bai, T., Li, Y., Ruan, X. and Li, H., 2013. Proteomic analysis of *Fusarium oxysporum* f. sp. *cubense* tropical race 4-inoculated response to Fusarium wilts in the banana root cells. *Proteome Science*, 11, 1–14.

Lindow, S.E., 1995. The use of reporter genes in the study of microbial ecology. *Molecular Ecology*, 4, 555–566.

López-Mondéjar, R., Kostovčík, M., Lladó, S., Carro, L. and García-Fraile, P., 2017. Exploring the plant microbiome through multi-omics approaches. In *Probiotics in Agroecosystem* (pp. 233–268). Springer, Berlin/Heidelberg, Germany.

Lowe, R.G., Cassin, A., Grandaubert, J., Clark, B.L., Van de Wouw, A.P., Rouxel, T. and Howlett, B.J., 2014. Genomes and transcriptomes of partners in plant–fungal interactions between canola (*Brassica napus*) and two *Leptosphaeria* species. *PloS One*, 9(7), e103098.

Lukumbuzya, M., Schmid, M., Pjevac, P. and Daims, H., 2019. A multicolor fluorescence in situ hybridization approach using an extended set of fluorophores to visualize microorganisms. *Frontiers in Microbiology*, 10, 1383.

Lynch, J., 1995. Root architecture and plant productivity. *Plant Physiology*, 109, 7–13.

Macho, A.P. and Zipfel, C., 2014. Plant PRRs and the activation of innate immune signaling. *Molecular Cell*, 54(2), 263–272.

Margaria, P., Abbà, S. and Palmano, S., 2013. Novel aspects of grapevine response to phytoplasma infection investigated by a proteomic and phospho-proteomic approach with data integration into functional networks. *BMC Genomics*, 14, 1–15.

Matz, M.V., Fradkov, A.F., Labas, Y.A., Savitsky, A.P., Zaraisky, A.G., Markelov, M.L. and Lukyanov, S.A., 1999. Fluorescent proteins from nonbioluminescent Anthozoa species. *Nature Biotechnology*, 17, 969–973.

Maxam, A.M. and Gilbert, W., 1977. A new method for sequencing DNA. *Proceedings of the National Academy of Sciences of the United States of America*, 72, 560e564.

McBratney, A., Field, D.J. and Koch, A., 2014. The dimensions of soil security. *Geoderma*, 213, 203–213.

Mehta, A., Brasileiro, A.C., Souza, D.S., Romano, E., Campos, M.A., Grossi-de-Sá, M.F., Silva, M.S., Franco, O.L., Fragoso, R.R., Bevitori, R. and Rocha, T.L., 2008. Plant–pathogen interactions: What is proteomics telling us? *The FEBS Journal*, 275(15), 3731–3746.

Miller, J., Ferreira, P. and LeJeune, J.T., 2022. Antimicrobial use and resistance in plant agriculture: A one health perspective. *Agriculture*, 12, 289.

Mishra, V., Ellouze, W. and Howard, R.J., 2018. Utility of arbuscular mycorrhizal fungi for improved production and disease mitigation in organic and hydroponic greenhouse crops. *Journal of Horticulture*, 5(03).

Misra, B.B. and Chaturvedi, R., 2015. When plants brace for the emerging pathogens. *Physiological and Molecular Plant Pathology*, 92, 181–185.

Monmeyran, A., Thomen, P., Jonquière, H., Sureau, F., Li, C., Plamont, M.A., Douarche, C., Casella, J.F., Gautier, A. and Henry, N., 2018. The inducible chemical-genetic fluorescent marker FAST outperforms classical fluorescent proteins in the quantitative reporting of bacterial biofilm dynamics. *Scientific Reports*, 8, 10336.

Mougel, C., Offre, P., Ranjard, L., Corberand, T., Gamalero, E., Robin, C. and Lemanceau, P., 2006. Dynamic of the genetic structure of bacterial and fungal communities at different developmental stages of *Medicago truncatula* Gaertn. cv. Jemalong line J5. *New Phytologist*, 170, 165–175.

Musat, N., Foster, R., Vagner, T., Adam, B. and Kuypers, M.M., 2012. Detecting metabolic activities in single cells, with emphasis on nanoSIMS. *FEMS Microbiology Reviews*, 36, 486–511.

Nakayasu, E.S., Nicora, C.D., Sims, A.C., Burnum-Johnson, K.E., Kim, Y.M., Kyle, J.E., Matzke, M.M., Shukla, A.K., Chu, R.K., Schepmoes, A.A. and Jacobs, J.M., 2016. MPLEx: A robust and universal protocol for single-sample integrative proteomic, metabolomic, and lipidomic analyses. *MSystems*, 1(3), 10–1128.

Narula, N., Remus, R., Deubel, A., Granse, A., Dudeja, S.S., Behl, R.K. and Merbach, W., 2007. Comparison of the effectiveness of wheat roots colonization by *Azotobacter chroococcum* and *Pantoea agglomerans* using serological techniques. *Plant Soil and Environment*, 53(4), 167.

Nescerecka, A., Hammes, F. and Juhna, T., 2016. A pipeline for developing and testing staining protocols for flow cytometry, demonstrated with SYBR Green I and propidium iodide viability staining. *Journal of Microbiological Methods*, 131, 172–180.

Newman, K.L., Almeida, R.P., Purcell, A.H. and Lindow, S.E., 2003. Use of a green fluorescent strain for analysis of *Xylella fastidiosa* colonization of *Vitis vinifera*. *Applied and Environmental Microbiology*, 69, 7319–7127.

Novello, G., Gamalero, E., Bona, E., Boatti, L., Mignone, F., Massa, N., Cesaro, P., Lingua, G. and Berta, G., 2017. The rhizosphere bacterial microbiota of *Vitis vinifera* cv. Pinot Noir in an integrated pest management vineyard. *Frontiers in Microbiology*, 8, 1528.

Nowicki, M., Foolad, M.R., Nowakowska, M. and Kozik, E.U., 2012. Potato and tomato late blight caused by *Phytophthora infestans*: An overview of pathology and resistance breeding. *Plant Disease*, 96(1), 4–17.

Pang, F., Li, Q., Solanki, M.K., Wang, Z., Xing, Y.X. and Dong, D.F., 2024. Soil phosphorus transformation and plant uptake are driven by phosphate-solubilizing microorganisms. *Frontiers in Microbiology*, 15, 1383813.

Parker, J., Koh, J., Yoo, M.J., Zhu, N., Feole, M., Yi, S. and Chen, S., 2013. Quantitative proteomics of tomato defense against *Pseudomonas syringae* infection. *Proteomics*, 13(12–13), 1934–1946.

Pasquet, J.C., Chaouch, S., Macadré, C., Balzergue, S., Huguet, S., Martin-Magniette, M.L., Bellvert, F., Deguercy, X., Thareau, V., Heintz, D. and Saindrenan, P., 2014. Differential gene expression and metabolomic analyses of *Brachypodium distachyon* infected by deoxynivalenol producing and non-producing strains of *Fusarium graminearum*. *BMC Genomics*, 15, 1–17.

Passos, M.A., de Cruz, V.O., Emediato, F.L., de Teixeira, C.C., Azevedo, V.C.R., Brasileiro, A.C., Amorim, E.P., Ferreira, C.F., Martins, N.F., Togawa, R.C. and Júnior, G.J.P., 2013. Analysis of the leaf transcriptome of *Musa acuminata* during interaction with *Mycosphaerella musicola*: Gene assembly, annotation and marker development. *BMC Genomics*, 14, 1–17.

Peck, M.C., Fisher, R.F. and Long, S.R., 2006. Diverse flavonoids stimulate NodD1 binding to nod gene promoters in *Sinorhizobium meliloti*. *Journal of Bacteriology*, 188(15), 5417–5427.

Peters, N.K. and Verma, D.P.S., 1990. Phenolic compounds as regulators of gene expression in plant–microbe interactions. *Molecular Plant–Microbe Interactions*, 3(4), 8.

Petriccione, M., Di Cecco, I., Arena, S., Scaloni, A. and Scortichini, M., 2013. Proteomic changes in *Actinidia chinensis* shoot during systemic infection with a pandemic *Pseudomonas syringae* pv. *actinidiae* strain. *Journal of Proteomics*, 78, 461–476.

Pierret, A., Gonkhamdee, S., Jourdan, C. and Maeght, J.L., 2013. IJ_Rhizo: An open-source software to measure scanned images of root samples. *Plant and Soil*, 373, 531–539.

Piovesan, A., Vancauwenberghe, V., Van De Looverbosch, T., Verboven, P. and Nicolaï, B., 2021. X-ray computed tomography for 3D plant imaging. *Trends in Plant Science*, 26(11), 1171–1185.

Qi, L.S., Larson, M.H., Gilbert, L.A., Doudna, J.A., Weissman, J.S., Arkin, A.P. and Lim, W.A., 2013. Repurposing CRISPR as an RNA-guided platform for sequence-specific control of gene expression. *Cell*, 5, 1173–1183.

Quirino, B.F., Candido, E.S., Campos, P.F., Franco, O.L. and Krüger, R.H., 2010. Proteomic approaches to study plant–pathogen interactions. *Phytochemistry*, 71(4), 351–362.

Raffaele, S. and Kamoun, S., 2012. Genome evolution in filamentous plant pathogens: Why bigger can be better. *Nature Reviews Microbiology*, 10(6), 417–430.

Ramlal, A., Rani, A., Nautiyal, A., Kalra, C., Kumari, R., Kumar, J., . . . Mishra, V., 2023. Importance of omics approaches in plant-microbe interaction for plant disease control. *Physiological and Molecular Plant Pathology*, 128, 102153.

Rochfort, S., 2005. Metabolomics reviewed: A new "omics" platform technology for systems biology and implications for natural products research. *Journal of Natural Products*, 68, 1813–1820.

Romano, I., Ventorino, V. and Pepe, O., 2020. Effectiveness of plant beneficial microbes: Overview of the methodological approaches for the assessment of root colonization and persistence. *Frontiers in Plant Science*, 11, 6.

Saito, K. and Matsuda, F., 2010. Metabolomics for functional genomics, systems biology, and biotechnology. *Annual Review of Plant Biology*, 61(1), 463–489.

Salas-González, I., Reyt, G., Flis, P., Custódio, V., Gopaulchan, D., Bakhoum, N., . . . Castrillo, G., 2021. Coordination between microbiota and root endodermis supports plant mineral nutrient homeostasis. *Science*, 371(6525), eabd0695.

Saletnik, A., Saletnik, B. and Puchalski, C., 2021. Overview of popular techniques of Raman spectroscopy and their potential in the study of plant tissues. *Molecules*, 26(6), 1537.

Sanger, F., Nicklen, S., Coulson, A.R., 1977. DNA sequencing with chain-terminating inhibitors. *Proceedings of the National Academy of Sciences of the United States of America*, 74, 5463e5467.

Sarim, K.M., Srivastava, R. and Ramteke, P.W., 2020. Next-generation omics technologies for exploring complex metabolic regulation during plant–microbe interaction. In *Microbial Services in Restoration Ecology* (pp. 123–138). Elsevier.

Scandiani, M.M., Luque, A.G., Razori, M.V., Ciancio Casalini, L., Aoki, T., O'Donnell, K., Cervigni, G.D. and Spampinato, C.P., 2015. Metabolic profiles of soybean roots during early stages of *Fusarium tucumaniae* infection. *Journal of Experimental Botany*, 66(1), 391–402.

Schenk, P.M., Carvalhais, L.C. and Kazan, K., 2012. Unraveling plant–microbe interactions: Can multi-species transcriptomics help? *Trends in Biotechnol*ogy, 30(3), 177–184.

Schmidt, P.E., Broughton, W.J. and Werner, D., 1994. Nod factors of *Bradyrhizobium japonicum* and *Rhizobium* sp. NGR234 induce flavonoid accumulation in soybean root exudate. *Plant Physiology*, 105(4), 1207–1214.

Schweizer, P., Polorny, J., Abderhalden, O. and Dudler, R., 1999. A transient assay system for the functional assessment of defense-related genes in wheat. *Molecular Plant–Microbe Interactions*, 12, 647–654.

Seethepalli, A., Dhakal, K., Griffiths, M., Guo, H., Freschet, G.T. and York, L.M., 2021. RhizoVision explorer: Open-source software for root image analysis and measurement standardization. *AoB Plants*, 13, plab056.

Seitaro Deguchi, S., Matsuda, Y., Takenaka, C., Hajime, Y.S., Ogata, Y., 2017. Proposal of a new estimation method of colonization rate of arbuscular mycorrhizal fungi in the roots of *Chengiopanax sciadophylloides*. *Mycobiology*, 45, 15–19.

Sharma, M., Sudheer, S., Usmani, Z., Rani, R. and Gupta, P., 2020. Deciphering the omics of plant–microbe interaction: Perspectives and new insights. *Current Genomics*, 21(5), 343–362.

Singh, J.S., Pandey, V.C. and Singh, D.P., 2011. Efficient soil microorganisms: A new dimension for sustainable agriculture and environmental development. *Agriculture, Ecosystems & Environment*, 140, 339e353.

Sørensen, J., Jensen, L.E. and Nybroe, O., 2011. Soil and rhizosphere as habitats for Pseudomonas inoculants: New knowledge on distribution, activity and physiological state derived from micro-scale and single-cell studies. *Plant and Soil*, 232, 97–108.

Sørensen, J. and Nybroe, O., 2006. Reporter genes in bacterial inoculants can monitor life conditions and functions in soil. In *Nucleic Acids and Proteins*, Vol. 8; Nannipieri, P. and Smalla, K., Eds. Springer, Berlin/Heidelberg, Germany.

Stöckel, S., Kirchhoff, J., Neugebauer, U., Rösch, P. and Popp, J., 2016. The application of Raman spectroscopy for the detection and identification of microorganisms. *Journal of Raman Spectroscopy*, 47(1), 89–109.

Stukenbrock, E.H., Banke, S., Javan-Nikkhah, M. and McDonald, B.A., 2007. Origin and domestication of the fungal wheat pathogen *Mycosphaerella graminicola* via sympatric speciation. *Molecular Biology and Evolution*, 24(2), 398–411.

Stukenbrock, E.H. and Bataillon, T., 2012. A population genomics perspective on the emergence and adaptation of new plant pathogens in agro-ecosystems. *Molecular Plant Pathology*, *13*(5), 424–437.

Sturite, I., Herkensen, T.M., Breland, T.A., 2005. Distinguishing between metabolically active and inactive roots by combined staining with 2,3,5 triphenultetrazolium chloride and image colour analysis. *Plant Soil, 271*, 75–82.

Swarupa, V., Pavitra, K., Shivashankara, K.S. and Ravishankar, K.V., 2016. Omics-driven approaches in plant–microbe interaction. In *Microbial Inoculants in Sustainable Agricultural Productivity*, Vol. 1, pp. 61–84. Springer, New Delhi, India.

Tajima, R. and Kato, Y., 2011. Comparison of threshold algorithms for automatic image processing of rice roots using freeware ImageJ. *Field Crops Research*, 121(3), 460–463.

Tennant, D., 1975. A test of modified line intersect method of estimating root length. *Journal of Ecology*, 63, 995–1001.

Thorn, K., 2016. A quick guide to light microscopy in cell biology. *Molecular Biology of the Cell*, 27(2), 219–222.

Thynne, E., McDonald, M.C. and Solomon, P.S., 2015. Phytopathogen emergence in the genomics era. *Trends in Plant Science*, 20(4), 246–255.

Tombolini, R., Unge, A., Davey, M.E., de Bruijn, F.J. and Jansson, J.K., 1997. Flow cytometric and microscopic analysis of GFP-tagged *Pseudomonas fluorescens* bacteria. *FEMS Microbiology Ecology*, 22, 17–28.

Tombolini, R., Van Der Gaag, D.J., Gerhardson, B. and Jansson, J.K., 1999. Colonization pattern of the biocontrol strain *Pseudomonas chlororaphis* MA 342 on barley seeds visualized by using green fluorescent protein. *Applied and Environmental Microbiology*, 65, 3674–3680.

Trivedi, P., Leach, J.E., Tringe, S.G., Sa, T. and Singh, B.K., 2020. Plant–microbiome interactions: From community assembly to plant health. *Nature Reviews Microbiology*, 18, 607–621.

Troxler, J., Zala, M., Natsch, A., Moënne-Loccoz, Y. and Défago, G., 1997. Autecology of the biocontrol strain *Pseudomonas fluorescens* CHA0 in the rhizosphere and inside roots at later stages of plant development. *FEMS Microbiology Ecology*, 23, 119–130.

Ul-Hamid, A., 2018. *A Beginners' Guide to Scanning Electron Microscopy* (Vol. 1, p. 402). Springer International Publishing, Cham.

Valette, M., Rey, M., Doré, J., Gerin, F. and Wisniewski-Dyé, F., 2020. Identification of a small set of genes commonly regulated in rice roots in response to beneficial rhizobacteria. *Physiology and Molecular Biology of Plants*, 26, 2537–2551.

van Noorden, G.E., Kerim, T., Goffard, N., Wiblin, R., Pellerone, F.I., Rolfe, B.G. and Mathesius, U., 2007. Overlap of proteome changes in *Medicago truncatula* in response to auxin and *Sinorhizobium meliloti*. *Plant Physiology*, 144(2), 1115–1131.

Vimal, S.R., Singh, J.S., Arora, N.K. and Singh, S., 2017. Soil–plant–microbe interactions in stressed agriculture management: A review. *Pedosphere*, 27, 177e192.

Wang, X., Zhang, X., Liu, L., Xiang, M., Wang, W., Sun, X., . . . Liu, X., 2015. Genomic and transcriptomic analysis of the endophytic fungus *Pestalotiopsis fici* reveals its lifestyle and high potential for synthesis of natural products. *BMC Genomics*, 16(1), 28.

Waśkiewicz, A., Irzykowska, L., Drzewiecka, K., Bocianowski, J., Dobosz, B., Weber, Z., Karolewski, Z., Krzyminiewski, R. and Goliński, P., 2013. Plant–pathogen interactions during infection process of asparagus with *Fusarium* spp. *Central European Journal of Biology*, 8, 1065–1076.

Wolinska, K.W., Vannier, N., Thiergart, T., Pickel, B., Gremmen, S., Piasecka, A., . . . Hacquard, S., 2021. Tryptophan metabolism and bacterial commensals prevent fungal dysbiosis in *Arabidopsis* roots. *Proceedings of the National Academy of Sciences*, 118(49), e2111521118.

Wood, K.V. and DeLuca, M., 1987. Photographic detection of luminescence in *Escherichia coli* containing the gene for firefly luciferase. *Analytical Biochemistry*, 161, 501–507.

Woods, A.E. and Stirling, J.W., 2018. Transmission electron microscopy. In *Bancroft's Theory and Practice of Histological Techniques*, 8th ed., pp. 434–475; Suvarna, S. K., Layton, C. and Bancroft, J. D., Eds. Elsevier, Philadelphia, PA.

Xie, H., Gu, Q., Chen, W., Meng, X., Guo, Z., Zhang, Y. and Li, H., 2024. Mitigation of oxidative stress and inflammatory factors, along with the antibrowning and antimicrobial effects of cassia seed microbial fermentation solution. *Frontiers in Microbiology*, 15, 1400505.

Yang, X., Varga, T., Liu, C. and Scheibe, T.D., 2017. What can we learn from in-soil imaging of a live plant: X-ray computed tomography and 3D numerical simulation of root–soil system. *Rhizosphere*, 3, 259–262.

Yi, Y., Li, Z., Song, C. and Kuipers, O.P., 2018. Exploring plant–microbe interactions of the rhizobacteria *Bacillus subtilis* and *Bacillus mycoides* by use of the CRISPR-Cas9 system. *Environmental Microbiology*, 20, 4245–4260.

Zhalnina, K., Louie, K.B., Hao, Z., Mansoori, N., Da Rocha, U.N., Shi, S., Cho, H., Karaoz, U., Loqué, D., Bowen, B.P. and Firestone, M.K., 2018. Dynamic root exudate chemistry and microbial substrate preferences drive patterns in rhizosphere microbial community assembly. *Nature Microbiology*, 3(4), 470–480.

Zhang, J., Subramanian, S., Stacey, G. and Yu, O., 2009. Flavones and flavonols play critical roles during nodulation of *Medicago truncatula* by *Sinorhizobium meliloti*. *The Plant Journal*, 57(1), 171–183.

Zhuang, X., McPhee, K.E., Coram, T.E., Peever, T.L. and Chilvers, M.I., 2012. Rapid transcriptome characterization and parsing of sequences in a non-model host–pathogen interaction; pea-*Sclerotinia sclerotiorum*. *BMC Genomics*, 13, 1–19.

3 Deciphering the Molecular Mechanisms of Symbiotic Nitrogen Fixation in Legumes

Manisha Yadav and Sabhyata Bhatia

3.1 INTRODUCTION

Among all the macronutrients required for proper plant growth and development, nitrogen is crucial in defining their occurrence in a habitat. It is the ability to fix atmospheric nitrogen that facilitates the leguminous plants to thrive in nitrogen-limited soils (Nutman 1997; Hughes, Ringelberg, and Bruneau 2025). Symbiotic nitrogen fixation (SNF) is thought to have originated about 100 million years ago (MYA) in angiosperms, followed by a pattern of "single gain and multiple losses" of the nodulation trait across species within the nitrogen-fixing clade (NFC). This clade comprises four orders, including, Fabales, Fagales, Cucurbitales, and Rosales (H.-L. Li et al. 2015; van Velzen, Doyle, and Geurts 2019). Additionally, approximately 65 MYA, distinct lineages of Caesalpinioideae and Papilionoid legumes acquired the ability to establish mutualistic associations with rhizobia (Zhao et al. 2021). Legume–rhizobia symbiosis occurs in approximately 88% of all legume species, occupying approximately 250 Mha area, and fixes about 90 Tg nitrogen (N) per year worldwide (de Faria et al. 1989; Sachs, Skophammer, and Regus 2011).

Root nodule symbiosis (RNS) represents a mutualistic association between plant roots and nitrogen-fixing rhizobia, where it performs this task in exchange for carbohydrates, thereby allowing the leguminous plants to grow with a competitive advantage compared to non-legumes (Ferguson et al. 2019). In nitrogen-limited conditions, root exudates containing flavonoid substances are released (Figure 3.1). These flavonoids are sensed by compatible rhizobial species, activating the transcription factor (TF) NodD (nodulation protein D), which further induces the expression of downstream nod genes (*nodA*, *nodB*, and *nodC*) responsible for Nod factor (NF) biosynthesis. NFs are lipochitooligosaccharide (LCO) molecules comprising a common core structure but different side groups, which confer specificity and are responsible for the selection of compatible rhizobial partners. The perception of rhizobia triggers a complex signaling cascade in the host plant, ultimately forming novel root structures called nodules (Ferguson et al. 2010). A nearly anoxic environment is maintained inside root nodules for the oxygen-labile nitrogenase enzyme to work and fix atmospheric nitrogen (Einsle and Rees 2020).

Based on the persistence of meristematic activity, nodules can be divided into determinate and indeterminate types. Irrespective of the types, further development, metabolism, and senescence of the nodule largely depend on the host plant (Andrews and Andrews 2017). Indeterminate nodules possess a persistent meristem at their apex, which keeps adding new cells in the proximal direction, resulting in an elongated nodule shape. A developmental gradient of cells is formed inside the nodule, where the youngest cells are located near the apical meristem and the oldest are closer to the root attachment site. Indeterminate nodules are organized into several zones, and the meristem comprises zone I. Next is the infection zone (zone II), where the rhizobia are released and subsequently undergo division and differentiation. Zone III, known as the fixation zone, is characterized by several changes, including the activation of *nif* genes and the peripheral accumulation of

DOI: 10.1201/9781003608080-3

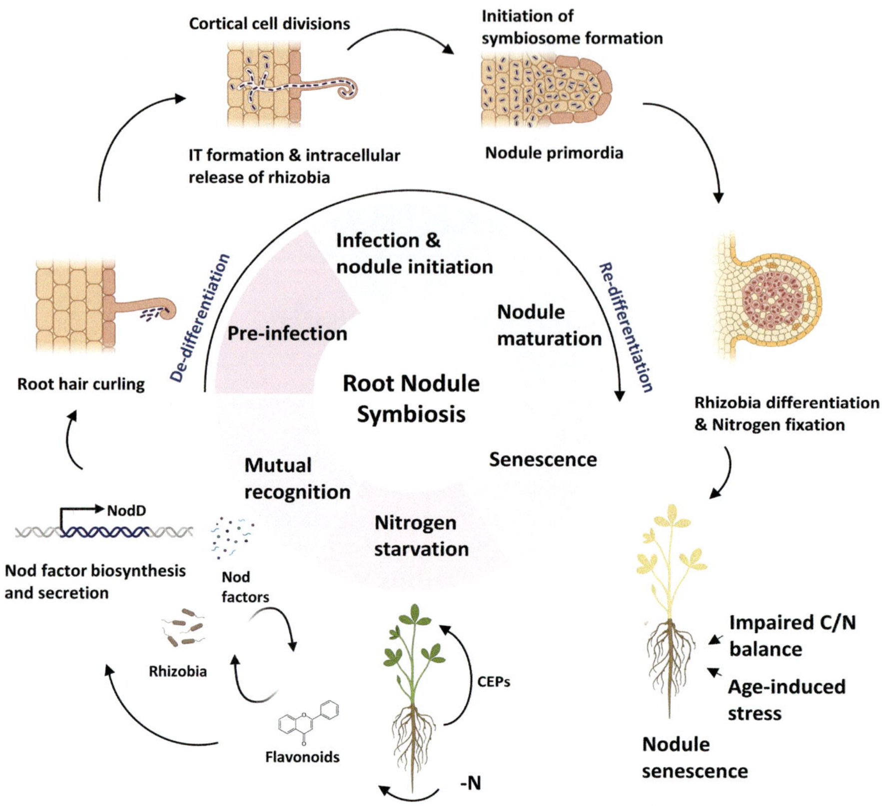

FIGURE 3.1 Step-by-step progression of root nodule symbiosis. N, nitrogen; CEPs, C-terminally encoded peptides; NodD, nodulation protein D; IT, infection thread.

amyloplasts in infected cells, which marks the start of the fixation zone. The oldest zone is zone IV (the senescence zone), where bacterial degradation occurs (Vasse et al. 1990). In contrast, determinate nodules possess a peripheral meristem that is only active in the early stages of development, resulting in spherical nodules (Shumilina et al. 2023).

The legume–rhizobia symbiosis has been rigorously investigated over the past several decades, resulting in a substantial body of knowledge. However, in-depth molecular mechanisms underlying the interactions between both partners remain partially understood and require further evaluation. Key pathways that require deeper exploration include the potential role of C-terminally encoded peptides (CEPs) in promoting flavonoid biosynthesis, flavonoid-mediated NodD activation, NF perception, regulation of infection thread (IT) development, bacteroid formation, and the regulation and coordination of nodule senescence, among others (Luo, Liu, and Xie 2023). Understanding the molecular mechanisms associated with each step of nodule development and function could enable us to engineer novel legume cultivars with improved crop productivity.

This chapter systematically reviews the signaling networks underlying nodule development: Mutual recognition, flavonoids as attractants for rhizobia, activation of rhizobial NFs and their perception by host plants, the symbiosis signaling pathway, terminal bacteroid differentiation, autoregulation of nodulation (AON), and the onset of nodule senescence and the environmental impact on it (Figure 3.1). Emphasis is given to the aspects that have been intensively addressed during the last few decades, representing the current frontiers in the study of RNS.

3.2 PLANT CUES AND ADAPTIVE RESPONSE IN RHIZOBIA

Plant perception of nitrogen limitation and the induction of starvation signals are linked to enhanced rhizobial attraction toward the roots. This step majorly relies on the biosynthesis and release of flavonoid substances, which act as chemoattractants, including sugars and various dicarboxylic acids such as succinate, malate, fumarate, and amino acids, viz., glutamate, proline, arginine, phenylalanine, and tryptophan (Zuluaga et al. 2020; Webb et al. 2014). Upon perception of these flavonoids, an array of genes required for NF synthesis is activated (Luo, Liu, and Xie 2023). In plants, flavonoids serve several functions, including defense against biotic and abiotic stress (e.g., flavones and flavonols), reproduction, and pigment formation (flavonols and anthocyanins). Their presence in the rhizosphere can be attributed to two main mechanisms. First, they may be passively released from the root sheath and border cells as they break down, or they may be actively secreted through membrane transporters upon sensing nitrogen limitation (Shaw, Morris, and Hooker 2006; Biała-Leonhard et al. 2021). For instance, genistein is released via ATP-binding cassette transporters (Sugiyama, Shitan, and Yazaki 2007). These transporters display higher affinity to glycosides than to their aglycone counterparts, thereby facilitating more efficient secretion of glycosides (Cesco et al. 2010; León-Barrios et al. 1993). This observation aligned with the chemical properties of flavonoid glycosides, which, due to their greater solubility, are more efficiently exported across membranes into the rhizosphere (Nouwen, Fardoux, and Giraud 2016). Once secreted, these glycosides are susceptible to hydrolysis by microbial glycosidases present in soil, leading to an increased presence of aglycons in root exudates compared to glycosides (Hungria, Joseph, and Phillips 1991). Some membrane transporters, such as multidrug and toxic compound extrusion type transporters, are essential for the release of the isoflavonoid genistein from roots. Under nitrogen and phosphate starvation, *LaMATE2* expression is upregulated, and silencing of *LaMATE2* leads to a reduction in genistein efflux, consequently impairing nodule formation (Biała-Leonhard et al. 2021). This suggests a positive role for *LaMATE2* in isoflavonoid release and nodule development. Thus, a decrease in "physiologically available nitrogen" induces flavonoid biosynthesis and the upregulation of genes associated with flavonoid transporters (Biała-Leonhard et al. 2021; Deng et al. 2019; Coronado et al. 1995). Despite plants synthesizing a broad range of flavonoids, only a few have been identified in root exudates, including aglycones of luteolin, genistein, naringenin, hesperetin, and daidzein, with a marginal contribution from other substances. Therefore, these may be considered as signaling molecules (Bag et al. 2022; Shumilina et al. 2023; Biała-Leonhard et al. 2021).

Flavonoids accumulate in actively growing root regions (Abdel-Lateif, Bogusz, and Hocher 2012), and supplementation with legume root exudates or pure naringenin in the growth medium results in a significant increase in rhizobial growth (*Bradyrhizobium* sp.) (Nouwen, Fardoux, and Giraud 2016). Nouwen *et al.* (2019) reported an approximately fourfold, dose-dependent increase in bacterial culture density upon supplementation of the medium with naringenin. This growth-promoting effect of flavonoids is independent of the regulators of the *nod* genes.

3.3 NODD ACTIVATION AND EARLY SIGNALING

A member of the LysR family of transcription activators, called NodD, works upstream of *nod* genes and regulates their expression. The members of this family are characterized by an N-terminal helix-turn-helix DNA-binding domain and are typically maintained in a repressed state, and only activated upon inducer binding. NodD has three conserved isoforms, NodD1, NodD2, and NodD3, showing a high level of homology with each other (Shang et al. 2022; Peck, Fisher, and Long 2006). Different species possess distinct numbers of *nod* genes. Consider the case of the *Sinorhizobium meliloti* genome, which contains all three isoforms with more than 77% similarity in their amino acid sequence. These isoforms differ in their physiological roles, and the activation of NodD1 and NodD2 requires plant-derived phenolic compounds, while NodD3 can be expressed even when flavonoid inducers are absent. For transactivation, NodD binds to a conserved 55-bp region within the promoter of *nod* genes, known as the nod box, thereby triggering the expression of *nod* genes (Peck, Fisher, and Long 2006; Györgypal, Iyer, and Kondorosi 1988). In addition to NodD proteins, a second mechanism for the recognition of isoflavonoids also exists, consisting of NodV and NodW in *Bradyrhizobium diazoefficiens* USDA110. NodW belongs to the two-component regulatory family that is crucial for mediating *nod* gene activation in response to isoflavonoid signals (Shang et al. 2022).

To date, we do not have evidence of a direct interaction between flavonoids and the NodD regulator. However, it has been reported that in *Sinorhizobium meliloti* culture, the affinity of NodD1 for the *nod* gene region increased several fold in the presence of GroEL protein and luteolin compared to the control. It has been hypothesized that a change in NodD1 conformation facilitates its binding to flavonoids (e.g., luteolin, eriodictyol, daidzein, 7-hydroxyflavone, and naringenin). This interaction is believed to be facilitated by GroEL and enhances NodD1's affinity for the promoter region of *nod* genes. Additionally, the study demonstrated that while a variety of flavonoids, such as naringenin, eriodictyol, luteolin, and daidzein, can bind NodD1, only luteolin effectively activates it to initiate *nod* gene transcription. Non-inducing flavonoids, in contrast, function as competitive inhibitors of luteolin, thus blocking NodD activation (Peck, Fisher, and Long 2006). However, various flavonoid ligands may trigger different conformational changes in the NodD protein (Wassem et al. 2017). Cooper et al. established a link between the structural characteristics of the polyphenols and their ability to either activate or suppress nodulation genes. It was established that hydroxyl groups positioned at the 7- and/or 4′- sites play a pivotal role in the activation of *nod* genes, independent of other ring substitutions. For example, in *Rhizobium leguminosarum*, *nod* gene expression significantly increased by flavonoids containing hydroxyl groups at the 4′-, 5-, and 7-positions (Cooper 2004).

Unfortunately, the mechanism of the transport of flavonoids into the rhizobial cell and further NodD activation has been minimally studied in rhizobial species besides *S. meliloti* (Ogawa and Long 1995). Expanding such mechanistic studies to other rhizobial species could help decipher the broader principles of NodD functioning and pave the way for future research in this direction.

3.4 ONSET OF SYMBIOSIS BY HOST–FLAVONOID-INDUCED NOD FACTOR SIGNALING

NF perception in the host induces developmental changes needed for the entry of rhizobia in legumes. The root hair tip curls up to entrap rhizobia within, from where they move into a plant-derived structure called IT. Parallelly, NFs also induce rapid cell divisions in the cortex and pericycle regions, marking the beginning of nodule formation (S. Roy et al. 2020). Some rhizobial strains induce nodulation more efficiently than others, and their success may be attributed to the individual NFs synthesized by them (Carlson et al. 1993; Sanjuan et al. 1992).

NFs are LCOs composed of β-1–4-linked *N*-acetyl glucosamine residues with a long acyl chain attached to the terminal glucosamine. Various bacteria encoded genes, e.g., *nodA* (acyl transferase), *nodB* (deacetylase), and *nodC* (*N*-acetyl glucosamine transferase), are required for the biosynthesis of a basic LCO structure (Shumilina et al. 2023). NF biosynthesis begins with the conversion of fructose-6-phosphate to glucosamine-6-phosphate, a reaction catalyzed by glucosamine synthase (NodM). This intermediate is subsequently acetylated to form N-acetylglucosamine, which is then oligomerized into chains composed of three to five N-acetylglucosamine monomers. In the subsequent step, a fatty acid moiety is attached to the oligosaccharide backbone by a specific acyltransferase. The fatty acid portion of NFs typically consists of mono- or polyunsaturated C16 or longer acyl chains (Spaink and Lugtenberg 1994; Spaink 1992). For instance, the NFs of *S. meliloti* contain C16:2 and C16:3 fatty acid moieties, while those from *R. leguminosarum* contain various C16, C18:2, C18:3, and C18:4 residues (Göttfert 1993; Gladchuk et al. 2020). This indicates that the species-specific fatty acids are involved in the NF synthesis.

The NFs are secreted through the membrane transporters of rhizobial cells. These NFs bind to the receptors present on the epidermal cells of the host, further triggering downstream signaling cascades comprising SYMRK, CCaMK, CYCLOPS, NIN, and other proteins (S. Roy et al. 2020).

The NF receptors are LysM-type serine/threonine kinases. These receptors were first identified in *Lotus japonicus* as NFR1 and NFR5, and in *Medicago truncatula* as LYK3 and pseudokinase NFP (Nod factor perception) (Luo, Liu, and Xie 2023). These NF receptors are integral proteins located within the plasma membrane of root epithelial cells. The signaling pathways activated upon receptor engagement have been partially elucidated. In the absence of symbiotic interactions, the malectin-like domain (MLD) remains dissociated from the leucine-rich repeat (LRR)-containing receptor kinase SYMRK. This SYMRK interacts with NFR5 and changes its conformation, thereby facilitating the attachment of a NF. After the formation of this complex, the kinase domain most likely triggers the reverse conformational change. In contrast to NFR1, NFR5 is a pseudo-kinase; i.e., it does not have an intracellular kinase domain. Therefore, depending upon the availability of NFR1 to interact with its activation partner, NFR5 may initiate two distinct signaling pathways. In the first scenario, NFR1, which possesses a kinase domain, forms a heterodimer with NFR5, thereby facilitating downstream signal transduction. As an alternative, in the absence of NFR1, NFR5 may interact with NFR5-interacting cytoplasmic kinase 4 (NiCK4) (Antolín-Llovera, Ried, and Parniske 2014). NiCK4 is known to phosphorylate the intracellular domain of NFR5. This phosphorylation event exposes the regulatory regions on NFR5, enabling it to recruit additional signaling partners and initiate downstream signaling (Shumilina et al. 2023; S. Roy et al. 2020). Interestingly, NFR1 can also phosphorylate NFR5 in a similar fashion. Activation of this pathway triggers the release of calcium ions from intracellular stores via calcium channels LjCASTOR and LjPOLLUX/MtDMI1, LjNUCLEOPORIN85 (LjNUP85) and LjNUP133, and MtCNGC15 (cyclic nucleotide gated channels) a/b/c (S. Roy et al. 2020). This calcium influx activates the calcium/calmodulin-dependent serine/threonine kinase (CCaMK), known as DMI3 (does not make infections 3) in *Medicago truncatula*. Activated CCaMK subsequently phosphorylates the TF CYCLOPS (referred to as IPD3, *Interacting Protein of DMI3* in *M. truncatula*). Furthermore, CYCLOPS facilitates the transcriptional activation of transcriptional regulator NIN (nodule inception), which activates downstream symbiosis genes (Figure 3.2). In *Medicago*, DELLA proteins facilitate the phosphorylation of IPD3

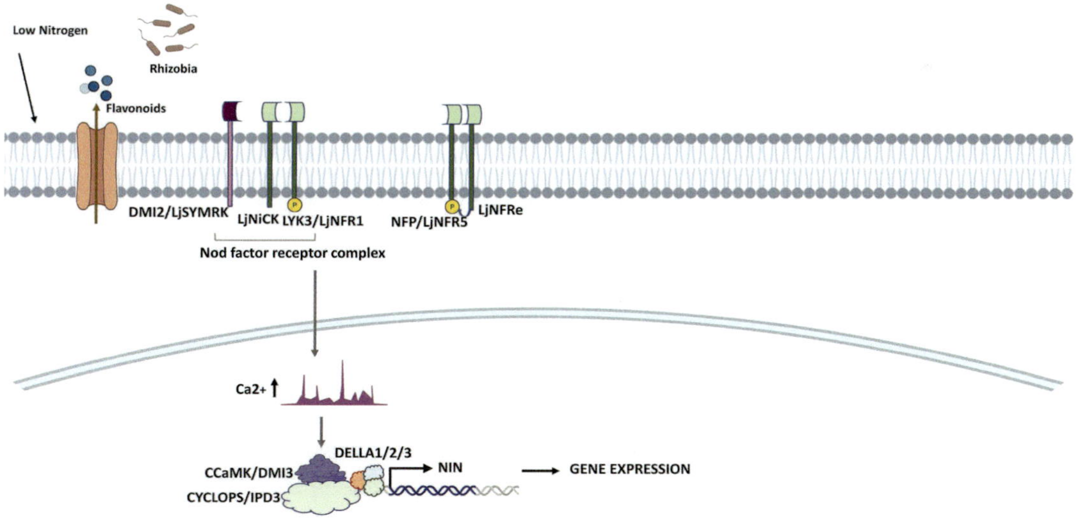

FIGURE 3.2 Molecular components involved in the early stage of nodulation. Flavonoids produced under low nitrogen conditions trigger NF biosynthesis and secretion. NFs are perceived by receptors present at the plasma membrane and subsequently activate LjSYMRK/MtDMI2 to initiate NF signaling. The calcium channel complex present on the nuclear membrane is activated, increasing calcium influx and resulting in Ca^{2+} spiking inside the nucleus. LjCCaMK/MtDMI3 decodes this Ca^{2+} spiking and phosphorylates LjCYCLOPS/MtIPD3, which, in association with DELLA1/2/3, activates the TF NIN, required for symbiotic gene expression.

and enhance its interaction with TFs NSP1 and NSP2 (nodulation signaling pathway proteins 1 and 2, respectively). This complex binds to the promoter region of the *NIN* gene (Figure 3.2), thereby promoting its transcriptional activation (Fonouni-Farde et al. 2016; Singh et al. 2014). It has been recently found that MtMYC2 positively regulates rhizobial infection by directly binding to the MtIPD3 promoter and activating its expression (Guo et al. 2024).

In addition to the classical NF-dependent pathway, alternative mechanisms also contribute to the establishment of legume–rhizobia symbiosis. Okazaki et al. reported that rhizobial strains lacking functional *nod* genes are still capable of initiating nodulation (Okazaki et al. 2013). This NF-independent symbiotic interaction appears to be partially mediated by the activity of type III and type IV secretion systems (T3SS and T4SS, respectively), which play essential roles in promoting nodule formation (Safronova et al. 2020; Okazaki et al. 2016). The involvement of the T3SS in nodule symbiosis has become increasingly known. Although the T3SS was first identified for its role in pathogenic bacterial infection of plant roots, its involvement in establishing symbiotic interactions has only been revealed in recent studies (Okazaki et al. 2016). Two additional components are crucial for the initiation of symbiosis through the T3SS pathway: (i) Activation of rhizobial infection by plant-secreted flavonoids and (ii) the presence of the NodD transcriptional regulator, which also activates the expression of the *nopL* gene required in T3SS assembly (Viprey et al. 1998; Marie et al. 2003). The resulting NopL protein is thought to initiate the expression of *T3SS* genes (Viprey et al. 1998). As this system relies on direct molecular interaction with host cells, it delivers specific effector proteins to the surface of root cells, known as nodulation outer proteins (Nops), enabling rhizobia to effectively modulate host cell metabolism by suppressing plant immune responses and promoting symbiotic development. Consequently, rhizobial mutants deficient in effector synthesis or defective in secretion machinery exhibit reduced nodulation efficiency (Staehelin and Krishnan 2015). For instance, *Sinorhizobium fredii* secretes eight characterized Nops—NopA–D, NopL, NopM, NopP, and NopX via T3SS in response to genistein, a flavonoid exuded by plant roots (Lorio, Kim, and Krishnan 2004; Jiménez-Guerrero et al. 2015; Deakin et al. 2005). Till now, about 30 Nops have been identified; however, the majority of them have been found based on *in silico* prediction and not yet confirmed at the functional protein level (Busset et al. 2021).

Recent studies have also shown that rhizobia capable of employing both NF-dependent and NF-independent pathways may gain a symbiotic advantage. One such example is the phenomenon of rhizobial synergy, where co-inoculation with multiple rhizobial strains leads to enhanced nodulation efficiency (Safronova et al. 2023).

3.5 MOLECULAR REGULATION OF NODULE INITIATION AND ORGANOGENESIS

Once the bacteria enter the root cells, they are surrounded by the plant-derived plasma membrane, forming a structure called infection chamber, which is eventually modified into IT (Coba de la Peña et al. 2017). The expansion of IT is facilitated by continuous biosynthesis and release of plant plasma membrane and cell wall components, involving an array of proteins including protein required for polar growth (RPG), actin rearrangement proteins, cystathionine-β-synthase-like 1 (CBS1), E3 ligase Lumpy infections (LIN)/CERBERUS, and Vapyrin (VPY) (Shumilina et al. 2023). The exact molecular processes underlying the expansion and polar growth of ITs remain unclear.

Among the key components of nodulation signaling are the regulatory pathways triggered by NIN. It induces the expression of *NF-YA1* and *NF-YB1* genes in *L. japonicus* roots (Soyano et al. 2013). These TFs are indispensable for nodulation. Their knockout prevents nodulation, and their overexpression results in increased nodule cell proliferation by causing increased local divisions in root tissue required for nodule establishment (Soyano et al. 2013). Additionally, an increased level of cytokinin increases NIN expression, and the NIN protein in turn promotes auxin accumulation in root cells, thereby promoting their proliferation (Kohlen et al. 2018; Tiwari et al. 2021).

Furthermore, NIN has been reported to upregulate the expression of CLAVATA3/embryo-surrounding region-related (CLE) peptides, which act as signaling molecules known to inhibit nodulation (Okamoto et al. 2013; Kereszt et al. 2018). Consequently, these peptides play a crucial role in modulating nodule development. In addition, NIN also controls the expression of nodulation pectate lyase (NPL), which plays a role in facilitating rhizobial entry into root cells by modulating cell wall permeability. Overall, this observation shows that NIN regulates various aspects of symbiosis, including nodule number, IT formation and expansion, and rhizobial entry in root cells.

3.6 TERMINAL DIFFERENTIATION OF SYMBIOTIC RHIZOBIA

Rhizobia are released into root cells via an infection droplet produced at the apex of the IT or at sites where the IT disrupts the host cell wall. At this interface, rhizobial cells encounter the host plasma membrane and are internalized into the cellular lumen through the process of endocytosis (S. Roy et al. 2020). The newly formed structure, called symbiosome, contains rhizobia that are differentiated into bacteroids and are enclosed by a plant cell membrane (Coba de la Peña et al. 2017). In determinate nodules, individual symbiosomes often undergo fusion or accommodate further division in bacteroids, resulting in symbiosomes containing multiple bacteroids (Coba de la Peña et al. 2017), whereas in indeterminate nodules, symbiosomes divide along with the bacteroids, usually resulting in one bacteroid per symbiosome (Prell and Poole 2006). The bacteroids in determinate nodules are spherical, while in indeterminate nodules they show complex branching, forming a Y-shaped structure (Kondorosi, Mergaert, and Kereszt 2013). Due to continuous DNA replication without subsequent cytokinesis, the bacteroid DNA content increases during differentiation. Indeterminate nodules display prominent morphological differentiation of bacteroids, which may be terminal (Peter Mergaert, Kereszt, and Kondorosi 2019). In contrast, the bacteroid differentiation in determinate nodules is less morphologically pronounced as in *L. japonicus* (Szczyglowski et al. 1998).

Nodule-specific cysteine-rich (NCR) peptides are the key regulators of terminal bacteroid differentiation. Mutations in the genes encoding for enzymes that regulate NCR peptide maturation result in defects in fixing nitrogen. *DNF1 (DEFECTIVE IN NITROGEN FIXATION1)* encodes a peptidase essential for cleaving the N-terminal signal peptide guiding the NCR peptides to the symbiosome. *MtDNF1* mutants are not capable of forming bacteroids (Van de Velde et al. 2010; D. Wang et al. 2010). Recently, it has been found that *DNF1* expression is positively regulated by MtMYC2 through direct binding to its promoter (Guo et al. 2024).

NCR peptides are species-specific and are essential for the viability of differentiated bacteroids (Yang, Gao, and Pan 2025). They are found to be present in species representing the inverted-repeat-lacking clade (IRLC), a distinct evolutionary lineage of the Faboideae (Papilionaceae) subfamily that includes several legume crops. NCR peptides have evolved independently in different species, and the size and composition of gene families determine the morphotype of bacteroids (Montiel et al. 2017; Downie and Kondorosi 2021). Usually, cationic, anionic, and neutral NCR peptides are known, and diversification of cationic peptides has resulted in the ability to impose major morphological changes on the endosymbionts (Montiel et al. 2017). The length of mature NCR peptides typically varies between 24 and 65 amino acids (Lima et al. 2020). They are characterized by a conserved N-terminal domain, and mature peptides form disulfide bonds due to the presence of four or six cysteine residues. Despite some similarities in structure, the sequences of NCR peptides are highly variable and heterogeneous. Across ten legume species representing a broad overview of the IRLC lineage, ranging from the most basal species *Glycyrrhiza uralensis* to representatives of the three major subclades: Hedysaroid (*Onobrychis viciifolia*), Astragalean (*Oxytropis lamberti* and *Astragalus canadensis*), and Vicioid (*Galega orientalis*, *Cicer arietinum*, *Ononis spinosa*, *Pisum sativum*, *Medicago sativa*, and *Medicago truncatula*), the number of NCR peptides ranges from as few as 7 in *Glycyrrhiza uralensis* to more than 600 in *M. truncatula*. These legumes show different bacteroid morphotypes, including spherical, elongated, swollen-pleomorphic, and elongated-branched forms (Peter Mergaert et al. 2003; Guefrachi et al. 2014).

The NCR peptides infiltrate bacterial cells via the plasma membrane and drive their terminal differentiation into bacteroids in the cytoplasm. They are hypothesized to do so by regulating the levels of cell cycle proteins: CtrA, DnaA, GcrA, and FtsZ (P. Roy et al. 2020; Farkas et al. 2014; P. Mergaert 2018). DnaA starts the replication of DNA and activates GcrA, which represses DnaA and upregulates CtrA. CtrA functions during the late stage of the bacterial cell cycle, where it represses the initiation of DNA replication and promotes cell division. NCR peptides modulate the activity of FtsZ, a key protein in Z-ring formation required for cytokinesis. Therefore, the promotion of DNA replication and inhibition of cell division is responsible for bacterial differentiation into bacteroids (P. Roy et al. 2020; Penterman et al. 2014). Some NCR peptides can act as positive regulators of nitrogen fixation in one legume and may negatively affect it in another, and this may be one of the contributing factors to NCR peptides being species-specific (Q. Wang et al. 2017).

3.7 SYSTEMIC REGULATION OF NODULE NUMBER

The regulation of nodule number may occur in two ways: Via the AON pathway or independently of the AON pathway. AON is a long-distance signaling pathway that involves signals from developing nodules to the shoot, where reverse signaling from the shoot to the root inhibits further nodulation (Caetano-Anollés and Gresshoff 1991; Krusell et al. 2002). Constitutive expression of a gene encoding nodulation-induced peptides, *MtCLE12/MtCLE13* (LjCLE-RS1/LjCLE-RS2), inhibits nodulation (Okamoto et al. 2009; Mortier et al. 2010), and LjHAR1 (*MtSUNN*) encoding HAR1(SUNN) is required for responses to these peptides (Figure 3.3) (Mortier et al. 2014). LjCLE-RS2 binds to LjHAR1 in the shoot and halts further nodulation in wild-type (WL) *L. japonicus* but not in a *har1* mutant, suggesting that the LjHAR1 receptor kinase is essential for regulating the AON pathway mediated by LjCLE (Okamoto et al. 2013).

CLE peptides are mobile root-to-shoot signals in the AON pathway, and the soil nitrogen content serves as a trigger for AON. The nitrate-induced expression of *LjCLE-RS1* and *LjCLE-RS2* is mediated by a NIN-like protein LjNRSYM1 (nitrate unresponsive symbiosis 1) (Nishida et al. 2018). In *Ljnrsym1* mutants, nodulation and IT development occur independently of nitrate levels. Studies on *Ljhar1–7 Ljnrsym1–1* double mutant revealed that while *LjHAR1* regulates nodule number under nitrogen-deficient conditions, it does not influence IT number, nodule cell size, or nitrogen fixation capacity (Nishida et al. 2018). Therefore, the regulation of SNF by *LjNRSYM1* under high-nitrogen conditions likely operates in parallel with the AON pathway. MtCLV2 is another component of the AON pathway that forms a complex with the receptor kinase MtSUNN and the pseudokinase MtCORYNE (CRN) (Crook, Schnabel, and Frugoli 2016). Knockdown and mutagenesis of *CLV2* showed a hypernodulation phenotype (Krusell et al. 2011). Like *Mtsunn* mutant, the effect of *Mtcrn* mutant is also shoot-derived, and *Mtcrn* mutants also do not respond to the overexpression of CLE peptides (Crook, Schnabel, and Frugoli 2016; Nowak, Schnabel, and Frugoli 2019). Together, these evidences suggest crosstalk between AON receptors in the shoot. *LjKLV* is another receptor-like kinase that works alongside *LjHAR1* in the AON pathway. MtCRA2 is another receptor kinase identified through mutation studies that systemically regulates the number of nodules from the shoot. It works independently of MtSUNN and helps limit nodulation when sufficient nitrate is present (Figure 3.3, represented by dotted line) (Laffont et al. 2019). Another pathway (pathway 2 in Figure 3.3) represents a relatively new class of peptides, the CEPs, that are required for the initiation of nodule formation. To date, more than 15 CEPs have been characterized. They work as signaling molecules in nitrogen-limited conditions (Figure 3.3) (Okamoto et al. 2009). CEPs bind to their receptors and induce the synthesis of C-terminally encoded peptide downstream (CEPD) types 1 and 2. These signaling peptides are synthesized in the shoot and transported to the root, where they increase the expression of nitrogen transporters in root cells (Figure 3.3, highlighted in red). CEPs bind to their receptor C-terminally encoded peptide receptor (CEPR) and induce the expression of miR2111, which migrates to the root and degrades its targets, *TML1/2* (Too Much Love 1/2, the inhibitors of nodulation), resulting in the induction of nodulation (Gautrat, Laffont, and Frugier

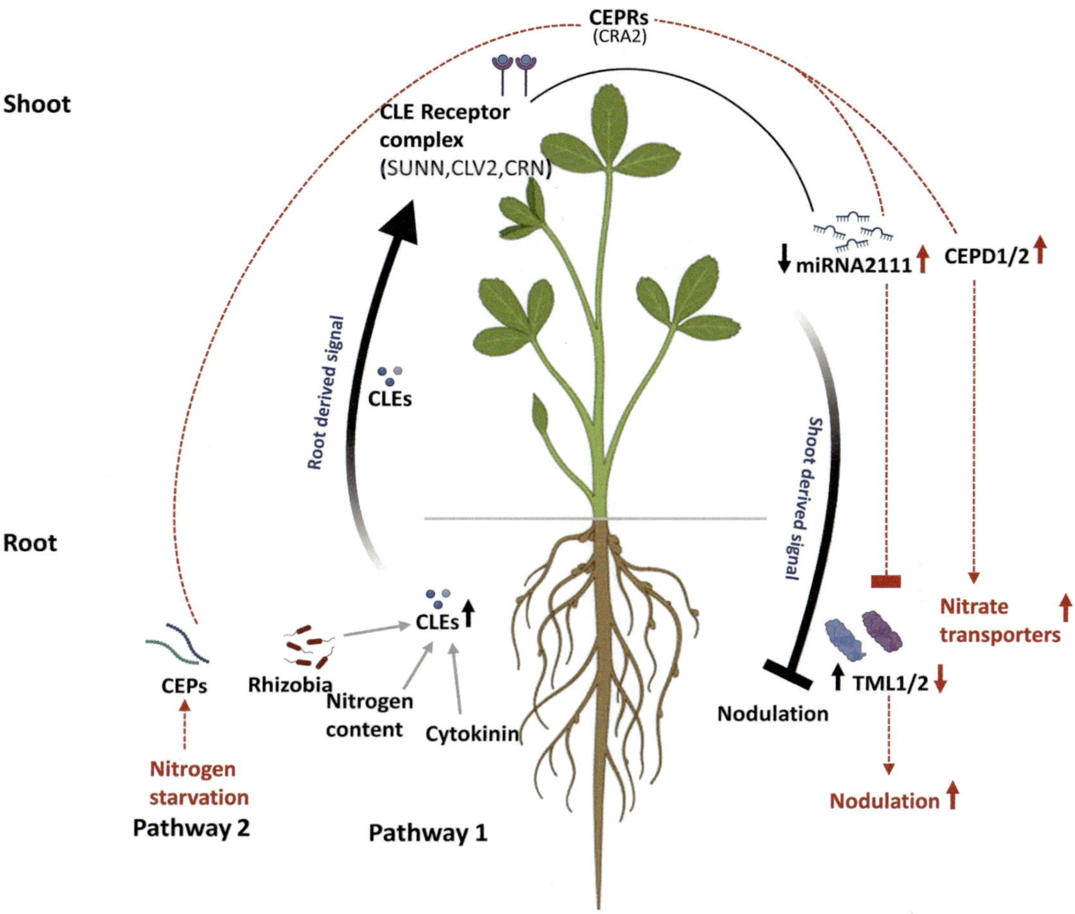

FIGURE 3.3 Two independent pathways regulate nodule number. Pathway 1 involves rhizobia, nitrogen, and cytokinin-induced expression of CLE peptides, which are transported to the shoot via the xylem. These CLE peptides are perceived by the SUNN receptor kinase in association with Clavata 2 (CLV2) and the pseudo-kinase CRN. This receptor complex suppresses miR2111 expression, leading to the accumulation of TML1/2, thereby inhibiting further nodulation. Pathway 2 involves Compact Root Architecture 2 (CRA2) mediated perception of CEPs in the shoot, which promotes the expression of nitrogen transporters and stimulates nodulation.

2020). Overexpression of MtCEP1 promotes nodulation (Imin et al. 2013), and CRA2 is needed for CEPs to work; however, the analysis of the *cra2 sunn* double mutant indicates that these nodule regulatory pathways function independently (Laffont et al. 2019; Nowak, Schnabel, and Frugoli 2019).

TML functions to control nodule number in the root after receiving a signal from the shoot (Magori and Kawaguchi 2009). In *L. japonicus*, *TML* encodes a Kelch-repeat F-box protein that performs the proteasomal degradation of its target proteins in the late stage of the AON pathway (Takahara et al. 2013). *Medicago* has two *TML* genes *MtTML1* and *MtTML2*, and silencing these genes through RNAi increases nodule formation, indicating their involvement in AON (Gautrat et al. 2021). In *L. japonicus*, the expression of miR2111 regulates *TML* through a shoot-dependent mechanism that requires *HAR1* (Tsikou et al. 2018).

Interestingly, *L. japonicus* mutant *Ljlhk1*, a cytokinin receptor mutant, exhibits reduced AON activity, resulting in enhanced nodulation (Murray et al. 2007), while mutation in the Medicago ortholog *MtCRE1* resulted in reduced nodule number. When this mutation occurs in the *Mtsunn* background, the nodule number becomes similar to that of WT. This indicates that two distinct pathways are employed in regulating nodule number (Gonzalez-Rizzo, Crespi, and Frugier 2006). Another gene is *CRE1*, which interacts with AON signals to locally regulate nodule number. Active MtCRE1 is required for the expression of *MtCLE12* and *MtCLE13* in the nodule meristem (Mortier, Holsters, and Goormachtig 2012). Similarly, LjCLE-RS1 and LjCLE-RS3 accumulate during the infection process and decline in *lhk1–1* mutant roots (Tsikou et al. 2018), indicating that early cytokinin signaling is essential for sending the AON signal. *LjLHK1* expression in cortical tissues further highlights that cytokinin signaling in cortical cells functions as a local mediator coordinating with the systemic AON mechanism (Miri et al. 2019).

3.8 NODULE SENESCENCE

Senescence is a precisely regulated process that relies on multiple factors such as nitrate levels and osmotic stress. It is the last stage of legume–rhizobia symbiosis (Kazmierczak et al. 2020; S. Roy et al. 2020) and is accompanied by a color change in the nodule from pink to green, resulting from the degradation of leghemoglobin (Roponen 1970). The membrane of the symbiosome bursts, releasing the enclosed molecules. Simultaneously, proteases, lipases, and transporters to transport the breakdown products get activated (Timmers et al. 2000; Pladys and Vance 1993; Schippers et al. 2015). In case of indeterminate nodules, a prominent senescence zone is formed, which migrates from the distal end to the apical part of the nodule, while in determinate nodules, it begins from the center and extends toward the periphery (Van de Velde et al. 2006).

Additionally, legumes can assess nodule performance and the quality of rhizobia, and if the gain from symbiosis is below a certain threshold, then the plant withdraws the allocation of resources to such a strain (Porter et al. 2024). Despite its functional relevance, comprehensive studies on this process are lacking; however, the transcriptomic studies on *Bradyrhizobium diazoefficiens* bacteroids revealed that, based on the developmental stage and the environmental factors, gene expression profiles of bacteroids varied and were majorly regulated by the host (Franck et al. 2018). The process of senescence usually begins when flowering is completed. However, environmental stresses can significantly accelerate its onset and intensify the associated changes in nodule morphology, anatomy, physiology, and metabolism (X. Li et al. 2018; de Zélicourt et al. 2012). This is accompanied by the generation of reactive oxygen species (ROS) in plant cells and thus the development of oxidative stress (Osmolovskaya et al. 2018; Huang et al. 2019).

Along with several factors, the nitric oxide (NO) levels also regulate nodule senescence (Fukudome et al). NO accumulation is reduced upon LjGlb1 (class 1 phytoglobin) overexpression in *L. japonicus*, resulting in prolonged nodule functioning. Contrastingly, treatment with 1-aminocyclopropane-1-carboxylic acid (ACC), ethylene precursor, resulted in the inhibition of nitrogenase activity and the induction of senescence genes (Fukudome et al. 2019). In *Medicago*, *MtNAC969* is induced in the nitrogen fixation zone upon nitrate treatment, and its knockdown caused early senescence by upregulating two senescence-induced cysteine proteases (CPs) (de Zélicourt et al. 2012). RNAi of the two CP genes, *MtCP6* and *MtVPE* (vacuolar processing enzyme), prolonged nodule life and nitrogen fixation. In functional nodules, reduced glutathione (GSH) is required in high amounts in tissues, whereas the oxidized glutathione (GSSG) correlates with nodule senescence (Ivanova et al. 2022).

Nodule senescence is accompanied by increased lipo- and glycoxidation. The age-related increase in oxidative stress is associated with high sugar concentrations present in root nodules. These molecules are known to accumulate in all parts of the plant, including root nodules (Shumilina et al. 2019, 2021; Soboleva et al. 2022). Importantly, the high glycation potential of regulatory sugars and sugar-related molecules, such as fructose, sucrose, and sugar phosphates, suggests a possible role for glycation in modulating plant stress responses (Bilova et al. 2016; Shumilina et al. 2019). Numerous genes linked to nodule senescence encode for CPs, TFs, regulatory peptides, and membrane proteins. The glycosylation process may alter the structural integrity of these proteins, thereby affecting their role in senescence.

Additionally, transcriptomic studies have deciphered age-related changes in the expression dynamics of senescence-related genes, including the upregulation of genes related to ethylene biosynthesis and ethylene-dependent TFs. Studies revealed that gibberellin inhibits the onset of nodule senescence (Serova et al. 2019). Together, these findings highlight the important role of phytohormones in regulating nodule senescence, but the exact regulatory mechanisms remain unclear.

3.9 CONCLUSION AND FUTURE PERSPECTIVES

RNS represents a highly specialized and evolutionarily significant plant–microbe interaction that has been extensively investigated over the past few decades. Despite substantial advancements in understanding the genetic and molecular framework governing this process, several regulatory components and key signaling pathways remain scarcely characterized, particularly those involved during the early stages of legume-rhizobia interaction. Understanding the precise molecular mechanisms and signaling cascades activated during the initial recognition and communication between rhizobia and host root cells is critical. A comprehensive understanding of these early events holds high potential for extending RNS to agriculturally important nonleguminous crop species. Furthermore, leveraging high-throughput genome sequencing across a diverse panel of genotypes within a species may facilitate the discovery of novel genes and allelic variants involved in modulating various aspects of nodule development and nitrogen fixation, thereby creating prospects for genetic improvement and sustainable agriculture.

REFERENCES

Abdel-Lateif, Khalid, Didier Bogusz, and Valérie Hocher. 2012. "The Role of Flavonoids in the Establishment of Plant Roots Endosymbioses with Arbuscular Mycorrhiza Fungi, Rhizobia and Frankia Bacteria." *Plant Signaling & Behavior* 7 (6): 636–41.

Andrews, Mitchell, and Morag E. Andrews. 2017. "Specificity in Legume–Rhizobia Symbioses." *International Journal of Molecular Sciences* 18 (4). https://doi.org/10.3390/ijms18040705.

Antolín-Llovera, Meritxell, Martina K. Ried, and Martin Parniske. 2014. "Cleavage of the Symbiosis Receptor-Like Kinase Ectodomain Promotes Complex Formation with Nod Factor Receptor 5." *Current Biology: CB* 24 (4): 422–27.

Bag, Sagar, Anupam Mondal, Anusha Majumder, Sunil Kanti Mondal, and Avishek Banik. 2022. "Flavonoid Mediated Selective Cross-Talk between Plants and Beneficial Soil Microbiome." *Phytochemistry Reviews: Proceedings of the Phytochemical Society of Europe* 21 (5): 1739–60.

Biała-Leonhard, Wanda, Laura Zanin, Stefano Gottardi, Rita de Brito Francisco, Silvia Venuti, Fabio Valentinuzzi, Tanja Mimmo, et al. 2021. "Identification of an Isoflavonoid Transporter Required for the Nodule Establishment of the Rhizobium–Fabaceae Symbiotic Interaction." *Frontiers in Plant Science* 12. https://www.frontiersin.org/journals/plant-science/articles/10.3389/fpls.2021.758213.

Bilova, Tatiana, Elena Lukasheva, Dominic Brauch, Uta Greifenhagen, Gagan Paudel, Elena Tarakhovskaya, Nadezhda Frolova, Juliane Mittasch, Gerd Ulrich Balcke, and Alain Tissier. 2016. "A Snapshot of the Plant Glycated Proteome: Structural, Functional, and Mechanistic Aspects." *The Journal of Biological Chemistry* 291 (14): 7621–36.

Busset, Nicolas, Djamel Gully, Albin Teulet, Joël Fardoux, Alicia Camuel, David Cornu, Dany Severac, Eric Giraud, and Peter Mergaert. 2021. "The Type III Effectome of the Symbiotic *Bradyrhizobium vignae* Strain ORS3257." *Biomolecules* 11 (11). https://doi.org/10.3390/biom11111592.

Caetano-Anollés, G., and P. M. Gresshoff. 1991. "Plant Genetic Control of Nodulation." *Annual Review of Microbiology* 45: 345–82.

Carlson, R. W., J. Sanjuan, U. R. Bhat, J. Glushka, H. P. Spaink, A. H. Wijfjes, A. A. van Brussel, T. J. Stokkermans, N. K. Peters, and G. Stacey. 1993. "The Structures and Biological Activities of the Lipo-Oligosaccharide Nodulation Signals Produced by Type I and II Strains of *Bradyrhizobium japonicum*." *The Journal of Biological Chemistry* 268 (24): 18372–81.

Cesco, Stefano, Guenter Neumann, Nicola Tomasi, Roberto Pinton, and Laure Weisskopf. 2010. "Release of Plant-Borne Flavonoids into the Rhizosphere and Their Role in Plant Nutrition." *Plant and Soil* 329 (1): 1–25.

Coba de la Peña, Teodoro, Elena Fedorova, José J. Pueyo, and M. Mercedes Lucas. 2017. "The Symbiosome: Legume and Rhizobia Co-Evolution toward a Nitrogen-Fixing Organelle?" *Frontiers in Plant Science* 8: 2229.

Cooper, James E. 2004. “Multiple Responses of Rhizobia to Flavonoids during Legume Root Infection.” In *Advances in Botanical Research*, 41:1–62. Academic Press.

Coronado, C., Jas Zuanazzi, C. Sallaud, J. C. Quirion, R. Esnault, H. P. Husson, A. Kondorosi, and P. Ratet. 1995. “Alfalfa Root Flavonoid Production Is Nitrogen Regulated.” *Plant Physiology* 108 (2): 533–42.

Crook, Ashley D., Elise L. Schnabel, and Julia A. Frugoli. 2016. “The Systemic Nodule Number Regulation Kinase SUNN in *Medicago truncatula* Interacts with MtCLV2 and MtCRN.” *The Plant Journal: For Cell and Molecular Biology* 88 (1): 108–19.

Deakin, William J., Corinne Marie, Maged M. Saad, Hari B. Krishnan, and William J. Broughton. 2005. “NopA Is Associated with Cell Surface Appendages Produced by the Type III Secretion System of *Rhizobium* sp. Strain NGR234.” *Molecular Plant–Microbe Interactions: MPMI* 18 (5): 499–507.

Deng, Bo, Yuanyuan Li, Dandan Xu, Qingqing Ye, and Guihua Liu. 2019. “Nitrogen Availability Alters Flavonoid Accumulation in *Cyclocarya paliurus* via the Effects on the Internal Carbon/Nitrogen Balance.” *Scientific Reports* 9 (1): 2370.

Downie, J. Allan, and Eva Kondorosi. 2021. “Why Should Nodule Cysteine-Rich (NCR) Peptides Be Absent from Nodules of Some Groups of Legumes but Essential for Symbiotic N-Fixation in Others?” *Frontiers in Agronomy* 3. https://www.frontiersin.org/journals/agronomy/articles/10.3389/fagro.2021.654576.

Einsle, Oliver, and Douglas C. Rees. 2020. “Structural Enzymology of Nitrogenase Enzymes.” *Chemical Reviews* 120 (12): 4969–5004.

Faria, S. M. de, G. P. Lewis, J. I. Sprent, and J. M. Sutherland. 1989. “Occurrence of Nodulation in the Leguminosae.” *The New Phytologist* 111 (4): 607–19.

Farkas, Attila, Gergely Maróti, Hajnalka Dürgő, Zoltán Györgypál, Rui M. Lima, Katalin F. Medzihradszky, Attila Kereszt, Peter Mergaert, and Éva Kondorosi. 2014. “*Medicago truncatula* Symbiotic Peptide NCR247 Contributes to Bacteroid Differentiation through Multiple Mechanisms.” *Proceedings of the National Academy of Sciences of the United States of America* 111 (14): 5183–88.

Ferguson, Brett J., Arief Indrasumunar, Satomi Hayashi, Meng-Han Lin, Yu-Hsiang Lin, Dugald E. Reid, and Peter M. Gresshoff. 2010. “Molecular Analysis of Legume Nodule Development and Autoregulation.” *Journal of Integrative Plant Biology* 52 (1): 61–76.

Ferguson, Brett J., Céline Mens, April H. Hastwell, Mengbai Zhang, Huanan Su, Candice H. Jones, Xitong Chu, and Peter M. Gresshoff. 2019. “Legume Nodulation: The Host Controls the Party.” *Plant, Cell & Environment* 42 (1): 41–51.

Fonouni-Farde, Camille, Sovanna Tan, Maël Baudin, Mathias Brault, Jiangqi Wen, Kirankumar S. Mysore, Andreas Niebel, Florian Frugier, and Anouck Diet. 2016. “DELLA-Mediated Gibberellin Signalling Regulates Nod Factor Signalling and Rhizobial Infection.” *Nature Communications* 7 (September): 12636.

Franck, Sooyoung, Kent N. Strodtman, Jing Qiu, and David W. Emerich. 2018. “Transcriptomic Characterization of *Bradyrhizobium diazoefficiens* Bacteroids Reveals a Post-Symbiotic, Hemibiotrophic-like Lifestyle of the Bacteria within Senescing Soybean Nodules.” *International Journal of Molecular Sciences* 19 (12): 3918.

Fukudome, Mitsutaka, Eri Watanabe, Ken-Ichi Osuki, Ryujiro Imaizumi, Toshio Aoki, Manuel Becana, and Toshiki Uchiumi. 2019. “Stably Transformed *Lotus japonicus* Plants Overexpressing Phytoglobin LjGlb1-1 Show Decreased Nitric Oxide Levels in Roots and Nodules as well as Delayed Nodule Senescence.” *Plant & Cell Physiology* 60 (4): 816–25.

Gautrat, Pierre, Carole Laffont, and Florian Frugier. 2020. “Compact Root Architecture 2 Promotes Root Competence for Nodulation through the miR2111 Systemic Effector.” *Current Biology: CB* 30 (7): 1339–45.

Gautrat, Pierre, Carole Laffont, Florian Frugier, and Sandrine Ruffel. 2021. “Nitrogen Systemic Signaling: From Symbiotic Nodulation to Root Acquisition.” *Trends in Plant Science* 26 (4): 392–406.

Gladchuk, Aleksey, Julia Shumilina, Alena Kusnetsova, Ksenia Bureiko, Susan Billig, Alexander Tsarev, Irina Alexandrova, et al. 2020. “High-Throughput Fingerprinting of Rhizobial Free Fatty Acids by Chemical Thin-Film Deposition and Matrix-Assisted Laser Desorption/Ionization Mass Spectrometry.” *Methods and Protocols* 3 (2). https://doi.org/10.3390/mps3020036.

Gonzalez-Rizzo, Silvina, Martin Crespi, and Florian Frugier. 2006. “The *Medicago truncatula* CRE1 Cytokinin Receptor Regulates Lateral Root Development and Early Symbiotic Interaction with *Sinorhizobium meliloti*.” *The Plant Cell* 18 (10): 2680–93.

Göttfert, M. 1993. “Regulation and Function of Rhizobial Nodulation Genes.” *FEMS Microbiology Reviews* 10 (1–2): 39–63.

Guefrachi, Ibtissem, Marianna Nagymihaly, Catalina I. Pislariu, Willem Van de Velde, Pascal Ratet, Mohamed Mars, Michael K. Udvardi, Eva Kondorosi, Peter Mergaert, and Benoît Alunni. 2014. “Extreme Specificity of NCR Gene Expression in *Medicago truncatula*.” *BMC Genomics* 15 (1): 712.

Guo, Da, Jingrui Li, Peng Liu, Yuzhan Wang, Na Cao, Xiangling Fang, Tao Wang, and Jiangli Dong. 2024. "The Jasmonate Pathway Promotes Nodule Symbiosis and Suppresses Host Plant Defense in *Medicago truncatula*." *Molecular Plant* 17 (8): 1183–203.

Györgypal, Zoltan, Narayan Iyer, and Adam Kondorosi. 1988. "Three Regulatory nodD Alleles of Diverged Flavonoid-Specificity Are Involved in Host-Dependent Nodulation by *Rhizobium meliloti*." *Molecular & General Genetics: MGG* 212 (1): 85–92.

Huang, Honglin, Farhan Ullah, Dao-Xiu Zhou, Ming Yi, and Yu Zhao. 2019. "Mechanisms of ROS Regulation of Plant Development and Stress Responses." *Frontiers in Plant Science* 10 (June): 800.

Hughes, Colin, Jens J. Ringelberg, and Anne Bruneau. 2025. "Legumes." *Current Biology: CB* 35 (9): R323–28.

Hungria, M., C. M. Joseph, and D. A. Phillips. 1991. "Rhizobium Nod Gene Inducers Exuded Naturally from Roots of Common Bean (*Phaseolus vulgaris* L.)." *Plant Physiology* 97 (2): 759–64.

Imin, Nijat, Nadiatul A. Mohd-Radzman, Huw A. Ogilvie, and Michael A. Djordjevic. 2013. "The Peptide-Encoding CEP1 Gene Modulates Lateral Root and Nodule Numbers in *Medicago truncatula*." *Journal of Experimental Botany* 64 (17): 5395–409.

Ivanova, Kira A., Ekaterina N. Chernova, Olga A. Kulaeva, Anna V. Tsyganova, Pyotr G. Kusakin, Iana V. Russkikh, Igor A. Tikhonovich, and Viktor E. Tsyganov. 2022. "The Regulation of Pea (*Pisum sativum* L.) Symbiotic Nodule Infection and Defense Responses by Glutathione, Homoglutathione, and Their Ratio." *Frontiers in Plant Science* 13 (March): 843565.

Jiménez-Guerrero, Irene, Francisco Pérez-Montaño, Carlos Medina, Francisco Javier Ollero, and Francisco Javier López-Baena. 2015. "NopC Is a *Rhizobium*-Specific Type 3 Secretion System Effector Secreted by *Sinorhizobium* (*Ensifer*) *fredii* HH103." *PloS One* 10 (11): e0142866.

Kazmierczak, Théophile, Li Yang, Eric Boncompagni, Eliane Meilhoc, Florian Frugier, Pierre Frendo, Claude Bruand, Véronique Gruber, and Renaud Brouquisse. 2020. "Legume Nodule Senescence: A Coordinated Death Mechanism between Bacteria and Plant Cells." In *Advances in Botanical Research*, 181–212. Elsevier.

Kereszt, Attila, Peter Mergaert, Jesús Montiel, Gabriella Endre, and Éva Kondorosi. 2018. "Impact of Plant Peptides on Symbiotic Nodule Development and Functioning." *Frontiers in Plant Science* 9. https://www.frontiersin.org/journals/plant-science/articles/10.3389/fpls.2018.01026.

Kohlen, Wouter, Jason Liang Pin Ng, Eva E. Deinum, and Ulrike Mathesius. 2018. "Auxin Transport, Metabolism, and Signalling during Nodule Initiation: Indeterminate and Determinate Nodules." *Journal of Experimental Botany* 69 (2): 229–44.

Kondorosi, Eva, Peter Mergaert, and Attila Kereszt. 2013. "A Paradigm for Endosymbiotic Life: Cell Differentiation of *Rhizobium* Bacteria Provoked by Host Plant Factors." *Annual Review of Microbiology* 67: 611–28.

Krusell, Lene, Lene H. Madsen, Shusei Sato, Grégoire Aubert, Aratz Genua, Krzysztof Szczyglowski, Gérard Duc, et al. 2002. "Shoot Control of Root Development and Nodulation Is Mediated by a Receptor-Like Kinase." *Nature* 420 (6914): 422–26.

Krusell, Lene, Naoto Sato, Izumi Fukuhara, Bjørn E. V. Koch, Christina Grossmann, Satoru Okamoto, Erika Oka-Kira, Yoko Otsubo, Grégoire Aubert, and Tomomi Nakagawa. 2011. "The *Clavata2* Genes of Pea and *Lotus japonicus* Affect Autoregulation of Nodulation." *The Plant Journal: For Cell and Molecular Biology* 65 (6): 861–71.

Laffont, Carole, Emeline Huault, Pierre Gautrat, Gabriella Endre, Peter Kalo, Virginie Bourion, Gérard Duc, and Florian Frugier. 2019. "Independent Regulation of Symbiotic Nodulation by the SUNN Negative and CRA2 Positive Systemic Pathways." *Plant Physiology* 180 (1): 559–70.

León-Barrios, M., F. D. Dakora, C. M. Joseph, and D. A. Phillips. 1993. "Isolation of *Rhizobium meliloti* Nod Gene Inducers from Alfalfa Rhizosphere Soil." *Applied and Environmental Microbiology* 59 (2): 636–39.

Li, Hong-Lei, Wei Wang, Peter E. Mortimer, Rui-Qi Li, De-Zhu Li, Kevin D. Hyde, Jian-Chu Xu, Douglas E. Soltis, and Zhi-Duan Chen. 2015. "Large-Scale Phylogenetic Analyses Reveal Multiple Gains of Actinorhizal Nitrogen-Fixing Symbioses in Angiosperms Associated with Climate Change." *Scientific Reports* 5 (1): 14023.

Li, Xin, Hao Feng, Jiangqi Wen, Jiangli Dong, and Tao Wang. 2018. "MtCAS31 Aids Symbiotic Nitrogen Fixation by Protecting the Leghemoglobin MtLb120-1 under Drought Stress in *Medicago truncatula*." *Frontiers in Plant Science* 9 (May). https://doi.org/10.3389/fpls.2018.00633.

Lima, R. M., S. Kylarová, P. Mergaert, and E. Kondorosi. 2020. "Unexplored Arsenals of Legume Peptides with Potential for Their Applications in Medicine and Agriculture." *Frontiers in Microbiology* 11: 1307.

Lorio, Julio C., Won Seok Kim, and Hari B. Krishnan. 2004. "NopB, a Soybean Cultivar-Specificity Protein from *Sinorhizobium fredii* USDA257, Is a Type III Secreted Protein." *Molecular Plant–Microbe Interactions: MPMI* 17 (11): 1259–68.

Luo, Zhenpeng, Haiyue Liu, and Fang Xie. 2023. "Cellular and Molecular Basis of Symbiotic Nodule Development." *Current Opinion in Plant Biology* 76 (102478): 102478.

Magori, Shimpei, and Masayoshi Kawaguchi. 2009. "Long-Distance Control of Nodulation: Molecules and Models." *Molecules and Cells* 27:129–34.

Marie, Corinne, William J. Deakin, Virginie Viprey, Joanna Kopcińska, Wladyslaw Golinowski, Hari B. Krishnan, Xavier Perret, and William J. Broughton. 2003. "Characterization of Nops, Nodulation Outer Proteins, Secreted via the Type III Secretion System of NGR234." *Molecular Plant–Microbe Interactions: MPMI* 16 (9): 743–51.

Mergaert, P. 2018. "Role of Antimicrobial Peptides in Controlling Symbiotic Bacterial Populations." *Natural Product Reports* 35 (4): 336–56.

Mergaert, Peter, Attila Kereszt, and Eva Kondorosi. 2019. "Gene Expression in Nitrogen-Fixing Symbiotic Nodule Cells in *Medicago truncatula* and Other Nodulating Plants." *The Plant Cell* 32 (1): 42–68.

Mergaert, Peter, Krisztina Nikovics, Zsolt Kelemen, Nicolas Maunoury, Danièle Vaubert, Adam Kondorosi, and Eva Kondorosi. 2003. "A Novel Family in *Medicago truncatula* Consisting of More than 300 Nodule-Specific Genes Coding for Small, Secreted Polypeptides with Conserved Cysteine Motifs." *Plant Physiology* 132 (1): 161–73.

Miri, Mandana, Preetam Janakirama, Terry Huebert, Loretta Ross, Tim McDowell, Kathleen Orosz, Katharina Markmann, and Krzysztof Szczyglowski. 2019. "Inside out: Root Cortex-localized LHK1 Cytokinin Receptor Limits Epidermal Infection of *Lotus japonicus* Roots by *Mesorhizobium loti*." *The New Phytologist* 222 (3): 1523–37.

Montiel, Jesús, J. Allan Downie, Attila Farkas, Péter Bihari, Róbert Herczeg, Balázs Bálint, Peter Mergaert, Attila Kereszt, and Éva Kondorosi. 2017. "Morphotype of Bacteroids in Different Legumes Correlates with the Number and Type of Symbiotic NCR Peptides." *Proceedings of the National Academy of Sciences of the United States of America* 114 (19): 5041–46.

Mortier, Virginie, Griet Den Herder, Ryan Whitford, Willem Van de Velde, Stephane Rombauts, Katrien D'haeseleer, Marcelle Holsters, and Sofie Goormachtig. 2010. "CLE Peptides Control *Medicago truncatula* Nodulation Locally and Systemically." *Plant Physiology* 153 (1): 222–37.

Mortier, Virginie, Marcelle Holsters, and Sofie Goormachtig. 2012. "Never Too Many? How Legumes Control Nodule Numbers." *Plant, Cell & Environment* 35 (2): 245–58.

Mortier, Virginie, Anton Wasson, Pavel Jaworek, Annick De Keyser, Martijn Decroos, Marcelle Holsters, Petr Tarkowski, Ulrike Mathesius, and Sofie Goormachtig. 2014. "Role of LONELY GUY Genes in Indeterminate Nodulation on *Medicago truncatula*." *The New Phytologist* 202 (2): 582–93.

Murray, Jeremy D., Bogumil J. Karas, Shusei Sato, Satoshi Tabata, Lisa Amyot, and Krzysztof Szczyglowski. 2007. "A Cytokinin Perception Mutant Colonized by *Rhizobium* in the Absence of Nodule Organogenesis." *Science (New York, N.Y.)* 315 (5808): 101–4.

Nishida, Hanna, Sachiko Tanaka, Yoshihiro Handa, Momoyo Ito, Yuki Sakamoto, Sachihiro Matsunaga, Shigeyuki Betsuyaku, Kenji Miura, Takashi Soyano, and Masayoshi Kawaguchi. 2018. "A NIN-LIKE PROTEIN Mediates Nitrate-Induced Control of Root Nodule Symbiosis in *Lotus japonicus*." *Nature Communications* 9 (1): 499.

Nouwen, Nico, Joel Fardoux, and Eric Giraud. 2016. "NodD1 and NodD2 Are Not Required for the Symbiotic Interaction of *Bradyrhizobium* ORS285 with Nod-Factor-Independent *Aeschynomene* Legumes." *PloS One* 11 (6): e0157888.

Nouwen, Nico, Daniel Gargani, and Eric Giraud. 2019. "The Modification of the Flavonoid Naringenin by *Bradyrhizobium* sp. Strain ORS285 Changes the Nod Genes Inducer Function to a Growth Stimulator." *Molecular Plant–Microbe Interactions: MPMI* 32 (11): 1517–25.

Nowak, Stephen, Elise Schnabel, and Julia Frugoli. 2019. "The *Medicago truncatula* CLAVATA3-LIKE CLE12/13 Signaling Peptides Regulate Nodule Number Depending on the CORYNE but Not the COMPACT ROOT ARCHITECTURE2 Receptor." *Plant Signaling & Behavior* 14 (6): 1598730.

Nutman, Phillip Sadler. 1997. "Genetics of Symbiosis and Nitrogen Fixation in Legumes." *Proceedings of the Royal Society of London. Series B, Containing Papers of a Biological Character. Royal Society (Great Britain)* 172 (1029): 417–37.

Ogawa, J., and S. R. Long. 1995. "The *Rhizobium meliloti* groELc Locus Is Required for Regulation of Early Nod Genes by the Transcription Activator NodD." *Genes & Development* 9 (6): 714–29.

Okamoto, Satoru, Erika Ohnishi, Shusei Sato, Hirokazu Takahashi, Mikio Nakazono, Satoshi Tabata, and Masayoshi Kawaguchi. 2009. "Nod Factor/Nitrate-Induced CLE Genes That Drive HAR1-Mediated Systemic Regulation of Nodulation." *Plant & Cell Physiology* 50 (1): 67–77.

Okamoto, Satoru, Hidefumi Shinohara, Tomoko Mori, Yoshikatsu Matsubayashi, and Masayoshi Kawaguchi. 2013. "Root-Derived CLE Glycopeptides Control Nodulation by Direct Binding to HAR1 Receptor Kinase." *Nature Communications* 4 (1): 2191.

Okazaki, Shin, Takakazu Kaneko, Shusei Sato, and Kazuhiko Saeki. 2013. "Hijacking of Leguminous Nodulation Signaling by the Rhizobial Type III Secretion System." *Proceedings of the National Academy of Sciences of the United States of America* 110 (42): 17131–36.

Okazaki, Shin, Panlada Tittabutr, Albin Teulet, Julien Thouin, Joël Fardoux, Clémence Chaintreuil, Djamel Gully, et al. 2016. "*Rhizobium*–Legume Symbiosis in the Absence of Nod Factors: Two Possible Scenarios with or without the *T3SS*." *The ISME Journal* 10 (1): 64–74.

Osmolovskaya, Natalia, Julia Shumilina, Ahyoung Kim, Anna Didio, Tatiana Grishina, Tatiana Bilova, Olga A. Keltsieva, et al. 2018. "Methodology of Drought Stress Research: Experimental Setup and Physiological Characterization." *International Journal of Molecular Sciences* 19 (12): 4089.

Peck, Melicent C., Robert F. Fisher, and Sharon R. Long. 2006. "Diverse Flavonoids Stimulate NodD1 Binding to Nod Gene Promoters in *Sinorhizobium meliloti*." *Journal of Bacteriology* 188 (15): 5417–27.

Penterman, Jon, Ryan P. Abo, Nicole J. De Nisco, Markus F. F. Arnold, Renato Longhi, Matteo Zanda, and Graham C. Walker. 2014. "Host Plant Peptides Elicit a Transcriptional Response to Control the *Sinorhizobium meliloti* Cell Cycle during Symbiosis." *Proceedings of the National Academy of Sciences of the United States of America* 111 (9): 3561–66.

Pladys, D., and C. P. Vance. 1993. "Proteolysis during Development and Senescence of Effective and Plant Gene-Controlled Ineffective Alfalfa Nodules." *Plant Physiology* 103 (2): 379–84.

Porter, Stephanie S., Simon E. Dupin, R. Ford Denison, E. Toby Kiers, and Joel L. Sachs. 2024. "Host-Imposed Control Mechanisms in Legume-Rhizobia Symbiosis." *Nature Microbiology* 9 (8): 1929–39.

Prell, Juergen, and Philip Poole. 2006. "Metabolic Changes of Rhizobia in Legume Nodules." *Trends in Microbiology* 14 (4): 161–68.

Roponen, Ilma. 1970. "The Effect of Darkness on the Leghemoglobin Content and Amino Acid Levels in the Root Nodules of Pea Plants." *Physiologia Plantarum* 23 (3): 452–60.

Roy, Proyash, Mingkee Achom, Helen Wilkinson, Beatriz Lagunas, and Miriam L. Gifford. 2020. "Symbiotic Outcome Modified by the Diversification from 7 to over 700 Nodule-Specific Cysteine-Rich Peptides." *Genes* 11 (4). https://doi.org/10.3390/genes11040348.

Roy, Sonali, Wei Liu, Raja Sekhar Nandety, Ashley Crook, Kirankumar S. Mysore, Catalina I. Pislariu, Julia Frugoli, Rebecca Dickstein, and Michael K. Udvardi. 2020. "Celebrating 20 Years of Genetic Discoveries in Legume Nodulation and Symbiotic Nitrogen Fixation." *The Plant Cell* 32 (1): 15–41.

Sachs, Joel L., Ryan G. Skophammer, and John U. Regus. 2011. "Evolutionary Transitions in Bacterial Symbiosis." *Proceedings of the National Academy of Sciences of the United States of America* 108 (Suppl 2): 10800–807.

Safronova, Vera I., Polina V. Guro, Anna L. Sazanova, Irina G. Kuznetsova, Andrey A. Belimov, Valentin V. Yakubov, Elizaveta R. Chirak, et al. 2020. "Rhizobial Microsymbionts of Kamchatka *Oxytropis* Species Possess Genes of the Type III and VI Secretion Systems, Which Can Affect the Development of Symbiosis." *Molecular Plant–Microbe Interactions: MPMI* 33 (10): 1232–41.

Safronova, Vera I., Anna Sazanova, Andrey Belimov, Polina Guro, Irina Kuznetsova, Denis Karlov, Elizaveta Chirak, et al. 2023. "Synergy between Rhizobial Co-Microsymbionts Leads to an Increase in the Efficiency of Plant–Microbe Interactions." *Microorganisms* 11 (5). https://doi.org/10.3390/microorganisms11051206.

Sanjuan, J., R. W. Carlson, H. P. Spaink, U. R. Bhat, W. M. Barbour, J. Glushka, and G. Stacey. 1992. "A 2-O-Methylfucose Moiety Is Present in the Lipo-Oligosaccharide Nodulation Signal of *Bradyrhizobium japonicum*." *Proceedings of the National Academy of Sciences of the United States of America* 89 (18): 8789–93.

Schippers, Jos H. M., Romy Schmidt, Carol Wagstaff, and Hai-Chun Jing. 2015. "Living to Die and Dying to Live: The Survival Strategy behind Leaf Senescence." *Plant Physiology* 169 (2): 914–30.

Serova, Tatiana A., Anna V. Tsyganova, Igor A. Tikhonovich, and Viktor E. Tsyganov. 2019. "Gibberellins Inhibit Nodule Senescence and Stimulate Nodule Meristem Bifurcation in Pea (*Pisum sativum* L.)." *Frontiers in Plant Science* 10 (March):285.

Shang, Jiao Ying, Pan Zhang, Yu Wen Jia, Yi Ning Lu, Yue Wu, Shuang Ji, La Chen, En Tao Wang, Wen Xin Chen, and Xin Hua Sui. 2022. "Coordinated Regulation of Symbiotic Adaptation by NodD Proteins and NolA in the Type I Peanut Bradyrhizobial Strain *Bradyrhizobium zhanjiangense* CCBAU51778." *Microbiological Research* 265 (September): 127188.

Shaw, Liz J., Phil Morris, and John E. Hooker. 2006. "Perception and Modification of Plant Flavonoid Signals by Rhizosphere Microorganisms." *Environmental Microbiology* 8 (11): 1867–80.

Shumilina, Julia, Daria Gorbach, Veronika Popova, Alexander Tsarev, Alena Kuznetsova, Maria Grashina, Mandy Dorn, et al. 2021. "Protein Glycation and Drought Response of Pea (*Pisum sativum* L.) Root Nodule Proteome: A Proteomics Approach." *Biological Communications* 66 (3). https://doi.org/10.21638/spbu03.2021.303.

Shumilina, Julia, Alena Kusnetsova, Alexander Tsarev, Henry C. Janse van Rensburg, Sergei Medvedev, Vadim Demidchik, Wim Van den Ende, and Andrej Frolov. 2019. "Glycation of Plant Proteins: Regulatory Roles and Interplay with Sugar Signalling?" *International Journal of Molecular Sciences* 20 (9): 2366.

Shumilina, Julia, Alena Soboleva, Evgeny Abakumov, Oksana Y. Shtark, Vladimir A. Zhukov, and Andrej Frolov. 2023. "Signaling in Legume–Rhizobia Symbiosis." *International Journal of Molecular Sciences* 24 (24). https://doi.org/10.3390/ijms242417397.

Singh, Sylvia, Katja Katzer, Jayne Lambert, Marion Cerri, and Martin Parniske. 2014. "CYCLOPS, a DNA-Binding Transcriptional Activator, Orchestrates Symbiotic Root Nodule Development." *Cell Host & Microbe* 15 (2): 139–52.

Soboleva, Alena, Nadezhda Frolova, Kseniia Bureiko, Julia Shumilina, Gerd U. Balcke, Vladimir A. Zhukov, Igor A. Tikhonovich, and Andrej Frolov. 2022. "Dynamics of Reactive Carbonyl Species in Pea Root Nodules in Response to Polyethylene Glycol (PEG)-Induced Osmotic Stress." *International Journal of Molecular Sciences* 23 (5): 2726.

Soyano, Takashi, Hiroshi Kouchi, Atsuko Hirota, and Makoto Hayashi. 2013. "Nodule Inception Directly Targets NF-Y Subunit Genes to Regulate Essential Processes of Root Nodule Development in *Lotus japonicus*." *PLoS Genetics* 9 (3): e1003352.

Spaink, H. P. 1992. "Rhizobial Lipo-Oligosaccharides: Answers and Questions." *Plant Molecular Biology* 20 (5): 977–86.

Spaink, H. P., and B. J. Lugtenberg. 1994. "Role of Rhizobial Lipo-Chitin Oligosaccharide Signal Molecules in Root Nodule Organogenesis." *Plant Molecular Biology* 26 (5): 1413–22.

Staehelin, Christian, and Hari B. Krishnan. 2015. "Nodulation Outer Proteins: Double-Edged Swords of Symbiotic Rhizobia." *The Biochemical Journal* 470 (3): 263–74.

Sugiyama, Akifumi, Nobukazu Shitan, and Kazufumi Yazaki. 2007. "Involvement of a Soybean ATP-Binding Cassette-Type Transporter in the Secretion of Genistein, a Signal Flavonoid in Legume–*Rhizobium* Symbiosis." *Plant Physiology* 144 (4): 2000–2008.

Szczyglowski, Krzysztof, Robert S. Shaw, Judith Wopereis, Sue Copeland, Dirk Hamburger, Beth Kasiborski, Frank B. Dazzo, and Frans J. de Bruijn. 1998. "Nodule Organogenesis and Symbiotic Mutants of the Model Legume *Lotus japonicus*." *Molecular Plant–Microbe Interactions: MPMI* 11 (7): 684–97.

Takahara, Masahiro, Shimpei Magori, Takashi Soyano, Satoru Okamoto, Chie Yoshida, Koji Yano, Shusei Sato, Satoshi Tabata, Katsushi Yamaguchi, and Shuji Shigenobu. 2013. "Too Much Love, a Novel Kelch Repeat-Containing F-Box Protein, Functions in the Long-Distance Regulation of the legume–*Rhizobium* Symbiosis." *Plant & Cell Physiology* 54 (4): 433–47.

Timmers, Antonius C. J., Eric Soupène, Marie-Christine Auriac, Françoise de Billy, Jacques Vasse, Pierre Boistard, and Georges Truchet. 2000. "Saprophytic Intracellular Rhizobia in Alfalfa Nodules." *Molecular Plant–Microbe Interactions: MPMI* 13 (11): 1204–13.

Tiwari, Manish, Manisha Yadav, Baljinder Singh, Vimal Pandey, Kashif Nawaz, and Sabhyata Bhatia. 2021. "Evolutionary and Functional Analysis of Two-Component System in Chickpea Reveals CaRR13, a TypeB RR, as Positive Regulator of Symbiosis." *Plant Biotechnology Journal* 19 (12): 2415–27.

Tsikou, Daniela, Zhe Yan, Dennis B. Holt, Nikolaj B. Abel, Dugald E. Reid, Lene H. Madsen, Hemal Bhasin, Moritz Sexauer, Jens Stougaard, and Katharina Markmann. 2018. "Systemic Control of Legume Susceptibility to Rhizobial Infection by a Mobile microRNA." *Science (New York, N.Y.)* 362 (6411): 233–36.

Van de Velde, Willem, Juan Carlos Pérez Guerra, Annick De Keyser, Riet De Rycke, Stéphane Rombauts, Nicolas Maunoury, Peter Mergaert, Eva Kondorosi, Marcelle Holsters, and Sofie Goormachtig. 2006. "Aging in Legume Symbiosis. A Molecular View on Nodule Senescence in *Medicago truncatula*." *Plant Physiology* 141 (2): 711–20.

Van de Velde, Willem, Grigor Zehirov, Agnes Szatmari, Monika Debreczeny, Hironobu Ishihara, Zoltan Kevei, Attila Farkas, et al. 2010. "Plant Peptides Govern Terminal Differentiation of Bacteria in Symbiosis." *Science (New York, N.Y.)* 327 (5969): 1122–26.

Vasse, J., F. de Billy, S. Camut, and G. Truchet. 1990. "Correlation Between Ultrastructural Differentiation of Bacteroids and Nitrogen Fixation in Alfalfa Nodules." *Journal of Bacteriology* 172 (8): 4295–306.

Velzen, Robin van, Jeff J. Doyle, and Rene Geurts. 2019. "A Resurrected Scenario: Single Gain and Massive Loss of Nitrogen-Fixing Nodulation." *Trends in Plant Science* 24 (1): 49–57.

Viprey, V., A. Del Greco, W. Golinowski, W. J. Broughton, and X. Perret. 1998. "Symbiotic Implications of Type III Protein Secretion Machinery in *Rhizobium*." *Molecular Microbiology* 28 (6): 1381–89.

Wang, Dong, Joel Griffitts, Colby Starker, Elena Fedorova, Erik Limpens, Sergey Ivanov, Ton Bisseling, and Sharon Long. 2010. "A Nodule-Specific Protein Secretory Pathway Required for Nitrogen-Fixing Symbiosis." *Science (New York, N.Y.)* 327 (5969): 1126–29.

Wang, Qi, Jinge Liu, Hua Li, Shengming Yang, Peter Körmöczi, Attila Kereszt, and Hongyan Zhu. 2017. "Nodule-Specific Cysteine-Rich Peptides Negatively Regulate Nitrogen-Fixing Symbiosis in a Strain-Specific Manner in *Medicago truncatula*." *Molecular Plant–Microbe Interactions: MPMI* 31 (2): 240–48.

Wassem, R., A. M. Marin, A. Daddaoua, R. A. Monteiro, L. S. Chubatsu, J. L. Ramos, W. J. Deakin, W. J. Broughton, F. O. Pedrosa, and E. M. Souza. 2017. "A NodD-like Protein Activates Transcription of Genes Involved with Naringenin Degradation in a Flavonoid-Dependent Manner in *Herbaspirillum seropedicae*." *Environmental Microbiology* 19 (3): 1030–40.

Webb, Benjamin A., Sherry Hildreth, Richard F. Helm, and Birgit E. Scharf. 2014. "*Sinorhizobium meliloti* Chemoreceptor McpU Mediates Chemotaxis toward Host Plant Exudates through Direct Proline Sensing." *Applied and Environmental Microbiology* 80 (11): 3404–15.

Yang, Jian, Fengzhan Gao, and Huairong Pan. 2025. "Essential Roles of Nodule Cysteine-Rich Peptides in Maintaining the Viability of Terminally Differentiated Bacteroids in Legume–Rhizobia Symbiosis." *Journal of Integrative Plant Biology* 67 (4): 1077–85.

Zélicourt, Axel de, Anouck Diet, Jessica Marion, Carole Laffont, Federico Ariel, Michaël Moison, Ons Zahaf, Martin Crespi, Véronique Gruber, and Florian Frugier. 2012. "Dual Involvement of a *Medicago truncatula* NAC Transcription Factor in Root Abiotic Stress Response and Symbiotic Nodule Senescence." *The Plant Journal: For Cell and Molecular Biology* 70 (2): 220–30.

Zhao, Yiyong, Rong Zhang, Kai-Wen Jiang, Ji Qi, Yi Hu, Jing Guo, Renbin Zhu, et al. 2021. "Nuclear Phylotranscriptomics and Phylogenomics Support Numerous Polyploidization Events and Hypotheses for the Evolution of Rhizobial Nitrogen-Fixing Symbiosis in Fabaceae." *Molecular Plant* 14 (5): 748–73.

Zuluaga, Ana M., Adal Mena-García, Ana C. Soria Monzón, Maite Rada-Mendoza, Diana M. Chito, Ana I. Ruiz-Matute, and Maria L. Sanz. 2020. "Microwave Assisted Extraction of Inositols for the Valorization of Legume By-Products." *Lebensmittel-Wissenschaft Und Technologie [Food Science and Technology]* 133 (November): 109971.

4 Engineering Nodule Development

Optimizing Symbiotic Efficiency for Enhanced Performance of Horticultural Crops

Akshara Kampurath, Afiya Rajan Selin, and Senthilkumar Shanmugasundram

4.1 INTRODUCTION

The interaction of higher plants with microbes, particularly legumes and bacteria such as Rhizobia, has been a cornerstone of agriculture for generations. In this mutually beneficial relationship, the rhizobia fixed themselves to the roots of the legume plants and developed into outgrowths known as nodules. In these nodules, the bacteria took nitrogen from the air and converted it to ammonia, which the plants could easily use. Since nitrogen is one of the vital macronutrients involved in various plant processes, such as the synthesis of proteins, enzymes, and chlorophyll—the green pigment required for photosynthesis—this natural process of nitrogen fixation is crucial for plant growth and health. Symbiotic nitrogen fixation has proven helpful for crops grown on nitrogen-deficient soils, where extending soil nitrogen fertility through conventional methods of fertilization may be less practical or cost-effective.

In horticulture, there is no better example of how this mutualism has been important, as numerous other plant and animal interactions demonstrate its significance. Other crops, such as legumes including peas, beans, and lentils, have shown that rhizobia make them compatible with nutrient-poor soils, particularly in terms of nitrogen. This biological nitrogen fixation reduced the reliance on synthetic nitrogen fertilizers, which are costly and have severe impacts on soil and water, leading to greenhouse gas emissions. The reduced use of chemical fertilizers led to environmentally friendly and cost-effective farming practices. Furthermore, legumes that benefit from nitrogen-fixing symbiotic associations are incorporated into crop rotation systems or used as cover crops. In that case, these legumes can rejuvenate soil fertility, thereby improving the overall production of different crops. This process has been beneficial in addressing growing concerns about environmental conservation and food supply worldwide.

Nevertheless, various factors have limited the overall potential of nitrogen fixation, and the development of nodules and symbiosis efficiency in horticultural crops. One of the main difficulties has been the fact that the effectiveness of different rhizobia strains when associated with the host plants was not constant. Some leguminous plants are not sensitive to rhizobia and nodule formation; other factors, such as soil type, pH, and temperature, can also impact nodule formation. Second, variability in nodule formation across different species of a plant, including those in the legume group, presents a significant challenge in enhancing nitrogen fixation. Drought, salinity, and nutrient stresses exacerbate these challenges, decreasing the efficiency of the partnership in achieving high crop yields.

 DOI: 10.1201/9781003608080-4

However, under certain conditions, the nitrogen fixation process is naturally slow, contributing to poor plant performance. For example, under conditions of low supply of other micronutrients, the efficiency of the interaction between rhizobia and plants decreases, which decreases the advantage of nodule formation. Such complications have therefore demanded higher research and development, and innovative approaches to improving the performance of this synergistic process. Engineering nodular development holds promise for enhancing the genetic, molecular, and environmental factors that characterize this interaction. New developments in the fields of biotechnology and synthetic biology have offered opportunities for enhancing symbiotic performance, improving nodule formation, and increasing nitrogen-fixing activity, leading to higher total horticultural yield.

Solving these challenges and optimizing the development of nodules with the help of science could significantly enhance crop productivity, decrease the use of chemical fertilizers, and build a more effective farming system. If applied successfully, all these innovations have the potential to transform horticultural practices and support farmers in feeding the ever-growing population without placing undue pressure on the environment due to changes in agricultural policies.

4.2 MOLECULAR MECHANISMS OF NODULE FORMATION

4.2.1 Early Signaling and Rhizobia Recognition

The perception of rhizobia by legume plants is a well-orchestrated biochemical and molecular event that is the basis of the symbiotic nitrogen fixation process. This complex interaction starts in the rhizosphere, where leguminous plants secrete flavonoids and other signaling molecules into the soil. These compounds serve as chemotactic attractants for the compatible rhizobacteria, which in turn synthesize lipochitin oligosaccharide molecules referred to as Nod factors, which, therefore, acted as essential alarm signals within the host plant, which, when perceived, led to the formation of nitrogen-fixing nodules.

When the Nod factors were detected, these molecules were perceived by the legume plant's root epidermal cells through receptor-like kinases (RLKs) present on the plasma membrane of plant cells. Several RLKs participated in this process, including Nod factor receptor 1 (NFR1) and Nod factor receptor 5 (NFR5) of legumes. These receptors were the key ones that identified the bacterial signals. These receptors were observed to interact with the Nod factors, leading to the activation of intracellular signal transduction. The same studies showed that this receptor-ligand binding creates calcium spiking in the nuclear region of the plant cell, known as calcium oscillation, which acts as a second messenger in the signaling process as well (Oldroyd, 2013).

Following the calcium oscillations, calcium- and calmodulin-dependent protein kinases (CCaMKs) were phosphorylated, and the phosphorylated transcription factors included nodulation signaling pathway 1 (NSP1) and NSP2. These transcription factors play a critical role in controlling early nodulation-related genes, such as root hair deformation and infection thread formation, through which rhizobia enter and penetrate the plant's root tissue (Wang et al., 2018). Other studies have demonstrated that during these developmental stages, symbiosis-related genes are transcribed, including the early nodulin (*ENOD*) genes, which are crucial for development.

At the same time, the host plant initiates a molecular signaling process, which also generates reactive oxygen species, such as hydrogen peroxide, and other defense compounds as part of a signal to the pathogen, thereby activating the defense mechanism. Usually, such defense mechanisms are masked to allow the plant to accommodate rather than attack and kill the rhizobia. Rhizobia, in turn, produce effector molecules via type III or type IV secretion systems that help suppress the plant's defense system, enabling the two partners to grow in harmony (Mamenko, 2021).

In all, legume recognition of rhizobia is mediated by the accurate swapping of molecular signals between both partners. This process is initiated by the release of flavonoids from the plant and the recognition of Nod factors by plant receptors, which in turn creates a series of calcium-dependent signal transduction events. It is this recognition mechanism that possibly leads to the formation of nodules, where nitrogen fixation can take place, highlighting the very high co-evolutionary relationship between legumes and rhizobia.

4.2.2 Development of the Infection Thread and Nodule Formation

The formation of nodules in leguminous plants requires a sequence of very controlled stages that commence with root hair curling and end with the development of nitrogen-fixing nodules. This process involves the accurate synchronization of cellular and molecular processes from the recognition between root hair cells of the plant and rhizobia (Figure 4.1).

4.2.2.1 Root Hair Curling

After the binding of rhizobial Nod factors, legume root hairs change their morphology and initiate the process of root hair deformation, a unique and systematized host response. Recognition of Nod factors by the plant RLKs causes cellular responses that bring about root hair curling. Research has shown that this deformation is crucial for containing rhizobia within the curl, thereby providing

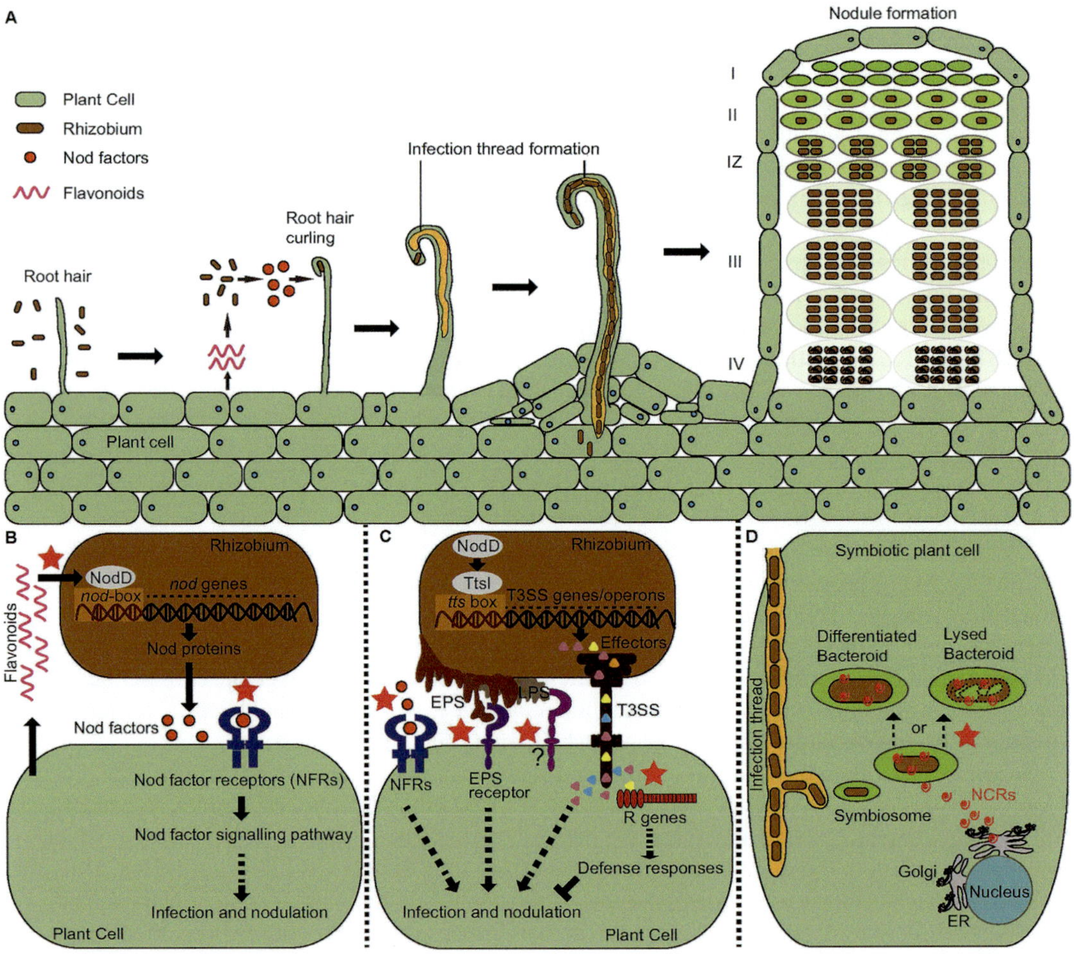

FIGURE 4.1 Symbiosis signaling and plant immunity involved in recognition specificity in the legume–rhizobia interactions (Wang et al., 2018).

a point of entry for bacteria (Oldroyd, 2013). At this stage, changes in the composition and texture of the cell wall in the root hairs are observed, enabling the formation of curls that could enfold around the rhizobial cells, thus trapping them within the curl.

4.2.2.2 Formation of the Infection Thread

After being confined in the curled root hair, the rhizobia embarked on constructing the infection thread, a pipe-like structure through which the bacteria invades the root tissues. There are small molecules of rhizobia, as well as other organic substances, including exopolysaccharides (EPS), which affect the plant's immune response. Consequently, the plant's full defense is not established, and therefore, the infection thread develops (Gage, 2004). Infection thread formation begins with the invagination of the plant cell wall and plasma membrane at the site of bacterial entrapment, and bacteria are incarcerated in a tubular structure. This thread stretches through the root hair cell up to the cortical cells of the root. Infection thread formation relies on the localized activity of the cytoskeleton, which consists of microtubules and actin filaments, controlling the advance of the infection thread (Timmers, 2008). This directional growth enables the rhizobia to penetrate further into the root tissues, where they must reach the cortical cells to form nodules.

4.2.2.3 Proliferation of Cortical Cells and Initiation of Nodule-P Primordium Formation

At the same time that infection thread formation occurs, cortical plant cells begin to divide in response to the Nod factor signal. The formation of the nodule primordium, the initial stage in nodule formation, is facilitated by these divisions. Cortical cell division is obtained through transcription factors NSP1 and NSP2, which control the early nodulin genes (Oldroyd et al., 2011). These form in the root cortex underneath the zone of infection and serve as the target of infection thread advancement. With the elongation of this specialized thread, it guides the rhizobia to penetrate the nodule tissue in the process.

4.2.2.4 Release and Nodule Formation

After the establishment of the infection thread, bacteria escapes into the plant cells within the nodule primordium. This release occurs through an endocytosis process, whereby the bacteria are engulfed by the plant cells and surrounded by a plant membrane to form an organelle-like structure referred to as the symbiosome (Wang et al., 2018). Within the symbiosome, rhizobia become committed to their nitrogen-fixing form, which is referred to as bacteroids. The plant regulates oxygen availability in the nodule by employing leghemoglobin, thereby maintaining an optimal oxygen range that is ideal for the nitrogenase enzyme of the rhizobium, which converts atmospheric nitrogen into ammonia.

4.2.2.5 The Progress and Development of the Nodule and Nitrogen Fixation

With the growth of the nodules, a quite clear tissue organization emerges based on the presence of specific zones. The central zone is occupied by infected cells with bacteroids, which are actively involved in nitrogen fixation, while the surrounding cells are healthy and contribute to nodule growth and development (Ferguson et al., 2010). In mature nodules, nitrogen fixation is at its most active stage. At the same time, ammonia is conducted to the host plant and converted into amino acids and other nitrogenous products vital for plant growth. This process minimizes the plant's reliance on external sources of nitrogen inputs to optimum levels. The ability of the infection thread and the formation of nitrogen-fixing nodules are well-developed and specialized, and the existence of this system highlights the co-evolution of the two organisms. From the processes that occur during nodule development, such as root hair curling, infection thread establishment, cortical cell division, rhizobial release, and nodule development, each process is well-coordinated to facilitate an efficient transfer of nitrogen to the host plant.

4.3 KEY GENETIC REGULATORS OF NODULE DEVELOPMENT

The process of formation and development of nitrogen-fixing nodules in legumes is known to be regulated by several genes, among which nodulation (*Nod*) genes and well-known transcription factors can be distinguished. Among them, NodD and NodL are involved in the initiation and control of legume–rhizobia interactions, as part of other gene products.

4.3.1 *NodD*

NodD is a transcriptional factor that plays a significant role in regulating the other nodulation genes present in rhizobia. The data obtained illustrate that NodD proteins function as transcriptional activators in response to plant-released flavonoids, which act as inducers of symbiosis (Oldroyd & Downie, 2008). Upon the interaction of specific flavonoid molecules in the rhizosphere, NodD initiates the expression of other *Nod* genes, including *nodABC*, which synthesize Nod factors that the host plant widely accepts. This process is crucial in initiating nodule formation, as it triggers the molecular signaling that is central to the establishment of symbiosis.

4.3.2 *NodL*

The *NodL* gene encodes an enzyme involved in attaching a specific group to Nod factors. For instance, *NodL* activity involves the acetylation of the Nod factor, a vital modification that determines the specificity of rhizobial signals that the host plant perceives. The presence of this O-acetyl group on the Nod factors determines whether the rhizobia are considered suitable partners by the legume species, as different legume species require different molecular modifications to the Nod factors for the formation of nodules (Gage, 2004). These mutations are observed to alter nodulation efficiency, demonstrating that the *NodL* gene is crucial for symbiosis specificity, as mutations to this gene commonly result in inefficient or completely impaired nodule formation.

4.3.3 Transcription Factors

Besides rhizobial Nod genes, nodule initiation and development involve the hosts' transcription factors, namely, NSP1 and NSP2. These transcription factors work downstream of Nod factor recognition and calcium signal in the root cells of the plant. The investigations establish that NSP1 and NSP2 control *ENOD* genes, which are associated with infection thread differentiation in the root cortex and cell division (Oldroyd et al., 2011). When there is no expression of NSP1 and NSP2, the root cortex is unable to initiate the cell division required for nodule organogenesis, and hence, nodule development is affected. Furthermore, one of the transcription factors, ethylene response factor required for nodulation (ERN), which belongs to the AP2/ERF family, is implicated in regulating the infection process. ERN controlled the root hair curling gene and the infection thread gene, and was connected to the early stage of rhizobial infection (Cerri et al., 2016). Defects in ERN lead to phenotypes that affect both the infection process and nodule development, confirming its role during symbiosis.

4.3.4 Other Key Regulators: NF-YA1 and NIN

The gene nodule inception (*NIN*) was assumed to control nearly all processes of nodule formation. It operates downstream of calcium signaling and is required for the initiation of cortical cell division and nodule primordium development. In addition, NIN is found to regulate genes involved in the processes of infection and nodule organogenesis, and its expression displays strict regulation for the formation of appropriate nodules (Soyano & Hayashi, 2014). Due to the

TABLE 4.1
Genes Involved in Nodule Development and Symbiotic Efficiency.

Gene	Function	Impact on Symbiotic Efficiency
Nod factor receptors (*NFR1*, *NFR5*)	Recognition of rhizobia	Essential for nodule formation
Ethylene response factor (*MtEFD*)	Control of nodule number and differentiation	Influences nodule efficiency and development
Nodule inception (*NIN*)	Initiation of nodule organogenesis	Critical for nodule formation

Source: Jardinaud et al., 2023.

overexpression of *NIN*, the plants develop more nodules than usual, while due to mutating *NIN*, the plants have no nodules at all.

Another essential regulator found to be involved in regulating nodule organogenesis is Nuclear Factor-YA1 (NF-YA1). NF-YA1 is a transcription factor that acts in the regulatory network for nodule-specific gene expression, programming the differentiated cells within the nodule (Laporte et al., 2014). Thus, NF-YA1 is proposed to orchestrate the actions of other transcriptional regulators, optimizing nodule development so that functional nitrogen-fixing structures are formed correctly. Various genes implicated in nodule development and symbiotic efficiency are summarized in Table 4.1.

4.4 ENGINEERING STRATEGIES TO OPTIMIZE NODULE DEVELOPMENT

4.4.1 Genetic Engineering Approaches to Enhance Symbiotic Efficiency

With the development of new technologies for genetic manipulation of plants, including clustered regularly interspaced short palindromic repeats (CRISPR), RNA interference (RNAi), and transcription activator-like effector nucleases (TALENs), it is now possible to enhance the efficiency of nitrogen fixation in legumes, thereby improving symbiotic relationships. These tools have been used to alter specific genes associated with nodule formation and nitrogen fixation, and then applied to improve the overall function of nitrogen-fixing symbiotic horticultural crops. Furthermore, recent reports on the importation of genes that promote symbiosis from other congeners or via synthetic biology have provided new opportunities to manage nodule size, number, and nitrogen-fixing capabilities.

4.4.1.1 CRISPR-Cas9 Technology

The development of CRISPR-Cas9 has opened up one of the most significant approaches for gene editing, which can be applied to increase symbiotic efficiency in legumes. CRISPR-Cas9 works by using a guide RNA to identify specific genomic targets, allowing it to cleave at these sites and edit genes. Nodulation genes have been targeted using CRISPR, and genes such as *NIN* and *NFR1* have been successfully disrupted to analyze the importance of both genes in nodulation and the plant–rhizobia symbiosis. Experiments carried out showed that silencing the negative regulators of nodulation resulted in the development of more nodules, and thus, more nitrogen fixation (Wang et al., 2016).

In one study, CRISPR-Cas9 was used to generate *Lotus japonicus* with a knockout mutation in the *NFR5* gene, which is a receptor kinase involved in Nod factor signaling. Reducing the expression of *NFR5* altered the plant's response to rhizobial signals, making modulation more efficient (Wang et al., 2016). Likewise, *NIN*, a master regulator of nodule formation, has been edited using CRISPR to produce plants with altered nodule organogenesis, making it easier for researchers to investigate and modify this complex process.

4.4.1.2 RNA Interference

Another strategy employed was the use of RNAi to silence genes that were deleterious to the efficiency of symbiosis. It was noted that mutations in genes responsible for inhibiting nodule formation or reducing nitrogen fixation improved symbiotic performance. For instance, the suppression of nuclear gene *ETR1*, which encodes an ethylene receptor inhibitory for nodule development, led to enhanced nodule formation in legumes, including soybeans, through RNAi (Chen et al., 2022). Ethylene is known to affect nodulation negatively, and the use of RNAi is seen as a positive sign for the occurrence of symbiosis.

RNAi has also been applied to the dissection of genes involved in regulating root architecture, e.g., *GmEXPA1*, which is involved in the regulation of root hairs and nodule formation in soybean. The increasing length of root hairs means that there is more area for the rhizobia to attach and infect the plant tissue, resulting in improved nodules and nitrogen fixation.

4.4.2 Introduction of Symbiosis-Enhancing Genes

4.4.2.1 Bringing over Beneficial Nodulation and Nitrogen Fixation Genes

To increase symbiotic efficiency, one tactic included transferring useful genes into the genetic background of leguminous crops, aiming to enhance nodule size, nodule number, or nitrogen fixation efficiency. To this end, the investigators aimed to identify genes involved in Nod factor signaling, nodule organogenesis, and nitrogen fixation. For instance, studies have reported that overexpression of the *NIN* gene enhances the number of nodules formed in legume roots, thereby increasing nitrogen uptake in plants (Soyano et al., 2014). The above approach enabled plants to fix nitrogen at higher rates and support better growth under nitrogen-starved conditions. Furthermore, genes for the reception of Nod factors, namely, *NFR1* and *NFR5*, were genetically manipulated to increase the plant's ability to respond to rhizobial signals, resulting in enhanced nodule initiation and maturation. Enhancing these genes enabled the engineering of plants to identify the optimal partners for rhizobia, resulting in increased nodule size and productivity (Madsen et al., 2003).

4.4.2.2 Creation of a Better Apparatus for Better Mycorrhization

The roles that could be assigned to root architecture include enhancing the plant's ability to communicate with rhizobia, thereby facilitating the development of nodules. Specific genes controlling root hairs, which elongate to increase the diameter of the root hairs, lateral root formation, and branching of the root system through the initiation of lateral roots, were manipulated to enhance the plant's ability to accommodate nitrogen-fixing bacteria. For instance, overexpression of the *LBD16* (lateral organ boundaries domain) gene in *Medicago truncatula* positively influenced lateral root emergence, thereby enhancing the number of sites that rhizobia could infect and develop nodules (Schiessl et al., 2019). It has been observed that these genetic modifications not only lead to an increase in the formation of nodules but also improve the effectiveness of nitrogen fixation in plants, thereby enhancing the overall growth of the plants.

Additionally, genes involved in promoting symbiosis with other species, specifically non-legume species, were transferred to legumes to enhance nitrogen fixation. For example, legume genes linked to specific phosphorus acquisition were genetically engineered into the organisms, as phosphorus can be a constraint to nitrogen fixation. When the *PTF1* gene from rice was successfully incorporated into the *Medicago* species, it subsequently increased PUE and had a consequent positive impact on nodule function and nitrogen fixation (Liang et al., 2014).

4.4.3 Synthetic Biology Approaches for Enhanced Nitrogen Fixation

Regarding nitrogen fixation, multiple genes and regulatory elements have been applied in synthetic biology to enhance symbiotic fixation. They achieved this through techniques such as establishing artificial routes for Nod factor synthesis or altering the plant's ability to manipulate the oxygen within nodules, which is vital for nitrogenase activities. Investigations were also conducted on the inoculation of plants with bacterial genes to optimize the relationship between the two organisms, including those that enhanced the efficiency of rhizobial nitrogenase or increased carbon availability for the bacteria (Rogers & Oldroyd, 2014).

4.4.3.1 Optimized Strains of Rhizobia through Synthetic Biology

To the best of the researchers' knowledge, the use of synthetic biology approaches to improve the efficiency of symbiosis has been primarily focused on the development of improved strains of the relevant rhizobia. The incorporation of synthetic gene circuits has enabled researchers to enhance rhizobia, allowing them to fix nitrogen or improve their ability to attach to plant roots. For example, by improving the modulation of the enzyme complex nitrogenase, which is involved in converting atmospheric nitrogen into a form that plants can utilize, certain rhizobia strains exhibit higher nitrogen-fixing capabilities under various environmental conditions (Mus et al., 2016). In the other study, synthetic promoters were constructed to optimize the levels of some of the genes encoding proteins involved in nitrogen fixation and nodule formation in *Sinorhizobium meliloti* (the microsymbiont of *Medicago truncatula*). By fine-tuning the synthesis of *Nif* genes in response to environmental cues, these engineered rhizobia improved the rate of nitrogen fixation in low-nitrogen fields, boosting symbiosis efficiency and sustainability (Ardi, 2023). Additionally, synthetic biology has helped introduce feedback control systems into rhizobia, enabling the bacteria to monitor the soil's nutrient status and adjust their nitrogen-fixing capacity, thereby reducing the need for fertilizers in plants.

In addition, researchers are also studying the ability to introduce synthetic pathways in the rhizobia to enhance their relationship with nonleguminous crops. In the past, when biologists transduced genes from associative nitrogen-fixing bacteria into rhizobial strains, they were able to develop strains that form symbiotic associations with other plant species, including cereals (Curatti & Rubio, 2014). These modified rhizobial strains enabled the capacity to spread nitrogen fixation not only in legume plants but also in other crops of agronomic importance, making the process more effective in increasing the sustainability of various farming systems.

4.4.3.2 Host Plant Modifications for Enhanced Symbiotic Traits

The use of synthetic biology to engineer rhizobia, in conjunction with efforts to alter host plants for enhanced adaptation to the bacteria and to improve symbiotic effectiveness, has been developed. Most of the host plant modifications were in the area of signaling, where the plants altered their release to attract more rhizobia or increased the nutrient exchange surface in the nodule.

4.4.3.2.1 Engineering Plant Signaling Pathways

Synthetic biology has been applied in manipulating plant signal transduction pathways that regulate the earliest stages of symbiosis, especially in the recognition of Nod factors synthesized by rhizobia. For instance, scientists have synthesized LysM receptor kinases, which are receptors that enable legumes to capture Nod factors, thereby improving their specificity and reliability (Kawaharada et al., 2015). This approach enabled the plants to deal with larger categories of these rhizobia and also escalated the possibility of forming mutualistic relations with them. Another focus of modification in the signaling cascade has also been discussed as a subject of synthetic biology in the downstream components. For instance, artificially developed molecules of calcium signaling proteins have been incorporated into nodules to improve their organogenesis, alongside enhancing nitrogen fixation output. These synthetic components enabled plants to gain control over the fabrication and maintenance of nodules, thereby minimizing energy expenditure on nodule formation and optimizing nitrogen absorption (Rogers & Oldroyd, 2014). Such changes optimized the mechanisms of intracellular signaling within the plant, thus increasing both the speed and efficiency of the mutual exchange.

4.4.3.2.2 *Enhancing the Secretion of Roots to Accommodate the Rhizobia*

It is essential to note that root exudates, a complex of secretions emitted by plant roots, play a significant role in attracting rhizobia to the root area and initiating the symbiotic processes. At other times, synthetic biology has enabled the enhancement of plants to produce more specific or potent exudates that attract a higher population of rhizobia or make the interaction more compatible. In one study, the exudation of flavonoids (substances that stimulate the production of Nod factors in rhizobia) was upregulated by synthetic gene networks, which in turn increased rhizobia colonization and nodule formation (Khandual, 2007). Researchers have modified flavonoids and other signaling molecules involved in the biosynthesis process to increase the root's efficiency in attracting high-performing rhizobia strains, thereby improving symbiotic efficiency. Furthermore, by altering the carbohydrate transporters in the root through transgenic modification, which prioritizes the export of sugars over biosynthesis, the plant effectively provides a greater energy source for the rhizobia and improves symbiosis (Das et al., 2023).

4.4.3.2.3 *Synthetic Engineering of Nutrient Exchange in Nodules*

The interactions between plants and rhizobia in the nodules, particularly the exchange of nutrients, are crucial for the establishment of symbiosis. Synthetic biology has advanced ways to improve this exchange. One of the focuses has been the release of synthetic transporters, which enables the shuttle of essential commodities like phosphorus and carbon between plant cells and rhizobia. By creating artificial versions of these transporters, plants are able to increase the rate of influx of the required nutrient into nodules, thereby increasing the rate of nitrogen fixation (Schulte et al., 2021). Moreover, synthetic biology strategies have been employed to enhance the oxygen supply within nodules. Nitrogenase, the enzyme that fixes nitrogen, is highly sensitive to oxygen. New synthetic gene circuits have been engineered to regulate oxygen in nodules, thereby providing higher control over nitrogenase. For example, the presence of synthetic hemoglobin within nodule cells was proposed as a means to regulate oxygen concentration, maintaining it optimal for nitrogenase, regardless of changes (Ramírez et al., 2016).

4.5 OPTIMIZING ENVIRONMENTAL CONDITIONS FOR ENHANCED SYMBIOTIC EFFICIENCY

Supplementary conditions improved the efficiency of biological symbiosis between legumes and rhizobia in the coculture system. These conditions affected the formation of nodules, nitrogen fixation, and plant health. To increase the nutrient supply and quality of horticultural crops, appropriate measures have been employed, including proper soil management, the use of biofertilizers, and the mitigation of abiotic stresses, to augment this mutual symbiotic relationship.

4.5.1 Soil and Nutrient Management

These include cultural factors such as pH, available nutrients, and levels of organic matter, which have a notable influence on the effectiveness of rhizobia and nodule formation. The most suitable pH for the majority of rhizobia is between 6 and 7, due to the high acidity or alkaline nature of the soil, which affects plant growth and rhizobial survival, and negatively impacts nodulation. One factor affecting nodulation in acidic soils is the toxicity of aluminum and manganese, which hindered the activity of rhizobia. In contrast, phosphorus availability was also a problem in alkaline soils (Mnalku, 2020). Soil pH can be adjusted through the use of lime when the soil is acidic, or sulfur when it is alkaline to enhance the symbiosis of beneficial bacteria. Concentrations of nutrients, including nitrogen, phosphorus, and calcium, also have a highly significant contribution to the results. High levels of nitrogen in the soil prevented the formation of nodules because the soil had easily accessible sources of nitrogen (Mwikali, 2018). On the other hand, phosphorus, which is involved in energy transfer during nodule formation, is needed; if there is a shortage of phosphorus, nodules develop poorly and are not very functional. An increase in the organic matter content have a positive impact, as it enhances the aggregation of soil particles, thereby providing a more favorable water regime and a suitable home for rhizobia, which in turn facilitates better colonization of the root system (Muthini, 2016). Other organic materials, such as compost and manure, are also used occasionally to enhance microbial activity and the rhizobial population at the site.

4.5.2 Application of Biofertilizers

Biofertilizers, such as rhizobia or other beneficial microbes, are employed to stimulate the growth of the soil microbiota and promote the formation of nodules. These appear to be effective as biofertilizers or microbial inoculants, providing plants with compatible and beneficial bacteria, mostly in situations where the initial range of native rhizobia is low (Dash & Gupta, 2011). Several researchers have reported that the use of rhizobial inoculants tends to enhance nodule number, nitrogen fixation rates, and biomass accumulation in plants (Htwe et al., 2019). It was also noted that the use of biofertilizers has a positive impact on plant performance, resulting in improved nutrient absorption, reduced chemical fertilizer use, and increased soil health over time. Apart from rhizobia, other soil microbes that enhance nodule formation and plant growth, including mycorrhizal fungi and plant growth-promoting rhizobacteria (PGPR), were also examined. These microbes play a role in the mobilization of nutrients and stress tolerance, thereby indirectly improving nitrogen-fixing capacity (Routray et al., 2021). Among P-solubilizing microorganisms, mycorrhizal fungi are found to significantly increase the level of phosphorus, which is very useful for the formation of nodules. Furthermore, researchers have also investigated the co-inoculation of rhizobia with PGPR, such as *Pseudomonas* and *Bacillus* species, which has led to an increased number of nodules and enhanced plant growth due to a positive interaction with beneficial microbes (Mesquita et al., 2023).

4.5.3 Abiotic Stress Mitigation

Drought, salinity, and temperature stress affect nodule formation and nitrogen fixation differently, as well as the renewable limitations of these stresses. The stress conditions affect plant root emergence and growth, adversely impacting the rhizobial endophytic ability and, consequently, nitrogen fixation ability (Zahran, 2009). These stresses constrain the function and survival of rhizobia through different strategies, which include the selective breeding of stress-tolerant rhizobial strains and the application of protective frames of microbial sizes.

4.5.3.1 Drought Stress

Water deficit stress negatively affects water relations between the host plant and rhizobia, ultimately resulting in reduced nodulation and early nodule senescence. Working on drought-tolerant rhizobial strains, it was found that these strains are even better adapted to grow in low water conditions and form nodules even under drought conditions (Saberi Riseh et al., 2021). Additionally, the drought-tolerant strain accumulates better osmotic potential in the cells, enabling it to perform various functions under conditions of water scarcity. One of these is the applicability of microbial coatings containing polysaccharides, which help increase the moisture content of rhizobia, thereby improving their chances of survival during periods of drought (Naseem et al., 2018).

4.5.3.2 Salinity Stress

Researchers found that high salinity has inhibitory effects on both plants and rhizobia, due to the disturbance of cellular osmotic balance and nutrient transport, which in turn results in poor nodule formation. K+-tolerant rhizobial strains have evolved to increase the concentration of osmoregulatory genes and ions in cells, including *betA*, which is responsible for synthesizing osmoprotectants such as glycine betaine (Saberi Riseh et al., 2021). These strains are capable of sustaining nitrogen-fixing activity under a saline environment, thus enhancing plant growth in saline soils. The next technical advancement supported bio-encapsulation techniques, in which rhizobia are encapsulated in a biodegradable polymer, thus enhancing resistance to salt stress and the efficiency of colonization.

4.5.3.3 Temperature Stress

Both high and low temperatures are found to have a detrimental effect on the rhizobia–legume symbiosis, inhibiting the rhizobia and impairing nodule function. The increased heat tolerance of rhizobial strains is through the selection of strains with tolerance to higher soil temperatures, which allows the heat shock proteins to protect the essential enzymes involved in nitrogen fixation (Bordeleau & Prévost, 1994). These heat-tolerant strains can nodulate and fix nitrogen more efficiently in temperatures that are not opportune for other rhizobial strains.

Employing these stress-tolerant rhizobial strains in conjunction with modern microbial seed coatings offers an implemental approach to counteract the impact of imposed abiotic stresses on nodule formation and nitrogen fixation. These approaches not only improve the ability of crops to withstand existing environmental conditions but also make farming practices more sustainable.

4.6 APPLICATIONS IN HORTICULTURAL CROPS

Engineering nodular development has been considered one of the most significant approaches to improving nitrogen fixation and yield in various leguminous horticultural crops, such as peas (*Pisum sativum*), beans (*Phaseolus vulgaris*), and lentils (*Lens culinaris*). Hence, the optimization of nodule formation and the efficiency of nitrogen fixation are sought to enhance crop output and efficiency in agricultural practices. Specifically, several approaches addressing these crops, including genetic engineering, bioinoculants, and soil management, are described through detailed case studies.

4.6.1 Peas (*Pisum sativum*)

In peas, the engineering process of nodule formation focuses on increasing the efficiency of symbiosis through genetic means or the use of microbes. Some of the expression changes made included *NodD* and *NIN*; these changes aimed to improve nodule initiation and thus increase nitrogen fixation potential. Baloglu et al. (2022) employed the CRISPR-Cas9 technique to remove negative regulatory genes that suppressed nodule formation. This study demonstrated that the overexpression of five major transcription factors enhanced nodulation strength and efficiency under favorable environmental conditions, both in the laboratory and in the field. In another method, microbial inoculation was used, involving the preferred strains of *Rhizobium leguminosarum*, which improved the size of the nodules and their efficiency. A reprint of an experimental study by Goyal et al. (2021) found that the virus infection of pea crops using an engineered strain of *R. leguminosarum* increased nitrogen fixation by a quarter more than that using commercial inoculants. This resulted in a proportional 20% increase in yield, primarily due to the use of engineered microbes for enhanced symbiosis. The better health of the soil and nutrients also played a significant role in making these interventions successful.

4.6.2 Beans (*Phaseolus vulgaris*)

Beans are also among the legumes that nodule engineering has greatly affected, particularly in terms of nitrogen fixation and crop yield. Because environmental stresses often reduce the symbiotic efficiency of beans, researchers have focused their efforts on enhancing nodulation by manipulating and identifying rhizobial strains with high levels of stress tolerance. There is the possibility of enhancing the heat tolerance level of *Rhizobium tropici* through strain improvement, which could improve nodule formation in high temperatures characteristic of the tropics and subtropics

(Bhadrecha et al., 2023). It also exhibited a 40% increase in nitrogen fixation—the ability of plants to convert atmospheric nitrogen into compounds required for growth—and a 25% enhancement in biomass when exposed to heat stress compared to the control groups, the wild-type strains. Other targets were also set based on the idea of enhancing the compatibility of the host plant with the rhizobia; this included genetically modifying the root structure of the host plant for effective root–rhizobia symbiosis. Another study by Estrada-Navarrete et al. (2006) also pointed out that overexpression of the Root Onset Lateral Branching (*ROLB*) gene in *Phaseolus vulgaris* resulted in enhanced root branching, leading to an increased ability of *Rhizobium* to colonize the nodules and, consequently, enhanced formation of nodules. This helps increase the efficiency of nitrogen fixation and results in a yield of beans grown under field conditions that is 15% higher than before. These were crucial to enhance productivity in places where environmental challenges affect the conventional mutual partnership.

4.6.3 Lentils (*Lens culinaris*)

Bioinoculants were also studied in the context of increasing the efficiency of symbiosis in lentils. The use of a coculture of rhizobia and mycorrhizal fungi has been shown to enhance nodule formation and nitrogen utilization, as well as plant biomass (Pandey et al., 2020). The two-inoculant approach, typical of the one used in this study, increased nitrogen fixation by 30% compared to by 20% in single-inoculant treatments and yield. This case study highlighted the potential for enhancing the effectiveness of symbiosis in lentils by leveraging phenomena related to the interaction of beneficial microbial species that can grow in marginal soils. Moreover, adaptation techniques that help plants avoid abiotic stresses are relevant in lentil crops in arid and semi-arid areas. The works done on drought-tolerant rhizobial strains also demonstrated positive effects in sustaining nodule performance under restricted water availability. An experiment conducted to study the drought tolerance of *Rhizobium leguminosarum* strains demonstrated that drought-tolerant strains, when inoculated in lentils, were able to maintain a higher nitrogen fixation capacity when grown under drought conditions and yielded 15% more than the control plants (Banjare et al., 2023). This particular case demonstrated that microbial engineering can be effectively applied to enhance the ability of plants to yield better crops under unfavorable conditions. Table 4.2 summarizes the crops and *Rhizobium* strains used for inoculation.

4.6.4 Expanding Nitrogen Fixation to Nonleguminous Crops

Nitrogen fixation in crops other than legumes, especially cereals and other horticultural crops, has been a focus area in molecular breeding in recent years. While the legume crops naturally undertake the process of symbiotic nitrogen fixation (SNF), the nonleguminous crops like the maize, rice, and wheat do not. Thus, engineering these crops to incorporate nitrogen-fixing bacteria makes them potentially decrease nitrogen fertilizer use and increase sustainability, a significant step forward. One of the experimental methodologies employed involved transferring *nif* (nitrogen fixation) genes from diazotrophic bacteria, such as *Klebsiella pneumoniae*, into plant genomes. Scientists developed transgenic rice through the genetic transformation of a group of *nif* genes that code for the nitrogenase enzyme, allowing these rice plants to convert nitrogen in the air. However, the problem was that nitrogenase was inhibited by oxygen, thereby limiting its applicability in plant cells. Initial experiments for evaluating nitrogenase were achievable at only a partial level, but it was understood that these activities needed to be optimized to guarantee nitrogen fixation in field conditions with higher and stable nitrogen levels (Rogers & Oldroyd, 2014).

Another promising experimental method involved the use of nitrogen-fixing bacteria to create a nonsymbiotic mutualism, which was then utilized as biofertilizers for nonleguminous plants. In rice, the use of biofertilizers with *Azospirillum* species resulted in a higher rate of nitrogen fixation in the rhizosphere. These nitrogen-fixing bacteria contributed to supplying nitrogen to plants

TABLE 4.2
Crops and *Rhizobium* Strains by Which They Are Inoculated.

Cross-Inoculation Groups	Leguminous Plant	*Rhizobium* Species
Clover group	*Trifolium* sp. (clovers and trefoil)	*Rhizobium trifolii*
Alfalfa group	*Medicago* sp. (alfalfa and burclover)	*Rhizobium trifolii*
Bean group	*Phaseolus* sp. (bean and wild bean)	*Rhizobium phaseoli*
Lupin group	*Ornithopus* sp. (lupines and serradella)	*Rhizobium lupine*
Pea group	Pea, sweet pea, lentil, and vetch	*Rhizobium leguminosarum*
Soybean group	*Glycine max* (soybean)	*Rhizobium japonicum*
Cowpea group	Cowpea, pigeonpea, lespedeza, groundnut, and kudzu	*Rhizobium* sp.

Source: Sermsathanaswadi, 2019.

and thus controlling the use of artificial fertilizers. Studies showed that high-rice and wheat grown with a biofertilizer treatment enhanced plant growth and yield under nitrogen stress conditions (Mus et al., 2016). This approach aimed to improve existing plant–microbe relationships without altering the genetic structure of the plant. It has been found that the possibility of expanding nitrogen fixation to nonleguminous horticultural plants has a very bright future. Still, more studies are needed to overcome the existing problems. This procedure is suspected to be complicated, and the continuous nitrogenase activity in plant tissues remains a target. Limitations with the use of nitrogenase, including oxygen sensitivity and the high energy requirement for nitrogen fixation, are considered critical technical challenges to be surmounted. Nevertheless, the listed problems are can be solved by the recent achievements in synthetic biology and genome editing, which aim to optimize gene expression regulation and create conditions that shield nitrogenase from negative impacts (Rogers & Oldroyd, 2014). Furthermore, microbial inoculants have also been applied as biofertilizers, as they act as natural stimulants to nitrogen-fixing crops, eliminating the need for genetic engineering. Bacterial strains capable of forming more effective symbiotic relationships with nonleguminous crops could serve as an environmentally friendly replacement for commercial nitrogen sources. Further field experiments are required to confirm that these biofertilizers can be effectively used in large-scale crops and under various environmental conditions.

4.7 SUSTAINABILITY AND PRODUCTIVITY GAINS

4.7.1 Higher Yield and Lower Use of Fertilizers

Enhancing grass nitrogen fixation and improving the nitrogen-fixing performance of nonconventional legumes are key to future efficiency in crop consumption. Studies have shown that nitrogen fixation efficiency improvement is one of the significant effects, as well as a decrease in the use of synthetic nitrogen fertilizers. Nitrogen fertilizers are costly to manufacture, involve the use of fossil fuels, and emit hazardous substances and pollutants into the environment through runoff and leaching. It may be possible to have crops that can automatically fix nitrogen or, better still, optimize their interaction with the nitrogen-fixing bacteria, thus reducing the use of synthetic fertilizers. In legumes, nitrogen fixation is a natural process of acquiring nitrogen, which, when enhanced through genetic engineering to promote nodule formation, significantly enhances nitrogen accumulation and crop yield. The accommodation of yield by optimizing nitrogen fixation through genetic manipulation, as demonstrated by experiments on soybean and pea, ranges between 20% and 30% (Oldroyd, 2013). These crops were even able to perform nitrogen fixation, fixing more nitrogen, especially in low-nitrogen soils, thus sparing the crop from the need for chemical fertilizers. In the case of nonleguminous crops, the possibility of extending nitrogen fixation is highly efficient because such crops are grown on a large scale. Maize, rice, and wheat are among the crops that demand the most nitrogen in the world today. If these crops could be made to fix nitrogen, either through regular crop breeding or through the enhancement of their microbial symbionts, already billions of hectares of land could be eliminated from using the destructive, costly, and environmentally unfriendly synthetic nitrogen fertilizer by tens and hundreds of millions of tons each year (Mus et al., 2016). This reduction would not only decrease the number of inputs required by an average farmer for production but also reduce the cost of production. Still, it would also decrease the polluting effects of nitrogen and the emission of greenhouse gases from fertilization processes.

4.7.2 Contribution to Sustainability

Enhancing nitrogen fixation is beneficial to the overall physical sustainability of farming, as well as the health of the soil and beyond. Nitrogen-fixing plants are those that help improve soil fertility and contribute to the soil's nitrogen content, thereby benefiting subsequent crops. On the same note, the application of nitrogen-based fertilizers takes it a step further and causes soil depletion, leading to the loss of beneficial soil qualities, such as acidity and biological activity. The utilization of biological nitrogen fixation thus lowers fertilizer use, resulting in better soil health in the long run (Mus et al., 2016). Moreover, increasing the efficiency of nitrogen fixation in horticultural systems can be beneficial for the development of sustainable agriculture worldwide, especially in regions with a lack of fertilizers or those with severe environmental pollution problems resulting from the overuse of fertilizers. That future, where the use of quality fertilizers is not as frequently relied upon as before, is not only environmentally healthy for agriculture but also affordable for smallholder farmers in developing nations. Adaptation of these nitrogen-fixing traits in nonleguminous crops could also make those crops more resilient to high and low prices of fertilizers or the unavailability of fertilizers, hence increasing food security (Rogers & Oldroyd, 2014). Therefore, the opportunity to increase N-fixation in nonleguminous crops through genetic engineering or a better association with microbial strains presents a significant opportunity to expand sustainability and productivity in horticultural crops. Such approaches are capable of reducing the proportion of synthetic fertilizer application by up to a third, leading to improved feasibility and sustainability of agricultural systems. However, more research and field tests are required to fully realize these benefits in the world.

4.8 FUTURE PROSPECTS AND CHALLENGES

4.8.1 Integrating Advances in Genomics and Biotechnology

There is a vast potential to expand nodule formation and efficiency for all leguminous crops, as well as for nonleguminous crops, through the application of the latest innovative technologies, such as genomics, bioinformatics, and precise breeding. It has been achieved through the use of these complex tools. Hence, it can be said that advanced tools enhance aspects of the plant–microbe relationship, particularly in the processes of nodule formation and nitrogen fixation, by identifying molecular signaling and genetic pathways.

4.8.2 Omics-Driven Approaches

The advanced "omics" techniques such as genomics, transcriptomics, proteomics, and metabolomics are instrumental in elucidating the biological mechanisms regulating the nodule formation process. These high-throughput techniques provide an overall picture of gene expression, protein–protein interactions, and metabolic maps during legume–rhizobia symbiosis. For instance, transcriptomics analysis presented data on the increased expression of particular gene families associated with nodulation, including *Nod* factors, *NIN*, and genes connected to nitrogenase activity, which promotes the nitrogen-fixing process. Through the use of multi-omics data, it has been possible to identify significant nodes that can be genetically modified to enhance nodule formation (Oldroyd, 2013). In addition, the innovation of more effective genomic editing tools, such as CRISPR-Cas9, enables the identification of genes to target within the nodules. For example, some genes that allow for low numbers of nodules or rhizobial colonization can be removed or downregulated to enhance nodule formation. Furthermore, some of the plant immune genes that could deter rhizobial infection may be silenced while the plant's immunity to pathogens is maintained. The precision breeding technology, such as CRISPR and other genome editing technologies, is thus opening ways to efficient nitrogen-fixing plants.

4.8.3 Bioinformatics and Systems Biology

A considerable amount of data accompanies omics; therefore, bioinformatics and systems biology methods are crucial for data handling. With the use of these tools, researchers can then map the necessary networks for nodule formation, thereby gaining insight into how genes and proteins relate to the process. The current advances in artificial intelligence are utilized in predicting gene function, selecting candidate genes for engineering, and determining the potential consequences of genetic manipulations. It has also been noted that such tools have enhanced the rate of discovery of desirable traits suitable for use in crop breeding.

Furthermore, bioinformatics tools have been utilized in the analysis of the rhizosphere microbiome. The overall goal of this research area is to identify and characterize the microbial communities associated with plant roots and the activities of symbiotic nitrogen fixation. In this way, the possibility of increasing the share of beneficial strains for further colonization of rhizobia and nodule formation, which could be important for nonleguminous crops, is considered through manipulations of the rhizosphere microbiome (Yang et al., 2024).

4.8.4 Precision Breeding

Precision breeding is the use of data from omics to enhance and optimize crop breeding strategies. The old ways of plant breeding primarily involved selecting plants with preferred characteristics and developing hybrids through successive generations; this is a slow and inefficient method of plant breeding. Precision breeding is the opposite of the strategy of using genomic data to select genes that enhance nodule formation and nitrogen fixation, such as root structure, rhizobial symbiotic signaling, and stress tolerance. This method of plant breeding accelerates the creation of potential offspring and positively filters the characteristics inherited by subsequent generations. For instance, marker-assisted selection (MAS) is currently used to select crops of the leguminous type with improved nitrogen fixation capabilities. MAS depends on techniques that establish linkages between DNA and presumably favorable traits; thus, MAS facilitates the screening of plants at a young age for higher nodule formation or nitrogenase activity. MAS has been used previously in crops such as soybean, in which breeding programs have managed to select plants with enhanced nitrogen-fixing traits.

4.8.5 Regulatory and Ecological Considerations

Altogether, the recent scientific achievements in engineering symbiotic systems are encouraging; however, several regulatory and ecological issues need to be addressed to utilize these technologies in agriculture effectively. Some of the advanced biotechnologies include genetically engineered crops, as well as synthetic biology and genome editing, the sale of which is subject to regulation in many countries. Legal and regulatory instruments provide determinations of the risks posed by genetically modified organisms (GMOs) in terms of their impact on human health, the environment, and the resultant effect on biological diversity. The approval process for such crops may be lengthy and complicated for innovations that involve new symbiotic systems, which have not undergone large-scale agricultural tests. For example, crops that have been bioengineered for improved nitrogen assimilation may require long-term studies to prove that they do not have adverse implications on nearby plants or any other ecological systems, or do not replace beneficial indigenous species. Regulatory agencies could also be concerned about the possibilities of horizontal gene transfer between engineered plants and other organisms beyond the target pests, as well as the long-term effects of introducing synthetic mutualisms (Gogarten & Townsend, 2005).

Considering these regulatory challenges, it is essential to enhance the partnership between scientists and policymakers to identify specific trends that can be applied once the engineered symbiotic systems are established. Other mitigation considerations that will be useful to consider include public participation and communication as a way of addressing GMO-related complaints and creating public confidence in such advancements.

4.8.6 Biosafety Concerns

Another factor of concern is the biosafety aspect of engineered nitrogen-fixing crops before they can be produced and released into the market. Therefore, one concern is the side effects of manipulating a specific process, such as nitrogen-fixing machinery in plants, for instance. For instance, hyperplasia of nodules may result in a wasteful allocation of plant resources, where more energy and nutrients are invested in the association for nitrogen fixation at the expense of other functions, such as growth and reproduction. Furthermore, the use of genetically modified rhizobia and their release into the environment could affect the native microbial population of the soil. These engineered bacteria could displace natural rhizobia, thus disrupting the balance in the soil ecosystem. Hence, regular biosafety measures must be implemented when it comes to these technologies to analyze their long-term impacts and avoid ecological disruptions (Gogarten & Townsend, 2005).

4.8.7 Ecosystem-Based Approaches

Researchers have recommended the introduction of engineered symbiotic systems into ecosystems. Unlike genetic engineering approaches to developing crop plants' resistance to pests and diseases, these approaches aim to improve the efficiency of natural plant–microbe associations using sustainable agricultural practices. For instance, bioinoculants that include nitrogen-fixing bacteria can be beneficially utilized to enrich crops without resorting to direct genetic engineering of the plant. This method has been well known to increase nitrogen fixation in leguminous and nonleguminous crops without promoting the effects of GMOs.

4.8.8 Adapting to Climate Change

Climate change poses a significant challenge to global agriculture, including the symbiotic systems involved in nitrogen fixation. Rising temperatures, shifting precipitation patterns, and increased incidence of extreme weather events can disrupt nodule development and reduce the efficiency of nitrogen fixation in leguminous crops.

4.8.9 Engineering for Climate Resilience

Engineering strategies aimed at enhancing plant resilience to climate change have become a priority. For example, researchers are exploring ways to modify plant signaling pathways to ensure robust nodulation even under stress. By introducing genes that promote stress tolerance, such as those regulating water-use efficiency or heat shock proteins, plants can maintain nodule formation and nitrogen fixation despite adverse environmental conditions.

4.9 CONCLUSION

Engineering advancements in nodule formation can be applied in horticulture development, particularly in nitrogen fixation processes, which offer possibilities for improving the performance and sustainability of horticultural crops. This advance is particularly relevant to leguminous crops, wherein the effectiveness of establishing nodules that contain nitrogen-fixing rhizobia can dramatically enhance nitrogen acquisition. Not only does it help improve the size and rate of growth of the nodule, but it also assists in increasing the overall yield while supporting the overall plant health, which in turn improves the resilience of the crops. Using modern biotechnological techniques, scientists have experimented with polymers related to nodule formation and function, employing CRISPR and RNAi. This has led to the development of varieties that possess more efficient and larger nodules, which are capable of fixing more nitrogen and thereby reducing the need for artificial fertilizers. In addition, extending such engineering to crops other than leguminous plants, such as cereals, is also a revolutionary opportunity. This is because it may be possible to harness the efficiency of nitrogen fixation by incorporating the nutrients in crops like maize and rice, thereby reducing the adverse effects of nitrogen fertilizer application while improving yields in resource-poor settings. In such areas, practice could be of special assistance to those associated with fertility, particularly in terms of soil and access to fertilizers, to feed the ever-growing global population.

Nevertheless, the primary premise of engineered nodule development remains to be achieved within the scope of these works, as well as with the help of cooperation between the branches of biotechnology and agronomy. The strengths are derived from the laboratory achievements; however, the shift to field applications is associated with numerous challenges. To this end, the multifaceted relationships among plants, microbes, and their environments must be comprehensively understood to ensure that these engineered traits are reliable in practical field crop production systems. Sustained above-ground field tests are required not only to determine the in-ground benefits of an engineered symbiotic system but also to assess its ecological implications. By civilizing interdisciplinary cooperation among scientists, agronomists, policymakers, and farmers, we can bridge the "from the lab to the farmland" gap. These announcements are clear indicators of an increasingly symbiotic strategy for improving food production, anchored in engineered nodule development to enhance agricultural performance and sustainability in the face of climate change and resource constraints.

REFERENCES

Ardi, A. (2023). *Improving Symbiotic Nitrogen Fixation in Sinorhizobium (Ensifer) Meliloti by Utilizing Synthetic Biology Approaches to Manipulate Key Regulatory Genes* (Master's thesis, North Dakota State University).

Baloglu, M. C., Celik Altunoglu, Y., Baloglu, P., Yildiz, A. B., Türkölmez, N., & Özden Çiftçi, Y. (2022). Gene-editing technologies and applications in legumes: Progress, evolution, and future prospects. *Frontiers in Genetics*, *13*, 859437.

Banjare, U., Patel, A. K., Pandey, A. K., Kumar, S., Singh, R. K., Masurkar, P., & Gupta, S. K. (2023). Biochemical and molecular evaluation of *Rhizobium* spp. and its growth promotion studies with lentil (*Lens culinaris* Medik. L.). *Journal of Pure and Applied Microbiology*, *17*(1), 155–166.

Bhadrecha, P., Singh, S., & Dwibedi, V. (2023). 'A plant's major strength in rhizosphere': The plant growth promoting rhizobacteria. *Archives of Microbiology*, *205*(5), 165.

Bordeleau, L. M., & Prévost, D. (1994). Nodulation and nitrogen fixation in extreme environments. *Plant and soil, 161*(1), 115–125.

Cerri, M. R., Frances, L., Kelner, A., Fournier, J., Middleton, P. H., Auriac, M. C., & de Carvalho-Niebel, F. (2016). The symbiosis-related ERN transcription factors act in concert to coordinate rhizobial host root infection. *Plant Physiology*, *171*(2), 1037–1054.

Chen, G., Li, J., Han, H., Du, R., & Wang, X. (2022). Physiological and molecular mechanisms of plant responses to copper stress. *International Journal of Molecular Sciences*, *23*(21), 12950.

Curatti, L., & Rubio, L. M. (2014). Challenges to develop nitrogen-fixing cereals by direct *nif*-gene transfer. *Plant Science*, *225*, 130–137.
Das, D., Tripathi, S., Mutyala, P., Srivastava, D. A., & Bücking, H. (2023). Development and resource exchange processes in root symbioses of legumes. *Symbiosis in Nature*.
Dash, S., & Gupta, N. (2011). Microbial bioinoculants and their role in plant growth and development. *International Journal of Biotechnology and Molecular Biology Research*, *2*(13), 232–251.
Estrada-Navarrete, G., Alvarado-Affantranger, X., Olivares, J. E., Díaz-Camino, C., Santana, O., Murillo, E.,. . . Sánchez, F. (2006). Agrobacterium rhizogenes transformation of the *Phaseolus* spp.: A tool for functional genomics. *Molecular Plant–Microbe Interactions*, *19*(12), 1385–1393.
Ferguson, B. J., Indrasumunar, A., Hayashi, S., Lin, M. H., Lin, Y. H., Reid, D. E., & Gresshoff, P. M. (2010). Molecular analysis of legume nodule development and autoregulation. *Journal of Integrative Plant Biology*, *52*(1), 61–76.
Gage, D. J. (2004). Infection and invasion of roots by symbiotic, nitrogen-fixing rhizobia during nodulation of temperate legumes. *Microbiology and Molecular Biology Reviews*, *68*(2), 280–300.
Gogarten, J. P., & Townsend, J. P. (2005). Horizontal gene transfer, genome innovation and evolution. *Nature Reviews Microbiology*, *3*(9), 679–687.
Goyal, R. K., Mattoo, A. K., & Schmidt, M. A. (2021). Rhizobia–host interactions and symbiotic nitrogen fixation in legume crops toward agriculture sustainability. *Frontiers in Microbiology*, *12*, 669404.
Htwe, A. Z., Moh, S. M., Soe, K. M., Moe, K., & Yamakawa, T. (2019). Effects of biofertilizer produced from *Bradyrhizobium* and *Streptomyces griseoflavus* on plant growth, nodulation, nitrogen fixation, nutrient uptake, and seed yield of mung bean, cowpea, and soybean. *Agronomy*, *9*(2), 77.
Jardinaud, M. F., Carrere, S., Gourion, B., & Gamas, P. (2023). Symbiotic nodule development and efficiency in the *Medicago truncatula Mtefd-1* mutant is highly dependent on *Sinorhizobium* strains. *Plant and Cell Physiology*, *64*(1), 27–42.
Kawaharada, Y., Kelly, S., Nielsen, M. W., Hjuler, C. T., Gysel, K., Muszyński, A., & Stougaard, J. (2015). Receptor-mediated exopolysaccharide perception controls bacterial infection. *Nature*, *523*(7560), 308–312.
Khandual, S. (2007). Flavonoids as signalling molecules and regulators of root nodule development. *Dynamic Soil, Dynamic Plant*, *1*(2), 83–94.
Laporte, P., Lepage, A., Fournier, J., Catrice, O., Moreau, S., Jardinaud, M. F., ... & Niebel, A. (2014). The CCAAT box-binding transcription factor NF-YA1 controls rhizobial infection. *Journal of experimental botany*, *65*(2), 481–494.
Liang, C., Wang, J., Zhao, J., Tian, J., & Liao, H. (2014). Control of phosphate homeostasis through gene regulation in crops. *Current Opinion in Plant Biology*, *21*, 59–66.
Madsen, E. B., Madsen, L. H., Radutoiu, S., Olbryt, M., Rakwalska, M., Szczyglowski, K., & Stougaard, J. (2003). A receptor kinase gene of the LysM type is involved in legume perception of rhizobial signals. *Nature*, *425*(6958), 637–640.
Mamenko, T. P. (2021). Regulation of legume-rhizobial symbiosis: Molecular genetic aspects and participation of reactive oxygen species. *Cytology and Genetics*, *55*, 447–459.
Mesquita, A. D. F. N., Bandeira, L. L., de Matos Neto, J. M., Martins, S. C. S., & Martins, C. M. (2023). Co-inoculation of rhizobia and other plant growth-promoting rhizobacteria as biofertilizers: A biotechnological overview, 14, 1–16. 10.5281/zenodo.7558384.
Mnalku, A. (2020). Intensification of symbiotic performances of grain legumes and rhizobia through joint use of agro-inputs. *Jima*, *29*, 32.
Mus, F., Crook, M. B., Garcia, K., Garcia Costas, A., Geddes, B. A., Kouri, E. D.,. . . Peters, J. W. (2016). Symbiotic nitrogen fixation and the challenges to its extension to nonlegumes. *Applied and Environmental Microbiology*, *82*(13), 3698–3710.
Muthini, M. O. R. R. I. S. (2016). Diversity, symbiotic efficiency and effect of water hyacinth compost on population of rhizobia nodulating *Phaseolus vulgaris* in Lake Victoria basin. *Ku. ac. ke*.
Mwikali, M. P. (2018). Influence of farmer-driven composting technology on below and above ground biology of common bean in Western Kenya.
Naseem, H., Ahsan, M., Shahid, M. A., & Khan, N. (2018). Exopolysaccharides producing rhizobacteria and their role in plant growth and drought tolerance. *Journal of Basic Microbiology*, *58*(12), 1009–1022.
Oldroyd, G. E. (2013). Speak, friend, and enter: Signalling systems that promote beneficial symbiotic associations in plants. *Nature Reviews Microbiology*, *11*(4), 252–263.
Oldroyd, G. E., & Downie, J. A. (2008). Coordinating nodule morphogenesis with rhizobial infection in legumes. *Annual Review of Plant Biology*, *59*(1), 519–546.
Oldroyd, G. E., Murray, J. D., Poole, P. S., & Downie, J. A. (2011). The rules of engagement in the legume–rhizobia symbiosis. *Annual Review of Genetics*, *45*(1), 119–144.

Pandey, A., Tripathi, P. H., Pandey, S. C., & Joshi, T. (2020). Multifaceted beneficial effects of plant growth promoting bacteria and rhizobium on legume production in hill agriculture. In *Recent Advancements in Microbial Diversity* (pp. 113–146). Academic Press.

Ramírez, M., Íñiguez, L. P., Guerrero, G., Sparvoli, F., & Hernández, G. (2016). *Rhizobium etli* bacteroids engineered for *Vitreoscilla* hemoglobin expression alleviate oxidative stress in common bean nodules that reprogramme global gene expression. *Plant Biotechnology Reports*, *10*, 463–474.

Rogers, C., & Oldroyd, G. E. (2014). Synthetic biology approaches to engineering the nitrogen symbiosis in cereals. *Journal of Experimental Botany*, *65*(8), 1939–1946.

Routray, S., Kumari, S., Borah, B., Shelat, H., & Veena Khanna, J. P. S. (2021). A review on rhizobia and PGPRs interactions in legumes. *Journal of Pharmaceutical Innovation*, *10*(7), 1448–57.

Saberi Riseh, R., Ebrahimi-Zarandi, M., Gholizadeh Vazvani, M., & Skorik, Y. A. (2021). Reducing drought stress in plants by encapsulating plant growth-promoting bacteria with polysaccharides. *International Journal of Molecular Sciences*, *22*(23), 12979.

Schiessl, K., Lilley, J. L., Lee, T., Tamvakis, I., Kohlen, W., Bailey, P. C., & Oldroyd, G. E. (2019). Nodule inception recruits the lateral root developmental program for symbiotic nodule organogenesis in *Medicago truncatula*. *Current Biology*, *29*(21), 3657–3668.

Schulte, C. C., Borah, K., Wheatley, R. M., Terpolilli, J. J., Saalbach, G., Crang, N.,. . . Poole, P. S. (2021). Metabolic control of nitrogen fixation in rhizobium–legume symbioses. *Science Advances*, *7*(31), eabh2433.

Sermsathanaswadi, J. (2019). Nitrogen-fixing bacteria and trends in agricultural applications. *Journal of Food Health and Bioenvironmental Science*, *12*(2), 49–56.

Soyano, T., & Hayashi, M. (2014). Transcriptional networks leading to symbiotic nodule organogenesis. *Current Opinion in Plant Biology*, *20*, 146–154.

Timmers, A. C. J. (2008). The role of the plant cytoskeleton in the interaction between legumes and rhizobia. *Journal of Microscopy*, *231*(2), 247–256.

Wang, L., Wang, L., Tan, Q., Fan, Q., Zhu, H., Hong, Z., & Duanmu, D. (2016). Efficient inactivation of symbiotic nitrogen fixation related genes in *Lotus japonicus* using CRISPR-Cas9. *Frontiers in Plant Science*, *7*, 1333.

Wang, Q., Liu, J., & Zhu, H. (2018). Genetic and molecular mechanisms underlying symbiotic specificity in legume–*Rhizobium* interactions. *Frontiers in Plant Science*, *9*, 313.

Yang, Y., Xu, N., Zhang, Z., Lei, C., Chen, B., Qin, G.,. . . Qian, H. (2024). Deciphering Microbial Community and Nitrogen Fixation in the Legume Rhizosphere. *Journal of Agricultural and Food Chemistry*, *72*(11), 5659–5670.

Zahran, H. H. (2009). Enhancement of rhizobia–legume symbioses and nitrogen fixation for crops productivity improvement. In *Microbial Strategies for Crop Improvement* (pp. 227–254). Berlin, Heidelberg: Springer Berlin Heidelberg.

5 Plant Immunity
Understanding Plant Defenses against Phytopathogens

Charanpreet Kaur and Vishnu Mishra

5.1 INTRODUCTION: GLOBAL FOOD INSECURITY AND ECONOMIC LOSSES

The Food and Agriculture Organization of the United Nations (FAO) predicts the world population to reach 10 billion by 2050 and the food demand to increase by 70%–100% (United Nations, 2017; Jones & Naidu, 2019; UN DESA, 2022). This surge in demand due to the growing population necessitates a significant increase in agricultural productivity. Pathogen attack significantly alters the physiological responses of plants compared to healthy ones. In healthy plants, physiological processes such as photosynthesis, respiration, water uptake, nutrient transport, and hormone regulation function normally, ensuring growth and productivity (Figure 5.1). However, when pathogens invade, they disrupt these processes, leading to visible and internal changes. Pathogens including bacteria, fungi, nematodes, and viruses are the main biotic factors limiting crop production (Suzuki et al., 2014; Zhu, 2016). Globally, pests or diseases may reduce crop yield by 20%–40%, leading to an economic loss of more than $200 billion annually (He & Creasey Krainer, 2020; Savary et al., 2019). Soybean cyst nematode alone causes loss of more than $1.5 billion annually in soybean production in the US (CNP, 2020). Climate change and shifts in agricultural practices further increase the risks of plant disease outbreaks with shifts in geographical distribution of pathogens (Chakraborty & Newton, 2011; Pautasso et al., 2012). Plant diseases have historically caused major food security crises. The Irish potato famine (1845–1852), caused by *Phytophthora infestans* resulted in over 2 million deaths and mass migration (Austin Bourke, 1964). Similarly, coffee rust outbreaks in Central America caused by *Hemileia vastatrix* reduced yields by over 50% (Avelino et al., 2015). These examples show the devastating impact of plant pathogens on food security, highlighting the need for sustainable solutions to mitigate crop losses.

Phytopathogens are broadly categorized as necrotrophs, biotrophs, and hemibiotrophs based on their modes of plant infection. Necrotrophs, such as *Botrytis cinerea* and *Alternaria alternata* (causing gray mold and leaf spot diseases, respectively, in a broad range of crops), kill its host and feed on the dead tissue for nutrition (Williamson et al., 2007). Pathogens that rely on living host cells for nutrition and ensure survival of their host to continue their life cycle are referred to as biotrophs (e.g., fungus *Blumeria graminis* causing powdery mildew in wheat and barley; *Xanthomonas oryzae* causing bacterial blight of rice) (Shen & Ronald, 2002; Geeta & Mishra, 2018). Pathogens that require a living host initially but kill it at a later stage of infection are described as hemibiotrophs (e.g., *Magnaporthe oryzae*, rice blast fungus) (Devanna et al., 2022). In general, bacteria and fungi are biotrophic or necrotrophic, whereas viruses are obligate biotrophs, although viral infection may lead to host cell death (Dou & Zhou, 2012; Guttman et al., 2014).

5.2 SUSTAINABILITY CHALLENGES OF CONVENTIONAL DISEASE CONTROL

Current disease management in agriculture primarily relies on synthetic pesticides, may pose risk to nontarget organisms, contaminate ecosystems, and threaten food safety due to pesticide residue

DOI: 10.1201/9781003608080-5

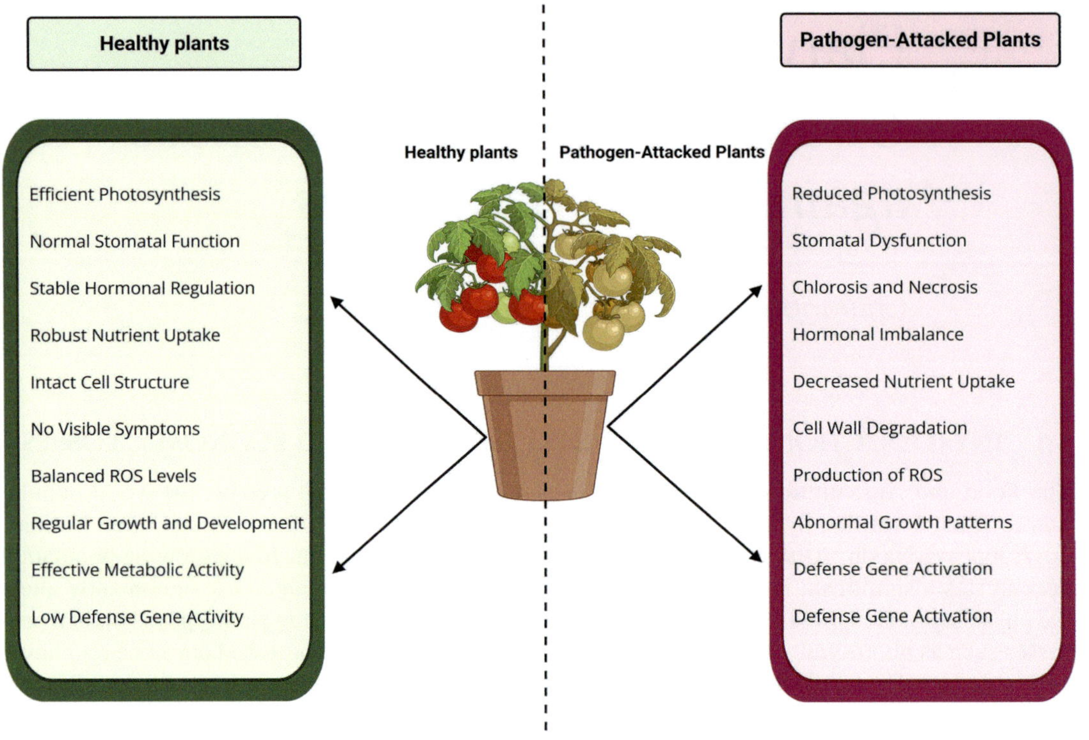

FIGURE 5.1 Comparison of physiological responses between healthy and pathogen-attacked plants highlighting photosynthesis, respiration, nutrient uptake, water balance, and hormone regulation.

(Damalas & Eleftherohorinos, 2011; Asghar et al., 2016). Pathogens develop resistance over time, leading to increased pesticide use, thus aggravating the problem. Although breeding resistant cultivars offers a long-term solution, it is time-intensive and often undermined by rapidly evolving pathogens. Transitioning to eco-friendly approaches, such as disease control through modulation of plant immunity, is a promising alternative.

Plants have an innate ability to recognize potential threats, triggering defense responses through a sophisticated surveillance system that monitors pathogen attack (Jones & Dangl, 2006). When pathogens attack, plants rapidly mobilize resources to activate immune responses, aiming to limit pathogen entry and colonization (Hawes et al., 2012; Melotto et al., 2006). This chapter provides an overview of plant immunity, explores the molecular mechanisms behind disease resistance, and discusses ways to use these concepts to enhance disease resistance in agricultural crops.

5.3 PLANT DEFENSE MECHANISMS AGAINST PATHOGENS

Plants defend themselves against pathogens through innate immunity, which includes both preformed and inducible defenses. Preformed defenses include structural barriers like the cuticle, cell walls, and stomata, which prevent pathogen entry. For instance, the waxy cuticle repels water and spores, while lignified cell walls resist fungal penetration by reinforcing cell integrity with cellulose, hemicellulose, and lignin (Thordal-Christensen, 2003). Inducible defenses activate upon pathogen recognition, involving both local and systemic responses. One key local response is the hypersensitive response (HR), which causes programmed cell death at infection sites, limiting pathogen spread. Additionally, plants produce pathogenesis-related (PR) proteins and antimicrobial compounds like phytoalexins that increase resistance to infection.

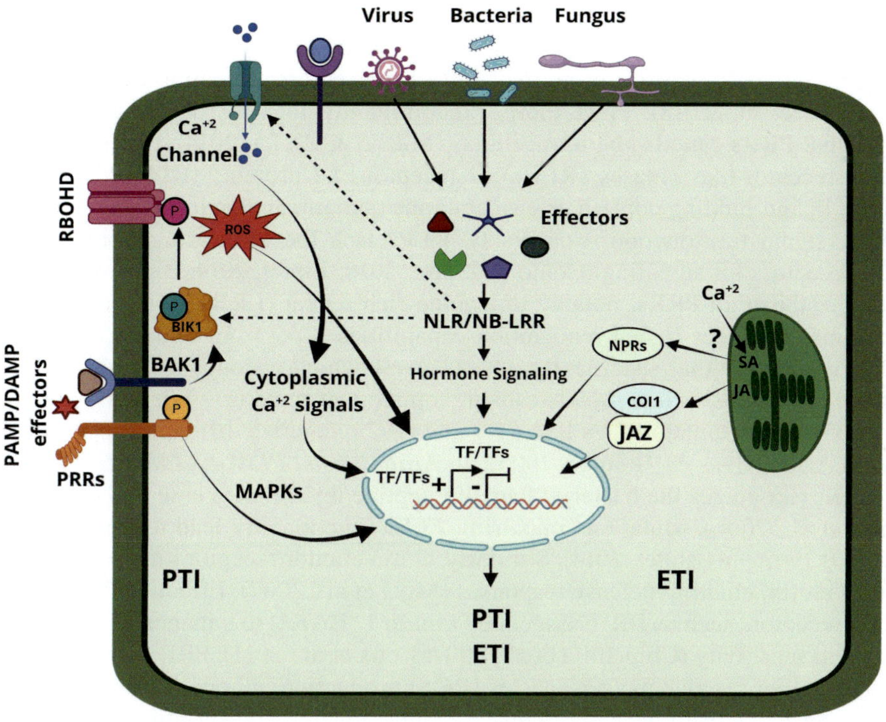

FIGURE 5.2 Schematic representation of PTI and ETI signaling under pathogen attack, highlighting PAMP/DAMP recognition, calcium influx, respiratory burst oxidase homolog B, BIK1, BAK1, and NLR-mediated responses.

5.4 PLANT IMMUNE SYSTEM: DUAL-TIERED DEFENSE MECHANISMS

Plants have a dual-tiered innate immune system comprising pattern-triggered immunity (PTI) and effector-triggered immunity (ETI) (Figure 5.2). These mechanisms work together, enabling plants to recognize and defend against microbial invaders and environmental threats. Plants rely on extracellular and intracellular immune receptors to detect molecular cues indicating microbial invasion or modified self (Van de Weyer et al., 2019; Zipfel, 2014). The first layer of defense, PTI involves surface-localized pattern recognition receptors (PRRs) that detect pathogen-associated molecular patterns (PAMPs) or other conserved molecular signals, whereas the second layer of defense, ETI, constitutes intracellular immune receptors known as resistance (R) proteins. These R proteins recognize pathogen-secreted effector molecules either directly or indirectly, activating a stronger defense response (Jones & Dangl, 2006; Macho & Zipfel, 2014).

5.5 MICROBIAL RECOGNITION

Microbial recognition in plants primarily involves two receptor types: Surface-localized PRRs and intracellular nucleotide-binding leucine-rich repeat (NLR) receptors (Ramírez-Zavaleta et al., 2022). PRRs initiate PTI, the plant's first defensive response, by detecting conserved microbial molecules termed molecular patterns (MPs). These patterns include PAMPs, microbe-associated molecular patterns (MAMPs), wound-induced molecular patterns (WIMPs), and damage-associated molecular patterns (DAMPs) (Macho & Zipfel, 2014). MPs typically enter the apoplast following pathogen invasion or mechanical injury, making their mutation-based evasion challenging for pathogens (Ngou et al., 2022). Examples of recognized MAMPs/PAMPs are bacterial flagellin, fungal chitin, and bacterial lipopolysaccharides (Couto & Zipfel, 2016). Plants encode numerous PRRs similar to animal Toll-like receptors, detecting extracellular PAMPs/DAMPs (Claus et al., 2018). Plant PRRs predominantly localize on the plasma membrane, whereas animal PRRs can also be intracellular (Macho & Zipfel, 2014). PRRs in plants are categorized as receptor-like kinases (RLKs) or receptor-like proteins (RLPs). RLKs contain an extracellular ligand-binding domain, transmembrane domain, and an intracellular kinase domain essential for signal transduction. Conversely, RLPs lack the intracellular kinase domain and require co-receptors for signaling (Couto & Zipfel, 2016; Zipfel, 2014; Zipfel & Oldroyd, 2017). Structural variability in PRRs, notably in leucine-rich repeat (LRR), lysine motif (LysM), and lectin domains, enhances ligand recognition capabilities. *Cf-9* from tomato, the first identified plant PRR, is an RLP that specifically detects *Avr9*, which encodes a fungal avirulence protein secreted by *Cladosporium fulvum* causing tomato leaf mold (De Wit et al., 2002; Huang & Joosten, 2025). Other tomato RLPs like *Cf-4* and *Cf-2* recognize different fungal effectors such as *Avr4* and *Avr2* (Rivas & Thomas, 2005). In *Arabidopsis*, *FLAGELLIN-SENSING 2* (*FLS2*), an LRR-RLK, recognizes the bacterial flagellin peptide flg22 that is essential for PTI activation (Chinchilla et al., 2006). Mutations impairing *FLS2* functionality lead to susceptibility toward *Pseudomonas syringae* (Zipfel, 2014). Similarly, chitin elicitor receptor kinase 1 (CERK1) recognizes fungal chitin, eliciting defense responses (Miya et al., 2007). Effective PRR function often requires co-receptors such as BRI1-associated kinase 1 (BAK1) to enhance pathogen recognition specificity and sensitivity (Chinchilla et al., 2007; Roux et al., 2011). PRR activation triggers PTI responses including reactive oxygen species (ROS) production, mitogen-activated protein kinase (MAPK) pathways activation, and extensive defense-related transcriptional changes, offering comprehensive protection against pathogens.

5.6 ROLE OF NLR RECEPTORS IN EFFECTOR RECOGNITION AND ETI

ETI is a precise and robust immune response initiated when plants detect pathogen-derived effector proteins intracellularly. These effectors, secreted by pathogens, manipulate host cellular processes to facilitate infection, often prompting a localized cell death known as the HR. ETI activation is mediated by intracellular immune receptors called NLRs, which are present across plants and animals (Jones et al., 2016). These receptors specifically recognize pathogen effectors, which are proteins secreted to undermine host immunity (Huang & Joosten, 2025). NLRs possess nucleotide-binding (NB) domains and LRR motifs critical for interaction with pathogen effectors or effector-altered host proteins.

NLRs activate ETI via two principal recognition models: The guard model and the decoy model. The guard model describes how NLRs monitor specific host proteins (guardees) targeted by pathogen effectors, activating defense upon detecting effector-induced modifications. Conversely, the decoy model involves host proteins mimicking effector targets to divert effector activity from genuine host proteins, allowing indirect effector detection by NLRs.

Pathogen effectors employ multiple strategies to undermine plant defenses, such as altering PAMPs, competing with PRRs, and disrupting immune signaling or hormone-mediated defenses (Dou & Zhou, 2012; Wang et al., 2022). Some bacterial pathogens, notably *Xanthomonas campestris* and *X. oryzae* pv. *oryzae* (Xoo), secrete transcription activator-like effectors (TALEs). These TALEs activate specific host susceptibility (S) genes by binding effector-binding elements (EBEs) in promoters, thereby promoting pathogen colonization. A prime example of an S gene is the Mildew Resistance Locus O (MLO), necessary for powdery mildew infection (Kusch & Panstruga, 2017). Together, PRRs and NLRs form an interconnected network of pathogen recognition mechanisms, enabling plants to mount multilayered and highly effective defense strategies against microbial threats.

5.7 PLANT IMMUNITY AGAINST VIRUSES

Plant antiviral immunity involves recognizing double-stranded RNAs (dsRNAs), produced during viral replication, as PAMPs. These dsRNAs activate RNA interference (RNAi), analogous to PTI, providing initial defense (Incarbone & Dunoyer, 2013). However, viruses can counter this through viral suppressors of RNAi (VSRs) like coat proteins, promoting effector-triggered susceptibility (ETS). Plants respond using R proteins to detect VSRs, activating ETI (Kørner et al., 2013; Mandadi & Scholthof, 2013). Together, RNAi-based PTI and R protein-mediated ETI constitute dual-layer antiviral defense (Mandadi & Scholthof, 2013; Kørner et al., 2013). Key discoveries include tobacco's *N* gene encoding a TIR-NB-LRR protein conferring resistance to tobacco mosaic virus (TMV) (Whitham et al., 1994). Additional *R* genes, such as *Rx1* and *Rx2* against potato virus X (PVX), and *Pvr4* against potato virus Y and pepper mottle virus, illustrate plant antiviral defenses predominantly through coiled-coil or TIR-NB-LRR classes (Gururani et al., 2012; Zhu et al., 2013; Kim et al., 2015).

5.8 SIGNAL TRANSDUCTION IN PLANT IMMUNITY

Signal transduction pathways in plant defense start with pathogen recognition by cell-surface PRRs. These receptors identify PAMPs, initiating PTI. Upon activation, a signaling cascade ensues, marked by calcium ion influx and ROS generation. Concurrently, crucial protein kinases such as calcium-dependent protein kinases (CDPKs) and MAPKs are activated, translating extracellular signals into intracellular responses. These cascades result in transcriptional reprogramming, enhancing the expression of various defense-related genes. Defense signaling also involves hormonal pathways, notably salicylic acid (SA), jasmonic acid (JA), and ethylene (ET), which fine-tune responses based on pathogen type. Additionally, protein modifications such as phosphorylation and ubiquitination play essential roles, influencing the stability and activity of key signaling proteins. Ultimately, these coordinated molecular events lead to the establishment of both localized defense and systemic acquired resistance (SAR), conferring broad-spectrum protection against subsequent infections.

5.9 PLANT SIGNAL TRANSDUCTION MOLECULES

5.9.1 Calcium ions (Ca^{2+})

Calcium ions play a key role in plant growth, reproduction, stress tolerance, and plant microbe interaction (Kudla et al., 2018; Tian et al., 2020). Ca^{2+} accumulates rapidly during pathogen attack to trigger immune responses. In ETI, calcium permeable channels like *NRG1.1* amplify defense signals, initiating downstream responses. CDPKs and calmodulins (CaMs) help translate Ca^{2+} signals into responses such as ROS production, nitric oxide (NO) synthesis, and SA accumulation. These processes activate defense genes and HR, localizing infections. Calcium also facilitates SAR by spreading signals to distant tissues. However, overactivation of CDPKs can impair resistance to necrotrophic pathogens, underscoring the importance of balanced Ca^{2+} signaling (Bi et al., 2021; Thor et al., 2020).

5.9.2 Reactive Oxygen Species

Plants produce ROS at different stages of both biotic and abiotic stresses, with each stress type triggering a distinct ROS response. Key ROS molecules such as hydrogen peroxide (H_2O_2) and superoxide, are rapidly produced during PTI and ETI via NADPH oxidases (e.g., RBOHD). ROS accumulate at infection sites to reinforce cell walls and inhibit pathogen spread. ROS also acts as secondary messengers, amplifying defense responses and coordinating with Ca^{2+} waves to transfer signals systemically. During HR, ROS mediates localized cell death to restrict biotrophic pathogens. Furthermore, ROS interacts with NO and SA pathways to enhance defense gene expression and SAR. However, at higher concentrations, ROS becomes toxic and can damage cellular components. Antioxidant systems, such as ascorbate and glutathione, help maintain ROS homeostasis, ensuring their dual roles in defense and signaling are balanced (Mittler, 2017).

5.9.3 Nitric Oxide

NO is a gaseous signaling molecule that enhances the production of phytoalexins, strengthens cell walls, and modulates protein S-nitrosation, which affects the activity of key defense proteins. It amplifies ROS production and contributes to HR-associated cell death. NO works synergistically with SA and ET to activate defense genes and establish SAR. Additionally, NO participates in signal transduction, facilitating communication between local and systemic tissues during pathogen attack. NO overaccumulation may cause nitrosative stress, underscoring the need for regulation through feedback and crosstalk mechanisms (Yu et al., 2014; Mur et al., 2013).

5.9.4 Mitogen-Activated Protein Kinase

Mitogen activated protein kinase (MAPK) cascades consist of three key kinases: MAP kinase kinase kinases (MAPKKKs), MAP kinase kinases (MAPKKs), and MAP kinases (MAPKs), which sequentially amplify signals through phosphorylation. MAPKKKs, located upstream, link external signals to downstream responses by activating MAPKKs, which in turn activate MAPKs through dual phosphorylation at Thr/Tyr sites. Activated MAPKs either stay in the cytoplasm to influence other kinases or move to the nucleus to regulate defense-related gene expression by phosphorylating transcription factors like WRKY. WRKY stands for a plant-specific transcription factor family, named after its highly conserved amino acid motif: These are the one-letter amino acid codes: W (Tryptophan), R (Arginine), K (Lysine), and Y (Tyrosine). This pathway is integral to both PTI and ETI, enhancing resistance mechanisms such as SA synthesis and PR gene expression. For instance, overexpression of MKK7 in *Arabidopsis* promotes SA-mediated SAR, while cotton GhNTF6 boosts resistance against *Verticillium dahliae.* Additionally, the reversibility of phosphorylation, regulated by protein phosphatases, ensures dynamic control of cellular responses, allowing efficient adaptation to diverse stress signals. The MAPK cascade exemplifies how plants integrate and magnify environmental cues to optimize defense strategies (Bi et al., 2021; Thor et al., 2020).

5.10 ROLE OF HORMONES IN SIGNAL TRANSDUCTION

5.10.1 Salicylic Acid

Salicylic acid is a key hormone in plant immunity (Peng et al., 2021). Upon infection, SA accumulates both locally and systemically, triggering SAR and inducing the expression of PR genes. *NPR1* acts as a transcription activator of SA-responsive genes. *NPR1* activity is promoted by elevated SA levels during defense (Fan & Dong, 2002; Rochon et al., 2006). SA also modulates ROS and NO production, strengthening defense responses. Its role extends to amplifying HR, which localizes infections through programmed cell death. SA interacts antagonistically with JA but acts synergistically with ROS and NO to tailor defenses against specific pathogens. However, excessive SA accumulation can divert resources from growth and reproduction, reflecting a trade-off between immunity and development, thus prioritizing immune signaling over growth under stress conditions. The dynamic regulation of SA ensures effective protection while maintaining plant fitness (Durrant & Dong, 2004).

5.10.2 Jasmonic Acid

JA is a central hormone in plant defense against necrotrophic pathogens and herbivorous insects. Upon detection of damage or infection, JA signaling activates genes encoding proteinase inhibitors, secondary metabolites, and other defense-related proteins. The JA receptor COI1 interacts with JAZ repressors, releasing transcription factors like MYC2 to drive transcriptional reprogramming. JA works synergistically with ET to counter necrotrophic pathogens while antagonizing SA signaling to prioritize defense over biotrophs. This crosstalk ensures resource allocation aligns with the nature of the threat. However, excessive JA signaling can suppress growth and SA-mediated immunity, highlighting the need for fine-tuned regulation. Its role in balancing immunity and development underscores its importance in plant survival (Wasternack & Song, 2017).

5.10.3 Ethylene

ET is a gaseous hormone that modulates plant defenses against necrotrophic pathogens and environmental stress. Pathogen attacks or wounding stimulates ET biosynthesis, which triggers transcriptional reprogramming through EIN2 and EIN3, the central regulators of ET signaling. ET works in concert with JA to activate defenses against necrotrophs, while its antagonistic interaction with SA modulates biotroph resistance. It also regulates ROS and NO production, enhancing HR and cell wall fortification. ET's role in systemic signaling ensures the activation of defenses in distant tissues, preparing the plant for future threats. However, excessive ET production can accelerate senescence and impair growth, necessitating precise control. By coordinating with other pathways, ET ensures a tailored and efficient defense response, balancing immunity with plant fitness (Broekgaarden et al., 2015).

Signal transduction pathways form the backbone of plant immune responses, translating pathogen recognition into effective defense strategies. The integration of secondary messengers, MAPK cascades, and hormonal signaling ensures a dynamic and tailored response to diverse threats. This intricate network enables plants to balance immunity with growth and development, underscoring the complexity and efficiency of plant defense systems.

5.11 EVASION OF PLANT IMMUNITY BY PATHOGENS

Plant defense mechanisms impose strong selection pressure on pathogens, prompting the evolution of sophisticated evasion strategies. These adaptations allow pathogens to suppress host immune responses and avoid detection, enabling them to cause diseases. The evasion of plant immunity occurs through several mechanisms, including modifications to PAMPs, suppression of immune signaling, and manipulation of plant receptors.

5.11.1 Evasion of PRR-Mediated Detection

Plant cell-surface PRRs identify conserved PAMPs, triggering immune responses. However, pathogens have evolved various mechanisms to evade PRR detection and suppress the activation of PTI. A common evasion strategy involves altering MAMP sequences to avoid recognition. For example, *Ralstonia solanacearum* modifies its flg22 peptide to escape detection by the plant *FLS2* receptor (Wei et al., 2020). Likewise, *X. oryzae* pv. *oryzae* (Xoo) and *oryzicola* (Xoc) introduce amino acid substitutions in their flg22 sequences, thereby reducing recognition by rice *FLS2* and avoiding immune activation (Wang et al., 2015). These modifications diminish the immunogenic potential of flg22, enabling the pathogen to bypass early immune responses. In a similar manner, fungal pathogens release polysaccharide deacetylase enzymes that chemically convert immunogenic chitin oligomers into chitosan, a molecule with much lower capacity to stimulate plant immune signaling (Gao et al., 2019; Martínez-Cruz et al., 2021).

To evade immune recognition, some pathogens release enzymes that degrade MAMPs. For instance, *Pseudomonas* species produce the alkaline protease ArpA, which digests both flagellin monomers and the immunogenic flg22 peptide, thereby suppressing the flg22-triggered immune response in plants (Bardoel et al., 2011). Likewise, *Podosphaera xanthii* secretes chitinase-active effectors that break down fungal chitin, helping the pathogen escape detection by plant chitin receptors (Martínez-Cruz et al., 2021). In addition to degrading MAMPs, some pathogens use effectors to shield their own structural components from degradation by host enzymes. For example, the effector Avr4 from *Cladosporium fulvum* binds to chitin in its cell walls, protecting it from enzymatic breakdown by plant chitinases (Van Den Burg et al., 2006; Van Esse et al., 2007). Another effector from *C. fulvum*, known as extracellular protein 6 (Ecp6), binds to free chitin oligomers and sequesters them, preventing their detection by the plant chitin receptor CERK1 (Van Den Burg et al., 2006; De Jonge et al., 2010). During infection, pathogens can minimize the production of MAMPs to evade plant immune detection. For example, *Pseudomonas syringae* utilizes the alternative sigma factor AlgU to suppress the expression of flagellin, a key MAMP recognized by plant receptors (Bao et al., 2020). Furthermore, plant-derived β-galactosidase 1 (BGAL1) cleaves the O-glycan on flagellin to release the immunogenic flg22 peptide (Buscaill et al., 2019). In response, certain bacterial pathogens have evolved mechanisms to either modify flagellin glycosylation or produce inhibitors that block BGAL1 activity, thereby preventing flg22 release and immune activation.

In addition to modulating MAMP levels, pathogens also interfere with PRR abundance and functionality. *Pseudomonas syringae*, e.g., delivers type III effectors (T3Es) such as HopU1, which disrupt the interaction between the RNA-binding protein GRP7 and FLS2 transcripts, leading to reduced accumulation of the FLS2 receptor during infection (Nicaise et al., 2013). Another effector, AvrPtoB, functions as an E3 ubiquitin ligase and facilitates the ubiquitin-mediated degradation of PRRs like FLS2 (Göhre et al., 2008). Similarly, the effector HopAO1 inhibits the tyrosine phosphorylation of FLS2, thereby blocking its activation and dampening the immune response (Macho & Zipfel, 2014).

5.11.2 Evasion of NLR-Mediated Detection

NLR receptors are crucial intracellular immune receptors that recognize pathogen-derived effectors and initiate defense responses in plants. To escape NLR-mediated surveillance, pathogens have developed a range of strategies including genetic rearrangements, molecular alterations, and the suppression of effector detection mechanisms. Effector genes are frequently located in dynamic, repeat-rich genomic regions, which facilitate rapid evolution through processes such as mutation, deletion, and the activity of transposable elements. For instance, effector genes in *Phytophthora infestans* are enriched in gene-sparse areas and exhibit strong diversifying selection—an adaptation strategy extensively discussed by Raffaele and Kamoun (2012). Similarly, the *Magnaporthe oryzae* AVR-Pita effector can be disrupted by Pot3 transposon insertions, as demonstrated by Zhou et al. (2007), impairing effector function and recognition. To further avoid immune detection, pathogens can introduce specific missense mutations into effector proteins. A single amino acid substitution in the *Hyaloperonospora arabidopsidis* effector ATR1, e.g., disrupts its recognition by the host NLR RPP1, effectively evading the plant's defense system (Ma et al., 2020). Likewise, *Phytophthora sojae* alters its Avr3c effector via a serine-to-glycine change, reducing its interaction with host targets and circumventing Rps3c-mediated immunity (Huang et al., 2019). Even with preserved host binding, extensive polymorphisms, such as those observed in the AVR2-like gene of *Phytophthora infestans* including 13 amino acid differences, can prevent the activation of potato receptor R2 (Gilroy et al., 2011). Beyond protein-level changes, pathogens may suppress effector gene expression to avoid triggering host defenses. In *Hyaloperonospora arabidopsidis*, the absence of *RxL103* expression correlates with successful evasion of RPP4-based resistance (Asai et al., 2018).

Regulatory disruptions such as promoter insertions (e.g., *Magnaporthe oryzae* AVR-Pita1) and structural rearrangements (*Phytophthora sojae* Avr3a/5) also block effector transcription (Asai et al., 2018; Zhou et al., 2007). Additionally, epigenetic modifications, particularly histone H3 methylation, have been linked to reduced expression of Avr1b in virulent *Phytophthora sojae* isolates (Wang et al., 2020). Pathogens may also deploy epistatic effectors that disrupt NLR recognition pathways. The *Pseudomonas syringae* effector HopZ3, for instance, acetylates the AvrB3–RPM1 complex, hindering its activation (Gilroy et al., 2011). In a similar manner, *Phytophthora infestans* IPIO4 binds to the potato receptor Rpiblb1, blocking its oligomerization, while other IPIO variants still induce resistance (Champouret et al., 2009; Zhao & Song, 2021). Finally, some pathogens target host proteins guarded by NLRs, such as RIN4, which is monitored by RPM1 and RPS2 in *Arabidopsis thaliana* (Deslandes & Rivas, 2012). The effector AvrRpt2 from *Pseudomonas syringae* cleaves RIN4 to trigger RPS2-mediated immunity yet simultaneously prevents RPM1 activation by interfering with phosphorylation dynamics (Mackey et al., 2003). These findings underscore the sophisticated strategies pathogens employ to manipulate and evade plant immune systems.

5.12 HOST–PATHOGEN CO-EVOLUTION: MECHANISTIC INSIGHTS

The ongoing evolutionary struggle between plants and pathogens fuels a cycle of mutual adaptation, as illustrated by the zig-zag model of plant immunity. In this dynamic interaction, pathogens release effector molecules specifically designed to interfere with and suppress plant immune responses, ultimately resulting in a state known as ETS (Dou & Zhou, 2012).

In response, plants evolve R proteins, such as NLRs, to detect these effectors and trigger ETI, often culminating in localized cell death to restrict pathogen spread. Pathogens counteract by mutating or acquiring new effectors, perpetuating a cycle of adaptation. This co-evolution diversifies pathogen effector repertoires and plant *R* genes, shaping the dynamic interplay in plant–pathogen interactions. This ongoing evolutionary battle between plant immune responses and pathogen effectors was conceptualized as the "zig-zag" model in a seminal review by Jones and Dangl (2006).

5.13 HARNESSING BENEFICIAL MICROBES TO ACTIVATE INDUCED SYSTEMIC RESISTANCE IN PLANTS

Plants host a diverse microbial community that forms beneficial, neutral, or harmful interactions with plants. These interactions influence plant growth, disease resistance, and stress tolerance. Beneficial microbes, such as mycorrhizal fungi and nitrogen-fixing bacteria, promote plant growth through enhanced nutrient uptake and improved resilience, whereas pathogenic microbes cause a range of plant diseases.

Beneficial microbes play a vital role in strengthening plant immunity by stimulating both PTI and induced systemic resistance (ISR). Recognition of MAMPs by PRRs initiate PTI, establishing a frontline defense (Boller & He, 2009). Certain beneficial microbes, such as *Pseudomonas fluorescens* and *Bacillus cereus*, can activate ISR through JA- and ET-mediated signaling pathways, providing broad-spectrum resistance against diverse pathogens (Pieterse et al., 2014). Transcription factors like WRKY8, WRKY33, MYB72, MYC2, and ERF1 regulate ISR via JA, ET, and SA pathways by activating defense genes in response to beneficial microbes (Sáenz-Mata et al., 2014; Jiang et al., 2016; Van der Ent et al., 2008).

In agricultural practice, beneficial microbes are introduced through seed treatments, foliar applications, or root inoculation to activate both PRR-mediated PTI and JA/ET-dependent ISR. The use of microbial consortia, alongside soil health management strategies, promotes a diverse rhizosphere microbiota that can prime and sustain robust immune responses. This approach aligns well with integrated pest management (IPM) and precision agriculture, linking molecular insights into PRRs, NLRs, and signaling networks with practical, field-level crop protection (Keswani et al., 2019).

Emerging evidence highlights that synthetic microbial communities (SynComs) deliberately assembled combinations of beneficial strains that can outperform single strains in promoting plant health and disease resistance (Santhanam et al., 2015; Durán et al., 2018). For example, SynComs applied to *Arabidopsis thaliana* effectively induced ISR against *Pseudomonas syringae* (Gonçalves et al., 2023), while *Bacillus*-based SynComs triggered ISR in maize (Ali et al., 2022). In wheat, SynComs enhanced host defense responses and suppressed *Rhizoctonia solani*, showcasing their promise in managing soil-borne diseases (Yin et al., 2022).

5.14 CONVENTIONAL BREEDING FOR DISEASE RESISTANCE

Conventional breeding approaches leverages from principles of plant immunity for identifying and introducing *R* genes and quantitative trait loci (QTLs) from wild relatives and landraces into cultivated crops (Pati et al., 2025). For instance, *R* genes such as *Xa21* in rice, *Lr24* in wheat, and *Rhg1* in soybean were introgressed to trigger immune responses against specific pathogens (Huang et al., 1997; Singh et al., 2011; Ruben et al., 2006). However, pathogens rapidly evolve to overcome single-gene resistance, therefore necessitating strategies like gene pyramiding to improve durability of resistance (Kaushik et al., 2016). Breeding methods such as backcrossing, pedigree selection, and recurrent selection have been widely applied based on crop type and reproduction mode (Miedaner, 2016). Marker-assisted selection (MAS) accelerates breeding by enabling precise tracking of *R* genes, reducing linkage drag, and improving efficiency (Hafeez et al., 2023; Jabran et al., 2023). Epigenetic mechanisms such as DNA methylation and histone modifications also control immune gene activity, offering new possibilities for improving disease resistance through epi-breeding (Xie & Duan, 2023). Despite these advancements, most commercial crops lack genetic diversity, making continuous breeding efforts essential to manage emerging diseases (Pati et al., 2025). Thus, integrating plant immunity with conventional and molecular breeding tools is critical for achieving a durable, broad-spectrum disease resistance.

5.15 ADVANCES IN PLANT IMMUNITY: MOLECULAR AND NANOTECHNOLOGICAL INTERVENTIONS

5.15.1 Nanotechnology-Mediated Modulation of Plant Immunity

Nanotechnology is gaining recognition as a valuable approach for strengthening plant immune responses and improving disease resistance. Engineered nanomaterials (ENMs) have demonstrated the ability to influence plant defense pathways by stimulating SAR and enhancing antioxidant activity. For instance, silica nanoparticles (SiNPs) have been shown to boost resistance against *Pseudomonas syringae* in *Arabidopsis thaliana* (El-Shetehy et al., 2021). Similarly, SA-coated iron nanoparticles (bio-FeNPs) effectively reduced *Fusarium* wilt incidence in watermelon (Noman et al., 2024). Zinc oxide nanoparticles (ZnO NPs) also contributed to the control of *Fusarium* wilt in tomatoes, achieving a 67.99% disease protection rate (Singh et al., 2024). In addition, carbon-based nanomaterials, including carbon nanotubes (CNTs), have been found to mitigate viral infections such as TMV in tobacco by triggering plant defense mechanisms (Adeel et al., 2021). Collectively, these studies underscore the promising role of nanomaterials in enhancing plant defense, although further research is necessary to refine their application and address potential safety concerns in agricultural systems.

5.15.2 RNAi for Disease Resistance in Crops

RNAi is a natural gene-silencing mechanism in plants that can be engineered to enhance disease resistance by targeting and silencing pathogen genes. This strategy has been successfully applied to combat plant viruses using hairpin RNAs or artificial microRNAs. For instance, transgenic barley expressing virus-targeting RNAi constructs has shown resistance to wheat dwarf virus (Kis et al., 2016). RNAi is also effective against fungal pathogens through host-induced gene silencing (HIGS), where plants express dsRNAs targeting essential fungal genes. For example, transgenic wheat expressing RNAi constructs against the *Fusarium graminearum* gene *Chs3b* showed strong resistance to *Fusarium* head and seedling blight (Cheng et al., 2015). These strategies are precise, can be multiplexed, and are especially useful when natural resistance is lacking.

5.15.3 CRISPR-Cas9-Mediated Modulation of Plant Immunity

Clustered regularly interspaced short palindromic repeats/CRISPR-associated protein 9 (CRISPR-Cas9) has emerged as a transformative genome editing technology in plant biotechnology. The system consists of two primary elements: Cas9 endonuclease and a guide RNA (gRNA). Acting like molecular scissors, Cas9 introduces precise double-stranded breaks at target genomic sites directed by the complementary sequence of the gRNA. This precise targeting capability has enabled researchers to modify *S* genes in crops to enhance resistance against various plant diseases. For example, in tomato, targeted mutagenesis of the SlPMR4 gene using CRISPR-Cas9 led to decreased susceptibility to powdery mildew and triggered a hypersensitive response-like cell death at infection sites, likely via SA signaling (Santillán Martínez et al., 2020). Editing of the nucleoredoxin genes (*SlNRX1* and *SlNRX2*) revealed that *SlNRX1* functions as a negative regulator of SA-mediated immunity, and its disruption increased resistance to both bacterial and fungal pathogens in tomato (Cha et al., 2023).

Likewise, targeted loss-of-function mutations in other well-characterized *S* genes, such as *MLO* in wheat and *LOB1* in citrus, conferred resistance to powdery mildew and citrus canker, respectively, by disrupting pathogen–host exploitation mechanisms (Peng et al., 2017; Wang et al., 2015). In the context of viral diseases, CRISPR-Cas9 has been successfully applied to edit *eIF4E* genes and their isoforms, which are critical for the replication of RNA potyviruses, in crops including *Arabidopsis*, cucumber, cassava, and rice. These cap-binding proteins interact with the viral VPg protein, and disrupting this interaction confers resistance to multiple viruses such as turnip mosaic virus (TuMV), zucchini yellow mosaic virus (ZYMV), clover yellow vein virus (ClYVV), and rice tungro spherical virus (Bastet et al., 2019; Chandrasekaran et al., 2016; Gomez et al., 2019; Macovei et al., 2018; Pyott et al., 2016). Collectively, these studies demonstrate the effectiveness of CRISPR-Cas9 in targeting a broad range of *S* genes, offering a powerful strategy for engineering durable, broad-spectrum disease resistance in crops. CRISPR-Cas9 modulates plant immunity and enhances disease management in crops.

5.15.4 Chemical Immunity Inducers

Plant immunity inducers are natural or synthetic compounds that activate defense responses without directly targeting pathogens. These inducers mimic endogenous immune signals and can trigger both local and systemic resistance SA, and its analogs, such as benzothiadiazole (BTH), isonicotinic acid derivatives like isonicotinic acid (INA) and 2,6-dichloroisonicotinic acid (NCI), and probenazole (PBZ), induce SAR by activating NPR1- and WRKY-dependent signaling pathways (Conrath et al., 2015; Dempsey & Klessig, 2017; Zhu et al., 2024). Similarly, jasmonate-based inducers (e.g., methyl jasmonate and coronatine mimics) stimulate defense responses against necrotrophic pathogens by activating JA/ET signaling and associated transcriptional networks (Wasternack & Song, 2017). Biopolymers such as chitin, chitosan, β-glucans, and oligogalacturonides (OGs) function as PAMPs or DAMPs, initiating defense responses through PRRs like CERK1 and PEPR1. These responses include rapid ROS production, calcium fluxes, MAPK activation, and upregulation of PR genes (Bigeard et al., 2015; De Kesel et al., 2021). A comprehensive review by Zhu et al. (2024) further highlights the central role of plant immunity inducers in sustainable disease control, documenting their mechanisms, efficacy, and integration into sustainable agricultural systems. Although these inducers present an eco-friendly alternative to conventional pesticides, their widespread use is limited by challenges in field scalability, formulation stability, and crop-specific efficacy.

5.16 CONCLUSION AND FUTURE PERSPECTIVES

Understanding and harnessing plant immunity has emerged as a cornerstone of sustainable agriculture in the face of growing food demand, climate change, and evolving pathogen threats. This chapter comprehensively reviewed the multilayered immune strategies employed by plants, highlighting both PTI and ETI, and the sophisticated roles of receptors such as PRRs and NLRs in pathogen recognition. It also explored the molecular signaling networks, including calcium flux, ROS, nitric oxide, and hormonal crosstalk, that orchestrate tailored immune responses. Additionally, the evasion tactics employed by pathogens underscore the ongoing co-evolutionary battle that shapes both host defense and pathogen virulence. The integration of cutting-edge technologies such as CRISPR-Cas9-mediated genome editing, RNAi, and nanotechnology offers unprecedented opportunities to engineer broad-spectrum and durable disease resistance. Furthermore, the use of beneficial microbes and SynComs to activate ISR provides an eco-friendly approach to crop protection. Chemical immunity inducers and biopolymer-based elicitors also hold potential for enhancing immunity without the adverse environmental effects associated with conventional pesticides. Looking forward, future research should focus on translating laboratory findings into field-ready solutions by optimizing delivery systems, enhancing the stability and specificity of immunity modulators, and integrating these tools into precision agriculture frameworks. Multi-omics approaches and machine learning can accelerate the identification of novel immune regulators and guide the design of more effective immune-enhancing strategies. Additionally, greater understanding of microbiome–plant interactions and epigenetic regulation of immunity can open new avenues for crop improvement. Ultimately, bridging molecular plant immunity with advanced biotechnological and ecological practices will enable the development of resilient crop systems, reduce chemical inputs, and ensure global food security in an environmentally responsible manner.

5.17 AUTHOR CONTRIBUTION

CPK: Conceptualization, original draft preparation, and editing. CPK and VM: Manuscript review, discussion, and editing.

ACKNOWLEDGMENTS

We acknowledge https://www.biorender.com/ for designing the figures. V.M. acknowledges the financial support from the Dirección General de Asuntos del Personal Académico (DGAPA), Universidad Nacional Autónoma de México (UNAM), through the UNAM Postdoctoral Program (POSDOC).

DECLARATION OF INTERESTS

All authors have reviewed and confirmed no conflicts of interest

REFERENCES

Adeel, M., Farooq, T., White, J. C., Hao, Y., He, Z., & Rui, Y. (2021). Carbon-based nanomaterials suppress tobacco mosaic virus (TMV) infection and induce resistance in *Nicotiana benthamiana*. *Journal of Hazardous Materials*, *404*, 124167. https://doi.org/10.1016/j.jhazmat.2020.124167

Ali, S. A. M., Sayyed, R. Z., Mir, M. I., Khan, M. Y., Hameeda, B., Alkhanani, M. F., Haque, S., Mohammad Al Tawaha, A. R., & Poczai, P. (2022). Induction of systemic resistance in maize and antibiofilm activity of surfactin from *Bacillus velezensis* MS20. *Frontiers in Microbiology*, *13*, 879739.

Asai, S., Furzer, O. J., Cevik, V., Kim, D. S., Ishaque, N., Goritschnig, S., Staskawicz, B. J., Shirasu, K., & Jones, J. D. G. (2018). A downy mildew effector evades recognition by polymorphism of expression and subcellular localization. *Nature Communications*, *9*(1). https://doi.org/10.1038/s41467-018-07469-3

Asghar, U., Malik, M. F., & Javed, A. (2016). Pesticide exposure and human health: A review. *Journal of Ecosystem & Ecography*, *5*(2).

Austin Bourke, P. M. (1964). Emergence of potato blight, 1843–46. *Nature*, *203*(4947), 805–808. https://doi.org/10.1038/203805a0

Avelino, J., Cristancho, M., Georgiou, S., Imbach, P., Aguilar, L., Bornemann, G., Läderach, P., Anzueto, F., Hruska, A. J., & Morales, C. (2015). The coffee rust crises in Colombia and Central America (2008–2013): Impacts, plausible causes and proposed solutions. *Food Security*, *7*(2), 303–321. https://doi.org/10.1007/s12571-015-0446-9

Bao, Z., Wei, H. L., Ma, X., & Swingle, B. (2020). *Pseudomonas syringae* AlgU downregulates flagellin gene expression, helping evade plant immunity. *Journal of Bacteriology*, *202*(4). https://doi.org/10.1128/JB.00418-19

Bardoel, B. W., van der Ent, S., Pel, M. J., Tommassen, J., Pieterse, C. M., van Kessel, K. P., & van Strijp, J. A. (2011). *Pseudomonas* evades immune recognition of flagellin in both mammals and plants. *PLoS Pathogens*, *7*(8), e1002206.

Bastet, A., Zafirov, D., Giovinazzo, N., Guyon-Debast, A., Nogué, F., Robaglia, C., & Gallois, J.-L. (2019). Mimicking natural polymorphism in eIF4E by CRISPR-Cas9 base editing is associated with resistance to potyviruses. *Plant Biotechnology Journal*, *17*(9), 1736–1750. https://doi.org/10.1111/pbi.13096

Bi, G., Su, M., Li, N., Liang, Y. U., Dang, S., Xu, J., Hu, M., Wang, J., Zou, M., Deng, Y., Li, Q., Huang, S., Li, J., Chai, J., He, K., Chen, Y. H., & Zhou, J. M. (2021). The ZAR1 resistosome is a calcium-permeable channel triggering plant immune signaling. *Cell*, *184*(13), 3528–3541.

Bigeard, J., Colcombet, J., & Hirt, H. (2015). Signaling mechanisms in pattern-triggered immunity (PTI). *Molecular Plant*, *8*(4), 521–539. https://doi.org/10.1016/j.molp.2014.12.022

Boller, T., & He, S. Y. (2009). Innate immunity in plants: An arms race between pattern recognition receptors in plants and effectors in microbial pathogens. *Science*, *324*(5928), 742–744.

Broekgaarden, C., Caarls, L., Vos, I. A., Pieterse, C. M., & Van Wees, S. C. (2015). Ethylene: Traffic controller on hormonal crossroads to defense. *Plant Physiology*, *169*(4), 2371–2379.

Buscaill, P., Chandrasekar, B., Sanguankiattichai, N., Kourelis, J., Kaschani, F., Thomas, E. L., Morimoto, K., Kaiser, M., Preston, G. M., Ichinose, Y., & Van Der Hoorn, R. A. L. (2019). Glycosidase and glycan polymorphism control hydrolytic release of immunogenic flagellin peptides. *Science*, *364*(6436). https://doi.org/10.1126/science.aav0748

Cha, J. Y., Uddin, S., Macoy, D. M., Shin, G. I., Jeong, S. Y., Ali, I., Hwang, J. W., Ji, M. G., Lee, S. C., Park, J. H., Sultana, M., Ryu, G. R., Ahn, G., Lee, S. Y., Kim, M. G., & Kim, W. Y. (2023). Nucleoredoxin gene *SlNRX1* negatively regulates tomato immunity by activating SA signaling pathway. *Plant Physiology and Biochemistry*, *200*, 107804.

Chakraborty, S., & Newton, A. C. (2011). Climate change, plant diseases and food security: An overview. *Plant Pathology*, *60*(1), 2–14. https://doi.org/10.1111/j.1365-3059.2010.02411.x

Champouret, N., Bouwmeester, K., Rietman, H., van der Lee, T., Maliepaard, C., Heupink, A., van de Vondervoort, P. J., Jacobsen, E., Visser, R. G., van der Vossen, E. A., Govers, F., & Vleeshouwers, V. G. (2009). *Phytophthora infestans* isolates lacking class I ipiO variants are virulent on Rpi-blb1 potato. *Molecular Plant–Microbe Interactions*, *22*, 1535–1545.

Chandrasekaran, J., Brumin, M., Wolf, D., Leibman, D., Klap, C., Pearlsman, M., Sherman, A., Arazi, T., & Gal-On, A. (2016). Development of broad virus resistance in non-transgenic cucumber using CRISPR/Cas9 technology. *Molecular Plant Pathology*, *17*(7), 1140–1153. https://doi.org/10.1111/mpp.12375

Cheng, W., Song, X.-S., Li, H.-P., Cao, L.-H., Sun, K., Qiu, X.-L., Xu, Y.-B., Yang, P., Huang, T., Zhang, J.-B., Qu, B., & Liao, Y.-C. (2015). Host-induced gene silencing of an essential chitin synthase gene confers durable resistance to *Fusarium* head blight and seedling blight in wheat. *Plant Biotechnology Journal*, *13*(9), 1335–1345. https://doi.org/10.1111/pbi.12352

Chinchilla, D., Bauer, Z., Regenass, M., Boller, T., & Felix, G. (2006). The *Arabidopsis* receptor kinase FLS2 binds flg22 and determines the specificity of flagellin perception. *The Plant Cell*, *18*, 465–476. https://doi.org/10.1105/tpc.105.036574

Chinchilla, D., Zipfel, C., Robatzek, S., Kemmerling, B., Nürnberger, T., Jones, J. D. G., Felix, G., & Boller, T. (2007). A flagellin-induced complex of the receptor FLS2 and BAK1 initiates plant defence. *Nature*, *448*(7152), 497–500. https://doi.org/10.1038/nature05999

Claus, L. A. N., Savatin, D. V, & Russinova, E. (2018). The crossroads of receptor-mediated signaling and endocytosis in plants. *Journal of Integrative Plant Biology*, *60*(9), 827–840. https://doi.org/10.1111/jipb.12672

Conrath, U., Beckers, G. J. M., Langenbach, C. J. G., & Jaskiewicz, M. R. (2015). Priming for enhanced defense. *Annual Review of Phytopathology*, *53*, 97–119. https://doi.org/10.1146/annurev-phyto-080614-120132

Couto, D., & Zipfel, C. (2016). Regulation of pattern recognition receptor signalling in plants. *Nature Reviews Immunology*, *16*(9), 537–552. https://doi.org/10.1038/nri.2016.77

Crop Protection Network (CPN). (2020). *Estimates of corn, soybean, and wheat yield losses due to disease: An online tool*. https://loss.cropprotectionnetwork.org/.

Damalas, C. A., & Eleftherohorinos, I. G. (2011). Pesticide exposure, safety issues, and risk assessment indicators. *International Journal of Environmental Research and Public Health*, *8*(5), 1402–1419. MDPI. https://doi.org/10.3390/ijerph8051402

De Jonge, R., Peter van Esse, H., Kombrink, A., Shinya, T., Desaki, Y., Bours, R., Van der Krol, S., Shibuya, N., Joosten, M. H., & Thomma, B. P. (2010). Conserved fungal LysM effector Ecp6 prevents chitin-triggered immunity in plants. *Science*, *329*(5994), 953–955.

De Kesel, J., Conrath, U., Flors, V., Luna, E., Mageroy, M. H., Mauch-Mani, B., Pastor, V., Pozo, M. J., Pieterse, C. M. J., Ton, J., & Kyndt, T. (2021). The induced resistance lexicon: Do's and don'ts. *Trends in Plant Science*, *26*(7), 685–691. https://doi.org/10.1016/j.tplants.2021.01.001

De Wit, P. J. G. M., Brandwagt, B. F., Van Den Burg, H. A., Cai, X., Van Der Hoorn, R. A. L., De Jong, C. F., Van 't Klooster, J., De Kock, M. J. D., Kruijt, M., Lindhout, W. H., Luderer, R., Takken, F. L. W., Westerink, N., Vervoort, J. J. M., & Joosten, M. H. A. J. (2002). The molecular basis of co-evolution between *Cladosporium fulvum* and tomato. *Antonie van Leeuwenhoek*, *81*.

Dempsey, D. A., & Klessig, D. F. (2017). How does the multifaceted plant hormone salicylic acid combat disease in plants and are similar mechanisms utilized in humans? *BMC Biology*, *15*(1), 23. https://doi.org/10.1186/s12915-017-0364-8

Desa, U. N. (2022). *World population prospects 2022: Summary of results*. United Nations Department of Economic and Social Affairs, Population Division, Tech. Rep. UN DESA/POP/2022/TR, (3)

Deslandes, L., & Rivas, S. (2012). Catch me if you can: Bacterial effectors and plant targets. *Trends in Plant Science*, *17*, 644–655.

Devanna, B. N., Jain, P., Solanke, A. U., Das, A., Thakur, S., Singh, P. K., Kumari, M., Dubey, H., Jaswal, R., Pawar, D., Kapoor, R., Singh, J., Arora, K., Saklani, B. K., AnilKumar, C., Maganti, S. M., Sonah, H., Deshmukh, R., Rathour, R., & Sharma, T. R. (2022). Understanding the dynamics of blast resistance in rice-*Magnaporthe oryzae* interactions. *Journal of Fungi*, *8*(6). https://doi.org/10.3390/jof8060584

Dou, D., & Zhou, J.-M. (2012). Phytopathogen effectors subverting host immunity: Different foes, similar battleground. *Cell Host & Microbe*, *12*(4), 484–495. https://doi.org/10.1016/j.chom.2012.09.003

Durán, P., Thiergart, T., Garrido-Oter, R., Agler, M., Kemen, E., Schulze-Lefert, P., & Hacquard, S. (2018). Microbial interkingdom interactions in roots promote *Arabidopsis* survival. *Cell*, *175*(4), 973–983. https://doi.org/10.1016/j.cell.2018.10.020

Durrant, W. E., & Dong, X. (2004). Systemic acquired resistance. *Annual Review of Phytopathology*, *42*, 185–209. https://doi.org/10.1146/annurev.phyto.42.040803.140421

El-Shetehy, M., Moradi, A., Maceroni, M., Reinhardt, D., Petri-Fink, A., Rothen-Rutishauser, B., Mauch, F., & Schwab, F. (2021). Silica nanoparticles enhance disease resistance in *Arabidopsis* plants. *Nature Nanotechnology*, *16*(3), 344–353.

Fan, W., & Dong, X. (2002). In vivo interaction between NPR1 and transcription factor TGA2 leads to salicylic acid–mediated gene activation in *Arabidopsis*. *The Plant Cell*, *14*(6), 1377–1389.

Gao, F., Zhang, B. Sen, Zhao, J. H., Huang, J. F., Jia, P. S., Wang, S., Zhang, J., Zhou, J. M., & Guo, H. S. (2019). Deacetylation of chitin oligomers increases virulence in soil-borne fungal pathogens. *Nature Plants*, *5*(11), 1167–1176. https://doi.org/10.1038/s41477-019-0527-4

Geeta, & Mishra, R. (2018). Fungal and bacterial biotrophy and necrotrophy. In Archana Singh & I. K. Singh (Eds.), *Molecular Aspects of Plant-Pathogen Interaction* (pp. 21–42). Springer Singapore. https://doi.org/10.1007/978-981-10-7371-7_2

Gilroy, E. M., Breen, S., Whisson, S. C., Squires, J., Hein, I., Kaczmarek, M., Turnbull, D., Boevink, P. C., Lokossou, A., Cano, L. M., Morales, J., Avrova, A. O., Pritchard, L., Randall, E., Lees, A., Govers, F., van West, P., Kamoun, S., Vleeshouwers, V. G. A. A., Cooke, D. E. L., & Birch, P. R. J. (2011). Presence/absence, differential expression and sequence polymorphisms between PiAVR2 and PiAVR2-like in *Phytophthora infestans* determine virulence on R2 plants. *New Phytologist*, *191*(3), 763–776. https://doi.org/10.1111/j.1469-8137.2011.03736.x

Göhre, V., Spallek, T., Häweker, H., Mersmann, S., Mentzel, T., Boller, T., de Torres, M., Mansfield, J. W., & Robatzek, S. (2008). Plant pattern-recognition receptor FLS2 is directed for degradation by the bacterial ubiquitin ligase AvrPtoB. *Current Biology*, *18*(23), 1824–1832.

Gomez, M. A., Lin, Z. D., Moll, T., Chauhan, R. D., Hayden, L., Renninger, K., Beyene, G., Taylor, N. J., Carrington, J. C., Staskawicz, B. J., & Bart, R. S. (2019). Simultaneous CRISPR/Cas9-mediated editing of cassava eIF4E isoforms nCBP-1 and nCBP-2 reduces cassava brown streak disease symptom severity and incidence. *Plant Biotechnology Journal*, *17*(2), 421–434. https://doi.org/10.1111/pbi.12987

Gonçalves, A. S., Leitão, M. M., Simões, M., & Borges, A. (2023). The action of phytochemicals in biofilm control. *Natural Product Reports*, *40*(3), 595–627. https://doi.org/10.1039/D2NP00053A

Gururani, M. A., Venkatesh, J., Upadhyaya, C. P., Nookaraju, A., Pandey, S. K., & Park, S. W. (2012). Plant disease resistance genes: Current status and future directions. *Physiological and Molecular Plant Pathology*, *78*, 51–65. https://doi.org/10.1016/j.pmpp.2012.01.002

Guttman, D. S., McHardy, A. C., & Schulze-Lefert, P. (2014). Microbial genome-enabled insights into plant–microorganism interactions. *Nature Reviews Genetics*, *15*(12), 797–813. Nature Publishing Group. https://doi.org/10.1038/nrg3748

Hafeez, U., Ali, M., Hassan, S. M., Akram, M. A., and Zafar, A. (2023). Advances in breeding and engineering climate-resilient crops: A comprehensive review. *International Journal of Research and Advances in Agricultural Sciences*, *2*, 85–99.

Hawes, M. C., Curlango-Rivera, G., Xiong, Z., & Kessler, J. O. (2012). Roles of root border cells in plant defense and regulation of rhizosphere microbial populations by extracellular DNA "trapping." *Plant and Soil*, *355*(1–2), 1–16. https://doi.org/10.1007/s11104-012-1218-3

He, S., & Creasey Krainer, K. M. (2020). Pandemics of people and plants: Which is the greater threat to food security? *Molecular Plant*, *13*(7), 933–934. https://doi.org/10.1016/j.molp.2020.06.007

Huang, J., Chen, L., Lu, X., Peng, Q., Zhang, Y., Yang, J., Zhang, B. Y., Yang, B., Waletich, J. R., Yin, W., Zheng, X., Wang, Y., & Dong, S. (2019). Natural allelic variations provide insights into host adaptation of *Phytophthora* avirulence effector PsAvr3c. *New Phytologist*, *221*(2), 1010–1022. https://doi.org/10.1111/nph.15414

Huang, N., Angeles, E. R., Domingo, J., Magpantay, G., Singh, S., Zhang, G., Kumaravadivel, N., Bennett, J., & Khush, G. S. (1997). Pyramiding bacterial blight resistance genes in rice: Marker-assisted selection using RFLP and PCR. *Theoretical and Applied Genetics*, *95*, 313–320. https://doi.org/10.1007/s001220050565

Huang, W. R. H., & Joosten, M. H. A. J. (2025). Immune signaling: Receptor-like proteins make the difference. *Trends in Plant Science*, *30*(1), 54–68. https://doi.org/10.1016/j.tplants.2024.03.012

Incarbone, M., & Dunoyer, P. (2013). RNA silencing and its suppression: Novel insights from *in planta* analyses. *Trends in Plant Science*, *18*(7), 382–392. https://doi.org/10.1016/j.tplants.2013.04.001

Jabran, M., Ali, M. A., Zahoor, A., Muhae-Ud-Din, G., Liu, T., Chen, W., & Gao., Li (2023). Intelligent reprogramming of wheat for enhancement of fungal and nematode disease resistance using advanced molecular techniques. *Frontiers in Plant Science*, *14*, 1132699. https://doi.org/10.3389/fpls.2023.1132699

Jiang, C. H., Huang, Z. Y., Xie, P., Gu, C., Li, K., Wang, D. C., Yu, Y. Y., Fan, Z. H., Wang, C. J., Wang, Y. P., Guo, Y. H., & Guo, J. H. (2016). Transcription factors WRKY70 and WRKY11 served as regulators in *Rhizobacterium Bacillus cereus* AR156-induced systemic resistance to *Pseudomonas syringae* pv. tomato DC3000 in Arabidopsis. *Journal of Experimental Botany*, *67*(1), 157–174.

Jones, J., & Dangl, J. (2006). The Plant immune system. *Nature*, *444*, 323–329. https://doi.org/10.1038/nature05286

Jones, J. D. G., Vance, R. E., & Dangl, J. L. (2016). Intracellular innate immune surveillance devices in plants and animals. *Science (New York, N.Y.)*, *354*(6316), aaf6395. https://doi.org/10.1126/science.aaf6395

Jones, R. A. C., & Naidu, R. A. (2019). Global dimensions of plant virus diseases: Current status and future perspectives. *Annual Review of Virology*, *6*, 387–409. https://doi.org/10.1146/annurev-virology-092818-015606

Kaushik, A., Kaswan, V., Sharma, S., & Kumar, R. (2016). Gene pyramiding: A novel approach for disease resistance. *Annals of Biology*, *32*, 13–16. https://doi.org/10.5555/20163192203

Keswani, C., Prakash, O., Bharti, N., Vílchez, J. I., Sansinenea, E., Lally, R. D., Borriss, R., Singh, S. P., Gupta, V. K., Fraceto, L. F., de Lima, R., & Singh, H. B. (2019). Re-addressing the biosafety issues of plant growth promoting rhizobacteria. *Science of the Total Environment*, *690*, 841–852.

Kim, S. B., Lee, H. Y., Seo, S., Lee, J. H., & Choi, D. (2015). RNA-dependent RNA polymerase (NIb) of the potyviruses is an avirulence factor for the broad-spectrum resistance gene Pvr4 in *Capsicum annuum* cv. CM334. *PLoS One*, *10*(3), e0119639.

Kis, A., Tholt, G., Ivanics, M., Várallyay, É., Jenes, B., & Havelda, Z. (2016). Polycistronic artificial miRNA-mediated resistance to wheat dwarf virus in barley is highly efficient at low temperature. *Molecular Plant Pathology*, *17*(3), 427–437. https://doi.org/10.1111/mpp.12291

Kørner, C. J., Klauser, D., Niehl, A., Domínguez-Ferreras, A., Chinchilla, D., Boller, T., Heinlein, M., & Hann, D. R. (2013). The immunity regulator BAK1 contributes to resistance against diverse RNA viruses. *Molecular Plant–Microbe Interactions®*, *26*(11), 1271–1280. https://doi.org/10.1094/MPMI-06-13-0179-R

Kudla, J., Becker, D., Grill, E., Hedrich, R., Hippler, M., Kummer, U., Parniske, M., Romeis, T., & Schumacher, K. (2018). Advances and current challenges in calcium signaling. *New Phytologist*, *218*(2), 414–431.

Kusch, S., & Panstruga, R. (2017). Mlo-based resistance: An apparently universal "weapon" to defeat powdery mildew disease. *Molecular Plant–Microbe Interactions*, *30*(3), 179–189. American Phytopathological Society. https://doi.org/10.1094/MPMI-12-16-0255-CR

Ma, S., Lapin, D., Liu, L. I., Sun, Y., Song, W., Zhang, X., Logemann, E., Yu, D., Wang, J., Jirschitzka, J., Han, Z., Schulze-Lefert, P., Parker, J. E., & Chai, J. (2020). Direct pathogen-induced assembly of an NLR immune receptor complex to form a holoenzyme. *Science*, *370*(6521), eabe3069.

Macho, A. P., & Zipfel, C. (2014). Plant PRRs and the activation of innate immune signaling. *Molecular Cell*, *54*(2), 263–272. Cell Press. https://doi.org/10.1016/j.molcel.2014.03.028

Mackey, D., Belkhadir, Y., Alonso, J. M., Ecker, J. R., & Dangl, J. L. (2003). *Arabidopsis* RIN4 is a target of the type III virulence effector AvrRpt2 and modulates RPS2-mediated resistance. *Cell*, *112*(3), 379–389.

Macovei, A., Sevilla, N. R., Cantos, C., Jonson, G. B., Slamet-Loedin, I., Čermák, T., Voytas, D. F., Choi, I.-R., & Chadha-Mohanty, P. (2018). Novel alleles of rice eIF4G generated by CRISPR/Cas9-targeted mutagenesis confer resistance to *Rice tungro* spherical virus. *Plant Biotechnology Journal*, *16*(11), 1918–1927. https://doi.org/10.1111/pbi.12927

Mandadi, K. K., & Scholthof, K.-B. G. (2013). Plant immune responses against viruses: How does a virus cause disease? *The Plant Cell*, *25*(5), 1489–1505. https://doi.org/10.1105/tpc.113.111658

Martínez-Cruz, J., Romero, D., Hierrezuelo, J., Thon, M., de Vicente, A., & Pérez-García, A. (2021). Effectors with chitinase activity (EWCAs), a family of conserved, secreted fungal chitinases that suppress chitin-triggered immunity. *Plant Cell*, *33*(4), 1319–1340. https://doi.org/10.1093/plcell/koab011

Melotto, M., Underwood, W., Koczan, J., Nomura, K., & He, S. Y. (2006). Plant stomata function in innate immunity against bacterial invasion. *Cell*, *126*(5), 969–980. https://doi.org/10.1016/j.cell.2006.06.054

Miedaner, T. (2016). Breeding strategies for improving plant resistance to diseases. In J. Al-Khayri, S. Jain and D. Johnson (Eds.), *Advances in Plant Breeding Strategies: Agronomic, Abiotic and Biotic Stress Traits* (pp. XX–XX). Springer, Cham. https://doi.org/10.1007/9783-319-22518-0_15

Mittler, R. (2017). ROS are good. *Trends in Plant Science*, *22*(1), 11–19. https://doi.org/10.1016/j.tplants.2016.08.002

Miya, A., Albert, P., Shinya, T., Desaki, Y., Ichimura, K., Shirasu, K., Narusaka, Y., Kawakami, N., Kaku, H., & Shibuya, N. (2007). CERK1, a LysM receptor kinase, is essential for chitin elicitor signaling in *Arabidopsis*. *Proceedings of the National Academy of Sciences*, *104*(49), 19613–19618. https://doi.org/10.1073/pnas.0705147104

Mur, L. A., Hebelstrup, K. H., & Gupta, K. J. (2013). Striking a balance: Does nitrate uptake and metabolism regulate both NO generation and scavenging? *Frontiers in Plant Science*, *4*, 288.

Ngou, B. P. M., Ding, P., & Jones, J. D. G. (2022). Thirty years of resistance: Zig-zag through the plant immune system. *Plant Cell*, *34*(5), 1447–1478. American Society of Plant Biologists. https://doi.org/10.1093/plcell/koac041

Nicaise, V., Joe, A., Jeong, B. R., Korneli, C., Boutrot, F., Westedt, I., Staiger, D., Alfano, J. R., & Zipfel, C. (2013). *Pseudomonas* HopU1 modulates plant immune receptor levels by blocking the interaction of their mRNAs with GRP7. *EMBO Journal*, *32*(5), 701–712. https://doi.org/10.1038/emboj.2013.15

Noman, M., Ahmed, T., Shahid, M., Nazir, M. M., Li, D., & Song, F. (2024). Salicylic acid-doped iron nano-biostimulants potentiate defense responses and suppress *Fusarium* wilt in watermelon. *Journal of Advanced Research*, *59*, 19–33. https://doi.org/10.1016/j.jare.2023.06.011

Pati, R., Sandhu, S., Kawadiwale, A., & Kaur, G. (2025). Unveiling the underlying complexities in breeding for disease resistance in crop plants. *Frontiers in Plant Science*, *16*, 1559751.

Pautasso, M., Döring, T., Garbelotto, M., Pellis, L., & Jeger, M. (2012). Impacts of climate change on plant disease—opinions and trend. *European Journal of Plant Pathology*. https://doi.org/10.1007/s10658-012-9936-1

Peng, A., Chen, S., Lei, T., Xu, L., He, Y., Wu, L., Yao, L., & Zou, X. (2017). Engineering canker-resistant plants through CRISPR/Cas9-targeted editing of the susceptibility gene CsLOB1 promoter in citrus. *Plant Biotechnology Journal*, *15*(12), 1509–1519. https://doi.org/10.1111/pbi.12733

Peng, Y., Yang, J., Li, X., & Zhang, Y. (2021). Salicylic acid: Biosynthesis and signaling. *Annual Review of Plant Biology*, *72*, 761–791. https://doi.org/10.1146/annurev-arplant-081320-092855

Pieterse, C. M. J., Zamioudis, C., Berendsen, R. L., Weller, D. M., Van Wees, S. C. M., & Bakker, P. A. H. M. (2014). Induced systemic resistance by beneficial microbes. *Annual Review of Phytopathology*, *52*, 347–375. https://doi.org/10.1146/annurev-phyto-082712-102340

Pyott, D. E., Sheehan, E., & Molnar, A. (2016). Engineering of CRISPR/Cas9-mediated potyvirus resistance in transgene-free *Arabidopsis* plants. *Molecular Plant Pathology*, *17*(8), 1276–1288. https://doi.org/10.1111/mpp.12417

Raffaele, S., & Kamoun, S. (2012). Genome evolution in filamentous plant pathogens: Why bigger can be better. *Nature Reviews Microbiology*, *10*(6), 417–430.

Ramírez-Zavaleta, C. Y., García-Barrera, L. J., Rodríguez-Verástegui, L. L., Arrieta-Flores, D., & Gregorio-Jorge, J. (2022). An overview of PRR- and NLR-mediated immunities: Conserved signaling components across the plant kingdom that communicate both pathways. *International Journal of Molecular Sciences*, *23*(21). MDPI. https://doi.org/10.3390/ijms232112974

Rivas, S., & Thomas, C. M. (2005). Molecular interactions between tomato and the leaf mold pathogen *Cladosporium fulvum*. *Annual Review of Phytopathology*, *43*, 395–436. https://doi.org/10.1146/annurev.phyto.43.040204.140224

Rochon, A., Boyle, P., Wignes, T., Fobert, P. R., & Després, C. (2006). The coactivator function of *Arabidopsis* NPR1 requires the core of its BTB/POZ domain and the oxidation of C-terminal cysteines. *The Plant Cell*, *18*(12), 3670–3685.

Roux, M., Schwessinger, B., Albrecht, C., Chinchilla, D., Jones, A., Holton, N., Malinovsky, F. G., Tör, M., de Vries, S., & Zipfel, C. (2011). The *Arabidopsis* leucine-rich repeat receptor-like kinases BAK1/SERK3 and BKK1/SERK4 are required for innate immunity to hemibiotrophic and biotrophic pathogens. *The Plant Cell*, *23*(6), 2440–2455. https://doi.org/10.1105/tpc.111.084301

Ruben, E., Jamai, A., Afzal, J., Njiti, V. N., Triwitayakorn, K., Iqbal, M. J., Yaegashi, S., Bashir, R., Kazi, S., Arelli, P., Town, C. D., Ishihara, H., Meksem, K., & Lightfoot, D. A (2006). Genomic analysis of the rhg1 locus: Candidate genes that underlie soybean resistance to the cyst nematode. *Molecular Genetics and Genomics*, *276*, 503–516. https://doi.org/10.1007/s00438-006-0150-8

Sáenz-Mata, J., Salazar-Badillo, F. B., & Jiménez-Bremont, J. F. (2014). Transcriptional regulation of *Arabidopsis thaliana* WRKY genes under interaction with beneficial fungus *Trichoderma atroviride*. *Acta Physiologiae Plantarum*, *36*(5), 1085–1093. https://doi.org/10.1007/s11738-013-1483-7

Santhanam, R., Luu, V. T., Weinhold, A., Goldberg, J., Oh, Y., & Baldwin, I. T. (2015). Native root-associated bacteria rescue a plant from a sudden-wilt disease that emerged during continuous cropping. *Proceedings of the National Academy of Sciences*, *112*(36), E5013–E5020. https://doi.org/10.1073/pnas.1505765112

Santillán Martínez, M. I., Bracuto, V., Koseoglou, E., Appiano, M., Jacobsen, E., Visser, R. G., Wolters, A. A., & Bai, Y. (2020). CRISPR/Cas9-targeted mutagenesis of the tomato susceptibility gene PMR4 for resistance against powdery mildew. *BMC Plant Biology*, 20(1), 284.

Savary, S., Willocquet, L., Pethybridge, S. J., Esker, P., McRoberts, N., & Nelson, A. (2019). The global burden of pathogens and pests on major food crops. *Nature Ecology & Evolution*, *3*(3), 430–439. https://doi.org/10.1038/s41559-018-0793-y

Shen, Y., & Ronald, P. (2002). *Molecular determinants of disease and resistance in interactions of Xanthomonas oryzae pv. oryzae and rice*. www.elsevier.com/locate/micinf

Singh, P. P., Chaturvedi, S., Bhatnagar, T., Das, S., & Debnath, N. (2024). Green synthesis of ZnO nanoparticles using Lantana camara leaf extract for the enhancement of plant growth. *Advances in Natural Sciences: Nanoscience and Nanotechnology*, *15*(3), 35010. https://doi.org/10.1088/2043-6262/ad6cc0

Singh, R. P., Huerta-Espino, J., & Rajaram, S. (2011). Achieving near-immunity to leaf rust and stripe rust in wheat by combining slow rusting resistance genes. *Crop Science*, *41*, 1144–1152. https://doi.org/10.2135/cropsci2001.1144

Suzuki, N., Rivero, R. M., Shulaev, V., Blumwald, E., & Mittler, R. (2014). Abiotic and biotic stress combinations. *New Phytologist*, *203*(1), 32–43. https://doi.org/10.1111/nph.12797

Thor, K., Jiang, S., Michard, E., George, J., Scherzer, S., Huang, S., Dindas, J., Derbyshire, P., Leitão, N., DeFalco, T. A., Köster, P., Hunter, K., Kimura, S., Gronnier, J., Stransfeld, L., Kadota, Y., Bücherl, C. A., Charpentier, M., Wrzaczek, M., MacLean, D., Oldroyd, G. E. D., Menke, F. L. H., Roelfsema, M. R. G., Hedrich, R., Feijo, J., & Zipfel, C. (2020). The calcium-permeable channel OSCA1. 3 regulates plant stomatal immunity. *Nature*, *585*(7826), 569–573.

Thordal-Christensen, H. (2003). Fresh insights into processes of nonhost resistance. *Current Opinion in Plant Biology*, *6*(4), 351–357. Elsevier Ltd. https://doi.org/10.1016/S1369-5266(03)00063-3

Tian, W., Wang, C., Gao, Q., Li, L., & Luan, S. (2020). Calcium spikes, waves and oscillations in plant development and biotic interactions. *Nature Plants*, *6*(7), 750–759.

United Nations (2017). *World Population Projected to Reach 9.8 Billion in 2050, and 11.2 Billion in 2100*. Department of Economic and Social Affairs.

Van de Weyer, A.-L., Monteiro, F., Furzer, O. J., Nishimura, M. T., Cevik, V., Witek, K., Jones, J. D. G., Dangl, J. L., Weigel, D., & Bemm, F. (2019). A species-wide inventory of NLR genes and alleles in *Arabidopsis thaliana*. *Cell*, *178*(5), 1260–1272.e14. https://doi.org/10.1016/j.cell.2019.07.038

Van Den Burg, H. A., Harrison, S. J., Joosten, M. H. A. J., Vervoort, J., & De Wit, P. J. G. M. (2006). *Cladosporium fulvum* Avr4 protects fungal cell walls against hydrolysis by plant chitinases accumulating during infection. *Molecular Plant–Microbe Interactions MPMI*, *19*(12), 1420–1430. https://doi.org/10.1094/MPMI

Van der Ent, S., Verhagen, B. W., Van Doorn, R., Bakker, D., Verlaan, M. G., Pel, M. J., Joosten, R. G., Proveniers, M. C., Van Loon, L. C., Ton, J., & Pieterse, C. M. (2008). MYB72 is required in early signaling steps of rhizobacteria-induced systemic resistance in *Arabidopsis*. *Plant Physiology*, *146*(3), 1293–1304.

Van Esse, H. P., Bolton, M. D., Stergiopoulos, I., de Wit, P. J., & Thomma, B. P. (2007). The chitin-binding *Cladosporium fulvum* effector protein Avr4 is a virulence factor. *Molecular Plant–Microbe Interactions*, *20*(9), 1092–1101.

Wang, S., Sun, Z., Wang, H., Liu, L., Lu, F., Yang, J., Zhang, M., Zhang, S., Guo, Z., Bent, A. F., & Sun, W. (2015). Rice OsFLS2-mediated perception of bacterial flagellins is evaded by *Xanthomonas oryzae* pvs. *oryzae* and *oryzicola*. *Molecular Plant*, *8*(7), 1024–1037. https://doi.org/10.1016/j.molp.2015.01.012

Wang, W., Feng, B., Zhou, J. M., & Tang, D. (2020). Plant immune signaling: Advancing on two frontiers. *Journal of Integrative Plant Biology*, *62*(1), 2–24. Blackwell Publishing Ltd. https://doi.org/10.1111/jipb.12898

Wang, Y., Pruitt, R. N., Nürnberger, T., & Wang, Y. (2022). Evasion of plant immunity by microbial pathogens. *Nature Reviews Microbiology*, *20*(8), 449–464. Nature Research. https://doi.org/10.1038/s41579-022-00710-3

Wasternack, C., & Song, S. (2017). Jasmonates: Biosynthesis, metabolism, and signaling by proteins activating and repressing transcription. *Journal of Experimental Botany*, *68*(6), 1303–1321.

Wei, Y., Balaceanu, A., Rufian, J. S., Segonzac, C., Zhao, A., Morcillo, R. J., & Macho, A. P. (2020). An immune receptor complex evolved in soybean to perceive a polymorphic bacterial flagellin. *Nature Communications*, *11*(1), 3763.

Whitham, S., Dinesh-Kumar, S. P., Choi, D., Hehl, R., Corr, C., & Baker, B. (1994). The product of the tobacco mosaic virus resistance gene N: Similarity to toll and the interleukin-1 receptor. *Cell*, *78*(6), 1101–1115.

Williamson, B., Tudzynski, B., Tudzynski, P., & van Kan, J. A. N. A. L. (2007). *Botrytis cinerea*: The cause of grey mould disease. *Molecular Plant Pathology*, *8*(5), 561–580. https://doi.org/10.1111/j.1364-3703.2007.00417.x

Xie, S. S., & Duan, C. G. (2023). Epigenetic regulation of plant immunity: From chromatin codes to plant disease resistance. *Abiotech*, *4*, 124–139. https://doi.org/10.1007/s42994-023-00101-z

Yin, C., Hagerty, C. H., & Paulitz, T. C. (2022). Synthetic microbial consortia derived from rhizosphere soil protect wheat against a soilborne fungal pathogen. *Frontiers in Microbiology*, *13*, 908981.

Yu, M., Lamattina, L., Spoel, S. H., & Loake, G. J. (2014). Nitric oxide function in plant biology: A redox cue in deconvolution. *New Phytologist*, *202*(4), 1142–1156. https://doi.org/10.1111/nph.12739

Zhao, J., & Song, J. (2021). NLR immune receptor RB is differentially targeted by two homologous but functionally distinct effector proteins. *Plant Communications*, *2*(6). https://doi.org/10.1016/j.xplc.2021.100236

Zhou, E., Jia, Y., Singh, P., Correll, J. C., & Lee, F. N. (2007). Instability of the *Magnaporthe oryzae* avirulence gene AVR-Pita alters virulence. *Fungal Genetics and Biology*, *44*(10), 1024–1034. https://doi.org/10.1016/j.fgb.2007.02.003

Zhu, F., Cao, M.-Y., Zhang, Q.-P., Mohan, R., Schar, J., Mitchell, M., Chen, H., Liu, F., Wang, D., & Fu, Z. Q. (2024). Join the green team: Inducers of plant immunity in the plant disease sustainable control toolbox. *Journal of Advanced Research*, *57*, 15–42. https://doi.org/10.1016/j.jare.2023.04.016

Zhu, J. K. (2016). Abiotic stress signaling and responses in plants. *Cell*, *167*(2), 313–324. Cell Press. https://doi.org/10.1016/j.cell.2016.08.029

Zhu, S., Jeong, R. D., Lim, G. H., Yu, K., Wang, C., Chandra-Shekara, A. C., Navarre, D., Klessig, D.F., & Kachroo, P. (2013). Double-stranded RNA-binding protein 4 is required for resistance signaling against viral and bacterial pathogens. *Cell Reports*, *4*(6), 1168–1184. https://doi.org/10.1016/j.celrep.2013.08.018

Zipfel, C. (2014). Plant pattern-recognition receptors. *Trends in Immunology*, *35*(7), 345–351. https://doi.org/10.1016/j.it.2014.05.004

Zipfel, C., & Oldroyd, G. (2017). Plant signalling in symbiosis and immunity. *Nature*, *543*, 328–336. https://doi.org/10.1038/nature22009

6 Enhancing Plant Disease Resistance

Strategies for Manipulating Soil Defense Responses

Bilal Ahmad Mir, Preedhi Kapoor, Ankita Thakur, Ritu Kumari, Zubair Ahmad Parrey, Muhammad Aamir Manzoor, Gurmeen Rakhra, and Zishan Ahmad

6.1 INTRODUCTION

Soil defense mechanisms refer to the natural processes and interactions in soil ecosystems that protect and maintain soil health, fertility, and structure (Tahat et al., 2020). These mechanisms are vital for promoting plant development, mitigating soil erosion, and maintaining ecological equilibrium (Albaladejo et al., 2021). Analyzing soil defense mechanisms entails studying the interrelated roles of soil organisms, the physical and chemical properties of soil, and their complex interactions (Rashid and Chung, 2017). Actinomycetes, fungi, and bacteria are examples of microorganisms that are essential for decomposing organic debris, recycling nutrients, and inhibiting soil-borne diseases (Aguilar-Paredes et al., 2023). Beneficial microbes suppress harmful pathogens by competing for resources, producing antibiotics, and triggering systemic resistance mechanisms in plants (Verma et al., 2017). Mycorrhizal fungi bind to plant roots to improve nutrition and water uptake and provide defense against root diseases (Khaliq et al., 2022). Soil organisms like nematodes and protozoa prey on bacteria and fungi, helping regulate microbial populations and recycle nutrients (Raj et al., 2019). Microorganisms, along with earthworms and insects, enhance soil structure, aeration, and nutrient availability through their burrowing activities and consumption of organic matter. This process aids in forming soil aggregates and improves water infiltration (Akhila and Entoori, 2022). Soil pH influences nutrient availability and microbial activity, making its regulation essential for maintaining soil health (Sharma et al., 2020). Allelopathy, where some plants release chemicals to prevent the growth of opposing plants or harmful microorganisms, is a natural defense mechanism (Cheng and Cheng 2015). Organic matter boosts soil fertility by steadily supplying nutrients, while humus enhances soil structure, water-holding capacity, and cation exchange efficiency (Bashir et al., 2021). Good soil structure, created by stable aggregates, enhances water infiltration, root penetration, and erosion resistance (Rabot et al., 2018). A well-structured soil with stable aggregates plays a crucial role in enhancing soil health, plant growth, and environmental sustainability (Rastogi et al., 2023). Soil structure refers to how individual soil particles, i.e., sand, silt, and clay, bind together to form aggregates that influence soil aeration, water movement, root growth, and microbial activity (Sharma and Kumar, 2023). Stable soil aggregates provide numerous benefits, particularly by improving water infiltration, root penetration, and erosion resistance, all of which contribute to the long-term fertility and productivity of the soil (Bodner et al., 2021). Stable soil aggregates create a network of pores that allow water to infiltrate efficiently rather than pooling on the surface and causing runoff. This prevents water loss, enhances groundwater recharge, and reduces soil erosion

DOI: 10.1201/9781003608080-6

caused by excessive surface water movement (Somasundaram et al., 2018). Well-aggregated soil retains moisture more effectively, ensuring that plants have a consistent water supply even during dry periods. This is especially beneficial in arid and semi-arid regions where water conservation is critical. Additionally, enhanced soil porosity reduces surface crusting, allowing plant roots to access moisture and nutrients more effectively (Abdelhak, 2022).

Good soil structure enables plant roots to penetrate deeply into the soil, facilitating access to water and nutrients from lower soil layers. In compacted or poorly aggregated soils, roots struggle to grow, limiting plant access to essential resources and ultimately affecting plant vigor (Aydin et al., 2025). A well-structured soil provides sufficient space for root expansion, leading to stronger root systems that improve plant stability and increase resilience to environmental stressors such as drought and wind (Hayat et al., 2025). Enhanced root growth also promotes better nutrient uptake, contributing to healthier and more productive plants. Erosion is a significant concern in agriculture and natural ecosystems, as it leads to nutrient loss, reduced soil fertility, and land degradation (Issaka and Ashraf, 2017). Stable aggregates help resist erosion by maintaining soil cohesion, reducing the impact of raindrops, and preventing the detachment of soil particles (Sadeghian et al., 2021). When soil aggregates are well-formed, water moves into the ground rather than washing away topsoil, which is rich in organic matter and nutrients essential for plant growth. This prevents sediment runoff into water bodies, reducing pollution and promoting sustainable land use practices (Álvarez et al., 2017).

A well-structured soil promotes proper aeration, ensuring an adequate oxygen supply to plant roots and soil microorganisms. Oxygen is essential for root respiration and microbial metabolism, both of which contribute to nutrient cycling and organic matter decomposition (Yu and Kuzyakov, 2021). Beneficial microbes, such as mycorrhizal fungi and nitrogen-fixing bacteria, thrive in well-aerated soil, enhancing nutrient availability and supporting plant growth. Conversely, poorly structured soils tend to become anaerobic, leading to the accumulation of toxic compounds that can hinder plant development and microbial activity (Hasnain et al., 2023). Stable soil aggregates also contribute to improved nutrient retention and availability. Organic matter and clay particles within aggregates help bind essential nutrients, preventing them from leaching away during heavy rainfall. This ensures a steady nutrient supply for plants, leading to healthier growth and higher crop yields (Radulov et al., 2024). Furthermore, well-aggregated soil supports diverse microbial communities that decompose organic matter and release nutrients in forms that plants can readily absorb. This natural process enhances soil fertility and reduces the need for synthetic fertilizers, promoting more sustainable agricultural practices (Tripathi et al., 2020).

In addition to its agronomic benefits, soil structure plays a key role in carbon sequestration, which is crucial for mitigating climate change. Well-aggregated soil helps store organic carbon, reducing the release of carbon dioxide into the atmosphere (Karbin et al., 2022). This contributes to sustainable soil management and helps combat the negative effects of greenhouse gas emissions. Maintaining soil structure through conservation practices, such as organic amendments, reduced tillage, and cover cropping, ensures that soil continues to function as a carbon sink while supporting long-term agricultural productivity (Francaviglia et al., 2023). A well-structured soil with stable aggregates is fundamental to maintaining soil health and agricultural productivity. By promoting water infiltration, root penetration, and erosion resistance, stable aggregates enhance soil fertility, plant resilience, and environmental sustainability (Raheem et al., 2025). Sustainable soil management practices, such as adding organic amendments, minimizing soil disturbance, and implementing cover cropping, can help maintain and improve soil structure over time. These strategies contribute to a more resilient agricultural system, ensuring food security and environmental conservation for future generations (Sarma et al., 2024).

Soil texture, determined by the proportions of sand, silt, and clay, affects drainage, aeration, and nutrient retention. Properly textured soils are better equipped to resist compaction and erosion (Wang et al., 2022). Mulches and cover crops protect the soil surface against erosion, decrease water evaporation, and inhibit weed development, while adding organic matter upon decomposition (Ngosong et al., 2019). Plant root exudates draw beneficial microbes while deterring harmful ones, aiding in nutrient absorption and providing disease protection (Olanrewaju et al., 2019). Biological control employs natural predators or antagonists to sustainably manage soil-borne pests and diseases (Iftikhar et al., 2020). A diverse soil ecosystem, resilient to disturbances and stresses, ensures functional redundancy, thereby maintaining overall soil health (Bhaduri et al., 2022). Understanding and managing complex soil ecosystem interactions can lead to enhanced agricultural productivity and environmental conservation. Implementing practices that preserve organic matter, protect soil structure, and support biodiversity is crucial for maintaining robust defense mechanisms.

6.2 SOIL'S CRITICAL IMPACT ON PLANT HEALTH AND PATHOGEN INTERACTIONS

Soil is vital for plant health and disease dynamics, acting as a medium that supports plant growth and influences the frequency and intensity of plant diseases (Lodha and Mawar, 2020). Nutrient-rich soil with abundant organic matter and beneficial microbes supports optimal plant growth by improving structure, enhancing water retention, and providing essential nutrients (Bhardwaj et al., 2014). However, soil can harbor harmful pathogens and pests, and the balance between beneficial and harmful organisms within the soil ecosystem significantly impacts plant health (Harman et al., 2021). Healthy soil improves nutrient availability, supplying essential elements like nitrogen, phosphorus, and potassium, which are vital for plant growth and development (Das et al., 2022). Soil microbes decompose organic materials, liberating nutrients that plants can assimilate. Mycorrhizal fungi expand root surface area, particularly enhancing phosphorus uptake for plants (Tunlid et al., 2022). Well-structured soil, with stable aggregates, promotes root penetration, water infiltration, and aeration, supporting beneficial microbial activity and reducing the risk of root diseases from waterlogging. Beneficial microorganisms like bacteria, fungi, and actinomycetes break down organic matter, recycle nutrients, and establish symbiotic relationships with plants (Bodner et al., 2021). They compete with harmful pathogens, produce antibiotics, and trigger systemic resistance in plants, thereby safeguarding them from diseases (Kumar and Dubey, 2020). Certain plants release allelopathic compounds into the soil that inhibit the growth of competing plants or harmful microorganisms, supporting the survival of specific plant species and assisting in the control of soil-borne diseases (Hussain and Farooqi, 2021). Soil health can be managed through biological control, by introducing natural predators or antagonists that control soil-borne pests and diseases, such as fungi that parasitize nematodes (Abd-Elgawad, 2020). Organic matter, including compost and decomposing plant material, improves soil fertility by providing a continuous source of nutrients. Humus, the stable form of organic matter, enhances soil structure, water retention, and cation exchange capacity, all of which are crucial for plant health and microbial activity (Biswas and Kole, 2017). Soil pH influences nutrient availability and microbial activity, making it essential to maintain an optimal pH for effective nutrient uptake by plants and the support of beneficial microbial processes (Msimbira and Smith, 2020). Soil texture, which is determined by the ratio of sand, silt, and clay, impacts drainage, aeration, and nutrient retention, thereby affecting overall soil health and the dynamics of disease (Shahane and Shivay, 2021). Soil–plant interactions include symbiotic partnerships between plants and soil organisms, such as mycorrhizal fungi and nitrogen-fixing bacteria, which enhance nutrient absorption and support disease resistance (Igiehon and Babalola, 2018). Beneficial microbes compete with pathogens for resources and space, reducing disease incidence, while some bacteria produce antibiotics that help suppress soil-borne pathogens (Meena et al., 2017). Natural predators or antagonists, like nematode-trapping fungi or predatory insects, help control pest populations and mitigate their detrimental effects on plants (Al-Ani et al., 2022). Allelopathic compounds secreted by plants can suppress the growth of competing plants and pathogens, affecting both plant health and soil ecology (Polyak and Sukcharevich, 2019). The function of soil in plant health and disease dynamics involves complex interactions among biological, chemical, and physical components (Dubey et al., 2019). Healthy soil ecosystems, supported by practices such as adding organic matter, maintaining optimal pH, employing crop rotation, and using cover crops, promote beneficial microbial activity and suppress soil-borne diseases (Jiang et al., 2022). Comprehending these interactions and mechanisms is crucial for effective soil management, leading to healthier plants, increased agricultural productivity, and sustainable farming practices (Das et al., 2022).

6.3 MICROBIAL INTERACTIONS IN SOIL AND PLANT DISEASE RESISTANCE

Microbial interactions in the soil play a vital role in enhancing plant disease resistance and supporting overall plant health (Vishwakarma et al., 2020). The soil microbiome, made up of a variety of microorganisms such as bacteria, fungi, actinomycetes, protozoa, and nematodes, interacts with plants in different ways, impacting their growth and disease resistance (Chauhan et al., 2023). These interactions can be beneficial, neutral, or detrimental, depending on the specific microbes and environmental conditions. Understanding them is essential for developing sustainable farming practices and effective soil management strategies (Das et al., 2022). Beneficial microbial interactions include symbiotic partnerships, such as mycorrhizal fungi forming mycorrhizae with plant roots to boost nutrient absorption and offer pathogen protection, and nitrogen-fixing bacteria like rhizobia forming nodules on leguminous plants to supply vital nutrients and support soil health (Huang, 2024). Competitive exclusion by beneficial microorganisms, including *Bacillus*, *Pseudomonas*, and *Trichoderma* species, helps diminish harmful pathogens by competing for nutrients and space (Wang et al., 2021a). Some soil microbes produce antimicrobial compounds, like *Pseudomonas fluorescens*, which suppress soil-borne pathogens using antibiotics like phenazine and pyoluteorin (Gade and Koche, 2022). Induced systemic resistance (ISR) is activated by beneficial bacteria such as *Bacillus subtilis* and *Pseudomonas* spp., boosting plant immune responses and increasing their resilience to infections (Yu et al., 2022). Microbial interactions contribute to nutrient cycling and soil fertility, with microbes decomposing organic matter to release essential nutrients, improving soil health (Bhattacharyya and Furtak, 2023). They also improve soil structure and health by forming and stabilizing soil aggregates, enhancing aeration and water retention (Sharma and Kumar, 2023). Allelopathy involves plants and microorganisms releasing compounds to inhibit the growth of competing plants or harmful microbes, protecting plants from diseases (Polyak and Sukcharevich, 2019). Biocontrol agents, such as the nematode-trapping fungus *Arthrobotrys oligospora* and the entomopathogenic fungus *Beauveria bassiana*, assist in managing soil-borne pests and pathogens, offering sustainable agricultural solutions (Rahman et al., 2023). Practical agricultural practices involve soil management strategies like crop rotation, cover cropping, and organic amendments to promote beneficial soil microbes, along with inoculating soils with helpful microbes to improve plant health and disease resistance (Kumar and Verma, 2019). Integrated pest management (IPM) integrates microbial biocontrol agents with other pest control methods, decreasing dependence on chemical pesticides and reducing environmental and health risks (Baker et al., 2020). By promoting beneficial microbial communities through sustainable soil management practices, plant resilience to diseases can be enhanced, soil fertility improved, and sustainable agricultural systems fostered, resulting in healthier plants, higher crop yields, and more eco-friendly farming practices (Roberts and Mattoo, 2018). Table 6.1 presents the microbes and their role in enhancing plant disease resistance and growth.

6.4 SOIL AMENDMENTS FOR ENHANCED PLANT DEFENSE

Enhancing plant defenses through soil amendments can effectively improve plant health and resilience to pests and diseases (Akanmu et al., 2021). Soil amendments improve soil structure, fertility, and microbial activity, promoting the growth of healthier and more resilient plants (Arif et al.,

TABLE 6.1
Beneficial Microbes and Their Roles in Enhancing Plant Disease Resistance and Growth.

Microbe	Interaction with Plants	Role in Disease Resistance	References
Rhizobia	Forms symbiotic relationships with legumes, fixes nitrogen	Encourages plant growth and nutrient absorption, while activating systemic resistance	Tonelli et al. (2020)
Mycorrhizae	Forms associations with plant roots	Boosts nutrient absorption and activates systemic resistance	Kadam et al. (2020)
Trichoderma	Soil fungus, biocontrol agent	Antagonizes plant pathogens, induces systemic resistance	Bisen et al. (2016)
Pseudomonas	Soil bacterium	Produces antibiotic, induces systemic resistance	Salomon et al. (2017)
Bacillus	Soil bacterium	Produces antimicrobial compounds, induces systemic resistance	Wu et al. (2018)
Azospirillum	Free-living nitrogen-fixing bacterium	Promotes plant growth, induces systemic resistance	Fukami et al. (2018)
Streptomyces	Soil actinomycete	Produces antibiotics, suppresses soil-borne pathogens	Gopalakrishnan and Srinivas (2019)
Actinomycetes	Soil bacteria, decomposers	Produce antimicrobial compounds, enhance nutrient cycling	Javed et al. (2021)
Penicillium	Soil fungus	Produces antibiotics, suppresses plant pathogens	Shen et al. (2020)
Bukholderia	Soil bacterium	Produces antibiotics, induces systemic resistance	Ramamoorthy et al. (2001)

2020) (Table 6.2). Organic amendments such as compost enhance soil structure, boost organic matter, and stimulate beneficial microbial activity, supporting microorganisms that outcompete pathogenic ones and thereby improving plant resistance to diseases (Neher et al., 2022). After discussing organic amendments, it is important to highlight the role of mineral amendments in enhancing plant defenses. This transition shifts the focus from organic amendments, such as compost, manure, and biochar, to mineral amendments like gypsum, lime, and rock phosphate (Altynbay et al., 2024). While organic amendments primarily improve soil fertility by adding organic matter and beneficial microorganisms, mineral amendments directly supply essential nutrients that enhance soil properties and plant health (Singh et al., 2020).

Mineral amendments play a crucial role in improving soil structure, optimizing nutrient availability, and enhancing plant stress tolerance. For instance, gypsum (calcium sulfate) enhances soil aggregation, reducing compaction and improving root penetration (Jing et al., 2024). This is particularly beneficial in clayey soils where water infiltration and aeration are poor. Lime (calcium carbonate) increases soil pH, making phosphorus, nitrogen, and potassium more available to plants, while gypsum supplies calcium and sulfur, both essential for plant growth (Shruthi et al., 2024). Rock phosphate acts as a slow-release phosphorus source, promoting root development and energy transfer. Additionally, calcium strengthens cell walls, helping plants resist diseases and physical damage, while sulfur contributes to the synthesis of key proteins and antioxidants that protect against oxidative stress (Kumari et al., 2022).

Regulating soil pH is essential for nutrient uptake, as acidic soils can limit nutrient availability. Lime is commonly used to raise soil pH, thereby improving nutrient accessibility, whereas gypsum does not alter pH but enhances soil structure and provides calcium (Minato et al., 2023). These amendments also ensure an adequate supply of key nutrients: Calcium reinforces cell walls,

TABLE 6.2
Soil Amendments: Composition, Mechanisms, and Applications for Enhancing Soil Health and Plant Growth.

Soil Amendment	Composition/Source	Mechanism of Action	Example Applications	References
Compost	Decomposed organic matter	Enhances soil structure, boosts microbial activity, and increases nutrient availability	General soil amendment in gardens and farms	De Corato (2020b)
Biochar	Charred organic material	Enhances nutrient retention, promotes beneficial microbes, and reduces soil-borne diseases	Incorporated into the soil to enhance soil health and increase plant resilience	Bolan et al. (2023)
Mycorrhizal fungi	Symbiotic fungi	Facilitates nutrient uptake, induces systemic resistance, and enhances drought tolerance	Inoculation at planting or transplanting	Bahadur et al. (2019)
Seaweed extracts	Algal extracts	Stimulates plant immune responses, enhances root growth, and improves stress tolerance	Foliar spray or soil drench	Shukla et al. (2021)
Plant growth-promoting rhizobacteria	Beneficial soil bacteria	Stimulates systemic resistance, encourages growth, and enhances nutrient absorption	Seed coating, soil inoculation	Khatoon et al. (2020)
Chitin	Derived from crustacean shells	Activates plant defense genes, enhances resistance to fungal pathogens	Soil amendment or foliar spray	Malerba and Cerana 2020)
Humic substances	Decomposed organic material	Enhances nutrient availability, improves soil structure, and increase stress tolerance	Soil conditioner or seed treatment	Canellas et al. (2020)
Silicon amendments	Silica-based materials	Strengthens cell walls, improving resistance to pests and diseases	Soil or foliar application	Verma et al. (2021)
Vermicompost	Worm-processed organic waste	Increases microbial activity, enhances nutrient cycling, and suppresses pathogens	Used in potting mixes or as a top dressing	Sivasabari et al. (2023)
Neem cake	Residue from neem oil extraction	Acts as a natural pesticide and nematicide, and enhances soil health	Mixed into soil to control pests and improve fertility	Pascoli et al. (2019)
Rock phosphate	Natural mineral phosphate	Provides slow-release phosphorus, enhances root development, and improves plant immunity	Applied as a phosphorus source	Bargaz et al. (2021)
Green manure	Decomposed cover crops	Improves soil fertility, adds organic matter, and suppresses diseases	Plowed under as a soil amendment	Das et al. (2020)
Gypsum	Calcium sulfate	Improves soil structure, reduces compaction, and helps manage salinity	Applied to soils with high sodium levels	Osman (2018)
Azolla	Aquatic ferns	Fixes atmospheric nitrogen, improves soil fertility, and suppresses weeds	Used in rice paddies or as a green manure	Abd El-Aal (2022)
Bone meal	Ground animal bones	Provides phosphorus and calcium, enhances root development, and improves plant vigor	Applied to the soil as a slow-release fertilizer	Balawejder et al. (2019)

reducing disease susceptibility and improving fruit quality; magnesium, a core component of chlorophyll, is vital for photosynthesis; and phosphorus supports root development, energy transfer (ATP production), and flowering (Sharma et al., 2023).

Moreover, gypsum helps break up compacted soils, increasing porosity and enhancing oxygen diffusion to plant roots. This improves root respiration and nutrient uptake, leading to healthier plant growth (Hasan et al., 2025). Mineral amendments also mitigate abiotic stresses such as salinity and drought by enhancing water retention and reducing harmful ion accumulation. High sodium levels in soil can be toxic to plants, but gypsum helps replace sodium ions with calcium, thereby reducing sodium toxicity and improving soil permeability (Bello et al., 2021). Additionally, amendments like zeolites and silica-based materials enhance soil water retention, reducing water loss during drought. Calcium and magnesium also facilitate water movement within plant cells, improving hydration and increasing tolerance to water stress. Excessive aluminum and sodium can be detrimental to plant growth (Ahanger et al., 2017). Lime neutralizes aluminum toxicity in acidic soils, while gypsum aids in leaching excess sodium from saline soils, preventing ion toxicity and promoting a healthier growing environment (Shruthi et al., 2024). The combined use of organic and mineral amendments leads to improved soil health, better plant growth, higher yields, and increased resistance to diseases and stress conditions (Shruthi et al., 2024). By integrating both types of amendments, farmers and researchers can develop sustainable soil management strategies that enhance plant resilience under adverse environmental conditions (De Corato, 2020a).

Manure enhances soil fertility by adding nutrients and organic matter, improving microbial diversity and activity, and supporting plant health and resilience (Bhunia et al., 2021). Biochar improves soil structure, enhances water retention, and boosts nutrient availability, creating an optimal environment for beneficial bacteria while reducing pathogen levels by adsorbing harmful compounds (Zhang et al., 2021). Mineral amendments like lime raise soil pH, improve nutrient availability, and supply calcium, strengthening cell walls and increasing structural integrity, thereby reducing plant susceptibility to pathogens (Gupta et al., 2017). Gypsum improves soil structure, particularly in heavy clay soils, and provides calcium and sulfur, with sulfur enhancing plant defense mechanisms and calcium strengthening cell walls (Shah et al., 2022). Biological supplements such as mycorrhizal fungi establish symbiotic relationships with plant roots, improving nutrient and water uptake and enhancing plant resistance to root infections and environmental stress by activating systemic defense mechanisms (Wahab et al., 2023). Beneficial bacteria like *Bacillus* spp. and *Pseudomonas* spp. promote plant growth and suppress soil-borne pathogens by producing antibiotics, siderophores, and other compounds that inhibit harmful microbes (Miljaković et al., 2020)

Nutrient amendments such as silicon fortify plant cell walls and enhance structural integrity, increasing resistance to both biotic stresses like pests and diseases and abiotic stresses such as drought (Bakhat et al., 2018). Micronutrients such as zinc, copper, and manganese are crucial for various physiological processes and enzymatic functions in plants, contributing to the production of defense compounds and the activation of defense mechanisms (Bhat et al., 2020). Organic matter-based amendments like seaweed extracts provide micronutrients, growth hormones, and biostimulants, enhancing plant stress tolerance and inducing systemic acquired resistance (SAR) (Ronga et al., 2021). Humic and fulvic acids enhance nutrient absorption and stimulate soil microbial activity, promoting root development and overall plant vitality, which improves resistance to pests and diseases (Shahrajabian and Sun, 2024). To effectively apply soil amendments, begin with soil testing to identify the soil's specific needs and prevent nutrient over-application. Consistently add organic matter, like compost, to maintain soil health and microbial diversity. Ensure plants receive balanced nutrition to lower their susceptibility to diseases and use microbial inoculants, such as mycorrhizal fungi and beneficial bacteria, to enhance soil biological activity and promote plant health (Alori and Babalola, 2018). By carefully selecting and applying these soil amendments, one can significantly improve plant defenses and overall health, leading to more robust and productive plants (Harman et al., 2021).

6.5 UTILIZING MICROBIAL INOCULANTS TO BOOST SOIL DEFENSES

Microbial inoculants, commonly known as biofertilizers or bio-stimulants, are essential for improving soil health and strengthening plant defense (Alori and Babalola, 2018). These products contain beneficial microorganisms that promote plant growth, aid in nutrient absorption, and protect plants from pests and diseases (Elnahal et al., 2022). Microbial inoculants include different types, such as mycorrhizal fungi like arbuscular mycorrhizal fungi (AMF) for many crops and ectomycorrhizal fungi for trees and shrubs. These fungi improve nutrient and water uptake, enhance soil structure, and increase plant resistance to pathogens and environmental stress (Wahab et al., 2023). Rhizobia, such as specific strains like *Rhizobium*, *Bradyrhizobium*, and *Sinorhizobium*, form root nodules in legumes to fix atmospheric nitrogen, which enhances soil fertility and supports plant growth. Plant growth-promoting rhizobacteria (PGPR), including *Bacillus* spp., *Pseudomonas* spp., and *Azospirillum* spp., boost plant growth by increasing nutrient availability, producing growth hormones, and defending plants against infections through the production of antibiotics and siderophores (de Andrade et al., 2023). Nitrogen-fixing bacteria, including *Azotobacter* and associative types like *Azospirillum*, convert atmospheric nitrogen into a form that plants can use, thereby improving soil fertility (Aasfar et al., 2021). Phosphate-solubilizing bacteria, including *Bacillus* and *Pseudomonas* species, enhance phosphorus availability for plants, which is crucial for root development and energy transfer. Moreover, beneficial fungi like *Trichoderma* spp. play a role in promoting plant health (Rawat et al., 2021). Soil-borne pathogens can be suppressed through competition, antibiosis, and mycoparasitism (Abdelaziz et al., 2023).

Microbial inoculants provide several advantages, such as improving nutrient availability and absorption, including nitrogen, phosphorus, and micronutrients. This results in healthier plants with improved growth and development (Parween et al., 2017). They also enhance soil structure, with fungi like mycorrhizae and bacteria secreting polysaccharides that help bind soil particles together, improving soil aeration, water retention, and root penetration (Wei et al., 2024). Another benefit is disease suppression, as beneficial microbes prevent pathogenic organisms by competing for space and resources, producing antibiotics, and stimulating systemic resistance in plants. This helps reduce soil-borne diseases and improves plant health (Chen et al., 2023). Additionally, microbial inoculants can help plants better withstand abiotic stresses such as drought, salinity, and heavy metal toxicity, enhancing overall plant vigor and resilience under harsh environmental conditions (Kumar et al., 2021).

Application strategies for microbial inoculants include seed inoculation, where seeds are coated with microbial inoculants before planting to promote early root colonization and immediate benefits from beneficial microbes (Rocha et al., 2019). Another approach is soil drenching, which involves applying a solution containing microbial inoculants directly to the soil or root zone to enhance nutrient uptake and suppress diseases (Elnahal et al., 2022). Foliar spray is also used, where microbial inoculants are sprayed on plant leaves to colonize the phyllosphere (leaf surface) and protect against foliar pathogens (Devarajan et al., 2021). Additionally, microbial inoculants can be added to compost, enhancing microbial activity and quality, which in turn improves soil health when the compost is applied (Greff et al., 2022).

Best practices for using microbial inoculants involve several key steps. The first is selecting the appropriate inoculant tailored to specific crops and soil conditions; this ensures maximum effectiveness (O'Callaghan et al., 2022). Second, proper storage and handling are crucial for maintaining viability and safeguarding against extreme temperatures and direct sunlight (Chandel and Agarwal, 2017). Timing the application during planting or early growth stages facilitates effective colonization and benefits (Backer et al., 2018). Ensuring compatibility with other agricultural inputs, such as fertilizers and pesticides, is also crucial, as these can affect microbial viability and activity (Santos et al., 2021). Finally, regular monitoring of soil and plant health allows for assessing effectiveness and making necessary adjustments (Rehman et al., 2022). By integrating microbial inoculants into soil management practices, significant improvements in defenses can enhance plant health and overall crop productivity (Alori and Babalola, 2018).

6.6 BIOCONTROL AGENTS AND THEIR IMPACT ON SOIL DEFENSE MECHANISMS

Biocontrol agents are helpful organisms employed to manage plant pathogens, pests, and weeds. They play a crucial role in strengthening soil defense systems, supporting plant health, and decreasing the need for chemical pesticides (Kumar and Verma, 2019). These agents include various types of bacteria, fungi, nematodes, predatory and parasitic insects, and viruses (Moosavi and Zare 2020) (Figure 6.1). Notable bacteria like *Bacillus* spp. and *Pseudomonas* spp. produce antibiotics and secondary metabolites that suppress plant pathogens and promote growth (Miljaković et al., 2020). Fungi like *Trichoderma* spp. and *Gliocladium* spp. fight soil-borne pathogens through mycoparasitism, competition, and promoting plant growth (Hariharan et al., 2022). Entomopathogenic nematodes and insects like ladybugs and parasitic wasps target and kill soil-dwelling insect pests, maintaining manageable pest populations (Balla et al., 2021). Viruses and bacteriophages, like baculoviruses and bacteriophages, infect and reduce populations of insect pests and bacterial pathogens, respectively (Gelaye and Negash, 2023).

Biocontrol agents impact soil defense mechanisms in multiple ways, including suppressing pathogens through antibiosis where antibiotics inhibit or kill pathogens and compete for nutrients and space, thereby reducing pathogen establishment (Syed Ab Rahman et al., 2018). They also activate plant defense mechanisms, such as ISR and SAR, which enhance plant resistance to various pathogens (Kamle et al., 2020). These agents also support soil microbial diversity, fostering a balanced microbial ecosystem that outcompetes harmful pathogens and works synergistically with beneficial microbes, ultimately enhancing soil health and plant resilience (Iqbal et al., 2023). Additionally,

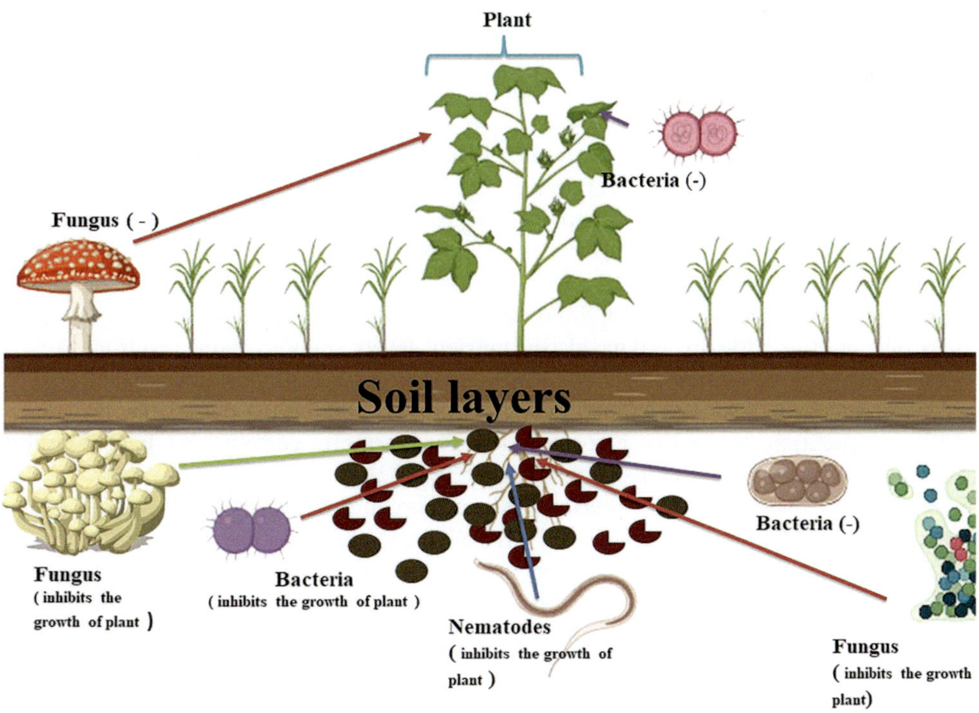

FIGURE 6.1 Soil microbiome and its interactions with plants and soil layers.

biocontrol agents enhance soil structure and health by promoting soil aggregation, which improves water infiltration and root growth, and by aiding in the decomposition of organic matter, which releases nutrients and boosts soil fertility (Wei et al., 2024). Using these agents also decreases the reliance on chemical pesticides, which helps reduce environmental impact and lowers the risk of pests and pathogens developing resistance (Tudi et al., 2021). Effective biocontrol agents include *Trichoderma* spp., which combat soil-borne fungal pathogens through mycoparasitism, competition, and the induction of plant defense responses when applied as a soil drench, seed treatment, or foliar spray (Al-Ani, 2018). *Bacillus subtilis*, another biocontrol agent, targets a wide range of bacterial and fungal pathogens by producing antibiotics and enzymes and inducing ISR when used as a soil amendment, seed treatment, or foliar spray (Khan et al., 2022). Similarly, *Pseudomonas fluorescens*, used as a soil inoculant or seed treatment, combats soil-borne and some foliar pathogens through antibiosis, competition, and ISR induction (Hussain et al., 2020). *Beauveria bassiana*, applied as a soil drench or foliar spray, targets soil-dwelling and foliar insect pests through pathogenicity, leading to pest mortality (Lee et al., 2017). For effective implementation, it is crucial to select biocontrol agents that are specific to the pathogens or pests present and suitable for your crop (Lahlali et al., 2022). These agents should be applied at the right time using appropriate methods, integrated with other sustainable agricultural practices, and regularly monitored to evaluate their effectiveness and make necessary adjustments (Lezoche et al., 2020). Using biocontrol agents can enhance soil defense mechanisms, promote sustainable agriculture, and achieve healthier, more resilient plants (Elnahal et al., 2022).

6.6.1 Influence of Soil Properties on Disease Resistance

Soil properties significantly influence disease resistance in plants through their physical, chemical, and biological characteristics (Larkin et al., 2021). Well-structured soils with good aeration and drainage prevent waterlogging, reducing root diseases like those caused by *Pythium* and *Phytophthora*, while poor structure and compaction promote conditions favorable for anaerobic pathogens (Dixon and Tilston, 2010). Sandy soils, with good drainage but low nutrient-holding capacity, can stress plants, making them more disease-prone, whereas clay soils retain water and nutrients but may hinder root penetration and promote root diseases due to poor drainage (Lu et al., 2020). Soil pH is also essential; a slightly acidic to neutral pH (6.0–7.0) maximizes nutrient availability and microbial activity, whereas extreme pH levels can stress plants and promote the growth of certain pathogens (Ali et al., 2022). High soil organic matter enhances nutrient supply, microbial diversity, and soil structure, all contributing to disease resistance (Dignam et al., 2019). Adequate soil moisture supports healthy plant growth and microbial activity, but excess moisture and drought stress can increase disease susceptibility (Ojuederie et al., 2019). Soil temperature affects root growth and microbial activity, with optimal temperatures promoting healthy roots and beneficial microbes (Wang et al., 2021b). Balanced soil nutrient levels support plant health and resistance, while imbalances can weaken plants and increase disease susceptibility (Gupta et al., 2017). A high degree of microbial diversity and activity in the soil aids in suppressing pathogens through mechanisms like competition, antibiosis, and predation. Furthermore, certain beneficial microbes can activate systemic resistance in plants (Jayaraman et al., 2021).

To improve soil properties for disease resistance, organic amendments such as compost and manure can enhance soil structure, nutrient availability, and microbial activity (Goldan et al., 2023). Cover crops enhance soil structure, contribute organic matter, and promote the growth of beneficial microbes (Martínez-García et al., 2018). Proper irrigation management avoids waterlogging and drought stress, maintaining optimal soil moisture levels (Mariani and Ferrante, 2017). Modifying soil pH using lime or sulfur sustains it within the ideal range for plant development and microbiological activity (Dhiman et al., 2019). Crop rotation prevents the build-up of specific pathogens and reduces disease pressure (Gamliel and Bruggen, 2016). Regular soil testing monitors nutrient levels, pH, and organic matter, allowing informed management decisions (Lobry de Bruyn and

Andrews, 2016). Understanding and managing these soil properties creates an environment that supports healthy plant growth and enhances disease resistance, leading to more resilient and productive crops (Shah and Wu, 2019).

6.6.2 The Microbiome's Role in Shaping Plant–Soil Interactions

The soil microbiome, which includes bacteria, fungi, archaea, viruses, and protozoa, is essential for shaping plant–soil interactions by creating intricate networks that impact plant health, growth, and resilience (Islam et al., 2020). Microorganisms break down organic matter into simpler substances, releasing essential nutrients such as nitrogen, phosphorus, and sulfur; making them available to plants; and ensuring a steady supply of nutrition (Biswas and Kole, 2017). Symbiotic bacteria, such as *Rhizobium* spp., fix atmospheric nitrogen, enhancing soil fertility, while phosphate-solubilizing microbes convert insoluble phosphorus to plant-available forms, supporting root development (Arif et al., 2017). Soil microorganisms produce phytohormones and establish mycorrhizal associations, which support root growth and enhance nutrient uptake (Liao et al., 2018). Beneficial microbes help suppress plant pathogens by competing for resources, producing antibiotics, and triggering ISR, which lowers the occurrence of diseases (Zhu et al., 2022). They also help plants cope with abiotic stresses like drought and salinity by producing compounds that aid in reducing stress (Iqbal et al., 2022). Microbial activity enhances soil structure and health through processes like soil aggregation and organic matter decomposition (Obalum et al., 2017). Chemical signaling between plants and microbes promotes plant growth and encourages mutualistic relationships, while microbial diversity contributes to ecosystem stability and ensures functional redundancy (Nizamani et al., 2024). Practical agricultural practices, including crop rotation, organic amendments, microbial inoculants, and soil health management, can support beneficial microbes, improving plant–soil interactions, plant health, yields, and overall sustainability (Tahat et al., 2020).

6.6.3 Integrating Soil Defense Strategies for Sustainable Agriculture

Integrating soil defense strategies into sustainable agriculture involves a holistic approach to managing soil health, enhancing plant resilience, and reducing dependency on chemical inputs (Roberts and Mattoo, 2018). This approach integrates different practices and technologies to build a more resilient and productive agricultural system (Altieri et al., 2015). Improving soil health with organic amendments like compost, manure, cover crops, and green manure boosts soil organic matter, microbial activity, and nutrient availability (Tully and McAskill, 2020). Conservation tillage methods, including reduced tillage and no-till agriculture, maintain soil structure, reduce erosion, and support microbial ecosystems (Somasundaram et al., 2020). Promoting beneficial microorganisms with microbial inoculants like mycorrhizal fungi, nitrogen-fixing bacteria, and biocontrol agents helps suppress soil-borne pathogens and improve plant health (Meena et al., 2017). Maintaining soil microbial diversity through crop rotation and intercropping disrupts pest cycles and promotes diverse microbial communities (Altieri et al., 2024). Nutrient management practices, such as routine soil testing, balanced fertilization, and the use of slow-release and organic fertilizers, help apply nutrients properly and improve soil structure and microbial activity (Selim, 2020). Efficient water management, utilizing drip irrigation, rainwater harvesting, mulching, and cover crops, optimizes water use and reduces conditions favorable for pathogens (El-beltagi et al., 2016). Integrated pest and disease management strategies incorporate cultural practices, biological control, and mechanical and physical controls to manage pests and diseases sustainably (Baker et al., 2020). Enhancing plant resilience by selecting resistant cultivars, utilizing advanced breeding techniques, and implementing healthy planting practices reduces the need for chemical controls (Roberts and Mattoo, 2018). Education and monitoring, including farmer training, workshops, regular soil testing, and pest surveillance, enable informed management decisions (Lamichhane et al., 2016). Policy support, government incentives, research, and community efforts further promote sustainable practices (Derks and Romijn, 2019). By integrating these soil defense strategies, farmers can achieve a sustainable agricultural system that benefits the environment, improves crop productivity, and enhances farm profitability (Kalogiannidis et al., 2022).

6.6.4 Successful Implementation of Soil Defense Manipulation

Successful implementation of soil defense strategies in sustainable agriculture requires integrated practices that enhance soil health, promote beneficial microbial activity, and improve plant resilience against pests and diseases (Tahat et al., 2020). Essential actions include evaluating and monitoring soil health through regular testing of pH, nutrient levels, organic matter content, and microbial activity, as well as adding organic amendments like compost, manure, green manures, and cover crops to boost organic matter and nutrient availability (Wang et al., 2003). Conservation tillage methods, including decreased tillage and no-till agriculture, preserve soil structure and microbial ecosystems. Promoting advantageous microorganisms via microbial inoculants, crop rotation, and intercropping enhances microbial diversity and reduces pathogen buildup (Koskey et al., 2021). Optimizing nutrient management with balanced fertilization based on soil tests and using organic fertilizers supports soil health, while efficient water management techniques like drip irrigation and mulching conserve soil moisture, regulate temperature, and suppress weeds (Demo and Bogale, 2024). Integrated pest and disease management strategies, including biological controls, cultural practices, and mechanical methods, help sustainably manage pests and diseases, and education and community engagement through farmer training and cooperative efforts facilitate knowledge sharing and implementation of sustainable practices. Key takeaways for successful implementation include adopting a holistic approach by integrating various soil management practices, tailoring practices to local conditions, continuously learning and adapting based on new research, engaging with communities and policymakers for support, and regularly monitoring soil health and crop performance. Future directions in soil-based plant disease management involve leveraging advanced technologies, enhancing the understanding of soil microbiomes, and promoting sustainable practices, with advanced microbial inoculants, precision agriculture, sustainable soil management practices, IPM, policy support, and climate. Real-world examples include the Rodale Institute in Pennsylvania, USA, showcasing the benefits of organic amendments, cover cropping, and reduced tillage (Mirsky et al., 2013). In India, the Andhra Pradesh Community Managed Natural Farming (APCNF) advocates for practices like using microbial inoculants, cover cropping, and organic mulching to improve soil health and increase crop yields (Rosenstock et al., 2024). The Long-Term Agroecosystem Research (LTAR) Network in the US studies sustainable practices, such as cover cropping and reduced tillage, that enhance soil microbial activity and water retention (Crézé and Horwath, 2021). Additionally, the MasHumus Project in Mexico focuses on sustainable soil management in maize and bean systems, enhancing soil structure and crop performance under stress conditions. Resilience strategies and biotechnological advances are critical areas of development for improving soil health and plant resilience (Sarker et al., 2021). Adopting these strategies and drawing insights from successful case studies can assist farmers in developing resilient agricultural systems that support soil health, strengthen plant defense mechanisms, and ensure long-term sustainability and productivity (Altieri et al., 2024).

6.7 CONCLUSION

The future of managing plant diseases through soil-based approaches relies on combining advanced technologies, sustainable practices, and innovative research. By utilizing precision agriculture, deepening our understanding of soil microbiomes, and adopting regenerative practices, we can build resilient agricultural systems that promote plant health, enhance soil fertility, and minimize the need for chemical inputs. Collaborative efforts among researchers, policymakers, and farmers are essential for driving these innovations and ensuring their effective implementation. Achieving sustainable food security is intricately linked to soil health, which underpins strong crop production, maintains environmental balance, and boosts resilience against climate change. Emphasizing soil health is vital for creating an agricultural system capable of meeting the world's increasing food demands while preserving natural resources for future generations. Higher crop productivity results from nutrient-rich soils that enhance plant growth and yields, as well as resilient crop systems supported by diverse microbial communities and organic matter. Sustainable farming practices, such as regenerative farming and using organic amendments, help improve soil structure, fertility, and biodiversity, reducing reliance on chemical inputs and supporting long-term productivity. Managing the soil microbiome, promoting beneficial microorganisms, and fostering microbial diversity enhance nutrient cycling, disease resistance, and plant growth. Integrated pest and disease management practices, like the use of biocontrol agents, crop rotation, and intercropping, support healthy soils and sustainable crop production by reducing chemical pesticide usage and breaking pest and disease cycles. Efficient water and nutrient management practices, such as drip irrigation and balanced fertilization, help conserve water, prevent soil degradation, and ensure nutrient availability without harming soil microorganisms. Climate resilience is improved through practices that boost soil organic matter and sequester carbon, such as agroforestry and cover cropping, and through climate-smart agricultural strategies like drought-resistant crops and better soil moisture management. Supportive policies and agricultural education play a crucial role in promoting the adoption of sustainable soil management practices and ensuring their effectiveness. Technological advances, including precision agriculture and biotechnological innovations, enable precise monitoring and management of soil health, optimizing farming practices and increasing crop yields. Investing in soil health is fundamental for sustainable agriculture and food security. By integrating soil management practices, enhancing microbial health, and using advanced technologies, we can establish resilient agricultural systems capable of sustaining global food production. Collaborative efforts among farmers, researchers, policymakers, and communities are critical to driving this change and ensuring a productive and sustainable agricultural future. Prioritizing soil health not only enhances crop productivity and resilience but also contributes to broader environmental sustainability, paving the way for a stronger and more equitable food system for future generations.

ACKNOWLEDGMENTS

The authors are grateful to Lovely Professional University for providing financial assistance.

CONFLICT OF INTEREST

There is no conflict of interest.

REFERENCES

Aasfar, A., Bargaz, A., Yaakoubi, K., Hilali, A., Bennis, I., Zeroual, Y., Meftah Kadmiri, I., 2021. Nitrogen fixing *Azotobacter* species as potential soil biological enhancers for crop nutrition and yield stability. *Front. Microbiol.* 12. https://doi.org/10.3389/fmicb.2021.628379

Abd El-Aal, A.A.M., 2022. *Anabaena-azollae*, significance and agriculture application: A case study for symbiotic *Cyanobacterium*, in: Pratap Singh, R., Manchanda, G., Bhattacharjee, K., Panosyan, H. (Eds.), *Microbial Syntrophy-Mediated Eco-Enterprising, Developments in Applied Microbiology and Biotechnology*. Academic Press, pp. 1–14. https://doi.org/10.1016/B978-0-323-99900-7.00006-7

Abdelaziz, A.M., Hashem, A.H., El-Sayyad, G.S., El-Wakil, D.A., Selim, S., Alkhalifah, D.H.M., Attia, M.S., 2023. Biocontrol of soil borne diseases by plant growth promoting rhizobacteria. *Trop. Plant Pathol.* 48, 105–127. https://doi.org/10.1007/s40858-022-00544-7

Abd-Elgawad, M.M.M., 2020. Plant-parasitic nematodes and their biocontrol agents: Current status and future vistas, in: Ansari, R.A., Rizvi, R., Mahmood, I. (Eds.), *Management of Phytonematodes: Recent Advances and Future Challenges*. Springer, Singapore, pp. 171–203. https://doi.org/10.1007/978-981-15-4087-5_8

Abdelhak, M., 2022. Chapter 4—Soil improvement in arid and semiarid regions for sustainable development, in: Jhariya, M.K., Meena, R.S., Banerjee, A., Meena, S.N. (Eds.), *Natural Resources Conservation and Advances for Sustainability*. Elsevier, pp. 73–90. https://doi.org/10.1016/B978-0-12-822976-7.00026-0

Aguilar-Paredes, A., Valdés, G., Araneda, N., Valdebenito, E., Hansen, F., Nuti, M., 2023. Microbial community in the composting process and its positive impact on the soil biota in sustainable agriculture. *Agronomy* 13, 542. https://doi.org/10.3390/agronomy13020542

Ahanger, M.A., Tomar, N.S., Tittal, M., Argal, S., Agarwal, R.M., 2017. Plant growth under water/salt stress: ROS production; antioxidants and significance of added potassium under such conditions. *Physiol. Mol. Biol. Plants* 23, 731–744. https://doi.org/10.1007/s12298-017-0462-7

Akanmu, A.O., Babalola, O.O., Venturi, V., Ayilara, M.S., Adeleke, B.S., Amoo, A.E., Sobowale, A.A., Fadiji, A.E., Glick, B.R., 2021. Plant disease management: Leveraging on the plant–microbe–soil interface in the biorational use of organic amendments. *Front. Plant Sci.* 12. https://doi.org/10.3389/fpls.2021.700507

Akhila, A., Entoori, K., 2022. Role of earthworms in soil fertility and its impact on agriculture: A review. *Int. J. Fauna Biol. Stud.* 9, 55–63. https://doi.org/10.22271/23940522.2022.v9.i3a.907

Al-Ani, L.K.T., 2018. Trichoderma: Beneficial role in sustainable agriculture by plant disease management, in: Egamberdieva, D., Ahmad, P. (Eds.), *Plant Microbiome: Stress Response*. Springer, Singapore, pp. 105–126. https://doi.org/10.1007/978-981-10-5514-0_5

Al-Ani, L.K.T., Soares, F.E. de F., Sharma, A., de los Santos-Villalobos, S., Valdivia-Padilla, A.V., Aguilar-Marcelino, L., 2022. Strategy of nematophagous fungi in determining the activity of plant parasitic nematodes and their prospective role in sustainable agriculture. *Front. Fungal Biol.* 3. https://doi.org/10.3389/ffunb.2022.863198

Albaladejo, J., Díaz-Pereira, E., de Vente, J., 2021. Eco-holistic soil conservation to support land degradation neutrality and the sustainable development goals. *CATENA* 196, 104823. https://doi.org/10.1016/j.catena.2020.104823

Ali, A., Li, M., Su, J., Li, Y., Wang, Z., Bai, Y., Ali, E.F., Shaheen, S.M., 2022. *Brevundimonas diminuta* isolated from mines polluted soil immobilized cadmium (Cd2+) and zinc (Zn2+) through calcium carbonate precipitation: Microscopic and spectroscopic investigations. *Sci. Total Environ.* 813, 152668. https://doi.org/10.1016/j.scitotenv.2021.152668

Alori, E.T., Babalola, O.O., 2018. Microbial inoculants for improving crop quality and human health in Africa. *Front. Microbiol.* 9. https://doi.org/10.3389/fmicb.2018.02213

Altieri, M.A., Nicholls, C.I., Dinelli, G., Negri, L., 2024. Towards an agroecological approach to crop health: Reducing pest incidence through synergies between plant diversity and soil microbial ecology. *npj Sustain. Agric.* 2, 1–6. https://doi.org/10.1038/s44264-024-00016-2

Altieri, M.A., Nicholls, C.I., Henao, A., Lana, M.A., 2015. Agroecology and the design of climate change-resilient farming systems. *Agron. Sustain. Dev.* 35, 869–890. https://doi.org/10.1007/s13593-015-0285-2

Altynbay, N., Tastambek, K., Akimbekov, N., Digel, I., Tagayev, K., Kamenov, B., Liu, X., 2024. A Comprehensive review on enhancing saline soil reclamation efficiency through bacterial solubilization of low-rank coal. *Eng. Sci.* 30 (August), 1147.

Álvarez, X., Valero, E., Santos, R.M.B., Varandas, S.G.P., Sanches Fernandes, L.F., Pacheco, F.A.L., 2017. Anthropogenic nutrients and eutrophication in multiple land use watersheds: Best management practices and policies for the protection of water resources. *Land Use Pol.* 69, 1–11. https://doi.org/10.1016/j.landusepol.2017.08.028

Arif, I., Batool, M., Schenk, P.M., 2020. Plant microbiome engineering: Expected benefits for improved crop growth and resilience. *Trends Biotechnol.* 38, 1385–1396. https://doi.org/10.1016/j.tibtech.2020.04.015

Arif, M.S., Shahzad, S.M., Yasmeen, T., Riaz, M., Ashraf, M., Ashraf, M.A., Mubarik, M.S., Kausar, R., 2017. Improving plant phosphorus (P) acquisition by phosphate-solubilizing bacteria, in: Naeem, M., Ansari, A.A., Gill, S.S. (Eds.), *Essential Plant Nutrients: Uptake, Use Efficiency, and Management*. Springer International Publishing, Cham, pp. 513–556. https://doi.org/10.1007/978-3-319-58841-4_21

Aydin, N., Sönmez, M.E., Güleç, T., Demir, B., Alipour, H., Türkoğlu, A., 2025. High-throughput phenotyping of wheat root angle and coleoptile length at different temperatures using 3D-printed equipment. *BMC Plant Biol.* 25, 112. https://doi.org/10.1186/s12870-025-06120-w

Backer, R., Rokem, J.S., Ilangumaran, G., Lamont, J., Praslickova, D., Ricci, E., Subramanian, S., Smith, D.L., 2018. Plant growth-promoting rhizobacteria: Context, mechanisms of action, and roadmap to commercialization of biostimulants for sustainable agriculture. *Front. Plant Sci.* 9. https://doi.org/10.3389/fpls.2018.01473

Bahadur, A., Batool, A., Nasir, F., Jiang, S., Mingsen, Q., Zhang, Q., Pan, J., Liu, Y., Feng, H., 2019. Mechanistic insights into arbuscular mycorrhizal fungi-mediated drought stress tolerance in plants. *Int. J. Mol. Sci.* 20, 4199. https://doi.org/10.3390/ijms20174199

Baker, B.P., Green, T.A., Loker, A.J., 2020. Biological control and integrated pest management in organic and conventional systems. *Biol. Control* 140, 104095. https://doi.org/10.1016/j.biocontrol.2019.104095

Bakhat, H.F., Bibi, N., Zia, Z., Abbas, S., Hammad, H.M., Fahad, S., Ashraf, M.R., Shah, G.M., Rabbani, F., Saeed, S., 2018. Silicon mitigates biotic stresses in crop plants: A review. *Crop Prot.* 104, 21–34. https://doi.org/10.1016/j.cropro.2017.10.008

Balawejder, M., Matłok, N., Gorzelany, J., Pieniążek, M., Antos, P., Witek, G., Szostek, M., 2019. Foliar fertilizer based on calcined bones, boron and molybdenum—a study on the development and potential effects on maize grain production. *Sustainability* 11, 5287. https://doi.org/10.3390/su11195287

Balla, A., Silini, A., Cherif-Silini, H., Chenari Bouket, A., Moser, W.K., Nowakowska, J.A., Oszako, T., Benia, F., Belbahri, L., 2021. The threat of pests and pathogens and the potential for biological control in forest ecosystems. *Forests* 12, 1579. https://doi.org/10.3390/f12111579

Bargaz, A., Elhaissoufi, W., Khourchi, S., Benmrid, B., Borden, K.A., Rchiad, Z., 2021. Benefits of phosphate solubilizing bacteria on belowground crop performance for improved crop acquisition of phosphorus. *Microbiol. Res.* 252, 126842. https://doi.org/10.1016/j.micres.2021.126842

Bashir, O., Ali, T., Baba, Z.A., Rather, G.H., Bangroo, S.A., Mukhtar, S.D., Naik, N., Mohiuddin, R., Bharati, V., Bhat, R.A., 2021. Soil organic matter and its impact on soil properties and nutrient status, in: Dar, G.H., Bhat, R.A., Mehmood, M.A., Hakeem, K.R. (Eds.), *Microbiota and Biofertilizers, Vol 2: Ecofriendly Tools for Reclamation of Degraded Soil Environs*. Springer International Publishing, Cham, pp. 129–159. https://doi.org/10.1007/978-3-030-61010-4_7

Bello, S.K., Alayafi, A.H., Al-Solaimani, S.G., Abo-Elyousr, K.A., 2021. Mitigating soil salinity stress with gypsum and bio-organic amendments: A review. *Agronomy* 11, 1735.

Bhaduri, D., Sihi, D., Bhowmik, A., Verma, B.C., Munda, S., Dari, B., 2022. A review on effective soil health bio-indicators for ecosystem restoration and sustainability. *Front. Microbiol.* 13. https://doi.org/10.3389/fmicb.2022.938481

Bhardwaj, D., Ansari, M.W., Sahoo, R.K., Tuteja, N., 2014. Biofertilizers function as key player in sustainable agriculture by improving soil fertility, plant tolerance and crop productivity. *Microb. Cell Fact.* 13, 66. https://doi.org/10.1186/1475-2859-13-66

Bhat, B.A., Islam, S.T., Ali, A., Sheikh, B.A., Tariq, L., Islam, S.U., Hassan Dar, T.U., 2020. Role of micronutrients in secondary metabolism of plants, in: Aftab, T., Hakeem, K.R. (Eds.), *Plant Micronutrients: Deficiency and Toxicity Management*. Springer International Publishing, Cham, pp. 311–329. https://doi.org/10.1007/978-3-030-49856-6_13

Bhattacharyya, S.S., Furtak, K., 2023. Soil–plant–microbe interactions determine soil biological fertility by altering rhizospheric nutrient cycling and biocrust formation. *Sustainability* 15, 625. https://doi.org/10.3390/su15010625

Bhunia, S., Bhowmik, A., Mallick, R., Mukherjee, J., 2021. Agronomic efficiency of animal-derived organic fertilizers and their effects on biology and fertility of soil: A review. *Agronomy* 11, 823. https://doi.org/10.3390/agronomy11050823

Bisen, K., Keswani, C., Patel, J.S., Sarma, B.K., Singh, H.B., 2016. Trichoderma spp.: Efficient inducers of systemic resistance in plants, in: Choudhary, D.K., Varma, A. (Eds.), *Microbial-Mediated Induced Systemic Resistance in Plants*. Springer Nature, Singapore, pp. 185–195. https://doi.org/10.1007/978-981-10-0388-2_12

Biswas, T., Kole, S.C., 2017. Soil organic matter and microbial role in plant productivity and soil fertility, in: Adhya, T.K., Mishra, B.B., Annapurna, K., Verma, D.K., Kumar, U. (Eds.), *Advances in Soil Microbiology: Recent Trends and Future Prospects: Volume 2: Soil-Microbe-Plant Interaction*. Springer, Singapore, pp. 219–238. https://doi.org/10.1007/978-981-10-7380-9_10

Bodner, G., Mentler, A., Keiblinger, K., 2021. Plant roots for sustainable soil structure management in cropping systems, in: *The Root Systems in Sustainable Agricultural Intensification*. John Wiley & Sons, Ltd, pp. 45–90. https://doi.org/10.1002/9781119525417.ch3

Bolan, S., Hou, D., Wang, L., Hale, L., Egamberdieva, D., Tammeorg, P., Li, R., Wang, B., Xu, J., Wang, T., Sun, H., Padhye, L.P., Wang, H., Siddique, K.H.M., Rinklebe, J., Kirkham, M.B., Bolan, N., 2023. The potential of biochar as a microbial carrier for agricultural and environmental applications. *Sci. Total Environ.* 886, 163968. https://doi.org/10.1016/j.scitotenv.2023.163968

Canellas, L.P., Canellas, N.O.A., da S. Irineu, L.E.S., Olivares, F.L., Piccolo, A., 2020. Plant chemical priming by humic acids. *Chem. Biol. Technol. Agric.* 7, 12. https://doi.org/10.1186/s40538-020-00178-4

Chandel, S.S., Agarwal, T., 2017. Review of current state of research on energy storage, toxicity, health hazards and commercialization of phase changing materials. *Renew. Sustain. Energy Rev.* 67, 581–596. https://doi.org/10.1016/j.rser.2016.09.070

Chauhan, P., Sharma, N., Tapwal, A., Kumar, A., Verma, G.S., Meena, M., Seth, C.S., Swapnil, P., 2023. Soil microbiome: Diversity, benefits and interactions with plants. *Sustainability* 15, 14643. https://doi.org/10.3390/su151914643

Chen, W., Modi, D., Picot, A., 2023. Soil and phytomicrobiome for plant disease suppression and management under climate change: A review. *Plants* 12, 2736. https://doi.org/10.3390/plants12142736

Cheng, F., Cheng, Z., 2015. Research progress on the use of plant allelopathy in agriculture and the physiological and ecological mechanisms of allelopathy. *Front. Plant* Sci. 6. https://doi.org/10.3389/fpls.2015.01020

Crézé, C.M., Horwath, W.R., 2021. Cover cropping: A malleable solution for sustainable agriculture? Meta-analysis of ecosystem service frameworks in perennial systems. *Agronomy* 11, 862. https://doi.org/10.3390/agronomy11050862

Das, K., Biswakarma, N., Zhiipao, R., Kumar, A., Ghasal, P.C., Pooniya, V., 2020. Significance and management of green manures, in: Giri, B., Varma, A. (Eds.), *Soil Health*. Springer International Publishing, Cham, pp. 197–217. https://doi.org/10.1007/978-3-030-44364-1_12

Das, P.P., Singh, K.R., Nagpure, G., Mansoori, A., Singh, R.P., Ghazi, I.A., Kumar, A., Singh, J., 2022. Plant-soil-microbes: A tripartite interaction for nutrient acquisition and better plant growth for sustainable agricultural practices. *Environ. Res.* 214, 113821. https://doi.org/10.1016/j.envres.2022.113821

de Andrade, L.A., Santos, C.H.B., Frezarin, E.T., Sales, L.R., Rigobelo, E.C., 2023. Plant growth-promoting rhizobacteria for sustainable agricultural production. *Microorganisms* 11, 1088. https://doi.org/10.3390/microorganisms11041088

De Corato, U., 2020a. Towards new soil management strategies for improving soil quality and ecosystem services in sustainable agriculture: Editorial overview. *Sustainability* 12, 9398. https://doi.org/10.3390/su12229398

De Corato, U., 2020b. Agricultural waste recycling in horticultural intensive farming systems by on-farm composting and compost-based tea application improves soil quality and plant health: A review under the perspective of a circular economy. *Sci. Total Environ.* 738, 139840. https://doi.org/10.1016/j.scitotenv.2020.139840

Demo, A.H., Asefa Bogale, G., 2024. Enhancing crop yield and conserving soil moisture through mulching practices in dryland agriculture. *Front. Agron.* 6. https://doi.org/10.3389/fagro.2024.1361697

Derks, M., Romijn, H., 2019. Sustainable performance challenges of rural microgrids: Analysis of incentives and policy framework in Indonesia. *Energy Sustain. Dev.* 53, 57–70. https://doi.org/10.1016/j.esd.2019.08.003

Devarajan, A.K., Muthukrishanan, G., Truu, J., Truu, M., Ostonen, I., Kizhaeral, S.S., Panneerselvam, P., Kuttalingam Gopalasubramanian, S., 2021. The foliar application of rice phyllosphere bacteria induces drought-stress tolerance in *Oryza sativa* (L.). *Plants* 10, 387. https://doi.org/10.3390/plants10020387

Dhiman, D., Sharma, R., Sankhyan, N.K., Sepehya, S., Sharma, S.K., Kumar, R., 2019. Effect of regular application of fertilizers, manure and lime on soil health and productivity of wheat in an acid alfisol. *J. Plant Nutr.* 42, 2507–2521. https://doi.org/10.1080/01904167.2019.1659317

Dignam, B.E.A., O'Callaghan, M., Condron, L.M., Raaijmakers, J.M., Kowalchuk, G.A., Wakelin, S.A., 2019. Impacts of long-term plant residue management on soil organic matter quality, *Pseudomonas* community structure and disease suppressiveness. *Soil Biol. Biochem.* 135, 396–406. https://doi.org/10.1016/j.soilbio.2019.05.020

Dixon, G.R., Tilston, E.L., 2010. Soil-borne pathogens and their interactions with the soil environment, in: Dixon, G.R., Tilston, E.L. (Eds.), *Soil Microbiology and Sustainable Crop Production*. Springer Netherlands, Dordrecht, pp. 197–271. https://doi.org/10.1007/978-90-481-9479-7_6

Dubey, A., Malla, M.A., Khan, F., Chowdhary, K., Yadav, S., Kumar, A., Sharma, S., Khare, P.K., Khan, M.L., 2019. Soil microbiome: A key player for conservation of soil health under changing climate. *Biodivers. Conserv.* 28, 2405–2429. https://doi.org/10.1007/s10531-019-01760-5

El-Beltagi, H.S., Ahmed, O.K., Hegazy, A.E., 2016. Protective effect of nitric oxide on high temperature induced oxidative stress in wheat (*Triticum aestivum*) callus culture. *Not. Sci. Biol.* 8, 192–198.

Elnahal, A.S., El-Saadony, M.T., Saad, A.M., Desoky, E.-S.M., El-Tahan, A.M., Rady, M.M., AbuQamar, S.F., El-Tarabily, K.A., 2022. The use of microbial inoculants for biological control, plant growth promotion, and sustainable agriculture: A review. *Eur. J. Plant Pathol.* 162, 759–792.

Francaviglia, R., Almagro, M., Vicente-Vicente, J.L., 2023. Conservation agriculture and soil organic carbon: Principles, processes, practices and policy options. *Soil Syst.* 7, 17. https://doi.org/10.3390/soilsystems7010017

Fukami, J., Cerezini, P., Hungria, M., 2018. *Azospirillum*: Benefits that go far beyond biological nitrogen fixation. *AMB Expr.* 8, 73. https://doi.org/10.1186/s13568-018-0608-1

Gade, R.M., Koche, M.D., 2022. Enhancing the growth and disease suppression ability of *Pseudomonas fluorescens*, in: Singh, H.B., Vaishnav, A. (Eds.), *New and Future Developments in Microbial Biotechnology and Bioengineering*. Elsevier, pp. 351–368. https://doi.org/10.1016/B978-0-323-85577-8.00012-3

Gamliel, A., van Bruggen, A.H.C., 2016. Maintaining soil health for crop production in organic greenhouses. *Sci. Hortic., Recent advances in organic horticulture technology and management—Part 1* 208, 120–130. https://doi.org/10.1016/j.scienta.2015.12.030

Gelaye, Y., Negash, B., 2023. The role of baculoviruses in controlling insect pests: A review. *Cogent Food Agric.* 9, 2254139. https://doi.org/10.1080/23311932.2023.2254139

Goldan, E., Nedeff, V., Barsan, N., Culea, M., Panainte-Lehadus, M., Mosnegutu, E., Tomozei, C., Chitimus, D., Irimia, O., 2023. Assessment of manure compost used as soil amendment—a review. *Processes* 11, 1167. https://doi.org/10.3390/pr11041167

Gopalakrishnan, S., Srinivas, V., 2019. Management of soil-borne diseases of grain legumes through broad-spectrum actinomycetes having plant growth-promoting and biocontrol traits, in: Varma, A., Tripathi, S., Prasad, R. (Eds.), *Plant Microbe Interface*. Springer International Publishing, Cham, pp. 129–144. https://doi.org/10.1007/978-3-030-19831-2_5

Greff, B., Szigeti, J., Nagy, Á., Lakatos, E., Varga, L., 2022. Influence of microbial inoculants on co-composting of lignocellulosic crop residues with farm animal manure: A review. *Journal of Environmental Management* 302, 114088. https://doi.org/10.1016/j.jenvman.2021.114088

Gupta, N., Debnath, S., Sharma, S., Sharma, P., Purohit, J., 2017. Role of nutrients in controlling the plant diseases in sustainable agriculture, in: Meena, V.S., Mishra, P.K., Bisht, J.K., Pattanayak, A. (Eds.), *Agriculturally Important Microbes for Sustainable Agriculture: Volume 2: Applications in Crop Production and Protection*. Springer, Singapore, pp. 217–262. https://doi.org/10.1007/978-981-10-5343-6_8

Hariharan, G., Rifnas, L.M., Prasannath, K., 2022. Role of *Trichoderma* spp. in biocontrol of plant diseases, in: Kumar, A. (Ed.), *Microbial Biocontrol: Food Security and Post Harvest Management: Volume 2*. Springer International Publishing, Cham, pp. 39–78. https://doi.org/10.1007/978-3-030-87289-2_3

Harman, G., Khadka, R., Doni, F., Uphoff, N., 2021. Benefits to plant health and productivity from enhancing plant microbial symbionts. *Front. Plant Sci.* 11. https://doi.org/10.3389/fpls.2020.610065

Hasan, Md.M., Liu, X.-D., Rahman, M.A., Hazzazi, Y., Wassem, ·Muhammad, Ghimire, S., Alabdallah, N.M., Alharbi, B.M., Kabir, A.H., Yao, G., Fang, X.-W., 2025. Plants breathing under pressure: Mechanistic insights into soil compaction-induced physiological, molecular and biochemical responses in plants. *Planta* 261, 52. https://doi.org/10.1007/s00425-025-04624-1

Hasnain, M., Munir, N., Abideen, Z., Zulfiqar, F., Koyro, H.W., El-Naggar, A., Caçador, I., Duarte, B., Rinklebe, J., Yong, J.W.H., 2023. Biochar-plant interaction and detoxification strategies under abiotic stresses for achieving agricultural resilience: A critical review. *Ecotoxicol. Environ. Saf.* 249, 114408. https://doi.org/10.1016/j.ecoenv.2022.114408

Hayat, M.K., Khan, A.R., Klutse, S., Rasool, M.W., Mohsin, M., 2025. Empowering crops: How organic, agronomic practices and technology can fortify our fields. *World News Nat. Sci.* 59, 358–383.

Huang, Q., 2024. Enhancing soil health and biodiversity through nitrogen fixation symbiosis in leguminous plants. *Mol. Microbiol. Res.* 14.

Hussain, M.M., Farooqi, Z.U.R., 2021. Allelopathic bacteria as an alternate weedicide: Progress and future standpoints, in: Hakeem, K.R., Dar, G.H., Mehmood, M.A., Bhat, R.A. (Eds.), *Microbiota and Biofertilizers: A Sustainable Continuum for Plant and Soil Health*. Springer International Publishing, Cham, pp. 211–230. https://doi.org/10.1007/978-3-030-48771-3_13

Hussain, T., Akthar, N., Aminedi, R., Danish, Mohd., Nishat, Y., Patel, S., 2020. Role of the potent microbial based bioagents and their emerging strategies for the ecofriendly management of agricultural phytopathogens, in: Singh, J., Yadav, A.N. (Eds.), *Natural Bioactive Products in Sustainable Agriculture*. Springer, Singapore, pp. 45–66. https://doi.org/10.1007/978-981-15-3024-1_4

Iftikhar, Y., Sajid, A., Shakeel, Q., Ahmad, Z., Ul Haq, Z., 2020. Biological antagonism: A safe and sustainable way to manage plant diseases, in: Ul Haq, I., Ijaz, S. (Eds.), *Plant Disease Management Strategies for Sustainable Agriculture Through Traditional and Modern Approaches*. Springer International Publishing, Cham, pp. 83–109. https://doi.org/10.1007/978-3-030-35955-3_5

Igiehon, N.O., Babalola, O.O., 2018. Below-ground-above-ground plant–microbial interactions: Focusing on soybean, rhizobacteria and mycorrhizal fungi. *Open Microbiol. J.* 12, 261–279. https://doi.org/10.2174/1874285801812010261

Iqbal, B., Li, G., Alabbosh, K.F., Hussain, H., Khan, I., Tariq, M., Javed, Q., Naeem, M., Ahmad, N., 2023. Advancing environmental sustainability through microbial reprogramming in growth improvement, stress alleviation, and phytoremediation. *Plant Stress* 10, 100283. https://doi.org/10.1016/j.stress.2023.100283

Iqbal, S., Wang, X., Mubeen, I., Kamran, M., Kanwal, I., Díaz, G.A., Abbas, A., Parveen, A., Atiq, M.N., Alshaya, H., Zin El-Abedin, T.K., Fahad, S., 2022. Phytohormones trigger drought tolerance in crop plants: Outlook and future perspectives. *Front. Plant Sci.* 12. https://doi.org/10.3389/fpls.2021.799318

Islam, W., Noman, A., Naveed, H., Huang, Z., Chen, H.Y.H., 2020. Role of environmental factors in shaping the soil microbiome. *Environ. Sci. Pollut. Res.* 27, 41225–41247. https://doi.org/10.1007/s11356-020-10471-2

Issaka, S., Ashraf, M.A., 2017. Impact of soil erosion and degradation on water quality: A review. *Geol. Ecol. Landsc.* 1, 1–11. https://doi.org/10.1080/24749508.2017.1301053

Javed, Z., Tripathi, G.D., Mishra, M., Dashora, K., 2021. Actinomycetes—the microbial machinery for the organic-cycling, plant growth, and sustainable soil health. *Biocatal. Agric. Biotechnol.* 31, 101893. https://doi.org/10.1016/j.bcab.2020.101893

Jayaraman, S., Naorem, A.K., Lal, R., Dalal, R.C., Sinha, N.K., Patra, A.K., Chaudhari, S.K., 2021. Disease-suppressive soils—beyond food production: A critical review. *J. Soil Sci. Plant Nutr.* 21, 1437–1465. https://doi.org/10.1007/s42729-021-00451-x

Jiang, Y., Zhang, J., Manuel, D.-B., Op de Beeck, M., Shahbaz, M., Chen, Y., Deng, X., Xu, Z., Li, J., Liu, Z., 2022. Rotation cropping and organic fertilizer jointly promote soil health and crop production. *J. Environ. Manage.* 315, 115190. https://doi.org/10.1016/j.jenvman.2022.115190

Jing, T., Li, J., He, Y., Shankar, A., Saxena, A., Tiwari, A., Maturi, K.C., Solanki, M.K., Singh, V., Eissa, M.A., Ding, Z., Xie, J., Awasthi, M.K., 2024. Role of calcium nutrition in plant physiology: Advances in research and insights into acidic soil conditions—a comprehensive review. *Plant Physiol. Biochem.* 210, 108602. https://doi.org/10.1016/j.plaphy.2024.108602

Kadam, S.B., Pable, A.A., Barvkar, V.T., 2020. Mycorrhiza induced resistance (MIR): A defence developed through synergistic engagement of phytohormones, metabolites and rhizosphere. *Functional Plant Biol.* 47, 880–890. https://doi.org/10.1071/FP20035

Kalogiannidis, S., Kalfas, D., Chatzitheodoridis, F., Papaevangelou, O., 2022. Role of crop-protection technologies in sustainable agricultural productivity and management. *Land* 11, 1680. https://doi.org/10.3390/land11101680

Kamle, M., Borah, R., Bora, H., Jaiswal, A.K., Singh, R.K., Kumar, P., 2020. Systemic acquired resistance (SAR) and induced systemic resistance (ISR): Role and mechanism of action against phytopathogens, in: Hesham, A.E.-L., Upadhyay, R.S., Sharma, G.D., Manoharachary, C., Gupta, V.K. (Eds.), *Fungal Biotechnology and Bioengineering*. Springer International Publishing, Cham, pp. 457–470. https://doi.org/10.1007/978-3-030-41870-0_20

Karbin, S., Kassam, A., Oza, A., Sawhney, T., Sahu, P., Mogare, B., Mitra, B., Viswakarma, S., Singh, J., Mahajan, R., Malviya, S., Badole, P., Patidar, N., 2022. Initiating conservation agriculture shows reduced soil CO_2 emissions and improved soil aggregate stability in the first season in rainfed cropping in India. *Int. J. Environ. Stud.* 79, 998–1014. https://doi.org/10.1080/00207233.2021.1987050

Khaliq, A., Perveen, S., Alamer, K.H., Zia Ul Haq, M., Rafique, Z., Alsudays, I.M., Althobaiti, A.T., Saleh, M.A., Hussain, S., Attia, H., 2022. Arbuscular mycorrhizal fungi symbiosis to enhance plant–soil interaction. *Sustainability* 14, 7840. https://doi.org/10.3390/su14137840

Khan, A.R., Mustafa, A., Hyder, S., Valipour, M., Rizvi, Z.F., Gondal, A.S., Yousuf, Z., Iqbal, R., Daraz, U., 2022. Bacillus spp. as bioagents: Uses and application for sustainable agriculture. *Biology* 11, 1763. https://doi.org/10.3390/biology11121763

Khatoon, Z., Huang, S., Rafique, M., Fakhar, A., Kamran, M.A., Santoyo, G., 2020. Unlocking the potential of plant growth-promoting rhizobacteria on soil health and the sustainability of agricultural systems. *J. Environ. Manage.* 273, 111118. https://doi.org/10.1016/j.jenvman.2020.111118

Koskey, G., Mburu, S.W., Awino, R., Njeru, E.M., Maingi, J.M., 2021. Potential use of beneficial microorganisms for soil amelioration, phytopathogen biocontrol, and sustainable crop production in smallholder agroecosystems. *Front. Sustain. Food Syst.* 5. https://doi.org/10.3389/fsufs.2021.606308

Kumar, A., Dubey, A., 2020. Rhizosphere microbiome: Engineering bacterial competitiveness for enhancing crop production. *J. Adv. Res.* 24, 337–352. https://doi.org/10.1016/j.jare.2020.04.014

Kumar, A., Verma, J.P., 2019. The role of microbes to improve crop productivity and soil health, in: Achal, V., Mukherjee, A. (Eds.), *Ecological Wisdom Inspired Restoration Engineering*. Springer, Singapore, pp. 249–265. https://doi.org/10.1007/978-981-13-0149-0_14

Kumar, M., Giri, V.P., Pandey, S., Gupta, A., Patel, M.K., Bajpai, A.B., Jenkins, S., Siddique, K.H.M., 2021. Plant-growth-promoting rhizobacteria emerging as an effective bioinoculant to improve the growth, production, and stress tolerance of vegetable crops. *Int. J. Mol. Sci.* 22, 12245. https://doi.org/10.3390/ijms222212245

Kumari, V.V., Banerjee, P., Verma, V.C., Sukumaran, S., Chandran, M.A.S., Gopinath, K.A., Venkatesh, G., Yadav, S.K., Singh, V.K., Awasthi, N.K., 2022. Plant nutrition: An effective way to alleviate abiotic stress in agricultural crops. *Int. J. Mol. Sci.* 23, 8519. https://doi.org/10.3390/ijms23158519

Lahlali, R., Ezrari, S., Radouane, N., Kenfaoui, J., Esmaeel, Q., El Hamss, H., Belabess, Z., Barka, E.A., 2022. Biological control of plant pathogens: A global perspective. *Microorganisms* 10, 596. https://doi.org/10.3390/microorganisms10030596

Lamichhane, J.R., Aubertot, J.-N., Begg, G., Birch, A.N.E., Boonekamp, P., Dachbrodt-Saaydeh, S., Hansen, J.G., Hovmøller, M.S., Jensen, J.E., Jørgensen, L.N., Kiss, J., Kudsk, P., Moonen, A.-C., Rasplus, J.-Y., Sattin, M., Streito, J.-C., Messéan, A., 2016. Networking of integrated pest management: A powerful approach to address common challenges in agriculture. *Crop Protect.* 89, 139–151. https://doi.org/10.1016/j.cropro.2016.07.011

Larkin, R.P., Griffin, T.S., Honeycutt, C.W., Olanya, O.M., He, Z., 2021. Potato cropping system management strategy impacts soil physical, chemical, and biological properties over time. *Soil Till. Res.* 213, 105148. https://doi.org/10.1016/j.still.2021.105148

Lee, S.J., Kim, S., Kim, J.C., Lee, M.R., Hossain, M.S., Shin, T.S., Kim, T.H., Kim, J.S., 2017. Entomopathogenic *Beauveria bassiana* granules to control soil-dwelling stage of western flower thrips, *Frankliniella occidentalis* (Thysanoptera: Thripidae). *BioControl* 62, 639–648. https://doi.org/10.1007/s10526-017-9818-8

Lezoche, M., Hernandez, J.E., Alemany Díaz, M. del M.E., Panetto, H., Kacprzyk, J., 2020. Agri-food 4.0: A survey of the supply chains and technologies for the future agriculture. *Comput. Ind.* 117, 103187. https://doi.org/10.1016/j.compind.2020.103187

Liao, D., Wang, S., Cui, M., Liu, J., Chen, A., Xu, G., 2018. Phytohormones regulate the development of arbuscular mycorrhizal symbiosis. *Int. J. Mol. Sci.* 19, 3146. https://doi.org/10.3390/ijms19103146

Lobry de Bruyn, L., Andrews, S., 2016. Are Australian and United States farmers using soil information for soil health management? *Sustainability* 8, 304. https://doi.org/10.3390/su8040304

Lodha, S., Mawar, R., 2020. Population dynamics of *Macrophomina phaseolina* in relation to disease management: A review. *J. Phytopathol.* 168, 1–17. https://doi.org/10.1111/jph.12854

Lu, J., Zhang, Q., Werner, A.D., Li, Y., Jiang, S., Tan, Z., 2020. Root-induced changes of soil hydraulic properties—a review. *J. Hydrol.* 589, 125203. https://doi.org/10.1016/j.jhydrol.2020.125203

Malerba, M., Cerana, R., 2020. Chitin- and Chitosan-based derivatives in plant protection against biotic and abiotic stresses and in recovery of contaminated soil and water. *Polysaccharides* 1, 21–30. https://doi.org/10.3390/polysaccharides1010003

Mariani, L., Ferrante, A., 2017. Agronomic management for enhancing plant tolerance to abiotic stresses—drought, salinity, hypoxia, and lodging. *Horticulturae* 3, 52. https://doi.org/10.3390/horticulturae3040052

Martínez-García, L.B., Korthals, G., Brussaard, L., Jørgensen, H.B., De Deyn, G.B., 2018. Organic management and cover crop species steer soil microbial community structure and functionality along with soil organic matter properties. *Agric. Ecosyst. Environ.* 263, 7–17. https://doi.org/10.1016/j.agee.2018.04.018

Meena, M., Swapnil, P., Zehra, A., Aamir, M., Dubey, M.K., Goutam, J., Upadhyay, R.S., 2017. Beneficial microbes for disease suppression and plant growth promotion, in: Singh, D.P., Singh, H.B., Prabha, R. (Eds.), *Plant–Microbe Interactions in Agro-Ecological Perspectives: Volume 2: Microbial Interactions and Agro-Ecological Impacts*. Springer, Singapore, pp. 395–432. https://doi.org/10.1007/978-981-10-6593-4_16

Miljaković, D., Marinković, J., Balešević-Tubić, S., 2020. The significance of *Bacillus* spp. in disease suppression and growth promotion of field and vegetable crops. *Microorganisms* 8, 1037. https://doi.org/10.3390/microorganisms8071037

Minato, E.A., Brignoli, F.M., Neto, M.E., Besen, M.R., Cassim, B.M.A.R., Lima, R.S., Tormena, C.A., Inoue, T.T., Batista, M.A., 2023. Lime and gypsum application to low-acidity soils: Changes in soil chemical properties, residual lime content and crop agronomic performance. *Soil Till. Res.* 234, 105860. https://doi.org/10.1016/j.still.2023.105860

Mirsky, S.B., Ryan, M.R., Teasdale, J.R., Curran, W.S., Reberg-Horton, C.S., Spargo, J.T., Wells, M.S., Keene, C.L., Moyer, J.W., 2013. Overcoming weed management challenges in cover crop–based organic rotational no-till soybean production in the Eastern United States. *Weed Technol.* 27, 193–203. https://doi.org/10.1614/WT-D-12-00078.1

Moosavi, M.R., Zare, R., 2020. Fungi as biological control agents of plant-parasitic nematodes, in: Mérillon, J.-M., Ramawat, K.G. (Eds.), *Plant Defence: Biological Control*. Springer International Publishing, Cham, pp. 333–384. https://doi.org/10.1007/978-3-030-51034-3_14

Msimbira, L.A., Smith, D.L., 2020. The roles of plant growth promoting microbes in enhancing plant tolerance to acidity and alkalinity stresses. *Front. Sustain. Food Syst.* 4. https://doi.org/10.3389/fsufs.2020.00106

Neher, D.A., Hoitink, H.A., Biala, J., Rynk, R., Black, G., 2022. Chapter 17—Compost use for plant disease suppression, in: Rynk, R. (Ed.), *The Composting Handbook*. Academic Press, pp. 847–878. https://doi.org/10.1016/B978-0-323-85602-7.00015-7

Ngosong, C., Okolle, J.N., Tening, A.S., 2019. Mulching: A sustainable option to improve soil health, in: Panpatte, D.G., Jhala, Y.K. (Eds.), *Soil Fertility Management for Sustainable Development*. Springer, Singapore, pp. 231–249. https://doi.org/10.1007/978-981-13-5904-0_11

Nizamani, M.M., Hughes, A.C., Qureshi, S., Zhang, Q., Tarafder, E., Das, D., Acharya, K., Wang, Y., Zhang, Z.-G., 2024. Microbial biodiversity and plant functional trait interactions in multifunctional ecosystems. *Applied Soil Ecology* 201, 105515. https://doi.org/10.1016/j.apsoil.2024.105515

Obalum, S.E., Chibuike, G.U., Peth, S., Ouyang, Y., 2017. Soil organic matter as sole indicator of soil degradation. *Environ Monit Assess* 189, 176. https://doi.org/10.1007/s10661-017-5881-y

O'Callaghan, M., Ballard, R.A., Wright, D., 2022. Soil microbial inoculants for sustainable agriculture: Limitations and opportunities. *Soil Use and Management* 38, 1340–1369. https://doi.org/10.1111/sum.12811

Ojuederie, O.B., Olanrewaju, O.S., Babalola, O.O., 2019. Plant growth promoting rhizobacterial mitigation of drought stress in crop plants: Implications for sustainable agriculture. *Agronomy* 9, 712. https://doi.org/10.3390/agronomy9110712

Olanrewaju, O.S., Ayangbenro, A.S., Glick, B.R., Babalola, O.O., 2019. Plant health: Feedback effect of root exudates–rhizobiome interactions. *Appl. Microbiol. Biotechnol.* 103, 1155–1166. https://doi.org/10.1007/s00253-018-9556-6

Osman, K.T., 2018. Saline and sodic soils, in: Osman, K.T. (Ed.), *Management of Soil Problems*. Springer International Publishing, Cham, pp. 255–298. https://doi.org/10.1007/978-3-319-75527-4_10

Parween, T., Bhandari, P., Jan, S., Mahmooduzzafar, Fatma, T., Raza, S.K., 2017. Role of bioinoculants as plant growth-promoting microbes for sustainable agriculture, in: Meena, V.S., Mishra, P.K., Bisht, J.K., Pattanayak, A. (Eds.), *Agriculturally Important Microbes for Sustainable Agriculture: Volume I: Plant-Soil-Microbe Nexus*. Springer, Singapore, pp. 183–206. https://doi.org/10.1007/978-981-10-5589-8_9

Pascoli, M., Jacques, M.T., Agarrayua, D.A., Avila, D.S., Lima, R., Fraceto, L.F., 2019. Neem oil based nanopesticide as an environmentally-friendly formulation for applications in sustainable agriculture: An ecotoxicological perspective. *Sci. Total Environ.* 677, 57–67. https://doi.org/10.1016/j.scitotenv.2019.04.345

Polyak, Y.M., Sukcharevich, V.I., 2019. Allelopathic interactions between plants and microorganisms in soil ecosystems. *Biol. Bull. Rev.* 9, 562–574. https://doi.org/10.1134/S2079086419060033

Rabot, E., Wiesmeier, M., Schlüter, S., Vogel, H.-J., 2018. Soil structure as an indicator of soil functions: A review. *Geoderma* 314, 122–137. https://doi.org/10.1016/j.geoderma.2017.11.009

Radulov, I., Berbecea, A., Radulov, I., Berbecea, A., 2024. Nutrient management for sustainable soil fertility. *IntechOpen*. https://doi.org/10.5772/intechopen.1006692

Raheem, A., Bankole, O.O., Danso, F., Musa, M.O., Adegbite, T.A., Simpson, V.B., 2025. Physical management strategies for enhancing soil resilience to climate change: Insights from Africa. *Eur. J. Soil Sci.* 76, e70030. https://doi.org/10.1111/ejss.70030

Rahman, M.U., Chen, P., Zhang, X., Fan, B., 2023. Predacious strategies of nematophagous fungi as biocontrol agents. *Agronomy* 13, 2685. https://doi.org/10.3390/agronomy13112685

Raj, A., Jhariya, M.K., Yadav, D.K., Banerjee, A., Meena, R.S., 2019. Soil for sustainable environment and ecosystems management, in: Jhariya, M.K., Banerjee, A., Meena, R.S., Yadav, D.K. (Eds.), *Sustainable Agriculture, Forest and Environmental Management*. Springer, Singapore, pp. 189–221. https://doi.org/10.1007/978-981-13-6830-1_6

Ramamoorthy, V., Viswanathan, R., Raguchander, T., Prakasam, V., Samiyappan, R., 2001. Induction of systemic resistance by plant growth promoting rhizobacteria in crop plants against pests and diseases. *Crop Protect.* 20, 1–11. https://doi.org/10.1016/S0261-2194(00)00056–9

Rashid, M.H.-O., Chung, Y.R., 2017. Induction of systemic resistance against insect herbivores in plants by beneficial soil microbes. *Front. Plant Sci.* 8. https://doi.org/10.3389/fpls.2017.01816

Rastogi, M., Verma, S., Kumar, S., Bharti, S., Kumar, G., Azam, K., Singh, V., 2023. Soil health and sustainability in the age of organic amendments: A review. *Int. J. Environ. Clim.* Change 13, 2088–2102. https://doi.org/10.9734/ijecc/2023/v13i102870

Rawat, P., Das, S., Shankhdhar, D., Shankhdhar, S.C., 2021. Phosphate-solubilizing microorganisms: Mechanism and their role in phosphate solubilization and uptake. *J. Soil Sci. Plant Nutr.* 21, 49–68. https://doi.org/10.1007/s42729-020-00342-7

Rehman, A., Saba, T., Kashif, M., Fati, S.M., Bahaj, S.A., Chaudhry, H., 2022. A revisit of internet of things technologies for monitoring and control strategies in smart agriculture. *Agronomy* 12, 127. https://doi.org/10.3390/agronomy12010127

Roberts, D.P., Mattoo, A.K., 2018. Sustainable agriculture—enhancing environmental benefits, food nutritional quality and building crop resilience to abiotic and biotic stresses. *Agriculture* 8, 8. https://doi.org/10.3390/agriculture8010008

Rocha, I., Ma, Y., Souza-Alonso, P., Vosátka, M., Freitas, H., Oliveira, R.S., 2019. Seed coating: A tool for delivering beneficial microbes to agricultural crops. *Front. Plant Sci.* 10. https://doi.org/10.3389/fpls.2019.01357

Ronga, D., Vitti, A., Zaccardelli, M., Pane, C., Caradonia, F., Cardarelli, M., Colla, G., Rouphael, Y., 2021. Root zone management for improving seedling quality of organically produced horticultural crops. *Agronomy* 11, 630. https://doi.org/10.3390/agronomy11040630

Rosenstock, T.S., Mayzelle, M., Namoi, N., Fantke, P., 2024. Climate impacts of natural farming: A cradle to gate comparison between conventional practice and Andhra Pradesh Community Natural Farming. *Res. Rev. J. Agri. Allied Sci.* https://doi.org/10.31220/agriRxiv.2020.00013

Sadeghian, N., Vaezi, A.R., Majnooniheris, A., Cerdà, A., 2021. Soil physical degradation and rill detachment by raindrop impact in semi-arid region. *CATENA* 207, 105603. https://doi.org/10.1016/j.catena.2021.105603

Salomon, M.V., Funes Pinter, I., Piccoli, P., Bottini, R., 2017. Use of plant growth-promoting rhizobacteria as biocontrol agents: Induced systemic resistance against biotic stress in plants, in: Kalia, V.C. (Ed.), *Microbial Applications Vol. 2: Biomedicine, Agriculture and Industry.* Springer International Publishing, Cham, pp. 133–152. https://doi.org/10.1007/978-3-319-52669-0_7

Santos, M.S., Rodrigues, T.F., Nogueira, M.A., Hungria, M., 2021. The challenge of combining high yields with environmentally friendly bioproducts: A review on the compatibility of pesticides with microbial inoculants. *Agronomy* 11, 870. https://doi.org/10.3390/agronomy11050870

Sarker, A., Ansary, M.W.R., Hossain, M.N., Islam, T., 2021. Prospect and challenges for sustainable management of climate change-associated stresses to soil and plant health by beneficial rhizobacteria. *Stresses* 1, 200–222. https://doi.org/10.3390/stresses1040015

Sarma, H.H., Paul, A., Kakoti, M., Talukdar, N., Hazarika, P., 2024. Climate resilient agricultural strategies for enhanced sustainability and food security: A review. *PA.* https://doi.org/10.51470/PLANTARCHIVES.2024.v24.no.1.108

Selim, M.M., 2020. Introduction to the integrated nutrient management strategies and their contribution to yield and soil properties. *Int. J. Agron.* 2020, 2821678. https://doi.org/10.1155/2020/2821678

Shah, F., Wu, W., 2019. Soil and crop management strategies to ensure higher crop productivity within sustainable environments. *Sustainability* 11, 1485. https://doi.org/10.3390/su11051485

Shah, S.H., Islam, S., Mohammad, F., 2022. Sulphur as a dynamic mineral element for plants: A review. *J. Soil Sci. Plant Nutr.* 22, 2118–2143. https://doi.org/10.1007/s42729-022-00798-9

Shahane, A.A., Shivay, Y.S., 2021. Soil health and its improvement through novel agronomic and innovative approaches. *Front. Agron.* 3. https://doi.org/10.3389/fagro.2021.680456

Shahrajabian, M.H., Sun, W., 2024. The importance of salicylic acid, humic acid and fulvic acid on crop production. *Lett. Drug Des. Discov.* 21, 1465–1480. https://doi.org/10.2174/1570180820666230411102209

Sharma, B., Tiwari, S., Kumawat, K.C., Cardinale, M., 2023. Nano-biofertilizers as bio-emerging strategies for sustainable agriculture development: Potentiality and their limitations. *Sci. Total Environ.* 860, 160476. https://doi.org/10.1016/j.scitotenv.2022.160476

Sharma, D., Gahtyari, N.C., Chhabra, R., Kumar, D., 2020. Role of microbes in improving plant growth and soil health for sustainable agriculture, in: Yadav, A.N., Rastegari, A.A., Yadav, N., Kour, D. (Eds.), *Advances in Plant Microbiome and Sustainable Agriculture: Diversity and Biotechnological Applications*. Springer, Singapore, pp. 207–256. https://doi.org/10.1007/978-981-15-3208-5_9

Sharma, P.K., Kumar, S., 2023. Soil structure and plant growth, in: Sharma, P.K., Kumar, S. (Eds.), *Soil Physical Environment and Plant Growth: Evaluation and Management*. Springer International Publishing, Cham, pp. 125–154. https://doi.org/10.1007/978-3-031-28057-3_5

Shen, N., Liang, Z., Liu, Q., Tu, C., Dong, K., Wang, C., Chen, M., 2020. Antifungal secondary metabolites isolated from mangrove rhizosphere soil-derived penicillium fungi. *J. Ocean Univ. China* 19, 717–721. https://doi.org/10.1007/s11802-020-4360-1

Shruthi, Prakash, N.B., Dhumgond, P., Goiba, P.K., Laxmanarayanan, M., 2024. The benefits of gypsum for sustainable management and utilization of acid soils. *Plant Soil* 504, 5–28. https://doi.org/10.1007/s11104-024-06907-0

Shukla, P.S., Borza, T., Critchley, A.T., Prithiviraj, B., 2021. Seaweed-based compounds and products for sustainable protection against plant pathogens. *Marine Drugs* 19, 59. https://doi.org/10.3390/md19020059

Singh, T.B., Ali, A., Prasad, M., Yadav, A., Shrivastav, P., Goyal, D., Dantu, P.K., 2020. Role of organic fertilizers in improving soil fertility, in: Naeem, M., Ansari, A.A., Gill, S.S. (Eds.), *Contaminants in Agriculture: Sources, Impacts and Management*. Springer International Publishing, Cham, pp. 61–77. https://doi.org/10.1007/978-3-030-41552-5_3

Sivasabari, K., Parthasarathy, S., Chandran, D., Sankaralingam, S., Ajaykumar, R., 2023. Vermicompost and vermi-leachate in pest and disease management, in: Mupambwa, H.A., Horn, L.N., Mnkeni, P.N.S. (Eds.), *Vermicomposting for Sustainable Food Systems in Africa*. Springer Nature, Singapore, pp. 279–304. https://doi.org/10.1007/978-981-19-8080-0_17

Somasundaram, J., Lal, R., Sinha, N.K., Dalal, R., Chitralekha, A., Chaudhary, R.S., Patra, A.K., 2018. Chapter three—Cracks and potholes in vertisols: Characteristics, occurrence, and management, in: Sparks, D.L. (Ed.), *Advances in Agronomy*. Academic Press, pp. 93–159. https://doi.org/10.1016/bs.agron.2018.01.001

Somasundaram, J., Sinha, N.K., Dalal, R.C., Lal, R., Mohanty, M., Naorem, A.K., Hati, K.M., Chaudhary, R.S., Biswas, A.K., Patra, A.K., Chaudhari, S.K., 2020. No-till farming and conservation agriculture in South Asia—issues, challenges, prospects and benefits. *Crit. Rev. Plant Sci.* 39, 236–279. https://doi.org/10.1080/07352689.2020.1782069

Syed Ab Rahman, S.F., Singh, E., Pieterse, C.M.J., Schenk, P.M., 2018. Emerging microbial biocontrol strategies for plant pathogens. *Plant Science* 267, 102–111. https://doi.org/10.1016/j.plantsci.2017.11.012

Tahat, M.M., Alananbeh, K.M., Othman, Y.A., Leskovar, D.I., 2020. Soil health and sustainable agriculture. *Sustainability* 12, 4859. https://doi.org/10.3390/su12124859

Tonelli, M.L., Figueredo, M.S., Rodríguez, J., Fabra, A., Ibañez, F., 2020. Induced systemic resistance-like responses elicited by rhizobia. *Plant Soil* 448, 1–14. https://doi.org/10.1007/s11104-020-04423-5

Tripathi, S., Srivastava, P., Devi, R.S., Bhadouria, R., 2020. Chapter 2—Influence of synthetic fertilizers and pesticides on soil health and soil microbiology, in: Prasad, M.N.V. (Ed.), *Agrochemicals Detection, Treatment and Remediation*. Butterworth-Heinemann, pp. 25–54. https://doi.org/10.1016/B978-0-08-103017-2.00002-7

Tudi, M., Daniel Ruan, H., Wang, L., Lyu, J., Sadler, R., Connell, D., Chu, C., Phung, D.T., 2021. Agriculture development, pesticide application and its impact on the environment. *Int. J. Environ. Res. Public Health* 18, 1112.

Tully, K.L., McAskill, C., 2020. Promoting soil health in organically managed systems: A review. *Org. Agr.* 10, 339–358. https://doi.org/10.1007/s13165-019-00275-1

Tunlid, A., Floudas, D., Op De Beeck, M., Wang, T., Persson, P., 2022. Decomposition of soil organic matter by ectomycorrhizal fungi: Mechanisms and consequences for organic nitrogen uptake and soil carbon stabilization. *Front. For. Glob. Change* 5. https://doi.org/10.3389/ffgc.2022.934409

Verma, K.K., Song, X.-P., Tian, D.-D., Guo, D.-J., Chen, Z.-L., Zhong, C.-S., Nikpay, A., Singh, M., Rajput, V.D., Singh, R.K., Minkina, T., Li, Y.-R., 2021. Influence of silicon on biocontrol strategies to manage biotic stress for crop protection, performance, and improvement. *Plants* 10, 2163. https://doi.org/10.3390/plants10102163

Verma, P., Yadav, A.N., Kumar, V., Singh, D.P., Saxena, A.K., 2017. Beneficial plant–microbes interactions: Biodiversity of microbes from diverse extreme environments and its impact for crop improvement, in: Singh, D.P., Singh, H.B., Prabha, R. (Eds.), *Plant–Microbe Interactions in Agro-Ecological Perspectives: Volume 2: Microbial Interactions and Agro-Ecological Impacts*. Springer, Singapore, pp. 543–580. https://doi.org/10.1007/978-981-10-6593-4_22

Vishwakarma, K., Kumar, N., Shandilya, C., Mohapatra, S., Bhayana, S., Varma, A., 2020. Revisiting plant–microbe interactions and microbial consortia application for enhancing sustainable agriculture: A review. *Front. Microbiol.* 11. https://doi.org/10.3389/fmicb.2020.560406

Wahab, A., Muhammad, M., Munir, A., Abdi, G., Zaman, W., Ayaz, A., Khizar, C., Reddy, S.P.P., 2023. Role of arbuscular mycorrhizal fungi in regulating growth, enhancing productivity, and potentially influencing ecosystems under abiotic and biotic stresses. *Plants* 12, 3102. https://doi.org/10.3390/plants12173102

Wang, H., Liu, R., You, M.P., Barbetti, M.J., Chen, Y., 2021a. Pathogen biocontrol using plant growth-promoting bacteria (PGPR): Role of bacterial diversity. *Microorganisms* 9, 1988. https://doi.org/10.3390/microorganisms9091988

Wang, L., He, Z., Zhao, W., Wang, C., Ma, D., 2022. Fine soil texture is conducive to crop productivity and nitrogen retention in irrigated cropland in a desert-oasis ecotone, Northwest China. *Agronomy* 12, 1509. https://doi.org/10.3390/agronomy12071509

Wang, Q., Chen, L., Xu, H., Ren, K., Xu, Z., Tang, Y., Xiao, J., 2021b. The effects of warming on root exudation and associated soil N transformation depend on soil nutrient availability. *Rhizosphere* 17, 100263. https://doi.org/10.1016/j.rhisph.2020.100263

Wang, Q., Li, Y., Klassen, W., 2003. Effects of soil amendments at a heavy loading rate associated with cover crops as green manures on the leaching of nutrients and heavy metals from a calcareous soil. *J. Environ. Sci. Health Part B* 38, 865–881. https://doi.org/10.1081/PFC-120025566

Wei, X., Xie, B., Wan, C., Song, R., Zhong, W., Xin, S., Song, K., 2024. Enhancing soil health and plant growth through microbial fertilizers: Mechanisms, benefits, and sustainable agricultural practices. *Agronomy* 14, 609. https://doi.org/10.3390/agronomy14030609

Wu, G., Liu, Y., Xu, Y., Zhang, G., Shen, Q., Zhang, R., 2018. Exploring elicitors of the beneficial *Rhizobacterium bacillus* amyloliquefaciens SQR9 to induce plant systemic resistance and their interactions with plant signaling pathways. *MPMI* 31, 560–567. https://doi.org/10.1094/MPMI-11-17-0273-R

Yu, G.-H., Kuzyakov, Y., 2021. Fenton chemistry and reactive oxygen species in soil: Abiotic mechanisms of biotic processes, controls and consequences for carbon and nutrient cycling. *Earth Sci. Rev*. 214, 103525. https://doi.org/10.1016/j.earscirev.2021.103525

Yu, Y., Gui, Y., Li, Z., Jiang, C., Guo, J., Niu, D., 2022. Induced systemic resistance for improving plant immunity by beneficial microbes. *Plants* 11, 386. https://doi.org/10.3390/plants11030386

Zhang, Y., Wang, Y., Wen, W., Shi, Z., Gu, Q., Ahammed, G.J., Cao, K., Shah Jahan, M., Shu, S., Wang, J., Sun, J., Guo, S., 2021. Hydrogen peroxide mediates spermidine-induced autophagy to alleviate salt stress in cucumber. *Autophagy* 17, 2876–2890. https://doi.org/10.1080/15548627.2020.1847797

Zhu, L., Huang, J., Lu, X., Zhou, C., 2022. Development of plant systemic resistance by beneficial rhizobacteria: Recognition, initiation, elicitation and regulation. *Front. Plant Sci.* 13. https://doi.org/10.3389/fpls.2022.952397

7 Exploring Microbial Interactions for Plant Pathogen Control

Strategies and Methods

Preedhi Kapoor, Bilal Ahmad Mir, Akanksha Shukla, Maridul Kundan, Gurmeen Rakhra, Muhammad Aamir Manzoor, and Zishan Ahmad

7.1 INTRODUCTION

Plant pathogens, such as fungi, bacteria, and viruses, pose a huge threat to the productivity of agriculture worldwide. Traditional chemical control measures, while effective in many cases, can have negative environmental and health implications (Singh et al. 2023). As a result, there is a huge demand for sustainable and environmentally friendly techniques for plant disease management. One promising strategy is to harness the power of microbial interactions to control plant pathogens (Iqbal et al. 2023). Microbial communities, both beneficial and pathogenic, exist in complex and dynamic relationships within and around plants. By understanding these interactions, researchers and farmers can manipulate microbial communities to promote plant health and suppress disease. This approach, known as microbial control or biocontrol, offers an array of advantages over traditional chemical techniques. It is often more selective and targets specific pathogens while minimizing harm to beneficial organisms. Additionally, microbial control agents can be renewable, reducing dependence on synthetic chemicals (Nadarajah and Abdul Rahman 2021). Given the critical need for effective methods to control plant pathogens and diseases, improving productivity is essential not only for food production but also for other plant-derived materials such as fiber, timber, oils, and medicines. Food output must increase by 70% by 2050 to meet the demands of a world population that is expanding at a rapid rate. This will assuage widespread worries about food security. It is crucial to adopt sustainable agricultural practices to mitigate biodiversity loss, currently at 60%, and reduce greenhouse gas emissions, which account for 25% of the total (Muhie 2022). Currently, most plant protection strategies rely on antibiotics and chemical treatments. Although these approaches offer immediate disease control, plants often develop resistance and the crop systems accumulate chemicals. These limitations highlight the pressing need for sustainable, eco-friendly disease management practices in agriculture (Pandit et al. 2022).

This chapter will delve into the various strategies and methods employed in microbial control of plant pathogens. Different types of beneficial microorganisms can be used for biocontrol, including bacteria, fungi, and viruses. We will also explore various biocontrol methods for combating plant diseases, along with emerging strategies that demonstrate enhanced potential for controlling a wide range of pathogens, which could contribute to achieving long-term sustainability objectives.

DOI: 10.1201/9781003608080-7

7.2 PLANT PATHOGEN TYPES

These microbes extract energy and nutrients directly from the cells of plants, frequently establishing close relationships with the host without triggering rapid cellular death (Pandit et al. 2022). These interactions, in turn, shape the plant's response to the pathogens (Panthapulakkal Narayanan et al. 2020).

7.2.1 Biotrophic Pathogens

Biotrophic pathogens usually feed on the living cells of host plants through intricate mechanisms that allow them to tap into the resources of plants. Some biotrophs have developed to the point where they are no longer able to thrive on artificial medium (nonliving), and as a result, they have become obligatory biotrophs. For example, *Uromyces fabae*, responsible for causing rusts, and *Blumeria* (*Erysiphe*) *graminis*, responsible for powdery mildews.

7.2.2 Necrotrophic Pathogens (Pathogens That Feed on Necrotic Tissue)

Unlike biotrophs, necrotrophic microorganisms appear to be opportunistic, nonspecialized pathogens that rapidly kill their hosts and consume their decomposing tissues (Pandit et al. 2022). These pathogens do not form haustoria but instead enter plants through natural wounds, releasing phytotoxins and lytic enzymes. They can easily be cultured on artificial media. These pathogens primarily attack plants that are young, weak, or damaged and can survive as saprotrophs. Both these fungal and bacterial necrotrophs follow a similar infection process involving host penetration, attachment, and eventual necrosis and plant tissue decay. For example, necrotrophs (fungal) include *Botrytis*, which leads to gray mold; *Cochliobolus*, which results in blight on corn leaves; and *Alternaria*, responsible for early blight in potatoes.

7.2.3 Hemibiotrophic Pathogens

Hemibiotrophic pathogens are unique as they exhibit traits of both necrotrophs and biotrophs and are known for transitioning between these two modes of infection (Pandit et al. 2022). They begin with an asymptomatic biotrophic phase and later shift to a destructive necrotrophic phase, during which they suppress the host's immune response at the appropriate moment, causing extensive damage that leads to the host's decay and death. Examples of hemibiotrophic fungi include *Pythium*, *Colletotrichum*, *Magnaporthe grisea*, *Venturia*, *Fusarium*, and *Phytophthora*, as well as the bacterium, *Pseudomonas syringae*, all of which start with a biotrophic interaction but ultimately switch to killing host cells in the necrotrophic phase.

7.3 PLANT–MICROBE INTERACTION TYPES

7.3.1 Mutualistic Interactions

Fungi that are mycorrhizal form symbiotic relationships with the roots of a plant to facilitate the exchange of nutrients. In return for the plant's carbohydrates, the fungi enhance the plant's uptake of water and nutrients, especially phosphorus and nitrogen (Bonfante, 2010). These fungi also colonize crops such as vegetables, fruits, ornamental plants, and cereals. In endomycorrhiza, the roots resemble normal plant roots in appearance, while the fungal hyphae penetrate the cortical cells of feeder roots, forming arbuscules (specialized feeding structures) and occasionally vesicles, which store food (Delavaux et al. 2019). In return for nutrients, fungi obtain sugars (carbon) produced by plants through photosynthesis. Ectomycorrhizae metabolize plant carbohydrates like mannitol and trehalose, which most plants cannot utilize (Adewole et al. 2010). Studies by Eccher et al. (2009) and Sokolovski et al. (2002) reported improved growth and amino acid uptake in *Calluna vulgaris* (heather) and *Vaccinium corymbosum* (highbush blueberry) due to the presence of *Hymenoscyphus ericae* mycorrhizae. This symbiotic relationship provides mutual benefits, supporting plant growth and potentially mitigating global warming (Bender et al. 2014). On the roots of leguminous plants, nitrogen-fixing bacteria belonging to the genus *Rhizobium* form nodules that transform atmospheric nitrogen into a form that plants can use (Wang et al. 2018). According to Vance (2001), interactions between species of *Rhizobium*, *Bradyrhizobium*, *Sinorhizobium*, *Azorhizobium*, *Mesorhizobium*, *Allorhizobium*, and leguminous plants contribute to about 80% of biologically fixed nitrogen. In this mutualistic relationship, bacteria supply nutrients to the plants, while the plants receive nitrogen in a reduced form from the microorganisms.

7.3.2 Pathogenic Interactions

Eight species of bacteria, 30 types of viruses, and 120 genera of fungi are responsible for the 11,000 known plant diseases (Nazarov et al. 2020). In contrast to the less common occurrence of fungal infections in animal diseases, fungi are responsible for a significant number of plant diseases. Numerous fungi can infect plants, leading to diseases that impair their growth, development, and productivity (Jain et al. 2019). Examples of diseases include rusts (*Puccinia triticina* affecting leaves), smuts (*Ustilago* species), and powdery mildew (genera *Erysiphe*, *Microsphaera*, *Phyllactinia*, *Podosphaera*, *Sphaerotheca*, and *Uncinula*). Bacteria can also harm plants, causing diseases such as fire blight, bacterial wilt, and bacterial blight. Plants can be infected by viruses, which may lead to symptoms like necrosis, stunted growth, and mosaic patterns. These small agents being infectious only replicate within living plant tissues and are typically composed of genetic material enclosed in a protein coat. Virus-infected plants often display various symptoms, including mosaic patterns on leaves, reduced growth, yellowing (chlorosis), and tissue death (necrosis). Understanding plant–virus interactions is significant for the discovery of strategies to protect crops and ensure food security.

7.3.3 Neutral or Commensalism Interactions

Certain bacteria inhabit plant surfaces without inflicting damage or providing significant benefits to the plant. These interactions may be impartial, exerting negligible influence on the plant's growth. Commensalism is a partnership in which one organism obtains benefit while the other stays apathetic, sustaining neither substantial harm nor advantage (Lewis et al. 2023). The bacterium receives a benefit from its relationship with the plant, but the plant experiences no significant beneficial or detrimental impacts. A neutral contact happens when the existence of a particular organism does not significantly impact the other's existence.

7.3.4 Endophytic Interactions

Endophytic connections denote beneficial interactions that are associated with plants and microbes, typically fungi or bacteria, residing within the tissues of plants without inflicting obvious damage. These endophytes can improve the growth of plants by increasing stress tolerance, nutrient absorption, and resistance to infections (Wippel 2023). Utilizing endophytes as biofertilizers, biopesticides, and promoters of plant stress tolerance offers an environmentally friendly alternative to chemical fertilizers. Furthermore, endophytes enhance ecosystem biodiversity and assist plants in adapting to diverse conditions (Sharma et al. 2023).

7.3.5 Allelopathy

When a plant exudes substances in the vicinity that affect the development, germination, and growth of other species of plants, this is known as allelopathy. Allelopathic substances may be exudated via volatilization, leaching, root exudation, or the breakdown of wastes of plants. Some plants release substances into the soil that can inhibit the growth of bacteria or other plants that compete with them. The surrounding vegetation may be impacted by this interaction in both positive and negative ways. Allelochemicals are secondary metabolites generated by plants that affect the growth, survival, and reproduction of nearby plants. These include phenolics, alkaloids, and terpenoids. Microorganisms and plants can create feedback loops through allelopathic interactions (Kong et al. 2019). For instance, certain plants may selectively attract or suppress specific microbial communities in their rhizosphere to enhance their allelopathic effects.

7.3.6 Phytoremediation

Phytoremediation refers to the use of plants and their associated microbes for environmental cleanup (Kafle et al. 2022). In this process, contaminants are absorbed, accumulated, or transformed by plants, frequently with the assistance of particular microbes. Phytoremediation is an effective method for addressing various organic and inorganic contaminants. Many organic pollutants in the environment are synthetic and regarded as xenobiotics by living organisms.

7.4 THE IMPACT OF CLIMATE CHANGE ON PLANT PATHOGENS AND THEIR INTERACTIONS WITH HOST PLANTS

7.4.1 Temperature

Physiological processes, plant growth, and development are all impacted by temperature, which also changes the variation, life cycle, and pathogenicity of plant diseases (Lahlali et al. 2024) (Figure 7.1). Higher temperatures generally encourage the proliferation of diseases associated with plants, especially those that are caused by bacteria and fungi, therefore weakening the plant's defenses (Mwangi et al. 2023). Given that many illnesses exhibit sensitivity to high temperatures, it is the need of the hour to understand that the optimal temperature range for pathogen–host interactions might vary (Porras et al. 2023; Skendžić et al. 2021). Research indicates that certain crop diseases, such as citrus canker (*Xanthomonas* spp.), rice blast (*Magnaporthe grisea*), potato late blight (*Phytophthora infestans*), and wheat rust (*Puccinia triticina*), can become more severe in response to rising temperatures (Angelotti et al. 2024; Lahlali et al. 2024).

7.4.2 Humidity

Humidity is a crucial environmental factor that influences plant illnesses. Mechanisms like transpiration, stomatal conductance, and evapotranspiration may be altered in host plants, as well as the emergence, illness, and persistence of plant infections (Velásquez et al. 2018). In general, increasing humidity speeds up the proliferation of the diseases associated with plants, especially those brought on by bacteria and fungi, while decreasing the efficiency of host defense systems and management strategies (Zhou et al. 2024) (Figure 7.1). The incidence and severity of foliar and fruit diseases may be exacerbated by higher humidity. Additionally, it may increase the likelihood of postharvest illnesses including soft rot (*Pandanus conoideus*) and gray mold, which could result in significant losses during storage and transit (Scortichini 2022). On the other hand, lower humidity may make diseases like rust and smut less common and less likely to spread, but it may also make the host species more susceptible to water stress and other diseases like canker and wilt. Additionally, lower

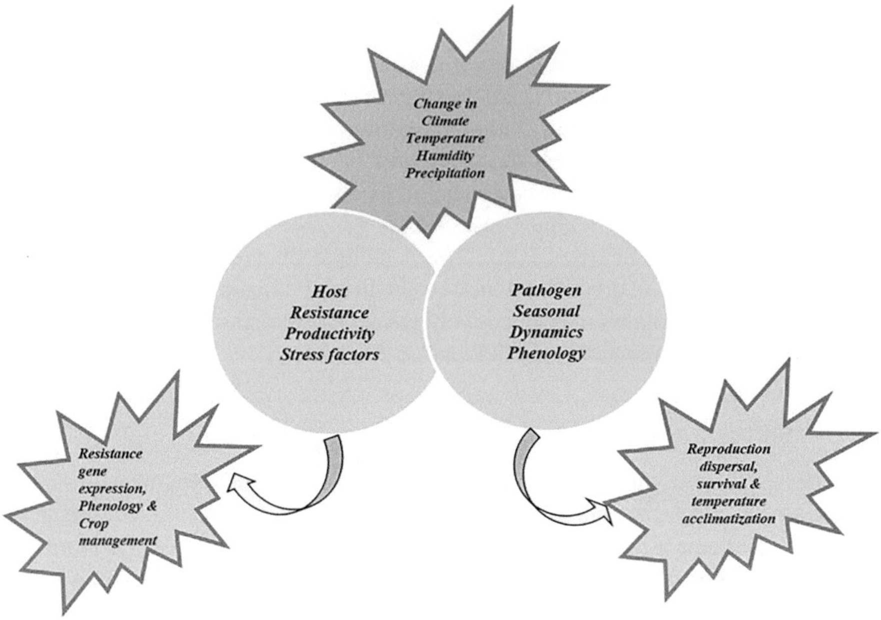

FIGURE 7.1 Impacts of climate change on the interactions between hosts and pathogens.

humidity may have an impact on the quantity and quality of nectar and pollen, which could impair host plant pollination and reproduction (Descamps et al. 2021).

7.4.3 Severe Weather Events

Severe weather events like physiological stress, physical damage, and altered biochemical reactions in host plants have a great effect on the diseases of plants. They also affect the diversity and population dynamics of pathogens of plants and their related vectors (Singh et al. 2023). These events, which are growing more persistently due to climate change, include storms, strong heat waves, freezing temperatures, and hailstones (Faranda et al. 2022) (Figure 7.1). Generally, extreme weather makes host plants more susceptible to diseases, which lowers the control strategies' efficacy (Gullino et al. 2022). Cold snaps and extreme temperatures cause oxidative stress and heat, which hinders respiration and photosynthesis and makes people more vulnerable to diseases like black rot, spots on the leaves, and powdered mildew (Zhang et al. 2023).

7.4.4 Precipitation

A crucial environmental component, precipitation affects the availability of water and nutrients, moisture of soil, pH levels, leaching of pesticides, and the development and dissemination of inoculum, all of which have an impact on plant diseases (Iqbal et al. 2023) (Figure 7.1). The timing, frequency, severity, and type of precipitation, and how each interacts with other meteorological features like humidity and temperature all affect how precipitation affects plant diseases (Lahlali et al. 2024). Root rot, damping-off, leaf spot, and wilt are among the soil-borne and foliar diseases that can become more prevalent and severe when precipitation levels are high (J. Zhou et al. 2024). On the other hand, less precipitation may make diseases like powdery mildew and downy mildew less common and less likely to spread, but it may also make host plants more susceptible to drought stress and other diseases like those caused by *Fusarium* and *Verticillium* (Hossain et al. 2024).

7.5 PEPTAIBIOTICS: HARNESSING MICROBIAL SECONDARY METABOLITES TO ADDRESS PLANT PATHOGENS

Peptaibiotics are a varied group of linear, amphipathic polypeptides made up of 5–21 amino acids, synthesized by the fungal non-ribosomal peptide synthetase (NRPS) pathway, exhibiting molecular weights between 500 and 2200 Da. (Pereira-Dias et al. 2023) (Table 7.1). Peptaibiotic sequences often exhibit a large amount of non-coded residues, such as Iva and Aib, with the latter being especially prevalent (De Zotti et al. 2012). This unusual composition underlies the helical 3D structure of peptaibols (e.g., α-helix and 310-helix) (Ségalas et al. 1999). Alterations to these conformations may arise in the presence of Aib–Pro motifs, resulting in a variant of the 310-helix, namely, the (Xaa–Yaa–Aib–Pro)-β-bend ribbon spiral (Christoffersen et al. 2015). These characteristics underpin the exceptional qualities of peptaibols concerning both antimicrobial activity and resistance to proteolysis (De Zotti et al. 2020). The N-terminal residue in peptaibols is often acetylated, while the C-terminal member is frequently an alcohol rather than a carboxyl, specifically an α-amino alcohol such as leucinol, phenylalaninol, or valinol, among others. In contrast, lipopeptaibols feature an N-terminal amino acid acylated with a short fatty acid (e.g., octanoic, decanoic, or cis-dec-4-enoic acids), while lipoaminopeptaibols generally have their N-terminal amino acid acylated by long-chain, α-methyl-branched fatty acids (Neumann et al. 2015).

Among peptaibiotics, peptaibols are the most extensive and well-researched category, including variants such as lipopeptaibols and lipoaminopeptaibols (Ramawat 2022). Peptaibiotics are categorized based on their amino acid sequence length: Long-sequence peptaibiotics have 17–21 residues (e.g., alamethicin, trichosporin, trichorzin, and chrysospermin), medium-sequence peptaibiotics have 12–16 residues (e.g., emerimicin, harzianins, and tylopetins), and short-sequence peptaibiotics have 4–11 residues (e.g., trichogin and trichodecenin) (Hou et al. 2022; Bortolus et al. 2015). Long-sequence peptaibiotics frequently feature central proline (Pro) and glutamine (Gln)

TABLE 7.1
Some Peptaibiotics Screened for Antimicrobial Activity against Phytopathogenic Bacteria, Fungi, and Viruses.

Peptaibiotics	Source Organisms	Target Organisms	Mode of Action	References
TvB I-II	*Trichoderma virens*	*Pseudomonas syringae pv Lachrymans*	Activation of plant defenses	Viterbo et al. (2007)
Trichokonins VI–VIII	*Trichoderma pseudokoningii*, *Trichoderma longibrachiatum*	*Erwinia carotovora, Xanthomonas oryzae pv. oryzae*; *Clavibacter michiganensis* ssp. *michiganensis*; *Pectobacterium carotovorum* ssp. *carotovorum*; *Ralstonia solanacearum*	Activation of plant defenses, rupture of the cell membrane	Zhang et al. (2022); Li et al. (2014)
Alamethicin F301	*Trichoderma* sp.	*Phoma destructiva*	Rupture of the cell membrane	Grigoriev et al. (2003)
Harzianins HC	*Trichoderma harzianum*	*Sclerotium cepivora*	Rupture of the cell membrane	Rebuffat (1995)
Texenomycin A	*Moszia lindtneri*	*Phoma destructiva*	Rupture of the cell membrane	Grigoriev et al. (2003)
Peptavirins A and B	*Apiocrea* sp.	Tobacco mosaic virus	Not known	Trojak-Goluch (2024)

residues near both ends, whereas medium-sequence peptaibiotics often contain an N-terminal Ac–Aib–Gln-segment or a similar structure with Gln replaced by asparagine (Asn). Both types commonly exhibit Aib–Pro motifs. In contrast, short peptaibiotics typically have a high glycine (Gly) content (Gavryushina et al. 2021).

Considering their antimicrobial properties, the mechanisms of action (MoAs) of peptaibiotics remain inadequately elucidated (Adam et al. 2018). According to the literature, peptaibiotics primarily affect microorganisms through two mechanisms of action: (i) Direct interaction with biological membranes (Bortolus et al. 2013; Khare et al. 2018) or (ii) intracellular interference by hindering or causing disruption to processes related to metabolism (e.g., nucleic acid synthesis, essential enzyme function, or other mechanisms) (Sella et al. 2021; Zhao et al. 2018), and by inducing systemic immune responses in plants (Li et al. 2014). The MOAs elucidate the powerful and rapid bioactivity of peptaibols, obstructing any efforts by pathogens to mitigate their effects (Adam et al. 2018).

7.6 THE SIGNIFICANCE OF PLANT MICROBIAL COMMUNITY IN AGRICULTURAL SUSTAINABILITY

7.6.1 Positive Interactions with Plant Microbiomes in Agriculture

Good plant microorganisms are essential for the development of plants, blossoming time, and yields of crops both directly and indirectly (Busby et al. 2017; Berg and Koskella 2018). Microbial reactions significantly influence the effects of environmental changes on the agricultural sector, leading to increased interest in utilizing plant-associated microbes to alleviate the detrimental effects of climate modifications on sustainable farming and the production of food (Singh et al. 2023). The outcome related to a particular plant–pathogen interaction depends upon its impact on plant overall health, which can be advantageous during certain conditions and detrimental under others. Notable instances of advantageous plant–microbe associations include the symbiotic relationships between legumes and nitrogen-fixing rhizobia, as well as arbuscular mycorrhizal (AM) fungal taxa, which assist host plants in acquiring nitrogen and phosphorus, respectively, in nutrient-deficient conditions (Cao et al. 2017).

7.6.2 Detrimental Plant–Microbe Interactions

Certain microbes adversely affect plants, leading to disease symptoms. Notable examples include *Pseudomonas syringae*, *Erwinia amylovora*, *Ralstonia solanacearum*, *Xanthomonas* spp., and *Xylella fastidiosa*. Pathogenic microbes of plants penetrate the tissues of the plants via natural openings or wounds to obtain nutrients and elicit immunity-related responses (Zoledowska et al. 2019). Multiple factors influence the results of plant–pathogen interactions, including the size of the population, host susceptibility, climate conditions, and biotic elements such as plant microbial communities (Brader et al. 2017).

7.7 RECENT METHODOLOGIES FOR INVESTIGATING PLANT–MICROBE INTERACTIONS

Comprehending the common processes of the specific plant–microbial community is an appropriate strategy for its application in the agricultural sector, given that the plant-related microbiota significantly affects the phenotype of hosts. Specifically, examining the plant–microbial genes implicated in the plant–microbe relationships is essential for the future utilization of the plant microbial community. Technologies associated with omics, and NGS technologies have greatly enhanced our awareness of plant microbes, encompassing both plant growth-promoting microorganisms and phytopathogens (Basu et al. 2018; Jansson and Hofmockel 2018). Genetic makeup changes in the beneficial microbiome population have been seen in plant–pathogen interactions (Fürnkranz et al. 2012). While certain metagenomic and proteogenomic studies have yielded assembly-level data on enhanced functional classifications, alternative methodologies are necessary to provide the functionality of genes or proteins. One study area regarding plant–microbe use in agriculture is on the utilization of microbial consortia, while another theme pertains to specific genetic modification (GM) of either plants or microbes. Historically, GM approaches, in conjunction with gene silencing, have been extensively employed to investigate the functions of genes or enhance traits. Transgenics offers a rapid solution; nevertheless, the incorporation of foreign genetic material restricts its broad application due to regulatory challenges (Muñoz et al. 2019). Genome editing (GE) tools are significant as they enable scientists to edit genomic sequences with greater precision without the incorporation of foreign genes. GE technology utilizes endonucleases that are engineered to induce a double-strand break (DSB), which is subsequently repaired by endogenous mechanisms, resulting in mutations (Gaj et al. 2013). Double-strand break (DSB) repair mechanisms are facilitated by two primary pathways: Nonhomologous end-joining (NHEJ) and homology-directed repair (HDR). HDR is a precise mechanism, although it is less prevalent than NHEJ and is used in specific contexts of donor-dependent gene replacement.

GMs may be achieved via various methods, with three types of meganucleases—transcription activator-like effector nucleases (TALENs), zinc finger nucleases (ZFNs), and the CRISPR-Cas system—being the most prevalent in recent applications. GE utilizing ZNFs and TALENs relies on the capacity of DNA-binding domains to specifically identify nearly any target DNA sequence. The gene editing capability of ZNFs/TALENs is primarily determined by the DNA-binding affinity and accuracy of the assembled zinc-finger and transcription activator-like effector (TALE) proteins. The CRISPR-Cas system offers advantages over ZFNs and TALENs regarding ease of design, versatility, cost-effectiveness, efficiency, multiplexing capabilities, and specificity (Brandt and Barrangou 2019). Since the introduction of CRISPR-Cas mechanisms to genetic engineering in eukaryotes, it has become the predominant tool in this field, often referred to as nature's toolbox and a "magical wand" of genome engineering.

7.8 EMERGING BIOCONTROL METHODS

7.8.1 Microbial Volatilome and Its Function in Biocontrol

Utilizing microorganisms as biocontrol agents (BCAs) is among the most effective and promising biocontrol techniques. The generation of volatile organic compounds (VOCs) is notably advantageous among the several microbiological techniques utilized by BCAs when direct connections between pathogens and their competitors are problematic (Pandit et al. 2022). All living creatures emit VOCs, which may be employed to manage plant pathogens. VOCs offer a sustainable alternative to chemical-based fungicides because of their simple means of usage, and minimum environmental and residue from plants (Tilocca et al. 2020). Tahir et al. (2017) argued that VOCs generated by *Bacillus* species are efficacious against the tobacco wilt disease, caused by *Ralstonia solanacearum*. This indicates that bacterial volatiles contribute to the biocontrol capabilities of *Bacillus*, possibly functioning through both direct and indirect processes, and that these VOCs may behave as complex signals influencing both host–pathogen interactions (Bailly and Weisskopf 2017).

7.8.2 Microbiome-Related Techniques for Plant Protection

Recent studies underscore considerable diversity in microbes among different plants, uncovering distinct bacteria capable of combating phytopathogens. Mosses, as some of the earliest terrestrial plants, possess a unique microbial population and harbor several hostile bacteria. Likewise, therapeutic and endemic plants provide potential reservoirs of unusual microbial biodiversity and antagonistic organisms. This distinctiveness originates from their particular metabolism, which affects the composition of their plant microbiome. Recent discoveries suggest that seed endophytes are likely to serve as significant sources of novel BCAs. Thus far, bacteria and fungi have predominantly been utilized for biocontrol. Microbial invasion can impact the network of microbes associated with plant life. Exploring emerging network models involving soil and plant microbial communities could give new prospects for managing diseases. While prior methods frequently relied on single species, which generated uneven outcomes, microbiome-based biocontrol tactics are now becoming practical.

In the years to come, microbial consortia and biocontrol products may be utilized to promote ecological diversity through microbiome engineering, targeting particular microbiome results. Crop-specific microbial consortia might be produced using a variety of BCAs similarly. By adopting a broad perspective and employing microbiome-based solutions, one can formulate targeted and predictive biocontrol techniques. Additionally, integrating breeding methods with biocontrol measures is vital for sustaining ecosystem variation and stability. These fundamental strategies are vital for minimizing the degradation of biodiversity while fostering sustainable agriculture methods (Pandit et al. 2022).

7.8.3 Phage Cocktail

Phage cocktails offer a promising technique for managing numerous diseases of plants, but attaining robust biological control requires overcoming certain technological barriers (Farooq et al. 2022). A comprehensive knowledge of the connections of crops, phages, and diseases is crucial, considering the distinctive and intricate characteristics of each plant's environment. This awareness may only be attained by lengthy field-related experiments, as tests in the lab may not adequately duplicate situations from the real world. Harnessing known phages from the plant's surroundings, such as those found in the phyllosphere, may boost defenses against local phytopathogens. The addition of light-absorbing compounds or protective formulations may help phages withstand UV degradation and increase their endurance in the phyllosphere. Because phytobacteria are so diverse, it is not possible to create a single universal phage cocktail. Creating assays to precisely identify pathogenic bacteria and their corresponding hostile phages could greatly enhance control measures. At present, there is no straightforward and economical method available for this goal.

7.8.4 Genetically Modified Biocontrol Agents

Genetic engineering approaches can be utilized across many organisms to boost the potency of BCAs. BCAs are naturally occurring organisms, including bacteria, fungi, viruses, and certain insects, employed to manage plant diseases, pests, or weeds by suppressing their growth, reproduction, or viability. They are seen as a safer and more eco-friendly option to synthetic pesticides, as they operate by natural processes such as competition, parasitism, predation, or the synthesis of antimicrobial chemicals (Haq et al. 2024). In the realm of genetic engineering, BCAs can be augmented to enhance their efficacy and specificity. Bacteria or fungi utilized as BCAs can be engineered to generate elevated concentrations of antimicrobial chemicals to augment their capacity to colonize plant roots or boost their stress tolerance, hence increasing their resilience and efficacy across diverse surroundings. Locating a gene that encodes the enzyme chitinase from *Serratia* to a *Pseudomonas* endophyte has demonstrated efficacy in controlling *Rhizoctonia solani* infections in beans. Likewise, the incorporation of a glucanase gene into *Trichoderma* has imparted resistance to pathogens such as *Pythium*, *Rhizoctonia*, and *Rhizopus*. Loper et al., (2007) effectively cloned the 2,4-diacetyl phloroglucinol (2,4-DAPG) biosynthetic gene cluster phlACBDE from strain CPF-10 into a mini-Tn5 transposon, subsequently integrating it into the chromosome of *Pseudomonas fluorescens* P32. Researchers ensure the sustained inheritance of desirable features across numerous generations by integrating the gene cluster into the chromosome with a mini-Tn5 transposon. The synthesis of 2,4-DAPG, a powerful antibiotic with significant antibacterial characteristics, enables the modified *Pseudomonas fluorescens* P32 to efficiently suppress a wide range of plant diseases, hence enhancing its effectiveness as a BCA. This genetic alteration enhances the strain's biocontrol efficacy and promotes sustainable agriculture practices by diminishing dependence on chemical pesticides. Moreover, the transfer of biosynthetic genes to other strains yields significant insights into gene activity and regulation, hence enhancing the creation of more effective and efficient BCAs for commercial applications.

This genetic modification enhanced the resistance of wheat to *Gaeumannomyces graminis* var. *tritici* and of tomato to *Ralstonia solanacearum* bacterial wilt. Clermont et al. (2011) used genome shuffling to develop enhanced biocontrol strains of *Streptomyces melanosporofaciens* EF76. After two rounds of genome shuffling, they obtained four strains with enhanced antagonistic activity against potato diseases caused by *Phytophthora infestans* and *Streptomyces scabiei.* Chemical mutagenesis can also improve the capacity for biological control. For example, nitrosoguanidine mutagenesis was used to create a strain of *Pseudomonas aurantiaca* B-162 with increased phenazine production and improved biocontrol activity (Feklistova and Maksimova 2008). Furthermore, genome editing can simplify the regulatory approval process, potentially aiding in the commercialization of BCAs.

7.8.5 Microbiome Engineering

Numerous researchers regard the microbiome as a "second genome", while others employ the term "holobiont" to characterize the varied microbial communities linked to animals and plant hosts (Zilber-Rosenberg and Rosenberg 2008).Microbiome engineering could profoundly alter agriculture, demanding the generation of microbiomes that have certain desirable features. Recent research indicates that specific strains that are endophytic can diminish the composition and species diversity inside the tissues of plants (Patel and Archana 2017). Practically, establishing microbial communities that are simultaneously resistant and resilient to stress could considerably boost productivity in agriculture. Microbiome engineering comprises the modification of microbial populations to enhance health or cure specific challenges, exemplified by engineered probiotics, fecal microbiota transplantation (FMT), and the engineering of soil microbiomes for agricultural purposes (Han and Yoshikuni 2022).

7.8.6 Mycoviruses as Agents for Biocontrol

Mycoviruses are pathogens that infect fungi, living as either double-stranded or single-stranded RNA or DNA inside the fungal cytoplasm. In contrast to most viruses, they inflict minimal damage on their fungal hosts, yet they can modify fungal physiology, including decreased virulence, positioning them as possible BCAs against fungal infections. Recently, mycoviruses infecting fungal pathogens have been recognized as potential biological agents of control for plant diseases. These viruses are known to produce hypovirulence, thereby lowering the infectious capacity of their fungal hosts (García-Pedrajas et al. 2019; Khalifa and Macdiarmid 2021). García-Pedrajas et al. (2019) indicated that the majority of mycoviruses in filamentous fungi possess double-stranded RNA (dsRNA) genomes or positive-sense single-stranded RNA (ssRNA) genomes featuring dsRNA replicative intermediates. They could possess real virion capsids or occasionally be capsids. Mycoviruses do not possess an extracellular transmission pathway; they are predominantly transferred through hyphal anastomosis or conidia during vertical transmission, enabling their transfer to progeny. Hypovirulence-associated mycoviruses diminish the potential for infection of their fungal hosts; however, their transmission among vegetatively incompatible groups poses significant challenges. This has made it challenging to create commercial mycovirus biocontrol techniques for fungi that are phytopathogenic (Khalifa and Macdiarmid 2021).

7.9 MICROBIAL NANOTECHNOLOGY ENHANCES AGRICULTURAL RESILIENCE TO CLIMATE CHANGE

Microbes are becoming pivotal in eco-friendly nanotechnology by offering a cost-effective and sustainable alternative to traditional methods that utilize hazardous chemicals and require high energy consumption. However, it is crucial to recognize that not all microbes are harmless; some may be pathogenic or generate harmful byproducts, necessitating meticulous selection and maximizing the efficiency for the safe and productive formation of nanomaterials (Kashyap et al. 2018). The capacity to produce nanoparticles (NPs) is peculiar to certain species and frequently varies by strain since only some bacteria, fungi, algae, and viruses have the necessary metabolic or enzymatic systems for NP production (Table 7.2). The microbial generation of NPs often encompasses bioreduction, biosorption, or precipitation processes. Certain bacteria and fungi synthesize enzymes or metabolites that may lower metal ions to their nanoscale elemental states, leading to the generation of NPs (Kashyap et al. 2013). The cell wall architecture of some microorganisms can act as nucleation sites for NP processing. They specialize in gathering and eliminating heavy metals via reductase enzymes, which can transform metal salts into NPs with accurate distribution of size and decreased polydispersity (Kashyap et al. 2013). The utilization of bacteria for NP formation is currently a fast-expanding topic in green nanotechnology, offering an environmentally friendly substitute to standard chemical and physical processes (Kashyap et al. 2018). By refining these techniques, scientists can rapidly and efficiently generate NPs with precise shapes and regulated sizes. The bacterial manufacture of metal NPs is acknowledged as an innovative, effective, and eco-friendly method. Psychosynthesis is a highly efficient process for generating stable and readily produced organic NPs. Fungi, possessing important metabolites, provide superior bioaccumulation capabilities and are readily cultivable,

TABLE 7.2
Biosynthesis of Nanoparticles Utilizing Microbiota.

Bacteria	Nanoparticles	References
Klebsiella pneumonia	Ag	Shahverdi et al. (2007)
Lactobacillus spp.	Gold	Nair and Pradeep (2002)
Clostridium thermoaceticum	Cadmium sulfide	Cunningham and Lundie (1993)
Nocardiopsis sp. MBRC-1	Ag	Manivasagan et al. (2013)
Serratia nematodiphila	Au	Malarkodi et al. (2013)
Fungi	Ag	Gajbhiye et al. (2009)
Alternaria alternata		
Alternaria fumigatus	ZnO	Raliya and Tarafdar (2013)
Candida albicans	Au	Chauhan et al. (2011)
Coriolus versicolor	Magnetite	Chen et al. (2011)
Fusarium oxysporum	Gold	Anitha and Palanivelu (2011)
Algae	Ag	Fathy (2016)
Acanthophora spicifera		
Shewanella sp.	Pt	Konishi et al. (2007)
Turbinaria conoides	Au	Vijayaraghavan et al. (2011)
Shewanella putrefaciens	Magnetite	Lovley et al. (1987)
Gelidiella acerosa	Ag	Vivek et al. (2011)
Virus	Ca	Bailey et al. (2013)
Bacteriophage		
Tobacco mosaic virus	Silica	Fernandes et al. (2014)
M13 bacteriophage	Ni	Oh et al. (2014)

rendering them optimal for economical NP synthesis with reduced subsequent processing. Fungi exhibit more metal absorption than bacteria, mostly owing to their elevated wall-binding potential for metal salts, facilitating high-yield NP production. Three primary mechanisms have been proposed to elucidate fungal NP synthesis: The activity of nitrate reductase, electron-shuttle quinones, or a combination of the two. Fungal enzymes, particularly nitrate reductase and NADPH-dependent reductases from *Penicillium* species and *Fusarium oxysporum*, are essential in the creation of NPs. *Fusarium oxysporum* has been thoroughly investigated for its capability to synthesize metallic NPs, especially those based on silver.

The extensive prevalence and accessibility of algae in nature render them an exceptional alternative for the synthesis of metallic NPs. The development of NPs is contingent upon dosage and varies according to the specific species of algae utilized. Various biomolecules, including polysaccharides, peptides, and pigments, engage in metal reduction. Proteins stabilize and cap metal NPs in aqueous solutions via amino acid sequences, residues of cysteine, and sulfated polysaccharides. Significantly, employing algae for NP manufacturing is more efficient in time than alternative biological approaches. Various seaweeds, notably *Sargassum wightii* and *Fucus vesiculosus*, have been exploited for producing silver nanoparticles (AgNPs) in an array of shapes and sizes (Mata et al. 2009). Viral nanoparticles (VNPs) are often tiny, monodisperse, stable, and resilient, facilitating production on a massive scale. Furthermore, they can be altered using GM or chemical biological conjugation methods. Furthermore, tobacco mosaic virus has effectively enabled the process of mineralization of lead sulfide (PbS) and cadmium sulfide (CdS) crystalline nanowires.

7.10 CONCLUSION

In conclusion, this chapter explores interactions between the microbes and offers a promising avenue for innovative strategies in plant pathogen control. Beneficial microorganisms, such as bacteria, fungi, viruses, and phages, participate in intricate and dynamic interactions with plants, frequently promoting plant development, immunity, and resistance against a wide array of diseases. Moreover, the integration of microbial biocontrol techniques with advanced technologies, including nanotechnology, provides novel opportunities for enhancing plant disease management. By harnessing the complex relationships between beneficial microbes and plants, researchers can develop sustainable and effective biocontrol methods that reduce reliance on chemical pesticides. Moreover, the integration of advanced techniques such as genomic analysis, metagenomics, and bioinformatics can enhance our understanding of these interactions, leading to the discovery of novel microbial agents with antagonistic properties against pathogens. Furthermore, genetic engineering and synthetic biology have promising opportunities to improve the potency and longevity of these agents, maximizing their function across different types of environments. As we continue to unravel the intricacies of microbial ecosystems, the potential for discovering new strategies for plant disease management becomes increasingly attainable, paving the way for resilient agricultural systems in the face of emerging plant pathogens.

REFERENCES

Adam, Catherine, Anna D. Peters, M. Giovanna Lizio, George F.S. Whitehead, Vincent Diemer, James A. Cooper, Scott L. Cockroft, Jonathan Clayden, and Simon J. Webb. 2018. "The Role of Terminal Functionality in the Membrane and Antibacterial Activity of Peptaibol-Mimetic Aib Foldamers." *Chemistry—A European Journal* 24 (9): 2249–56. https://doi.org/10.1002/chem.201705299.

Adewole, M.B., O.O. Awotoye, M.O. Ohiembor, and A.O. Salami. 2010. "Influence of Mycorrhizal Fungi on Phytoremediation Potential and Yield of Sunflower in Cd and Pb Polluted Soils." *Journal of Agricultural Sciences, Belgrade* 55 (1): 17–28. https://doi.org/10.2298/jas1001017a.

Angelotti, Francislene, Emília Hamada, and Wagner Bettiol. 2024. "A Comprehensive Review of Climate Change and Plant Diseases in Brazil." *Plants* 13 (17): 2447. https://doi.org/10.3390/plants13172447.

Anitha, Thoomatti Sudharsan, and Peramchi Palanivelu. 2011. "Synthesis and Structural Characterization of Polydisperse Silver and Multishaped Gold Nanoparticles Using *Fusarium oxysporum* MTCC 284." *Digest Journal of Nanomaterials and Biostructures* 6 (4): 1587–95.

Bailey, Jordan, Anthony Oliveri, and Edward Levin. 2013. "基因的改变NIH Public Access." *Bone* 23 (1): 1–7. https://doi.org/10.1021/bm200274r.Self-assembly.

Bailly, A., & Weisskopf, L. 2017. Mining the volatilomes of plant-associated microbiota for new biocontrol solutions. *Frontiers in Microbiology*, 8, 1638. https://doi.org/10.3389/fmicb.2017.01638.

Basu, Supratim, Roel C. Rabara, Sangeeta Negi, and Pratyoosh Shukla. 2018. "Engineering PGPMOs Through Gene Editing and Systems Biology: A Solution for Phytoremediation?" *Trends in Biotechnology* 36 (5). Elsevier Ltd.: 499–510. https://doi.org/10.1016/j.tibtech.2018.01.011.

Bender, S. Franz, Faline Plantenga, Albrecht Neftel, Markus Jocher, Hans Rudolf Oberholzer, Luise Köhl, Madeline Giles, Tim J. Daniell, and Marcel G.A. Van Der Heijden. 2014. "Symbiotic Relationships between Soil Fungi and Plants Reduce N_2O Emissions from Soil." *ISME Journal* 8 (6): 1336–45. https://doi.org/10.1038/ismej.2013.224.

Berg, Maureen, and Britt Koskella. 2018. "Nutrient- and Dose-Dependent Microbiome-Mediated Protection against a Plant Pathogen." *Current Biology* 28 (15). Elsevier Ltd.: 2487–2492.e3. https://doi.org/10.1016/j.cub.2018.05.085.

Bonfante, P., & Genre, A. (2010). Mechanisms underlying beneficial plant–fungus interactions in mycorrhizal symbiosis. *Nature Communications, 1,* 48. https://doi.org/10.1038/ncomms1046.

Bortolus, Marco, Annalisa Dalzini, Fernando Formaggio, Claudio Toniolo, Marina Gobbo, and Anna Lisa Maniero. 2015. "An EPR Study of Ampullosporin A, a Medium-Length Peptaibiotic, in Bicelles and Vesicles." *Physical Chemistry Chemical Physics* 18 (2). Royal Society of Chemistry: 749–60. https://doi.org/10.1039/c5cp04136h.

Bortolus, Marco, Marta De Zotti, Fernando Formaggio, and Anna Lisa Maniero. 2013. "Alamethicin in Bicelles: Orientation, Aggregation, and Bilayer Modification as a Function of Peptide Concentration." *Biochimica et Biophysica Acta—Biomembranes* 1828 (11). Elsevier B.V.: 2620–27. https://doi.org/10.1016/j.bbamem.2013.07.007.

Brader, Günter, Stéphane Compant, Kathryn Vescio, Birgit Mitter, Friederike Trognitz, Li Jun Ma, and Angela Sessitsch. 2017. "Ecology and Genomic Insights into Plant-Pathogenic and Plant-Nonpathogenic Endophytes." *Annual Review of Phytopathology* 55: 61–83. https://doi.org/10.1146/annurev-phyto-080516-035641.

Brandt, Katelyn, and Rodolphe Barrangou. 2019. "Annual Review of Food Science and Technology Applications of CRISPR Technologies across the Food Supply Chain." https://doi.org/10.1146/annurev-food-032818-.

Busby, Posy E., Chinmay Soman, Maggie R. Wagner, Maren L. Friesen, James Kremer, Alison Bennett, Mustafa Morsy, Jonathan A. Eisen, Jan E. Leach, and Jeffery L. Dangl. 2017. "Research Priorities for Harnessing Plant Microbiomes in Sustainable Agriculture." *PLoS Biology* 15 (3): 1–14. https://doi.org/10.1371/journal.pbio.2001793.

Cao, Yangrong, Morgan K. Halane, Walter Gassmann, and Gary Stacey. 2017. "The Role of Plant Innate Immunity in the Legume–Rhizobium Symbiosis." *Annual Review of Plant Biology* 68: 535–61. https://doi.org/10.1146/annurev-arplant-042916-041030.

Chauhan, Arun, Swaleha Zubair, Saba Tufail, Asif Sherwani, Mohammad Sajid, Suri C. Raman, Amir Azam, and Mohammad Owais. 2011. "Fungus-Mediated Biological Synthesis of Gold Nanoparticles: Potential in Detection of Liver Cancer." *International Journal of Nanomedicine* 6: 2305–19. https://doi.org/10.2147/ijn.s23195.

Chen, Gui Qiu, Zheng Jun Zou, Guang Ming Zeng, Ming Yan, Jia Qi Fan, An Wei Chen, Fan Yang, Wen Juan Zhang, and Liang Wang. 2011. "Coarsening of Extracellularly Biosynthesized Cadmium Crystal Particles Induced by Thioacetamide in Solution." *Chemosphere* 83 (9). Elsevier Ltd.: 1201–7. https://doi.org/10.1016/j.chemosphere.2011.03.063.

Christoffersen, Heidi F., Sara K. Hansen, Brian S. Vad, Erik H. Nielsen, Jakob T. Nielsen, Thomas Vosegaard, Troels Skrydstrup, and Daniel E. Otzen. 2015. "The Natural, Peptaibolic Peptide SPF-5506-A4 Adopts a β-Bend Spiral Structure, Shows Low Hemolytic Activity and Targets Membranes through Formation of Large Pores." *Biochimica et Biophysica Acta—Proteins and Proteomics* 1854 (8). Elsevier B.V.: 882–89. https://doi.org/10.1016/j.bbapap.2015.03.003.

Clermont, N., S. Lerat, and Carole Beaulieu. 2011. "Genome Shuffling Enhances Biocontrol Abilities of *Streptomyces* Strains against Two Potato Pathogens." *Journal of Applied Microbiology* 111 (3): 671–82. https://doi.org/10.1111/j.1365-2672.2011.05078.x.

Cunningham, D.P., and L.L. Lundie. 1993. "Precipitation of Cadmium by *Clostridium thermoaceticum*." *Applied and Environmental Microbiology* 59 (1): 7–14. https://doi.org/10.1128/aem.59.1.7-14.1993.

Delavaux, Camille S., Patrick Weigelt, Wayne Dawson, Jessica Duchicela, Franz Essl, Mark van Kleunen, Christian König, et al. 2019. "Mycorrhizal Fungi Influence Global Plant Biogeography." *Nature Ecology and Evolution* 3 (3): 424–29. https://doi.org/10.1038/s41559-019-0823-4.

Descamps, Charlotte, Muriel Quinet, and Anne Laure Jacquemart. 2021. "Climate Change–Induced Stress Reduce Quantity and Alter Composition of Nectar and Pollen from a Bee-Pollinated Species (*Borago officinalis*, Boraginaceae)." *Frontiers in Plant Science* 12 (October): 1–12. https://doi.org/10.3389/fpls.2021.755843.

Eccher, T., M. Bacchetta, and G. Granelli. 2009. "Long-Term Effects of Ericoid Endomycorrhizae on the Growth of Micropropagated Plants of *Vaccinium corymbosum* L. in the Field." *Acta Horticulturae* 810: 657–64. https://doi.org/10.17660/ActaHortic.2009.810.87.

Faranda, D., Messori, G., & Pons, F. M. E. 2022. A climate-change attribution retrospective of some impactful extreme events. *Weather and Climate Dynamics*, 3, 1311–1330. https://doi.org/10.5194/wcd-3-1311-2022.

Farooq, T., Hussain, M. D., Shakeel, M. T., Tariqjaveed, M., Aslam, M. N., Naqvi, S. A. H., Amjad, R., Tang, Y., She, X., & He, Z. 2022. Deploying viruses against phytobacteria: potential use of phage cocktails as a multifaceted approach to combat resistant bacterial plant pathogens. *Viruses*, 14(2), 171. https://doi.org/10.3390/v14020171.

Fathy, W.A. 2016. "Green Biosynthesis of Silver Nanoparticles Using Marine Red Algae *Acanthophora specifera* and Its Antibacterial Activity." *Journal of Nanomedicine & Nanotechnology* 07 (06): 7–10. https://doi.org/10.4172/2157-7439.1000409.

Feklistova, I.N., and N.P. Maksimova. 2008. "Obtaining *Pseudomonas aurantiaca* Strains Capable of Overproduction of Phenazine Antibiotics." *Microbiology* 77 (2): 176–80. https://doi.org/10.1134/S0026261708020094.

Fernandes, Francisco M., Thibaud Coradin, and Carole Aimé. 2014. "Self-Assembly in Biosilicification and Biotemplated Silica Materials." *Nanomaterials* 4 (3): 792–812. https://doi.org/10.3390/nano4030792.

Fürnkranz, Michael, Birgit Lukesch, Henry Müller, Herbert Huss, Martin Grube, and Gabriele Berg. 2012. "Microbial Diversity Inside Pumpkins: Microhabitat-Specific Communities Display a High Antagonistic Potential against Phytopathogens." *Microbial Ecology* 63 (2): 418–28. https://doi.org/10.1007/s00248-011-9942-4.

Gaj, Thomas, Charles A. Gersbach, and Carlos F. Barbas. 2013. "ZFN, TALEN, and CRISPR/Cas-Based Methods for Genome Engineering." *Trends in Biotechnology* 31 (7). Elsevier Ltd.: 397–405. https://doi.org/10.1016/j.tibtech.2013.04.004.

Gajbhiye, Monali, Jayendra Kesharwani, Avinash Ingle, Aniket Gade, and Mahendra Rai. 2009. "Fungus-Mediated Synthesis of Silver Nanoparticles and Their Activity against Pathogenic Fungi in Combination with Fluconazole." *Nanomedicine: Nanotechnology, Biology, and Medicine* 5 (4). Elsevier Inc.: 382–86. https://doi.org/10.1016/j.nano.2009.06.005.

García-Pedrajas, M. D., Cañizares, M. C., Sarmiento-Villamil, J. L., Jacquat, A. G., & Dambolena, J. S. 2019. Mycoviruses in biological control: From basic research to field implementation. *Phytopathology*, 109 (11), 1828–1839. https://doi.org/10.1094/PHYTO-05-19-0166-RVW.

Gavryushina, I.A., M.L. Georgieva, A.E. Kuvarina, and V.S. Sadykova. 2021. "Peptaibols as Potential Antifungal and Anticancer Antibiotics: Current and Foreseeable Development (Review)." *Applied Biochemistry and Microbiology* 57 (5): 556–63. https://doi.org/10.1134/S0003683821050070.

Grigoriev, Pavel A., Brigitte Schlegel, Matthias Kronen, Albrecht Berg, Albert Härtl, and Udo Gräfe. 2003. "Differences in Membrane Pore Formation by Peptaibols." *Journal of Peptide Science* 9 (11–12): 763–68. https://doi.org/10.1002/psc.502.

Gullino, M. L., Albajes, R., Al-Jboory, I., Angelotti, F., et al. 2022. Climate change and pathways used by pests as challenges to plant health in agriculture and forestry. *Sustainability*, 14 (19), 12421. https://doi.org/10.3390/su141912421.

Han, Sang Woo, and Yasuo Yoshikuni. 2022. "Microbiome Engineering for Sustainable Agriculture: Using Synthetic Biology to Enhance Nitrogen Metabolism in Plant-Associated Microbes." *Current Opinion in Microbiology* 68. Elsevier: 102172. https://doi.org/10.1016/j.mib.2022.102172.

Haq, Ihtisham Ul, Kashif Rahim, Galal Yahya, Bushra Ijaz, Sajida Maryam, and Najeeba Parre Paker. 2024. "Eco-Smart Biocontrol Strategies Utilizing Potent Microbes for Sustainable Management of Phytopathogenic Diseases," *Biotechnology Reports* 44 (July). Elsevier B.V.: e00859. https://doi.org/10.1016/j.btre.2024.e00859.

Hossain, Motaher, Farjana Sultana, Mahabuba Mostafa, Humayra Ferdus, and Mridula Rahman. 2024. *Plant Disease Dynamics in a Changing Climate: Impacts, Molecular Mechanisms, and Climate – Informed Strategies for Sustainable Management. Discover Agriculture*. Springer International Publishing. https://doi.org/10.1007/s44279-024-00144-w.

Hou, Xuewen, Ruonan Sun, Yanyan Feng, Runfang Zhang, Tianjiao Zhu, Qian Che, Guojian Zhang, and Dehai Li. 2022. "Peptaibols: Diversity, Bioactivity, and Biosynthesis." *Engineering Microbiology* 2 (3). Elsevier B.V.: 100026. https://doi.org/10.1016/j.engmic.2022.100026.

Iqbal, Babar, Guanlin Li, Khulood Fahad Alabbosh, Hamad Hussain, Ismail Khan, Muhammad Tariq, Qaiser Javed, Muhammad Naeem, and Naveed Ahmad. 2023. "Advancing Environmental Sustainability through Microbial Reprogramming in Growth Improvement, Stress Alleviation, and Phytoremediation." *Plant Stress* 10 (August). Elsevier B.V.: 100283. https://doi.org/10.1016/j.stress.2023.100283.

Jain, Archana, Surendra Sarsaiya, Qin Wu, Yuanfu Lu, and Jingshan Shi. 2019. "A Review of Plant Leaf Fungal Diseases and Its Environment Speciation." *Bioengineered* 10 (1). Taylor & Francis: 409–24. https://doi.org/10.1080/21655979.2019.1649520.

Jansson, Janet K., and Kirsten S. Hofmockel. 2018. "The Soil Microbiome—from Metagenomics to Metaphenomics." *Current Opinion in Microbiology* 43. The Authors: 162–68. https://doi.org/10.1016/j.mib.2018.01.013.

Kafle, Arjun, Anil Timilsina, Asmita Gautam, Kaushik Adhikari, Anukul Bhattarai, and Niroj Aryal. 2022. "Phytoremediation: Mechanisms, Plant Selection and Enhancement by Natural and Synthetic Agents." *Environmental Advances* 8 (February). Elsevier Ltd: 100203. https://doi.org/10.1016/j.envadv.2022.100203.

Kashyap, P. L., Kumar, S., Srivastava, A. K., & Sharma, A. K. 2013. Myconanotechnology in agriculture: A perspective. *World Journal of Microbiology and Biotechnology*, 29, 191–207.

Kashyap, P. L., Kumar, S., & Srivastava, A. K. 2018. Nanobiotechnology in agriculture: Prospects and challenges. *Journal of Environmental Management*, 217, 590–602. https://doi.org/10.1016/j.jenvman.2018.03.073.

Khalifa, M. E., & MacDiarmid, R. M. 2021. A new frontier for biological control of plant diseases: Mycoviruses and their potential role in integrated pest management. *Phytopathology*, 111 (12), 2151–2163. https://doi.org/10.1094/PHYTO-03-21-0106-RVW.

Khare, Ekta, Sanoj Kumar, and Kangmin Kim. 2018. "Role of Peptaibols and Lytic Enzymes of *Trichoderma cerinum* Gur1 in Biocontrol of *Fusarium oxysporum* and Chickpea Wilt." *Environmental Sustainability* 1 (1). Springer Singapore: 39–47. https://doi.org/10.1007/s42398-018-0001-7.

Kong, Chui Hua, Tran Dang Xuan, Tran Dang Khanh, Hoang Dung Tran, and Nguyen Thanh Trung. 2019. "Allelochemicals and Signaling Chemicals in Plants." *Molecules (Basel, Switzerland)* 24 (15): 1–19. https://doi.org/10.3390/molecules24152737.

Konishi, Yasuhiro, Kaori Ohno, Norizoh Saitoh, Toshiyuki Nomura, Shinsuke Nagamine, Hajime Hishida, Yoshio Takahashi, and Tomoya Uruga. 2007. "Bioreductive Deposition of Platinum Nanoparticles on the Bacterium *Shewanella* Algae." *Journal of Biotechnology* 128 (3): 648–53. https://doi.org/10.1016/j.jbiotec.2006.11.014.

Lahlali, Rachid, Mohammed Taoussi, Salah Eddine Laasli, Grace Gachara, Rachid Ezzouggari, Zineb Belabess, Kamal Aberkani, et al. 2024. "Effects of Climate Change on Plant Pathogens and Host–Pathogen Interactions." *Crop and Environment* 3 (3). The Author(s): 159–70. https://doi.org/10.1016/j.crope.2024.05.003.

Lewis, Cecil M., Mercy Y. Akinyi, Sharon N. DeWitte, and Anne C. Stone. 2023. "Ancient Pathogens Provide a Window into Health and Well-Being." *Proceedings of the National Academy of Sciences of the United States of America* 120 (4): 1–8. https://doi.org/10.1073/pnas.2209476119.

Li, Hai Yun, Yan Luo, Xiu Sheng Zhang, Wei Ling Shi, Zhi Ting Gong, Mei Shi, Lei Lei Chen, Xiu Lan Chen, Yu Zhong Zhang, and Xiao Yan Song. 2014. "Trichokonins from *Trichoderma pseudokoningii* SMF2 Induce Resistance against Gram-Negative *Pectobacterium carotovorum* Subsp. *carotovorum* in Chinese Cabbage." *FEMS Microbiology Letters* 354 (1): 75–82. https://doi.org/10.1111/1574-6968.12427.

Loper, J. E., Henkels, M. D., Shaffer, B. T., Valeriote, F. A., & Gross, D. C. 2007. Isolation and characterization of a 2,4-diacetylphloroglucinol-biosynthetic gene cluster from Pseudomonas fluorescens Q2-87. *Molecular Plant-Microbe Interactions*, 20 (12), 1531–1541. https://doi.org/10.1094/MPMI-20-12-1531.

Lovley, Derek R., John F. Stolz, Gordon L. Nord, and Elizabeth J.P. Phillips. 1987. "Anaerobic Production of Magnetite by a Dissimilatory Iron-Reducing Microorganism." *Nature* 330 (6145): 252–54. https://doi.org/10.1038/330252a0.

Malarkodi, Chelladurai, Shunmugam Rajeshkumar, Kanniah Paulkumar, Mahendran Vanaja, Gnana Dhas Gnana Jobitha, and Gurusamy Annadurai. 2013. "Bactericidal Activity of Bio Mediated Silver Nanoparticles Synthesized by *Serratia nematodiphila*." *Drug Invention Today* 5 (2). Elsevier: 119–25. https://doi.org/10.1016/j.dit.2013.05.005.

Manivasagan, Panchanathan, Jayachandran Venkatesan, Kalimuthu Senthilkumar, Kannan Sivakumar, and Se Kwon Kim. 2013. "Biosynthesis, Antimicrobial and Cytotoxic Effect of Silver Nanoparticles Using a Novel *Nocardiopsis* sp. MBRC-1." *BioMed Research International* 2013. https://doi.org/10.1155/2013/287638.

Mata, Y., Torres, E., Blázquez, M., Ballester, A., González, F., & Muñoz, J. 2009. Gold(III) biosorption and bioreduction with the brown alga Fucus vesiculosus. *Journal of Hazardous Materials*, 166 (2–3), 612–618. https://doi.org/10.1016/j.jhazmat.2008.11.064.

Muhie, Seid Hussen. 2022. "Novel Approaches and Practices to Sustainable Agriculture." *Journal of Agriculture and Food Research* 10 (August). Elsevier B.V.: 100446. https://doi.org/10.1016/j.jafr.2022.100446.

Muñoz, Isabel Vicente, Sabrina Sarrocco, Luca Malfatti, Riccardo Baroncelli, and Giovanni Vannacci. 2019. "CRISPR-CAS for Fungal Genome Editing: A New Tool for the Management of Plant Diseases." *Frontiers in Plant Science* 10 (February): 1–5. https://doi.org/10.3389/fpls.2019.00135.

Mwangi, Ruth W., Mohammed Mustafa, K. Charles, Isabel W. Wagara, and Noemi Kappel. 2023. "Selected Emerging and Reemerging Plant Pathogens Affecting the Food Basket: A Threat to Food Security." *Journal of Agriculture and Food Research* 14 (August). Elsevier B.V.: 100827. https://doi.org/10.1016/j.jafr.2023.100827.

Nadarajah, Kalaivani, and Nur Sabrina Natasha Abdul Rahman. 2021. "Plant–Microbe Interaction: Aboveground to Belowground, from the Good to the Bad." *International Journal of Molecular Sciences* 22 (19). https://doi.org/10.3390/ijms221910388.

Nair, Binoj, and T. Pradeep. 2002. "Coalescence of Nanoclusters and Formation of Submicron Crystallites Assisted by *Lactobacillus* Strains." *Crystal Growth and Design* 2 (4): 293–98. https://doi.org/10.1021/cg0255164.

Nazarov, P.A., D.N. Baleev, M.I. Ivanova, L.M. Sokolova, and M.V. Karakozova. 2020. "Infectious Plant Diseases: Etiology, Current Status, Problems and Prospects in Plant Protection." *Acta Naturae* 12 (3): 46–59. https://doi.org/10.32607/actanaturae.11026.

Neumann, Nora K.N., Norbert Stoppacher, Susanne Zeilinger, Thomas Degenkolb, Hans Brückner, and Rainer Schuhmacher. 2015. "The Peptaibiotics Database—a Comprehensive Online Resource." *Chemistry and Biodiversity* 12 (5): 743–51. https://doi.org/10.1002/cbdv.201400393.

Oh, Dahyun, Jifa Qi, Binghong Han, Geran Zhang, Thomas J. Carney, Jacqueline Ohmura, Yong Zhang, Yang Shao-Horn, and Angela M. Belcher. 2014. "M13 Virus-Directed Synthesis of Nanostructured Metal Oxides for Lithium-Oxygen Batteries." *Nano Letters* 14 (8): 4837–45. https://doi.org/10.1021/nl502078m.

Pandit, Manisha Arora, Jitendra Kumar, Saloni Gulati, Neeru Bhandari, Poonam Mehta, Roma Katyal, Charu Dogra Rawat, Vachaspati Mishra, and Jasleen Kaur. 2022. "Major Biological Control Strategies for Plant Pathogens." *Pathogens* 11 (2): 1–21. https://doi.org/10.3390/pathogens11020273.

Panthapulakkal Narayanan, Saritha, Shiu Cheung Lung, Pan Liao, Clive Lo, and Mee Len Chye. 2020. "The Overexpression of OsACBP5 Protects Transgenic Rice against Necrotrophic, Hemibiotrophic and Biotrophic Pathogens." *Scientific Reports* 10 (1). Nature Publishing Group UK: 1–19. https://doi.org/10.1038/s41598-020-71851-9.

Patel, J. S., & Archana, G. (2017). Diverse endophytic bacteria colonization in root tissues of plants and their role in plant growth promotion. *3 Biotech*, 7, 1–12. https://doi.org/10.1007/s13205-017-0909-4.

Pereira-Dias, Leandro, Paulo R. Oliveira-Pinto, Juliana O. Fernandes, Laura Regalado, Rafael Mendes, Cátia Teixeira, Nuno Mariz-Ponte, Paula Gomes, and Conceição Santos. 2023. "Peptaibiotics: Harnessing the Potential of Microbial Secondary Metabolites for Mitigation of Plant Pathogens." *Biotechnology Advances* 68 (August). https://doi.org/10.1016/j.biotechadv.2023.108223.

Porras, M.-F., Navas, C. A., Agudelo-Cantero, G. A., Santiago-Martínez, M. G., Loeschcke, V., Sørensen, J. G., Crandall, S., Biddinger, D., & Rajotte, E. G. 2023. Extreme heat alters the performance of hosts and pathogen. *Frontiers in Ecology and Evolution*, 11, Article 1186452. https://doi.org/10.3389/fevo.2023.1186452.

Raliya, Ramesh, and J.C. Tarafdar. 2013. "ZnO Nanoparticle Biosynthesis and Its Effect on Phosphorous-Mobilizing Enzyme Secretion and Gum Contents in Clusterbean (*Cyamopsis tetragonoloba* L.)." *Agricultural Research* 2 (1): 48–57. https://doi.org/10.1007/s40003-012-0049-z.

Ramawat, Kishan Gopal. 2022. *Parthenium Hysterophorus (Parthenium Weed). PlantwisePlus Knowledge Bank*. Vol. Species Pa. https://doi.org/10.1079/pwkb.species.45573.

Rebuffat, S., Goulard, C., Bodo, B., & Roquebert, M. F. 1995. The peptaibols from Trichoderma spp.: Structure, biological activity, and membrane interactions. *Journal of Industrial Microbiology*, 14, 134–141. https://doi.org/10.1007/BF01569895.

Scortichini, Marco. 2022. "Sustainable Management of Diseases in Horticulture: Conventional and New Options." *Horticulturae* 8 (6). https://doi.org/10.3390/horticulturae8060517.

Ségalas, Isabelle, Yann Prigent, Daniel Davoust, Bernard Bodo, and Sylvie Rebuffat. 1999. "Characterization of a Type of β-Bend Ribbon Spiral Generated by the Repeating (Xaa-Yaa-Aib-Pro) Motif: The Solution Structure of Harzianin HC IX, a 14-Residue Peptaibol Forming Voltage-Dependent Ion Channels." *Biopolymers* 50 (1): 71–85. https://doi.org/10.1002/(SICI)1097-0282(199907)50:1<71::AID-BIP7>3.0.CO;2–5.

Sella, Luca, Rakshita Govind, Rocco Caracciolo, Alessandra Quarantin, Van V. Vu, Silvio Tundo, Hung Minh Nguyen, Francesco Favaron, Rita Musetti, and Marta De Zotti. 2021. "Transcriptomic and Ultrastructural Analyses of *Pyricularia oryzae* Treated with Fungicidal Peptaibol Analogs of *Trichoderma trichogin*." *Frontiers in Microbiology* 12 (October): 1–20. https://doi.org/10.3389/fmicb.2021.753202.

Shahverdi, Ahmad R., Sara Minaeian, Hamid Reza Shahverdi, Hossein Jamalifar, and Ashraf Asadat Nohi. 2007. "Rapid Synthesis of Silver Nanoparticles Using Culture Supernatants of Enterobacteria: A Novel Biological Approach." *Process Biochemistry* 42 (5): 919–23. https://doi.org/10.1016/j.procbio.2007.02.005.

Sharma, Itika, Ashish Raina, Malvi Choudhary, Apra, Sanjana Kaul, and Manoj K. Dhar. 2023. "Fungal Endophyte Bioinoculants as a Green Alternative towards Sustainable Agriculture." *Heliyon* 9 (9). Elsevier Ltd: e19487. https://doi.org/10.1016/j.heliyon.2023.e19487.

Singh, Brajesh K., Manuel Delgado-Baquerizo, Eleonora Egidi, Emilio Guirado, Jan E. Leach, Hongwei Liu, and Pankaj Trivedi. 2023. "Climate Change Impacts on Plant Pathogens, Food Security and Paths Forward." *Nature Reviews Microbiology* 21 (10). Springer US: 640–56. https://doi.org/10.1038/s41579-023-00900-7.

Skendžić, S., Zovko, M., Pajač Živković, I., Lešić, V., & Lemić, D. 2021. The Impact of Climate Change on Agricultural Insect Pests. *Insects*, 12 (5), 440. https://doi.org/10.3390/insects12050440.

Sokolovski, Sergei G., Andy A. Meharg, and Frans J.M. Maathuis. 2002. "*Calluna vulgaris* Root Cells Show Increased Capacity for Amino Acid Uptake When Colonized with the Mycorrhizal Fungus *Hymenoscyphus ericae*." *New Phytologist* 155 (3): 525–30. https://doi.org/10.1046/j.1469-8137.2002.00485.x.

Tahir, H. A. S., Gu, Q., Wu, H., Niu, Y., Huo, R., & Gao, X. 2017. Bacillus volatiles adversely affect the physiology and ultra-structure of Ralstonia solanacearum and induce systemic resistance in tobacco against bacterial wilt. *Scientific Reports*, 7, 40481. https://doi.org/10.1038/srep40481.

Tilocca, B., Cao, A., & Migheli, Q. 2020. Scent of a killer: Microbial volatilome and its role in the biological control of plant pathogens. *Frontiers in Microbiology*, 11, 41. https://doi.org/10.3389/fmicb.2020.00041.

Trojak-Goluch, Anna. 2024. "The Use of Bacteria, Actinomycetes and Fungi in the Bioprotection of Solanaceous Crops against Tobacco Mosaic Virus (TMV)." *Agriculture (Switzerland)* 14 (8). https://doi.org/10.3390/agriculture14081220.

Vance, C. P. (2001). Symbiotic nitrogen fixation and phosphorus acquisition. *Plant Physiology*, *127*(2), 390–397. https://doi.org/10.1104/pp.010331.

Velásquez, André C., Christian Danve M. Castroverde, and Sheng Yang He. 2018. "Plant–Pathogen Warfare under Changing Climate Conditions." *Current Biology* 28 (10): R619–34. https://doi.org/10.1016/j.cub.2018.03.054.

Vijayaraghavan, K., A. Mahadevan, M. Sathishkumar, S. Pavagadhi, and R. Balasubramanian. 2011. "Biosynthesis of Au(0) from Au(III) via Biosorption and Bioreduction Using Brown Marine Alga *Turbinaria conoides*." *Chemical Engineering Journal* 167 (1). Elsevier B.V.: 223–27. https://doi.org/10.1016/j.cej.2010.12.027.

Viterbo, Ada, Aric Wiest, Yariv Brotman, Ilan Chet, and Charles Kenerley. 2007. "The 18mer Peptaibols from *Trichoderma virens* Elicit Plant Defence Responses." *Molecular Plant Pathology* 8 (6): 737–46. https://doi.org/10.1111/j.1364-3703.2007.00430.x.

Vivek, Marimuthu, Palanisamy Senthil Kumar, Sesurajan Steffi, and Sellappa Sudha. 2011. "Biogenic Silver Nanoparticles by *Gelidiella acerosa* Extract and Their Antifungal Effects." *Avicenna Journal of Medical Biotechnology* 3 (3): 143–48.

Wang, Qi, Jinge Liu, and Hongyan Zhu. 2018. "Genetic and Molecular Mechanisms Underlying Symbiotic Specificity in Legume–Rhizobium Interactions." *Frontiers in Plant Science* 9 (March): 1–8. https://doi.org/10.3389/fpls.2018.00313.

Wippel, Kathrin. 2023. "Plant and Microbial Features Governing an Endophytic Lifestyle." *Current Opinion in Plant Biology* 76. The Author: 102483. https://doi.org/10.1016/j.pbi.2023.102483.

Zhang, Yu Qiang, Shan Zhang, Mei Ling Sun, Hai Nan Su, Hao Yang Li, Kun-Liu, Yu Zhong Zhang, Xiu Lan Chen, Hai Yan Cao, and Xiao Yan Song. 2022. "Antibacterial Activity of Peptaibols from *Trichoderma longibrachiatum* SMF2 against Gram-Negative *Xanthomonas oryzae* pv. *oryzae*, the Causal Agent of Bacterial Leaf Blight on Rice." *Frontiers in Microbiology* 13 (October): 1–13. https://doi.org/10.3389/fmicb.2022.1034779.

Zhang, Y., Li, B., & Wang, Z. 2023. Plants' response to abiotic stress: mechanisms and strategies. *International Journal of Molecular Sciences*, 24 (14), 11002. https://doi.org/10.3390/ijms241411002.

Zhao, P., A. Ren, P. Dong, Y. Sheng, and D. Li. 2018. "Antimicrobial Peptaibols, Trichokonins, Inhibit Mycelial Growth and Sporulation and Induce Cell Apoptosis in the Pathogenic Fungus *Botrytis cinerea*." *Applied Biochemistry and Microbiology* 54 (4): 396–403. https://doi.org/10.1134/S0003683818040154.

Zhou, Jien, Xueyan Zhang, Zheng Qu, Chenchen Zhang, Feng Wang, Tongguo Gao, Yanpo Yao, and Junfeng Liang. 2024. "Progress in Research on Prevention and Control of Crop Fungal Diseases in the Context of Climate Change." *Agriculture (Switzerland)* 14 (7): 1–19. https://doi.org/10.3390/agriculture14071108.

Zilber-Rosenberg, I., & Rosenberg, E. 2008. Role of microorganisms in the evolution of animals and plants: the hologenome theory of evolution. *FEMS Microbiology Reviews*, 32 (5), 723–735. https://doi.org/10.1111/j.1574-6976.2008.00123.x.

Zoledowska, Sabina, Luana Presta, Marco Fondi, Francesca Decorosi, Luciana Giovannetti, Alessio Mengoni, and Ewa Lojkowska. 2019. "Metabolic Modeling of *Pectobacterium parmentieri* Scc3193 Provides Insights into Metabolic Pathways of Plant Pathogenic Bacteria." *Microorganisms* 7 (4). https://doi.org/10.3390/microorganisms7040101.

Zotti, Marta De, Barbara Biondi, Yoonkyung Park, Kyung Soo Hahm, Marco Crisma, Claudio Toniolo, and Fernando Formaggio. 2012. "Antimicrobial Lipopeptaibol Trichogin GA IV: Role of the Three Aib Residues on Conformation and Bioactivity." *Amino Acids* 43 (4): 1761–77. https://doi.org/10.1007/s00726-012-1261-7.

Zotti, Marta De, Luca Sella, Angela Bolzonello, Laura Gabbatore, Cristina Peggion, Alessandro Bortolotto, Ibrahim Elmaghraby, Silvio Tundo, and Francesco Favaron. 2020. "Targeted Amino Acid Substitutions in a *Trichoderma* Peptaibol Confer Activity against Fungal Plant Pathogens and Protect Host Tissues from *Botrytis cinerea* Infection." *International Journal of Molecular Sciences* 21 (20): 1–19. https://doi.org/10.3390/ijms21207521.

8 The Hidden Hand
Exploring miRNA-Mediated Signaling in Plant–Microbe Synergy

Priya Jaiswal, Virevol Thakro, and Pinky Agarwal

8.1 INTRODUCTION

MicroRNAs (miRNAs) have emerged as key players in regulating plant adaption to changing environmental conditions. These small, noncoding RNA molecules, approximately 20–24 nucleotides long, play crucial roles in transcriptional or post-transcriptional gene regulation (Dong et al., 2022). Widely expressed across tissues and developmental stages, miRNAs orchestrate various complex biological processes, including plant growth, development, hormone signaling, stress responses, and interactions with beneficial microbes and pathogens (Navarro et al., 2006; Mishra et al., 2018; Ding et al., 2024). The discovery of the first miRNA, lin-4, in *Caenorhabditis elegans* in 1993 revealed its role in developmental timing via suppression of the nuclear protein LIN-14 (Lee et al., 1993). Since then, research on miRNAs has grown exponentially, with the miRBase database (https://www.mirbase.org) cataloging 48,885 miRNAs across 271 species, often found as single, multiple, or clustered gene copies (Kozomara et al., 2019; Ding et al., 2024). In plants, miRNAs demonstrate evolutionary flexibility by adapting rapidly to environmental changes and co-evolving with microbes, where they not only conserve critical functions across species but also diversify to regulate species-specific plant–microbe interactions (Jones-Rhoades et al., 2006; Middleton et al., 2021). This evolutionary versatility allows plants to modulate their responses to microbial signals, enhancing adaptability and resilience to changing environmental conditions (Middleton et al., 2021).

In plants, miRNAs are encoded by *MIRNA* genes and regulate gene expression via an RNA silencing mechanism (Dong et al., 2022). This fine-tuning enables plants to dynamically respond to diverse biotic and abiotic signals, promoting growth, development, and survival (Navarro et al., 2006; Mishra et al., 2018; Ding et al., 2024). miRNAs, thus, serve as central components in signaling networks, enhancing plant immunity and adaptability to various stressors (Middleton et al., 2021). Another key aspect of miRNA-mediated defense involves the regulation of phytohormone signaling pathways (Ding et al., 2024). Phytohormones like jasmonic acid (JA), salicylic acid (SA), and abscisic acid (ABA) are pivotal in modulating plant immune responses (Ding et al., 2024). Specific miRNAs, such as miR160 and miR393, regulate auxin signaling components, which interact with SA and JA pathways to balance growth and defense (Pinweha et al., 2015; Natarajan et al., 2018; Jiang et al., 2022).

Plants interact with diverse microbiota, including bacteria and fungi, which significantly influence their health and development. Beneficial microbes, such as rhizobia and mycorrhizal fungi, contribute to nutrient acquisition, such as nitrogen fixation and phosphate uptake, and enhance plant disease resistance. Plants in turn supply these microbes with photosynthetically derived carbon (Bais et al., 2006). During pathogen attack, plants can trigger miRNA-mediated immune responses (Yadav & Ramakrishna, 2024). miRNAs are crucial in managing these complex relationships by modulating symbiotic interactions with beneficial microbes such as rhizobia and mycorrhizal fungi, and activating immune responses against pathogens (bacteria, fungi, and viruses) (Hoang et al., 2020; Yadav & Ramakrishna, 2024). For instance, miRNAs regulate nodulation in legumes during

DOI: 10.1201/9781003608080-8

rhizobial symbiosis and mediate immune responses to pathogens, balancing the need for growth with defense. Recent studies also revealed that miRNAs can move between plant and pathogen cells, with plant-derived miRNAs suppressing pathogenic genes, and pathogen-derived miRNAs interfering with plant immunity, adding complexity to plant–pathogen interactions (Chi et al., 2023). This cross-kingdom communication makes miRNAs a critical component of the plant–microbe signaling network, which provides new avenues for enhancing crop resilience. Advances in miRNA research have led to the discovery of numerous miRNAs, enhancing our understanding of miRNA regulatory mechanisms and shedding light on their potential for developing disease-resistant crops (Gualtieri et al., 2020). In agricultural systems, leveraging miRNA-mediated regulation holds immense potential for reducing reliance on chemical inputs, improving soil health, and enhancing resilience to environmental stresses.

This chapter provides an overview of the current understanding of miRNAs in regulating symbiotic and plant–pathogen interactions in plants, integrating recent discoveries on miRNA–microbe synergies. It also discusses potential applications for enhancing crop disease resistance through targeted breeding, transgenic approaches, and gene editing techniques.

8.2 BIOGENESIS AND MECHANISMS OF MIRNA ACTION

miRNAs are produced through a highly regulated multistep process, involving transcription, precursor processing, methylation, and assembly into silencing complexes (Xie et al., 2005). miRNA (*MIR*) genes are first transcribed by RNA polymerase II to generate primary miRNA (pri-miRNA), which are single-stranded molecules and several hundred nucleotides long. The Dicer-like protein 1 (DCL1) then recognizes these pri-miRNAs in the nucleus, which sequentially cleaves them into precursor miRNAs (pre-miRNAs) and further processes them into mature miRNA/miRNA* duplexes (Chen, 2005; Bologna et al., 2018). These duplexes then undergo methylation and are transported to the cytoplasm, where they interact with Argonaute (AGO) proteins and integrate into the RNA-induced silencing complex (RISC). The miRNA then guides the complex to target specific mRNAs through sequence complementarity, leading to either degradation of mRNAs or inhibition of translation (Bin & Wang, 2010; Wang et al., 2019). miRNAs can also regulate gene expression by modulating histone modifications and DNA methylation, further adding to the complexity of miRNA-mediated gene regulation (Khraiwesh et al., 2010). The precise regulation and targeting mechanisms of miRNAs enable plants to dynamically adjust their gene expression in response to environmental cues and developmental signals. By targeting a wide array of mRNAs, miRNAs play a pivotal role in processes ranging from growth and development to responses to biotic and abiotic stresses, establishing miRNAs as central components in plant regulatory networks.

8.3 ROLE OF MIRNA IN PLANT DEFENSE MECHANISMS AGAINST PATHOGENS

Being sessile, plants are frequently exposed to various pathogens, such as bacteria, fungi, and viruses. To resist pathogen attacks, plants have evolved various defense mechanisms that include two primary defense strategies: Pathogen-associated molecular pattern (PAMP)-triggered immunity (PTI) and effector-triggered immunity (ETI) (Zipfel, 2008). PTI, the plant's first line of defense, is activated when the plant detects PAMPs on pathogen surfaces, triggering a broad-spectrum immune response involving MAP kinase, calcium, and reactive oxygen species (ROS) signaling pathways. In response, pathogens deploy a diverse array of virulence factors or effectors to suppress host immune responses, which led plants to evolve ETI as a second line of defense. ETI is induced when plant resistance (R) proteins, encoded by *R* genes, recognize these pathogen effectors, triggering a more targeted immune response, often resulting in localized cell death (Zhang and Zhou, 2010). Research has shown that miRNAs are involved in modulating plant defense responses to pathogens by fine-tuning gene expression through multiple defense signaling pathways. They influence defense mechanisms by targeting disease-resistance genes, modulating hormone signaling, and regulating

ROS production. Such regulatory processes are pivotal in the evolutionary arms race between plants and pathogens, shaping plant resistance mechanisms.

8.3.1 miRNA-Mediated Regulations of Disease Resistance Genes

Plants have evolved intricate complex defense mechanisms to detect and respond to various pathogen attacks. A key target of miRNA is the nucleotide-binding site-leucine-rich repeat (NBS-LRR) gene family, which encodes R proteins, crucial for recognizing pathogen effectors and triggering immune responses through ETI (Su et al., 2018). miRNAs act as master regulators of *NBS-LRR* genes, helping prevent unnecessary immune activation and optimize energy allocation between defense and growth. Table 8.1 lists the differential regulation of plant–microbe interactions by miRNAs. For example, the miR482/2118 family has been shown to target *NLR* genes, which fine-tunes the immune response to conserve energy for growth and development (Liao et al., 2022). In potatoes, the *R3a* resistance gene is cleaved by miR482 to produce phasiRNAs (Wei et al., 2014). In contrast, its homolog, *I2*, in tomatoes is targeted by miR6024, indicating the evolutionary divergence of resistance genes and their distinct miRNA-mediated regulation (Wei et al., 2014).

Variations in miRNA levels can significantly impact disease resistance (Zhang et al., 2023). For instance, miR472 downregulates the NBS-LRR gene *RPS5*, which recognizes effector protein c, and triggers ETI against *Pseudomonas syringae* infection. Similarly, miR482 and miR5300 confer resistance to curl virus in tomatoes by suppressing nucleotide-binding (NB)-domain genes, and miR482 expression in cotton enhances resistance to *Verticillium dahliae* infection (Ouyang et al., 2014). In poplar, miR482 and miR472a reduce NBS-LRR transcripts and provide defense against fungal pathogens *Colletotrichum gloeosporioides* and *Cytospora chrysosperma* (Su et al., 2018). In rice, Osa-miR535 targets *SPL14*, which regulates *DEP1* and *WRKY45*, enhances yield, but reduces resistance to *Magnoporthe oryzae*, indicating a negative regulatory effect (Zhang et al., 2022). In *Arabidopsis*, lectin receptor-like kinases (LecRLKs) are known to recognize bacterial lipopolysaccharides, activating immune responses (Ding et al., 2024). miR393 targets LecRLKs to regulate these immune responses (Djami-Tchatchou and Dubery, 2019). In apples, Md-miR156ab and Md-miR395 target WRKY transcription factors (TFs) WRKYN1 and WRKY26, respectively, which enhances resistance by upregulating pathogenesis-related genes (Zhang et al., 2017). These findings highlight the dual roles of miRNA in the positive and negative regulation of disease-resistance genes, underscoring their role in fine-tuning plant immune responses.

8.3.2 miRNA-Mediated Regulation of Phytohormone Signaling for Plant Immunity

miRNAs are known to play a key role in mediating plant defense responses via phytohormone signaling pathways, including auxin, JA, SA, and ethylene. Their role in plant hormone signaling was first reported in *Arabidopsis*, where the perception of bacterial flagellin (flg22) during *Pseudomonas syringae* infection increases miR393 levels, leading to the repression of F-box auxin receptors TIR1, AFB2, and AFB3 (Navarro et al., 2006). This repression enhances resistance to *Pseudomonas syringae* pv tomato DC3000. Similarly, miR393 and miR160 protect crops against fungal infection by targeting auxin-response factors (ARFs), ARF10 and TIR1, to suppress auxin signaling and prevent anthracnose disease caused by *Colletotrichum gloeosporioides* (Pinweha et al., 2015). In potatoes, miR160 targets ARF10 and ARF16, providing resistance to *Phytophthora infestans* (Natarajan et al., 2018). Interestingly, miR160 overexpression and knockdown lines showed increased susceptibility, suggesting that balanced miR160 levels are essential for effective defense (Natarajan et al., 2018). Beyond auxin signaling, miR168 in *Malus hupehensis* targets ARGONAUTE1 (AGO1), part of the RISC complex, which enhances salicylic acid-mediated signaling during *Botryosphaeria dothidea* infection (Yu et al., 2017). JA signaling is similarly influenced by miRNAs; in tomatoes, miR319 binds to *Teosinte branched1/Cycloidea/Proliferating cell factor 4* (*TCP4*), leading to the suppression of JA synthesis genes in response to root-knot nematode infection (Zhao et al., 2015).

TABLE 8.1
Different Mechanisms of miRNA Regulation in Plants.

S. No.	miRNAs	Plants	Pathogens	Functions	Mechanisms	References
1.	Can-miRn371	Chilli	*Colletotrichum truncatum*	Hormonal signaling	ET	Mishra et al. (2018)
2.	miR393	Arabidopsis	*Pseudomonas syringae*	Hormonal signaling	IAA	Navarro et al. (2006)
3.	miR393, miR160	Cassava	*Colletotrichum gloeosporioides*	Hormonal signaling	IAA	Pinweha et al. (2015)
4.	miR167d	Rice	*Magnoporthe oryzae*	Hormonal signaling	IAA	Zhao et al. (2020)
5.	miR319	Tomato	Root-knot nematode	Hormonal signaling	JA	Zhao et al. (2015)
6.	miR394	Tomato	*Phythopthora infestans*	Hormonal signaling	JA	Zhang et al. (2021)
7.	miR159a	Poplar tree	*Cytospora chrysosperma*	Hormonal signaling	JA/ET	Yang et al. (2023)
8.	miR168	*Malus hupehensis*	*Botryosphaeria dothidea*	Hormonal signaling	SA	Yu et al. (2017)
9.	miR393	Arabidopsis	Bacterial lipopolysaccharide	Plant disease resistance proteins	LecRLKs	Djami-Tchatchou and Dubery (2019)
10.	miR482, miR5300	Tomato	*Fusarium oxysporum*	Plant disease resistance proteins	NB-LRR	Ouyang et al. (2014)
11.	miR156ab and miR395	Apple	Leaf spot fungus	Plant disease resistance proteins	PR	Zhang et al. (2017)
12.	miR396	Rice	*Aspergillus oryzae*	Reactive oxygen and plant secondary metabolism	Flavonoids	Chandran et al. (2019); Dai et al. (2019)
13.	miR528	Rice	Rice stripe virus	Reactive oxygen and plant secondary metabolism	AO, SPL9	Yao et al. (2019)
14.	miR397b	*Malus hupehensis*	*Botryosphaeria dothidea*	Reactive oxygen and plant secondary metabolism	Laccase	Yu et al. (2020)
15.	miR482d, miR397	Potato	Potato virus A	Reactive oxygen and plant secondary metabolism	Laccase	Li et al. (2017a)
16.	miR168, miR528, miR159, miR397, miR827	Corn	Sugarcane mosaic virus	Reactive oxygen and plant secondary metabolism	Laccase	Xia et al. (2018)

Conversely, miR394 has been shown to increase susceptibility to *Phytopthora infestans* infection by suppressing the JA synthesis genes in tomatoes (Zhang et al., 2021). Moreover, miR159a enhances poplar resistance to fungal infection caused by *Cytospora chrysosperma* by regulating key genes involved in JA/ET crosstalk (Yang et al., 2023). Ethylene signaling also contributes to defense, with miRNAs modulating key genes. For example, Can-miRn371 suppresses ethylene response factors (ERFs), providing immunity against anthracnose pathogen *Colletotrichum truncatum* (Mishra et al., 2018). Additionally, Osa-miR167d targets *ARF12* in rice, downregulating its expression to help balance between indole-3-acetic acid (IAA) and JA signaling during *Magnoporthe oryzae* infection, activating defense genes contributing to enhance immunity (Zhao et al., 2020). Several other miRNAs, such as miR472, miR482, miR2109, and miR2118, are known for their regulation of resistant proteins, particularly through the regulation of *NBS-LRR* genes. Together, these findings underscore miRNAs as key modulators of hormone-mediated immunity, reinforcing their role in regulating responses to pathogen attacks.

8.3.3 miRNAs Involved in ROS to Mediate Plant Immunity

ROS accumulation is a key defense mechanism plants use against various biotic and abiotic stresses (Saha, 2024). miRNAs play crucial roles in maintaining cellular redox balance and mediating immune responses by regulating ROS-responsive genes. For example, miR398 targets superoxide dismutase (SOD) genes in *Arabidopsis thaliana*, enhancing resistance to the fungal pathogen *Magnoporthe oryzae* (Li et al., 2019). Similarly, in rice, miR398b upregulates SOD activity, leading to increased hydrogen peroxide (H_2O_2) levels to promote resistance to *Magnoporthe oryzae* (Li et al., 2019). Furthermore, miR528 targets L-ascorbate oxidase (AO) and impacts ROS levels to regulate viral resistance in rice (Zhu et al., 2020). In poplar, transgenic lines overexpressing miR472a exhibit reduced H_2O_2 accumulation compared to both non-transgenic plants and those expressing STTM472a after pathogen exposure (Su et al., 2018). In tomato, miR6024 targets *NLR* genes, downregulating its expression, leading to increased ROS production, and enhancing resistance to *Alternaria solani* infection (Dey et al., 2022). Viral infections can suppress miRNA28, which consequently increases AO levels and lowers ROS, weakening plant defenses. The SPL9 TF modulates this response by upregulating miR528 expression, which in turn downregulates *AO* gene expression. This SPL9-miR528-AO regulatory network can be targeted to enhance plant viral resistance (Yao et al., 2019).

While some miRNAs enhance resistance by promoting ROS accumulation, others like miR482 reduce ROS, which can compromise immunity, as observed in tomato (Ouyang et al., 2014). In addition to ROS regulation, miRNAs also influence secondary metabolite production, including lignins, flavonoids, and glucosinolates, which help neutralize ROS (Anjali et al., 2023). For example, Mh-miR397b targets the laccase gene *LAC7* involved in lignin biosynthesis and weakens *Malus hupehensis* resistance to *Botryosphaeria dothidea* (Yu et al., 2020; Sun et al., 2022). Other miRNAs, including miR168, miR397, miR482d, miR528, miR159, miR397, and miR827, also target laccase genes to regulate lignification of cell wall (Li et al., 2017a; Xia et al., 2018). Additionally, miR396 negatively regulates resistance to rice blast by targeting growth-regulating factor (*OsGRF*) genes linked to flavonoid biosynthesis genes, such as *flavanone 3-hydroxylase* (*F3H*) (Chandran et al., 2019; Dai et al., 2019). Through the coordinated regulation of ROS, hormone signaling, and secondary metabolism, miRNAs play a vital role in orchestrating complex immune responses to pathogens, suggesting their potential as targets for enhancing stress tolerance in crops.

8.4 CROSS-KINGDOM MOVEMENT OF MIRNAS DURING PLANT–PATHOGEN INTERACTION

Beyond their role in internal gene regulation, miRNAs act as intercellular signaling molecules, crossing species barriers and even kingdom boundaries to facilitate interaction between plants, microbes, and other organisms (Chi et al., 2023). This cross-kingdom movement of miRNAs adds complexity to the plant–pathogen interactions, allowing organisms to influence each other's gene expression. Through these exchanges, plants can deploy miRNAs to silence pathogen virulence genes, while pathogens can also transfer miRNAs to manipulate plant defenses, enabling a dynamic, reciprocal regulation that shapes infection outcomes. miRNAs are transported between organisms via extracellular vesicles or through direct cell-to-cell contact, using specialized structures such as fungal haustoria and plant plasmodesmata (Cai et al., 2018). Some of the miRNAs involved in the cross-kingdom movement from plants to pathogens are listed in Table 8.2.

TABLE 8.2
Cross-Kingdom Transfer of Representative miRNAs from Plants to Pathogens.

S. No.	Plants	miRNAs	Cross-Kingdom Species	Targets	References
1.	Arabidopsis	miR166	*Verticillium dahliae*	Virulence genes	Cai et al. (2018)
2.	Arabidopsis	miR173	*Botrytis cinerea*	Target mRNAs	Cai et al. (2018)
3.	Cotton	miR166 and miR159	*Verticillium dahliae*	*Clp-1* and *HiC-15*	Zhang et al. (2016)
4.	Honeysuckle	miR2911	Influenza A viruses (IAVs)	*PB2* and *NS1*	Zhou et al. (2015)
5.	Tomato	miR1001	*Botrytis cinerea*	*Bcin03g02170.1* and *Bcin10g01400.1*	Meng et al. (2020)
6.	Wheat	miR1023	*Fusarium graminearum*	*FGSG_03101*	Jiao and Peng (2018)

For instance, in *Arabidopsis*, miR159a and novel-7703-5p have been shown to move into the cabbage moth, targeting pathogen genes and hampering their development (Yang et al., 2023). Similarly, miR166 and miR173 in *Arabidopsis* target genes in *Verticillium dahliae* and *Botrytis cinerea*, respectively, to enhance plant immunity (Cai et al., 2018). In cotton, miR159 and miR166 are induced in response to *Verticillium dahliae* infection and exported to the pathogen to silence virulence-related genes *Clp-1* and *HiC-15*, respectively, thereby reducing the pathogen's virulence (Zhang et al., 2016). The reach of miRNAs extends even further across biological kingdoms. For example, miR2911 from honeysuckle (*Lonicera japonica*) inhibits the replication of influenza virus proteins PB2 and NS1, demonstrating the broad utility of plant miRNAs in cross-kingdom gene regulation (Zhou et al., 2015). This inter-organismal miRNA communication illustrates how plants respond to pathogens, turning the tables on microbial invaders.

Pathogens have evolved their miRNA-based strategies to counter plant defenses. The parasitic plant *Cuscuta campestris* delivers miR393 into its host, suppressing pathogenesis-related genes to weaken the host's defense response (Shahid et al., 2018; Hudzik et al., 2023). Similarly, *Fusarium oxysporum* uses fol-milR1 to target host protein kinases at the post-transcriptional level to enhance resistance in tomatoes (Ji et al., 2021). Other examples include miR236 from *Magnoporthe oryzae*, which targets an acetyltransferase gene, affecting appressorium formation and enhancing pathogenicity (Li et al., 2020), and milR7 from *Penicillium italicum*, which enhances resistance in *Citrus sinensis* (Yin et al., 2020). The rust pathogen *Puccinia striiformis* deploys Pst-milR1 to bind and suppress plant defense gene *PR2*, weakening the host's immune response (Wang et al., 2017). These studies demonstrate that pathogen-derived miRNA can selectively target host defense mechanisms to facilitate infection. Some of the other miRNAs that transfer from pathogens to plants are listed in Table 8.3.

These exchanges of miRNAs exemplify a molecular "arms race" between plants and pathogens, where each continually evolves strategies to outmaneuver the other. This cross-kingdom movement of miRNAs exemplifies the complex interplay of signaling molecules during plant–microbe interaction and underscores the significance of plant–microbe synergy in regulating immune responses. By targeting these inter-organismal miRNA networks, researchers may develop innovative strategies to manipulate plant immune responses, providing a promising approach to bolster crop resistance against an ever-evolving landscape of microbial threats.

8.5 CROSS-KINGDOM MOVEMENT OF MIRNAS IN PLANT–MICROBE SYMBIOSIS

In addition to their roles in pathogen defense, miRNAs play vital roles in symbiotic relationships, by enabling cross-kingdom communication between plants and their microbial

TABLE 8.3
Cross-Kingdom Transfer of Representative miRNAs from Pathogens to Plants.

S. No.	Pathogens	miRNA	Cross-Kingdom Species	Targets	References
1.	*Cuscuta campestris*	miR393	*Arabidopsis thaliana*	*TIR1*, *AFB2/3*, *HSFB4*, and *BIK1*	Shahid et al. (2018); Hudzik et al. (2023)
2.	*Fusarium oxysporum*	Fol-milR1	*Solanum lycopersicum*	*FRG-4*	Ji et al. (2021)
3.	*Magnaporthe oryzae*	milR236	*Oryza sativa*	*MoHAT1*	Li et al. (2020)
4.	*Penicillium italicum*	milR7	*Citrus sinensis*	*AP2/B3*	Yin et al. (2020)
5.	*Pisolithus microcarpus*	Pmic_miR-8	*Eucalyptus grandis*	*NB-ARC*	Wong-Bajracharya et al. (2022)
6.	*Puccinia striiformis*	pst-milR1	*Triticum aestivum*	*PR2*	Wang et al. (2017)
7.	*Valsa mali*	MilR16	*Malus domestica*	*VmSNF1*, *VmDODA*, and *VmHy1*	Xu et al. (2020)
8.	*Verticillium dahliae*	VdmilR1	*Gossypium herbaceum*	*VdHy1*	Jin et al. (2019)

symbionts. This exchange facilitates gene expression regulation in both plant hosts and symbiotic microbes, optimizing nutrient acquisition, growth, and stress tolerance resilience (Gabriel et al., 2019). This bidirectional exchange, as summarized in Figure 8.1, allows for a dynamic interaction, enhancing adaptability for both partners in response to environmental challenges (Gualtieri et al., 2020).

In symbiotic plant species, miRNA expression patterns differ significantly from those in non-symbiotic species. For example, miRNAs transferred from microbes to plants alter metabolic and defense processes in the roots of *Medicago* species, enhancing the plant's response to the environment.

8.5.1 Rhizobia–Legume Nitrogen-Fixing Symbiosis

Rhizobia–legume symbiosis is a model of plant–microbe synergy, where legumes benefit from a unique mutualistic relationship with nitrogen-fixing rhizobia. During this interaction, nitrogen-fixing *Rhizobium* bacteria convert atmospheric nitrogen (N_2) into ammonia to support the host plant's growth, while the host provides the bacteria with essential nutrients. This nitrogen-fixation process is initiated when legumes secrete flavonoids, which attract rhizobia to the root surfaces. Following root hair curling and bacterial infection, specialized root structures known as nodules form, where nitrogen fixation occurs (Yang et al., 2022).

Several miRNAs play crucial regulatory roles in these symbiotic interactions. miR172 silences *APETALA2* (*AP2*) TFs for the development of nodulation primordium in several legumes like common beans, soybeans, *Medicago truncatula*, and *Lonicera japonica* (Wang et al., 2014). Overexpression of miR172 has been shown to increase both nodule numbers and nitrogenase activity, enhancing nitrogen fixation efficiency (Yan et al., 2013; Nova-Franco et al., 2015; Lelandais-Brière et al., 2016). In soybean, miRNAs, such as miR482 and miR1515, also contribute to enhancing nodule development (Li et al., 2010). In *Lotus japonicus*, miR171c and miR397 are involved in nodule development, with miR171c highly expressed in younger nodules and miR397 more abundant in mature nodules (De Luis et al., 2012). In *Medicago truncatula*, miRNAs such as Mtr-MIR2586 and Mtr-sRNA107 are expressed in the meristematic zone, while miR172 and miR398 accumulate in the differentiation zone of the root nodules (Lelandais-Brière et al., 2009). Collectively, these miRNAs regulate different stages of nodule formation, ensuring efficient nitrogen fixation.

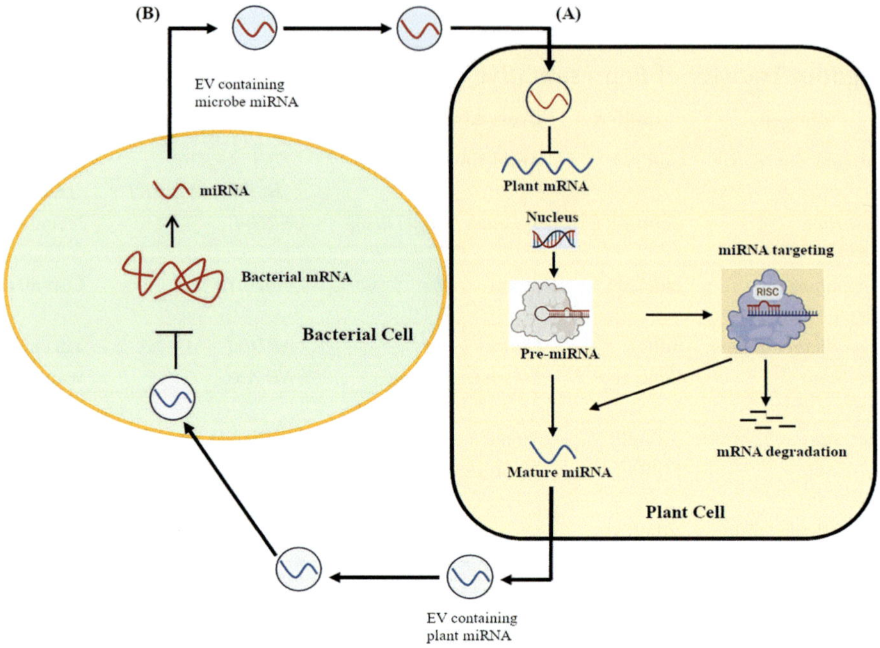

FIGURE 8.1 Overview of miRNA biogenesis and cross-kingdom movement of miRNA. (A) Biogenesis of miRNAs: *MIR* genes are transcribed by RNA polymerase II in the nucleus to produce pri-miRNAs. The pri-miRNAs are processed by Dicer-like enzymes into pre-miRNAs, which are cleaved into mature miRNAs. The mature miRNAs are incorporated into the RISC, guiding mRNA degradation or translational repression. (B) Cross-kingdom movement of miRNAs between microbes and plants: Microbial miRNAs are packaged in extracellular vesicles (EVs) and transferred into plant cells. Once inside, they target plant mRNAs, suppressing the expression of genes related to plant immunity, including *R* genes facilitating pathogen virulence. Conversely, plant-derived mature miRNAs are packaged in EVs and transferred into microbial cells. These miRNAs target complementary microbial mRNAs, silencing virulence factors or other key microbial genes, influencing microbial pathogenicity or promoting beneficial symbiotic behaviors. Plant miRNA and microbial miRNA are represented in blue and red, respectively.

miRNAs also mediate the crosstalk between phytohormones and rhizobia in root nodulation (Liu et al., 2018). Auxin-responsive miRNAs, like miR164, target NAC TF MtNAC1 in *Medicago truncatula*, influencing cell division and auxin signaling during nodule initiation (D'haeseleer et al., 2011). Additionally, miR160 regulates *ARF* genes, *ARF10*, *16*, and *17*, and influences plant development. Silencing *ARF17* regulates the transcript level of *GH3*, which is auxin-inducible, and ultimately impacts plant developmental responses (Mallory et al., 2005). In soybean, miR167c targets *ARF6* and *ARF8* and regulates the number of nodules (Wang et al., 2015). Cytokinin, another important hormone in nodule formation, influences miR171h expression, which targets, *MtNSP2*, a key component of the nodule signaling pathway in *Medicago truncatula* (Ariel et al., 2012). Similar to auxin and cytokinin, strigolactones also positively regulate nodulation, while other phytohormones like gibberellic acid (GA), ethylene, JA, brassinosteroids, and ABA act as negative regulators (Liu et al., 2018). This coordinated miRNA–phytohormone regulation underscores a finely tuned regulatory system to optimize rhizobia–legume symbiosis. The complex interplay of miRNAs and phytohormones in these interactions highlights the importance of plant–microbe synergy in optimizing legume productivity and ecosystem nitrogen cycling.

8.5.2 Plant–Fungi Symbiosis

The mutualistic association between plant roots and arbuscular mycorrhizal fungi (AMF) represents a significant symbiotic relationship. In this symbiosis, AMF rely on host plants for photosynthetic products, while plants benefit from improved nutrient acquisition, especially phosphorus, and resistance to biotic and abiotic stresses (Montesinos-Navarro et al., 2012). This symbiosis primarily involves flowering plants and fungi from the phylum Glomeromycota (Wipf et al., 2019). During AMF colonization, the fungi form arbuscles within the root cortical cells, where nutrient exchange takes place. These arbuscules are surrounded by a periarbuscular membrane, a specialized plant-derived structure that facilitates nutrient transfer (Lanfranco et al., 2016). The colonization of AMF triggers significant reprogramming of root cells to support and maintain this endosymbiotic relationship.

The plant–AMF symbiosis exemplifies a coordinated balance between nutrient acquisition and immune response, acting either antagonistically or synergistically. For instance, in *Oryza sativa–Magnaporthe oryzae* interaction, high phosphate and nitrogen levels suppress immunity to fungal infection, while increased iron levels promote resistance (Li et al., 2017b; Sánchez-Sanuy et al., 2019; Yang et al., 2022; Julian Maywald et al., 2023). *Magnaporthe oryzae* can also suppress plant immunity by reducing potassium ion influx through the potassium channel, underscoring the role of ion balance in plant defense (Shi et al., 2018). Additionally, nutrient transporters like ammonium transporters (AMT) facilitate ammonium uptake, which promotes good pathogen growth. AMT downregulation in wheat has been linked to increased resistance to mildew, while *AMT* mutants in cucumber exhibit resistance to fungal infections (Jiang et al., 2022). Reduced ammonium levels, in turn, activate defense responses. This highlights the importance of nutrient homeostasis in plant immunity.

miRNAs are increasingly recognized as key regulators in AMF symbiosis, mediating key pathways in nutrient management and immune response. Notably, miR399 accumulates in the roots of *Medicago truncatula* and *Nicotiana tabacum* during AMF colonization and regulates phosphorus homeostasis (Wu et al., 2016). miR827 in rice regulates phosphorus levels by targeting the phosphate transporter gene *OsSPX-MFS*, ensuring controlled phosphorus levels in plants (Puga et al., 2017). The wheat stripe rust pathogen (*Puccinia striiformis*) introduces its miRNA, pst-milR1, into wheat cells to suppress immune responses by targeting immune-related genes such as *SM638* and *PR* genes (Wang et al., 2017).

8.5.3 Plant Growth-Promoting Rhizobacteria Symbiosis

The rhizosphere, the soil region surrounding plant roots, hosts a diverse microbiome, including bacteria, fungi, and other microorganisms. These microbes can be transferred from the environment or directly inherited through the seed of the parent plant (Qu et al., 2020). Plants secrete phenolic compounds and other organic materials to attract rhizobacteria, while rhizobacteria produce phytohormones, secondary metabolites, to enhance plant tolerance to biotic and abiotic stresses (Middleton et al., 2021). Rhizobacteria play a crucial role in stress mitigation by inducing antioxidant enzyme activity and stimulating the production of metabolites like proline and sugars, which increase plant resilience. Additionally, rhizobacteria ameliorate biotic stress by activating immune responses, often through the production of elicitors that primed plant defenses (Khan et al., 2021). Rhizobacteria also promote plant growth by releasing phytohormones such as auxins and gibberellins, which modulate root architecture, nutrient uptake, and overall plant development (Saeed et al., 2021). The composition and function of rhizospheric microbiota are shaped by root exudates, which guide the colonization and structuring of microbial community (Middleton et al., 2021).

miRNAs are integral to plant–rhizobacteria symbiosis, enabling cross-kingdom communication through EVs, which allow miRNA transfer from plant to microbe and vice versa. A notable example is the interaction between ginger and *Lactobacillus*, where plant miRNAs target the bacterial monooxygenase genes, thereby regulating bacterial growth (Teng et al., 2018). This cross-kingdom exchange facilitates mutual benefits by fine-tuning gene expression on both sides, thus reinforcing the synergy between plants and rhizobacteria.

Both plant- and rhizobacteria-derived EVs play distinct roles in this interspecies communication. In plants, EVs originate from the budding of the plasma membrane, while in rhizobacteria, they are derived from the outer bacterial membrane. Within these vesicles, miRNAs are sorted and transported to interact with specific genes in other organisms (Choi et al., 2017). This exchange not only regulates essential processes like nutrient acquisition and growth but also modulates immune responses. miRNAs from pathogens can target virulence genes and suppress the immune response, while plant miRNAs can target these virulence-related genes in pathogens to counteract these effects (Gabriel et al., 2019). The role of miRNAs in both intracellular communication over short distances and intercellular communication over long distances underlines their importance in maintaining plant–microbe symbiosis (Middleton et al., 2021). These miRNA-mediated mechanisms offer a fine-tuned regulatory framework that supports plant growth, enhances stress tolerance, and mitigates biotic stresses, demonstrating the critical role of rhizobacteria in the broader scope of plant–microbe synergy.

8.6 ENGINEERING MIRNA FOR ENHANCING PLANT–MICROBE SYNERGY AND FOOD SECURITY

Plant pathogens significantly impact crop yields and present a major challenge to global agricultural productivity (Ristaino et al., 2021). While pesticides help control disease spread, they pose environmental and health risks. miRNAs present an eco-friendly alternative for enhancing pathogen resistance and promoting synergistic plant–microbe interactions. Advances in miRNA research have opened up new avenues for miRNA-based approaches in plant breeding, particularly in developing disease-resistant crops. Overexpression of rice-specific miRNA osa-miR7695 enhances resistance against the fungal pathogen *Magnoporthe oryzae* (Sánchez-Sanuy et al., 2019).

Advances in high-throughput sequencing and molecular tools, like small RNA sequencing and computational analysis, have enabled the identification of both conserved and novel miRNAs mediating plant–microbe signaling (Zhou et al., 2022a). Single-cell RNA sequencing (scRNA-seq) has further advanced the field by providing a high-resolution view of miRNA expression at the cellular level. This technology reveals the cell-specific roles of miRNAs in regulating plant–microbe interactions across cellular niches (Wu et al., 2016). The role of miRNA in plant–microbe symbiosis and its application in crop improvement are summarized in Figure 8.2.

Artificial miRNA (amiRNA) technology has emerged as a powerful tool for precise silencing of genes (Franco-Zorrilla et al., 2007). Unlike RNA interference (RNAi), which can sometimes yield off-target effects, amiRNAs are specifically engineered to precisely target and manipulate specific genes, especially against rapidly mutating viral pathogens (Thakur et al., 2015). Engineered amiRNAs that target conserved viral genome regions have provided broad-spectrum resistance across multiple viral strains. For example, amiRNAs designed to inhibit the ATP/GTP binding domain of the viral *AC1* gene conferred resistance against tomato leaf curl New Delhi virus (ToLCNDV) in tomatoes and resistance against rice stripe virus (Sharma et al., 2020; Zhou et al., 2022b). amiRNAs have also been successful in reducing infections by potato virus Y (PVY) and tobacco etch virus (TEV) (Song et al., 2024).

One of the most promising areas in miRNA application is engineering miRNAs to enhance plant–microbe symbiosis. miRNAs can be genetically manipulated to improve nitrogen fixation, nutrient exchange, and stress tolerance in plants. For example, engineering miRNAs that regulate nitrogen-fixing genes can improve legume–rhizobia symbiosis, reducing the need for synthetic fertilizers (Zhang et al., 2019). Similarly, engineering miRNAs to suppress genes that negatively affect symbiosis can enhance the mutual benefits between plants and microbes. Additionally, miRNAs

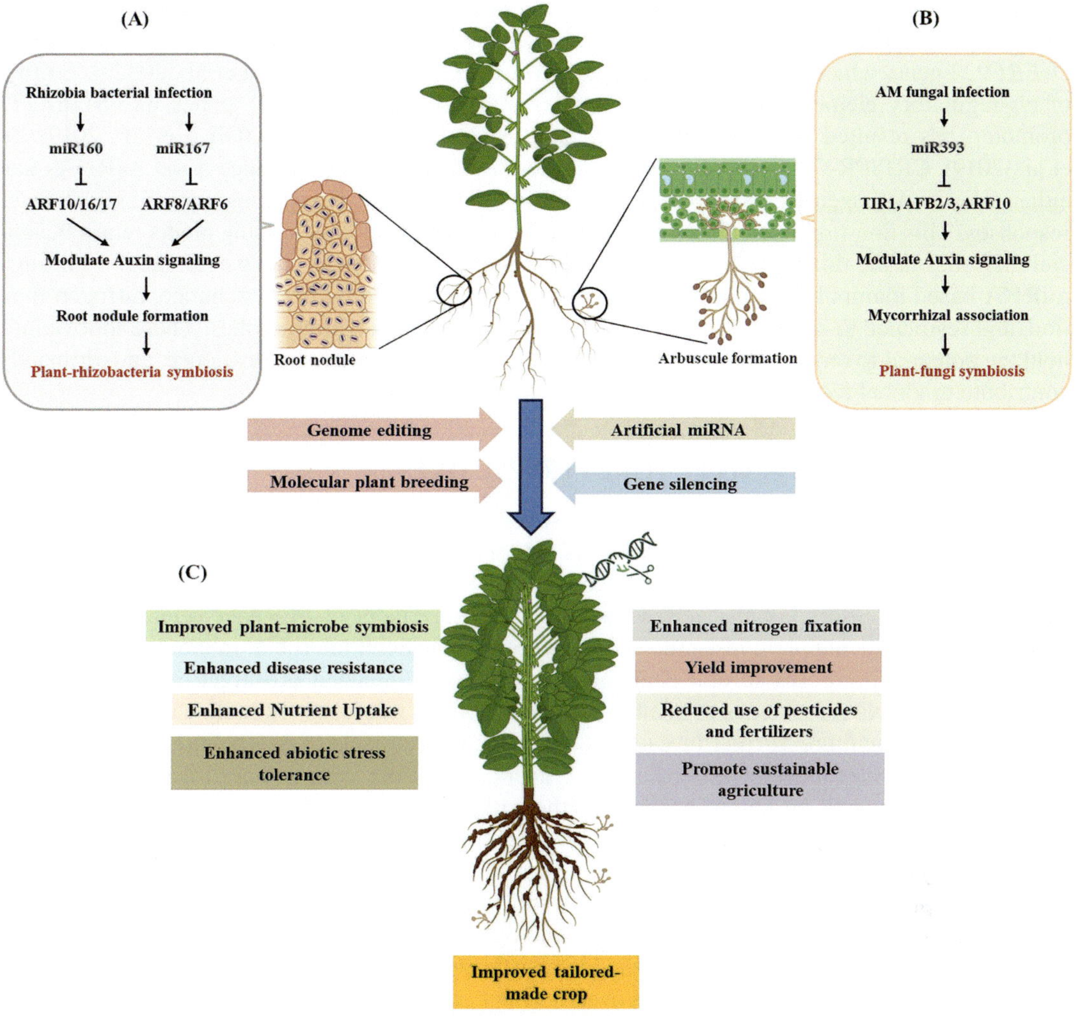

FIGURE 8.2 miRNA-mediated regulation of plant–microbe symbiosis and applications for crop improvement. (A) Depiction of root nodule formation in legumes, where miR160 and miR167 jointly regulate auxin signaling during rhizobial infection. After rhizobial bacteria attaches to root hairs, it triggers miR160-mediated repression of *ARF10/16/17* and miR167 represses *ARF8/ARF6* genes. Both pathways modulate auxin signaling to promote root nodule formation. This coordinated regulation ensures effective nodulation in response to rhizobial infection. (B) Mycorrhizal association: AMF attach to plant roots, forming a symbiotic association with the plant. During this interaction, miR393 regulates auxin signaling by repressing *TIR1*, *AFB2/3*, and *ARF10*, facilitating mycorrhizal colonization. (C) Visual representation of crop improvement strategies, including genome editing, artificial miRNA, plant breeding, and gene silencing, used to enhance traits such as disease resistance, nutrient uptake, nitrogen fixation, and phosphorus uptake. These innovations can lead to healthier, improved plants with better agronomic performance.

can be designed to enhance the secretion of root exudates, improve beneficial microbial colonization, improve phosphate uptake, and increase stress resilience (Yogindran and Rajam, 2021). The CRISPR-Cas9 system has become a powerful tool for precise editing of miRNA or their target sites in both plants and pathogens, to modulate host–pathogen interactions effectively (Mishra et al., 2018). By facilitating the design of synthetic miRNAs that specifically target pathogen virulence genes, this technology offers a safer, and more targeted alternative to traditional transgenic methods, helping alleviate public safety concerns associated with the introduction of foreign DNA. This tool can be used to directly disrupt disease-causing genes, as shown by knockouts like the rice *OsERF922* gene, which enhanced resistance to *Magnaporthe oryzae* (Zhou et al., 2022a). Editing of *cis*-regulatory elements, such as the effector-binding element (EBE) in genes like *SWEET11* promoter, has resulted in enhanced resistance to bacterial blight in rice (Oliva et al., 2019; Xu et al., 2019). CRISPR-Cas9 can be used to engineer miRNA genes to produce novel miRNAs that either silence pathogen targets or introduce silent mutations in host genes, optimizing defense responses. This fine-tuning can improve a balanced immune response, helping plants retain beneficial microbes in the rhizosphere while warding off harmful pathogens (Jiang et al., 2017). Through miRNA-based manipulation, plants can be tailored for enhanced nutrient exchange, nitrogen fixation, and resilience to stress, reinforcing symbiotic relationships with microbes. These innovations hold the potential to reduce dependency on chemical inputs, promote crop resilience, and ultimately contribute to global food security.

8.7 CONCLUSIONS AND FUTURE PERSPECTIVES

As global temperatures continue to rise, biotic stresses, including bacteria, fungi, and insects, pose an increasing threat to food security. miRNAs play a crucial role in plant growth, stress response, hormone signaling, and defense against pathogens by regulating gene expression at the transcriptional and post-transcriptional levels. Cross-kingdom movement of miRNAs offers a promising approach for enhancing plant resistance and promoting beneficial microbial interactions. However, miRNA engineering poses several challenges, including unintended off-target effects that could disrupt essential plant processes or symbiosis with beneficial microbes. To realize the full potential of miRNA in agriculture, further investigation into the mechanisms of miRNA transport, specificity, and cross-species movement is essential. Understanding the signals that direct miRNA selection and transport will be key to harnessing their regulatory functions. Research should explore whether miRNA transport mechanisms vary by pathogen type and how miRNA influences target gene expression at both transcriptional and post-transcriptional levels. Integrating CRISPR-Cas9 and high-throughput technologies, like scRNA-seq, with miRNA research holds significant promise for addressing these challenges. These technologies open up exciting opportunities to develop crops that not only exhibit enhanced pathogen resistance but also optimize symbiotic relationships with beneficial microbes, leading to improved nutrient acquisition and stress resilience. As the field advances, miRNAs may become central players in next-generation agricultural practices, leading to the development of climate-resilient and high-yielding crops for global food security amid climate change and population challenges.

ACKNOWLEDGMENTS

The authors acknowledge the financial support provided for this study by a research grant from the Department of Biotechnology (DBT), government of India, and the Science and Engineering Research Board (SERB). P.J. acknowledges support from the Research Associateship awarded by SERB, and V.T. acknowledges the DBT-Research Associateship (DBT-RA) Program.

REFERENCES

Anjali, K.S., Korra, T., Thakur, R., Arutselvan, R., Kashyap, A.S., Nehela, Y., Chaplygin, V., Minkina, T., Keswani, C. (2023) Role of plant secondary metabolites in defense and transcriptional regulation in response to biotic stress. *Plant Stress* 8: 100154.

Ariel, F., Brault-Hernandez, M., Laffont, C. (2012) Two direct targets of cytokinin signaling regulate symbiotic nodulation in *Medicago truncatula*. *The Plant Cell* 24: 3838–3852.

Bais, H.P., Weir, T.L., Perry, L.G., Gilroy, S., Vivanco, J.M. (2006) The role of root exudates in rhizosphere interactions with plants and other organisms. *Annual Reviews of Plant Biology* 57, 233–266.

Bin, Y., Wang, H. (2010) Translational inhibition by microRNAs in plants. *Progress in Molecular and Subcellular Biology* 50: 41–57.

Bologna, N.G., Iselin, R., Abriata, L.A, Sarazin, A., Pumplin, N., Jay, F., Grentzinger, T., Dal Peraro, M., Voinnet, O. (2018) Nucleo-cytosolic shuttling of ARGONAUTE1 prompts a revised model of the plant microRNA pathway. *Molecular Cell* 69: 709–719.

Cai, Q., Qiao, L., Wang, M., He, B., Lin, F.M., Palmquist, J., Huang, S.D., Jin, H. (2018) Plants send small RNAs in extracellular vesicles to fungal pathogen to silence virulence genes. *Science* 360: 1126–1129.

Chandran, V., Wang, H., Gao, F., Cao, X.L., Chen, Y.P., Li, G.B., Zhu, Y., Yang, X.M., Zhang, L.L., Zhao, Z.X., Zhao, J.H., Wang, Y.G., Li, S., Fan, J., Li, Y., Zhao, J.Q., Li, S.Q., Wang, W.M. (2019) miR396-OsGRFs module balances growth and rice blast disease-resistance. *Frontiers in Plant Science* 9: 1999.

Chen, X. (2005) MicroRNA biogenesis and function in plants. *FEBS Letters* 579: 5923–5931.

Chi, X., Wang, Z., Wang, Y., Liu, Z., Wang, H., Xu, B. (2023) Cross-kingdom regulation of plant-derived miRNAs in modulating insect development. *International Journal of Molecular Sciences* 24: 7978.

Choi, J.W., Um, J.H., Cho, J.H., Lee, H.J. (2017) Tiny RNAs and their voyage via extracellular vesicles: Secretion of bacterial small RNA and eukaryotic microRNA. *Experimental Biology and Medicine* 242: 1475–1481.

Dai, Z., Tan, J., Zhou, C., Yang, X., Yang, F., Zhang, S., Sun, S., Miao, X., Shi, Z. (2019) The OsmiR396-OsGRF8-OsF3H-flavonoid pathway mediates resistance to the brown planthopper in rice (*Oryza sativa*). *Plant Biotechnology Journal* 17: 1657–1669.

De Luis, A., Markmann, K., Cognat, V., Holt, D.B., Charpentier, M., Parniske, M., Stougaard, J., Voinnet, O. (2012) Two microRNAs linked to nodule infection and nitrogen-fixing ability in the legume *Lotus japonicus*. *Plant Physiology* 160: 2137–2154.

Dey, S., Sarkar, A., Chowdhury, S., Singh, R., Mukherjee, A., Ghosh, Z., Kundu, P. (2022) Heightened miR6024-NLR interactions facilitate necrotrophic pathogenesis in tomato. *Plant Molecular Biology* 109: 717–739.

D'haeseleer, K., Den Herder, G., Laffont, C. (2011) Transcriptional and post-transcriptional regulation of a NAC1 transcription factor in *Medicago truncatula* roots. *New Phytologist* 191: 647–661.

Ding, T., Li, W., Li, F., Ren, M., Wang, W. (2024) microRNAs: Key regulators in plant responses to abiotic and biotic stresses via endogenous and cross-kingdom mechanisms. *International Journal of Molecular Sciences* 25: 1154.

Djami-Tchatchou, A.T., Dubery, I.A. (2019) miR393 regulation of lectin receptor-like kinases associated with LPS perception in *Arabidopsis thaliana. Biochemical and Biophysical Research Communications* 513: 88–92.

Dong, Q., Hu, B., Zhang, C. (2022) microRNAs and their roles in plant development. *Frontiers in Plant Science* 13: 824240.

Gabriel, A.F., Costa, M.C., Enguita, F.J., Leitão, A.L. (2019) Si vis pacem para bellum: A prospective in silico analysis of miRNA-based plant defenses against fungal infections. *Plant Science* 288: 110241.

Gualtieri, C., Leonetti, P., Macovei, A. (2020) Plant miRNA cross-kingdom transfer targeting parasitic and mutualistic organisms as a tool to advance modern agriculture. *Frontiers in Plant Science* 11: 930.

Franco-Zorrilla, J.M., Valli, A., Todesco, M., Mateos, I., Puga, M.I., Rubio-Somoza, I., Leyva, A., Weigel, D., García, J.A., Paz-Ares, J. (2007) Target mimicry provides a new mechanism for regulation of microRNA activity. *Nature Genetics* 39: 1033–1037.

Hoang, N.T., Tóth, K., Stacey, G. (2020) The role of microRNAs in the legume–*Rhizobium* nitrogen-fixing symbiosis. *Journal of Experimental Botany* 71(5): 1668–1680.

Hudzik, C., Maguire, S., Guan, S., Held, J., Axtell, M.J. (2023) Trans-species microRNA loci in the parasitic plant *Cuscuta campestris* have a U6-like snRNA promoter. *The Plant Cell* 35: 1834–1847.

Ji, H.M., Mao, H.Y., Li, S.J., Feng, T., Zhang, Z.Y., Cheng, L., Luo, S.J., Borkovich Katherine, A., Ouyang, S.Q. (2021) Fol-milR1, a pathogenicity factor of *Fusarium oxysporum*, confers tomato wilt disease resistance by impairing host immune responses. *New Phytologist* 232: 705–718.

Jiang, J., Zhum, H., Li, N., Batley, J., Wang, Y. (2022) The miR393-target module regulates plant development and responses to biotic and abiotic stresses. *International Journal of Molecular Sciences* 23: 9477.

Jiang, S., Wu, H., Liu, H., Zheng, J., Lin, Y., Chen, H. (2017) The overexpression of insect endogenous small RNAs in transgenic rice inhibits growth and delays pupation of striped stem borer (*Chilo suppressalis*). *Pest Management Science* 73: 1453–1461.

Jiao, J., Peng, D. (2018) Wheat microRNA1023 suppresses invasion of *Fusarium graminearum* via targeting and silencing FGSG_03101. *Journal of Plant Interactions* 13: 514–521.

Jin, Y., Zhao, J.H., Zhao, P., Zhang, T., Wang, S., Guo, H.S. (2019) A fungal milRNA mediates epigenetic repression of a virulence gene in *Verticillium dahliae*. *Philosophical Transactions of the Royal Society B: Biological Sciences* 374: 20180309.

Jones-Rhoades, M.W., Bartel, D.P., Bartel, B. (2006) MicroRNAs and their regulatory roles in plants. *Annual Review of Plant Biology* 57: 19–53.

Julian Maywald, N., Francioli, D., Mang, M., Ludewig, U. (2023) Role of mineral nitrogen nutrition in fungal plant diseases of cereal crops. *Critical Reviews in Plant Sciences* 42: 93–123.

Khan, N., Ali, S., Shahid, M.A., Mustafa, A., Sayyed, R.Z., Curá, J.A. (2021) Insights into the Interactions among roots, rhizosphere, and rhizobacteria for improving plant growth and tolerance to abiotic stresses: A review. *Cells* 10: 1551.

Khraiwesh, B., Arif, M.A., Seumel, G.I., Ossowski, S., Weigel, D., Reski, R., Frank, W. (2010) Transcriptional control of gene expression by MicroRNAs. *Cells* 140: 111–122.

Kozomara, A., Birgaoanu, M., Griffiths-Jones, S. (2019) miRBase: From microRNA sequences to function. *Nucleic Acids Research* 47: D155–D162.

Lanfranco, L., Bonfante, P., Genre, A. (2016) The mutualistic interaction between plants and arbuscular mycorrhizal fungi. *Microbiology Spectrum* 4. https://doi.org/10.1128/microbiolspec.funk-0012-2016

Lee, R.C., Feinbaum, R.L., Ambros, V. (1993) The *C. elegans* heterochronic gene *lin-4* encodes small RNAs with antisense complementarity to lin–14. *Cell* 75: 843–54.

Lelandais-Brière, C., Moreau, J., Hartmann, C., Crespi, M. (2016) Noncoding RNAs, emerging regulators in root endosymbioses. *Molecular Plant–Microbe Interactions* 29: 170–180.

Lelandais-Brière, C., Naya, L., Sallet, E., Calenge, F., Frugier, F., Hartmann, C., Gouzy, J., Crespi, M. (2009) Genome-wide *Medicago truncatula* small RNA analysis revealed novel microRNAs and isoforms differentially regulated in roots and nodules. *The Plant Cell* 21: 2780–2796.

Li, H., Deng, Y., Wu, T., Subramanian, S., Yu, O. (2010) Misexpression of miR482, miR1512, and miR1515 increases soybean nodulation. *Plant Physiology* 153: 1759–1770.

Li, Y., Cao, X.L., Zhu, Y., Yang, X.M., Zhang, K.N., Xiao. Z.Y., Wang, H., Zhao, J.H., Zhang, L.L., Li, G.B., Zheng, Y.P., Fan, J., Wang, J., Chen, X.Q., Wu, X.J., Zhao, J.Q., Dong, O.X., Chen, X.W., Chern, M., Wang, W.M. (2019) Osa-miR398b boosts H_2O_2 production and rice blast disease-resistance via multiple superoxide dismutases. *New Phytologist* 222: 1507–1522.

Li, Y., Hu, X., Chen, J., Wang, W., Xiong, X., He, C. (2017a) Integrated mRNA and microRNA transcriptome analysis reveals miRNA regulation in response to PVA in potato. *Scientific Reports* 7: 16925.

Li, Y., Liu, X., Liu, M., Wang, Y., Zou, Y., You, Y., Yang, L., Hu, J., Zhang, H., Zheng, X., Wang, P., Zhang, Z. (2020) *Magnaporthe oryzae* auxiliary activity protein MoAa91 functions as chitin-binding protein to induce appressorium formation on artificial inductive surfaces and suppress plant immunity. *Microbiology* 11. https://doi.org/10.1128/mbio.03304-03319

Li, Y., Zhao, S.L., Li, J.L., Hu, X.H., Wang, H., Cao, X.L., Xu, Y.J., Zhao, Z.X., Xiao, Z.Y., Yang, N., Fan, J. (2017b) Osa-miR169 negatively regulates rice immunity against the blast fungus *Magnaporthe oryzae*. *Frontiers in Plant Science* 8: 2.

Liao, L., Xie, B., Guan, P., Jiang, N., Cui, J. (2022) New insight into the molecular mechanism of miR482/2118 during plant resistance to pathogens. *Frontiers in Plant Science* 13: 1026762.

Liu, H., Zhang, C., Yang, J., Yu, N., Wang, E. (2018) Hormone modulation of legume–rhizobial symbiosis. *Journal of Integrative Plant Biology* 60: 632–648.

Mallory, A.C., Bartel, D.P., Bartel, B. (2005) MicroRNA-directed regulation of Arabidopsis AUXIN RESPONSE FACTOR17 is essential for proper development and modulates expression of early auxin response genes. *The Plant Cell* 17: 1360–1375.

Meng, X., Jin, W., Wu, F. (2020) Novel tomato miRNA miR1001 initiates cross-species regulation to suppress the conidiospore germination and infection virulence of *Botrytis cinerea* in vitro. *Gene* 759: 145002.

Middleton, H., Yergeau, É., Monard, C., Combier, J.P., El Amrani, A. (2021) Rhizospheric plant–microbe interactions: miRNAs as a key mediator. *Trends in Plant Science* 26: 132–141.

Mishra, R., Mohanty, J.N., Chand, S.K., Joshi, R.K. (2018) Can-miRn37a mediated suppression of ethylene response factors enhances the resistance of chilli against anthracnose pathogen *Colletotrichum truncatum* L. *Plant Science* 267: 135–147.

Montesinos-Navarro, A., Segarra-Moragues, J.G., Valiente-Banuet, A., Verdú, M. (2012) The network structure of plant–arbuscular mycorrhizal fungi. *New Phytologist* 194: 536–547.

Natarajan, B., Kalsi, H.S., Godbole, P., Malankar, N., Thiagarayaselvam, A., Siddappa, S., Thulasiram, H.V., Chakrabarti, S.K., Banerjee, A.K. (2018) MiRNA160 is associated with local defense and systemic acquired resistance against *Phytophthora infestans* infection in potato. *Journal of Experimental Botany* 69: 2023–2036.

Navarro, L., Dunoyer, P., Jay, F., Arnold, B., Dharmasiri, N., Estelle, M., Voinnet, O., Jones, J.D.G. (2006) A plant miRNA contributes to antibacterial resistance by repressing auxin signaling. *Science* 312: 436–439.

Nova-Franco, B., Íñiguez, L.P., Valdés-López, O. (2015) The micro-RNA72c-APETALA2-1 node as a key regulator of the common bean–*Rhizobium etli* nitrogen fixation symbiosis. *Plant Physiology* 168: 273–291.

Oliva, R., Ji, C., Atienza-Grande, G., Huguet-Tapia, J.C., Perez-Quintero, A., Li, T., Eom, J.S., Li, C., Nguyen, H., Liu, B., Auguy, F., Sciallano, C., Luu, V.T., Dossa, G.S., Cunnac, S., Schmidt, S.M., Slamet-Loedin, I.H., Cruz, C.V., Szurek, B., Frommer, W.B., White, F.F, Yang, B. (2019) Broad-spectrum resistance to bacterial blight in rice using genome editing. *Nature Biotechnology* 37: 1344–1350.

Ouyang, S., Park, G., Atamian, H.S., Han, C.S., Stajich, J.E, Kaloshian, I., Borkovich, K.A. (2014) MicroRNAs suppress NB domain genes in tomato that confer resistance to *Fusarium oxysporum*. *PLOS Pathogens* 10: e1004464.

Pinweha, N., Asvarak, T., Viboonjun, U., Narangajavana, J. (2015) Involvement of miR160/miR393 and their targets in cassava responses to anthracnose disease. *Journal of Plant Physiology* 174: 26–35.

Puga, M.I., Rojas-Triana, M., de Lorenzo, L., Leyva, A., Rubio, V., Paz-Ares, J. (2017) Novel signals in the regulation of Pi starvation responses in plants: Facts and promises. *Current Opinion in Plant Biology* 39: 40–49.

Qu, Q., Zhang, Z., Peijnenburg, W., Liu, W., Lu, T., Hu, B., Chen, J., Chen, J., Lin, Z., Qian, H. (2020) Rhizosphere microbiome assembly and its impact on plant growth. *Journal of Agricultural and Food Chemistry* 68: 5024–5038.

Ristaino, J.B., Anderson, P.K., Bebber, D.P., Brauman, K.A., Cunniffe, N.J., Fedoroff, N.V., Finegold, C., Garrett, K.A., Gilligan, C.A., Jones, C.M., Martin, M.D., MacDonald, G.K., Neenan, P., Records, A., Schmale, D.G., Tateosian, L., Wei, Q. (2021) The persistent threat of emerging plant disease pandemics to global food security. *Proceedings of the National Academy of Sciences* 118: e2022239118.

Saeed, Q., Xiukang, W., Haider, F.U., Kučerik, J., Mumtaz, M.Z., Holatko, J., Naseem, M., Kintl, A., Ejaz, M., Naveed, M., Brtnicky, M., Mustafa, A. (2021) Rhizosphere bacteria in plant growth promotion, biocontrol, and bioremediation of contaminated sites: A comprehensive review of effects and mechanisms. *International Journal of Molecular Sciences* 22: 10529.

Saha, S. (2024) Role of microRNA in oxidative stress. *Stresses* 4: 269–281.

Sánchez-Sanuy, F., Peris-Peris, C., Tomiyama, S., Okada, K., Hsing, Y.I., San Segundo, B., Campo, S. (2019) Osa-miR7695 enhances transcriptional priming in defense responses against the rice blast fungus. *BMC Plant Biology* 19(1): 563.

Shahid, S., Kim, G., Johnson, N.R., Wafula, E., Wang, F., Coruh, C., Bernal-Galeano, V., Phifer, T., dePamphilis, C.W., Westwood, J.H., Axtell, M.J. (2018) MicroRNAs from the parasitic plant *Cuscuta campestris* target host messenger RNAs. *Nature* 553: 82–85.

Sharma, N., Prasad, M. (2020) Silencing *AC1* of Tomato leaf curl virus using artificial microRNA confers resistance to leaf curl disease in transgenic tomato. *Plant Cell Reports* 39(11): 1565–1579.

Shi, X., Long, Y., He, F., Zhang, C., Wang, R., Zhang, T., Wu, W., Hao, Z., Wang, Y. I., Wang, G. L., Ning, Y. (2018) The fungal pathogen *Magnaporthe oryzae* suppresses innate immunity by modulating a host potassium channel. *PLoS Pathogens* 14: e1006878.

Song, H., Gao, X., Song, L., Jiao, Y., Shen, L., Yang, J., Li, C., Shang, J., Wang, H., Zhang, S., Li, Y. (2024) Unraveling the regulatory network of miRNA expression in potato Y virus-infected of *Nicotiana benthamiana* using integrated small RNA and transcriptome sequencing. *Frontiers in Genetics*, 14, 1290466.

Su, Y., Li, H.G., Wan, Y., Li, S., Wang, H.L., Yu, L., He, F., Yang, Y., Feng, C.H., Shuai, P., Liu, C., Yin, W., Xia, X. (2018) Poplar miR472a targeting NBS-LRRs is involved in effective defense against the necrotrophic fungus *Cytospora chrysosperma*. *Journal of Experimental Botany* 69: 5519–5530.

Sun, Z., Zhou, Y., Hu, Y., Jiang, N., Hu, S., Li, L., Li, T. (2022) Identification of wheat LACCASEs in response to *Fusarium graminearum* as potential deoxynivalenol trappers. *Frontiers in Plant Science* 13: 832800.

Teng, Y., Ren, Y., Sayed, M., Hu, X., Lei, C., Kumar, A., Hutchins, E., Mu, J., Deng, Z., Luo, C., Sundaram, K., Sriwastva, M.K., Zhang, L., Hsieh, M., Reiman, R., Haribabu, B., Yan, J., Jala, V.R., Miller, D.M., Van Keuren-Jensen, K., Merchant, M.L, McClain, C.J, Park, J.W., Egilmez, N.K., Zhang, H.G. (2018) Plant-derived exosomal microRNAs shape the gut microbiota. *Cell Host & Microbe* 24: 637–652.e638.

Thakur, A., Sanju, S., Siddappa, S., Srivastava, N., Shukla, P.K., Pattanayak, D., Sharma. S., Singh, B. (2015) Artificial microRNA mediated gene silencing of *Phytophthora infestans* single effector *Avr3a* gene imparts moderate type of late blight resistance in potato. *Plant Pathology Journal* 14: 1–12.

Wang, B., Sun, Y., Song, N., Zhao, M., Liu, R., Feng, H., Wang, X., Kang, Z. (2017) *Puccinia striiformis f. sp. tritici* microRNA-like RNA 1 (Pst-milR1), an important pathogenicity factor of Pst, impairs wheat resistance to Pst by suppressing the wheat pathogenesis-related 2 gene. *New Phytologist* 215: 338–350.

Wang, J., Mei, J., Ren, G. (2019) Plant microRNAs: Biogenesis, homeostasis, and degradation. *Frontiers in Plant Science* 10: 360.

Wang, Y., Li, K., Chen, L. (2015) MicroRNA167-directed regulation of the auxin response factors GmARF8a and GmARF8b is required for soybean nodulation and lateral root development. *Plant Physiology* 168: 984–999.

Wang, Y., Wang, L., Zou, Y. (2014) Soybean miR172c targets the repressive AP2 transcription factor NNC1 to activate ENOD40 expression and regulate nodule initiation. *The Plant Cell* 26: 4782–4801.

Wei, C., Kuang, H., Li, F., Chen, J. (2014) The *I2* resistance gene homologues in *Solanum* have complex evolutionary patterns and are targeted by miRNAs. *BMC Genomics* 15: 743.

Wipf, D., Krajinski, F., van Tuinen, D., Recorbet, G., Courty, P.E. (2019) Trading on the arbuscular mycorrhiza market: From arbuscules to common mycorrhizal networks. *New Phytologist* 223: 1127–1142.

Wong-Bajracharya, J., Singan, V.R., Monti, R., Plett, K.L., Ng, V., Grigoriev, I.V., Martin, F.M., Anderson, I.C., Plett, J.M. (2022) The ectomycorrhizal fungus *Pisolithus microcarpus* encodes a microRNA involved in cross-kingdom gene silencing during symbiosis. *Proceedings of the National Academy of Sciences* 119: e2103527119.

Wu, P., Wu, Y., Liu, C.C., Liu, L.W., Ma, F.F., Wu, X.Y., Wu, M., Hang, Y.Y., Chen, J.Q., Shao, Z.Q., Wang, B. (2016) Identification of arbuscular mycorrhiza (AM)-responsive microRNAs in tomato. *Frontiers in Plant Science* 7: 429.

Xia, Z., Zhao, Z., Li, M., Chen, L., Jiao, Z., Wu, Y., Zhou, T., Yu, W., Fan, Z. (2018) Identification of miRNAs and their targets in maize in response to Sugarcane mosaic virus infection. *Plant Physiology and Biochemistry* 125: 143–152.

Xie, Z., Allen, E., Fahlgren, N., Calamar, A., Givan, S.A., Carrington, J.C. (2005) Expression of Arabidopsis miRNA genes. *Plant Physiology* 138: 2145–2154.

Xu, M., Guo, Y., Tian, R., Gao, C., Guo, F., Voegele, R.T., Bao, J., Li, C., Jia, C., Feng, H., Huang, L. (2020) Adaptive regulation of virulence genes by microRNA-like RNAs in *Valsa mali*. *New Phytologist* 227: 899–913.

Xu, Z., Xu, X., Gong, Q., Li, Z., Li, Y., Wang, S., Yang, Y., Ma, W., Liu, L., Zhu, B., Zou, L., Chen, G. (2019) Engineering broad-spectrum bacterial blight resistance by simultaneously disrupting variable TALE-binding elements of multiple susceptibility genes in rice. *Molecular Plant* 12: 1434–1446.

Yadav, R., Ramakrishna, W. (2024) MicroRNAs involved in nutritional regulation during plant–microbe symbiotic and pathogenic interactions with rice as a model. *Molecular Biotechnology* 66: 1754–1771.

Yan, Z., Hossain, M.S., Wang, J., Valdés-López, O., Liang, Y., Libault, M., Qiu, L., Stacey, G. (2013) miR172 regulates soybean nodulation. *Molecular Plant–Microbe Interactions* 26: 1371–1377.

Yang, J., Lan, L., Jin, Y., Yu, N., Wang, D, Wang, E. (2022) Mechanisms underlying legume–rhizobium symbioses. *Journal of Integrative Plant Biology* 64: 244–267.

Yang, X., Fu, T., Yu, R., Zhang, L., Yang, Y, Xiao, D., Wang, Y., Wang, Y., Wang, Y. (2023) miR159a modulates poplar resistance against different fungi and bacteria. *Plant Physiology and Biochemistry* 201: 107899.

Yao, S., Yang, Z., Yang, R., Huang, Y., Guo, G., Kong, X., Lan, Y., Zhou, T., Wang, H., Wang, W., Cao, X., Wu, J., Li, Y. (2019) Transcriptional regulation of miR528 by OsSPL9 orchestrates antiviral response in rice. *Molecular Plant* 12: 1114–1122.

Yin, C., Zhu, H., Jiang, Y., Shan, Y., Gong, L. (2020) Silencing dicer-like genes reduces virulence and sRNA generation in *Penicillium italicum*, the cause of citrus blue mold. *Cells* 9: 363.

Yogindran, S., Rajam, M.V. (2021) Host-derived artificial miRNA-mediated silencing of ecdysone receptor gene provides enhanced resistance to *Helicoverpa armigera* in tomato. *Genomics* 113: 736–747.

Yu, X., Gong, H., Cao, L., Hou, Y., Qu, S. (2020) MicroRNA397b negatively regulates resistance of *Malus hupehensis* to *Botryosphaeria dothidea* by modulating *MhLAC7* involved in lignin biosynthesis. *Plant Science* 292: 110390.

Yu, X., Hou, Y., Chen, W., Wang, S., Wang, P., Qu, S. (2017) *Malus hupehensis* miR168 targets to ARGONAUTE1 and contributes to the resistance against *Botryosphaeria dothidea* infection by altering defense responses. *Plant and Cell Physiology* 58: 1541–1557.

Zipfel, C. (2008) Pattern-recognition receptors in plant innate immunity. *Current Opinion in Immunology* 20: 10–16.

Zhang, A., Zhang, S., Wang, F., Meng, X, Ma, Y., Guan, J., Zhang, F. (2023) The roles of microRNAs in horticultural plant disease resistance. *Frontiers in Genetics* 14: 1137471.

Zhang, J., Zhou, J.M. (2010) Plant Immunity triggered by microbial molecular signatures. *Molecular Plant* 3: 783–793.

Zhang, L.L., Huang, Y.Y., Zheng, Y.P, Liu, X.X., Zhou, S.X., Yang, X.M., Liu, S.L., Li, Y., Li, J.L., Zhao, S.L., Wang, H., Ji, Y.P., Zhang, J.W., Pu, M., Zhao, Z.X., Fan, J., Wang, W.A-O. (2022) Osa-miR535 targets SQUAMOSA promoter binding protein-like 4 to regulate blast disease resistance in rice. *Plant Journal* 110: 166–178.

Zhang, L.-L., Jing. X.-D., Chen, W., Wang, Y., Lin, J.-H., Zheng, L., Dong, Y.-H., Zhou, L., Li, F.-F., Yang, F.-Y., Peng, L., Vasseur, L., He, W.-Y., You, M.-S. (2019) Host plant-derived miRNAs potentially modulate the development of a cosmopolitan insect pest, *Plutella xylostella*. *Biomolecules* 9: 602.

Zhang, Q., Li, Y., Zhang, Y., Wu, C., Wang, S., Hao, L., Wang, S., Li, T. (2017) Md-miR156ab and Md-miR395 target WRKY transcription factors to influence apple resistance to leaf spot disease. *Frontiers in Plant Science* 8: 526.

Zhang, T., Zhao, Y.-L., Zhao, J.-H., Wang, S., Jin, Y., Chen, Z.-Q., Fang, Y.-Y., Hua, C.-L., Ding, S.-W., Guo, H.-S. (2016) Cotton plants export microRNAs to inhibit virulence gene expression in a fungal pathogen. *Nature Plants* 2: 16153.

Zhang, Y.-Y., Hong, Y.-H., Liu, Y.-R., Cui, J., Luan, Y.-S. (2021) Function identification of miR394 in tomato resistance to *Phytophthora infestans*. *Plant Cell Reports* 40: 1831–1844.

Zhao, W., Li, Z., Fan, J., Hu, C., Yang, R., Qi, X., Chen, H., Zhao, F., Wang, S. (2015) Identification of jasmonic acid-associated microRNAs and characterization of the regulatory roles of the miR319/TCP4 module under root-knot nematode stress in tomato. *Journal of Experimental Botany* 66: 4653–4667.

Zhao, Z.X, Feng, Q., Cao, X.L., Zhu, Y., Wang, H., Chandran, V., Fan, J., Zhao, J.Q., Pu, M., Li, Y., Wang, W.M. (2020) Osa-miR167d facilitates infection of *Magnaporthe oryzae* in rice. *Journal of Integrative Plant Biology* 62: 702–715.

Zhou, L., Yuan, Q., Ai, X., Chen, J., Lu, Y., Yan, F. (2022b) Transgenic rice plants expressing artificial miRNA targeting the rice stripe virus MP gene are highly resistant to the virus. *Biology* 11: 332.

Zhou, Y., Xu, S., Jiang, N., Zhao, X., Bai, Z., Liu, J., Yao, W., Tang, Q., Xiao, G., Lv, C., Wang, K., Hu, X., Tan, J., Yang, Y. (2022a) Engineering of rice varieties with enhanced resistance to both blast and bacterial blight diseases via CRISPR/Cas9. *Plant Biotechnology Journal* 20(5): 876–885.

Zhou, Z., Li, X., Liu, J., Dong, L., Chen, Q., Liu, J., Kong, H., Zhang, Q., Qi, X., Hou, D., Zhang, L., Zhang, G., Liu, Y., Zhang, Y., Li, J., Wang, J., Chen, X., Wang, H., Zhang, J., Chen, H., Zen, K., Zhang, C.-Y. (2015) Honeysuckle-encoded atypical microRNA2911 directly targets influenza A viruses. *Cell Research* 25: 39–49.

Zhu, H., Chen, C., Zeng, J., Yun, Z., Liu, Y., Qu, H., Jiang, Y., Duan, X., Xia, R. (2020) MicroRNA528, a hub regulator modulating ROS homeostasis via targeting of a diverse set of genes encoding copper-containing proteins in monocots. *New Phytologist* 225: 385–399.

9 Long non-coding RNAs
Emerging Players in Plant–Microbe Interaction

Jaishri Rubina Das and Joyati Das

9.1 INTRODUCTION

Plants persistently experience various biotic and abiotic stress conditions throughout their life cycle. To counter these stress conditions, plants have developed complex and tightly regulated mechanisms that include a plethora of critical regulatory molecules. For a long time, only protein-coding genes were primarily considered as regulatory molecules to modulate the gene expression in response to stress. However, in the past two decades, the constant and rapid advancement in sequencing technologies and transcriptomic analysis have highlighted the regulation by noncoding RNAs (ncRNAs), a class of RNAs that do not code for protein yet control the level of gene expression. The ncRNAs are classified into two main groups: (i) Housekeeping ncRNAs, which include transfer RNAs (tRNAs), small nuclear RNAs (snRNAs), ribosomal RNAs (rRNAs), small nucleolar RNAs (snoRNAs), and telomerase RNAs, and (ii) regulatory ncRNAs, which comprise microRNAs (miRNAs), short interfering RNAs (siRNAs), piwi-interacting RNAs (piRNAs), and long noncoding RNAs (lncRNAs) (Wang *et al.*, 2017; Waititu *et al.*, 2020; Traubenik *et al.*, 2024). Notably, regulatory ncRNAs vary in size and therefore can be subdivided as (i) small regulatory ncRNAs, which are typically 21–24 nucleotides in size, such as miRNAs, siRNAs, and piRNAs, and (ii) large regulatory ncRNAs, which can be 200 nucleotides to several kilobases long, such as lncRNAs (Traubenik *et al.*, 2024). Small regulatory ncRNAs have been extensively studied and reports suggests their critical role in plant development and stress response (Yu *et al.*, 2019; Waititu *et al.*, 2020; Nie *et al.*, 2023). On the other hand, lncRNAs discovered through large-scale genome sequencing still require further comprehensive studies to fully understand their mechanistic regulation in plants.

Similar to small ncRNAs, lncRNAs are also considered as a versatile regulator due to their involvement in plant development, molecular processes, and abiotic and biotic stress responses (Wang and Chekanova, 2017; Jin *et al.*, 2024; Gonzales *et al.*, 2024). For instance, they are reported to play a crucial role in developmental stages such as pollen development, fertilization, fruit ripening, and sexual differentiation in crop species like *Brassica rapa*, tomato, cucumber, etc. (Zhu *et al.*, 2015; Li *et al.*, 2018; Cho *et al.*, 2005; Jin *et al.*, 2024). The interaction of plant lncRNAs with other molecular regulators like DNA, RNA, protein, and other ncRNAs like miRNAs consequently leads them to perform critical functions in molecular processes such as transcriptional and epigenetic regulations, chromatin remodeling, posttranscriptional processing, alternative splicing, and R-loop formation (Gonzales *et al.*, 2024). lncRNAs also govern abiotic stress responses including thermotolerance, cold tolerance, drought tolerance, and salt tolerance (Li *et al.*, 2023; Jin *et al.*, 2024; Traubenik *et al.*, 2024). Similarly, lncRNAs display a pivotal role in plant immune response against biotic stress from perception to signaling (Li *et al.*, 2023; Huang *et al.*, 2023; Jiang *et al.*, 2023; Yajnik *et al.*, 2024).

The present chapter is focused on highlighting the emerging role of lncRNAs in plant defense response during plant–microbe interactions. The key findings will be discussed in detail with the aim to provide timely and precise information about the function of lncRNAs in governing plant

 DOI: 10.1201/9781003608080-9

immunity. Areas that require subsequent research and any potential gaps in this field are also identified in this chapter.

9.2 BIOGENESIS, CLASSIFICATION, AND MODE OF ACTION OF PLANT LNCRNAS

The biogenesis of lncRNAs begins with transcription primarily by RNA polymerase II (RNA Pol II) followed by subsequent processing such as splicing, 5′ capping, and 3′polyadenylation, similar to any protein-coding gene (Liu *et al.*, 2012; Wierzbicki *et al.*, 2021; Patra *et al.*, 2023; Fahad *et al.*, 2024). Additionally, lncRNAs are also reported to be transcribed by RNA Pol III, IV, and V, which further decides the fate of lncRNAs (Wierzbicki *et al.*, 2021; Patra *et al.*, 2023; Fahad *et al.*, 2024). For instance, lncRNAs AtR8 and AtR18 are transcribed by RNA Pol III and play a predominant role during hypoxia stress conditions (Wu *et al.*, 2012). On the other hand, RNA pol IV and V transcribed lncRNAs have been reported to be involved in siRNA-mediated pathways, act as scaffold RNAs, and recognize and silence transposable elements (TEs) (He *et al.*, 2014; Wierzbicki *et al.*, 2021). LncRNAs can be classified into six types based on their transcription from diverse genomic locations and orientation: (i) Sense lncRNAs, which are transcribed from the sense strand of DNA; (ii) antisense lncRNAs, also known as natural antisense transcripts (NATs), which are transcribed from the antisense strand of DNA; (iii) intronic lncRNAs (incRNAs), which originate from the intronic regions; (iv) long intergenic lncRNAs (linclncRNAs), which originate from the intergenic regions; (v) bidirectional lncRNAs, which are transcribed from the same promoter of a protein-coding gene but in the opposite direction; and (vi) eRNAs, which are derived from the enhancer region (Wu *et al.*, 2020; Traubenik *et al.*, 2024) (Figure 9.1).

So far, the studies on lncRNAs suggest that they do not have a conserved mode of action due to their diverse roles in numerous biological processes. lncRNAs can be *cis*-acting and *trans*-acting to regulate the expression of proximal and distant genes, respectively (Wierzbicki *et al.*, 2021; Fahad *et al.*, 2024). *Cis*-acting lncRNAs primarily affect the transcription and epigenetics of nearby genes, while *trans*-acting lncRNAs majorly act as scaffolds and alter the stability and function of proteins. The regulation of *FLC* (*FLOWERING LOCUS C*) gene expression by three lncRNAs, *COOLAIR*,

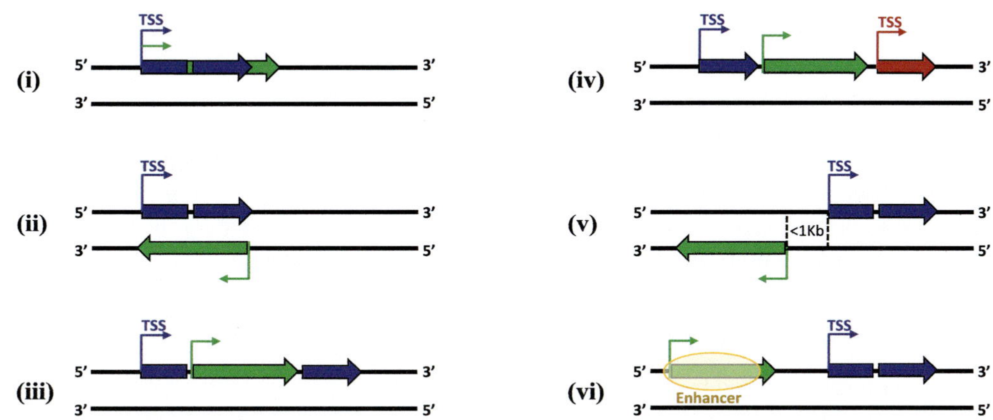

FIGURE 9.1 Classification of lncRNAs according to their genomic positioning and orientation. lncRNAs can be classified into six types based on their transcription from diverse genomic locations and orientation: (i) Sense lncRNAs; (ii) antisense lncRNAs; (iii) incRNAs; (iv) linclncRNAs; (v) bidirectional lncRNAs; and (v) eRNAs, which are derived from the enhancer region. Black solid lines represent two strands (sense and antisense) of DNA; blue box and arrow represent exons of gene *A*; green arrow represents lncRNA; red arrow represents gene *B*; and the little arrows show the direction of transcription.

COLDAIR, and *COLDWRAP*, in *Arabidopsis thaliana* during the vernalization process is a classic example of *cis*-acting mode of action of lncRNAs. Cold-induced *COOLAIR* (cold-induced long antisense intragenic RNA) is transcribed from the antisense strand of *FLC*, which is recruited by PRC2 (Polycomb Repressive Complex 2), ultimately leading to the silencing of *FLC* (Swiezewski *et al.*, 2009); *COLDAIR* (cold-assisted intronic noncoding RNA) is transcribed from the intron of *FLC*, which physically interacts with PRC2, causing the stable repression of *FLC* (Heo and Sung 2011); and *COLDWRAP* is also a PRC2-interacting lncRNA, transcribed from the repressed promoter of *FLC*. Both COLDAIR and COLDWRAP form an intragenic chromatin loop at the *FLC* locus, which is required for the stable epigenetic suppression of *FLC* during the vernalization process (Kim and Sung, 2017). Similar to *cis*-acting lncRNAs, several studies have confirmed the *trans*-acting mode of action of lncRNAs. For instance, lncRNA *APOLO* (Auxin-Regulated Promoter Loop) has been well reported to create DNA–RNA duplexes (R-loop) at distantly located loci in *Arabidopsis* genome, further altering the 3D conformation of chromatin at these loci, which eventually leads to their altered expression (Ariel *et al.*, 2020; Moison *et al.*, 2021; Fonouni-Farde *et al.*, 2022). The *APOLO* lncRNA controls the lateral root development and root hair growth and development in *Arabidopsis* through R-loop formation (Ariel *et al.*, 2020; Moison *et al.*, 2021). At low temperature conditions, *APOLO* interacts with transcription factor WRKY42 and enhances the expression of root hair master regulator, ROOT HAIR DEFECTIVE 6 (RHD6), which results in enhanced root hair growth and development in *Arabidopsis* (Ariel *et al.*, 2020; Moison *et al.*, 2021). In a similar way, *APOLO* interacts with LIKE HETEROCHROMATIN PROTEIN 1 (LHP1), which is a PRC1-component, and VARIANT IN METHYLATION 1 (VIM1), which is a DNA methylation regulator (Fonouni-Farde *et al.*, 2022). Under high temperature conditions, the *APOLO*–LHP1–VIM1 complex induces epigenetic modification at the promoter of an auxin biosynthesis gene *YUCCA2*, which further leads to enhanced auxin synthesis, thus causing elongated hypocotyl in *Arabidopsis* (Fonouni-Farde *et al.*, 2022).

lncRNAs often play a regulatory role by interconnecting with other ncRNAs like miRNAs and siRNAs either by physically interacting with them to induce the ncRNA pathways or by sponging them to regulate the gene expression. In order to inhibit the miRNA action, lncRNAs compete with their target mRNAs for binding, thereby sponging these miRNAs to block their action. These lncRNAs are referred as endogenous target mimics (eTMs); for instance, lncRNA IPS1 (Induced by Phosphate Starvation 1) acts as an eTM of miR399, which causes the accumulation of miR399 target mRNA PHO2 during the phosphate starvation condition in *Arabidopsis* (Franco-Zorrilla *et al.*, 2007). Likewise, heat stress-responsive lncRNAs (TCONS_00048391) act as eTMs of miR164, which leads to the inhibition of miR164-mediated silencing of NAC1 in Chinese cabbage during heat stress conditions (Wang *et al.*, 2019). These studies confirm the distinct mode of action of lncRNAs in various developmental and stress conditions in plants. The central role of lncRNAs in abiotic stress conditions like cold, heat, salt, drought, and nutrient deficiency has been well-documented in many review articles (Li *et al.*, 2023; Patra *et al.*, 2023; Jin *et al.*, 2024; Traubenik *et al.*, 2024). To provide a comprehensive understanding of the field, the emerging role of lncRNAs in plant–microbe interactions will be thoroughly covered in this chapter.

9.3 REGULATORY ROLE OF LNCRNAS IN PLANT IMMUNITY

Plants are constantly attacked by pathogens and pests, resulting in crop yield and economical losses globally. To counter these attacks, plants have developed complex and effective immune responses against the pathogens and pests. The plant immune response can be broadly classified in two major types of immunity: (i) Pathogen-associated molecular pattern (PAMP)-triggered immunity (PTI) and (ii) effector-triggered immunity (ETI) (Glazebrook, 2005; Jones and Dangl, 2006; Schwessinger and Zipfel, 2008). In PTI, membrane-bound receptors, pattern recognition receptors (PRRs) and surface receptors like kinases, recognize the PAMPs, such as glycoprotein, chitin, and flagellin, to further induce the defense responses including the expression of pathogenesis-related (PR) gene expression, salicylic acid (SA) accumulation, and callose deposition (Glazebrook, 2005; Jones and Dangl, 2006; Schwessinger and Zipfel, 2008). ETI primarily acts inside plant cells through plant disease resistance (*R*) genes encoding polymorphic protein having nucleotide binding (NB) and leucine-rich repeat (LRR) (Kong *et al.*, 2017).

LncRNAs play a critically important role in plant immunity from perception to defense response against pathogens. In tomato, lcRNA16397 reduced the ROS accumulation and lncRNA33732 induced the expression of respiratory burst oxidase (RBOH) against the attack of *Phytophthora infestans* (Cui *et al.*, 2017, 2019). In *Arabidopsis*, lncRNA SABC1 (salicylic acid biogenesis controller 1) suppresses the immunity and induces growth in healthy plants, while upon pathogen infection the SABC1 level is reduced due to calcium influx and ROS burst, further causing enhanced expression of NAC3, which in turn activates the expression of ICS1 (isochorismate synthase 1), which is a key SA biosynthesis enzyme (Liu *et al.*, 2022). lncRNAs regulate the expression of defense-related gene expression either by targeting or by interacting with transcription factors, epigenetic factors, splicing factors, and/or proteins to modulate downstream signaling pathways during pathogen attack. The ALTERNATIVE SPLICING COMPETITOR (ASCO) lncRNA interacts with multiple spliceosome components PRP8 and SmD1b, and alternative splicing regulator NSRs, which further affects the alternative splicing of several genes, thereby altering global transcriptome in the response of *Arabidopsis* to flagellin (Rigo *et al.*, 2020). In *Arabidopsis*, lncRNA ELF18-INDUCED LONG-NONCODING RNA1 (ELENA1) was identified as a positive regulator to establish increased resistance against *Pseudomonas syringe* pv *tomato* DC3000 (Seo *et al.*, 2017). The authors showed that the ELENA1 interacts with MED19a (Mediator subunit 19a), which leads to the enhanced expression of *PR1* (*PATHOGENESIS-RELATED GENE1*). Additionally, the overexpression lines of ELENA1 showed pathogen resistance phenotype, whereas the knockdown plants displayed susceptibility (Seo *et al.*, 2017). Moreover, a subsequent study also revealed that ELENA1 can interact with fibrillarin-2 (FIB2), which is a negative regulator of PR1 (Seo *et al.*, 2019). The genetic and biochemical analysis by the authors confirmed that ELENA1 evicts FIB2 from MED19a, ultimately resulting the higher expression of PR1 genes and thereby enhancing pathogen defense response (Seo *et al.*, 2019). The lncRNAs can also induce the defence-related hormones and hormone pathways in plants, for instance lncRNA SABC1, suppression of SABC1 leads to induce the SA biosynthesis in *Arabidopsis* (mentioned earlier in this section). Similarly, *Verticillium dahliae* or *Botrytis cinerea* induced two lncRNAs *GhlncNAT-ANX2* and *GhlncNAT-RLP7* which repress the expression of *LOX1* (lipoxygenase 1) and *LOX2* (lipoxygenase 2), which are JA pathway genes. This regulation causes attenuated resistance against fungi in cotton (Zhang *et al.*, 2018).

9.4 EMERGING ROLE OF LNCRNAS IN PLANT–MICROBE INTERACTION

9.4.1 lncRNAs and Plant–Bacterial Interaction

Various studies have highlighted the role of lncRNAs in regulating plant immune responses to bacterial infections. Plants deploy lncRNA-dependent immune-mediated defense mechanisms against various bacterial pathogens. The bacterial pathogen *Xanthomonas oryzae* pv. *oryzae*, which infects rice, induces the expression of lncRNA ALEX1, leading to enhanced resistance by activating the jasmonic acid (JA) signaling pathway (Yu *et al.*, 2020). The lncRNA AtR8, induced by low levels of SA, has a role in SA-mediated resistance to *Pseudomonas syringae* infection (Li *et al.*, 2020). SA also induces U3 snoRNA-related lincRNA (SUNA1), another lncRNA that provides resistance to *Arabidopsis* against *Pseudomonas syringae* pv. tomato DC3000 infection. It has been suggested that SUNA1 interacts with fibrillarin, which improves the translational and processing efficiency of pre-rRNA for particular defense-related genes that has a positive regulatory role in plant immunity (Kong *et al.*, 2022). A recent study showed an interaction among lncRNAs, miRNAs, and mRNAs (eTM482e-3p-1-Sly-miR482e-3p-NBS-LRRs network) in conferring resistance against *Ralstonia solanacearum*, which causes the destructive wilting disease in tomato (Si *et al.*, 2023). Another such interacting module is reported in mulberry, wherein MuLncl forms a regulatory network with mulmiR3954, siRNAs, and mRNAs to enhance resistance against *Pseudomonas syringae* pv. tomato DC3000 (Gai *et al.*, 2018). In this case, the cleavage of MuLncl by mulmiR3954 results in si161579, which cleaves the calmodulin-like protein encoding gene transcript (*CML27*), a key calcium sensor playing a crucial role in plant defense mechanism. Transcriptome analysis during *Xanthomonas oryzae* pv. *oryzae* (*Xoo*)–rice interaction identified 1,130 lncRNAs, wherein it is validated that downregulation of some lncRNAs resulted in defense gene modulation, such as genes encoding arginase, agmatine deiminase, and linoleate 13-s-lipoxygenase, which are key components in polyamine and JA biosynthesis (Selvamani *et al.*, 2025). In *Arabidopsis*, the lncRNA SABC1 is reported to regulate the balance between plant immunity and growth during both bacterial and viral infection (Liu *et al.*, 2022).

9.4.2 lncRNAs and Plant–Fungi Interaction

In a comparative study between resistant and susceptible tomato plants against *Phytophthora*, the pathogen causing tomato late blight revealed that lncRNA16397 enhances *SlGRX22* expression, providing resistance to late blight. Cotton plants exhibited enhanced resistance against the fungal diseases caused by *Verticillium dahliae* and *Botrytis cinerea* when the lncRNAs GhlncNAT-ANX2 and GhlncNAT-RLP7 were silenced, due to increased expression of *LOX1* and *LOX2* (Zhang *et al.*, 2015). The role of lncRNAs in the growth and development of fungal pathogens has been elucidated in many studies; for instance, it has been demonstrated that lncRNAs are essential for the development of *Fusarium graminearum*, a phytopathogenic fungus that causes head blight in many economically important crops (Kim *et al.*, 2018). Moreover, the expression profiles of lncRNAs during infection and development in *Ustilaginoidea virens*, which causes false smut disease in rice, have been highlighted (Tang *et al.*, 2021). Additionally, the lncRNAs have regulatory roles in many phytopathogenic fungi. Notably, in *Pyricularia oryzae*, the causative agent of rice blast disease, a specific lncRNA regulates a neighboring gene required for hyphal growth, highlighting the role of lncRNAs as crucial regulators of fungal growth (Li *et al.*, 2021b). In *F. graminearum*, GzmetE, encoding a homoserine O-acetyltransferase and having role in sexual development and plant infection, is regulated by a new antisense lncRNA (Wang *et al.*, 2021a). Using transcriptome sequencing, Zhao *et al.* (2023) elucidated how lncRNAs are differentially regulated in Kentucky bluegrass during rust infection. In order to demonstrate their role in plant hormone signal transduction and plant–pathogen interaction pathways, the study identified 15,711 differently expressed genes and 105 differentially expressed lncRNAs, which could be potential targets for creating rust-resistant Kentucky bluegrass cultivars.

9.4.3 lncRNAs and Plant–Virus Interaction

Certain viral pathogens encode own lncRNAs, which play crucial roles in regulating infection in plants by disrupting the host immune pathways. Some of them were shown to interfere with RNA silencing, the major antiviral defense mechanism in plants. Some lncRNAs are reported to disrupt RNA silencing, the primary antiviral defense mechanism present in plants. For instance, in cauliflower mosaic pararetrovirus, noncoding 8S RNA encoded from the viral genome acts as a sponge for some of the components of the host RNA silencing machinery, thereby preventing them from targeting viral RNAs (Blevins *et al.*, 2011). Another noncoding subgenomic RNA (ncRNA30) of beet necrotic yellow vein virus, which acts as a suppressor of antiviral RNA silencing, is required for the systemic spread of virus (Flobinus *et al.*, 2016). SR1f, a noncoding subgenomic RNA of red clover necrotic mosaic virus, supports viral infection by regulating host and viral gene expression, independent of RNA silencing suppression (Kanodia and Miller, 2022). The lncRNA low-molecular-weight tristeza 1 (LMT1), expressed by citrus tristeza virus (CTV), is essential for successful infection in citrus (Kang *et al.*, 2019) and for maintaining an optimal balance between

CTV and citrus host by interacting with p33, a viral protein that triggers plant immune response, enabling the virus to persist throughout the tree's lifespan (Folimonova and Sun, 2022).

9.4.4 lncRNAs and Plant–Herbivore Interaction

The defense responses of plants against insect pests is a key area of research, and plants have complex mechanisms to combat these pests, including the regulatory roles of lncRNAs. However, the role of lncRNAs as key regulators in plant–herbivore resistance is relatively underexplored. A study in *Nicotiana attenuata*, a model wild tobacco species with well-characterized herbivore interactions, identified 1,290 lncRNAs to be significantly modulated in response to herbivore elicitation by *Manduca sexta* larvae (Li *et al.*, 2021a). Another study demonstrated the role of lncRNAs in armyworm-induced defense responses in rice, wherein antisense lncRNA was found to be transcribed from jasmonate ZIM-domain (*JAZ10*) gene, a key regulator in JA-mediated defense signaling, highlighting the potential role of lncRNAs in modulating JA signaling pathways during herbivory (Wang *et al.*, 2021b). Transcriptomic analysis of tea plants (*Camellia sinensis*) in response to *Helopeltis theivora*, one of the deadly pests of tea, highlighted the role of lncRNAs as potential regulators of primary and secondary metabolic pathways in the host plant (Bordoloi *et al.*, 2023). Transcriptomic studies during cotton (*Gossypium hirsutum*)–aphids (*Aphis gossypii*) interaction identified several lncRNAs, and functional analysis of selected lncRNAs revealed them to be linked to stress pathways and predicted them to mimic cotton miRNAs (Zhang *et al.*, 2019) (Figure 9.2).

9.5 CONCLUSION AND FUTURE PERSPECTIVES

The rapid advancement in next-generation sequencing technologies have led to the identification of thousands of lncRNAs in plant growth, development, and various stress responses. However, only few lncRNAs have been validated functionally due to the technical challenges in studying lncRNAs. Nevertheless, constant efforts are underway for their functional characterization in plants. Here, we focus on providing a comprehensive insight about the role of lncRNAs in plant immunity

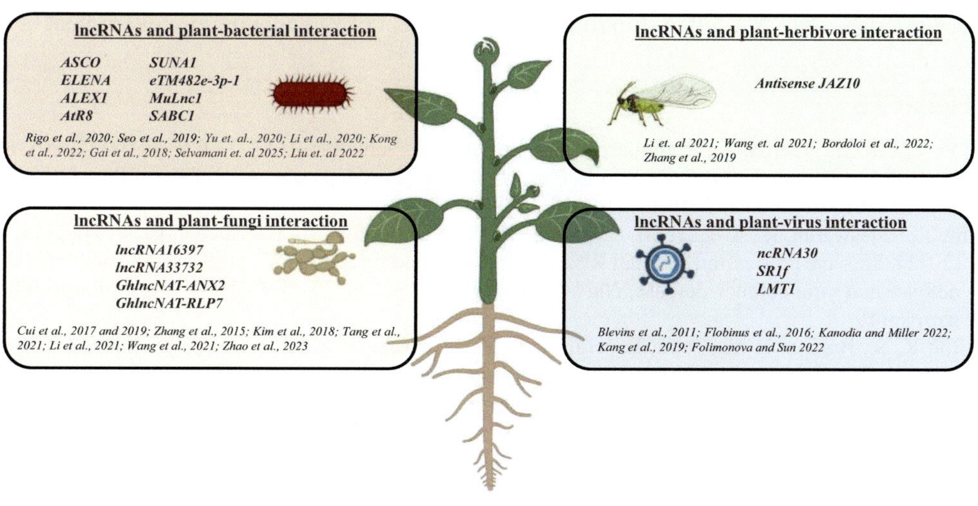

FIGURE 9.2 lncRNAs and plant–microbe interaction. The interaction of plant and microbes like bacteria, fungi, herbivore, and virus leads to the altered expression of several lncRNAs. These alterations ultimately influence the expression of essential defense-related genes, thereby facilitating the plant's response to stress. Some illustrations were created in https://BioRender.com.

and plant–microbe interaction. Studies suggest that lncRNAs act as critical players during plant–microbe interaction, from perceiving the stress upon interaction with microbes to activating the defense mechanism at the molecular level by inducing stress-responsive gene expression. However, in-depth analysis is still required to determine their exact molecular and biological function in plant–microbe interaction. Along with a reverse genetics approach, genome editing tools such as CRISPR-Cas9 (Clustered Regularly Interspaced Short Palindromic Repeats) and Transcription Activator-Like Effector Nucleases (TALENs) are being widely used to decipher the regulatory mechanism of lncRNAs in plant immunity. Moreover, the secondary structure of lncRNAs is critically important in deciding their regulatory roles in diverse biological processes. A more precise and accurate version of secondary structure prediction tools might improve our understanding on the structure of lncRNAs. Similarly, the databases for plant lncRNAs are relatively less as compared to the databases available for mRNA, protein and sRNA. It is interesting to study the network involving mRNAs, lncRNAs, sRNAs, and protein-coding genes, which can be achieved by well-developed databases and bioinformatic tools. The lncRNAs are potential candidates to study plant immunity due to their involvement at almost every step during plant–microbe interaction. Studies on lncRNAs may unfold underlying mechanistic regulation during biotic stress responses, which will ultimately improve our understanding of plant immune systems.

9.6 AUTHOR CONTRIBUTIONS

Both authors JRD and JD equally contributed to the writing and editing of the manuscript and preparation of the figures.

9.7 FUNDING

The study was financed by a postdoctoral fellowship to JRD (ANR-21-CE20-0034-03). The Institute Jean-Pierre Bourgin (IJPB) for Plant Sciences also benefits from the support of Saclay Plant Sciences (SPS) (ANR-17-EUR-0007).

ACKNOWLEDGMENT

JD acknowledges DBT-RA fellowship from the Department of Biotechnology, government of India.

REFERENCES

Ariel, F., Lucero, L., Christ, A., Mammarella, M. F., Jegu, T., Veluchamy, A.,. . . Crespi, M. (2020). R-loop mediated trans action of the APOLO long noncoding RNA. *Molecular Cell*, *77*(5), 1055–1065. https://doi.org/10.1016/j.molcel.2019.12.015

Blevins, T., Rajeswaran, R., Aregger, M., Borah, B. K., Schepetilnikov, M., Baerlocher, L.,. . . Pooggin, M. M. (2011). Massive production of small RNAs from a non-coding region of cauliflower mosaic virus in plant defense and viral counter-defense. *Nucleic Acids Research*, *39*(12), 5003–5014. https://doi.org/10.1093/nar/gkr119

Bordoloi, K. S., Baruah, P. M., Tanti, B., Gill, S. S., & Agarwala, N. (2023). *Helopeltis theivora* responsive transcriptomic reprogramming uncovers long non-coding RNAs as possible regulators of primary and secondary metabolism in tea plant. *Journal of Plant Growth Regulation*, *42*(10), 6523–6548. https://doi.org/10.1007/s00344-022-10893-x

Cho, J., Koo, D. H., Nam, Y. W., Han, C. T., Lim, H. T., Bang, J. W., & Hur, Y. (2005). Isolation and characterization of cDNA clones expressed under male sex expression conditions in a monoecious cucumber plant (*Cucumis sativus* L. cv. Winter Long). *Euphytica*, *146*, 271–281. https://doi.org/10.1007/s10681-005-9023-1

Cui, J., Jiang, N., Meng, J., Yang, G., Liu, W., Zhou, X.,. . . Luan, Y. (2019). LncRNA33732-respiratory burst oxidase module associated with WRKY1 in tomato–*Phytophthora infestans* interactions. *The Plant Journal*, *97*(5), 933–946. https://doi.org/10.1111/tpj.14173

Cui, J., Luan, Y., Jiang, N., Bao, H., & Meng, J. (2017). Comparative transcriptome analysis between resistant and susceptible tomato allows the identification of lnc RNA 16397 conferring resistance to *Phytophthora infestans* by co-expressing glutaredoxin. *The Plant Journal*, *89*(3), 577–589. https://doi.org/10.1111/tpj.13408

Fahad, M., Tariq, L., Muhammad, S., & Wu, L. (2024). Underground communication: Long noncoding RNA signaling in the plant rhizosphere. *Plant Communications*. https://doi.org/10.1016/j.xplc.2024.100927

Flobinus, A., Hleibieh, K., Klein, E., Ratti, C., Bouzoubaa, S., & Gilmer, D. (2016). A viral noncoding RNA complements a weakened viral RNA silencing suppressor and promotes efficient systemic host infection. *Viruses*, *8*(10), 272. https://doi.org/10.3390/v8100272

Folimonova, S. Y., & Sun, Y. D. (2022). Citrus tristeza virus: From pathogen to panacea. *Annual Review of Virology*, *9*(1), 417–435. https://doi.org/10.1146/annurev-virology-100520-114412

Fonouni-Farde, C., Christ, A., Blein, T., Legascue, M. F., Ferrero, L., Moison, M.,. . . Ariel, F. (2022). The Arabidopsis APOLO and human UPAT sequence-unrelated long noncoding RNAs can modulate DNA and histone methylation machineries in plants. *Genome Biology*, *23*(1), 181. https://doi.org/10.1186/s13059-022-02750-7

Franco-Zorrilla, J. M., Valli, A., Todesco, M., Mateos, I., Puga, M. I., Rubio-Somoza, I.,. . . Paz-Ares, J. (2007). Target mimicry provides a new mechanism for regulation of microRNA activity. *Nature Genetics*, *39*(8), 1033–1037. https://doi.org/10.1038/ng2079

Gai, Y. P., Yuan, S. S., Zhao, Y. N., Zhao, H. N., Zhang, H. L., & Ji, X. L. (2018). A novel LncRNA, MuLnc1, associated with environmental stress in mulberry (*Morus multicaulis*). *Frontiers in Plant Science*, *9*, 669. https://doi.org/10.3389/fpls.2018.00669

Glazebrook, J. (2005). Contrasting mechanisms of defense against biotrophic and necrotrophic pathogens. *Annual Review of Phytopathology*, *43*(1), 205–227. https://doi.org/10.1146/annurev.phyto.43.040204.135923

Gonzales, L. R., Blom, S., Henriques, R., Bachem, C. W., & Immink, R. G. (2024). LncRNAs: The art of being influential without protein. *Trends in Plant Science*. https://doi.org/10.1016/j.tplants.2024.01.006

He, X. J., Ma, Z. Y., & Liu, Z. W. (2014). Non-coding RNA transcription and RNA-directed DNA methylation in *Arabidopsis*. *Molecular Plant*, *7*(9), 1406–1414. https://doi.org/10.1093/mp/ssu075

Heo, J. B., & Sung, S. (2011). Vernalization-mediated epigenetic silencing by a long intronic noncoding RNA. *Science*, *331*(6013), 76–79. https://doi.org/10.1126/science.1197349

Huang, J., Zhou, W., Zhang, X., & Li, Y. (2023). Roles of long non-coding RNAs in plant immunity. *PLoS Pathogens*, *19*(5), e1011340. https://doi.org/10.1371/journal.ppat.1011340

Jiang, C. H., Li, Z. J., Zheng, L. Y., Yu, Y. Y., & Niu, D. D. (2023). Small RNAs: Efficient and miraculous effectors that play key roles in plant–microbe interactions. *Molecular Plant Pathology*, *24*(8), 999–1013. https://doi.org/10.1111/mpp.13329

Jin, X., Wang, Z., Li, X., Ai, Q., Wong, D. C. J., Zhang, F.,. . . Si, H. (2024). Current perspectives of lncRNAs in abiotic and biotic stress tolerance in plants. *Frontiers in Plant Science*, *14*, 1334620. https://doi.org/10.3389/fpls.2023.1334620

Jones, J. D., & Dangl, J. L. (2006). The plant immune system. *Nature*, *444*(7117), 323–329. https://doi.org/10.1038/nature05286

Kang, S. H., Sun, Y. D., Atallah, O. O., Huguet-Tapia, J. C., Noble, J. D., & Folimonova, S. Y. (2019). A long non-coding RNA of citrus tristeza virus: Role in the virus interplay with the host immunity. *Viruses*, *11*(5), 436. https://doi.org/10.3390/v11050436

Kanodia, P., & Miller, W. A. (2022). Effects of the noncoding subgenomic RNA of red clover necrotic mosaic virus in virus infection. *Journal of Virology*, *96*(3), e01815–e01821. https://doi.org/10.1128/jvi.01815-21

Kim, D. H., & Sung, S. (2017). Vernalization-triggered intragenic chromatin loop formation by long noncoding RNAs. *Developmental Cell*, *40*(3), 302–312. https://doi.org/10.1016/j.devcel.2016.12.021

Kim, W., Miguel-Rojas, C., Wang, J., Townsend, J. P., & Trail, F. (2018). Developmental dynamics of long noncoding RNA expression during sexual fruiting body formation in *Fusarium graminearum*. *MBio*, *9*(4), 10–1128. https://doi.org/10.1128/mbio.01292-18

Kong, L., Qiu, X., Kang, J., Wang, Y., Chen, H., Huang, J.,. . . Wang, Y. (2017). A Phytophthora effector manipulates host histone acetylation and reprograms defense gene expression to promote infection. *Current Biology*, *27*(7), 981–991. https://doi.org/10.1016/j.cub.2017.02.044

Kong, X., Wang, H., Zhang, M., Chen, X., Fang, R., & Yan, Y. (2022). A SA-regulated lincRNA promotes *Arabidopsis* disease resistance by modulating pre-rRNA processing. *Plant Science*, *316*, 111178. https://doi.org/10.1016/j.plantsci.2022.111178

Li, Q., Shen, H., Yuan, S., Dai, X., & Yang, C. (2023). miRNAs and lncRNAs in tomato: Roles in biotic and abiotic stress responses. *Frontiers in Plant Science*, *13*, 1094459. https://doi.org/10.3389/fpls.2022.1094459

Li, R., Fu, D., Zhu, B., Luo, Y., & Zhu, H. (2018). CRISPR/Cas9-mediated mutagenesis of lncRNA1459 alters tomato fruit ripening. *The Plant Journal*, *94*(3), 513–524. https://doi.org/10.1111/tpj.13872

Li, R., Jin, J., Xu, J., Wang, L., Li, J., Lou, Y., & Baldwin, I. T. (2021a). Long non-coding RNAs associate with jasmonate-mediated plant defence against herbivores. *Plant, Cell & Environment*, *44*(3), 982–994. https://doi.org/10.1111/pce.13952

Li, S., Nayar, S., Jia, H., Kapoor, S., Wu, J., & Yukawa, Y. (2020). The *Arabidopsis* hypoxia inducible AtR8 long non-coding RNA also contributes to plant defense and root elongation coordinating with WRKY genes under low levels of salicylic acid. *Non-Coding RNA*, *6*(1), 8. https://doi.org/10.3390/ncrna6010008

Li, Z., Yang, J., Peng, J., Cheng, Z., Liu, X., Zhang, Z.,. . . Peng, Y. L. (2021b). Transcriptional landscapes of long non-coding RNAs and alternative splicing in *Pyricularia oryzae* revealed by RNA-Seq. *Frontiers in Plant Science*, *12*, 723636. https://doi.org/10.3389/fpls.2021.723636

Liu, J., Jung, C., Xu, J., Wang, H., Deng, S., Bernad, L.,. . . Chua, N. H. (2012). Genome-wide analysis uncovers regulation of long intergenic noncoding RNAs in *Arabidopsis*. *The Plant Cell*, *24*(11), 4333–4345. https://doi.org/10.1105/tpc.112.102855

Liu, N., Xu, Y., Li, Q., Cao, Y., Yang, D., Liu, S.,. . . Zhang, X. (2022). A lncRNA fine-tunes salicylic acid biosynthesis to balance plant immunity and growth. *Cell Host & Microbe*, *30*(8), 1124–1138. https://doi.org/10.1016/j.chom.2022.07.001

Moison, M., Pacheco, J. M., Lucero, L., Fonouni-Farde, C., Rodríguez-Melo, J., Mansilla, N.,. . . Ariel, F. (2021). The lncRNA APOLO interacts with the transcription factor WRKY42 to trigger root hair cell expansion in response to cold. *Molecular Plant*, *14*(6), 937–948. https://doi.org/10.1016/j.molp.2021.03.008

Nie, H., Cheng, C., Kong, J., Li, H., & Hua, J. (2023). Plant Non-coding RNAs Function in Pollen Development and Male Sterility. *Frontiers in Plant Science*, *14*, 291. https://doi.org/10.3389/fpls.2023.1109941

Patra, G. K., Gupta, D., Rout, G. R., & Panda, S. K. (2023). Role of long non-coding RNA in plants under abiotic and biotic stresses. *Plant Physiology and Biochemistry*, *194*, 96–110. https://doi.org/10.1016/j.plaphy.2022.10.030

Rigo, R., Bazin, J., Romero-Barrios, N., Moison, M., Lucero, L., Christ, A.,. . . Ariel, F. (2020). The *Arabidopsis* lncRNA ASCO modulates the transcriptome through interaction with splicing factors. *EMBO Reports*, *21*(5), e48977. https://doi.org/10.15252/embr.201948977

Schwessinger, B., & Zipfel, C. (2008). News from the frontline: Recent insights into PAMP-triggered immunity in plants. *Current Opinion in Plant Biology*, *11*(4), 389–395. https://doi.org/10.1016/j.pbi.2008.06.001

Selvamani, S. B., Nallusamy, S., Kamalanathan, V., Ravichandran, A., Singaravelu, R., Selvapandian, U.,. . . Natesan, S. (2025). Understanding the mechanisms of long noncoding RNAs in response to bacterial blight in rice. *Physiological and Molecular Plant Pathology*, *135*, 102486. https://doi.org/10.1016/j.pmpp.2024.102486

Seo, J. S., Diloknawarit, P., Park, B. S., & Chua, N. H. (2019). ELF18-induced long noncoding RNA 1 evicts fibrillarin from mediator subunit to enhance pathogenesis-related gene 1 (*PR1*) expression. *New Phytologist*, *221*(4), 2067–2079. https://doi.org/10.1111/nph.15530

Seo, J. S., Sun, H. X., Park, B. S., Huang, C. H., Yeh, S. D., Jung, C., & Chua, N. H. (2017). ELF18-INDUCED LONG-NONCODING RNA associates with mediator to enhance expression of innate immune response genes in *Arabidopsis*. *The Plant Cell*, *29*(5), 1024–1038. https://doi.org/10.1105/tpc.16.00886

Si, X., Liu, H., Cheng, X., Xu, C., Han, Z., Dai, Z.,. . . Lu, G. (2023). Integrative transcriptomic analysis unveils lncRNA-miRNA-mRNA interplay in tomato plants responding to *Ralstonia solanacearum*. *International Journal of Biological Macromolecules*, *253*, 126891. https://doi.org/10.1016/j.ijbiomac.2023.126891

Swiezewski, S., Liu, F., Magusin, A., & Dean, C. (2009). Cold-induced silencing by long antisense transcripts of an *Arabidopsis* Polycomb target. *Nature*, *462*(7274), 799–802. http://www.nature.com/doifinder/10.1038/nature08618

Tang, J., Chen, X., Yan, Y., Huang, J., Luo, C., Tom, H., & Zheng, L. (2021). Comprehensive transcriptome profiling reveals abundant long non-coding RNAs associated with development of the rice false smut fungus, *Ustilaginoidea virens*. *Environmental Microbiology*, *23*(9), 4998–5013. https://doi.org/10.1111/1462-2920.15432

Traubenik, S., Charon, C., & Blein, T. (2024). From environmental responses to adaptation: The roles of plant lncRNAs. *Plant Physiology*, *195*(1), 232–244. https://doi.org/10.1093/plphys/kiae034

Waititu, J. K., Zhang, C., Liu, J., & Wang, H. (2020). Plant non-coding RNAs: Origin, biogenesis, mode of action and their roles in abiotic stress. *International Journal of Molecular Sciences*, *21*(21), 8401. https://doi.org/10.3390/ijms21218401

Wang, A., Hu, J., Gao, C., Chen, G., Wang, B., Lin, C.,. . . Zhou, G. (2019). Genome-wide analysis of long non-coding RNAs unveils the regulatory roles in the heat tolerance of Chinese cabbage (Brassica rapa ssp. chinensis). *Scientific Reports*, *9*(1), 5002. https://doi.org/10.1038/s41598-019-41428-2

Wang, H. L. V., & Chekanova, J. A. (2017). Long noncoding RNAs in plants. *Long Non Coding RNA Biology*, 133–154. https://doi.org/10.1007/978-981-10-5203-3_5

Wang, J., Meng, X., Dobrovolskaya, O. B., Orlov, Y. L., & Chen, M. (2017). Non-coding RNAs and their roles in stress response in plants. *Genomics, Proteomics & Bioinformatics*, *15*(5), 301–312. https://doi.org/10.1016/j.gpb.2017.01.007

Wang, J., Zeng, W., Xie, J., Fu, Y., Jiang, D., Lin, Y.,. . . Cheng, J. (2021a). A novel antisense long non-coding RNA participates in asexual and sexual reproduction by regulating the expression of *GzmetE* in *Fusarium graminearum*. *Environmental Microbiology*, *23*(9), 4939–4955. https://doi.org/10.1111/1462-2920.15399

Wang, L., Wu, S., Jin, J., & Li, R. (2021b). Identification of herbivore-elicited long non-coding RNAs in rice. *Plant Signaling & Behavior*, *16*(7), 1916702. https://doi.org/10.1080/15592324.2021.1916702

Wierzbicki, A. T., Blevins, T., & Swiezewski, S. (2021). Long noncoding RNAs in plants. *Annual Review of Plant Biology*, *72*(1), 245–271. https://doi.org/10.1146/annurev-arplant-093020-035446

Wu, J., Okada, T., Fukushima, T., Tsudzuki, T., Sugiura, M., & Yukawa, Y. (2012). A novel hypoxic stress-responsive long non-coding RNA transcribed by RNA polymerase III in *Arabidopsis*. *RNA Biology*, *9*(3), 302–313. https://doi.org/10.4161/rna.19101

Wu, L., Liu, S., Qi, H., Cai, H., & Xu, M. (2020). Research progress on plant long non-coding RNA. *Plants*, *9*(4), 408. https://doi.org/10.3390/plants9040408

Yajnik, K. N., Singh, I. K., & Singh, A. (2024). lncRNAs and epigenetics regulate plant's resilience against biotic stresses. *Plant Physiology and Biochemistry*, 108892. https://doi.org/10.1016/j.plaphy.2024.108892

Yu, W., Wang, L., Zhao, R., Sheng, J., Zhang, S., Li, R., & Shen, L. (2019). Knockout of SlMAPK3 enhances tolerance to heat stress involving ROS homeostasis in tomato plants. *BMC Plant Biology*, *19*(1), 1–13. https://doi.org/10.1186/s12870-019-1939-z

Yu, Y., Zhou, Y. F., Feng, Y. Z., He, H., Lian, J. P., Yang, Y. W.,. . .& Chen, Y. Q. (2020). Transcriptional landscape of pathogen-responsive lncRNAs in rice unveils the role of ALEX 1 in jasmonate pathway and disease resistance. *Plant Biotechnology Journal*, *18*(3), 679–690. https://doi.org/10.1111/pbi.13234

Zhang, J., Yang, Z., Feng, P., Zhong, X., Ma, Q., Su, Q.,. . . Yang, Y. (2019). Identification and the potential roles of long non-coding RNAs in cotton leaves damaged by *Aphis gossypii*. *Plant Growth Regulation*, *88*, 215–225. https://doi.org/10.1007/s10725-019-00500-7

Zhang, L., Wang, M., Li, N., Wang, H., Qiu, P., Pei, L.,. . . Zhang, X. (2018). Long noncoding RNA s involve in resistance to *Verticillium dahliae*, a fungal disease in cotton. *Plant Biotechnology Journal*, *16*(6), 1172–1185. https://doi.org/10.1111/pbi.12861

Zhang, Y., Wang, W., Chen, J., Liu, J., Xia, M., & Shen, F. (2015). Identification of miRNAs and their targets in cotton inoculated with *Verticillium dahliae* by high-throughput sequencing and degradome analysis. *International Journal of Molecular Sciences*, *16*(7), 14749–14768. https://doi.org/10.3390/ijms160714749

Zhao, X., Sun, X., Chen, Y., Wu, H., Liu, Y., Jiang, Y.,. . . Chen, Y. (2023). Mining of long non-coding RNAs with target genes in response to rust based on full-length transcriptome in Kentucky bluegrass. *Frontiers in Plant Science*, *14*, 1158035. https://doi.org/10.3389/fpls.2023.1158035

Zhu, B., Yang, Y., Li, R., Fu, D., Wen, L., Luo, Y., & Zhu, H. (2015). RNA sequencing and functional analysis implicate the regulatory role of long non-coding RNAs in tomato fruit ripening. *Journal of Experimental Botany*, *66*(15), 4483–4495. https://doi.org/10.1093/jxb/erv203

10 Role of Microorganisms in Tackling Heavy Metal Stress in Crops

Avantika Pandey, Yogesh Kumar Pandey, and Shashi Bhushan Agrawal

10.1 INTRODUCTION

Heavy metal pollution is one of the significant global issues that affect the agricultural soil and the productivity of crops worldwide. Increased anthropogenic activities like rapid industrialization, prolonged use of untreated wastewater, and intense agricultural activities are the major causes of heavy metal pollution in agricultural soil. A recent study carried out by Rani et al. (2023) depicts the effect of industrialization on the contamination of heavy metals in agricultural soil in Haryana state of India. Increased accumulation of these heavy metals in agricultural soil leads to enhanced absorption and bioaccumulation by crops, which consequently affects the food chain (Narayanan and Ma 2023). Although some heavy metals (Cu, Zn, Ni, Co, Mn, and Mo) are required in small quantities and are essential for the normal growth and development of crops, they become toxic at higher levels. On the other hand, some of the heavy metals, such as Se, Cr, Hg, Cd, Pb, and As, have no known role in the metabolism of crops and are toxic even at low levels (Tiwari and Lata 2018). Increased bioaccumulation of heavy metals causes growth, developmental, physiological, biochemical, genetic, and molecular changes that consequently affect the overall production of crops. Redox-active metals (Cu, Cr, Fe, and Mn) directly cause oxidative stress due to elevated production of reactive oxygen species (ROS) that results in disturbance in cell homeostasis and damage to DNA, proteins, lipid membranes, and pigments. However, non-redox-active metals (Zn, Ni, Cd, Hg, and Al) indirectly cause oxidative stress via stimulating the activity of ROS-inducing enzymes (NADPH oxidase), binding to the sulfhydryl group of proteins, depletion of nonenzymatic antioxidants (glutathione), and inhibition of antioxidative enzymes (Emamverdian et al. 2015). El-Ballat et al. (2023) reported that Cr stress significantly reduces the growth of *Cicer arietinum* by inducing the production of ROS and oxidative stress. To deal with stress caused by heavy metals, crops have several defense mechanisms; however, these mechanisms become ineffective beyond a certain threshold, jeopardizing the plant's survival (Narayanan and Ma 2023). The use of beneficial microorganisms has emerged as a key alternative solution to enhance the effectiveness of plant defense against heavy metal stress because these microorganisms offer an efficient, sustainable, and environment-friendly approach (El-Ballat et al. 2023).

10.2 MICROORGANISM-MEDIATED ALLEVIATION OF HEAVY METAL STRESS IN CROPS

The rhizosphere is an intricate and dynamic environment where countless interactions among the roots, organic compounds, minerals, solutes, gases, and microorganisms play a key role in driving the biogeochemical cycling of metals (Oburger and Schmidt 2016). Rhizospheres are colonized by a variety of microorganisms such as bacteria, fungi, algae, and cyanobacteria. Moreover, interactions

DOI: 10.1201/9781003608080-10

among microorganisms and crops can be utilized to help the crops adapt to metal-contaminated environments and offer the potential for enhancing metal tolerance in crops (Nazli et al. 2020). This interaction enhances the production of secondary metabolites, phytohormones, growth-promoting potential, the expression of genes and proteins, the activation of antioxidant molecules, and ultimately the stress tolerance in crops (Shaffique et al. 2023). However, not all microbes are capable of carrying out the alleviation of heavy metal toxicity and enhancing the tolerance of crops, as the effectiveness of microorganisms to alleviate heavy metal stress can vary from species to species and metal to metal (Singh and Narzary 2022). A study performed by Li et al. (2024) in a rice farmland in China reported that Zn, As, Cd, Sb, and Pb were the main pollutants of the farmland, and due to contamination of these heavy metals, the diversity of bacteria was found to be reduced. However, some of the bacteria, such as of the order Chlorofexales, Gemmatimonadales, Myxococcales, and Solibacterales, showed strong tolerance to heavy metals and might reduce the accumulation of Pb and As in rice grains. Therefore, a crucial need for employing microorganisms in heavy metal-contaminated environments to decrease the mobility and thereby toxicity of heavy metals to crops is the tolerance or resistance of microorganisms against the heavy metals. In microorganisms, both chromosomal and plasmid-encoded metal resistance is reported; however, the majority of the resistance mechanisms are plasmid-encoded and are very specific to an anion or cation (Etesami 2018). Additionally, there is a diverse group of beneficial microorganisms called plant growth-promoting microorganisms (PGPMs) that colonize the rhizosphere and promote the growth of plants by ameliorating various environmental stresses, including heavy metal stress. There are generally two types of PGPMs: (i) Plant growth-promoting rhizobacteria (PGPR) and (ii) plant growth-promoting fungi (PGPF).

In general, the strategies utilized by microorganisms to cope with heavy metal toxicity may involve one or more (Figure 10.1) of the following mechanisms: (i) Heavy metals can be absorbed at the cell surface of microbes through processes such as precipitation, adsorption, ion exchange, or

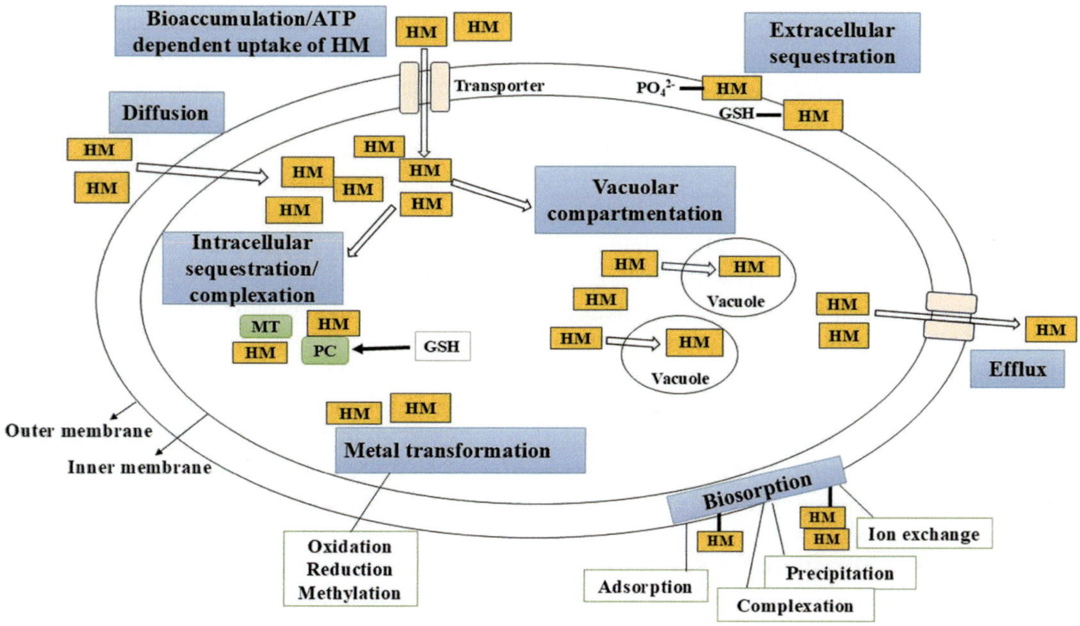

FIGURE 10.1 Mechanisms employed by microorganisms under heavy metal stress, adapted from Singh and Narzary (2022). Abbreviations: HM: heavy metals; GSH: glutathione; PC: phytochelatins; MT: metallothionein; PO_4^{2-}: phosphate.

complexation. Wang et al. (2020) reported that two strains of urease-producing bacteria (*Bacillus megaterium* HD8 and *Enterobacter bugandensis* TJ6) protect *Lactuca sativa* from Cd and Pb toxicity by immobilization and extracellular sequestration of Pb and Cd, leading to lower absorption of Pb and Cd by the root of *Lactuca sativa*. (ii) In the periplasmic space of microbes, heavy metals are complexed with metabolites and form a heavy metal–metabolite complex (referred to as extracellular sequestration), hence reducing heavy metal toxicity. (iii) If heavy metals enter cells, it involves chelation or complexation with various metabolites, enzymes, or regulatory molecules (referred to as intracellular sequestration). (iv) The transformation of heavy metals or organo-metal compounds inside cells takes place by various methods such as oxidation, reduction, methylation, and dealkylation. Two bacterial strains, *Raoultella planticola* (C9) and *Acinetobacter beijerinckii* (C5), were found to reduce the toxic forms of Cr and As to the less toxic and stable forms in the rhizosphere, thus making them unavailable to *Glycine max* and reducing the accumulation of As and Cr by the root of *G. max* (Husna et al. 2021). (v) The exclusion of heavy metal complexes or heavy metals from the cell occurs either through volatilization or efflux mechanism (Singh and Narzary 2022).

10.3 ROLE OF BACTERIA IN THE ALLEVIATION OF HEAVY METAL STRESS

Soil inhabits a wide range of bacterial populations. Among these bacterial populations, a class of beneficial bacteria called plant growth-promoting rhizobacteria (PGPR) promotes the growth of plants. Besides improving plant growth, these PGPR have been found to induce various mechanisms in crops against heavy metal toxicity (Table 10.1). These PGPR either live intracellularly in symbiotic association with plants or live freely. Examples of intracellular symbiotic PGPR include *Rhizobium*, *Azorhizobium*, and *Allorhizobium*, while extracellular PGPR include *Bacillus*, *Azotobacter*, *Burkholderia*, *Pseudomonas*, and *Azospirillum*.

Through either direct or indirect means, PGPR improves the growth of crops under heavy metal stress. The primary way that PGPR directly promotes plant growth is by either secreting beneficial substances (siderophores, phytohormones, and aminocyclopropane-1-carboxylate (ACC) deaminase) to plants or by making it easier for them to absorb specific nutrients from their surroundings. Plant growth is indirectly promoted by PGPR via the activation of stress-responsive mechanisms (stimulation of physiological and antioxidant systems) in crops (Figure 10.2).

Production of phytohormones: PGPR produces several phytohormones such as auxin (indole acetic acid (IAA)), gibberellins, and cytokinins. Among these phytohormones, IAA is one of the most significant plant hormones that PGPR produce. One of the traits of encouraging the growth of crops is the production of IAA by PGPR, which directly contributes to the growth of crops under heavy metal stress. Cu- and Pb-tolerant and IAA-producing *Azotobacter chroococcum* CAZ3 was found to greatly enhance growth by stimulating cell division, enlargement, and initiation of root elongation in *Zea mays* growing under Cu and Pb stress (Rizvi and Khan 2018).

Enhances nitrogen fixation: Various heavy metals are known to severely affect nitrogen fixation and thereby productivity of crops. However, studies reported the effective role of PGPR in enhancing nitrogen fixation in crops under heavy metal stress conditions. In a pot study, it was found that Cu and Zn at an elevated level (400 mg/kg) reduced plant growth, nodule number, and nitrogenase activity (a key enzyme of biological nitrogen fixation) in *Medicago lupulina*. However, co-inoculation of *M. lupina* with *Sinorhizobium meliloti* CCNWSX0020 and *Agrobacterium tumefaciens* CCNWGS0286 significantly enhances the number of nodules and nitrogenase activity under Cu and Zn stress (Jian et al. 2019).

Phosphate solubilization: Phosphorus is an essential macronutrient for plant metabolism and nutrient cycling. Although soil contains a lot of phosphorus in both organic and inorganic forms, it frequently bonds with calcium, iron, aluminum, and other ions, making it difficult for plants to absorb and use phosphorus efficiently. To convert insoluble phosphorus into a form that plants can utilize, PGPR primarily secrete organic acids (oxalic, citric, and gluconic acids), chelates metal ions to create soluble complexes (such as iron, calcium, and aluminum phosphates), and produces

TABLE 10.1
Role of PGPR in Growth Promotion and Heavy Metal Stress Alleviation in Crops.

Crops	Heavy Metals	PGPR	PGPR-Mediated Heavy Metal Stress Alleviation and Growth promotion	References
Triticum aestivum L.	Cr (25, 50, and 100 mg/kg soil)	*Bacillus subtilis*	Reduced the total Cr, MDA, and H_2O_2 content and increased the shoot and root dry weight, leaf area, chlorophyll, carotenoid content, and APX activity	Seleiman et al. (2020)
Medicago sativa L.	Cr (10 mg/kg soil)	*Pseudomonas* sp.	Increased the shoot and root dry weight and chlorophyll content and decreased the levels of stress markers (MDA, H_2O_2 and proline)	Tirry et al. (2021)
Lactuca sativa L.	Cd (3 mg/L) and Pb (5 mg/L)	*Enterobacter bugandensis* TJ6 and *Bacillus megaterium* HD8	Increased the biomass, soluble protein, and vitamin C content and reduced the Cd and Pb contents of shoots	Wang et al. (2020)
Triticum aestivum cv. Akbar and Dilkash	Cd (0, 40, and 80 mg/kg)	*Bacillus subtilis* NA2 and *Aspergillus niger* PMI-118	Improvement in plant biomass, shoot length, root length, chlorophyll, total sugar, total protein, and ascorbic acid content	Bashir et al. (2022)
Brassica napus L.	Sb (608.8 mg/L)	*Cupriavidus* sp. S-8-2	Increased fresh and dry weight of root and decreased MDA content	Zheng et al. (2023)
Cicer arietinum L	Cr (130 and 260 μM)	*Azospirillum brasilense* EMCC1454	Alleviate the oxidative stress markers (MDA and H_2O_2)	El-Ballat et al. (2023)
Sorghum bicolor L.	Cu (100, 200, and 400 mg/kg), Cd (10, 20, and 40 mg/kg), Zn (250, 500, and 1,000 mg/kg), and Pb (200, 400, and 1,000 mg/kg)	*Alcaligenes faecalis* MG257493.1, *Bacillus cereus* MG257494.1, and *Alcaligenes faecalis* MG966440.1	Increased the photosynthetic pigments, plant height, shoot, and root fresh weight, and shoot and root dry weight	El-Meihy et al. (2019)
Lycopersicon esculentum L.	Cd (0.4 mM)	*Pseudomonas aeruginosa* and *Burkholderia gladioli*	Enhanced the root length, shoot length, and fresh weight chlorophyll, carotenoid, and xanthophyll content	Khanna et al. (2019)
Oryza sativa L.	Cd and Ni (0.5 and 1.0 mM)	*Enterobacter ludwigii* SAK5 and *Exiguobacterium indicum* SA22	Enhancement of biomass and growth attributes such as the root length, shoot length, root fresh weight, shoot fresh weight, and chlorophyll content	Jan et al. (2019)
Zea mays L.	Cu (669, 1,838, and 2,007 mg/kg) and Pb (195, 390, and 585 mg/kg)	*Azotobacter chroococcum* CAZ3	Dry biomass of roots, and number, yield, and protein of kernels were increased, while the levels of proline, MDA, and antioxidant enzymes decreased	Rizvi and Khan (2018)

(Continued)

TABLE 10.1 ***(Continued)***
Role of PGPR in Growth Promotion and Heavy Metal Stress Alleviation in Crops.

Crops	Heavy Metals	PGPR	PGPR-Mediated Heavy Metal Stress Alleviation and Growth promotion	References
Hordeum vulgare L.	Cr, Zn, and Ni (0.5, 1, and 2 mM)	*Bacillus glycinifermentans* IS-2T strain	Improved the growth parameters	Belhassan et al. (2024)
Oryza sativa L.	Cd (200 μg/mL)	*Klebsiella pneumoniae* MCC 3091	Enhanced seed germination, root and shoot length, and root and shoot fresh weight. Improved antioxidant enzyme activity, proline, α-amylase, protease, total sugar, total protein, chlorophyll content, and reduced Cd accumulation	Pramanik et al. (2017)
Hordeum vulgare L.	Cr (50 and 100 μM)	PGPR (HAS31)	Enhanced plant growth, biomass, photosynthetic pigments, antioxidant enzymes, and mineral uptake	Zhu et al. (2023)

Abbreviations: MDA: malondialdehyde; H_2O_2: hydrogen peroxide; APX: ascorbate peroxidase.

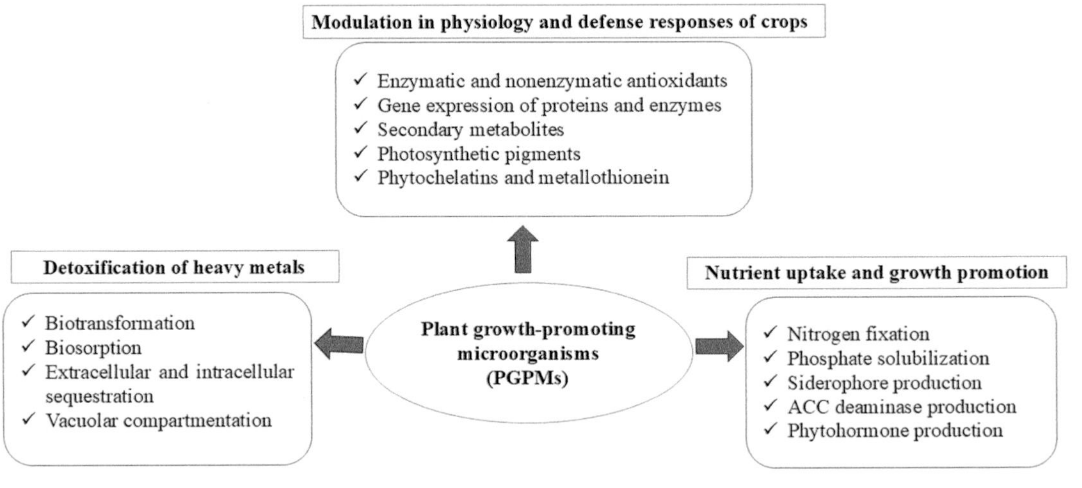

FIGURE 10.2 Mechanisms utilized by plant growth-promoting microorganisms (PGPMs) in the alleviation of heavy metal stress and growth promotion of crops.

enzymes (phytase and pyridoxal phosphatase) that hydrolyze organic phosphorus in soil into inorganic forms (Qin et al. 2024). As heavy metals are known to disturb the uptake of essential nutrients, Husna et al. (2021) reported that inoculation of PGPR strains *Raoultella planticola* (C9) and *Acinetobacter beijerinckii* (C5) enhanced the availability of nutrients to soybean seedlings through phosphate solubilization.

Siderophore production: Iron, although abundantly dispersed throughout the Earth's crust, is usually found in an insoluble oxidation state that hinders its absorption and utilization by plant roots. Certain microbes and crops are capable of producing a siderophore, which is a low-molecular-weight organic molecule that has a high affinity for iron, thus making it bioavailable for plants. Siderophores can form complexes with insoluble forms of iron (Fe^{3+}) and convert them into soluble forms (Fe^{2+}) for efficient plant uptake. Besides, siderophores also chelate with heavy metals, thereby reducing their uptake and consequently the toxicity of heavy metals to crops. Wang et al. (2020) reported that *Bacillus megaterium* HD8 and *Enterobacter bugandensis* TJ6 reduce the availability of Pb and Cd in soil by chelating the Pb and Cd ions with siderophore, which results in reduced accumulation of these heavy metals by *Lactuca sativa*.

ACC deaminase production: Ethylene is a gaseous hormone that plays a crucial role in controlling plant growth. However, elevated levels of ethylene are found to prevent plants from growing normally and encourage senescence. Under heavy metal stress conditions, it was reported that the level of ethylene increased. Moreover, ACC deaminase) can degrade ACC (precursor of ethylene) into ammonia and α-ketobutyrate and hence lower the level of ethylene under the condition of heavy metal stress. Zainab et al. (2020) reported that ACC deaminase-producing PGPR *Bacillus gibsonii* (PM11) and *Bacillus xiamenensis* (PM14) stimulate plant growth by reducing the stress caused by heavy metals, as observed by enhanced root and shoot length, dry weight, and enzymatic antioxidant activity of *Linum usitatissimum*.

Induction of physiological and antioxidant systems of crops: PGPMs also have the ability to indirectly boost the defense mechanisms of crops against heavy metal stress. *Cicer arietinum* exposed to two levels of Cr (130 and 260 μM) showed a reduction in physiological attributes. However, *Cicer arietinum* inoculated with *Azospirillum brasilense* EMCC1454 showed improvement in physiological traits as observed by enhanced carotenoid and chlorophyll content, photosynthetic rate, relative water content, and stomatal conductance compared to uninoculated stressed plants (El-Ballat et al. 2023). The application of Cr-resistant *B. subtilis* reduced the total Cr in the shoot and root of wheat. Furthermore, *B. subtilis* was also found to reduce oxidative stress and Cr toxicity by converting the highly toxic form of Cr (Cr^{6+}) to the less toxic form of Cr (Cr^{3+}) in wheat grown under Cr toxicity, as observed by reduced oxidative stress markers such as hydrogen peroxide (H_2O_2), malondialdehyde (MDA) content, and increased activity of enzymatic antioxidants (Seleiman et al. 2020). Inoculation of *Brassica juncea* with *Cupriavidus* sp. S-8-2 enhanced the fresh and dry weight of root by four- and twofold, respectively, under Sb stress compared to non-inoculated seedlings. Furthermore, the PGPR strain was also found to reduce oxidative stress (as observed by lower MDA content) by reducing the absorption of Sb by *Brassica juncea* (Zheng et al. 2023). Similarly, in a study, Rizvi and Khan (2018) reported that metal-tolerant PGPR *Azotobacter chroococcum* CAZ3 enhanced the number, yield, and protein of the kernel of *Z. mays* even at higher concentrations of Cu (2,007 mg/kg) by 10%, 45%, and 6%, respectively. The enhancement in yield attributes of *Z. mays* due to the inoculation of *Azotobacter chroococcum* CAZ3 could be due to the amelioration of the toxic effects of Cu as observed by the reduced uptake of Cu and improvement in nutrient uptake as well as reduced oxidative stress (reduced MDA, proline, and antioxidative enzyme activity).

Besides, some other compounds secreted by bacteria also have a prominent role in alleviating heavy metal toxicity. Melanins are polymeric, heterogenous, amorphous compounds secreted by several microorganisms formed by the oxidation of amino acid tyrosine and can chelate heavy metals and thereby protect microorganisms from heavy metal toxicity. Rizvi and Khan (2018) reported the metal-binding property of melanins extracted from *Azotobacter chroococcum* CAZ3 under excess Cu and Pb and suggested the role of *Azotobacter chroococcum* CAZ3 in Cu and Pb toxicity.

10.4 ROLE OF FUNGI IN THE ALLEVIATION OF HEAVY METAL STRESS

Fungi are eukaryotic heterotrophic microorganisms that have a wide range of distributions. Among these, a group of beneficial fungi known as plant growth-promoting fungi (PGPF) holds great potential for enhancing plant growth under heavy metal stress (Table 10.2). These PGPF either live in symbiotic association with plant roots and are called mycorrhiza or are free-living (*Trichoderma* species) (Gopalan et al. 2023). Mycorrhiza is generally of two types, i.e., endomycorrhiza and ectomycorrhiza. Endomycorrhiza includes arbuscular mycorrhizal fungi (AMF), in which fungal hyphae penetrate the cortical cell of roots and form a vesicle- or arbuscle-like structure. In ectomycorrhiza, fungal hyphae are extracellular and form a thick sheath on the surface of generally perennial woody plants (Rosling et al. 2009).

PGPF also enhance the growth of crops and alleviate the heavy metal toxicity to crops by employing several mechanisms, such as enhancing the nutrient uptake (through phosphate solubilization and siderophore production), converting the toxic heavy metals to less toxic forms, reducing the bioavailability of heavy metals to crops (through biosorption, sequestration, and complexation of heavy metals), and activating the physiological and antioxidant systems of crops (Figure 10.2). The negative impact of Cd on the physiology of *Solanum lycopersicum* was found to be mitigated by the inoculation of AMF (*Rhizophagus intraradices*, *Funneliformis mosseae*, and *Claroideoglomus etunicatum*), as observed by increased chlorophyll and carotenoid content compared to that in Cd-stressed plants. Hashem et al. (2016) suggested that enhancement of chlorophyll content in AMF-inoculated *Solanum lycopersicum* might be due to enhanced uptake of nutrients (Mg) under Cd stress. Additionally, AMF inoculation also reduced the oxidative stress caused by Cd to *Solanum lycopersicum* by strengthening the nonenzymatic (phenol content) and enzymatic antioxidants (peroxidase (POD), superoxide dismutase (SOD), catalase (CAT), ascorbate peroxidase (APX), and glutathione reductase (GR)). Husna et al. (2021) reported that the endophytic fungus *Aspergillus welwitschiae* is tolerant to Cr and As (up to 1,200 μg/mL) stress and has several growth-promoting potentials, i.e., showed enhanced phytohormone production, solubilized the inorganic phosphate, and produced stress-related metabolites. Furthermore, when *Aspergillus welwitschiae* was inoculated into *Glycine max*, it was found to reduce Cr and As toxicity and improve the growth of *G. max* by reducing the Cr and As uptake by *G. max*, and by alleviating the oxidative stress by stimulating the antioxidative system of *G. max*.

In a study by Sun et al. (2021), it was found that the uptake of Se from the soil to root reduced when *Z. mays* was inoculated with AMF (*Funneliformis mosseae and Claroideoglomus etunicatum*). Se is not essential for plant metabolism; it is transported via phosphorus, sulfate, and silicon transporters and is known to affect the growth of crops adversely at higher concentrations. Sun et al. (2021) further suggested that the reduced uptake of Se by *Z. mays* inoculated with AFM was due to the downregulation of the gene expression of Se uptake-related transporters (ZmNIP2;1, ZmPht2, and ZmSultr1;3). A pot experiment was conducted by Zhang et al. (2019) to observe the role of AMF and the mechanism employed by AMF (*Rhizophagus intraradices* and *Glomus versiforme*) in the amelioration of Cd stress and the growth promotion of *Z. mays*. Zhang et al. (2019) found several-fold increases in phytochelatins and glutathione content in AMF-inoculated Cd-stressed *Z. mays*. Phytochelatins and glutathione are very well known to form complexes with heavy metals that further sequester inside the vacuole, thereby reducing the toxicity of heavy metals. Zhang et al. (2019) suggested that AMF inoculation reduces Cd toxicity by enhancing the level of phytochelatins and glutathione that form complexes with Cd that are consequently sequestered inside the vacuoles (as observed by the decreased distribution proportion of the active form and enhanced proportion of the inactive form of Cd).

Some studies reported that inoculation of a mixture of AMF was more efficient in heavy metal stress alleviation than the inoculation of single-species AMF. Crossay et al. (2020) reported that mixtures of AFM with different taxa are more efficient in reducing the metal content in the aerial parts of *Sorghum vulgare* than the single AMF.

TABLE 10.2
Role of PGPF in Growth Promotion and Heavy Metal Stress Alleviation in Crops.

Crops	Heavy Metals	PGPF	PGPF-Mediated Heavy Metal Stress Alleviation and Growth Promotion	References
Solanum lycopersicum L.	Cd (50 μM)	*Funneliformis mosseae*, *Rhizophagus intraradices*, and *Claroideoglomus etunicatum*	Enhanced the shoot and root length, chlorophylls, carotenoids, phenol content, and the activity of enzymatic antioxidants CAT, POD, SOD, APX, and GR	Hashem et al. (2016)
Glycine max (L.) Merr.	Cu (100 and 300 mg/kg), Pb (100 and 300 mg/kg), and Zn (300 and 600 mg/kg)	*Rhizophagus intraradices*	Enhanced growth and phosphorus uptake	Adeyemi et al. (2021)
Zea mays L.	Se (2, 5, and 10 mg/kg)	*Funneliformis mosseae* and *Claroideoglomus etunicatum*	Improved the root morphology, nitrogen and phosphorus nutrition, and shoot growth. Inhibited the transport of Se from soil to roots	Sun et al. (2021)
Triticum aestivum L.	Pb (25 and 75 ppm) and Cu (100 and 200 ppm)	*Aspergillus terreus*	Reduced the toxicity of Pb and Cu, and increased the shoot and root length, fresh and dry weight, protein, and sugar content	Seleiman et al. (2020)
Triticum aestivum L.	Pb (25 and 75 ppm) and Cu (100 and 200 ppm)	*Aspergillus terreus*	Increased the shoot and root length, fresh and dry weight of shoot and root, protein, and sugar content	Dilawar et al. (2024)
Glycine max (L.) Merr.	Cr and As	*Aspergillus welwitschiae*	Increased shoot and root length, and CAT and POD activity	Husna et al. (2021)
Hordeum vulgare L.	Cd (10, 30, and 60 mg/kg sand)	*Microdochium bolleyi* strain B26	Enhanced the shoot and root length, and dry biomass of shoot and root	Shadmani et al. (2021)
Hordeum vulgare L.	Cd (150 mg/kg soil)	*Trichoderma harzianum*	Increased tolerance index, shoot dry weight, and chlorophyll content	Taghavi et al. (2022)
Zea mays L.	Cd	*Rhizophagus intraradices* and *Glomus versiforme*	Decreased Cd toxicity and enhanced the phytochelatin and glutathione content	Zhang et al. (2019)

Abbreviations: CAT: catalase; POD: peroxidase; SOD: superoxide dismutase; APX: ascorbate peroxidase; GR: glutathione reductase.

10.5 ROLE OF ALGAE AND CYANOBACTERIA IN ALLEVIATION OF HEAVY METAL STRESS

Algae and cyanobacteria also constitute major parts of soil microorganisms and have the great ability to tolerate and bioaccumulate high concentrations of toxic heavy metals. Several studies have reported the beneficial role of algae and cyanobacteria in the remediation of toxic heavy metals from soil and water (Xu et al. 2024; Yeheyo et al. 2024). Besides bioremediation, these microorganisms also have plant growth-promoting potential and enhance crop yield. Cyanobacteria can stimulate plant growth by producing bioactive compounds such as phytohormones, amino acids, and polysaccharides. The cyanobacteria *Anabaenopsis* and *Anabaena* were reported to enhance seed propagation and plant growth by producing phytohormones. Several growth, physiological, and yield attributes of important crops such as wheat, rice, and maize improved due to the inoculation of cyanobacteria (Nawaz et al. 2024). Although algae and cyanobacteria are efficiently utilized in the bioremediation of soil and water and have growth-promoting potential for crops, studies related to the role of algae and cyanobacteria in the alleviation of heavy metal toxicity to crops are scarce.

10.6 COMPARATIVE EFFICIENCY OF MICROORGANISMS IN THE ALLEVIATION OF HEAVY METAL STRESS

Microorganisms including PGPR and PGPF are very well known to alleviate heavy metal stress. However, only a small number of studies have compared the efficiency of these microorganisms under heavy metal stress conditions as these microorganisms utilized specific mechanisms to alleviate heavy metal stress depending on various other factors. Bashir et al. (2022) suggested that *Bacillus subtilis* is more effective than *Aspergillus niger* in mitigating the Cd stress (40 and 80 mg/kg soil) and promoting the growth of wheat in heavy metal-contaminated soils. However, the study does not shed light on the exact mechanisms that are responsible for the better efficiency of *Bacillus subtilis* than *Aspergillus niger* in heavy metal stress alleviation, although some studies nowadays have utilized the combined potential of PGPR and PGPF in alleviating the heavy metal toxicity of plants. In a study by Bilal et al. (2018), it was found that co-inoculation of *Glycine max* with endophytic fungus (*Paecilomyces formosus* LHL10) and bacteria (*Sphingomonas* sp. LK11) under Al and Zn stress conditions showed significantly higher plant growth attributes compared to plants inoculated individually or non-inoculated. Recently, Hasan et al. (2024) reported that co-inoculation of a PGPR (*Rhizobium*) and PGPF (*Funneliformis caledonius* and *Glomus bagyarajii*) synergistically reduced the heavy metal uptake with more efficiency in the roots and shoots of *Lens culinaris* grown in sewage sludge-amended soil.

10.7 CONCLUSIONS AND FUTURE PROSPECTS

Due to increased anthropogenic activities, heavy metal pollution has become a significant concern that affects the production of crops. There are several approaches to alleviate the effects of heavy metal stress on crops. However, the use of microorganisms nowadays has become very popular as it offers an eco-friendly and sustainable way to reduce the toxicity of heavy metals to crops. Additionally, PGPMs promote the growth of crops besides alleviating heavy metal toxicity. These PGPMs utilize several direct and indirect mechanisms such as enhanced phytohormone production, nutrient uptake (through phosphate solubilization and siderophore production), conversion of toxic heavy metals to less toxic forms, reduction in the bioavailability of heavy metals (through biosorption, sequestration, and complexation), and activation of stress-responsive mechanisms (alleviation of stress markers, stimulation of secondary metabolites, and antioxidant system) in crops. Sometimes co-inoculation of plants with two different microorganisms is also preferred as it showed a synergistic effect on plant growth promotion and heavy metal stress alleviation. Despite the upsurge in the use of microorganisms for heavy metal stress alleviation and growth promotion of crops, studies related to their wide use in natural field conditions are scarce, as most of the studies are restricted to lab conditions. Additionally, there is a need to utilize the bioremediation and plant growth-promoting potential of algae and cyanobacteria in heavy metal stress alleviation for crops.

ACKNOWLEDGMENTS

Avantika Pandey is grateful to the Head, Department of Botany, University of Lucknow, for providing all the necessary facilities. Prof. S.B. Agrawal is also grateful to the National Academy of Sciences (NASI) for Senior Scientist position.

REFERENCES

Adeyemi, Nurudeen Olatunbosun, Mufutau Olaoye Atayese, Olalekan Suleiman Sakariyawo, Jamiu Oladipupo Azeez, Soremi Paul Abayomi Sobowale, Adebanke Olubode, Ridwan Mudathir, Rukayat Adebayo, and Samuel Adeoye. (2021). Alleviation of heavy metal stress by arbuscular mycorrhizal symbiosis in *Glycine max* (L.) grown in copper, lead and zinc contaminated soils. *Rhizosphere* 18: 100325. https://doi.org/10.1016/j.rhisph.2021.100325

Bashir, Sarmad, Sadia Javed, Khalid Mashay Al-Anazi, Mohammad Abul Farah, and Sajad Ali. (2022). Bioremediation of cadmium toxicity in wheat (*Triticum aestivum* L.) plants primed with L-proline, *Bacillus subtilis* and *Aspergillus niger*. *International Journal of Environmental Research and Public Health* 19 (19): 12683. https://doi.org/10.3390/ijerph191912683

Belhassan, Mayssa, Ameny Farhat, Hanen El Abed, Zayneb Chaabeen, Fériel Bouzid, Amine Elleuch, Imen Fendri, and Bassem Khemakhem. (2024). Isolation and identification of a new *Bacillus glycinifermentans* strain from date palm rhizosphere and its effect on barley seeds under heavy metal stress. *Brazilian Journal of Microbiology* 55 (1): 843–854. https://doi.org/10.1007/s42770-024-01263-8

Bilal, Saqib, Raheem Shahzad, Abdul L. Khan, Sang-Mo Kang, Qari M. Imran, Ahmed Al-Harrasi, Byung-Wook Yun, and In-Jung Lee. (2018). Endophytic microbial consortia of phytohormones-producing fungus *Paecilomyces formosus* LHL10 and bacteria *Sphingomonas* sp. LK11 to *Glycine max* L. regulates physio-hormonal changes to attenuate aluminum and zinc stresses. *Frontiers in Plant Science* 9: 1273. https://doi.org/10.3389/fpls.2018.01273

Crossay, Thomas, Yvon Cavaloc, Clarisse Majorel, Dirk Redecker, Valérie Medevielle, and Hamid Amir. (2020). Combinations of different arbuscular mycorrhizal fungi improve fitness and metal tolerance of sorghum in ultramafic soil. *Rhizosphere* 14: 100204. https://doi.org/10.1016/j.rhisph.2020.100204

Dilawar, Naveen, Muhammad Hamayun, Amjad Iqbal, Bokyung Lee, Sajid Ali, Ayaz Ahmad, Abdulwahed Fahad Alrefaei, Turki Kh Faraj, Ho-Youn Kim, and Anwar Hussain. (2024). Rhizofungus *Aspergillus terreus* mitigates heavy metal stress-associated damage in *Triticum aestivum* L. *Plants* 13 (18): 2643. https://doi.org/10.3390/plants13182643

El-Ballat, Enas M., Sobhy E. Elsilk, Hayssam M. Ali, Hamada E. Ali, Christophe Hano, and Mohamed A. El-Esawi. (2023). Metal-resistant PGPR strain *Azospirillum brasilense* EMCC1454 enhances growth and chromium stress tolerance of chickpea (*Cicer arietinum* L.) by modulating redox potential, osmolytes, antioxidants, and stress-related gene expression. *Plants* 12 (11): 2110. https://doi.org/10.3390/plants12112110

El-Meihy, Rasha M., Hamed E. Abou-Aly, Ahmed M. Youssef, Taha A. Tewfike, and Eman A. El-Alkshar. (2019). Efficiency of heavy metals-tolerant plant growth promoting bacteria for alleviating heavy metals toxicity on sorghum. *Environmental and Experimental Botany* 162: 295–301. https://doi.org/10.1016/j.envexpbot.2019.03.005

Emamverdian, Abolghassem, Yulong Ding, Farzad Mokhberdoran, and Yinfeng Xie. (2015). Heavy metal stress and some mechanisms of plant defense response. *The Scientific World Journal* 2015 (1): 756120. https://doi.org/10.1155/2015/756120

Etesami, Hassan. (2018). Bacterial mediated alleviation of heavy metal stress and decreased accumulation of metals in plant tissues: Mechanisms and future prospects. *Ecotoxicology and Environmental Safety* 147: 175–191. https://doi.org/10.1016/j.ecoenv.2017.08.032

Gopalan, N. S. Raja, P. T. Nikhil, Raunak Sharma, and Sridev Mohapatra. (2023). The use of microbes as a combative strategy for alleviation of abiotic and biotic stresses. In *Unravelling Plant–Microbe Synergy*, pp. 175–193. Academic Press. https://doi.org/10.1016/B978-0-323-99896-3.00010-2

Hasan, Mudassara, Fauzia Naushin, Mohd Irfan Naikoo, and Fareed Ahmad Khan. (2024). Co-applications of Rhizobium and arbuscular mycorrhizal fungi synergistically alleviate heavy metal uptake and toxicity in lentil plants (*Lens culinaris* Medik.) grown in sewage sludge amended soil. *South African Journal of Botany* 165: 349–366. https://doi.org/10.1016/j.sajb.2023.12.045

Hashem, Abeer, E. F. Abd-Allah, A. A. Alqarawi, Asma A. Al Huqail, D. Egamberdieva, and S. Wirth. (2016). Alleviation of cadmium stress in *Solanum lycopersicum* L. by arbuscular mycorrhizal fungi via induction of acquired systemic tolerance. *Saudi Journal of Biological Sciences* 23 (2): 272–281. https://doi.org/10.1016/j.sjbs.2015.11.002

Husna, Anwar Hussain, Mohib Shah, Muhammad Hamayun, Muhammad Qadir, and Amjad Iqbal. (2021). Heavy metal tolerant endophytic fungi *Aspergillus welwitschiae* improves growth, ceasing metal uptake and strengthening antioxidant system in *Glycine max* L. *Environmental Science and Pollution Research*: 1–15. https://doi.org/10.1007/s11356-021-16640-1

Jan, Rahmatullah, Muhammad Aaqil Khan, Sajjad Asaf, Lubna, In-Jung Lee, and Kyung Min Kim. (2019). Metal resistant endophytic bacteria reduces cadmium, nickel toxicity, and enhances expression of metal stress related genes with improved growth of *Oryza sativa*, via regulating its antioxidant machinery and endogenous hormones. *Plants* 8 (10): 363. https://doi.org/10.3390/plants8100363

Jian, Liru, Xiaoli Bai, Hui Zhang, Xiuyong Song, and Zhefei Li. (2019). Promotion of growth and metal accumulation of alfalfa by coinoculation with *Sinorhizobium* and *Agrobacterium* under copper and zinc stress. *PeerJ* 7: e6875. https://doi.org/10.7717/peerj.6875

Khanna, Kanika, Vijay Lakshmi Jamwal, Sumit G. Gandhi, Puja Ohri, and Renu Bhardwaj. (2019). Metal resistant PGPR lowered Cd uptake and expression of metal transporter genes with improved growth and photosynthetic pigments in *Lycopersicon esculentum* under metal toxicity. *Scientific Reports* 9 (1): 5855. https://doi.org/10.1038/s41598-019-41899-3

Li, Chuanzhang, He Huang, Xinyi Gu, Kai Zhong, Juan Yin, Jun Mao, Junxian Chen, and Chaolan Zhang. (2024). Accumulation of heavy metals in rice and the microbial response in a contaminated paddy field. *Journal of Soils and Sediments* 24 (2): 644–656. https://doi.org/10.1007/s11368-023-03643-3

Narayanan, Mathiyazhagan, and Ying Ma. (2023). Mitigation of heavy metal stress in the soil through optimized interaction between plants and microbes. *Journal of Environmental Management* 345: 118732. https://doi.org/10.1016/j.jenvman.2023.118732

Nawaz, Taufiq, Shah Saud, Liping Gu, Imran Khan, Shah Fahad, and Ruanbao Zhou. (2024). Cyanobacteria: Harnessing the power of microorganisms for plant growth promotion, stress alleviation, and phytoremediation in the era of sustainable agriculture. *Plant Stress* 11: 100399. https://doi.org/10.1016/j.stress.2024.100399

Nazli, Farheen, Adnan Mustafa, Maqshoof Ahmad, Azhar Hussain, Moazzam Jamil, Xiukang Wang, Qaiser Shakeel, Muhammad Imtiaz, and Mohamed A. El-Esawi. (2020). A review on practical application and potentials of phytohormone-producing plant growth-promoting rhizobacteria for inducing heavy metal tolerance in crops. *Sustainability* 12 (21): 9056. https://doi.org/10.3390/su12219056

Oburger, Eva, and Hannes Schmidt. (2016). New methods to unravel rhizosphere processes. *Trends in Plant Science* 21 (3): 243–255. https://doi.org/10.1016/j.tplants.2015.12.005

Pramanik, Krishnendu, Soumik Mitra, Anumita Sarkar, Tithi Soren, and Tushar Kanti Maiti. (2017). Characterization of cadmium-resistant *Klebsiella pneumoniae* MCC 3091 promoted rice seedling growth by alleviating phytotoxicity of cadmium. *Environmental Science and Pollution Research* 24: 24419–24437. https://doi.org/10.1007/s11356-017-0033-z

Qin, Haichen, Zixiao Wang, Wenya Sha, Shuhong Song, Fenju Qin, and Wenchao Zhang. (2024). Role of plant-growth-promoting rhizobacteria in plant machinery for soil heavy metal detoxification. *Microorganisms* 12 (4): 700. https://doi.org/10.3390/microorganisms12040700

Rani, Jyoti, Sudesh Chaudhary, and Tripti Agarwal. (2023). Impact of industrialization on heavy metals contamination in agricultural soils of Sonepat, a satellite township of New Delhi, India. *Sustainability, Agri, Food and Environmental Research-Discontinued* 12 (1). https://doi.org/10.7770/safer-V12N1-art2436

Rizvi, Asfa, and Mohd Saghir Khan. (2018). Heavy metal induced oxidative damage and root morphology alterations of maize (*Zea mays* L.) plants and stress mitigation by metal tolerant nitrogen fixing *Azotobacter chroococcum*. *Ecotoxicology and Environmental safety* 157: 9–20. https://doi.org/10.1016/j.ecoenv.2018.03.063

Rosling, Anna, Tiina Roose, Anke M. Herrmann, Fordyce A. Davidson, Roger D. Finlay, and Geoffrey M. Gadd. (2009). Approaches to modelling mineral weathering by fungi. *Fungal Biology Reviews* 23 (4): 138–144. https://doi.org/10.1016/j.fbr.2009.09.003

Seleiman, Mahmoud F., Shafaqat Ali, Yahya Refay, Muhammad Rizwan, Bushra Ahmed Alhammad, and Salah E. El-Hendawy. (2020). Chromium resistant microbes and melatonin reduced Cr uptake and toxicity, improved physio-biochemical traits and yield of wheat in contaminated soil. *Chemosphere* 250: 126239. https://doi.org/10.1016/j.chemosphere.2020.126239

Shadmani, Leila, Samad Jamali, and Akram Fatemi. (2021). Effects of root endophytic fungus, *Microdochium bolleyi* on cadmium uptake, translocation and tolerance by *Hordeum vulgare* L. *Biologia* 76: 711–719. https://doi.org/10.2478/s11756-020-00598-5

Shaffique, Shifa, Saddam Hussain, Sang-Mo Kang, Muhammad Imran, Eun-Hae Kwon, Muhammad Aaqil Khan, and In-Jung Lee. (2023). Recent progress on the microbial mitigation of heavy metal stress in soybean: Overview and implications. *Frontiers in Plant Science* 14: 1188856. https://doi.org/10.3389/fpls.2023.1188856

Singh, Khomdram Niren, and Diganta Narzary. (2022). Harnessing the Power of Microbes to Overcome Heavy Metal Stress in Crop Plants. In *Plant Stress: Challenges and Management in the New Decade*, pp. 251–275. Cham: Springer International Publishing. https://doi.org/10.1007/978-3-030-95365-2_16

Sun, Chenyu, Yisen Yang, Muhammad Zeeshan, Shengfeng Qin, Junqing Ma, Lu Liu, Juan Yang, Xunbo Zhou, and Jinghua Huang. (2021). Arbuscular mycorrhizal fungi reverse selenium stress in *Zea mays* seedlings by improving plant and soil characteristics. *Ecotoxicology and Environmental Safety* 228: 113000. https://doi.org/10.1016/j.ecoenv.2021.113000

Taghavi Ghasemkheili, Fatemeh, Flemming Ekelund, Jesper Liengaard Johansen, Hemmatollah Pirdashti, Seyede Roghie Ghadirnezhad Shiade, Amin Fathi, and Rasmus Kjøller. (2022). Ameliorative effects of *Trichoderma harzianum* and rhizosphere soil microbes on cadmium biosorption of barley (*Hordeum vulgare* L.) in Cd-polluted soil. *Journal of Soil Science and Plant Nutrition*: 1–13. https://doi.org/10.1007/s42729-021-00666-y

Tirry, Nabil, Aziza Kouchou, Bouchra El Omari, Mohamed Ferioun, and Naïma El Ghachtouli. (2021). Improved chromium tolerance of *Medicago sativa* by plant growth-promoting rhizobacteria (PGPR). *Journal of Genetic Engineering and Biotechnology* 19: 1–14. https://doi.org/10.1186/s43141-021-00254-8

Tiwari, Shalini, and Charu Lata. (2018). Heavy metal stress, signaling, and tolerance due to plant-associated microbes: An overview. *Frontiers in Plant Science* 9: 452. https://doi.org/10.3389/fpls.2018.00452

Wang, Tiejun, Shilin Wang, Xingchun Tang, Xianpeng Fan, Sheng Yang, Lunguang Yao, Yadong Li, and Hui Han. (2020). Isolation of urease-producing bacteria and their effects on reducing Cd and Pb accumulation in lettuce (*Lactuca sativa* L.). *Environmental Science and Pollution Research* 27: 8707–8718. https://doi.org/10.1007/s11356-019-06957-3

Xu, Ke, Ming Du, Rao Yao, Jing Luo, Zixi Chen, Chao Li, Anping Lei, and Jiangxin Wang. (2024). Microalgae-mediated heavy metal removal in wastewater treatment: Mechanisms, influencing factors, and novel techniques. *Algal Research*: 103645. https://doi.org/10.1016/j.algal.2024.103645

Yeheyo, Hillary Agaba, Anu Mary Ealias, Giphin George, and Umamaheswari Jagannathan. (2024). Bioremediation potential of microalgae for sustainable soil treatment in India: A comprehensive review on heavy metal and pesticide contaminant removal. *Journal of Environmental Management* 363: 121409. https://doi.org/10.1016/j.jenvman.2024.121409

Zainab, Nida, Bashir Ud Din, Muhammad Tariq Javed, Muhammad Siddique Afridi, Tehmeena Mukhtar, Muhammad Aqeel Kamran, Amir Abdullah Khan et al. (2020). Deciphering metal toxicity responses of flax (*Linum usitatissimum* L.) with exopolysaccharide and ACC-deaminase producing bacteria in industrially contaminated soils. *Plant Physiology and Biochemistry* 152: 90–99. https://doi.org/10.1016/j.plaphy.2020.04.039

Zhang, Xiao-Feng, Zun-He Hu, Ting-Xiu Yan, Rui-Rui Lu, Chang-Lian Peng, Shao-Shan Li, and Yuan-Xiao Jing. (2019). Arbuscular mycorrhizal fungi alleviate Cd phytotoxicity by altering Cd subcellular distribution and chemical forms in *Zea mays*. *Ecotoxicology and Environmental Safety* 171: 352–360. https://doi.org/10.1016/j.ecoenv.2018.12.097

Zheng, Yu, Jianquan Tang, Can Liu, Xinlin Liu, Zihan Luo, Di Zou, Guohong Xiang et al. (2023). Alleviation of metal stress in rape seedlings (*Brassica napus* L.) using the antimony-resistant plant growth-promoting rhizobacteria *Cupriavidus* sp. S-8-2. *Science of the Total Environment* 858: 159955. https://doi.org/10.1016/j.scitotenv.2022.159955

Zhu, Yanfeng, Liping Wang, Jing Ma, Yuhang Li, Fu Chen, and Willie Peijnenburg. (2023). Comparative physiological and metabolomics analyses using Ag–NPs and HAS31 (PGPR) to alleviate Cr stress in barley (*Hordeum vulgare* L.). *Environmental Pollution* 333: 122010. https://doi.org/10.1016/j.envpol.2023.122010

11 Symbiotic Strategies
Plant–Microbe Partnerships for Abiotic Stress Resilience

Lydia Pramitha, G. Jeeva, and M. Jincy

11.1 INTRODUCTION

Abiotic stressors affect agricultural productivity by restricting plant growth and development and cause irreversible damage to the plant cell (Jincy et al., 2019). In the future, abiotic stresses, *viz.*, drought, high temperature, salt, and cold and heavy metal stress, are predicted to intensify due to climate change. Plants are sessile; therefore, they often adapt to extreme environmental changes during their growth and development. The extreme change in environmental conditions is one of the main threats to the increasing population. The accumulation of reactive oxygen species (ROS) in plant cells, *viz.*, superoxide (O^{2-}), hydroxyl radical (OH^-), and hydrogen peroxide (H_2O_2), damages the plant cell membrane and collapses the membrane integrity, leading to oxidative stress and cell death under stress conditions (Miller et al., 2008; Choudhury et al., 2013; Jincy et al., 2017). Plants use a combination of morphological and physiological adaptation strategies regulated by various signaling pathways that kindle molecular and biochemical mechanisms to withstand harsh environmental conditions. The adaptation strategies include antioxidant activities, osmotic adjustments, and maintaining the integrity of cell membrane under stress (Villalobos-López et al., 2022). To overcome the stress, plants activate multiple mechanisms such as enzymatic and nonenzymatic antioxidant defense mechanisms; accumulation of osmolytes to maintain the homeostasis of cells and to protect the plants against oxidative damage; inhibition of shoot growth to avoid water loss; biosynthesis and assimilation of osmolytes, *viz.*, betaines, prolines, polyols, amino acids, and inositol for maintaining the osmotic balance; and synthesis of antioxidants such as ascorbic acid, glutathione (GSH), superoxide dismutase (SOD), peroxidase (POD), and catalase (CAT) to quench the ROS burst under oxidative stress (Djanaguiraman et al., 2018; Sharma et al., 2012; Sandrini et al., 2022).

Many mitigation strategies can be used to cope with such impacts, but eco-friendly and low-cost methods should be developed to withstand under abiotic stress. In addition, these few studies showed that certain microbes can also improve the tolerance of plants to abiotic stresses, especially rhizospheric microbes. To increase plant tolerance against abiotic stresses and to ensure agricultural sustainability, microbes play an essential role in various symbiotic adaptation strategies. Naturally, plants and microbes can adapt to the stress conditions by developing new tolerant communities *via* regulatory processes involving many genes. Most of the microbes can survive under stress in an inactive state; once exposed to favorable conditions, the microbes' activity gets restored. The number of microbes in the soil may be affected by prolonged exposure to stress, but it may not affect the metabolic activity of microbes (Milošević et al., 2012). Plant microbial populations from harsh environments such as endophytes, plant growth-promoting rhizobacteria (PGPR) and arbuscular mycorrhizal fungi (AMF) induce plant growth and also enhance soil fertility (Umesha et al., 2018; Verma et al., 2020). The other mechanism behind the stress tolerance in plant–microbe interactions is the triggering of the growth of root hairs and root area *via* the synthesis of indole acetic acid

DOI: 10.1201/9781003608080-11

(IAA), gibberellin, abscisic acid (ABA), and other growth-promoting substances, which in turn facilitate water and nutrient uptake by the roots, promote plant growth, and increase the tolerance against abiotic stresses (Yang et al., 2009). AMF regulate proline accumulation and osmoregulation, especially under drought and salt stress conditions, and increase tolerance in plants; in addition, dual symbiotic mechanisms tend to alleviate the effect of abiotic stress on plants (Grover et al., 2011). This chapter highlights how microbial symbiosis tackles the impact of abiotic stresses and enables plants to survive under hostile regimes.

11.2 MICROBIAL SYMBIOSIS REGULATING CROP RESILIENCE TO ABIOTIC STRESS

Under natural conditions, plants are not alone: They are growing along with many microbes like bacteria, fungi, protists, viruses, etc. These microbes cohabit with the plants in different tissues: phyllosphere, where microbes live on stem and leaf surfaces; endosphere, where microbes live within the plant in internal tissues; and rhizosphere, where microbes live on the root surfaces and in the soil around them (Sharma et al., 2020). Many studies concluded that beneficial microbes such as *Bacillus*, *Rhizobium*, *Glomus*, *Azospirillum*, *Pseudomonas*, and *Aspergillus* species, plays a major role in improving plant growth and enhancing tolerance in plants against abiotic stress conditions and also with great potential in sustainable agriculture (Figure 11.1) (Sandrini et al., 2022).

11.3 MICROBIAL SYMBIOSIS CONFERRING DROUGHT STRESS TOLERANCE

In arid and semi-arid regions, drought stress severely influences crop growth and productivity, mainly due to climate change. According to the FAO 2013 report, worldwide the intensity and frequency of drought have increased, and it adversely affects plant growth and development by influencing the morphology, physiology, uptake of nutrients, and metabolism of plants (Evelin et al., 2009). Drought stress influences the photosynthetic rate by affecting the expansion of the leaf, closure of stomata, disruption in photosynthetic machinery and respiration, change in photosynthetic pigments, and leaf senescence: This in turn influences the yield factors in plants (Wahid et al., 2007; Jincy et al., 2022). The initial changes for tackling the uptake of water begin from modifications

FIGURE 11.1 Impact of abiotic stresses and microbial symbiosis conferring abiotic stress tolerance.

to the morphology of roots to the activation of the antioxidant defense systems in crops. However, severe effects of water deprivation resulted in reduced seed yield, flowering, and seed set in crops. In legumes, many rhizobial symbionts confer drought tolerance by improving the photosynthetic rate *via* an increased number of nodules and by maintaining the water potential under drought conditions (Yang et al., 2013). To sustain the stress, the crops seek the help of endophytes such as *Pseudomonas* and *Bacillus*, which enhanced the synthesis of 1-aminocyclopropane-1-carboxylic acid (ACC) deaminase, IAA, and proline, which helps the plants improve growth under stress conditions; offers them support for crop sustenance through exopolysaccharide (EPS) production; enhances the circulation of water in the plants, osmoregulation, nutrient acquisition, phytohormones, and biofilm formation; and alters plant morphology and genetic regulation (Table 11.1) (Grover et al., 2011; Yadav et al., 2020).

TABLE 11.1
List of Microbes Conferring Drought Stress Tolerance.

S. No.	Host Plant	Microbe	Drought Tolerance Mechanism	References
1	*Glycine max*	*Sinorhizobium fredii* strain SMH12	Improved number of nodules, maintained the water potentials	Kibido et al. (2020)
		Bacillus diazoefficiens	Higher glutamine synthetase activity Maintained N metabolism under drought	Cerezini et al. (2020)
2	*Astragalus sinicus* L.	*Mesorhizobium huakuii* 7653R	Increased the level of proline, arginine, alanine, glutamate, putrescine, spermidine, and spermine Improved nitrogen fixation and ammonium assimilation	Liu et al. (2022a)
3	*Medicago truncatula*	*Sinorhizobium medicae* or *Sinorhizobium meliloti*	Delayed drought-induced leaf senescence, enhanced allocation of reserves to osmolytes	Staudinger et al. (2016)
4	*Vigna mungo*, *Vigna radiata*	*Rhizobium meliloti*	Increased the number of nodules, improved photosynthesis	Siddiqui et al. (2021)
5	*Medicago sativa*	*Rhizobium meliloti*	Enhanced antioxidants	Yang et al. (2013)
6	*Vicia faba*	*Rhizobium leguminosarum*	Enhanced the production of osmoprotectants	Amine-Khodja et al. (2022)
7	*Phaseolus vulgaris*	*Rhizobium tropici* CIAT 899	Increased leaf ABA content	Figueiredo et al. (2008)
8	*Medicago sativa*	IAA-overproducing *Ensifer meliloti* 1021 (Ms-RD64)	Enhanced the production of osmolytes	Defez et al. (2017)
		Sinorhizobium meliloti	Overexpressed cytokinin and antioxidant enzymes	Xu et al. (2012)
9	*Vigna radiata*	*Bradyrhizobium* sp. SUTN9-2	Enhanced ACC deaminase activity	Sarapat et al. (2020)
10	*Phaseolus vulgaris*	*Rhizobium etli*	Overexpressed trehalose-6-phosphate synthase Enhanced the expression of cytochrome cbb3 oxidases	Suárez et al. (2008)Talbi et al. (2012)

TABLE 11.1 (*Continued*)
List of Microbes Conferring Drought Stress Tolerance.

S. No.	Host Plant	Microbe	Drought Tolerance Mechanism	References
11	*Triticum aestivum*	*Rhizophagus intraradices*, *Funneliformis geosporum*	Increased relative water content (RWC), photochemistry, higher chlorophyll content	Mathur et al. (2018)
		Bacillus subtilis (LDR2)	Increased photosynthetic efficiency	Barnawal et al. (2017)
		Glomus fasciculatum, *Glomus mosseae*, *Gigaspora decipiens*	Enhanced plant growth Increased total chlorophyll content	Pal and Pandey (2016)
12	*Cicer arietinum* L.	*Bacillus megaterium* *Bacillus thuringiensis*	Increased proline content in leaf Enhanced enzymatic antioxidant (CAT, POX, POD, and SOD)	Khan et al. (2019)
13	*Solanum tuberosum*	*Rhizobacteria*	Increased ROS-scavenging enzyme activity and enhanced the photosynthesis	Gururani et al. (2013)
14	*Zea mays*	*Azospirillum* sp.	Increased osmolyte accumulation Enhanced growth characteristics	García et al. (2017)
		Bacillus subtilis	Increased plant growth, gas exchange parameters through the synthesis of siderophores, ABA, IAA, and salicylic acid (SA)	Lima et al. (2019); Khan et al. (2019)
		Pseudomonas putida	Increased proline content Maintained RWC and leaf water potential Increased biomass	Sandhya et al. (2010)
		Pseudomonas fluorescens	Increased proline content Improved the production of ABA, auxin, cytokinin, and gibberellin	Ansary et al. (2012)
		Glomus versiforme	Increased chlorophyll content Enhanced uptake of mineral Increased the accumulation of compatible solutes such as proline, sugars, and free amino acids	Begum et al. (2019)
		Trichoderma atroviride	Enhanced the antioxidant system Decreased the level of oxidants (H_2O_2)	Guler et al. (2016)
		Cladosporium oxysporum, *Embellisia chlamydospore*, and *Paraphoma* sp.	Increased the biomass	Li et al. (2019)
15	*Helianthus annuus*	*Pseudomonas putida* GAP-P45	Produced EPSs and improved the stable soil aggregates	Sandhya et al. (2009)
		Pseudomonas lurida EOO26	Enhanced ACC deaminase, siderophore Increased IAA	Kumar et al. (2021)

(Continued)

TABLE 11.1 (*Continued*)
List of Microbes Conferring Drought Stress Tolerance.

S. No.	Host Plant	Microbe	Drought Tolerance Mechanism	References
16	*Citrullus lanatus*	*Glomus versiforme*	Increased proline content Increased antioxidant enzyme activity Increased soluble sugar	Mo et al. (2016)
17	*Poncirus trifoliata*	*Paraglomus occultum*	Increased length of hyphae Improved the rate of water absorption and maintained the water potential	Zhang et al. (2018a)
18	*Solanum lycopersicum*	*Ampelomyces* sp.	Increased growth and yield	Morsy et al. (2020)
19	*Vicia sativa*	*Rhizobium leguminosarum*	Differential expression of genes for N_2 fixation and increase of amino acid transporters like glutamate synthases, dehydrogenases, glutamine synthases, or nitrate and nitrite reductases	Álvarez-Aragón et al. (2023)
20	*Pisum sativum*	*Rhizobium leguminosarum*	Enhanced osmolytes production Increased antioxidants Decreased malonaldehyde and H_2O_2	Álvarez-Aragón et al. (2023)

Source: Goyal and Habtewold, 2023; Inbaraj, 2021; Soni et al., 2023.

11.3.1 Symbiont-Induced Phytohormonal Signaling under Drought Stress

The synthesis of signaling molecules during stress is an initial gateway for adaptation under drought. Under abiotic stress conditions, many bacteria enhance the production of the ACC deaminase enzyme. This bacterial enzyme facilitates plant growth by inhibiting plant ACC (ethylene precursor) activity and thereby leads to the reduction of ethylene levels in plants (Saleem et al., 2007). The phytohormonal modifications and ACC deaminase activity also alter the morphology of root and cell walls and inhibit oxidative damage by the production of glycine betaine and proline, the compatible solutes that assist the process of osmoregulation (Chukwuneme et al., 2020). *Achromobacter piechaudii* ARV8 in tomato and pepper, and *Variovorax paradoxus* and *Pseudomonas* sp. in pea increased the synthesis of ACC deaminase, which resulted in a reduction in the ethylene level and increased the resistance against drought stress (Grover et al., 2011). The ACC deaminase containing rhizobacteria also induces root elongation in plants and thereby increases the uptake of water from the deeper layer of soil, thus facilitating crop growth and development under drought conditions. The inoculation of *Bacillus* in velvet beans and chickpeas and *Enterobacter* in maize was found to synthesize ACC deaminase under drought to reduce the production of ethylene, which inhibits germination and root development (Poudel et al., 2021). AMF maintain the water status under drought by increasing hydraulic conductivity, stomatal conductance, relative water content (RWC), water use efficiency (WUE), and biomass content. AMF also improve the water status of the plant by changing the expression of aquaporins in the plasma membrane and tonoplast (Goyal et al., 2020). In maize, AMF have also been found to induce the synthesis of mRNA for aquaporins to facilitate the passive efflux of water (García, 2020). In leguminous crops, the accumulation of ABA was triggered by phytosymbionts. ABA acts as a hormonal stress signal and modulates the transpiration rate and expression of aquaporins. *Bacillius subtilis* GOT9 in *Arabidopsis* has been reported to trigger endogenous ABA levels in plants (Woo et al., 2020). Moreover, rice inoculated with bacterial symbionts was observed to upregulate OSM34, which is a key regulator of the ABA response pathway. *Piriformospora indica* inoculated in rice upregulates microRNAs like mir159 and mir396 for ABA, and redox potential propelled the same function of phytohormonal regulation in drought (Álvarez-Aragón et al., 2023). Similar induction of IAA in wheat treated with *Rhizobium leguminosarum*, *Mesorhizobium ciceri*, and *Rhizobium phaseoli* promoted the synthesis of auxin components for higher growth and biomass (Hussain et al., 2014). Furthermore, soybean inoculated with *Pseudomonas putida* was found to enhance the secretion of gibberellins (Vurukonda et al., 2016). Thus, it could be understood that endophytes act as a major elicitor in stimulating stress regulatory mechanisms in crops. In addition to this, salicylic acid (SA) is yet another crucial growth hormone that regulates stress acclimatization, and the differential expression of genes related to SA in nodulated *Arabidopsis* and *Triticum durum* was observed to mitigate drought stress (Poudel et al., 2021).

11.3.2 Symbiont-Induced Activation of Osmoprotectants under Drought Stress

The production of osmolytes in plants is a key factor for the uptake of water and minerals; however, this has been the major mechanism adapted by crops to survive drought stress. Studies revealed that this could be accelerated through symbionts. *Bacillus subtilis* and *Bacillus pumilis* ameliorate water deprivation by modifying the plant's osmolyte accumulation (Xie et al., 2019). The impact on endophytes in plants under drought enhances the production of osmolytes such as proline, glycine betaine, soluble sugars, and organic acids for maintaining the turgor potential. This has been evident with the enhanced expression of P5CCS, which acts as the precursor for proline synthesis (Nataraja et al., 2022). Few studies showed that *Rhizobium* also induced the accumulation of osmoprotectants such as proline and soluble sugars under drought stress (Yang et al., 2013; Amine-Khodja et al., 2022). *Bacillus megaterium* and *Glomus* sp. increased the level of IAA and proline content in clover and improved the plant's resistance against drought stress (Grover et al., 2011). Consecutively, in drought-stressed maize, endophytes (*Pirifomospora indica*) were found to upregulate the native stress-responsive genes, *viz.*, *ANAC072*, *DREB2A*, *CIPK3*, *ERD1*, *CBL1*, *HAT*, and *PLDδ*, for osmolyte production. Furthermore, chickpea, maize, sorghum, potato, and cucumber inoculated with *Bacillus* were found to comprise elevated proline contents. Moreover, the accumulation of proline, choline, and glycine betaine was found to be in connection with *Klebsiella variicola*, *Pseudomonas fluorescens*, and *Raoultella planticola* in maize, sorghum, and tomato (Tang et al., 2022). Under drought conditions, more than 70% of vascular plants symbiotically associate with AMF for enzymatic antioxidant activity and osmotic adjustment. This also increases the photosynthetic rate and biomass accumulation by enhanced nutrient uptake. In osmotic adjustment, compatible solutes such as sugar, glycine betaine, proline, polyamine, and organic acids accumulate in plant cells to maintain the turgor pressure and plant growth (Wu and Zou, 2017; Li et al., 2020). In addition to osmolytes, mannitol and sorbitol; soluble sugars such as sucrose and fructans; and amino acids such as asparagine, glutamic acid, GABA, and glutamine were found in higher folds in the shoots and roots of colonized stressed plants (Gagné-Bourque et al., 2016). To conclude, AMF has been reported to embrace higher synthesis of osmolytes, phenols, secondary metabolites, glomalin, and other soluble sugars to maintain higher WUE and RWC stature in crops under drought (Tang et al., 2022).

11.3.3 Symbiont-Induced Exopolysaccharide Production under Drought Stress

One of the unique mechanisms endorsed by bacterial and fungal symbionts under drought is the support offered for water retention in crops. Bacterial symbionts secrete EPSs, which are components that form a biofilm to support the hydrophilic system in the rhizosphere. The symbionts form a hydrophilic biofilm around the roots to protect them from desiccation. This polysaccharide film maintains the soil structure and its aggregation properties under moisture stress. AMF possesses unique properties to influence higher plants in response to drought stress *via* the synthesis of EPSs and biofilm formation, and the production of osmoprotectants and heat shock proteins (HSP), which can enhance tolerance in plants to drought stress (Milošević et al., 2012). *Bacillus amyloliquefaciens* improves the rhizosphere soil structure due to the formation of macroaggregates by the production of EPSs. Macroaggregates are important for soil fertility; this will maintain the aerobic and anaerobic conditions and also regulate the uptake of nutrients from the soil reserves. Therefore, it increases the tolerance of plants to drought stress (Milošević et al., 2002). Similarly, *Pantoea agglomerans* in wheat, and *Rhizobium* sp. and *Pseudomonas putida* P45 in sunflower improve the rhizosphere soil structure by the production of EPS. Under drought stress, *Azospirillum* sp. in wheat and AMF in sorghum increased the circulation of water (Grover et al., 2011). Common beans inoculated with *Bacillus amyloliquefaciens* produce more EPSs than non-inoculated plants. Furthermore, maize seeds treated with EPS-producing bacterial strains such as *Proteus penneri*, *Alcaligenes faecalis*, and *Pseudomonas aeruginosa* improved soil moisture retention, leading to higher plant biomass, and root, shoot, and leaf area (Naseem and Bano, 2014). Improved drought mitigation in okra using treatment with rhizobia, in wheat with *Rhizobium leguminosarum*, in chickpea with *Mesorhizobium ciceri*, and in sunflower seedlings with rhizobium enhanced macroporosity through EPSs under water-deprived conditions (Vurukonda et al., 2016).

11.3.4 Symbiont-Induced Antioxidant Mechanism under Drought Stress

Detoxification under stress conditions is an important pathway for maintaining homeostasis. Concerning this, the symbionts under drought were found to play an important role in activating the radical scavenging systems in various crops. Detoxification processes in nodulated plants using thioredoxin and peroxidases were higher than that in the control. Maize plants inoculated with *Bacillus* species had higher ROS protection conferred by ascorbate peroxidase (APX) and glutathione peroxidase (Vurukonda et al., 2016). The antioxidants induced by rhizobacteria scavenge and nullify the toxic effects of ROS and minimize membrane damage and cell death during drought stress (Milošević et al., 2012). *Pseudomonas mendocina* and *Glomus intraradices* in lettuce increased the antioxidant content during drought stress (Grover et al., 2011). Moreover, a similar increase in antioxidant defense mechanisms in maize and soybean with *Bacillus subtilis* under drought showed the reason beneath the biomass and WUE (Lima et al., 2019). Drought tolerance deliberated by glutathione reductase (GR) in guinea grass and catalase in maize as well as in wheat endows the major stimulation of ROS defense in plants by symbionts (Gagné-Bourque et al., 2016). Furthermore, AMF inoculation in brassica and sesbania was found to improve the ROS scavenging pathway through enhanced SOD activity (Allah et al., 2015). Apart from this, AMF has also been found to stimulate phenolic compounds, which play a major role in ROS scavenging. AMF inoculation has been observed to improve phenylalanine ammonia-lyase (PAL) and flavonoids under drought stress (Tang et al., 2022). Therefore, AMF inoculation has been found to maintain the redox balance under stress, limiting the negative effects on the photosynthetic potential of crops under drought.

11.4 MICROBIAL SYMBIOSIS CONFERRING HIGH-TEMPERATURE STRESS TOLERANCE

Increasing global average temperature has an alarming impact on the distribution of crops and crop production (Porter, 2005). High-temperature stress is the increase in temperature beyond the threshold level for a while that is sufficient to cause irreversible damage to plant growth and development, thus affecting the production of the crop (Wahid et al., 2007). Even a minor increase or decrease in temperature for a short period affects the biochemical and physiological mechanisms of crop plants (Kaur et al., 2015). The impacts due to climate change have been witnessed globally in different ways. The day-by-day increase in ambient temperature is alarming, and the crops being cultivated worldwide are being affected daily. It was predicted that the atmospheric temperature may increase by 1.8–3.6 °C by 2100. The increasing temperature will affect the current and future mean temperature, reduce the soil water content. and lead to drought stress in the future (Ma, 2019; Yan et al., 2015). High-temperature stress is sensitive at all growth stages of plants; this results in severe yield loss; it may also change the adaptation of plants or even cause species extinction (Howarth, 2005). High-temperature stress affects plant growth and yield by the alteration of the plasma membrane and cell division, and disruption of the photosynthetic apparatus and its function (Quinn and Williams, 1985). ROS burst under high-temperature stress collapses the structural integrity of the cells by changing the ratio of biomolecules, which leads to lipid peroxidation, membrane damage, and denaturation of proteins, and affects protein synthesis, enzyme activity, root growth, and uptake of water and nutrients, thereby resulting in localized cell and tissue death in plants (Djanaguiraman et al., 2018; Jincy et al., 2022). Microbial association in plants have been found to boost the survival percentage of plants under high-temperature stress (Table 11.2). Microbes in plants act as triggering molecules for synthesizing pigments, secondary metabolites, and hormones compared to the control (Reva et al., 2021; Dastogeer et al., 2022). Microbes that live in extreme climates use various defense mechanisms to overcome stress, e.g., most of the rhizobacteria produce osmoprotectants such as

TABLE 11.2
List of Microbes Conferring High-Temperature Stress Tolerance.

S. No.	Host Plant	Microbe	High-Temperature Tolerance Mechanism	References
1	*Triticum aestivum*	*Bacillus amyloliquefaciens*	Upregulation of HSP transcription factors Decreased ROS level	Abd El-Daim et al. (2014)
		Bacillus Velezensis	Enhanced metabolite production and enzymatic production for antioxidant systems	Abd El-Daim et al. (2019)
		Pseudomonas putida strain AKMP7	Increased dry weight Increased tillers, and root and shoot length Improved metabolites Enhanced the grain and spikelet formation	Ali et al. (2011)
		Funneliformis geosporum *Rhizophagus irregularis*	Improved the allocation of nutrients Increased nutrient composition in roots Increased grain number	Cabral et al. (2016)
2	*Brassica campestris* spp. *pekinensis*	*Bacillus siamensis* H30-3 *Bacillus aryabhattai* H26-2	Increased ABA content and decreased the stomatal opening	Yoo et al. (2019b)
3	*Solanum lycopersicum*	*Paraburkholderia phytofirmans* strain PsJN	Enhanced photosystem II (PSII) activity Increased chlorophyll content Enhanced the accumulation of proline and total amino acids	Issa et al. (2018)
		Septoglomus deserticola *Septoglomus constrictum*	Enhanced the antioxidant level Decreased lipid peroxidation and H_2O_2	Duc et al. (2018)
4	*Sorghum bicolor*	*Bacillus cereus* *Myroides odoratimimus*	Improved plant growth Enhanced antioxidant enzyme activity Decreased MDA content	Bruno et al. (2020)

(Continued)

TABLE 11.2 (*Continued*)
List of Microbes Conferring High-Temperature Stress Tolerance.

S. No.	Host Plant	Microbe	High-Temperature Tolerance Mechanism	References
5	*Glycine max*	*Bacillus cereus* SA1	Increased the biomass and chlorophyll content Enhanced antioxidant enzyme activity Activation of HSPs	Khan et al. (2019)
		Bacillus aryabhattai	Improved the plant growth and modulation of phytohormones	Park et al. (2017)
		Paecilomyces formosus	Maintained the uptake of macronutrients Enhanced antioxidants such as GSH, CAT, and SOD activities Decreased lipid peroxidation Activation of drought-related and HSP genes	Bilal et al. (2020)
6	*Arabidopsis thaliana*	*Paraphaeosphaeria quadriseptata*	HSP101 and HSP70 expression	McLellan et al. (2007)
7	*Oryza sativa*	*Chaetomium sp.*	Increased shoot and root growth Improved survival percentage	Sangamesh et al. (2018)
		Paecilomyces formosus LWL1	Increased protein and chlorophyll content Increased plant height	Waqas et al. (2015)
8	*Poncirus trifoliata*	*Glomus mosseae*	Increased biomass of shoot and root	Wu (2011)
9	*Zea mays*	*Glomus geosporum* *Rhizophagus intraradices*	Enhanced plant growth Increased chlorophyll content Increased the quantum efficiency of PSII	Mathur et al. (2018)
10	*Helianthus annuus*	*Aspergillus violaceofuscus*	Increased chlorophyll content Decreased ROS level	Ismail et al. (2020a)
11	*Cucumis sativus*	*Thermomyces* sp.	Enhanced quantum efficiency of PSII Increased root length Increased WUE	Ali et al. (2018)
12	*Lolium perenne*	*Aspergillus aculeatus*	Improved efficiency of thw photosynthetic apparatus	Li et al. (2021)

Source: Soni et al., 2023.

proline, K+, trehalose, glutamate, glycine, and polysaccharides. In sorghum seedlings, *Pseudomonas* sp. strain NBRI0987 increased the tolerance against high-temperature stress by synthesizing high-molecular-weight proteins in leaves and increased the plant biomass (Milošević et al., 2012).

11.4.1 Symbiont-Induced Plant Growth under High-Temperature Stress

Microbial studies have been found to stimulate the biomass, including the root and shoot length, in normal and high temperatures. There were significant improvements in root length during high-temperature stress. However, the colonized plants had the same content of relative water potential in both stress and nonstress conditions. The colonized plants were found to have a higher chlorophyll index in thermal stress. The key findings included the inoculation of *Pseudomonas putida* in wheat. In this study, higher root and shoot lengths with higher chlorophyll content, sugar, proline, starch, and amino acids were found to offer a higher dry weight in wheat subjected to heat stress (Ali et al. 2011). Yeasmin et al. (2019) reported a higher growth rate in *Asparagus officinalis* colonized with *Glomus intraradices* due to higher nutrient acquisition under high temperatures. Improved survival percentage, and root and shoot biomass in rice treated with *Chaetomium* sp. was observed by Sangamesh et al. (2018). Similar studies in pepper inoculated with *Glomus intraradices* and *Penicillium resedanum* were found to improve the crop growth rate and survival at higher temperatures (Dastogeer et al., 2022). The vigor for biomass, leaf area, and crop growth rate was also enhanced by *Paecilomyces formosus* and *Exophiala* sp. in cucumber (Khan et al., 2012a, 2012b). Thereby, it is evident that the endophytes play a major role in providing additional support for crop growth under high-temperature stress.

11.4.2 Maintenance of Optimal Photosynthetic Potential under High-Temperature Stress

The microbial colonization under high-temperature stress was found to significantly increase the net photosynthetic efficiency and photochemical efficiency through enhanced production and protection of photosynthetic pigments like chlorophyll, carotenoids, flavonoids, and phenolics. The improved biomass of plants under high-temperature stress is an effect of the higher photosynthetic potential. High-temperature stress has a major effect on photosynthesis and gaseous exchange parameters. However, there is augmented maintenance of chlorophyll and carotenoid pigments in mycorrhizal plants under high-temperature stress (Reva et al., 2021). The symbiotic association of AMF increased the tolerance in maize plants by increasing the active centers of photosystem II (PSII), PSII quantum efficiency, trapping of excitation energy, and net photosynthetic rate and thereby protecting the oxygen-evolving complex (OEC) (Mathur et al., 2018). Cucumber inoculated with *Thermomyces* sp. mediates heat tolerance through the maintenance of maximum quantum efficiency in PSII. Thus, there is not much deterioration in the photosynthetic rate and WUE in mycorrhizal plants under high temperature when compared with normal plants in nonstress conditions (Ali et al., 2018; Mathur et al., 2021). Improved WUE and maximum quantum efficiency in PSII with higher hydrothermal time were observed by Hubbard et al. (2012) in wheat inoculated with endophytic ascomycetous mitosporic fungi. Moreover, heat-sensitive crops like tomatoes when inoculated with *Paraburkholderia phytofirmans* had enhanced chlorophyll content, gas exchange parameters, amino acids, sugars, and proline (Issa et al., 2018). Regulation and augmentation of PSII resulting in higher stomatal conductance and optimum transpiration were also found in maize treated with *Glomus* sp. and *Glomus etunicatum* under thermal stress (Zhu et al., 2011; Mathur and Jajoo, 2020). Thus, symbionts are found to have a major role in stimulating the protection of photosystems under high-temperature stress.

11.4.3 Symbiont-Induced Production of Osmolytes and Plant Growth Hormones under High-Temperature Stress

Organic compounds like osmolytes and hormones play a major role in tolerating high-temperature stress. Although there are a variety of components produced for tackling stress, the major components include sugars, proline, glycine betaine, and sulfonium compounds (Kaur and Asthir, 2015). The accumulation of proline helps in the osmotic balance under high-temperature stress, and this minimizes the stress injury of the plants. Moreover, other osmolytes like mannitol, sorbitol, soluble sugars, and GABA are found to accumulate in mycorrhizal plants to confer a chain of events for stress tolerance (Dastogeer et al., 2022). Sorghum with *Pseudomonas* sp., wheat with *Pseudomonas putida*, and pepper with *Penicillium resedanum* portrayed higher proline, sugars, and amino acids with improved biomass under high temperature, which provides an insight for stimulating a chain of events for defense against high temperature (Ali et al., 2009; Ali et al., 2011; Khan et al., 2013).

In addition to osmolytes, the growth hormones also played an important role in conferring stress tolerance to plants. Major phytohormones like ABA, jasmonic acid (JA), and SA play a major role in conferring high-temperature tolerance. ABA is a major thermotolerant hormone that is crucial for stomatal opening and initiation of heat shock proteins (HSP70). SA is known for stabilizing heat shock elements, and JA plays an important role in the synthesis of defensive secondary metabolites and proteins. Microbial symbiosis tends to induce signaling events for the synthesis of these hormones, and these microbes play an important role in conferring heat tolerance (Dastogeer et al., 2022). In addition to the above, gibberellins and cytokinin are also found to confer high-temperature tolerance in crops *via Penicillium resedanum* in pepper (Khan et al., 2015). Enhanced synthesis of ABA in sunflower and soybean was reported by *Aspergillus flavus* under high-temperature stress. Higher concentrations of SA and IAA were found in sunflower in soybeans with *Rhizopus oryzae* (Ismail et al., 2020a, 2020b). Tomato was also found with higher SA and ABA on inoculation with *Bacillus cereus* (Khan et al., 2020a). Additionally, the same endophyte in soybean was found to induce higher ABA, gibberellic acid (GA), IAA, and SA to reduce the negative impacts of heat stress (Khan et al., 2020b). In addition to this, rice with *Paecilomyces formosus* has also been found to have higher stimulated phytohormone signaling and secondary metabolite production, showing amenable results for using these endophytes in conveying sustainable cropping patterns (Waqas et al., 2015).

11.4.4 Symbiont-Induced Antioxidant Mechanisms under High-Temperature Stress

ROS are detrimental to the physiological metabolism in crops, and during high-temperature stress, the scavenging systems are very important to protect the photosynthetic apparatus (Xie et al., 2014). Both enzymatic and nonenzymatic antioxidant systems are found to be stimulated through symbiosis under stress. The decrease in malondialdehyde (MDA) and hydrogen peroxide with an efficient antioxidant mechanism was found in sorghum inoculated with *Bacillus cereus, Providencia rettgeri*, and *Myroides odoratimimus* (Bruno et al., 2020). In sunflower and soybean, it was observed that an enhanced production of phenols, flavonoids, catalase, ascorbic acid oxidase, SOD, and POD when inoculated with *Aspergillus flavus* and *Aspergillus niger* (Ismail et al., 2020a, 2020b). In the soybean plant, the fungus *Paecilomyces formosus* increased the activity of antioxidants such as catalase, SOD, and GSH and decreased lipid peroxidation, enhancing the uptake of macronutrients and upregulating the HSP genes (Bilal et al., 2020). Furthermore, the inoculation of ryegrass with *Aspergillus aculeatus* (Li et al., 2021) and asparagus with *Glomus intraradices* is evident for higher ROS scavenging systems, which enable these plants to sustain higher temperatures. Thus, inoculation of these microbial symbionts as fungal bioprimers would be a crucial measure to offer additional support to crops in acclimatization.

11.5 MICROBIAL SYMBIOSIS CONFERRING SALT STRESS TOLERANCE

One of the accelerating problems in edaphic conditions for plant growth is increased salinity in soils due to urbanization and saltwater intrusion. Due to the leaching of the salts beneath the water table, the irrigation water is transitioning toward salinity, and this results in the increase in saline soils. Around the globe, 20% of the cultivated land is under salinity, and this tends to gradually increase in the upcoming years (Ahmad et al., 2015). Salinity in soils has a great influence on the vegetative and reproductive growth of plants. This leads to a decrease in the germination and biomass of the crops. The increase in saline conditions leads to a physiological drought, inhibiting the uptake of water due to unbalanced osmotic potential and leading to significant impairment in respiration and photosynthesis. The stress of deprived water and nutritional intake leads to excessive synthesis of ROS (Evelin et al., 2019). However, saline tolerance in plants is a multifaceted mechanism, and symbiotic relationships offer a better scope in improving saline environments (Table 11.3). Plant symbiosis for saline tolerance is a multifaceted mechanism that plays a major role in enhancing nutrient uptake, osmotic balance, and hormonal regulation; improving water relations; and alleviating oxidative stress.

TABLE 11.3
List of Microbes Conferring Salt Stress Tolerance.

S. No.	Host Plant	Microbe	Salt Tolerance Mechanism	References
1	*Triticum aestivum*	*Rhizophagus irregularis*	Increase in leaf area and stomatal density Enhanced stomatal conductance and maintained leaf water potential Increased stomatal conductance, stomatal density, and size Increased soluble sugar concentration, K^+/Na^+ ratio, C assimilation, and N uptake	Dar et al. (2023); Oljira et al. (2020)
		Piriformospora indica	Increased total biomass, photosynthetic pigments, and soluble solutes	Gupta et al. (2021a)
		Glomus mosseae	Elevated total phenol and proline content Increased leaf area, shoot, and root dry weights	Evelin et al. (2019)
2	*Triticum durum*	*Rhizophagus irregularis* and *Funneliformis mosseae*	Upregulation of genes related to trehalose metabolism, RNA processing, vesicle trafficking, cell wall organization, and signal transduction	Puccio et al. (2023)
3	*Oryza sativa*	*Piriformospora indica*	Improved antioxidant enzyme system and osmoregulation ability Increased the concentration of K^+ and inhibited the transfer of Na^+ to rice leaves Overexpression of genes related to salt tolerance, photosynthesis, and ion homeostasis	Gupta et al. (2021a); Zhang et al. (2023)
		Claroideoglomus etunicatum	Enhanced uptake of P and K in shoots with reduced Na^+ in shoots	Evelin et al. (2019)
4	*Cucumis sativus*	*Funneliformis mosseae*	Improved nutrient uptake, plant biomass, germination percentage, and photosynthetic rate	Dar et al. (2023); Evelin et al. (2019)
		Glomus etunicatum, *Glomus intraradices*	Increased total phenol and ascorbic acid, SOD, CAT, APX, and GR Reduced H_2O_2 and MAD contents	Dar et al. (2023)
		Glomus versiforme	Reduction in the uptake of Na^+ and increase in antioxidant enzymes (SOD, POD, and CAT)	Gupta et al. (2021a)
5	*Ocimum basilicum*	*Glomus deserticola*	Increased WUE, chlorophyll content, and K, P, and Ca contents	Evelin et al. (2019)
		Glomus mosseae and *Glomus Intraradices*	Increased proline accumulation	Gupta et al. (2021a)
6	*Trigonella foenum-graecum*	*Glomus monosporum*	Elevated growth attributes and increased proline antioxidant enzymes and phosphatase	Dar et al. (2023); Evelin et al. (2019)
		Glomus intraradices	Increased ascorbic acid, α-tocopherol, GSH (γ-glutamyl-cysteinyl-glycine), carotenoid, SOD, APX, POD, GR, and CAT	Dar et al. (2023); Evelin et al. (2019)

(Continued)

TABLE 11.3 (*Continued*)
List of Microbes Conferring Salt Stress Tolerance.

S. No.	Host Plant	Microbe	Salt Tolerance Mechanism	References
7	*Medicago sativa*	*Funneliformis mosseae*	Higher in K^+, K^+/Na^+, and Ca^{2+}/Na^+ ratios	Dar et al. (2023); Evelin et al. (2019)
8	*Capsicum annuum*	*Glomus irradicans*	Higher total fresh and dry weights, chlorophyll a and b concentrations, and RWC with decreased membrane permeability	Dar et al. (2023)
9	*Cajanus cajan*	*Funneliformis mosseae* *Rhizophagus irregularis*	Upgraded plant biomass, reduced oxidative radicals due to increased antioxidant enzymatic activities and total soluble sugars	
		Funneliformis mosseae	Increased GSH/GSSG ratio, AsA (ascorbic acid)/DHA (Dehydroascorbic acid) ratio, and GR, APX, and SOD activities, and decreased MDA content	
10	*Arachis hypogea*	*Rhizophagus irregularis*	Increased photosynthetic rate, RWC, plant height, osmolyte accumulation, SOD, POD, CAT, APX, and lower MAD	Evelin et al. (2019)
11	*Solanum lycopersicum*	*Rhizophagus irregularis* and *Piriformospora indica*	Increased K^+, cytokinin, indoleacetic acid, ABA concentrations, and K^+/Na^+ ratio, and enhanced proline content	Dar et al. (2023); Evelin et al. (2019); Gupta et al. (2021a)
12	*Cucurbita pepo*	*Glomus intraradices*	Enhanced RWC	Dar et al. (2023)
13	*Acacia nilotica*	*Glomus deserticola*	Elevated uptake and solubilization of P	Dar et al. (2023)
14	*Lycopersicon esculentum*	*Glomus mosseae*	Increased antioxidant enzyme activities, decreased MAD	Gupta et al. (2021a)
15	*Suaeda salsa*	*Glomus mosseae*	Higher SOD and CAT with decreased MAD	
16	*Sesbania sesban*	*Glomus etunicatum*, *Glomus intraradices*, and *Glomus mosseae*	Increased activities of enzymatic and non-enzymatic antioxidants	
17	*Digitaria eriantha*	*Rhizophagus irregularis*	Increased root and shoot SOD, CAT, GSH, and APX	
18	*Cicer arietinum*	*Glomus Mosseae*	Elevated CAT, GPOD (guaiacol peroxidase), APX, SOD, MDHAR (Monodehydroascorbate reductase), DHAR (dehydroascorbate reductase), GR, GSH, GSSG, and total GSH contents	Evelin et al. (2019)
19	*Lupinus termis*	*Glomus etunicatum*, *Glomus intraradices*, *Glomus mosseae*	Increased activities of key antioxidant enzymes (SOD, CAT, APX, and POD)	Dar et al. (2023)

TABLE 11.3 (*Continued*)
List of Microbes Conferring Salt Stress Tolerance.

S. No.	Host Plant	Microbe	Salt Tolerance Mechanism	References
20	Lettuce	*Rhizophagus irregularis*	Increased ABA concentrations under salinity	Dar et al. (2023)
21	*Brassica juncea*	*Trichoderma harzianum*	Elevated SOD, POD, APX, GR, GST, GPX, GSH, and GSSGi	Ahmad et al. (2015); Boamah et al. (2021)
22	*Medicago sativa*	*Glomus viscosum*	Increased K and Mg content in shoots and decreased Na, Ca, and Cl content in shoots	Dar et al. (2023)
23	*Pennisetum glaucum*	*Glomus fasciculatum*	Increased proline in shoot and roots	Evelin et al. (2019)
24	*Capsicum annum*	*Glomus intraradices*	Increased leaf and root proline	Evelin et al. (2019)
25	*Zea mays*	*Rhizophagus intraradices, Septoglomus constrictum*, and *Claroideoglomus etunicatum*	Elevated root and shoot proline content	Evelin et al. (2019)
		Funneliformis mosseae	Enhanced H^+/K^+ ratio, higher Na^+ efflux upregulation of transporter genes	Wang et al. (2021)
		Yarrowia lipolytica	Increased shoot proline, total flavonoid, and total phenolics	Gupta et al. (2021a)
26	*Panicum turgidum*	*Funneliformis mosseae, Rhizophagus intraradices*, and *Claroideoglomus etunicatum*	Increased proline and glycine betaine content	Evelin et al. (2019)
27	*Glycine max*	*Fusarium verticillioides*	Increased production of ABA, SA, and lipid peroxidation activities	Gupta et al. (2021a)

11.5.1 Symbiont-Induced Nutrient Uptake under Salt Stress

The initial support offered by the endophytes is the enhanced nutrient uptake through roots with increased biomass. The effect of salt stress in plants impacts severe reduction in biomass and yield. The decrease in plant height during salt stress of untreated plants was about 33.79% at 200 mM NaCl. However, on the contrary, there was an increase of 20%–23% in the root and 13%–15% in the shoot length of *Trichoderma*-treated plants at 100 mM and 200 mM NaCl in mustard seedlings (Ahmad et al., 2015). Reports on increased root and shoot length of pretreated fungal cultures, *viz.*, *Alternaria chlamydospora*, *Embellisia phragmospora*, *Phoma betae*, *Chaetomium coarctatum*, *Fusarium equiseti*, and *Fusarium graminearum*, in barley seedlings under salinity projected the prospects of utilizing this strategy for mitigating stress-affected plants in saline regions (Kouadria et al., 2023). Detailing of the mechanism behind this increase in vegetative growth was well explored with AMF. The rhizosphere in a saline region is prone to reduced cell division and growth. However, the AMF was found to improve this ability by altering the root architecture. AMF colonization in these saline soils was observed to increase the root length, surface area, and volume in citrus, jatropha, and curcas plants (Evelin et al., 2019). This alteration in the root architecture resulted in improved nutrition acquisition by plants. Chickpea plants inoculated with *Mesorhizobium ciceri* were observed to have higher nitrogen, potassium, phosphorus, and magnesium content in their roots and shoots than the control (Egamberdieva et al., 2017). Similar results of enhanced nutrition uptake were observed from treatments of *Trichoderma* and *Bacillus* in saline conditions (Oljira et al., 2020). During salinity, an excess competitive environment for nutrient uptake exists due to the presence of Na^+ and Cl^- ions. Thus, there is an imbalance in the ratio of ions impacting physiological growth. However, the colonization of endophytes influences the supply of N, P, K, Mg, and Ca. AMF colonization enhances the synthesis of organic acids and polyamines in plants to lower the electrical conductivity of the soil. Thereby, this helps the plants imbibe nutrients and water in an ionic homeostasis (Evelin et al., 2012). The enhanced P acquisition in plants by AMF is affected by the secretion of acid and alkaline phosphatases from the AMF hyphae. This liberates the P from its complex and helps in the intrinsic maintenance of P at a lower threshold in soils. Simultaneously, in roots, the expression of *GvPT*, *GiPT*, and *GmosPT* genes increases the affinity for the transport of P (Evelin et al., 2019).

Concerning K^+, there is a stiff competence with Na^+ in the rhizosphere due to their similar chemical nature. Mycorrhizal plants have shown an increased $K^+ : Na^+$ ratio than their counterparts. This was observed due to the ability of the mycorrhizal plants to promote the sequester of Na^+ ions into the vacuole and exclude them through the cytosol. The mycorrhizal plants were found to retract the Na^+ ions through the xylem and divert them away from the photosynthetic tissues to the roots. Thereby, an increase in the efflux of Na^+ / Na^+ exchanger through the salt overly sensitive (SOS) pathway and vacuolar Na^+ / H^+ (NHX) compartmentalization is found to be accelerated through symbionts in plants. As examples in mainstream cereals, *NHX1, NHX2*, and *NHX3* in maize are stimulated by *Bacillus amyloliquefaciens* by producing spermidine, and with AMF, *Funneliformis mosseae* had upregulations in *ZmSOS1*, *ZmHKT1*, and *ZmNHX* for ion homeostasis (Wang et al., 2021; Zhang et al., 2023). In addition, mycorrhizal colonization has an added advantage with the accumulation of glomalin for WinRHIZO™ balance in saline tolerance. Glomalin is a glycoprotein and a homolog of HSP60 that reduces the cytosolic damage due to Na^+ -mediated protein misfolding. Mycorrhiza-treated rice was found to compartmentalize more Na^+ than the normal plants in the vacuole, and deportation of excess Na^+ for maintaining ionic balance was found to be more beneficial in K^+ uptake (Zhang et al., 2023). Pertinent to this, HKT1 is a key determinant in plants that is expressed in shoots for K^+ balance in saline tolerance, and this is found to be expressed in plants treated with *Bacillus subtilis* and *Puccinia indica* (Zhang et al., 2018b).

Similarly, a mechanism for the uptake of nitrogen exists in symbiosis. Higher nitrate reductase activity for enhanced mobilization of nitrate and ammonia by AMF has been reported (Talaat and Shawky, 2014). Higher expression of nitrate and ammonium transporter genes such as *NRT 1.1*, *NAR 2.2*, *AMT 1.1*, and *AMT 1.2* in wheat colonized with *Rhizophagus irregularis* and *Funneliformis mosseae* was reported by Fileccia et al. (2017). Further additional reports on enhanced Mg^{2+} and Ca^{2+} uptake of mycorrhizal plants under salinity with balanced hydraulic conductivity have been observed (Evelin et al., 2012). The mechanism beneath the uptake of calcium and magnesium in plants under salinity needs to be further explored. However, enhanced expression of genes for micronutrient uptake through *MtZIP2* for the uptake of zinc in plants treated with AMF under salinity was reported by Burleigh et al. (2003). Although a mechanism for major nutrient acquisition has been studied with AMF, further deeper insights on the key determinants expressed in plants through symbiosis to overcome saline tolerance in macro- and micronutrient uptake with other symbionts provide a higher thrust in developing efficient fungal and bacterial biopriming techniques for ameliorating the saline regions.

11.5.2 Symbiont-Induced Osmotic Balance under Salt Stress

The water potential of soil during salinity tends to be extra negative due to excessive salt accumulation. Thus, plants should decrease their water potential by effective osmoregulation for incessant water flow into roots. This kind of osmotic adjustment is enabled by plants through the secretion of organic ions and solutes such as proline, betaines, soluble sugars, and polyamines. This is beneficial in maintaining cell turgor, opening of stomata, cellular expansion, and photosynthesis. Among all, proline is the most crucial solute produced by plants during salinity. Several studies with AMF have emphasized the increased synthesis of proline in the roots of treated plants (Klinsukon et al., 2021; Dar et al., 2023). Inoculation of chickpea with *Mesorhizobium ciceri* and *Bacillus subtilis* was observed to significantly increase proline by 29%, and when it was inoculated only with *Mesorhizobium ciceri*, it resulted in an increase of proline by 11% (Egamberdieva et al., 2017). Similarly, treatments of *Trichoderma* resulted in a 56.25% increase in proline at 100 mM and in 70.37% at 200 mM of NaCl in mustard seedlings (Ahmad et al., 2015). The higher folds of proline accumulation in mycorrhizal plants are attributed to the increased expression of *P5CS*. This gene enhances the activity of glutamate dehydrogenase for synthesizing glutamate, which is the precursor of proline (Evelin et al., 2019). Trehalose is yet another disaccharide for carbohydrate metabolism secreted for maintaining the $K^+ : Na^+$ ratio and scavenging the ROS. AMF symbiosis has been found to boost this osmolyte by increasing the activity of trehalose-6-phosphate synthase (TPS) and trehalose-6-phosphate phosphatase (TPP) while lowering trehalase (TRE), which degenerates trehalose (Garg and Pandey, 2016). Furthermore, *Bacillus subtilis* was found to induce glycine betaine in *Arabidopsis* through an ABA-dependent pathway. In addition to this, organic acids like acetic acid, citric, fumaric, malic and oxalic acids in maize were found to be accumulated through AMF colonization. Although deeper mechanisms for organic acids are yet to be explored, it has been ventured that AMF colonization protects the enzymes involved in the biosynthesis of organic acids (Evelin et al., 2019).

Moreover, polyamines play an important role in osmoregulation under salinity, and AMF has been found to enhance this polyamine pool in plants under stress. Polyamines play a vital role in ionic channels for regulating Ca^{2+} ions in the cytoplasm to facilitate stomatal opening, and this has been triggered by AMF colonization (Gupta et al., 2021a). *Trigonella foenum-graceaum* plants under AMF colonization reported higher spermidine: spermine ratio than non-mycorrhizal plants. Furthermore, AMF has also been known to play an important role in accelerating total soluble solids like alpha- and beta-amylases, and this has been observed in AMF inoculated in chickpeas under salt stress (Evelin et al., 2019; Gupta et al., 2021a). Thus, it is understood that AMF colonization in plants regulates the multiple networks in plants for realizing the stress and balancing them to maintain optimum photosynthetic efficiency.

11.5.3 Impact of Microbial Symbiosis in Hormonal Regulation under Salt Stress

One of the crucial stress responses of plants under salinity is the regulation of stomatal movement and water potential, which is often mitigated by hormones, *viz.*, ABA, auxin, JA, and SA. Considering all, ABA is the signature response of plants during salt stress. It has been observed that ABA accumulation is often enhanced upon microbial inoculation for restoration of osmotic balance (Liu et al., 2022b). Studies reported that *Bacillus licheniformis*, *Paenibacillus yonginensis*, and *Bacillus mesonae* increased the ABA contents in chrysanthemum, Panax ginseng, and tomato under saline conditions (Yoo et al., 2019a; Sukweenadhi et al., 2018). During AMF symbiosis, ABA, auxin, JA, SA, and strigolactones coordinate in root growth and architecture based on the availability of nutrients in the soil. Suppression of secondary branches in shoots and alternation of internodal lengths occur on crosstalk with auxin, which is observed to be triggered by AMF symbiosis. Dual AMF colonization with *Pseudomonas* and *Bradyrhizobium* showed an enhanced N_2 fixation under salt stress by reducing the expression of ethylene in soybeans (Win et al., 2023). Furthermore, crops like *Sesbania cannabina*, under salinity, exhibited a positive correlation between strigolactones and ABA, which are mediated by AMF (Evelin et al., 2019). AMF colonization has also been found to enhance the concentrations of GA and SA under salinity. Moreover, the symbiosis of *Digitaria eriantha* with *Rhizophagus irregularis* enhanced the concentration of JA in imparting saline tolerance (Pedranzani et al., 2016). Several reports on enhancing GA, ABA, and IAA by *Trichoderma asperellum* and other major endophytes projected the importance of symbiosis in stimulating the interplay of hormones during salinity (Gupta et al., 2021b).

11.5.4 Symbiont-Induced Antioxidant Mechanisms under Salt Stress

Salinity in soil imposes secondary stress on plants due to the disturbance in the equilibrium between the ROS and antioxidant scavenging mechanism. The accumulation of ROS in stress depends upon the balance between its production and scavenging systems. Plants under salinity have both enzymatic (CAT, SOD, POD, APX, GR, and glutathione peroxidase) and nonenzymatic antioxidant systems (ascorbate, carotenoids, phenolics, GSH, and tocopherols), which are triggered by AMF (Puccio et al., 2023). It has been observed that the AMF-colonized plants show increased SOD, CAT, and APX activities in *Olea europaea* and *Retama sphaerocarpa*. Apart from this, the varied species of *Glomus* and *Rhizophagus* in tomato, cucumber, sesbania, trigonella, chickpea, red gram, citrus, and lupinus were found to show enhanced levels of antioxidant enzymes for imparting saline tolerance. In addition, AMF-inoculated plants like marjoram, lemon balm, and marigold were reported to produce increased polyphenols under salinity (Ahmad et al., 2015; Dar et al., 2023). Hashem et al. (2016) reported the role of *Bacillus subtilis* in *Arabidopsis* with reduced H_2O_2 synthesis. Furthermore, SOD, GR, CAT, guaiacol peroxidase, APX, and POD were found to be increased under salinity in *Trichoderma*-treated mustard seedlings (Ahmad et al., 2015). Studies on the implications of this increase in metalloenzymes suggest that their increased activity in mycorrhizal plants might be due to their higher uptake of Zn, Cu, Mn, and Fe during salinity. Since they are the major co-factors, their uptake and activation help plants sustain the stress through an efficient antioxidant defense mechanism (Evelin et al., 2019; Dar et al., 2023). Beyond this, AMF was also found to cater for the nonenzymatic antioxidant pathway with enhanced α-tocopherols to prevent the auto-oxidation of lipids. Thus, the pool of antioxidant systems, such as GR, GSH, ascorbate, SOD, APX, CAT, POD, and tocopherols, were observed to be accelerated due to the symbiosis of AMF to protect the photosynthetic electron transport in plants (Evelin et al., 2019; Liu et al., 2022b).

11.5.5 Symbiont-Induced Water Relation under Salt Stress

Endophytes have an innate ability to positively affect plant growth, and they tend to enter the crops *via* roots, stomata, lenticels, hydathodes, and wounds by suppressing the pathogenic response in the plants. This relationship is being established in a mutually beneficial nature. Thereby, the endophytes play a key role in managing the water potential of the interface between the plant and soil. Based on the concentration of nutrients in the soil, the roots interact with the endophytes that it sustained under stressful situations (Gupta et al., 2021a). Under salinity, the symbionts play a major role in signaling to prevent ROS and sustain the optimum root-induced respiration in the rhizosphere. The increase in respiration in roots limits C assimilation by plants. However, it was noticed by Rewald et al. (2015) that endophytic colonization increased C assimilation with reduced respiration in roots, which had the potential for optimum photosynthesis. Moreover, endophytic colonization was also found to impart lower osmotic potential and increase leaf turgidity in *Malus hupehensis* mycorrhizal plants (Evelin et al., 2019). *Robinia pseudoacacia* treated with mycorrhizal colonies were found to establish an improved root architecture with enhanced hydraulic activity under salt stress. The desirable root profile in mycorrhizal plants for water uptake under salt stress is also found to be affected by the secretion of aquaporins in roots and leaves. Similar reports were found in tomato and *Lactuca sativa* which exhibited an upregulation of aquaporin and SOS genes in the leaves and roots (Liu et al., 2023). Thus, further deeper insights into the relationship of roots and endophytes for desirable modifications in root architecture lay the foundation in developing sustainable bioagents that could modify the root traits for enhanced water uptake.

11.6 MICROBIAL SYMBIOSIS CONFERRING COLD STRESS TOLERANCE

Subzero temperatures pose a major challenge to plant cultivation and significantly impact plant fitness and survival. Around two-thirds of the earth's surface is being exposed to chilling stress, and enabling plants to survive under chilling stress is a key factor in ensuring food security. Cold stress in plants has a severe impact on crop growth and regulation. The first impact of cold stress is the hardening of cell membranes, leading to morphological changes like leaf curl, discoloration, tissue damage, and stem cracking. This also results in affecting the photosystems due to restricted root surface area, and imbalanced nutrient and ionic balance. Moreover, freezing temperatures can result in altered lipid composition, protoplasmic streaming, and cytoskeleton rearrangement, leading to impaired cellular regulation mechanisms (Dastogeer et al., 2022). However, there are functional roles of endosymbionts in enhancing the cold stress tolerance mechanisms and several reports on antioxidant activity, hormonal signaling, and osmotic balance (Table 11.4).

TABLE 11.4
List of Microbes Conferring Cold Stress Tolerance.

S. No.	Host Plant	Microbe	Cold Tolerance Mechanism	References
1	*Triticum aestivum*	*Glomus mosseae*	Increased chlorophyll content under cold stress	Acuña-Rodríguez et al. (2020)
2	*Phaseolus vulgaris*	*Glomus intraradices*	Enhanced production of aquaporins in roots for tackling cold stress	
3	*Zea mays*	*Glomus etunicatum*	Decreased transpiration rate and lipid peroxidation with increased photosynthesis and biomass production, and antioxidant activity	
4	*Solanum lycopersicum*	*Glomus mosseae*	Increased leaf sugar, photosynthetic pigments, and soluble protein with higher antioxidant (SOD, CAT, POD, and APX) activities leading to enhanced biomass	
		Pseudomonas vancouverensis, *Pseudomonas fredericksbergensis*	The plants exhibited high chilling (10–12 °C) tolerance with reduced cell membrane damage and ROS Upregulation of cold acclimation genes (*LeCBF1* and *LeCBF3*)	Acuña-Rodríguez et al. (2020)
		Trichoderma harzianum	Improved root, leaf weights, and photochemical PSII efficiency	Acuña-Rodríguez et al. (2020)

(Continued)

TABLE 11.4 (*Continued*)
List of Microbes Conferring Cold Stress Tolerance.

S. No.	Host Plant	Microbe	Cold Tolerance Mechanism	References
5	*Oryza sativa*	*Curvularia protuberata*	Higher germination and crop establishment at 5–10 °C	Dastogeer et al. (2022)
		Bacillus amyloliquefaciens and *Brevibacillus laterosporus*	Increased foliar proline and chlorophyll contents at 0–5 °C Higher antioxidant (SOD and CAT) activity	Ganie et al. (2022)
		Glomus intraradices	Increased uptake of water by roots at 15 °C and 25 °C Enhanced production of aquaporins at 15 °C with higher trehalose synthesis in AMF-colonized roots	Ganie et al. (2022)
6	*Glycine max*	*Galactomyces geotrichum*	Harmonized the imbalance of ABA, GA, and sugar metabolism under cold stress	Cui et al. (2024)
7	*Vitis Vinifera*	*Burkholderia phytofirmans*	Altered carbohydrate metabolism at 4 °C	Acuña-Rodríguez et al. (2020)
8	*Cucumis sativus*	*Funneliformis mosseae*	Improvement in biomass, phenols, flavonoids, lignin, and other secondary metabolites	
9	*Hordeum vulgare*	*Chaetomium globosum*, *Epicoccum nigrum*, *Piriformospora indica*	Increased biomass and grain weight under cold stress	
10	*Digitaria eriantha*	*Rhizophagus irregularis*	Reduced lipid peroxidation, increased antioxidant activity, and higher JA concentrations at 4 °C	Cui et al. (2024)
11	*Citrus tangerina*	*Glomus mosseae*	Enhanced root length and root Ca^{2+} concentration for signaling	Kamran et al. (2022)
12	*Arabidopsis thaliana*	*Burkholderia phytofirmans*	Increased cell wall strength, photosynthetic efficiency due to high chlorophyll contents at 1 °C	
13	*Vitis vinifera*	*Burkholderia phytofirmans*	Higher cellular concentrations of soluble carbohydrates like raffinose, trehalose, or starch	
14	*Musa acuminata*	*Piriformospora indica*	Reduced MDA and hydrogen peroxide with an increase in SOD, CAT, soluble sugars, and proline content	Begum et al. (2019)

11.6.1 Symbiont-Induced Antioxidant Mechanisms under Cold Stress

ROS is an unavoidable product during stress, and this has been found to have detrimental effects on cellular levels at low temperatures. The main role of balancing ROS is performed by their scavenging systems enabled through enzymes like CAT, POD, and SOD. These enzymes were found to reduce electrolyte leakage and MAD production in cells under freezing conditions. Inoculation of maize with *Glomus etunicatum* and rice with *Bacillus amyloliquefaciens* and *Brevibacillus laterosporus* has reported an increase in CAT, SOD, and POD under subzero temperatures (Zhu et al., 2010; Ganie et al., 2022). A similar increase in the antioxidant levels in shoots and roots of *Digitaria eriantha* treated with *Rhizophagus irregularis* under 4°C projected the beneficial role of symbionts in embarking the scavenging systems in cold tolerance (Pedranzani et al., 2016). In bananas, cold tolerance is triggered on treatment with *Piriformospora indica* in roots, which reduced MAD and H_2O_2 concentrations with increased SOD. Thus, it is understood that microbial symbionts play an important role in signaling and accelerating the production of antioxidants. One such example was studied in a polyculture of symbionts in lettuce, chard, and spinach with *Penicillium fuscoglaucum* and *Penicillium glabrum*. This resulted in the enhanced production of flavonoids, and there was an increased uptake of nutrients in subzero conditions (below 4 °C) (Morales-Vargas et al., 2024). Although the acceleration of antioxidants has been reported, the genetic determinants expressed in plants and microbes have to be further explored in investigations to boost the development of biopriming agents for cold tolerance.

11.6.2 Symbiont-Induced Hormonal Signaling under Cold Stress

Hormonal signaling plays a major role in connecting the environment and the plant's response under a stress situation. Thus, the concentrations of hormonal secretion vary from high to low depending upon the mechanism adapted by the plants for sustaining the stress. Research on microbial symbionts has revealed their significant role in tolerating freezing conditions. There were higher folds of expression for ACC deaminase in *Phaseolus vulgaris* treated with six bacterial symbionts. These treated plants with microbial symbionts had been found to have reduced ice nucleation and freezing injury than the control (Tiryaki et al., 2019). The major role of the ACC deaminase was to hydrolyze an ethylene precursor to ammonia and α-ketobutyrate. This was beneficial in reducing the harmful effects of ethylene on crop growth under freezing temperatures (Singh et al., 2015). Nevertheless, this mechanism of enhancement and repression of the ethylene pathways was found to be beneficial in minimizing the injury, and the potential support for growth and survival was not reported (Acuña-Rodríguez et al., 2020).

Inclusively, Kakar et al. (2016) detected the enhanced hormonal profile in rice treated with a consortium of bacterial symbionts, *viz.*, *Bacillus laterosporus* and *Bacillus amyloliquefaciens* at 0–5 °C. In-depth analysis of their mechanism exhibited higher folds of elicitors like SA and β-aminobutyric acid. Moreover, the upregulation of C-repeat/DREB (dehydration-responsive element-binding) binding factors like *OsDREB*1A and *OsMYBR2*, *OsCDPK13*, and *OsDIL* through symbiosis was known to be pivotal in cold tolerance by triggering the calcium-dependent signals, and the synthesis of IAA was intriguing. Similar findings from AMF symbiosis in the enhancement of jasmonates and IAA in *Digitaria eriantha* and *Ipomoea batatas* portrayed further evidence of the beneficial impacts of microbial symbionts under chilling stress (Pedranzani et al., 2016). Despite this, the treatment of rice seeds with the fungal endophyte, *viz.*, *Curvularia protuberata*, exhibited an increase in the germination percentage at 5–10 °C (Redman et al., 2002).

11.6.3 Symbiont-Induced Osmoprotectants under Cold Stress

The reduction in the temperatures of the soil has a huge impact on the access to water uptake. However, microbial symbionts have a crucial role in the maintenance of osmotic potential at low temperatures through the synthesis of metabolites. The increase in the acclimation of grapevine, namely, *Vitis vinifera*, at 4°C with *Burkholderia phytofirmans* resulted in elevated production of trehalose, raffinose, and starch (Fernandez et al., 2012). These soluble sugars in plants act as osmoprotectants to prevent the dehydration of cells under freezing temperatures (Kamran et al., 2022). Tomatoes incubated with *Trichoderma harzianum* at 8 °C enhanced the synthesis of proline and improved biomass, owing to the higher efficiency of PSII (Ghorbanpour et al., 2018). Proline ensured the protection of cellular membranes and reduced electrolyte leakage in rice treated with *Bacillus laterosporus* and *Brevibacillus amyloliquefaciens* (Kakar et al., 2016). In addition, aquaporins were also found to be improved *via* symbiosis in plants like *Phaseolus vulgaris* and rice when inoculated with *Glomus intraradices*. The higher folds of expression of *PvPIP1; 3* and *OsPIP* under cold stress conditions (15 °C) unravelled the beneficial effects of AMF across crops (Javot et al., 2003; Liu et al., 2014). Furthermore, the upregulation of molecular chaperons like HSP70 by the gene *DnaK* in *Alicyclobacillus acidocaldarius* during heat and cold stresses laid the basis for improving the host plant's ability under temperature stress (Kamran et al., 2022). This was similar in wheat when treated with *Bacillus velezensis*, which showed higher proteins and osmotic metabolites (Dastogeer et al., 2022).

Despite ensuring the synthesis of metabolites, the microbial symbionts also have a profound effect on water relations. AMF-inoculated crops were found to have a higher WUE and conservation than the control under cold stress. Their main role in acclimatization involves the retainment of moisture in crops by enhancing the production of secondary metabolites like phenols and proteins to combat the freezing stress (Begum et al., 2019). This sort of osmotic adjustments and water uptake is a result of an efficient signaling process triggered by the symbionts at an early sensation of stress in the soil. It was observed that the altered lipid composition and membrane fluidity of cells under stress were regulated by Ca^{2+} flow turned on by symbiosis. The studies in *Arabidopsis* with *Piriformospora indica* increased the expression of C-repeat binding factors and cold regulator genes for calcium-mediated signaling in plants (Morales-Vargas et al., 2024). Thus, we could understand that the frequency of these beneficial symbionts at lower temperatures is low, and strategies of inoculating symbionts to lower freezing injuries in crops should be manifested in the upcoming years.

11.6.4 Symbiont-Induced Nutrient Acquisition under Cold Stress

The presence of AMF in colder regions is rare, and the array of symbionts that infect arctic regions varies from those in tropical zones. Instead of AMF, a higher frequency of dark septate endophytes was reported. However, inoculation of *Glomus claroideum* in cudweed showed enhanced N utilization, and *Cucumis sativus* with *Rhizophagus intraradices* showed improved P acquisition at 15 and 10 °C (Acuña-Rodríguez et al., 2020). *Phialocephala fortinii* inoculated in alpine sedge showed higher shoot P concentrations and improved seed germination under 15 °C. *Piriformospora indica* in barley at 8°C also reported higher N utilization and improved biomass than the control (Murphy et al., 2014). Although studies are emphasizing cold tolerance in plants, the major role of these symbionts in natural environments is yet to be explored. Thus, further studies on various aspects of cold tolerance through symbionts need to be conducted in the future (Acuña-Rodríguez et al., 2020).

11.7 MICROBIAL SYMBIOSIS CONFERRING HEAVY METAL STRESS TOLERANCE

This is a stress conferred by severe human activities like industrialization, urbanization, and mining. These activities result in the deposition of heavy metals like lead, arsenic, cobalt, manganese, copper, mercury, nickel, and cadmium. These elements are present in the soils of cultivable regions and have a severe impact on physiology, metabolism, respiration, ion efflux, and photosynthesis. The presence of heavy metals in soil impacts key enzymes like PEP (phosphoenolpyruvate) carboxylase and RuBisCO. This causes a decline in the photosynthetic efficiency of the crops, affecting the plant's survival. The presence of heavy metals initially disrupts the REDOX potential of plants by synthesizing excessive H_2O_2, OH^-, and O^{2-} ions. Therefore, to survive the heavy metal stress, around 20%–30% of the plants have developed detoxification mechanisms like chelation and subcellular compartmentalization. Possible mechanisms of promoting stress tolerance in plants through symbiosis have also been beneficial, and these microbes mitigate ionic toxicity by binding to the ions and transporting them into the cell (Table 11.5). Thus, there are several ways in which microbial symbiosis supports the plants to sustain under a toxic environment, therby ameliorating soil health (Swamy et al., 2019).

TABLE 11.5
List of Microbes Conferring Heavy Metal Stress Tolerance.

S. No.	Host Plant	Microbe	Heavy Metal	Tolerance Mechanism	References
1	*Trigonella foenum-graecum*	*Glomus monosporum*, *Glomus clarum*, *Gigaspora nigra*, and *Acaulospora laevis*	Cd	Increased growth attributes, chlorophyll, and protein contents with reduced MDA at higher concentrations	Abdelhameed and Metwally (2019)
2	*Brassica juncea L.*	*Bacillus mucilaginosus*	Cd	Enhanced production of IAA and gibberellins	Devi et al. (2022)
		Rhizobium leguminosarum	Zn	Improved chelation of heavy metals through organic acids in rhizosphere	
		Psychrobacter sp.	Ni	Higher production of ACC deaminase, IAA, and enzymes for P solubilization	
3	*Brassica napus*	*Enterobacter cloacae*	As	Higher synthesis of IAA, ACC deaminase, and proline	
		Pseudomonas fluorescens	Pb	Improved production of IAA and ACC deaminase	
		Pseudomonas tolaasii	Cd	Synthesis of ACC deaminase, siderophores, and IAA	
4	*Lycopersicum esculentum*	*Burkholderia* sp.	Pb	Increased phosphate solubilization and the synthesis of IAA and ACC	
		Glomus intraradices	Zn	Genetic regulation and expression of *LeNramP1* and *Lemt2* for heavy metal transporter	Devi et al. (2022)
5	*Mesorhizobium huakuii*	*Astragalus sinicus*	Cd	Elevated production of metallothionein	
6	*Solanum nigrum*	*Pseudomonas* sp.	Cd	Development of biosurfactants and siderophores for ion mobilization	
7	*Triticum aestivum*	*Pseudomonas putida*	Cd	Synthesis of phytochelatins	
		Pseudomonas aeruginosa	Zn	Enhanced uptake of N and P, and increase in total soluble protein, leaf chlorophyll, and biomass	
8	*Helianthus annus*	*Micrococcus* sp.	Cd	Increased synthesis of IAA and ACC deaminase	
		Pseudomonas putida	Cd	Synthesis of metal-binding peptides	

(Continued)

TABLE 11.5 (*Continued*)
List of Microbes Conferring Heavy Metal Stress Tolerance.

S. No.	Host Plant	Microbe	Heavy Metal	Tolerance Mechanism	References
9	*Amaranthus*	*Rahnella* sp.	Cd	Synthesis of IAA, siderophores, and ACC deaminase	
10	*Ricinus communis*	*Pseudomonas jessenii*	Zn	Enhanced biosorption of heavy metals and increase in ACC deaminase and IAA	
11	*Oudemansiella radicata*	*Bacillus amyloliquefaciens*	As, Pb, Al, and Cd	Increased production of iron chelating agent and siderophores	
12	*Oryza sativa*	*Brevundimonas diminuta*	As	Increased biomass and chlorophyll and reduced MDA content	
13	*Vigna mungo*	*Klebsiella pneumoniae*	Cd	Increased synthesis of siderophores and IAA	
14	*Medicago sativa*	*Rhizobium* and *Piriformospora indica*	Cd	Elevated siderophores and IAA in alfaalfa	
15	*Cicer arietinum*	*Trichoderma harzianum*	Cd	Increased solubilization of P	
16	*Glycine max*	*Funneliformis mosseae*	Cu, Pb, and Zn	Enhanced phosphorus uptake and its mobilization	Adeyemi et al. (2021)
17	*Eucalyptus globulus*	*Pisolithus albus*	Ni	The symbiont acts as a barrier in the plant to uptake heavy metals from soil	Chot and Reddy (2022)
18	*Eucalyptus tereticornis*	*Pisolithus albus*	Cd and Cu	Heavy metal immobilization in the soil through fungal mycelium due to the synthesis of metallothionein	Chot and Reddy (2022); Tiwari and Lata (2018)
19	*Eucalyptus* and *Pinus massoniana*	*Pisolithus tinctorius* and *Suillus luteus*	Mn	Impact on heavy metal dilution and reduction in heavy metal induced oxidative stress	Chot and Reddy (2022)
20	*Pinus thunbergii*	*Pisolithus* sp.	Cr	Reduced accumulation of Cr due to reduced bioavailability	
21	*Populus canescens*	*Paxillus involutus*	Pb	Heavy metal concentration is biofiltered by ectomycorrhizae (ECM) fungi	

11.7.1 Symbiont-Induced Secretion of Phytochelatins under Heavy Metal Stress

The exudates or the secretions in the rhizosphere from the plants include oxalic acid, SA, malic acid, chitin, citric acid, and formic acid to reduce the pH in the rhizosphere for mineral uptake by plants. These acids also play an important role in the chelation and detoxification of heavy metals in the soil. Trivalent ions such as Fe^{3+} and Al^{3+} in the soil could easily chelate with these organic acids to form citrates, oxalates, and malates, which reduces the toxicity in the soil. Understanding these features of the plants, the ectomycorrhizal symbiosis in host plants induces the secretion of these root exudates like oxalic acid in larger quantities. Higher oxalic acid was synthesized in mycorrhiza-treated *Pinus sylvestris* in aluminum-bound toxic soil (Chot and Reddy, 2022). However, a similar study by Fernández-Fuego et al. (2017) portrayed a decline in the organic acids produced by mycorrhizal plants under nickel stress. This exhibited the metabolic costs to be paid by plants for survival under heavy metal stress. The plants have to spend their maximum energy to accumulate and mitigate the toxic metals through sequestration. Thus, the role of metallothioneins (MTs) in heavy metal sequestration in cells was found to reduce the organic acids produced by mycorrhizal plants. In-depth analysis of the mycorrhizal plants under heavy metal stress projected fluctuations in the organic acids, which were also reported to be dependent on the fungal species involved in the symbiosis (Chot and Reddy, 2022). Thus, it could be understood that the exudation of organic acids in the intra- and extra-radical hyphae helps in the immobilization of the heavy metals accumulated in the rhizosphere (Shi et al., 2019; Adeyemi et al., 2021). In addition, the exudation of organic acids and the association of AMF in plants under heavy metal stress also enhance the production of glomalin, a glycoprotein from the mycelium. This secretion from the AMF helps in enhancing the heavy metal-binding sites and subsequently reduces the accumulation of heavy metals in the soil (Bano and Ashfaq, 2013; Swamy et al., 2019). Mycorrhizal inoculation, *viz.*, *Rhizophagus intraradices*, *Glomus versiforme*, and *Funneliformis mosseae*, in crops like soybean and maize exhibited a reduced concentration of Cd, Pb, Zn, and Cu due to the release of glomalin (Adeyemi et al., 2021).

11.7.2 Symbiont-Induced Modifications in Roots under Heavy Metal Stress

Roots inoculated with symbiotic organisms have been found to play a vital role in defense in tolerating and avoiding the accumulation of heavy metals in the crop system. *Pinus sylvestris* with *Paxillus involutus* demonstrated a higher Pb aggregation in its epidermis and cortex. Surprisingly, this Pb was not found in the shoots and endodermis of the crop. Thus, it is understood that the root associated with the fungal mycelium increases the electronegative sites for binding. This prevents these toxic metals from entering the crop system, and thus they remain as an unavailable element in the root-hyphal regions. Several reports on heavy metal binding on hyphae have been reported (Chot and Reddy, 2022). These heavy metals are loaded from the soil through the Hartig net of the mycelium to the roots, from where their entry is restricted by the apoplastic pathway. This kind of pattern has been observed only in the ectomycorrhizal connections in plants (Fernández-Fuego et al., 2017). In addition to this, another modification for enhancing the root uptake of heavy metals is the secretion of siderophores. Numerous filamentous fungi like *Penicillium*, *Mucor*, *Trichoderma*, and *Aspergillus* establish the synthesis of siderophores for tolerating heavy metal stress (Tiwari and Lata, 2018). These help in the enhanced uptake of P in the plants, and P mobilization in cells enhances the size of the roots and hyphal regions, giving more room for heavy metal accumulation in the roots and cortex (Adeyemi et al., 2021). Thus, it could be understood that there is an effective nutrient uptake mechanism enabled by mycorrhizal symbiosis in heavy metal-infested soils that enhances the concentrations of essential minerals like N, P, K, Ca, and Mg in shoots (Chot and Reddy, 2022).

The genus *Rhizobium* is also one of the effective P solubilizers, and this also enhances the production of siderophores for enhanced Fe uptake and other micronutrient availability. Rhizobia symbiosis also secretes components like acetoin and 2,3-butanediol to attract the plant growth-promoting bacteria under heavy metal stress for nutrient mobilization. This phytoprotection mechanism has also been observed in alkaline and calcareous soils. Thus, the legume–rhizobium symbiotic system is efficient in the phytoremediation of heavy metal soils (Jach et al., 2022). Furthermore, the structural modification in the root for the crystallization of heavy metals is another kind of adaptation established with hyphal regions. AMF encouraged Zn tolerance *via* crystallizing in the cortical region. Similar accumulation of Mn, Ni, Cd, Cr, and As in roots and mycelium in association with symbionts like *Aspergillus*, *Ganoderma*, and *Trichoderma* have been found to play an important role in preventing the entry of these toxic metals into the apoplastic system of plants (Bano and Ashfaq, 2013; Swamy et al., 2019).

11.7.3 Symbiont-Induced Synthesis of Enzymes and Growth Regulators under Heavy Metal Stress

Under heavy metal stress, endophytes play a major role in sustaining plant health through a cascade of events. These include the synthesis of phytohormones, ROS scavenging, and genetic expression in plants for acclimatization (Abdelhameed and Metwally, 2019). Soybean was treated with two fungal endosymbionts, namely, *Paecilomyces formosus* LHL10 and *Sphingomonas* sp. LK11, and revealed the enhanced production of ABA, SA, and gibberellins than the control under zinc and aluminum toxicity. Elevation in these hormones was found to be a cofounded result of the overexpression of *GmARI1* and downregulation of *GmHMA18*, *GmHMA13*, *GmHMA19*, and *GmPHA1*. Similar experiments for Cr tolerance in soybean with endophytic fungi showed higher production of IAA, ABA, SA, CAT, SOD, and POD activities (Shaffique et al., 2023). Overexpression of *GmHMA13*, *GmHMA14*, and *GmHMA18* for enhanced production of ABA and JA under Ni, Cd, and Al stress in soybean with two endophytic strains also confirmed the importance of symbiosis in heavy metal tolerance. The genetic expression studies revealed the importance of mitigating the oxidative stress caused by heavy metals in soils, which resulted in an increased production of GSH, SOD, APX, and CAT. Similar reports with nitrogen-fixing bacteria for elevated IAA in the presence of Cu and Zn were found to be beneficial for N assimilation in *Medicago sativa*. IAA under heavy metal stress enhances the activity of enzymes like citrate synthase and ketoglutarate dehydrogenase in the tricarboxylic acid (TCA) cycle, which is necessary for N fixation. Nitrogen-fixing bacteria also impart heavy metal stress tolerance by increasing the production of thiol, EPS, GSH, proline, and urease under Cd, Cu, Zn, and Ni (Jach et al., 2022). Enhanced synthesis of GSH to prevent oxidative stress under symbiosis is another factor that prevents the cells from oxidative stress and xenobiotic detoxification. This is evident in the upregulation of fungus-associated genes with GSH genes, which enhances the sequestration of heavy metal–GSH complexes in vacuoles (Chot and Reddy, 2022).

11.7.4 Symbiont-Induced Heavy Metal Sequestration under Heavy Metal Stress

MTs are low-molecular-weight crystal-rich proteins that can bind metal ions through the thiol group of cysteine residues for sequestration and detoxification. Several reports on the differential expression of MT under heavy metal stress have been analyzed for metal detoxification. Studies on *Oryza sativa OsMT2c* and *OsMT1b* under Cr stress and *OsMT1e* under Cd stress; *Physcomitrella patens PpMT2* under Cu and Cd stress; and *Phytolacca americana PaMT3-1* under Cd stress are some of the examples on heavy metal detoxification. *PaMT1* in *Pisolithus albus* under excess Cu and Cd, *HmMT3* in *Hebeloma mesophaeum* under Cd stress, and *SuiMT1* and *SuiMT2* in *Suillus indicus* under Cu stress have been found to play a key role in metal detoxification. The above reports analyzed the importance of increased MT under ectomycorrhizal connections through mycelium with AMF (Chot and Reddy, 2022). The expression of MTs in *Arabidopsis* for *MTL4* with *Astragalus sinicus* under Cd and *PsMT1* and *PsMT2* in *Pisum sativum* with *Rhizobium leguminosarum* under Cd further suggests the ectomycorrhizal mechanisms for enhanced MT in heavy metal stress tolerance (Jach et al., 2022). The overexpression of the 90-kD shock protein MT related

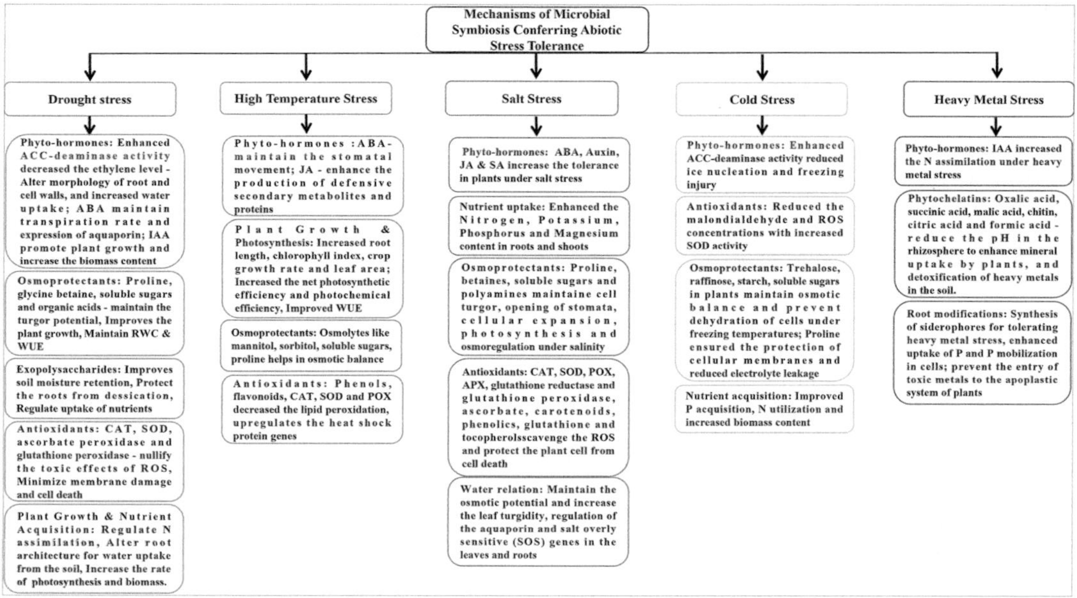

FIGURE 11.2 Mechanisms of plant–microbe interaction for abiotic stress tolerance.

to glutathione-S-transferase genes in arbuscular fungi plants for the immobilization of heavy metals (Bano and Ashfaq, 2013) shows the multifaceted role of MTs in plants under heavy metal stress. Thus, it is understood that heavy metal stress is sustained by ectomycorrhizal plants than normal plants, making them highly effective for bioamelioration (Devi et al., 2022).

Collectively, all these processes enhanced water balance, hormonal regulation, ROS detoxification, and nutrient supply *via* a unified mechanism through which microbes enable plants to withstand drought, salinity, and other environmental stresses (Figure 11.2).

11.8 CONCLUSION

Drastic climate change increases the incidence of abiotic stresses, severely affecting crop productivity, posing a significant threat to global food security. Plant–microbe relationships can alleviate susceptibility to various abiotic stresses in plants. A few mechanisms behind tolerance to abiotic stresses are maintenance of water relation through enhanced production of osmolytes, improved soil moisture retention through enhanced production of exopolysaccharides, morphological adaptation through altered phytohormonal signaling, countering oxidative stress through antioxidant activity, enhanced nutrient acquisition, and enhanced heavy metal sequestration through the production of MT. In addition to tackling climate change by developing climate-smart crops through plant breeding, transgenics, genome editing, and identification of beneficial crop variety-based and/or agro-climatic condition-based plant–microbial relationships/microbial strains for various crop plants will help in achieving food security in a sustainable way.

REFERENCES

Abd El-Daim, I.A., Bejai, S., & Meijer, J. (2014). Improved heat stress tolerance of wheat seedlings by bacterial seed treatment. *Plant and Soil*, **379**, 337–350. https://doi.org/10.1007/s11104-014-2063-3

Abd El-Daim, I.A., Bejai, S., & Meijer, J. (2019). *Bacillus velezensis* 5113 induced metabolic and molecular reprogramming during abiotic stress tolerance in wheat. *Scientific Reports*, **9(1)**, 16282. https://doi.org/10.1038/s41598-019-52567-x.

Abdelhameed, R.E., & Metwally, R.A. (2019). Alleviation of cadmium stress by arbuscular mycorrhizal symbiosis. *International Journal of Phytoremediation*, **21(7)**, 663–671. https://doi.org/10.1016/j.sjbs.2015.03.004

Acuña-Rodríguez, I.S., Newsham, K.K., Gundel, P.E., Torres-Díaz, C., & Molina-Montenegro, M.A. (2020). Functional roles of microbial symbionts in plant cold tolerance. *Ecology Letters*, **23(6)**, 1034–1048. https://doi.org/10.1111/ele.13502

Adeyemi, N.O., Atayese, M.O., Sakariyawo, O.S., Azeez, J.O., Sobowale, S.P.A., Olubode, A., Mudathir, R., Adebayo, R., & Adeoye, S. (2021). Alleviation of heavy metal stress by arbuscular mycorrhizal symbiosis in *Glycine max* (L.) grown in copper, lead and zinc contaminated soils. *Rhizosphere*, **18**, 100325. https://doi.org/10.1016/j.rhisph.2021.100325

Ahmad, P., Hashem, A., Abd-Allah, E.F., Alqarawi, A.A., John, R., Egamberdieva, D., & Gucel, S. (2015). Role of *Trichoderma harzianum* in mitigating NaCl stress in Indian mustard (*Brassica juncea* L) through antioxidative defense system. *Frontiers in Plant Science*, **6**, 868. https://doi.org/10.3389/fpls.2015.00868

Ali, A.H., Abdelrahman, M., Radwan, U., El-Zayat, S., & El-Sayed, M.A. (2018). Effect of *Thermomyces* fungal endophyte isolated from extreme hot desert-adapted plant on heat stress tolerance of cucumber. *Applied Soil Ecology*, **124**, 155–162. https://doi.org/10.1016/j.apsoil.2017.11.004

Ali, S.Z., Sandhya, V., Grover, M., Kishore, N., Rao, L.V., & Venkateswarlu, B. (2009). Pseudomonas sp. strain AKM-P6 enhances tolerance of sorghum seedlings to elevated temperatures. *Biology and Fertility of Soils*, **46**, 45–55. https://doi.org/10.1007/s00374-009-0404-9

Ali, S.Z., Sandhya, V., Grover, M., Linga, V.R., & Bandi, V. (2011). Effect of inoculation with a thermotolerant plant growth promoting *Pseudomonas putida* strain AKMP7 on growth of wheat (*Triticum* spp.) under heat stress. *Journal of Plant Interactions*, **6(4)**, 239–246. https://doi.org/10.1080/17429145.2010.545147

Allah, E.F., Hashem, A., Alqarawi, A.A., Bahkali, A.H., & Alwhibi, M.S. (2015). Enhancing growth performance and systemic acquired resistance of medicinal plant *Sesbania sesban* (L.) Merr. using arbuscular mycorrhizal fungi under salt stress. *Saudi Journal of Biological Sciences*, **22(3)**, 274–283. https://doi.org/10.1016/j.sjbs.2015.03.004

Álvarez-Aragón, R., Palacios, J.M., & Ramírez-Parra, E. (2023). Rhizobial symbiosis promotes drought tolerance in *Vicia sativa* and *Pisum sativum*. *Environmental and Experimental Botany*, **208**, 105268. https://doi.org/10.1016/j.envexpbot.2023.105268

Amine-Khodja, I.R., Boscari, A., Riah, N., Kechid, M., Maougal, R.T., Belbekri, N., & Djekoun, A. (2022). Impact of two strains of *Rhizobium leguminosarum* on the adaptation to terminal water deficit of two cultivars *Vicia faba*. *Plants*, **11(4)**, 515. https://doi.org/10.3390/plants11040515

Ansary, M.H., Rahmani, H.A., Ardakani, M.R., Paknejad, F., Habibi, D., & Mafakheri, S. (2012). Effect of *Pseudomonas fluorescent* on proline and phytohormonal status of maize (*Zea mays* L.) under water deficit stress. *Annals of Biological Research*, **3(2)**, 1054–1062.

Bano, S.A., & Ashfaq, D. (2013). Role of mycorrhiza to reduce heavy metal stress. *Natural Science*, 2013. https://doi.org/10.4236/ns.2013.512A003

Barnawal, D., Bharti, N., Pandey, S.S., Pandey, A., Chanotiya, C.S., & Kalra, A. (2017). Plant growth-promoting rhizobacteria enhance wheat salt and drought stress tolerance by altering endogenous phytohormone levels and TaCTR1/TaDREB2 expression. *Physiologia Plantarum*, **161(4)**, 502–514. https://doi.org/10.1111/ppl.12614

Begum, N., Qin, C., Ahanger, M.A., Raza, S., Khan, M.I., Ashraf, M., Ahmed, N., & Zhang, L. (2019). Role of arbuscular mycorrhizal fungi in plant growth regulation: Implications in abiotic stress tolerance. *Frontiers in Plant Science*, **10**, 1068. https://doi.org/10.3389/fpls.2019.01068

Bilal, S., Shahzad, R., Imran, M., Jan, R., Kim, K.M., & Lee, I.J. (2020). Synergistic association of endophytic fungi enhances *Glycine max* L. resilience to combined abiotic stresses: Heavy metals, high temperature and drought stress. *Industrial Crops and Products*, **143**, 111931. https://doi.org/10.1016/j.indcrop.2019.111931

Boamah, S., Zhang, S., Xu, B., Tong, L.I., Inayat, R., & Calderón-Urrea, A. (2021). The role of *Trichoderma* species in plants response to salt stress. *Asian Journal of Research in Crop Science*, **6(2)**, 28–43.

Bruno, L.B., Karthik, C., Ma, Y., Kadirvelu, K., Freitas, H., & Rajkumar, M. (2020). Amelioration of chromium and heat stresses in *Sorghum bicolor* by Cr^{6+} reducing-thermotolerant plant growth promoting bacteria. *Chemosphere*, **244**, 125521.

Burleigh, S.H., Kristensen, B.K., & Bechmann, I.E. (2003). A plasma membrane zinc transporter from *Medicago truncatula* is up-regulated in roots by Zn fertilization, yet down-regulated by arbuscular mycorrhizal colonization. *Plant Molecular Biology*, **52**, 1077–1088. https://doi.org/10.1023/A:1025479701246

Cabral, C., Ravnskov, S., Tringovska, I., & Wollenweber, B. (2016). Arbuscular mycorrhizal fungi modify nutrient allocation and composition in wheat (*Triticum aestivum* L.) subjected to heat-stress. *Plant and Soil*, **408**, 385–399. https://doi.org/10.1007/s11104-016-2942-x

Cerezini, P., Kuwano, B.H., Grunvald, A.K., Hungria, M., & Nogueira, M.A. (2020). Soybean tolerance to drought depends on the associated *Bradyrhizobium* strain. *Brazilian Journal of Microbiology*, **51(4)**, 1977–1986. https://doi.org/10.1007/s42770-020-00375-1

Chot, E., & Reddy, M.S. (2022). Role of ectomycorrhizal symbiosis behind the host plants ameliorated tolerance against heavy metal stress. *Frontiers in Microbiology*, **13**, 855473. https://doi.org/10.3389/fmicb.2022.855473

Choudhury, S., Panda, P., Sahoo, L., & Panda, S.K. (2013). Reactive oxygen species signaling in plants under abiotic stress. *Plant Signaling & Behavior*, **8(4)**, 23681. https://doi.org/10.4161/psb.23681

Chukwuneme, C.F., Babalola, O.O., Kutu, F.R., & Ojuederie, O.B. (2020). Characterization of actinomycetes isolates for plant growth promoting traits and their effects on drought tolerance in maize. *Journal of Plant Interactions*, **15(1)**, 93–105. https://doi.org/10.1080/17429145.2020.1752833

Cui, J., Nie, F., Zhao, Y., Zhang, D., Zhou, D., Wu, J., Qu, L., Xiao, L.,& Liu, L. (2024). A review on plant endophytes in response to abiotic stress. *Environmental Pollutants and Bioavailability*, **36(1)**, 2323123. https://doi.org/10.1080/26395940.2024.2323123

Dar, M., Razvi, H., Singh, S.M., Mushtaq, N., Shahnawaz, A., D.A.R., & Hussain, S. (2023). Arbuscular mycorrhizal fungi for salinity stress: Anti-stress role and mechanisms. *Pedosphere*, **33(1)**, 212–224. https://doi.org/10.1016/j.pedsph.2022.06.027

Dastogeer, K.M., Zahan, M.I., Rhaman, M.S., Sarker, M.S., & Chakraborty, A. (2022). Microbe-mediated thermotolerance in plants and pertinent mechanisms-a meta-analysis and review. *Frontiers in Microbiology*, **13**, 833566. https://doi.org/10.3389/fmicb.2022.833566

Defez, R., Andreozzi, A., Dickinson, M., Charlton, A., Tadini, L., Pesaresi, P., & Bianco, C. (2017). Improved drought stress response in alfalfa plants nodulated by an IAA over-producing *Rhizobium* strain. *Frontiers in Microbiology*, **8**, 2466. https://doi.org/10.3389/fmicb.2017.02466

Devi, R., Kaur, T., Kour, D., Hricovec, M., Mohan, R., Yadav, N., Rai, P.K., Rai, A.K., Yadav, A., Kumar, M., & Yadav, A.N. (2022). Microbes-mediated alleviation of heavy metal stress in crops. *Current Research and Future Challenges*, **10**, 25–37. https://doi.org/10.7324/JABB.2022.10s203

Djanaguiraman, M., Boyle, D.L., Welti, R., Jagadish, S.V.K., & Prasad, P.V.V. (2018). Decreased photosynthetic rate under high temperature in wheat is due to lipid desaturation, oxidation, acylation, and damage of organelles. *BMC Plant Biology*, **18**, 1–17. https://doi.org/10.1186/s12870-018-1263-z

Duc, N.H., Csintalan, Z., & Posta, K. (2018). Arbuscular mycorrhizal fungi mitigate negative effects of combined drought and heat stress on tomato plants. *Plant Physiology and Biochemistry*, **132**, 297–307. https://doi.org/10.1016/j.plaphy.2018.09.011

Egamberdieva, D., Wirth, S.J., Shurigin, V.V., Hashem, A., & Abd-Alla, E.F. (2017). Endophytic bacteria improve plant growth, symbiotic performance of chickpea (*Cicer arietinum* L.) and induce suppression of root rot caused by *Fusarium solani* under salt stress. *Frontiers in Microbiology*, **8**, 1887. https://doi.org/10.3389/fmicb.2017.01887

Evelin, H., Devi, T.S., Gupta, S., & Kapoor, R. (2019). Mitigation of salinity stress in plants by arbuscular mycorrhizal symbiosis: Current understanding and new challenges. *Frontiers in Plant Science*, **10**, 470.

Evelin, H., Giri, B., & Kapoor, R. (2012). Contribution of *Glomus intraradices* inoculation to nutrient acquisition and mitigation of ionic imbalance in NaCl-stressed *Trigonella foenum-graecum*. *Mycorrhiza*, **22**, 203–217. https://doi.org/10.1007/s00572-011-0392-0

Evelin, H., Kapoor, R., & Giri, B. (2009). Arbuscular mycorrhizal fungi in alleviation of salt stress: A review. *Annals of Botany*, **104(7)**, 1263–1280. https://doi.org/10.1093/aob/mcp251Fernandez, O., Theocharis, A., Bordiec, S., Feil, R., Acquens, J.L., Clément Fontaine, C.F., & Barka, E.A. (2012). *Burkholderia phytofirmans* PsJN acclimates grapevine to cold by modulating carbohydrate metabolism. *Molecular Plant–Microbe Interactions*, **25(4)**, 496–504. https://doi.org/10.1094/MPMI-09-11-0245

Fernández-Fuego, D., Bertrand, A., & González, A. (2017). Metal accumulation and detoxification mechanisms in mycorrhizal *Betula pubescens*. *Environmental Pollution*, **231**, 1153–1162. https://doi.org/10.1016/j.envpol.2017.07.072

Figueiredo, M.V., Burity, H.A., Martinez, C.R., & Chanway, C.P. (2008). Alleviation of drought stress in the common bean (*Phaseolus vulgaris* L.) by co-inoculation with *Paenibacillus polymyxa* and *Rhizobium tropici*. *Applied Soil Ecology*, **40(1)**, 182–188. https://doi.org/10.1016/j.apsoil.2008.04.005

Fileccia, V., Ruisi, P., Ingraffia, R., Giambalvo, D., Frenda, A.S., & Martinelli, F. (2017). Arbuscular mycorrhizal symbiosis mitigates the negative effects of salinity on durum wheat. *PloS One*, **12(9)**, e0184158. https://doi.org/10.1371/journal.pone.0184158

Gagné-Bourque, F., Bertrand, A., Claessens, A., Aliferis, K.A., & Jabaji, S. (2016). Alleviation of drought stress and metabolic changes in timothy (*Phleum pratense* L.) colonized with *Bacillus subtilis* B26. *Frontiers in Plant Science*, **7**, 584. https://doi.org/10.3389/fpls.2016.00584

Ganie, S.A., Bhat, J.A., & Devoto, A. (2022). The influence of endophytes on rice fitness under environmental stresses. *Plant Molecular Biology*, **109(4)**, 447–467. https://doi.org/10.1007/s11103-021-01219-8

García, J.E., Maroniche, Gcreus, C., Suárez-Rodríguez, R., Ramirez-Trujillo, J.A., & Groppa, M.D. (2017). In vitro PGPR properties and osmotic tolerance of different *Azospirillum* native strains and their effects on growth of maize under drought stress. *Microbiological Research*, **202**, 21–29. https://doi.org/10.1016/j.micres.2017.04.007

García, Q.G. (2020). *Deciphering the Contribution of Maize Aquaporins Regulated by Arbuscular Mycorrhizae to the Transport in Planta of Water and/or Solutes of Physiological Importance under Drought*. University De Granada. ISBN: 9788413064369.

Garg, N., & Pandey, R. (2016). High effectiveness of exotic arbuscular mycorrhizal fungi is reflected in improved rhizobial symbiosis and trehalose turnover in *Cajanus cajan* genotypes grown under salinity stress. *Fungal Ecology*, **21**, 57–67. https://doi.org/10.1016/j.funeco.2016.04.001

Ghorbanpour, A., Salimi, A., Ghanbary, M.A.T., Pirdashti, H., & Dehestani, A. (2018). The effect of *Trichoderma harzianum* in mitigating low temperature stress in tomato (*Solanum lycopersicum* L.) plants. *Scientia Horticulturae*, **230**, 134–141. https://doi.org/10.1016/j.scienta.2017.11.028

Goyal, D., Mishra, S., & Dantu, P.K. (2020). Arbuscular mycorrhizae associations and role in mitigation of drought stress in plants. In *Agriculturally Important Fungi for Sustainable Agriculture: Volume 2: Functional Annotation for Crop Protection*, 81–100. https://doi.org/10.1007/978-3-030-48474-3_3

Goyal, R.K., & Habtewold, J.Z. (2023). Evaluation of legume–rhizobial symbiotic interactions beyond nitrogen fixation that help the host survival and diversification in hostile environments. *Microorganisms*, **11(6)**, 1454. https://doi.org/10.3390/microorganisms11061454

Grover, M., Ali, S.Z., Sandhya, V., Rasul, A., & Venkateswarlu, B. (2011). Role of microorganisms in adaptation of agriculture crops to abiotic stresses. *World Journal of Microbiology and Biotechnology*, **27**, 1231–1240. https://doi.org/10.1007/s11274-010-0572-7

Guler, N.S., Pehlivan, N., Karaoglu, S.A., Guzel, S., & Bozdeveci, A. (2016). *Trichoderma atroviride* ID20G inoculation ameliorates drought stress-induced damages by improving antioxidant defence in maize seedlings. *Acta Physiologiae Plantarum*, **38**, 1–9. https://doi.org/10.1007/s11738-016-2153-3

Gupta, S., Schillaci, M., Walker, R., Smith, P.M., Watt, M., & Roessner, U. (2021a). Alleviation of salinity stress in plants by endophytic plant–fungal symbiosis: Current knowledge, perspectives and future directions. *Plant and Soil*, **461**, 219–244. https://doi.org/10.1007/s11104-020-04618-w

Gupta, S., Smith, P.M., Boughton, B.A., Rupasinghe, T.W., Natera, S.H., & Roessner, U. (2021b). Inoculation of barley with *Trichoderma harzianum* T-22 modifies lipids and metabolites to improve salt tolerance. *Journal of Experimental Botany*, **72(20)**, 7229–7246. https://doi.org/10.1093/jxb/erab335

Gururani, M.A., Upadhyaya, C.P., Baskar, V., Venkatesh, J., Nookaraju, A., & Park, S.W. (2013). Plant growth-promoting rhizobacteria enhance abiotic stress tolerance in *Solanum tuberosum* through inducing changes in the expression of ROS-scavenging enzymes and improved photosynthetic performance. *Journal of Plant Growth Regulation*, **32**, 245–258. https://doi.org/10.1007/s00344-012-9292-6

Hashem, A., Abd-Allah, E.F., Alqarawi, A.A., Al-Huqail, A.A., Wirth, S., & Egamberdieva, D. (2016). The interaction between arbuscular mycorrhizal fungi and endophytic bacteria enhances plant growth of *Acacia gerrardii* under salt stress. *Frontiers in Microbiology*, **7**, 1089. https://doi.org/10.3389/fmicb.2016.01089

Howarth, C.J. (2005). Genetic improvements of tolerance to high temperature. In *Abiotic Stresses: Plant Resistance through Breeding and Molecular Approaches*, 277–300. Howarth Press Inc., New York.

Hubbard, M.J., Germida., & Vujanovic, V. (2012). Fungal endophytes improve wheat seed germination under heat and drought stress. *Botany*, **90(2)**, 137–149.

Hussain, M.B., Zahir, Z.A., Asghar, H.N., & Asgher, M. (2014). Can catalase and exopolysaccharides producing rhizobia ameliorate drought stress in wheat. *International Journal Agriculture and Biology*, **16(1)**, 3–13.

Inbaraj, M.P. (2021). Plant–microbe interactions in alleviating abiotic stress—a mini review. *Frontiers in Agronomy*, **3**, 667903.

Ismail, M.H., Hussain, A., Khan, S.A., Iqbal, A., & Lee, I.J. (2020a). An endophytic fungus *Aspergillus violaceofuscus* can be used as heat stress adaptive tool for *Glycine max* L. and *Helianthus annuus* L. *Journal of Applied Botany & Food Quality*, **93**. https://doi.org/10.5073/JABFQ.2020.093.014

Ismail, M.H., Hussain, A., Iqbal, A., Khan, S.A., & Lee, I.J. (2020b). *Aspergillus niger* boosted heat stress tolerance in sunflower and soybean via regulating their metabolic and antioxidant system. *Journal of Plant Interactions*, **15(1)**, 223–232. https://doi.org/10.1080/17429145.2020.1771444

Issa, A., Esmaeel, Q., Sanchez, L., Courteaux, B., Guise, J.F., Gibon, Y., Ballias, P., Clément, C., Jacquard, C., Vaillant-Gaveau, N., & Aït Barka, E. (2018). Impacts of *Paraburkholderia phytofirmans* strain PsJN on tomato (*Lycopersicon esculentum* L.) under high temperature. *Frontiers in Plant Science*, **9**, 1397. https://doi.org/10.3389/fpls.2018.01397

Jach, M.E., Sajnaga, E., & Ziaja, M. (2022). Utilization of legume-nodule bacterial symbiosis in phytoremediation of heavy metal-contaminated soils. *Biology*, **11(5)**, 676. https://doi.org/10.3390/biology11050676

Javot, H., Lauvergeat, V., Santoni, V., Martin-Laurent, F., Güçlü, J., Vinh, J., Heyes, J., Franck, K.I., Schaffner, A.R., Bouchez, D., & Maurel, C. (2003). Role of a single aquaporin isoform in root water uptake. *The Plant Cell*, **15(2)**, 509–522. https://doi.org/10.1105/tpc.008888

Jincy, M., Djanaguiraman, M., Jeyakumar, P., Subramanian, K.S., Jayasankar, S., & Paliyath, G. (2017). Inhibition of phospholipase D enzyme activity through hexanal leads to delayed mango (*Mangifera indica* L.) fruit ripening through changes in oxidants and antioxidant enzymes activity. *Scientia Horticulturae*, **218**, 316–325. https://doi.org/10.1016/j.scienta.2017.02.026

Jincy, M., Jeyakumar, P., Boominathan, P., Manivannan, N., Varanavasiappan, S., & Rajendraprasad, V.B. (2019). Evaluation of greengram genotypes for high temperature tolerance at seedling stage by temperature induction response (TIR). *Environment and Ecology*, **37**, 1007–1013.

Jincy, M., Prasad, V. Senthil, A., Jeyakumar, P., & Manivannan, N. (2022). Physiological divergence in green gram [*Vigna radiata* (L.) Wilczek] genotypes for drought and high temperature stress tolerance during flowering phase. *Legume Research-An International Journal*, **45(8)**, 960–967.

Kakar, K.U., Ren, X.L., Nawaz, Z., Cui, Z.Q., Li, B., Xie, G.L., Hassan, M.A., Ali, E., & Sun, G.C. (2016). A consortium of rhizobacterial strains and biochemical growth elicitors improve cold and drought stress tolerance in rice (*Oryza sativa* L.). *Plant Biology*, **18(3)**, 471–483. https://doi.org/10.1111/plb.12427

Kamran, M., Imran, Q.M., Ahmed, M.B., Falak, N., Khatoon, A., & Yun, B.W. (2022). Endophyte-mediated stress tolerance in plants: A sustainable strategy to enhance resilience and assist crop improvement. *Cells*, **11(20)**, 3292. https://doi.org/10.3390/cells11203292

Kaur, G., & Asthir, B.J.B.P. (2015). Proline: A key player in plant abiotic stress tolerance. *Biologia Plantarum*, **59**, 609–619. https://doi.org/10.1007/s10535-015-0549-3

Kaur, R., Bains, T.S., Bindumadhava, H., & Nayyar, H. (2015). Responses of mungbean (*Vigna radiata* L.) genotypes to heat stress: Effects on reproductive biology, leaf function and yield traits. *Scientia Horticulturae*, **197**, 527–541. https://doi.org/10.1016/j.scienta.2015.10.015

Khan, A.L., Hamayun, M., Radhakrishnan, R., Waqas, M., Kang, S.M., Kim, Y.H., Shin, J.H., Choo, Y.S., Kim, J.G., & Lee, I.J. (2012a). Mutualistic association of *Paecilomyces formosus* LHL10 offers thermotolerance to *Cucumis sativus*. *Antonie Van Leeuwenhoek*, **101**, 267–279. https://doi.org/10.1007/s10482-011-9630-x

Khan, A.L., Hamayun, M., Waqas, M., Kang, S.M., Kim, Y.H., Kim, D.H, & Lee, I.J. (2012b). *Exophiala* sp. LHL08 association gives heat stress tolerance by avoiding oxidative damage to cucumber plants. *Biology and Fertility of Soils*, **48**, 519–529. https://doi.org/10.1007/s00374-011-0649-y

Khan, A.L., Kang, S.M., Dhakal, K.H., Hussain, J., Adnan, M., Kim, J.G., & Lee, I.J. (2013). Flavonoids and amino acid regulation in *Capsicum annuum* L. by endophytic fungi under different heat stress regimes. *Scientia Horticulturae*, **155**, 1–7. https://doi.org/10.1016/j.scienta.2013.02.028

Khan, M.A., Asaf, S., Khan, A.L., Jan, R., Kang, S.M. Kim, K.M., & Lee, I.J. (2020a). Extending thermotolerance to tomato seedlings by inoculation with SA1 isolate of *Bacillus cereus* and comparison with exogenous humic acid application. *PLoS One*, **15(4)**, e0232228. https://doi.org/10.1371/journal.pone.0232228

Khan, M.A., Asaf, S., Khan, A.L., Jan, R., Kang, S.M., Kim, K.M., & Lee, I.J. (2020b). Thermotolerance effect of plant growth-promoting *Bacillus cereus* SA1 on soybean during heat stress. *BMC Microbiology*, **20**, 1–14. https://doi.org/10.1186/s12866-020-01822-7

Khan, M.I.R., Fatma, M., Per, T.S., Anjum, N.A., & Khan, N.A. (2015). Salicylic acid-induced abiotic stress tolerance and underlying mechanisms in plants. *Frontiers in Plant Science*, **6**, 462. https://doi.org/10.3389/fpls.2015.00462

Khan, N., Bano, A. Rahman, M.A., Guo, J., Kang, Z., & Babar, M.A. (2019). Comparative physiological and metabolic analysis reveals a complex mechanism involved in drought tolerance in chickpea (*Cicer arietinum* L.) induced by PGPR and PGRs. *Scientific Reports*, **9(1)**, 2097. https://doi.org/10.1038/s41598-019-38702-8

Kibido, T., Kunert, K., Makgopa, M., Greve, M., & Vorster, J. (2020). Improvement of rhizobium–soybean symbiosis and nitrogen fixation under drought. *Food and Energy Security*, **9(1)**, 177. https://doi.org/10.1002/fes3.177

Klinsukon, C., Lumyong, S., Kuyper, T.W., & Boonlue, S. (2021). Colonization by arbuscular mycorrhizal fungi improves salinity tolerance of eucalyptus (*Eucalyptus camaldulensis*) seedlings. *Scientific Reports*, **11(1)**, 4362. https://doi.org/10.1038/s41598-021-84002-5

Kouadria, R., Bouzouina, M., & Lotmani, B. (2023). Endophytic fungi from salt adapted plants confer salt tolerance in barley. *Abiotic Stress*, **18(2)**.

Kumar, A., Voropaeva, O., Maleva, M., Panikovskaya, K., Borisova, G., Rajkumar, M., & Bruno, L.B. (2021). Bioaugmentation with copper tolerant endophyte *Pseudomonas lurida* strain EOO26 for improved plant growth and copper phytoremediation by *Helianthus annuus*. *Chemosphere*, **266**, 128983. https://doi.org/10.1016/j.chemosphere.2020.128983

Li, X., He, C., He, X., Su, F., Hou, L., Ren, Y., & Hou, Y. (2019). Dark septate endophytes improve the growth of host and non-host plants under drought stress through altered root development. *Plant and Soil*, **439**, 259–272. https://doi.org/10.1007/s11104-019-04057-2

Li, X., Zhao, C., Zhang, T., Wang, G., Amombo, E., Xie, Y., & Fu, J. (2021). Exogenous *Aspergillus aculeatus* enhances drought and heat tolerance of perennial ryegrass. *Frontiers in Microbiology*, **12**, 593722. https://doi.org/10.3389/fmicb.2021.593722

Li, Z., Wu, N., Meng, S., Wu, F., & Liu, T. (2020). Arbuscular mycorrhizal fungi (AMF) enhance the tolerance of *Euonymus maackii* Rupr. at a moderate level of salinity. *Plos One*, **15(4)**, 0231497. https://doi.org/10.1371/journal.pone.0231497

Lima, B.C., Moro, A.L., Santos, A.C.P., Bonifacio, A., Araujo, A.S.F., & de Araujo, F.F. (2019). *Bacillus subtilis* ameliorates water stress tolerance in maize and common bean. *Journal of Plant Interactions*, **14(1)**, 432–439. https://doi.org/10.1080/17429145.2019.1645896

Liu, M.Y., Li, Q.S., Ding, W.Y., Dong, L.W., Deng, M., Chen, J.H., Tian, X., Hashem, A., Al-Arjani, A.B.F., Alenazi, M.M., & Abd-Allah, E.F. (2023). Arbuscular mycorrhizal fungi inoculation impacts expression of aquaporins and salt overly sensitive genes and enhances tolerance of salt stress in tomato. *Chemical and Biological Technologies in Agriculture*, **10(1)**, 5.

Liu, Y., Guo, Z., & Shi, H. (2022a). *Rhizobium* symbiosis leads to increased drought tolerance in Chinese milk vetch (*Astragalus sinicus* L.). *Agronomy*, **12(3)**, 725. https://doi.org/10.3390/agronomy12030725

Liu, Y., Xun, W., Chen, L., Xu, Z., Zhang, N., Feng, H., Zhang, Q., & Zhang, R. (2022b). Rhizosphere microbes enhance plant salt tolerance: Toward crop production in saline soil. *Computational and Structural Biotechnology Journal*, **20**, 6543–6551. https://doi.org/10.1016/j.csbj.2022.11.046

Liu, Z., Ma, L., He, X., & Tian, C. (2014). Water strategy of mycorrhizal rice at low temperature through the regulation of PIP aquaporins with the involvement of trehalose. *Applied Soil Ecology*, **84**, 185–191. https://doi.org/10.1016/j.apsoil.2014.07.010

Ma, Y. (2019). Biotechnological potential of plant–microbe interactions in environmental decontamination. *Frontiers in Plant Science*, **10**, 1519. https://doi.org/10.3389/fpls.2019.01519

Mathur, S., Agnihotri, R., Sharma, M.P., Reddy, V.R., & Jajoo, A. (2021). Effect of high-temperature stress on plant physiological traits and mycorrhizal symbiosis in maize plants. *Journal of Fungi*, **7(10)**, 867. https://doi.org/10.3390/jof7100867

Mathur, S., & Jajoo, A. (2020). Arbuscular mycorrhizal fungi protects maize plants from high temperature stress by regulating photosystem II heterogeneity. *Industrial Crops and Products*, **143**, 111934. https://doi.org/10.1016/j.indcrop.2019.111934

Mathur, S., Sharma, M.P., & Jajoo, A. (2018). Improved photosynthetic efficacy of maize (*Zea mays*) plants with arbuscular mycorrhizal fungi (AMF) under high temperature stress. *Journal of Photochemistry and Photobiology B: Biology*, **180**, 149–154. https://doi.org/10.1016/j.jphotobiol.2018.02.002McLel lan, C.A., Turbyville, T.J., Wijeratne, E.K., Kerschen, A., Vierling, E., Queitsch, C., Whitesell, L., & Gunatilaka, A.L. (2007). A rhizosphere fungus enhances *Arabidopsis* thermotolerance through production of an HSP90 inhibitor. *Plant Physiology*, **145(1)**, 174–182. https://doi.org/10.1104/pp.107.101808

Miller, G., Shulaev, V., & Mittler, R. (2008). Reactive oxygen signaling and abiotic stress. *Physiologia Plantarum*, **133(3)**, 481–489. https://doi.org/10.1111/j.1399-3054.2008.01090.x

Milošević, N., Govedarica, M., Kastori, R., & Petrović, N. (2002). Effect of nickel on wheat plants, soil microorganisms and enzymes. *Studia Universitatis Babeș-Bolyai, Biologia*, **47**, 79–84.

Milošević, N.A., Marinković, J.B., & Tintor, B.B. (2012). Mitigating abiotic stress in crop plants by microorganisms. *Zbornik Matice srpske za prirodne nauke*, **123**, 17–26.

Mo, Y., Wang, Y., Yang, R., Zheng, J., Liu, C., Li, H., Ma, J., Zhang, Y., Wei, C., & Zhang, X. (2016). Regulation of plant growth, photosynthesis, antioxidation and osmosis by an arbuscular mycorrhizal fungus in watermelon seedlings under well-watered and drought conditions. *Frontiers in Plant Science*, **7**, 644. https://doi.org/10.3389/fpls.2016.00644

Morales-Vargas, A.T., López-Ramírez, C., Álvarez-Mejía & Vázquez-Martínez, J. (2024). Endophytic fungi for crops adaptation to abiotic stresses. *Microorganisms*, **12(7)**, 1357. https://doi.org/10.3390/microorganisms12071357

Morsy, M., Cleckler, B., & Armuelles-Millican, H. (2020). Fungal endophytes promote tomato growth and enhance drought and salt tolerance. *Plants*, **9(7)**, 877. https://doi.org/10.3390/plants9070877

Murphy, B.R., Doohan, F.M., & Hodkinson, T.R. (2014). Yield increase induced by the fungal root endophyte *Piriformospora indica* in barley grown at low temperature is nutrient limited. *Symbiosis*, **62**, 29–39. https://doi.org/10.1007/s13199-014-0268-0

Naseem, H., & Bano, A. (2014). Role of plant growth-promoting rhizobacteria and their exopolysaccharide in drought tolerance of maize. *Journal of Plant Interactions*, **9(1)**, 689–701. https://doi.org/10.1080/17429145.2014.902125

Nataraja, K.N., Dhanyalakshmi, K.H., Govind, G., & Oelmüller, R. (2022). Activation of drought tolerant traits in crops: Endophytes as elicitors. *Plant Signaling & Behavior*, **17(1)**, 2120300. https://doi.org/10.1080/15592324.2022.2120300

Oljira, A.M., Hussain, T., Waghmode, T.R., Zhao, H., Sun, H.X., Liu, X. Wang & Liu, B. (2020). *Trichoderma* enhances net photosynthesis, water use efficiency, and growth of wheat (*Triticum aestivum* L.) under salt stress. *Microorganisms*, **8(10)**, 1565. https://doi.org/10.3390/microorganisms8101565

Pal, A., & Pandey, S., (2016). Role of arbuscular mycorrhizal fungi on plant growth and reclamation of barren soil with wheat (*Triticum aestivum* L.) crop. *International Journal for Soil Science*, **12(1)**, 25–31.

Park, Y.G., Mun, B.G., Kang, S.M., Hussain, A., Shahzad, R., Seo, C.W., Kim, A.Y., Lee, S.U., Oh, K.Y., Lee, D.Y., & Lee, I.J. (2017). *Bacillus aryabhattai* SRB02 tolerates oxidative and nitrosative stress and promotes the growth of soybean by modulating the production of phytohormones. *PloS One*, **12(3)**, 0173203. https://doi.org/10.1371/journal.pone.0173203

Pedranzani, H., Rodríguez-Rivera, M., Gutiérrez, M., Porcel, R., Hause, B., & Ruiz-Lozano, J.M. (2016). Arbuscular mycorrhizal symbiosis regulates physiology and performance of *Digitaria eriantha* plants subjected to abiotic stresses by modulating antioxidant and jasmonate levels. *Mycorrhiza*, **26**, 141–152. https://doi.org/10.1007/s00572-015-0653-4

Porter, J.R. (2005). Rising temperatures are likely to reduce crop yields. *Nature*, **436(7048)**, 174–174. https://doi.org/10.1038/436174b

Poudel, M., Mendes, R., Costa, L.A., Bueno, C.G., Meng, Y., Folimonova, S.Y., Garrett, K.A., & Martins, S.J. (2021). The role of plant-associated bacteria, fungi, and viruses in drought stress mitigation. *Frontiers in Microbiology*, **12**, 743512. https://doi.org/10.3389/fmicb.2021.743512

Puccio, G., Ingraffia, R., Mercati, F., Amato, G., Giambalvo, D., Martinelli, F., Sunseri, F., & Frenda, A.S. (2023). Transcriptome changes induced by Arbuscular mycorrhizal symbiosis in leaves of durum wheat (*Triticum durum* Desf.) promote higher salt tolerance. *Scientific Reports*, **13(1)**, 116. https://doi.org/10.1038/s41598-022-26903-7

Quinn, P.J., & Williams, W.P., (1985). Environmentally induced changes in chloroplast membranes and their effects on photosynthetic function. *Photosynthetic Mechanisms and the Environment*, **6**, 1–47.

Redman, R.S., Sheehan, K.B., Stout, R.G., Rodriguez, R.J., & Henson, J.M. (2002). Thermotolerance generated by plant/fungal symbiosis. *Science*, **298(5598)**, 1581–1581. https://doi.org/10.1126/science.1078055

Reva, M., Cano, C., Herrera, M.A., & Bago, A. (2021). Arbuscular mycorrhizal inoculation enhances endurance to severe heat stress in three horticultural crops. *HortScience*, **56(4)**, 396–406.

Rewald, B., Holzer, L., & Göransson, H. (2015). Arbuscular mycorrhiza inoculum reduces root respiration and improves biomass accumulation of salt-stressed *Ulmus glabra* seedlings. *Urban Forestry & Urban Greening*, **14(2)**, 432–437. https://doi.org/10.1016/j.ufug.2015.04.011

Saleem, M., Arshad, M., Hussain, S., & Bhatti, A.S. (2007). Perspective of plant growth promoting rhizobacteria (PGPR) containing ACC deaminase in stress agriculture. *Journal of Industrial Microbiology and Biotechnology*, **34(10)**, 635–648. https://doi.org/10.1007/s10295-007-0240-6

Sandhya, V.S.K.Z., Sk, Z.A., Grover, M., Reddy, G., & Venkateswarlu, B. (2010). Effect of plant growth promoting *Pseudomonas* spp. on compatible solutes, antioxidant status and plant growth of maize under drought stress. *Plant Growth Regulation*, **62**, 21–30. https://doi.org/10.1007/s10725-010-9479-4

Sandhya, V.Z.A.S., Sk, Z.A., Grover, M., Reddy, G., & Venkateswarlu, B.S.S.S. (2009). Alleviation of drought stress effects in sunflower seedlings by the exopolysaccharides producing *Pseudomonas putida* strain GAP-P45. *Biology and Fertility of Soils*, **46**, 17–26. https://doi.org/10.1007/s00374-009-0401-z

Sandrini, M., Nerva, L., Sillo, F., Balestrini, R., Chitarra, W., & Zampieri, E. (2022). Abiotic stress and below-ground microbiome: The potential of omics approaches. *International Journal of Molecular Sciences*, **23(3)**, 1091. https://doi.org/10.3390/ijms23031091

Sangamesh, M.B., Jambagi, S., Vasanthakumari, M.M., Shetty, N.J., Kolte, H., Ravikanth, G., Nataraja, K.N., & Uma Shaanker, R. (2018). Thermotolerance of fungal endophytes isolated from plants adapted to the Thar Desert, India. *Symbiosis*, **75**, 135–147. https://doi.org/10.1007/s13199-017-0527-y

Sarapat, S., Songwattana, P., Longtonglang, A., Umnajkitikorn, K., Girdthai, T., Tittabutr, P., Boonkerd, N., & Teaumroong. (2020). Effects of increased 1-aminocyclopropane-1-carboxylate (ACC) deaminase activity in *Bradyrhizobium* sp. SUTN9-2 on mung bean symbiosis under water deficit conditions. *Microbes and Environments*, **35(3)**, 20024. https://doi.org/10.1264/jsme2.ME20024

Shaffique, S., Hussain, S., Kang, S.M., Imran, M., Kwon, E.H., Khan, M.A., & Lee, I.J. (2023). Recent progress on the microbial mitigation of heavy metal stress in soybean: Overview and implications. *Frontiers in Plant Science*, **14**, 1188856. https://doi.org/10.3389/fpls.2023.1188856

Sharma, M., Sudheer, S., Usmani, Z., Rani, R., & Gupta, P. (2020). Deciphering the omics of plant–microbe interaction: Perspectives and new insights. *Current Genomics*, **21(5)**, 343–362. https://doi.org/10.2174/1389202921999200515140420

Sharma, P., Jha, A.B., Dubey, R.S., & Pessarakli, M. (2012). Reactive oxygen species, oxidative damage, and antioxidative defense mechanism in plants under stressful conditions. *Journal of Botany*, **2012(1)**, 217037. https://doi.org/10.1155/2012/217037

Shi, L., Deng, X., Yang, Y., Jia, Q., Wang, C., Shen, Z., & Chen, Y. (2019). A Cr (VI)-tolerant strain, *Pisolithus sp1*, with a high accumulation capacity of Cr in mycelium and highly efficient assisting *Pinus thunbergii* for phytoremediation. *Chemosphere*, **224**, 862–872. https://doi.org/10.1016/j.chemosphere.2019.03.015

Siddiqui, Z.S., Ali, F., & Uddin, Z. (2021). Sustainable effect of a symbiotic nitrogen-fixing bacterium *Sinorhizobium meliloti* on nodulation and photosynthetic traits of four leguminous plants under low moisture stress environment. *Letters in Applied Microbiology*, **72(6)**, 714–724. https://doi.org/10.1111/lam.13463

Singh, R.P., Shelke, G.M., Kumar, A., & Jha, P.N. (2015). Biochemistry and genetics of ACC deaminase: A weapon to "stress ethylene" produced in plants. *Frontiers in Microbiology*, **6**, 937. https://doi.org/10.3389/fmicb.2015.00937

Soni, R., Prakash, G., Sharma, S., Sinha, D., & Mishra, R. (2023). Role of microbes in alleviating abiotic stress in plants. *Plant Science Today*, **10(3)**, 160–187. https://doi.org/10.14719/pst.2215

Staudinger, C., Mehmeti-Tershani, V., Gil-Quintana, E., Gonzalez, E.M., Hofhansl, F., Bachmann, G., & Wienkoop, S. (2016). Evidence for a rhizobia-induced drought stress response strategy in *Medicago truncatula*. *Journal of Proteomics*, **136**, 202–213. https://doi.org/10.1016/j.jprot.2016.01.006

Suárez, R., Wong, A. Ramírez, M., Barraza, A., Orozco, M.D.C., Cevallos, M.A., Lara, M., Hernández, G., & Iturriaga, G. (2008). Improvement of drought tolerance and grain yield in common bean by overexpressing trehalose-6-phosphate synthase in rhizobia. *Molecular Plant–Microbe Interactions*, **21(7)**, 958–966. https://doi.org/10.1094/MPMI-21-7-0958

Sukweenadhi, J., Balusamy, S.R., Kim, Y.J., Lee, C.H., Kim, Y.J., Koh, S.C., & Yang, D.C. (2018). A growth-promoting bacteria, *Paenibacillus yonginensis* DCY84T enhanced salt stress tolerance by activating defense-related systems in *Panax ginseng*. *Frontiers in Plant Science*, **9**, 813. https://doi.org/10.3389/fpls.2018.00813

Swamy, M.K., Nalina, N., Nalina, D., Akhtar, M.S., & Purushotham, B. (2019). Heavy metal stress and tolerance in plants mediated by rhizospheric microbes. In *Salt Stress, Microbes, and Plant Interactions: Causes and Solution*, 181–198. https://doi.org/10.1007/978-981-13-8801-9_8

Talaat, N.B., & Shawky, B.T. (2014). Protective effects of arbuscular mycorrhizal fungi on wheat (*Triticum aestivum* L.) plants exposed to salinity. *Environmental and Experimental Botany*, **98**, 20–31. https://doi.org/10.1016/j.envexpbot.2013.10.005

Talbi, C., Sanchez, C., Hidalgo-Garcia, A., González, E.M., Arrese-Igor, C., Girard, L., Bedmar, E.J., & Delgado, M.J. (2012). Enhanced expression of *Rhizobium etli* cbb 3 oxidase improves drought tolerance of common bean symbiotic nitrogen fixation. *Journal of Experimental Botany*, **63(14)**, 5035–5043. https://doi.org/10.1093/jxb/ers101

Tang, H., Hassan, M.U., Feng, L., Nawaz, M., Shah, A.N., Qari, S.H., Liu, Y., & Miao, J. (2022). The critical role of arbuscular mycorrhizal fungi to improve drought tolerance and nitrogen use efficiency in crops. *Frontiers in Plant Science*, **13**, 919166. https://doi.org/10.3389/fpls.2022.919166

Tiryaki, D., Aydınand, İ., &Atıcı, Ö. (2019). Psychrotolerant bacteria isolated from the leaf apoplast of cold-adapted wild plants improve the cold resistance of bean (*Phaseolus vulgaris* L.) under low temperature. *Cryobiology*, **86**, 111–119. https://doi.org/10.1016/j.cryobiol.2018.11.001

Tiwari, S., & Lata, C. (2018). Heavy metal stress, signaling, and tolerance due to plant-associated microbes: An overview. *Frontiers in Plant Science*, **9**, 452. https://doi.org/10.3389/fpls.2018.00452

Umesha, S., Singh, P.K., & Singh, R.P. (2018). Microbial biotechnology and sustainable agriculture. In *Biotechnology for Sustainable Agriculture*, 185–205. https://doi.org/10.1016/B978-0-12-812160-3.00006-4

Verma, K.K., Song, X.P., Li, D.M., Singh, M., Rajput, V.D., Malviya, M.K., Minkina, T., Singh, R.K., Singh, P., & Li., Y.R. (2020). Interactive role of silicon and plant–rhizobacteria mitigating abiotic stresses: A new approach for sustainable agriculture and climate change. *Plants*, **9(9)**, 1055.

Villalobos-López, M.A., Arroyo-Becerra, A., Quintero-Jiménez, A., & Iturriaga, G. (2022). Biotechnological advances to improve abiotic stress tolerance in crops. *International Journal of Molecular Sciences*, **23(19)**, 12053. https://doi.org/10.3390/ijms231912053

Vurukonda, S.S.K.P., Vardharajula, S., Shrivastava, M., & Sk, Z.A. (2016). Enhancement of drought stress tolerance in crops by plant growth promoting rhizobacteria. *Microbiological Research*, **184**, 13–24. https://doi.org/10.1016/j.micres.2015.12.003

Wahid, A., Gelani, S., Ashraf, M., & Foolad, M.R. (2007). Heat tolerance in plants: An overview. *Environmental and Experimental Botany*, **61(3)**, 199–223. https://doi.org/10.1016/j.envexpbot.2007.05.011

Wang, H., An, T., Huang, D., Liu, R., Xu, B., Zhang, S., Deng, X., Siddique, K.H., & Chen, Y. (2021). Arbuscular mycorrhizal symbioses alleviating salt stress in maize is associated with a decline in root-to-leaf gradient of Na+/K+ ratio. *BMC Plant Biology*, **21**, 1–15. https://doi.org/10.1186/s12870-021-03237-6

Waqas, M., Khan, A.L., Shahzad, R., Ullah, I., Khan, A.R., & Lee, I.J. (2015). Mutualistic fungal endophytes produce phytohormones and organic acids that promote japonica rice plant growth under prolonged heat stress. *Journal of Zhejiang University Science*, **16(12)**, 1011. https://doi.org/10.1631/jzus.B1500081

Win, K.T., Wasai-Hara, S., Tanaka, F., Oo, A.Z., Minamisawa, K., Shimoda, Y., & Imaizumi-Anraku, H. (2023). Synergistic N_2 -fixation and salt stress mitigation in soybean through dual inoculation of ACC deaminase-producing *Pseudomonas* and *Bradyrhizobium*. *Scientific Reports*, **13(1)**, 17050. https://doi.org/10.1038/s41598-023-43891-4

Woo, O.G., Kim, H., Kim, J.S., Keum, H.L., Lee, K.C. Sul, W.J., & Lee, J.H. (2020). *Bacillus subtilis* strain GOT9 confers enhanced tolerance to drought and salt stresses in *Arabidopsis thaliana* and *Brassica campestris*. *Plant Physiology and Biochemistry*, **148**, 359–367. https://doi.org/10.1016/j.plaphy.2020.01.032

Wu, Q.S. (2011). Mycorrhizal efficacy of trifoliate orange seedlings on alleviating temperature stress. *Plant, Soil and Environment*, **57(10)**, 459–464.

Wu, Q.S., & Zou, Y.N. (2017). Arbuscular mycorrhizal fungi and tolerance of drought stress in plants. In *Arbuscular Mycorrhizas and Stress Tolerance of Plants*, 25–41. https://doi.org/10.1007/978-981-10-4115-0_2

Xie, H.T., Wan, L., Li, Y.Z.S., & Zhang, Y. (2014). Spatiotemporal production of reactive oxygen species by NADPH oxidase is critical for tapetal programmed cell death and pollen development in *Arabidopsis*. *The Plant Cell*, **26(5)**, 2007–2023. https://doi.org/10.1105/tpc.114.125427

Xie, Z., Chu, Y., Zhang, W., Lang, D., & Zhang, X. (2019). *Bacillus pumilus* alleviates drought stress and increases metabolite accumulation in *Glycyrrhiza uralensis* Fisch. *Environmental and Experimental Botany*, **158**, 99–106. https://doi.org/10.1016/j.envexpbot.2018.11.021

Xu, J., Li, X.L., & Luo, L. (2012). Effects of engineered *Sinorhizobium meliloti* on cytokinin synthesis and tolerance of alfalfa to extreme drought stress. *Applied and Environmental Microbiology*, **78(22)**, 8056–8061. https://doi.org/10.1128/AEM.01276-12

Yadav, V.K., Raghav, M., Sharma, S.K., & Bhagat, N. (2020). Rhizobacteriome: Promising candidate for conferring drought tolerance in crops. *Journal of Pure and Applied Microbiology*, **14(1)**, 73–92. https://doi.org/10.22207/JPAM.14.1.10

Yan, N., Marschner, P., Cao, W., Zuo, C., & Qin, W. (2015). Influence of salinity and water content on soil microorganisms. *International Soil and Water Conservation Research*, **3(4)**, 316–323. https://doi.org/10.1016/j.iswcr.2015.11.003

Yang, J., Kloepper, J.W., & Ryu, C.M. (2009). Rhizosphere bacteria help plants tolerate abiotic stress. *Trends in Plant Science*, **14(1)**, 1–4. https://doi.org/10.1016/j.tplants.2008.10.004

Yang, P., Zhang, P., Li, B., & Hu, T. (2013). Effect of nodules on dehydration response in alfalfa (*Medicago sativa* L.). *Environmental and Experimental Botany*, **86**, 29–34. https://doi.org/10.1016/j.envexpbot.2011.05.012

Yeasmin, R., Bonser, S.P., Motoki, S., & Nishihara, E. (2019). Arbuscular mycorrhiza influences growth and nutrient uptake of asparagus (*Asparagus officinalis* L.) under heat stress. *HortScience*, **54(5)**, 846–850. https://doi.org/10.21273/HORTSCI13587–18

Yoo, S.J., Hong, J.K., Weon, H.YSong, J., & Sang, M.K. (2019b). Effect of *Bacillus aryabhattai* H26-2 and *B. siamensis* H30-3 on growth promotion and alleviation of heat and drought stresses in Chinese cabbage. *The Plant Pathology Journal*, **35(2)**, 178. https://doi.org/10.5423/PPJ.NT.08.2018.0159

Yoo, S.J., Weon, H.Y., Song, J., & Sang, M.K. (2019a). Induced tolerance to salinity stress by halotolerant bacteria *Bacillus aryabhattai* H19-1 and *B. mesonae* H20-5 in tomato plants. *Journal of Microbiology and Biotechnology*, **29(7)**, 1124–1136.

Zhang, B., Shi, F., Zheng, X., Pan, H., Wen, Y., & Song, F. (2023). Effects of AMF compound inoculants on growth, ion homeostasis, and salt tolerance-related gene expression in *Oryza sativa* L. under salt treatments. *Rice*, **16(1)**, 18. https://doi.org/10.1186/s12284-023-00635-2

Zhang, F., Jia-Dong, H.E., Qiu-Dan, N.I., Qiang-Sheng, W.U., & Ying-Ning, Z.O.U. (2018a). Enhancement of drought tolerance in trifoliate orange by mycorrhiza: Changes in root sucrose and proline metabolisms. *Notulae Botanicae Horti Agrobotanici Cluj-Napoca*, **46(1)**, 270–276. https://doi.org/10.15835/nbha46110983

Zhang, W., Wang, C., Lu, T., & Zheng, Y. (2018b). Cooperation between arbuscular mycorrhizal fungi and earthworms promotes the physiological adaptation of maize under a high salt stress. *Plant and Soil*, **423**, 125–140. https://doi.org/10.1007/s11104-017-3481-9

Zhu, X.C., Song, F.B., Liu, S.Q., & Liu, T.D. (2011). Effects of arbuscular mycorrhizal fungus on photosynthesis and water status of maize under high temperature stress. *Plant and Soil*, **346**, 189–199. https://doi.org/10.1007/s11104-011-0809-8

Zhu, X.C., Song, F.B., Liu, T.D., & Liu, S.Q. (2010). Arbuscular mycorrhizae reducing water loss in maize plants under low temperature stress. *Plant Signaling & Behavior*, **5(5)**, 591–593. https://doi.org/10.4161/psb.11498

12 Harnessing the Power of Beneficial Plant–Microbe Synergy in Sustainable Agriculture

Jay Hind Nishad and Bipin Maurya

12.1 INTRODUCTION

Agriculture is the backbone of human civilization, providing the food, fiber, and fuel necessary to sustain global populations. However, modern agricultural practices, characterized by high-input systems reliant on chemical fertilizers, pesticides, and intensive tillage, have led to significant environmental challenges. These include soil degradation, water pollution, loss of biodiversity, and greenhouse gas emissions (Foley, 2011). With the global population projected to reach 9.7 billion by 2050, the challenge of achieving food security while maintaining environmental sustainability has never been more pressing (FAO, 2021).

In response to these challenges, sustainable agriculture has emerged as a paradigm that seeks to harmonize productivity with environmental health. A key aspect of this approach involves leveraging natural processes, such as the interactions between plants and their associated microbes, to enhance agricultural productivity and resilience. Beneficial plant–microbe interactions, particularly in the rhizosphere, the soil zone influenced by root exudates, play a critical role in nutrient cycling, disease suppression, stress tolerance, and soil structure improvement (Berg et al., 2017). By harnessing these interactions, it is possible to reduce dependence on synthetic inputs, thereby minimizing environmental impacts and fostering ecological balance.

Plant-associated microbes encompass a diverse array of organisms, including bacteria, fungi, archaea, and viruses. These microbes establish complex relationships with their host plants, ranging from mutualistic symbioses to commensal and even parasitic interactions. Among the most well-known beneficial interactions are those involving nitrogen-fixing bacteria, mycorrhizal fungi, and plant growth-promoting rhizobacteria (PGPR) (Jacoby et al., 2017). For example, nitrogen-fixing bacteria such as *Rhizobium* form symbiotic associations with leguminous plants, converting atmospheric nitrogen into ammonia, which is a bioavailable form of nitrogen for plants (Mus et al., 2016). Similarly, mycorrhizal fungi enhance nutrient and water uptake by extending the root network, while PGPR promote plant growth through mechanisms such as hormone modulation and pathogen suppression (Vessey, 2003).

Recent advances in microbial ecology, genomics, and biotechnology have shed light on the vast potential of plant-associated microbiomes for agricultural innovation. High-throughput sequencing techniques have revealed the complexity and functional diversity of microbial communities associated with plants, while synthetic biology has enabled the engineering of microbial strains with enhanced capabilities (Bulgarelli et al., 2013; Trivedi et al., 2020). These advancements pave the way for the development of bioinoculants, biofertilizers, and biopesticides that can be integrated into sustainable farming systems.

DOI: 10.1201/9781003608080-12

Despite their promise, translating laboratory findings on plant–microbe interactions into field-scale applications presents several challenges. The performance of microbial inoculants can vary significantly across different environmental conditions, soil types, and crop systems (Compant et al., 2010). Additionally, the long-term stability and persistence of introduced microbes in soil ecosystems remain critical concerns. Addressing these challenges requires interdisciplinary research, farmer engagement, and supportive policy frameworks.

This chapter explores the potential of beneficial plant–microbe interactions to transform agriculture into a more sustainable and resilient system. It provides an overview of the types of beneficial interactions, the mechanisms underlying their effects, and their practical applications in sustainable agriculture. Furthermore, it discusses recent advancements in microbial biotechnology and highlights challenges and future directions for integrating these natural partnerships into agricultural practices. By harnessing the power of plant–microbe interactions, we can build a foundation for sustainable agriculture that supports both productivity and environmental health.

12.2 PLANT–MICROBE INTERACTIONS: AN OVERVIEW

Beneficial plant–microbe interactions are diverse and multifaceted, spanning symbiotic relationships, commensal associations, and mutualistic partnerships. These interactions occur primarily in the rhizosphere but also extend to endophytic and phyllosphere communities. These plant–microbe interactions play a key role in supporting sustainable agriculture (Table 12.1).

12.2.1 Mycorrhizal Association

Mycorrhizal fungi establish symbiotic relationships with over 80% of terrestrial plants. Arbuscular mycorrhizal fungi (AMF), in particular, enhance plant nutrient acquisition, especially phosphorus, while receiving carbon from the host plant. Recent studies have shown that AMF also play a role in improving soil structure and carbon sequestration (Field et al., 2020).

12.2.2 Nitrogen-Fixing Microorganisms

Nitrogen fixation by microbes such as *Rhizobium*, *Frankia*, and cyanobacteria contributes to sustainable nitrogen management. These microbes convert atmospheric nitrogen into ammonium, which plants can assimilate, reducing their dependence on synthetic nitrogen fertilizers (Mus et al., 2016).

12.2.3 Plant Growth-Promoting Rhizobacteria

PGPR are a diverse group of bacteria that colonize the rhizosphere and enhance plant growth. Species such as *Bacillus subtilis* and *Pseudomonas fluorescens* produce phytohormones, solubilize nutrients, and suppress pathogens. PGPR-based biofertilizers are increasingly being adopted in organic farming systems (Bhattacharyya & Jha, 2012).

12.2.4 Endophytes

Endophytic microbes, which reside within plant tissues, contribute to host fitness by promoting growth and providing resistance to biotic and abiotic stresses. For example, certain fungal endophytes enhance drought tolerance by modulating the host plant's stress response pathways (Rodriguez et al., 2008).

The increasing demand for sustainable agricultural practices highlights the importance of understanding and harnessing the power of beneficial plant–microbe interactions. These interactions significantly influence plant health, productivity, and resilience, offering promising alternatives to conventional agricultural inputs such as chemical fertilizers and pesticides. The following

TABLE 12.1
Key Plant–Microbe Interactions Supporting Sustainable Agriculture.

S. No.	Interaction Type	Representative Microbes	Primary Mechanisms	Plant Benefits	Contribution to Sustainable Agriculture	References
1	Biological nitrogen fixation	*Rhizobium leguminosarum*, *Frankia*	Nitrogenase-mediated atmospheric nitrogen fixation	Improved nitrogen availability, enhanced plant growth	Reduces synthetic nitrogen use, lowers pollution	Kaur and Pandove (2023)
2	Mycorrhizal aymbiosis	*Rhizophagus irregularis*, *Glomus intraradices*	Hyphal network extension, nutrient exchange	Increased uptake of phosphorus, zinc, and water	Enhances soil fertility and drought tolerance	Karunarathna (2024)
3	Phosphate solubilization	*Pseudomonas fluorescens*, *Bacillus megaterium*	Organic acid and enzyme-mediated phosphorus solubilization	Converts unavailable phosphorus into plant-usable form	Reduces dependence on chemical phosphorus fertilizers	Fasusi et al. (2023)
4	Plant growth promotion (PGPR)	*Azospirillum brasilense*, *Pseudomonas putida*	Production of IAA, siderophores, and ACC deaminase	Promotes root elongation, nutrient uptake, stress resilience	Enhances crop yield with lower agrochemical input	Ehinmitan et al. (2024)
5	Biocontrol of pathogens	*Trichoderma harzianum*, *Bacillus subtilis*	Antibiosis, enzyme production, niche competition	Suppresses fungal pathogens like *Fusarium* and *Pythium*	Reduces pesticide usage, promotes ecological pest control	Jambhulakar et al. (2015)
6	Induced systemic resistance (ISR)	*Pseudomonas fluorescens*, *Bacillus amyloliquefaciens*	Activation of jasmonic acid/ ethylene pathways	Enhances plant immunity without direct pathogen contact	Natural alternative to chemical-based plant protection	Kaur and Pandove (2023)
7	Endophytic association	*Herbaspirillum seropedicae*, *Enterobacter cloacae*	Internal colonization, modulation of stress-related hormones	Enhances stress tolerance, growth under adverse conditions	Improves resilience in marginal environments	Jain et al. (2024)
8	Silicon mobilization	*Bacillus* spp., *Penicillium chrysogenum*	Solubilization of mineral-bound silicon	Improves structural integrity, stress resistance	Aids climate-resilient agriculture	Ghosh et al. (2023)
9	Volatile organic compound (VOC)-mediated interactions	*Trichoderma virens*, *Bacillus subtilis*	Release of VOCs like acetoin, butanediol	Stimulates plant growth and systemic resistance from a distance	Offers noncontact bio-stimulation and natural defense priming	Lee et al. (2016)

mechanisms illustrate how beneficial microbes support plants in acquiring nutrients, suppressing diseases, tolerating abiotic stress, modulating hormonal pathways, and enhancing resistance to pathogens.

12.3 MECHANISMS OF BENEFICIAL INTERACTIONS

12.3.1 Nutrient Acquisition and Solubilization

One of the primary benefits of plant–microbe interactions is the enhancement of nutrient acquisition. Many plant-associated microbes, such as nitrogen-fixing bacteria and phosphate-solubilizing microbes, play a crucial role in improving nutrient availability to plants, especially in nutrient-poor soils.

Rhizobium and *Azospirillum* species are well-known for their ability to fix atmospheric nitrogen into forms that plants can use. *Azospirillum*, e.g., has been shown to improve nitrogen uptake and promote root growth in various crops, increasing plant productivity (Trivedi et al., 2020). Similarly, phosphate-solubilizing bacteria such as *Bacillus* spp. convert insoluble phosphorus compounds into bioavailable forms, making phosphorus more accessible to plants, which is particularly important in phosphorus-deficient soils (Rawat et al., 2021).

Moreover, mycorrhizal fungi form symbiotic relationships with plant roots, where the fungal hyphae extend into the soil, significantly increasing the plant's surface area for nutrient absorption, particularly for phosphorus, water, and trace minerals. These fungal networks are vital in nutrient cycling, making them a cornerstone of sustainable agriculture (Choudhary et al. 2022).

12.3.2 Disease Suppression and Biocontrol

Beneficial microbes play an essential role in protecting plants from various pathogens, offering a natural alternative to chemical pesticides. These microbes suppress diseases through a variety of mechanisms, including competition, the production of antimicrobial compounds, and inducing plant defense responses.

Microorganisms like *Trichoderma* spp. have been extensively studied for their biocontrol properties. *Trichoderma* spp. produces a wide array of secondary metabolites, including enzymes that degrade fungal cell walls, and they also produce volatile organic compounds (VOCs) that inhibit pathogen growth (Rajani et al., 2021). For example, *Trichoderma koningiopsis* has been shown to control *Verticillium dahliae*—a soil-borne pathogen responsible for wilt disease—by interfering with the pathogen's spore formation and disrupting its virulence factors (Kong et al., 2022).

Other microbes, such as *Bacillus amyloliquefaciens*, not only produce antimicrobial compounds but also trigger induced systemic resistance (ISR) in plants. This resistance enhances the plant's ability to defend itself against pathogens through the activation of its own immune system, without direct pathogen interaction. For instance, *Bacillus amyloliquefaciens* effectively reduces *Botrytis cinerea* infection in crops like strawberries by stimulating plant defense responses (Yang et al., 2024).

12.3.3 Abiotic Stress Tolerance

Abiotic stresses such as drought, salinity, and extreme temperatures significantly affect crop yields and quality. Beneficial microbes contribute to mitigating these stresses, thereby enhancing plant resilience and promoting sustainable agriculture.

Bacterial strains like *Bacillus amyloliquefaciens* and *Pseudomonas fluorescens* are known to improve drought tolerance in plants. These microbes can produce exopolysaccharides that protect plant roots by maintaining moisture under drought conditions and by modulating plant growth-promoting hormones (Rawat et al., 2021). Similarly, microbes can alleviate salt stress by modulating ion balance and producing compounds that help plants maintain cellular turgor under saline conditions (Choudhary et al., 2022).

Mycorrhizal fungi also contribute to abiotic stress tolerance by enhancing water uptake during drought and protecting plants from salt stress by limiting sodium ion uptake. The symbiotic relationship between mycorrhizal fungi and plants is particularly critical in arid and semi-arid regions where water availability is a major limiting factor for crop production (Trivedi et al., 2020).

12.3.4 Modulation of Plant Hormonal Pathways

One of the most profound effects of plant–microbe interactions is the modulation of plant hormonal pathways, which are central to plant growth, development, and stress responses. Certain beneficial microbes produce plant hormones, such as auxins, cytokinins, and gibberellins, that influence various physiological processes in plants.

For instance, *Azospirillum* spp. are known to produce auxins, which promote root elongation and branching, facilitating better nutrient and water absorption. These bacteria have been used in agricultural practices to improve crop yields by stimulating root growth and enhancing nutrient uptake (Ramlal et al., 2023). Similarly, *Bacillus* spp. can synthesize gibberellins that stimulate seed germination and root growth, helping plants overcome stress and improve productivity.

Modulation of ethylene production is another key mechanism by which beneficial microbes enhance plant stress tolerance. Ethylene, a plant hormone, is often associated with plant stress responses, but its production must be carefully regulated. Microbial inoculants can modulate ethylene levels to optimize plant responses to both biotic and abiotic stresses (Shoaib et al., 2023). This has been demonstrated in the case of *Pseudomonas* species, which can produce acetoin, a compound that reduces ethylene production, improving plant tolerance to stresses such as salinity and drought.

12.3.5 Induced Systemic Resistance

ISR refers to a plant's ability to mount a broad-spectrum defense response against a range of pathogens following exposure to beneficial microbes. Unlike direct resistance, which is activated locally at the site of infection, ISR provides protection throughout the entire plant.

Microorganisms like *Bacillus amyloliquefaciens* and *Pseudomonas fluorescens* are well-known for their ability to express ISR in plants. These microbes interact with the plant's immune system to trigger the production of defense-related proteins, such as pathogenesis-related (PR) proteins, and activate signaling pathways involving jasmonic acid, ethylene, and salicylic acid (Pieterse et al., 2014). For example, when *Bacillus amyloliquefaciens* was applied to tomato plants, it triggered ISR, enhancing resistance to the foliar pathogen *Botrytis cinerea* without the need for direct exposure to the pathogen itself.

ISR is particularly important for managing plant diseases in organic farming systems, where chemical pesticides are avoided. By inducing systemic resistance, beneficial microbes reduce the reliance on chemical controls, making farming more environmentally sustainable and economically viable.

The growing interest in sustainable agricultural practices has led to the integration of beneficial plant–microbe interactions as a cornerstone of modern farming. These interactions not only enhance plant growth but also improve soil health, reduce dependency on chemical fertilizers and pesticides, and promote organic farming systems. Below, we explore several key applications of beneficial plant–microbe interactions in sustainable agriculture.

12.4 APPLICATIONS IN SUSTAINABLE AGRICULTURE

12.4.1 Biofertilizers: Innovations and Field Applications

Biofertilizers are substances containing living microorganisms that, when applied to plants or soils, improve plant growth by enhancing nutrient availability and uptake. The use of biofertilizers has been increasingly recognized as an eco-friendly alternative to chemical fertilizers, addressing concerns about soil degradation, water contamination, and unsustainable farming practices. Recent innovations have focused on microbial strains that can fix nitrogen, solubilize phosphates, and enhance potassium availability, all essential nutrients for plant growth (Singh et al., 2024).

For instance, nitrogen-fixing bacteria, such as *Rhizobium* and *Azotobacter*, have been extensively used as biofertilizers. These bacteria form symbiotic relationships with plants, especially legumes, to convert atmospheric nitrogen into a form that plants can use (Bashan et al., 1998). Similarly, phosphate-solubilizing bacteria like *Pseudomonas* spp. and *Bacillus* spp. release organic acids that break down insoluble phosphates in the soil, making them available to plants (Meena et al., 2017). Field trials have shown significant improvements in crop yields, particularly for cereals and legumes, when biofertilizers are used in combination with organic amendments (Khoshru et al., 2025).

12.4.2 Biopesticides and Disease Control

Biopesticides, derived from natural organisms such as bacteria, fungi, and viruses, offer an effective solution to managing pests and plant diseases without harming the environment. Beneficial microbes, such as *Bacillus thuringiensis* (Bt) and *Trichoderma* spp., are widely used in biocontrol applications. *Bt*, e.g., produces toxins that are lethal to specific insect pests but are nontoxic to humans, animals, and beneficial insects (Ehis-Eriakha et al., 2025).

Recent advancements in the use of biopesticides have focused on their integration into integrated pest management (IPM) strategies. For instance, *Trichoderma* fungi not only combat soil-borne pathogens like *Fusarium* but also promote plant growth through the production of plant growth hormones and by enhancing root development (Mandal et al., 2019). The adoption of biopesticides in both organic and conventional farming systems has led to reduced reliance on chemical pesticides, thus mitigating the risks of resistance buildup and environmental pollution (Prasad et. al., 2024).

12.4.3 Role in Organic Farming Systems

Organic farming systems emphasize the use of natural processes and inputs to maintain soil fertility and plant health, aligning with the goals of sustainability and biodiversity preservation. Beneficial plant–microbe interactions play a crucial role in organic farming by improving nutrient cycling, promoting disease resistance, and reducing the need for synthetic chemicals.

In organic farming, biofertilizers and biopesticides are essential tools. Studies have shown that organic farming systems, which often rely on microbe-based amendments, foster a diverse microbiome in the soil. This diversity supports nutrient cycling and enhances plant resilience to stressors, such as pests and diseases (Mohanram and Kumar, 2019). The use of beneficial microbes in organic fields also helps mitigate the harmful effects of monocropping, promoting crop rotation and biodiversity, which are key principles in organic agriculture (Khoshru et al., 2025).

For example, the use of *Azospirillum* spp., which promotes nitrogen fixation, and mycorrhizal fungi, which assist in phosphorus uptake, has been shown to boost crop productivity and soil health in organic systems (Andreozzi et al., 2019). Additionally, organic potato fields, when treated with organic amendments, exhibited stronger defense responses against pests, demonstrating that microbial interactions can enhance plant resistance and improve crop yields even under organic management (Lalitha, 2017).

12.4.4 Contributions to Soil Health and Fertility

One of the most significant contributions of plant–microbe interactions to sustainable agriculture is their positive impact on soil health and fertility. Microbial communities in the soil play essential roles in nutrient cycling, organic matter decomposition, and soil structure formation, all of which are vital for maintaining soil fertility.

Microorganisms such as nitrogen-fixing bacteria, phosphorus-solubilizing microbes, and mycorrhizal fungi contribute directly to nutrient availability, helping plants access otherwise inaccessible nutrients. The use of biofertilizers, which introduce or enhance the activity of these microorganisms, can significantly improve soil fertility. For instance, the use of *Rhizobium* in leguminous crops enhances nitrogen fixation, thereby reducing the need for synthetic nitrogen fertilizers (Bashan et al., 1998). Similarly, mycorrhizal fungi form symbiotic relationships with plant roots, facilitating the uptake of water and nutrients, particularly in nutrient-poor soils (Gupta et al., 2021).

In addition to nutrient cycling, beneficial microbes also improve soil structure. Fungi, bacteria, and actinomycetes contribute to the formation of soil aggregates, which enhances soil aeration, water retention, and drainage. These microbes also help suppress soil-borne pathogens, reducing the need for chemical interventions and promoting a healthy soil ecosystem (Mohanram and Kumar, 2019). For example, *Trichoderma* spp. and other beneficial fungi have been shown to improve soil microbial diversity and protect plants from diseases like root rot and wilt (Ehis-Eriakha et al., 2025).

Furthermore, plant–microbe interactions enhance the resilience of soil ecosystems to various stressors. They help mitigate the impact of soil erosion, nutrient depletion, and contamination by heavy metals, which are common challenges in conventional agricultural systems (Singh et al., 2024). This makes microbe-based approaches not only environmentally friendly but also economically viable, as they reduce the long-term costs of soil degradation and improve crop productivity.

12.5 BIOTECHNOLOGICAL ADVANCES AND INNOVATIONS

The application of biotechnological innovations has accelerated the development of beneficial plant–microbe interactions that support sustainable agricultural practices. With advances in metagenomics, genome editing, synthetic biology, and microbial consortia, the potential to enhance crop production and soil health has increased significantly. Below, we explore some of the key advancements that are shaping the future of microbial technologies in agriculture.

12.5.1 Metagenomics and Microbial Diversity

Metagenomics, the study of genetic material recovered directly from environmental samples, has revolutionized the understanding of microbial diversity in the soil and rhizosphere. By sequencing the DNA of entire microbial communities, metagenomics provides insights into the complex interactions between plants and microbes that are often missed by traditional culturing methods (Schlatter et al., 2015). This has allowed researchers to uncover previously unknown microbial species with potential beneficial effects on plant growth, nutrient cycling, and disease suppression.

For example, metagenomic approaches have revealed how specific microbial taxa in the rhizosphere can enhance soil fertility and improve crop yields. One study by Thotakuri et al. (2024) identified unique soil bacteria that significantly promoted phosphorus uptake in maize, which is essential for enhancing crop productivity in phosphorus-deficient soils. Moreover, the identification of microbial consortia, or groups of microbes working synergistically, has opened new avenues for developing biofertilizers that combine multiple beneficial species to improve nutrient availability and stress resistance in plants (Hassani et al., 2025).

The integration of metagenomics with other high-throughput techniques, such as transcriptomics and proteomics, is further advancing the understanding of plant–microbe interactions. These technologies help in deciphering the molecular mechanisms through which microbes influence plant health, allowing for more precise interventions in agricultural practices (Sharma et al., 2022).

12.5.2 Genome Editing and Synthetic Biology

Genome editing technologies, such as CRISPR-Cas9, have paved the way for the targeted modification of both plant and microbial genomes. These tools allow for the precise manipulation of microbial strains to enhance their beneficial properties, such as increased nitrogen fixation, enhanced pathogen resistance, or improved plant growth-promoting activities (Malik, 2024). Genome editing can also be applied to plants to enhance their responsiveness to beneficial microbes, optimizing their ability to interact with beneficial microorganisms in the soil.

A significant example of this is the editing of plant receptor genes to promote better communication between plants and their beneficial microbial partners. Studies have shown that modifying plant genes related to immune receptors can enhance plant responses to beneficial rhizobacteria, thereby increasing the plants' resistance to pathogens and improving overall growth (Pankaj et al., 2020). Similarly, microbial genome editing allows for the development of custom-designed strains that can be more effective in specific environments, such as those with nutrient deficiencies or high pathogen pressure (Malik, 2024).

Moreover, synthetic biology, which involves the construction of new biological parts, devices, and systems, is being explored to create novel microbes with engineered functions tailored to agricultural needs. These microbes can be designed to interact more effectively with plants, perform specific biochemical transformations, or synthesize compounds that enhance plant resistance to abiotic and biotic stresses (Kimotho et al., 2024).

12.5.3 Microbial Consortia and Formulation Technologies

Microbial consortia, composed of multiple beneficial microorganisms working together, are gaining attention as a means to improve plant health and productivity. These consortia can provide a broad spectrum of benefits, such as enhanced nutrient cycling, pathogen suppression, and increased tolerance to environmental stresses (Khoshru et al., 2025). By combining different microbial species, researchers aim to harness synergistic effects that lead to better outcomes than using single strains alone.

The formulation of microbial consortia is a challenging but crucial step in ensuring the effectiveness of these biological products. Innovations in formulation technologies are focusing on improving the stability, shelf-life, and delivery methods of microbial inoculants. Advances in encapsulation technologies, e.g., allow for the protection of microbes from environmental stressors during storage and application, ensuring that they remain viable and effective when introduced into the soil (Khoshru et al., 2025). Additionally, biotechnological approaches to optimizing the delivery of microbial consortia to the plant roots are also being explored, with techniques such as root coating and hydrogel formulations enhancing microbial colonization and persistence in the rhizosphere (Sharma & Kumawat, 2022).

Case studies have demonstrated the efficacy of microbial consortia in field applications. For instance, a consortium of *Bacillus*, *Pseudomonas*, and *Trichoderma* spp. was found to significantly improve crop yields and reduce the incidence of soil-borne diseases in tomato and pepper fields (Khoshru et al., 2025). Similarly, consortia of nitrogen-fixing bacteria combined with phosphate-solubilizing microbes have been shown to improve nutrient uptake and growth in rice and wheat, especially in nutrient-poor soils (Wu et al., 2024).

12.5.4 Case Studies of Successful Microbial Inoculants

Several case studies highlight the successful application of microbial inoculants in agricultural settings, demonstrating their potential for enhancing crop productivity and sustainability. One notable example is the use of *Azospirillum brasilense* in maize production. In a field trial conducted by Dos Santos et al. (2020), maize inoculated with *Azospirillum brasilense* showed a significant increase in nitrogen content and biomass compared to untreated control plants. This beneficial bacterium promotes nitrogen fixation, reducing the need for chemical fertilizers and contributing to more sustainable farming practices.

Another successful case is the use of mycorrhizal fungi in improving plant drought tolerance. Studies by Singh et al. (2022) demonstrated that inoculating crops like wheat and sorghum with *Glomus* spp. mycorrhizal fungi led to improved water uptake and better survival rates under drought conditions. The symbiotic relationship between the fungi and plant roots enhances nutrient and water absorption, making these crops more resilient to abiotic stressors, particularly in arid regions.

Furthermore, a combination of biocontrol agents such as *Trichoderma harzianum* and *Bacillus subtilis* has been successfully used in organic farming systems to suppress soil-borne pathogens. In a case study by Sharma and Singh (2025), this combination significantly reduced the incidence of *Fusarium* wilt in tomato plants, demonstrating the potential of microbial inoculants to replace chemical fungicides in sustainable agriculture.

12.6 CHALLENGES AND LIMITATIONS

While the potential benefits of utilizing plant–microbe interactions in sustainable agriculture are immense, there are several challenges and limitations that need to be addressed before these technologies can be widely adopted. Issues such as variability in field performance, environmental factors affecting efficacy, and regulatory hurdles present significant obstacles. Understanding and overcoming these challenges are critical for the successful implementation of microbial solutions in agriculture.

12.6.1 Variability in Field Performance

One of the primary challenges in the application of beneficial plant–microbe interactions is the variability in their performance across different environmental conditions and field settings. While laboratory and greenhouse studies often show promising results, the effectiveness of microbial inoculants can vary significantly when transferred to large-scale field trials. This variability is influenced by numerous factors, including soil type, crop species, climate, and management practices (Schlatter et al., 2015).

For example, a study by Khoshru et al. (2025) found that *Azospirillum* inoculants, which are known to enhance nitrogen fixation in laboratory conditions, exhibited inconsistent results in field trials. In some locations, inoculated crops showed a marked improvement in growth, while in others, there was little to no benefit. Similarly, the use of mycorrhizal fungi for phosphorus uptake enhancement has yielded mixed results, with some studies showing significant growth improvements in certain crops and others showing no effect (Duan et al., 2024).

This variability can be attributed to differences in microbial communities and environmental conditions that affect microbial activity. Soil pH, moisture levels, temperature, and nutrient availability are just a few of the factors that can influence the success of microbial inoculants (Bashan et al., 1998). The interaction between the introduced microbes and native soil microbiota also plays a critical role, as native microorganisms may compete with or inhibit the beneficial microbes.

12.6.2 Environmental Factors Affecting Efficacy

Environmental factors such as temperature, humidity, soil texture, and moisture significantly impact the efficacy of microbial inoculants in the field. Microorganisms are sensitive to changes in these factors, which can affect their survival, activity, and ability to colonize plant roots. For instance, the survival of nitrogen-fixing bacteria like *Rhizobium* can be compromised in soils with extreme temperatures or fluctuating moisture levels (Ehis-Eriakha et al., 2025).

Additionally, the application of microbial inoculants in arid or semi-arid regions, where water availability is limited, may result in poor colonization and reduced effectiveness. A study by Zapata-Sarmiento et al. (2025) on the use of *Trichoderma* spp. for disease suppression found that while the inoculant was effective in regions with moderate temperatures and consistent rainfall, its performance was significantly reduced in drought-prone areas, where soil moisture fluctuated. The microbial inoculants failed to establish a stable population in dry conditions, undermining their intended benefits.

Moreover, the introduction of microbial inoculants into agricultural systems can sometimes lead to unintended ecological consequences. For example, the use of a single-strain inoculant may disrupt the natural balance of microbial communities, potentially causing an imbalance that could harm plant health or even contribute to disease outbreaks. Understanding the relationship between microbial inoculants and the broader ecosystem is crucial to minimizing such risks and ensuring the long-term sustainability of these technologies (Kimotho & Maina, 2024).

12.6.3 Regulatory Hurdles and Public Perception

In addition to the technical and environmental challenges, the commercialization of microbial inoculants faces significant regulatory hurdles and public perception issues. Many countries have stringent regulations governing the approval and use of microbial products, which often involve lengthy and expensive approval processes. These regulations are designed to ensure the safety and efficacy of the products but can also delay the market introduction of potentially beneficial technologies (Bashan et al., 1998).

For instance, the approval process for a new microbial product in the European Union or the United States can take several years and requires extensive testing to demonstrate its safety, effectiveness, and environmental impact. In some cases, regulatory authorities may require large-scale field trials, which can be costly and time-consuming (Prasad et al., 2024). These regulatory delays can be a major barrier to the widespread adoption of microbial inoculants, especially for small and medium-sized enterprises that lack the financial resources to navigate complex regulatory landscapes.

Public perception of genetically modified organisms (GMOs) and microbe-based agricultural products is another challenge. While beneficial microbes are generally considered safe, there is still some skepticism among farmers and consumers regarding the use of genetically engineered microbes in agriculture. This is particularly true in regions where the public is sensitive to GMOs, such as in parts of Europe and Africa. Even though synthetic biology and genome editing can help create more efficient microbial strains, concerns about the potential for unintended ecological consequences or the spread of modified genes into wild populations may hinder the acceptance of these technologies (Malik, 2024).

Furthermore, lack of awareness and understanding of the benefits of microbial inoculants among farmers can also pose a challenge. Farmers may be hesitant to adopt these technologies if they are unsure of their effectiveness or the potential return on investment. Educational efforts and demonstrations of successful field trials are essential to build trust and encourage the widespread use of microbial products (Bashan et al., 1998).

12.7 FUTURE PROSPECTS AND RECOMMENDATIONS

The potential for beneficial plant–microbe interactions to revolutionize agriculture is enormous, but this requires continuous advancements in science, technology, policy, and farmer education to fully realize their benefits. Below are some key future prospects and recommendations for maximizing the application of microbial technologies in sustainable farming.

12.7.1 Integrating Microbiomes into Precision Agriculture

Precision agriculture, which uses data-driven techniques to optimize crop production, is a natural fit for integrating microbial management. Advances in microbial genomics, metagenomics, and high-throughput sequencing technologies have allowed for a deeper understanding of plant microbiomes and their roles in nutrient cycling, disease suppression, and stress tolerance. Integrating microbiome-based strategies into precision agriculture could enhance the efficiency of microbial inoculants and allow farmers to apply tailored microbial products based on specific soil conditions and crop requirements (Bashan et al., 1998; Schlatter et al., 2015).

For instance, with the aid of sensors and microbiome profiling, it may be possible to precisely monitor microbial populations in real time, allowing for the application of microbial treatments only when needed, reducing costs, and enhancing efficacy (Sharma & Kumawat, 2022). Moreover, the integration of microbial management into digital farming platforms could help farmers optimize the application of both chemical and biological inputs, fostering more sustainable agricultural practices. This approach could significantly improve the precision with which microbes are applied to crops, leading to more consistent field performance and better yields under varying environmental conditions.

12.7.2 Research Priorities for Enhanced Microbial Applications

Future research should focus on understanding the dynamic interactions between plants and microbes in different agricultural ecosystems. A major research priority is to improve the consistency of microbial performance across diverse environments. This includes understanding the complex interplay between environmental factors such as soil type, moisture, and temperature, and how these factors influence microbial colonization, survival, and activity (Malik, 2024). Furthermore, the development of more robust and effective microbial consortia, which combine different species of microbes for synergistic effects, holds promise for improving the overall performance of microbial inoculants in various agricultural contexts (Hassani et al., 2021).

Another critical area for research is the exploration of microbial biostimulants that can enhance plant growth and resilience without relying on GMOs. Non-GMO microbial products may be more widely accepted by farmers and consumers, addressing some of the regulatory and public perception concerns surrounding genetically engineered products (Kimotho & Maina, 2024; Maurya et al., 2024). Additionally, there is a need for research on how microbial inoculants can improve soil health over the long term, particularly in terms of soil structure, nutrient cycling, and organic matter content (Duan et al., 2024).

12.7.3 Policy and Farmer Education

For the widespread adoption of microbial technologies, supportive policies and farmer education are essential. Policymakers must work toward creating regulatory frameworks that facilitate the approval and commercialization of microbial products while ensuring their safety and efficacy. Streamlining approval processes and providing clear guidelines for the registration of microbial inoculants would help accelerate their availability on the market (Bashan et al., 1998). Governments should also consider offering financial incentives to farmers who adopt microbial solutions, which would encourage more sustainable practices and reduce the reliance on chemical inputs.

Farmer education is equally important to increase the acceptance and adoption of microbial technologies. Many farmers are still unfamiliar with the benefits of microbial inoculants and may be hesitant to change from conventional farming practices. Extension programs, field demonstrations, and farmer-to-farmer networks can be instrumental in promoting microbe-based practices. Moreover, educating farmers about the long-term benefits of soil health and the role of microbes in sustaining agricultural productivity is essential for fostering a broader understanding of sustainable farming systems (Schlatter et al., 2015).

12.8 CONCLUSION AND FUTURE PERSPECTIVES

The intricate relationship between plants and beneficial microbes offers a transformative avenue for enhancing crop productivity and sustainability in agriculture. Beneficial microbes, including rhizobacteria, endophytes, mycorrhizal fungi, and nitrogen-fixing bacteria, significantly contribute to plant health by improving nutrient uptake, enhancing stress tolerance, and suppressing pathogens. The use of microbial consortia and bioinoculants presents a promising alternative to chemical fertilizers and pesticides, minimizing environmental degradation and fostering soil health.

Advancements in genomics, metagenomics, and systems biology have deepened our understanding of plant–microbe interactions, revealing the molecular dialogs and signaling networks involved. The integration of multi-omics technologies and synthetic biology approaches holds potential to design next-generation microbial formulations with tailored functionalities. However, the field still faces challenges related to the inconsistency of microbial performance in field conditions, host specificity, and complex soil microbiome dynamics. Future research should focus on unraveling the context-dependent mechanisms that govern plant–microbe symbioses under diverse agro-climatic conditions. Emphasis should also be placed on developing robust, scalable microbial technologies that are resilient to abiotic stressors and adaptable to various cropping systems. A multidisciplinary approach combining microbiology, plant science, agronomy, and data-driven modeling will be key to translating laboratory successes into field-level applications.

Furthermore, regulatory frameworks and public awareness must be strengthened to facilitate the safe and widespread adoption of microbial technologies. Collaborations between academia, industry, and policymakers are essential to accelerate innovation and commercialization in this domain. Overall, harnessing beneficial plant–microbe interactions represent a sustainable, eco-friendly strategy to meet global food demands while preserving natural resources. Strategic investments in research, infrastructure, and education are critical to fully realize the potential of plant-associated microbiomes in shaping the future of agriculture.

12.9 AUTHOR CONTRIBUTION

JHN: Conceptualization, original draft preparation, editing, and drafting of tables and figures.

JHN and BM: Review, discussion, and editing. BM: Provided scientific feedback, edited the manuscript (MS), and contributed valuable insights for improving the MS.

12.10 DECLARATION OF INTERESTS

All authors have reviewed and confirmed no conflicts of interest.

REFERENCES

Ali, Q., Khan, A. R., Raza, W., Bilal, M. S., Khalid, S., Ayaz, M., . . . & Mundra, S. (2025). Mechanisms of microbial VOC-mediated communication in plant ecosystems and agricultural applications. *Journal of Sustainable Agriculture and Environment*, 4(1), e70044.

Andreozzi, A., Prieto, P., Mercado-Blanco, J., Monaco, S., Zampieri, E., Romano, S., & Bianco, C. (2019). Efficient colonization of the endophytes Herbaspirillum huttiense RCA24 and Enterobacter cloacae RCA25 influences the physiological parameters of Oryza sativa L. cv. baldo rice. *Environmental Microbiology*, 21(9), 3489–3504

Bashan, Y., Puente, M. E., Myrold, D. D., & Toledo, G. (1998). In vitro transfer of fixed nitrogen from diazotrophic filamentous cyanobacteria to black mangrove seedlings. *FEMS Microbiology Ecology*, 26(3), 165–170. https://doi.org/10.1111/j.1574-6941.1998.tb00502.x

Berg, G., Grube, M., Schloter, M., & Smalla, K. (2017). Unraveling the plant microbiome: Looking back and future perspectives. *Frontiers in Microbiology*, 8, 161.

Bhattacharyya, P. N., & Jha, D. K. (2012). Plant growth-promoting rhizobacteria (PGPR): Emergence in agriculture. *World Journal of Microbiology and Biotechnology*, 28(4), 1327–1350. https://doi.org/10.1007/s11274-011-0979-9

Bulgarelli, D., Rott, M., & Schlaeppi, K. (2013). Revealing structure and assembly of the root-associated microbiome. *Nature*, 488(7409), 91–95.

Choudhary, M., Chandra, P., Dixit, B., Nehra, V., Choudhary, U., & Choudhary, S. (2022). Plant growth-promoting microbes: Role and prospective in amelioration of salt stress. *Communications in Soil Science and Plant Analysis*, 53(13), 1692–1711.

Compant, S., Clément, C., & Sessitsch, A. (2010). Plant growth-promoting bacteria in the rhizo- and endosphere of plants: Their role in plant health. *Plant and Soil*, 321(1–2), 341–361.

Dos Santos, A. C., Kandasamy, S., & Rigobelo, E. C. (2020). *Bacillus cereus*, *Lactobacillus acidophilus* and *Succinovibrio dextrinosolvens* promoting the growth of maize and soybean plants. *African Journal of Microbiology Research*, 14(5), 189–197. https://doi.org/10.5897/AJMR2020.9322

Duan, H. X., Luo, C. L., Wang, X., Cheng, Y. S., Abrar, M., & Batool, A. (2024). Responses of legumes to rhizobia and arbuscular mycorrhizal fungi under abiotic stresses: A global meta-analysis. *Agronomy*, 14(11), 2597. https://doi.org/10.3390/agronomy14112597

Ehinmitan, E., Losenge, T., Mamati, E., Ngumi, V., Juma, P., & Siamalube, B. (2024). BioSolutions for green agriculture: Unveiling the diverse roles of plant growth-promoting Rhizobacteria. *International Journal of Microbiology*, 2024(1), 6181491.

Ehis-Eriakha, C. B., Oleghe, P. O., Akemu, S. E., Ajuzieogu, C. A., & Oboh, E. J. (2025). Microbial inoculant as a viable green technology for combating food insecurity in Africa. *European Journal of Plant Pathology*, 1–16. https://doi.org/10.1007/s10658-025-03020-5

FAO. (2021). *The state of food security and nutrition in the world 2021*. Food and Agriculture Organization of the United Nations.

Fasusi, O. A., Babalola, O. O., & Adejumo, T. O. (2023). Harnessing of plant growth-promoting rhizobacteria and arbuscular mycorrhizal fungi in agroecosystem sustainability. *CABI Agriculture and Bioscience*, *4*(1), 26.

Field, K. J., Pressel, S., Duckett, J. G., Rimington, W. R., & Bidartondo, M. I. (2020). Symbiotic options for the conquest of land. *Trends in Ecology & Evolution*, 35(7), 492–504. https://doi.org/10.1016/j.tree.2020.02.004

Foley, J. A., Foley, J., Ramankutty, N., Brauman, K., Cassidy, E., Gerber, J., Johnston, M., Mueller, N., O'Connell, C., Ray, D., West, P., Balzer, C., Bennett, E., Carpenter, S., Hill, J., Monfreda, C., Polasky, S., Rockström, J., Sheehan, J., Siebert, S., & Zaks, D. (2011). Solutions for a cultivated planet. *Nature*, 478(7369), 337–342.

Ghosh, S., Mondal, S., Banerjee, S., Mukherjee, A., & Bhattacharyya, P. (2023). Temporal dynamics of potassium release from waste mica as influenced by potassium mobilizing bacteria. *Journal of Pure and Applied Microbiology*, 17(1), 273–288.

Gupta, R., Anand, G., Gaur, R., & Yadav, D. (2021). Plant–microbiome interactions for sustainable agriculture: A review. *Physiology and Molecular Biology of Plants*, 27(1), 165–179.

Hassani, L. M. I., & Jauregui, I. (2025). Harnessing the power of microbial. In *Sustainable bioeconomy development in the global south. Volume I. Status and perspectives* (p. 261). Springer Nature.

Jacoby, R., Peukert, M., Succurro, A., Koprivova, A., & Kopriva, S. (2017). The role of soil microorganisms in plant mineral nutrition—current knowledge and future directions. *Frontiers in Plant Science*, 8, 1617.

Jain, A., Sarsaiya, S., Singh, R., Gong, Q., Wu, Q., & Shi, J. (2024). Omics approaches in understanding the benefits of plant–microbe interactions. *Frontiers in Microbiology*, 15, 1391059. https://doi.org/10.3389/fmicb.2024.1391059

Jambhulkar, P. P., Sharma, M., Lakshman, D., & Sharma, P. (2015). Natural mechanisms of soil suppressiveness against diseases caused by Fusarium, Rhizoctonia, Pythium, and Phytophthora. In *Organic amendments and soil suppressiveness in plant disease management* (pp. 95–123). Cham: Springer International Publishing.

Kaur, J., & Pandove, G. (2023). Understanding the beneficial interaction of plant growth promoting rhizobacteria and endophytic bacteria for sustainable agriculture: A bio-revolution approach. *Journal of Plant Nutrition*, 46(14), 3569–3597.

Khoshru, B., Fallah Nosratabad, A., Mahjenabadi, V. A. J., Knežević, M., Hinojosa, A. C., Fadiji, A. E., & Mitra, D. (2025). Multidimensional role of *Pseudomonas*: From biofertilizers to bioremediation and soil ecology to sustainable agriculture. *Journal of Plant Nutrition*, 48(6), 1016–1042. https://doi.org/10.1080/01904167.2024.2416078

Kimotho, R. N., & Maina, S. (2024). Unraveling plant–microbe interactions: Can integrated omics approaches offer concrete answers?. *Journal of Experimental Botany*, 75(5), 1289–1313. https://doi.org/10.1093/jxb/erad448

Kong, W. L., Ni, H., Wang, W. Y., & Wu, X. Q. (2022). Antifungal effects of volatile organic compounds produced by Trichoderma koningiopsis T2 against Verticillium dahliae. *Frontiers in Microbiology*, 13, 1013468.

Lalitha, S. (2017). Plant growth–promoting microbes: A boon for sustainable agriculture. In *Sustainable agriculture towards food security* (pp. 125–158). Singapore: Springer Singapore.

Lee, S., Yap, M., Behringer, G., Hung, R., & Bennett, J. W. (2016). Volatile organic compounds emitted by Trichoderma species mediate plant growth. *Fungal Biology and Biotechnology*, 3(1), 7.

Malik, S. (2024). The intersection of biotechnology and genetics in agricultural biotechnology. *Frontiers in Biotechnology and Genetics*, 1(3), 244–261. https://doi.org/10.3389/fbioe.2024.1350722

Mandal, S. K. (2019). Impact of pest control chemicals on biological activity of biocontrol agents. In *The souvenir of national seminar on agro-chemical inputs and its extension approaches towards food-security and bio-safety: Prospects and challenges* (pp. 80–84). State Agricultural Management and Extension Training Institute – West Bengal.

Maurya, B., Mishra, V., Divyanshu, K., & Nishad, J. H. (2024). Pros and cons of genetic engineering technologies in reference to biofortified crops. In *Harnessing crop biofortification for sustainable agriculture* (pp. 349–370). Singapore: Springer Nature Singapore. https://doi.org/10.1007/978-981-97-3438-2_18

Meena, M., Swapnil, P., Zehra, A., Aamir, M., Dubey, M. K., Goutam, J., & Upadhyay, R. S. (2017). Beneficial microbes for disease suppression and plant growth promotion. In *Plant–microbe interactions in agro-ecological perspectives: Volume 2: Microbial interactions and agro-ecological impacts* (pp. 395–432). Singapore: Springer Singapore.

Mohanram, S., & Kumar, P. (2019). Rhizosphere microbiome: Revisiting the synergy of plant-microbe interactions. *Annals of Microbiology*, 69(4), 307–320.

Mus, F., Crook, M. B., Garcia, K., Costas, A. G., Geddes, B. A., Kouri, E. D., Paramasivan, P., Ryu, M. H., Oldroyd, G. E. D., Poole, P. S., Udvardi, M. K., Voigt, C. A., Ane, J. M., & Peters, J. W. (2016). Symbiotic nitrogen fixation and the challenges to its extension to non-legumes. *Applied and Environmental Microbiology*, 82(13), 3698–3710. https://doi.org/10.1128/AEM.01055-16

Pankaj, U., & Dubey, N. (2020). Application of genetically engineered microbes for sustainable development of agro-ecosystem. In *Microbes in agriculture and environmental development* (pp. 209–234). CRC Press.

Pieterse C. M., Zamioudis C., Does D. V., Van Wees S. (2014). Signalling networks involved in induced resistance. In *Induced resistance for plant defense: A sustainable approach to crop protection* (pp. 58–80). Wiley & Sons Ltd.

Prasad, K., Rajasekaran, S., Pandey, R. K., & Rajat, K. (2024). Multiple functions of microbes to improve food and soil nutrition in support of environment management to maintain global agricultural practices. *Journal of Energy and Environmental Science*, 2(1), 1–18. https://doi.org/10.23880/jeesc-16000112

Rajani, P., Rajasekaran, C., Vasanthakumari, M. M., Olsson, S. B., Ravikanth, G., & Shaanker, R. U. (2021). Inhibition of plant pathogenic fungi by endophytic Trichoderma spp. through mycoparasitism and volatile organic compounds. *Microbiological Research*, 242, 126595.

Ramlal, A., Rani, A., Nautiyal, A., Kalra, C., Kumari, R., Kumar, J., & Mishra, V. (2023). Importance of omics approaches in plant–microbe interaction for plant disease control. *Physiological and Molecular Plant Pathology*, 128, 102153. https://doi.org/10.1016/j.pmpp.2023.102153

Rawat, P., Das, S., Shankhdhar, D., & Shankhdhar, S. C. (2021). Phosphate-solubilizing microorganisms: Mechanism and their role in phosphate solubilization and uptake. *Journal of Soil Science and Plant Nutrition*, 21(1), 49–68.

Rodriguez, R. J., Woodward, C. J., Redman, R. S., & Freeman, D. C. (2008). Fungal endophytes: Diversity and functional roles. *New Phytologist*, 180(1), 1–2. https://doi.org/10.1111/j.1469-8137.2008.02672.x

Schlatter, D. C., Bakker, M. G., Bradeen, J. M., & Kinkel, L. L. (2015). Plant community richness and microbial interaction's structure bacterial communities in soil. *Ecology*, 96(1), 134–142.

Sharma, D., & Singh, S. (2025). Encouraging Probiotics as Plant Stimulants for Sustainable Agriculture Through Microbial Consortium. *Microbial Consortia in Plant Science and Sustainable Agriculture*, 223–238.

Sharma, S., & Kumawat, K. C. (2022). Role of rhizospheric microbiome in enhancing plant attributes and soil health for sustainable agriculture. In *Core microbiome: Improving crop quality and productivity* (pp. 139–162). Technology & Engineering.

Shoaib, M., Zhang, H., Hussain, S., Arshad, M., Li, G., Fan, Y., Brestic, M., Yang, X., Zhang, S., Liu, X. and Jiao, J., Hu, F., Wu, J., & Li, H. (2024). Adaptive mechanisms of Maize and Soybean against nanoplastic-induced stress: A multistage defense mechanism with metabolic and transcriptomic alterations. Available at SSRN: https://ssrn.com/abstract=4957813 or http://dx.doi.org/10.2139/ssrn.4957813

Singh, N. K., Sachan, K., Bp, M., Panotra, N., & Katiyar, D. (2024). Building soil health and fertility through organic amendments and practices: A review. *Asian Journal of Soil Science and Plant Nutrition*, 10(1), 175–197.

Thotakuri, G., Angidi, S., & Athelly, A. (2024). Soil carbon pool as influenced by soil microbial activity—an overview. *American Journal of Climate Change*, 13(2), 175–193. https://https://doi.org/10.4236/ajcc.2024.132010

Vessey, J. K. (2003). Plant growth promoting rhizobacteria as biofertilizers. *Plant and Soil*, 255(2), 571–586.

Wu, Q., Yang, L., Liang, H., Liu, M., Chen, Y., Chen, D., & Shen, P. (2024). Impacts of soil compaction and phosphorus levels on the dynamics of phosphate-solubilizing and nitrogen-fixing bacteria in the peanut rhizosphere. *Agronomy*, 14(9), 1971.

Yang, P., Yuan, P., Liu, W., Zhao, Z., Bernier, M. C., Zhang, C., . . . & Xia, Y. (2024). Plant growth promotion and plant disease suppression induced by Bacillus amyloliquefaciens strain GD4a. *Plants*, 13(5), 672.

Zapata-Sarmiento, D. H., Rodríguez-Hernández, A. A., Sepúlveda-Jiménez, G., & Rodríguez-Monroy, M. (2025). Tolerance and antioxidant response to heavy metals are differentially activated in *Trichoderma asperellum* and *Trichoderma longibrachiatum*. *PeerJ*, 13, e19016.

13 *Rhizobium*–Legume Symbiosis
Fueling Crop Yield through Nitrogen Fixation

Ankita Thakur, Bilal Ahmad Mir, Ranjna Kumari, Simran Kauts, Simarjot Singh, and Zishan Ahmad

13.1 INTRODUCTION

Nitrogen (N) is an essential macronutrient required for plant growth and development. Although nitrogen constitutes about 78% of the Earth's atmosphere, most plants cannot directly utilize atmospheric nitrogen (N_2) for their metabolic needs. This limitation often results in nitrogen deficiency, which significantly reduces crop yields, as nitrogen is a critical component of proteins, nucleic acids, and other vital biomolecules.

Nitrogen fixation, the process of converting atmospheric nitrogen into forms usable by plants, occurs through physical, chemical, and biological means. Among these, biological nitrogen fixation (BNF) is the most environmentally sustainable approach. Nitrogen-fixing bacteria, particularly those belonging to the *Rhizobium* genus, play a crucial role in this process by converting atmospheric nitrogen into ammonia (NH_3), which plants can readily absorb (Griesmann et al., 2018). BNF not only contributes significantly to global food security but also maintains the natural nitrogen cycle.

Leguminous plants have developed a specialized symbiotic relationship with *Rhizobium* species, facilitating nitrogen fixation within root nodules (Liu et al., 2021). Phylogenetic studies suggest that the capacity for nodulation in angiosperms evolved approximately 100 million years ago (Zhao et al., 2021). About 65 million years ago, certain legume species acquired the ability to form mutualistic associations with *Rhizobium* bacteria—a trait still prevalent among modern legumes (Sprent & James, 2007).

In this symbiosis, *Rhizobium* bacteria colonize the roots of legume plants and form nodules, where nitrogenase enzymes catalyze the conversion of atmospheric nitrogen into ammonia. The host plant utilizes this ammonia as a nitrogen source, while providing the bacteria with carbohydrates derived from photosynthesis (Oldroyd et al., 2011; Lindström & Mousavi, 2019). This mutualistic interaction has been widely utilized in sustainable agriculture to reduce dependency on chemical fertilizers and improve soil fertility.

13.1.1 Limitations and Challenges in Large-Scale Application

Despite its potential, the large-scale application of BNF faces several challenges. The efficiency of nitrogen fixation varies significantly among different *Rhizobium* strains and legume species. Environmental factors such as soil pH, temperature, nutrient availability, and competition with native soil microbes can adversely affect nodulation and nitrogen fixation efficiency. Moreover, introducing high-performing *Rhizobium* strains into natural ecosystems may not always guarantee successful colonization or nitrogen fixation.

Recent advancements, including the development of genetically improved *Rhizobium* strains, genome editing of legumes, and improved soil management practices, hold promise in overcoming

DOI: 10.1201/9781003608080-13

these challenges (Garrido-Oter et al., 2021). However, further research is required to optimize these strategies for field-level application and long-term sustainability.

13.1.2 Importance of Nitrogen in Agriculture

Nitrogen is essential for plant growth, particularly because it forms the backbone of amino acids, the building blocks of proteins (Table 13.1). Nitrogen is also a crucial component of chlorophyll, the molecule responsible for photosynthesis. Its availability directly influences plant vigor, yield, and quality. However, synthetic nitrogen fertilizers, while effective, contribute to environmental degradation, such as water pollution and greenhouse gas emissions.

Nitrogen (N) is one of the most essential macronutrients for plant growth and crop production. It plays a fundamental role in the synthesis of key biomolecules such as proteins, nucleic acids, chlorophyll, and enzymes, directly influencing plant productivity and yield. Although nitrogen is abundantly available in the atmosphere, accounting for about 78% of air, its molecular form (N_2) is inert and unavailable for direct uptake by most plants. Consequently, nitrogen deficiency is commonly recognized as a critical limiting factor in agricultural systems worldwide (Griesmann et al., 2018).

To meet the increasing food demand driven by global population growth, external nitrogen inputs have become necessary in modern agriculture. However, the excessive use of synthetic nitrogen fertilizers has led to severe environmental challenges, including soil degradation, water pollution, and greenhouse gas emissions (Ribaudo et al., 2011). Therefore, enhancing BNF represents a sustainable alternative to reduce reliance on chemical fertilizers while maintaining high crop yields.

Long-term experiments, such as those conducted at Rothamsted Research in the United Kingdom, have demonstrated that nitrogen fertilization can increase the yields of cereal crops like wheat and maize by two to three times compared to crops grown without nitrogen supplementation (Dobermann & Cassman, 2002, 2004). In this context, BNF, primarily mediated by nitrogen-fixing

TABLE 13.1
Role of Nitrogen in Plant Growth along with Bacteria Involved, Their Roles, and Benefits to Plants.

S. No.	Component	Bacteria Responsible	Role in Nitrogen Cycle	Benefits to Plants
1	Nitrogen fixation	*Rhizobium*, *Azotobacter*, *Frankia*	Convert atmospheric N_2 into NH_3	Provides usable nitrogen (NH_3) for plant growth
2	Nitrification	*Nitrosomonas*, *Nitrobacter*	Convert ammonia (NH_3) to nitrite (NO_2^-), and then nitrate (NO_3^-)	Provides nitrate (NO_3^-), the primary nitrogen source for plants
3	Ammonification	*Bacillus*, *Clostridium*	Decomposition of organic matter into ammonia	Recycles nitrogen in organic matter for plant uptake
4	Denitrification	*Pseudomonas*, *Clostridium*	Convert nitrate (NO_3^-) back to N_2 gas	Maintains nitrogen balance in the ecosystem, prevents excess accumulation of nitrates
5	Symbiotic nitrogen fixation	*Rhizobium*, *Bradyrhizobium*	Form symbiotic relationships with legume roots, fixing nitrogen	Enhanced plant growth, higher crop yields, and improved soil fertility
6	Nonsymbiotic nitrogen fixation	*Azospirillum*, *Azotobacter*	Free-living bacteria that fix nitrogen independently	Improve soil nitrogen content, supporting plant nutrition

bacteria such as *Rhizobium* spp., offers an ecologically viable solution. These bacteria possess the ability to convert atmospheric nitrogen into ammonia (NH_3), a form readily assimilated by plants, thereby promoting sustainable crop production.

13.1.3 Overview of Symbiotic Relationships in Nature

Symbiotic relationships are specialized associations between two different organisms, which can be classified into mutualism, commensalism, and parasitism. Among these, mutualistic symbiosis provides mutual benefits to both partners, often enhancing their survival, growth, or reproduction (Table 13.2). A prominent example of such an interaction in nature is the symbiotic association between plants and soil microorganisms, including nitrogen-fixing bacteria and mycorrhizal fungi (Wang et al., 2012).

The symbiotic interaction between legumes and *Rhizobium* bacteria serves as a well-established model of mutualism in sustainable agriculture. In this relationship, *Rhizobium* species colonize the roots of leguminous plants, leading to the formation of specialized structures called root nodules. Within these nodules, nitrogenase enzymes produced by the bacteria facilitate the conversion of atmospheric nitrogen (N_2) into ammonia (NH_3), which is utilized by the host plant. In exchange, the bacteria receive carbon compounds derived from plant photosynthesis, supporting their growth and metabolism (Oldroyd et al., 2011).

This symbiotic process is initiated when legume roots secrete flavonoids into the rhizosphere, acting as chemical signals that attract compatible *Rhizobium* strains. In response, the bacteria produce Nod factors (NFs), signaling molecules essential for the initiation of nodule development. These NFs trigger the activation of specific plant signaling pathways, including the common symbiosis signaling pathway (CSSP), which regulates both nitrogen-fixing symbiosis and arbuscular mycorrhizal (AM) symbiosis (Oldroyd, 2013).

Notably, genes involved in nodulation, often carried on symbiotic plasmids, can be horizontally transferred among *Rhizobium* species, enhancing their adaptability and competitiveness in diverse soil environments (Remigi et al., 2016). Additionally, the *noeI* gene, initially considered nonessential for symbiosis, has been found to significantly influence nitrogen fixation efficiency in soybean root nodules (Liu et al., 2018).

Recent research highlights the complexity of the *Rhizobium*–legume symbiosis, emphasizing the roles of molecular signaling, root development programs, and environmental factors in successful nodule formation and nitrogen fixation (Suzaki & Kawaguchi, 2014; Poole et al., 2018; Soyano et al., 2021). Variations in plant *CSSP* genes can also influence the composition of the root microbiome, further affecting symbiotic efficiency (Zgadzaj et al., 2016).

Overall, the *Rhizobium*–legume symbiotic relationship exemplifies a highly evolved mutualistic interaction with profound implications for sustainable agriculture. Harnessing this natural nitrogen-fixing system offers great potential to reduce chemical fertilizer use, improve soil health, and promote environmentally friendly farming practices (Table 13.2).

13.1.4 Significance of *Rhizobium*–Legume Symbiosis

The symbiotic association between legumes and *Rhizobium* species represents a fundamental natural mechanism that facilitates BNF, significantly contributing to sustainable agriculture. This interaction enables legumes to form specialized root nodules, within which atmospheric N_2 is biologically converted into NH_3, an accessible nitrogen source for plant growth. In exchange, legumes supply essential carbohydrates and a favorable environment for the proliferation of *Rhizobium* bacteria (Verma et al., 2020).

Globally, BNF is estimated to contribute approximately 139–175 million tons of fixed nitrogen annually, with *Rhizobium*–legume symbiosis accounting for nearly 35–44 million tons (Verma et al., 2020). This substantial nitrogen input reduces the dependence on synthetic nitrogen

TABLE 13.2
Symbiotic Relationships in Nature.

S. No.	Symbiotic Relationship	Organisms Involved	Type of Symbiosis	Benefit to Each Organism
1	*Rhizobium*–legume symbiosis	*Rhizobium* bacteria and legumes	Mutualism	Bacteria fix nitrogen for plant; plant provides nutrients
2	Mycorrhizal association	Fungi and plant roots	Mutualism	Fungi enhance nutrient absorption; plants supply carbohydrates
3	Lichen	Algae (or cyanobacteria) and fungi	Mutualism	Algae provide food through photosynthesis; fungi offer protection
4	Oxpecker and large herbivores	Oxpecker birds and herbivores	Mutualism/ parasitism	Birds eat parasites off herbivores; sometimes feed on their blood

fertilizers, lowering production costs and minimizing environmental hazards associated with chemical fertilization.

Beyond nitrogen fixation, *Rhizobium* species also exhibit multiple plant growth-promoting (PGP) traits. These include the production of phytohormones such as indole-3-acetic acid (IAA), gibberellic acid (GA), and cytokinins, which stimulate plant growth and development. Additionally, *Rhizobium* enhances phosphate solubilization, alleviates ethylene-induced stress through 1-aminocyclopropane-1-carboxylate (ACC) deaminase activity, and improves soil health (Wani et al., 2016). These multifaceted attributes not only benefit plant nutrition but also promote agro-ecosystem sustainability.

Legume crops, including soybean (*Glycine max*), lentil (*Lens culinaris*), and alfalfa (*Medicago sativa*), are commonly integrated into crop rotation systems to enhance soil fertility. Thus, *Rhizobium*–legume symbiosis forms the ecological backbone of sustainable agricultural practices, contributing to food security and environmental conservation.

13.2 BIOLOGY OF *RHIZOBIUM* BACTERIA

The biology of *Rhizobium* is central to understanding its symbiotic efficiency and nitrogen-fixing ability. This section focuses on the classification, morphological characteristics, life cycle, and genetic mechanisms that underpin the symbiosis between *Rhizobium* and legumes.

13.2.1 Classification and Characteristics

Rhizobium species are Gram-negative, rod-shaped, motile bacteria belonging to the family Rhizobiaceae within the class Alphaproteobacteria. These bacteria are characterized by their ability to induce root nodule formation on leguminous plants, facilitating nitrogen fixation under symbiotic conditions (Masson-Boivin et al., 2009).

The classification of *Rhizobium* is primarily based on their host range specificity, nodulation capability, and nitrogen-fixing efficiency. Table 13.3 summarizes the classification of *Rhizobium* species based on their legume hosts and symbiotic characteristics.

13.2.1.1 Morphological and Physiological Traits

Rhizobium bacteria are motile *via* polar flagella and thrive in diverse environments using aerobic respiration. They tolerate variations in soil pH and temperature, making them adaptable to different ecosystems.

TABLE 13.3
Classification of *Rhizobium*.

Taxonomic Rank	Information
Kingdom	Bacteria
Phylum	Proteobacteria
Class	Alphaproteobacteria
Order	Rhizobiales
Family	Rhizobiaceae
Genus	*Rhizobium*
Species	*Rhizobium leguminosarum*, *R. etli*, *R. phaseoli*, etc.

TABLE 13.4
Genetic Elements in *Rhizobium*–Legume Symbiosis.

Genetic Element	Function
Nod genes	Synthesize NFs for nodule formation
Fix genes	Regulate nitrogenase activity for N fixation
Sym plasmids	Harbor nodulation and nitrogen fixation genes

13.2.2 Life Cycle and Growth Conditions

The life cycle of *Rhizobium* begins in the soil, where free-living cells detect plant-secreted flavonoids, triggering NF production. This signaling initiates root colonization and nodule formation.

Key Lifecycle Stages:

1. **Free-living Stage:** Bacteria survive in the soil.
2. **Root Colonization:** *Rhizobium* attaches to root hairs, inducing curling and infection thread formation.
3. **Nodule Formation:** Bacteria enter the root cortex, stimulating nodule development.
4. **Nitrogen Fixation:** Within nodules, *Rhizobium* differentiates into bacteroids that fix atmospheric nitrogen into ammonia.

Optimal growth conditions include neutral to slightly alkaline soils (pH ~6.5–7.5) and temperatures between 25°C and 30°C, with root exudates serving as a primary carbon source.

13.2.3 Genetic Mechanisms of Symbiosis

The *Rhizobium*–legume symbiosis is governed by genetic elements that regulate nodulation and nitrogen fixation (Table 13.4).

Key Genetic Components:

- **Nod Genes:** Encode NFs that trigger nodule formation, activated by NodD proteins in response to plant flavonoids.

The establishment of symbiotic associations between host plants and microbial partners such as rhizobia and mycorrhizal fungi involves intricate signaling mechanisms regulated by overlapping genes and pathways (Oldroyd, 2013). Both types of symbionts release specialized signaling molecules known as lipo-chitooligosaccharides (LCOs). These are referred to as NFs in rhizobia and Myc factors in mycorrhizal fungi, which are recognized by specific LysM-type receptor kinases present on host plant roots (Chabaud et al., 2002).

Mycorrhizal symbiosis represents an ancient form of plant–microbe interaction, primarily involving fungi from the phylum Glomeromycota (Oldroyd, 2013). In contrast, rhizobial symbiosis is a relatively recent evolutionary adaptation, restricted to leguminous plants and a few species such as *Parasponia* (Madsen et al., 2003).

The structural and functional similarity between Myc and NF signaling pathways suggests that NF signaling may have evolved as a modification of ancestral Myc factor pathways (Streng et al., 2011). NFs are composed of 3–5 β(1–4)-linked N-acetylglucosamine residues, often modified with a fatty acid chain and various functional groups (Mylona et al., 1995). These structures closely resemble chitin, a component of fungal cell walls that typically triggers plant immune responses (Pusztahelyi, 2018). However, NFs are distinguished by their shorter oligosaccharide chains. In the well-studied *Sinorhizobium meliloti–Medicago truncatula* symbiosis, NFs are perceived by specific LysM receptor kinases such as MtNFP and MtLYK3, which mediate symbiotic signal transduction (Oldroyd, 2013).

13.2.4 Role of Reactive Oxygen Species in Nodulation

NF signaling not only initiates symbiotic interactions but also induces defense-like responses in host plants, including the production of reactive oxygen species (ROS) such as hydrogen peroxide (H_2O_2). This transient ROS generation has been observed particularly around the root hair tips of *Medicago truncatula* following NF exposure (Cardenas et al., 2008). However, excessive accumulation of ROS may hinder symbiosis. Mutants of *S. meliloti* that overexpress catalase, an ROS-detoxifying enzyme, showed delayed nodulation and abnormal infection thread development, indicating the necessity of maintaining optimal ROS levels for effective nodulation (Jamet et al., 2007). This highlights the dual role of ROS in symbiosis: While initially acting as stress signals, controlled ROS production facilitates infection thread stabilization and successful bacterial entry.

13.2.5 Nod Factor Recognition and Signal Perception

The establishment of *Rhizobium*–legume symbiosis is primarily initiated by the secretion of flavonoids from the roots of legume plants into the rhizosphere. These flavonoids, which are secondary metabolites synthesized *via* the phenylpropanoid pathway, act as key signaling molecules, particularly under nitrogen-deficient conditions, to stimulate the production of NFs (LCOs) in rhizobia (Subramanian et al., 2007). The interaction of flavonoids with NodD, a LysR-type transcriptional regulator in rhizobia, activates the expression of nodulation genes such as *nodA*, *nodB*, and *nodC*, which subsequently leads to the synthesis of NFs (Long, 1996).

In model leguminous species like *Medicago truncatula* and *Lotus japonicus*, NFs are perceived by specific receptor complexes comprising NF perception proteins (MtNFP or LjNFR5) and LysM domain receptor-like kinases (MtLYK3 or LjNFR1) (Limpens et al., 2003). These receptors are characterized by extracellular LysM motifs, a single transmembrane domain, and intracellular kinase domains. Although MtNFP and LjNFR5 function as pseudokinases with inactive kinase domains, the kinase activity of MtLYK3 and LjNFR1 is essential for the transduction of NF signals, ultimately leading to successful nodule formation.

13.2.5.1 Engineering Symbiotic Traits

Recent advancements in receptor engineering have provided new opportunities to extend symbiotic traits to non-legume species. For example, the integration of extracellular domains from MtNFP and MtLYK3 into rice LysM receptors (OsMYR1 and OsCERK1) successfully restored nodulation phenotypes in *M. truncatula* mutants, demonstrating the feasibility of engineering nitrogen-fixing abilities in cereal crops (He et al., 2019). Moreover, receptor-like cytoplasmic kinases (RLCKs) have emerged as essential components in NF signaling. In *Lotus japonicus*, RLCK LjNiCK4 interacts with and phosphorylates NF receptors LjNFR1 and LjNFR5, linking NF detection to downstream signaling pathways that facilitate nodule organogenesis (Wong et al., 2019).

13.2.5.2 Alternative Symbiosis Mechanisms

Although NFs play a central role in most legume–*Rhizobium* symbioses, certain legume species have evolved NF-independent mechanisms. For instance, *Aeschynomene indica* establishes symbiosis with photosynthetic *Bradyrhizobium* strains lacking typical nodulation genes. In this case, alternative signaling molecules such as cytokinin-like compounds are proposed to trigger nodule formation (Giraud et al., 2007). Additionally, the type III secretion system (T3SS) in *Bradyrhizobium* USDA61 enables the direct activation of the NF signaling cascade in soybean, bypassing the requirement of NFs (Okazaki et al., 2013). Interestingly, in *Aeschynomene evenia*, a cysteine-rich receptor-like kinase (AeCRK) has been identified as critical for NF-independent symbiosis (Quilbé et al., 2021).

These insights into diverse mechanisms of symbiosis highlight the potential for developing novel strategies to enhance BNF in nonleguminous crops, thus contributing to sustainable agriculture and improving soil health.

13.3 MECHANISM OF SYMBIOSIS

The symbiotic relationship between *Rhizobium* species and legume plants represents a well-characterized and ecologically significant interaction that benefits both partners. Legumes gain access to a reliable nitrogen source essential for their growth and productivity, while *Rhizobium* bacteria receive carbohydrates and a protective niche within plant tissues (Zhu et al., 2022). The complex sequence of events in this symbiotic association involves multiple stages, including host recognition, bacterial infection, root nodule development, nitrogen fixation, and the involvement of diverse signaling molecules (Li et al., 2023). A schematic representation of this mechanism is illustrated in Figure 13.1.

13.3.1 Host Recognition and Infection Process

The symbiotic association between *Rhizobium* bacteria and leguminous plants begins with a sophisticated mechanism of recognition and communication. Initially, legume roots secrete specific flavonoids into the rhizosphere, which act as chemical attractants for *Rhizobium* strains present in the surrounding soil (Wang et al., 2021). These flavonoid signals vary across different legume species, thereby promoting selective interaction with compatible *Rhizobium* strains (Zhang et al., 2023).

Upon sensing these flavonoids, *Rhizobium* bacteria activate their nodulation (*nod*) genes, leading to the synthesis of NFs, which are LCO signaling molecules. These NFs are essential in initiating the infection process by inducing morphological and physiological changes in the host plant. They promote root hair curling, localized degradation of the cell wall, and formation of infection threads—tube-like structures that facilitate bacterial entry into the root tissue (Xu et al., 2022). Through these infection threads, *Rhizobium* bacteria migrate toward the root cortex, where they begin colonization and eventually induce nodule formation (Liu et al., 2023). This highly regulated infection process ensures a mutualistic and nonpathogenic relationship between the host plant and *Rhizobium* (Chen et al., 2022).

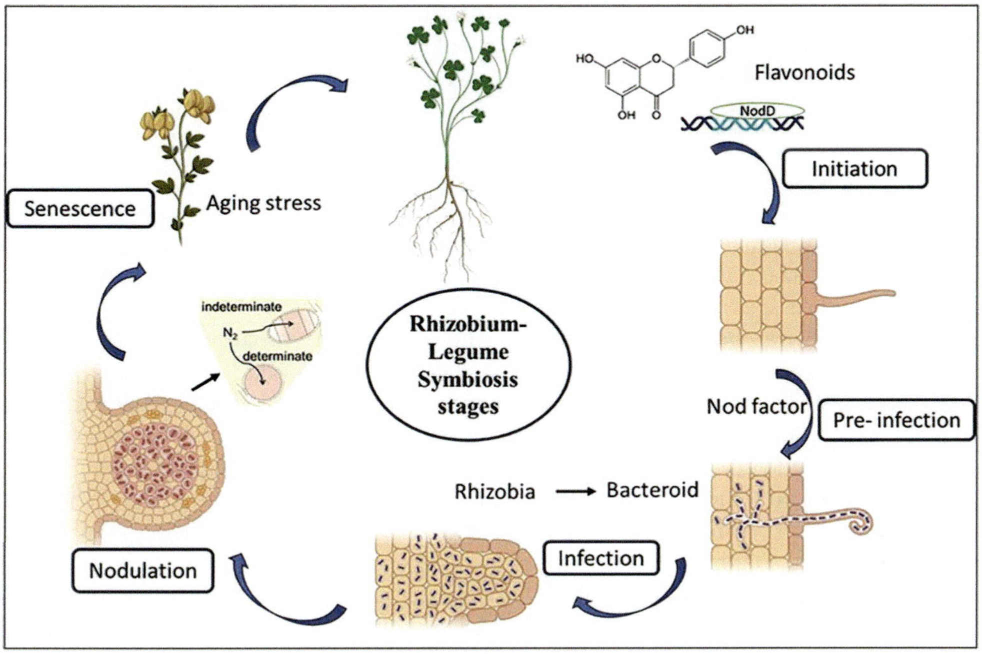

FIGURE 13.1 Stages of *Rhizobium*–legume symbiosis.

13.3.2 Formation of Root Nodules

The formation of root nodules is a complex developmental process that requires precise coordination between the plant and *Rhizobium* bacteria. Following bacterial entry into the root cortex *via* infection threads, the plant responds by initiating localized cell divisions, leading to the formation of a nodule primordium—the early developmental stage of a root nodule (Tian et al., 2021).

Within the developing nodules, specialized plant cells create a microenvironment conducive to *Rhizobium* proliferation. The bacteria are encapsulated within membrane-bound structures called symbiosomes, where they differentiate into bacteroids capable of fixing atmospheric nitrogen (Liu et al., 2023). The development and functioning of nodules are regulated by the coordinated expression of genes from both symbiotic partners. Plant hormones such as auxins, cytokinins, and ethylene play crucial roles in modulating nodule organogenesis. Moreover, *Rhizobium*-derived signaling molecules activate plant-specific transcription factors like NIN (Nodule Inception), which further govern the structural and functional development of nodules.

Despite these advances, challenges like strain specificity, environmental stress conditions, and competition with native soil microbiota often limit the efficiency of nodulation under field conditions.

13.3.3 Nitrogen Fixation Process: Biochemical Pathways

Nitrogen fixation is the most vital process within the *Rhizobium*–legume symbiosis. Although atmospheric nitrogen (N_2) is abundant, most plants cannot utilize it directly. However, *Rhizobium* bacteria, with the help of the nitrogenase enzyme complex, convert atmospheric N_2 into ammonia (NH_3), which plants can readily assimilate for their growth and metabolism (Wagner et al., 2023).

The nitrogenase enzyme complex consists of two major proteins: The iron (Fe) protein and the molybdenum-iron (MoFe) protein. This ATP-dependent process also requires reducing power, generally provided by ferredoxin or flavodoxin (Zhou et al., 2021).

A significant challenge in nitrogen fixation is that the nitrogenase enzyme is extremely sensitive to oxygen. Therefore, the plant produces leghemoglobin—an oxygen-binding protein—that regulates oxygen concentration within the nodule, maintaining conditions optimal for nitrogenase activity while allowing respiration (Gupta et al., 2021).

13.3.4 Role of Nod Factors and Signaling Molecules

NFs represent crucial signaling molecules that mediate the early stages of symbiotic interaction between *Rhizobium* bacteria and leguminous plants. These molecules are LCOs synthesized and secreted by *Rhizobium* species in response to specific flavonoids exuded by legume roots (García et al., 2022). NFs play a pivotal role in host-specific recognition, initiating a series of signaling events that ultimately lead to successful infection and nodule formation.

The perception of NFs by leguminous plants occurs through LysM domain-containing receptor-like kinases (RLKs) located on the root epidermal cells. Upon binding of NFs, these receptors activate downstream calcium spiking and the subsequent activation of symbiosis-specific transcription factors, including NIN and ERN (Ethylene Responsive Factor Required for Nodulation), which govern gene expression for infection thread formation and cortical cell division (Wu & Zhang, 2023; Singh & Kumar, 2023).

Additionally, plant hormones, particularly auxins, cytokinins, and ethylene, regulate various stages of nodulation. Cytokinins promote cell division within the root cortex to facilitate nodule primordium formation, while auxins contribute to root hair deformation and infection thread development. Conversely, ethylene often acts as a negative regulator, restricting excess nodule formation and maintaining symbiotic balance (Muller et al., 2021; Zhou et al., 2022).

In addition to NFs, other bacteria-derived molecules, such as exopolysaccharides (EPS) and lipopolysaccharides (LPS), play supportive roles in modulating plant immune responses. These molecules assist in suppressing plant defense mechanisms, enabling successful colonization and establishment of a stable symbiotic relationship (Muller et al., 2021; Yang et al., 2023).

However, the efficiency of NF signaling can be influenced by environmental conditions such as soil pH, temperature, nutrient availability, and the presence of incompatible microbial strains, which may hinder optimal symbiotic performance (Khan et al., 2022).

The *Rhizobium*–legume symbiosis thus represents a highly coordinated and dynamic molecular dialog essential for sustainable nitrogen fixation. This interaction not only enhances plant growth but also contributes to environmentally friendly agricultural practices by reducing dependency on chemical nitrogen fertilizers (Miyamoto & Fukuhara, 2021). Consequently, understanding the intricacies of NF signaling and plant–microbe interactions is vital for improving BNF and advancing sustainable agriculture (Qiu et al., 2021).

13.4 IMPACT ON LEGUME PHYSIOLOGY AND GROWTH

13.4.1 Nutrient Uptake and Utilization

A critical aspect of legume productivity is the symbiotic relationship between *Rhizobium* bacteria and leguminous plants, which significantly enhances nitrogen uptake and utilization. This symbiotic association facilitates BNF, wherein *Rhizobium* bacteria colonize the root nodules of legumes and convert atmospheric nitrogen (N_2) into ammonia (NH_3), a form readily assimilable by plants (Santos et al., 2023). This natural process substantially reduces the dependence on synthetic nitrogen fertilizers, promoting environmentally sustainable agricultural practices.

Nitrogen fixation commences when *Rhizobium* bacteria infect the root hairs of legumes, initiating nodule formation (Nguyen et al., 2022). Within these nodules, nitrogen is reduced to ammonium (NH_4^+), which is then utilized by plants for the synthesis of essential biomolecules, including amino acids, nucleotides, and proteins (Zhao & Hu, 2021). The enhanced availability of nitrogen significantly improves the nutritional profile of legumes, supporting their optimum growth and development (Liu et al., 2021).

Moreover, this symbiotic association promotes the absorption of other vital nutrients, such as phosphorus, potassium, and micronutrients. The increased nitrogen supply stimulates extensive root development, enhancing the plant's ability to explore a larger soil volume for nutrient acquisition (Singh & Verma, 2023). In addition, *Rhizobium*–legume symbiosis can upregulate nutrient transporter genes, thereby improving nutrient uptake efficiency (Patel et al., 2023).

This interaction is particularly beneficial in nutrient-deficient soils, enabling legumes to thrive under suboptimal conditions while contributing to long-term soil fertility improvement (Xie & Ma, 2021). Furthermore, the decomposition of root nodules and associated organic matter enhances soil nutrient cycling, supporting the growth of subsequent crops (Li et al., 2022).

13.4.2 Enhanced Plant Growth and Development

In addition to nutrient acquisition, the *Rhizobium*–legume symbiosis profoundly influences overall plant growth and development. The increased nitrogen availability ensures the synthesis of critical biomolecules such as proteins, enzymes, and chlorophyll, which are essential for plant physiological processes and growth.

One of the most evident outcomes of this symbiosis is enhanced biomass production. Symbiotic legumes often exhibit greater shoot and root biomass compared to their nonsymbiotic counterparts due to the efficient nitrogen supply (Singh & Verma, 2023). Additionally, increased leaf area and chlorophyll content lead to improved photosynthetic capacity, supplying the energy required for accelerated plant growth.

Rhizobial interaction also positively affects reproductive parameters, including flowering and seed development. Symbiotic legumes tend to flower earlier and produce more abundant floral structures, leading to improved yield quality and quantity—an essential trait in both natural ecosystems and agricultural production systems (Nguyen et al., 2022).

The symbiosis further promotes the development of a well-branched and extensive root architecture. The formation of root nodules not only facilitates nitrogen fixation but also enhances water and nutrient uptake, improves root anchorage, and provides greater resilience against environmental challenges.

Moreover, the *Rhizobium*–legume interaction modulates the plant hormonal profile by elevating the levels of key phytohormones such as gibberellins, cytokinins, and auxins. These hormones regulate critical physiological processes, including cell division, elongation, and differentiation, thereby promoting plant vigor, enhanced growth, and higher yield potential.

13.4.3 Resistance to Environmental Stresses

One of the most significant advantages of *Rhizobium*–legume symbiosis is its contribution to enhanced resistance to various environmental stresses. With the increasing global challenges like climate change, water scarcity, and soil degradation, the ability of leguminous plants to withstand such stresses is important for ensuring food security and ecological stability (Sharma & Verma, 2022). One of the most frequent environmental pressures on legume crops is drought. It has been demonstrated that symbiotic legumes are more drought-tolerant than nonsymbiotic plants (Singh et al., 2021). Legumes are able to obtain deeper soil moisture, thanks to the improved root system that results from *Rhizobium* infection, which acts as a buffer against water stress (Liu et al., 2023). Furthermore, even in situations of severe drought, nitrogen fixation persists, guaranteeing that the plant has a constant supply of nitrogen—a necessity for survival and recovery both during and after drought periods (Kumar et al., 2022). Another environmental stressor that seriously hinders the growth of legumes is salinity, especially in dry and semi-arid areas. Legumes that contain *Rhizobium* can lessen the negative effects of salt in the soil (Patel & Singh, 2023). *Rhizobium*–legume symbiosis enhances ion homeostasis in the plant, regulating the uptake and transport of harmful ions such as sodium (Na^+) and chloride (Cl^-), which can disrupt cellular function (Chen et al., 2022). Furthermore, symbiotic plants are better equipped to maintain osmotic balance, reducing the physiological stress caused by high salinity levels (Khan & Ali, 2023). Temperature extremes, both high and low, can also negatively impact plant growth and productivity. Symbiotic legumes are more resilient to temperature fluctuations (Zhou & Li, 2022). The presence of *Rhizobium* helps stabilize cellular functions under heat stress by maintaining enzyme activity and protecting against protein denaturation. Similarly, under cold stress, *Rhizobium*-infected legumes exhibit better root growth and nitrogen uptake, which helps sustain metabolic processes despite low temperatures (Xie & Liu, 2021).

Rhizobium–legume symbiosis confers tolerance not only against abiotic challenges but also against biotic pressures like infections and pests. *Rhizobium* enhances plant defense mechanisms by promoting systemic resistance, a state where the plant's immune system is primed for a stronger response to pathogens (Jain & Gupta, 2022). Certain *Rhizobium* strains also produce antimicrobial compounds that inhibit harmful soil microorganisms. This symbiotic relationship profoundly influences legume physiology and productivity by improving nutrient uptake, fostering growth, and increasing resilience to environmental stresses (Patel et al., 2023). Leveraging this natural partnership can significantly contribute to sustainable agriculture, enabling farming systems to adapt to changing environmental conditions (Chen & Wang, 2023).

13.5 AGRONOMIC BENEFITS AND CROP YIELD

The symbiotic association between legumes and *Rhizobium* bacteria plays a crucial role in sustainable agriculture by enabling BNF, a natural process that converts atmospheric nitrogen into a plant-usable form. This interaction not only enhances the growth and productivity of legume crops but also provides significant agronomic benefits, such as improving soil fertility, reducing dependency on synthetic nitrogen fertilizers, and increasing the yield potential of subsequent crops in crop rotation systems (Peoples et al., 2009).

13.5.1 Improvement in Soil Fertility

Soil fertility is a critical factor influencing agricultural productivity, as it determines the soil's ability to supply essential nutrients to plants. In legume–*Rhizobium* symbiosis, the nitrogen-fixing capability of *Rhizobium* bacteria significantly enhances soil nitrogen content, particularly in nitrogen-deficient soils. This natural enrichment of soil nitrogen reduces the need for chemical fertilizers, supporting eco-friendly and cost-effective farming practices (Shah et al., 2021).

13.5.1.1 Biological Nitrogen Fixation Process

The process of BNF begins with the formation of specialized root nodules in legumes, where *Rhizobium* bacteria reside. These bacteria utilize the nitrogenase enzyme to convert inert atmospheric nitrogen gas (N_2) into ammonia (NH_3), a form that plants can readily assimilate (Telles et al., 2023) (Figure 13.2). The fixed nitrogen is then incorporated into essential organic compounds like amino acids, proteins, and nucleotides, facilitating the growth and development of leguminous plants.

In addition to benefiting the host legume, the BNF process contributes to soil nitrogen enrichment. When legume residues decompose after harvesting, the organic matter returns nitrogen to the soil, increasing its fertility for subsequent crops (Table 13.5). This makes legumes valuable components in crop rotation and intercropping systems, where they enhance the nitrogen status of the soil, benefiting nonleguminous crops grown afterward.

13.5.1.2 Soil Fertility Enhancement through Residual Nitrogen

One of the significant agronomic benefits of nitrogen-fixing legumes is the enhancement of soil fertility through the residual nitrogen left in the soil after the harvest of legume crops. This residual nitrogen, present in microbial biomass, decomposed plant residues, and root exudates, becomes

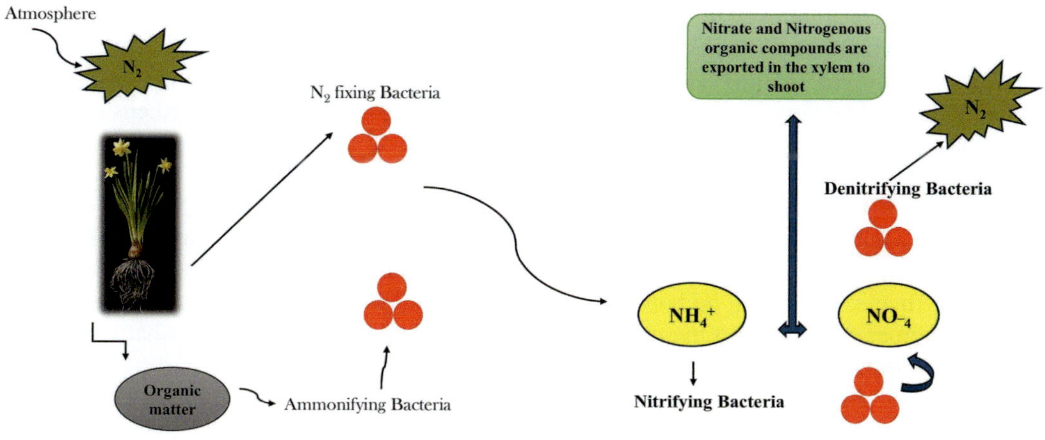

FIGURE 13.2 BNF process.

TABLE 13.5
Contribution of Nitrogen Fixation to Soil Fertility.

Crop	Estimated N Fixed (kg/ha/year)	Residual N Contribution to Soil (kg/ha)
Soybean	60–150	30–50
Pea	50–125	25–45
Alfalfa	150–250	80–100
Chickpea	40–100	20–30

available for subsequent crops grown in rotation. Nonleguminous crops particularly benefit from this residual nitrogen, reducing the need for external nitrogen fertilizer applications and supporting more sustainable nutrient management practices (Yuvaraj et al., 2020).

13.5.1.3 Contribution to Soil Organic Matter

In addition to nitrogen enrichment, legumes also play a vital role in increasing the organic matter content of the soil. The decomposition of their roots and above-ground biomass releases organic substances that improve soil structure, enhance moisture retention, and promote the proliferation of beneficial soil organisms, including earthworms, bacteria, and fungi. This increase in microbial biomass and activity facilitates nutrient cycling processes and contributes to maintaining long-term soil fertility and productivity (Shabtai et al., 2023).

13.5.1.4 Reduction in Soil Acidification

A key environmental advantage of utilizing *Rhizobium*–legume symbiosis for nitrogen fixation is the mitigation of soil acidification. Unlike synthetic nitrogen fertilizers such as ammonium sulfate and urea, which often lower soil pH and lead to soil acidification over time, BNF generates nitrogen in a manner that minimizes pH reduction. This natural nitrogen input decreases the dependence on liming materials typically required to neutralize soil acidity resulting from prolonged chemical fertilizer use (Zhao et al., 2021).

13.5.1.5 Benefits of Nutrient Cycling and Microbial Diversity

Incorporating legumes into agricultural systems significantly enhances nutrient cycling by improving the availability of nitrogen and other essential nutrients like phosphorus (P) and potassium (K).

Legume roots interact with soil microorganisms to solubilize phosphorus, which is often present in unavailable forms in the soil. Additionally, legumes foster microbial diversity by creating favorable conditions for nitrogen-fixing bacteria, mycorrhizal fungi, and other beneficial microorganisms. This diverse microbial population not only enhances nutrient uptake but also strengthens soil health and resilience (Yadav et al., 2021) (Table 13.6).

13.5.1.6 Sustainable Agricultural Systems

In sustainable agriculture, the role of *Rhizobium*–legume symbiosis in improving soil fertility is crucial. Nitrogen-fixing legumes provide a natural source of nitrogen, reducing dependence on synthetic fertilizers, which can be both costly and environmentally harmful. By replenishing soil nutrients, legumes contribute to the creation of more resilient cropping systems that can sustain productivity without depleting the soil (Mnalku et al., 2020).

13.5.2 Increased Yield in Legume Crops

By giving the plants a consistent source of nitrogen, legumes' symbiotic association with *Rhizobium* bacteria allows them to fix nitrogen, which has a direct effect on crop yields (Table 13.7). Increased biomass and seed production are the results of this nitrogen's support for the synthesis of proteins, chlorophyll, and other essential growth factors in legume crops (Goyal et al., 2021).

13.5.2.1 Impact of Nitrogen Fixation on Legume Growth

Nitrogen is a vital nutrient required for the synthesis of amino acids, proteins, enzymes, and nucleic acids, all essential for plant growth and development. Unlike nonleguminous crops that depend on

TABLE 13.6
Impact of *Rhizobium*–Legume Symbiosis on Soil Fertility.

Aspect of Soil Fertility	Impact of *Rhizobium*–Legume Symbiosis
Nitrogen availability	Increases nitrogen levels through BNF, reducing the need for synthetic fertilizers
Organic matter	Enhances soil organic content through root biomass and plant residue decomposition
Soil structure	Improves soil aggregation and reduces compaction
Microbial activity	Promotes the growth of beneficial microorganisms such as nitrogen-fixers and fungi
Soil pH	Reduces soil acidification compared to chemical nitrogen fertilizers
Nutrient cycling	Enhances phosphorus solubilization and overall nutrient cycling

TABLE 13.7
Yield Increases in Legume Crops due to *Rhizobium*-Induced Nitrogen Fixation.

Legume Crop	Yield without Inoculation (kg/ha)	Yield with Inoculation (kg/ha)	% Increase in Yield
Soybean	2,500	3,100	24%
Chickpea	2,800	2,200	22%
Lentil	1,200	1,500	25%
Faba bean	3,000	1,600	20%

soil nitrogen or fertilizers, legumes benefit from their symbiotic relationship with *Rhizobium* bacteria, enabling them to fix atmospheric nitrogen. This continuous nitrogen supply throughout the growing season offers several significant advantages for legume crops:

- **Enhanced Vegetative Growth:** Nitrogen plays a key role in producing proteins and enzymes that stimulate cell division and elongation, contributing to robust vegetative growth.
- **Improved Chlorophyll Content:** As a fundamental component of chlorophyll, nitrogen increases the chlorophyll concentration in leaves, enhancing photosynthesis and biomass production.
- **Increased Flowering and Pod Formation:** Adequate nitrogen supply promotes reproductive development, resulting in a higher number of flowers, pods, and seeds.
- **Higher Seed Protein Content:** Nitrogen fixation improves the protein content of legume seeds, making crops like soybean, chickpea, and lentil valuable protein sources for human and animal consumption.

13.5.2.2 Factors Influencing Yield in Legume Crops

The yield of legume crops through *Rhizobium*-mediated nitrogen fixation depends on multiple factors, which collectively determine the efficiency of nitrogen fixation and overall crop productivity:

- **Efficiency of *Rhizobium* Strains:** The nitrogen-fixing ability of *Rhizobium* varies with strain type. Utilizing highly efficient strains through seed inoculation can significantly enhance nodulation, nitrogen fixation, and ultimately crop yield, particularly in soils deficient in native *Rhizobium* populations.
- **Soil Conditions:** Soil pH, texture, aeration, and organic matter content greatly influence the success of nitrogen fixation. Optimal conditions promote better nodule formation. Conversely, acidic or compacted soil can hinder the symbiotic process, reducing nitrogen fixation and yield potential.
- **Nodulation:** The quantity and health of nodules on legume roots directly correlate with nitrogen fixation. More nodules typically translate into greater nitrogen availability and higher yields. Seed inoculation with compatible *Rhizobium* strains ensures optimal nodulation, especially in soils lacking native populations.
- **Environmental Conditions:** Factors like temperature, soil moisture, rainfall, and sunlight intensity affect the nitrogen fixation process. Extreme environmental stress, such as drought or high temperatures, can impair nodulation and symbiotic efficiency, leading to reduced nitrogen fixation and lower yields (Barbosa et al., 2022).

These yield improvements, supported by *Rhizobium*-mediated nitrogen fixation, not only enhance plant growth but also reduce the need for chemical nitrogen fertilizers, resulting in lower production costs and higher economic returns for farmers (Goyal et al., 2021).

Through proper management, including the use of rhizobial inoculants in nitrogen-poor soils, farmers can achieve significant yield improvements in legume crops. The economic savings from reduced fertilizer inputs and increased yield make this symbiosis particularly valuable in low-input farming systems.

13.5.3 Benefits to Subsequent Crops in Crop Rotation

A major advantage of *Rhizobium*–legume symbiosis is that its benefits extend beyond the immediate legume crop. The residual nitrogen left in the soil after the legume crop is harvested helps improve the growth and productivity of subsequent crops, particularly nitrogen-demanding cereals like wheat, maize, and rice.

13.5.3.1 Residual Nitrogen Contribution

The nitrogen that *Rhizobium* fixes in legume crops after harvesting is found in the soil as microbial biomass, decaying plant tissues, and root nodules. Through mineralization, this nitrogen progressively becomes available to crops that come after it. According to research, the next crop can utilize 30%–50% of the nitrogen fixed by legumes, negating the requirement for synthetic nitrogen fertilizers.

Key components of residual nitrogen:

- **Root Nodules:** After the legume roots die, the nitrogen stored in nodules becomes available for the next crop.
- **Decomposing Plant Biomass:** Plant residues decompose in the field, contributing organic nitrogen to the soil.
- **Microbial Nitrogen Pools:** Soil microbes, which use nitrogen during the legume growth cycle, also contribute to nitrogen cycling and availability as they decompose (Sanzovo et al., 2023).

13.5.3.2 Nitrogen Demand of Subsequent Crops

Wheat, maize, and rice are examples of crops with high nitrogen needs. Legumes supply nitrogen through residual nitrogen left in the soil, which helps crop rotations meet these nitrogen requirements. By increasing the natural nitrogen supply through the development, production, and nutrient intake of succeeding crops, the need for artificial nitrogen fertilizers can be reduced. This method is particularly useful in agricultural systems when fertilizer access is costly or restricted.

Crops that benefit the most from legume residues include:

- Wheat: Studies show that wheat planted after a legume crop has significantly higher yields than wheat planted after non-legume crops.
- Maize: Residual nitrogen from legumes helps support the growth of maize, often leading to substantial yield improvements.
- Rice: Legumes, such as mung beans, rotated with rice, improve soil fertility and rice yields (Table 13.8).

13.5.3.3 Improved Soil Structure and Microbial Activity

Legume crops not only provide nitrogen but also improve soil health. As legume residues decompose, they enhance the soil's organic matter content, improving its structure, water retention, and aeration, which supports healthy root growth in subsequent crops. Moreover, organic matter stimulates microbial activity, which is vital for nutrient cycling. As soil microbes break down organic matter, nitrogen and other nutrients become available for uptake by future crops. Sanyal et al. (2020)

TABLE 13.8
Yield Increases in Subsequent Crops due to Legume Rotation.

Crop	Yield without Legume Rotation (kg/ha)	Yield with Legume Rotation (kg/ha)	% Increase in Yield
Wheat	3,000	3,600	20%
Maize	5,000	6,500	30%
Rice	4,500	5,300	18%

highlighted that increased microbial activity fosters a more resilient soil ecosystem, reducing the risk of disease outbreaks and promoting long-term soil health.

13.5.3.4 Reduction in Nitrogen Fertilizer Use

The residual nitrogen left by legumes can greatly reduce the need for synthetic nitrogen fertilizers in subsequent crops. This reduction provides multiple advantages:

- **Economic Savings:** Farmers save on input costs by reducing the amount of nitrogen fertilizer required, which is particularly beneficial in low-input farming systems or regions where fertilizers are costly.
- **Environmental Sustainability:** Lower fertilizer use reduces the environmental impact of farming by minimizing nitrate leaching, soil acidification, and greenhouse gas emissions (such as nitrous oxide, a potent greenhouse gas produced by excessive nitrogen fertilizer use).
- **Improved Crop Quality:** Using less chemical fertilizer can improve crop quality by encouraging more efficient uptake of naturally available nitrogen, resulting in better nutrient balance.

13.5.3.5 Enhanced Pest and Disease Management

Rotating legumes with other crops can help break pest and disease cycles. Monocropping frequently results in the build-up of particular pathogens, pests, and soil-borne illnesses. Legumes interfere with these pests' and diseases' life cycles, decreasing their frequency and the necessity for chemical management methods. Because of this, crop rotations based on legumes are essential to integrated pest management (IPM) techniques.

These examples illustrate how the inclusion of legumes in crop rotations contributes to higher yields in subsequent crops due to the enhanced nitrogen supply, improved soil health, and better pest management.

13.5.3.6 Sustainability and Long-Term Benefits

Incorporating legumes into crop rotations not only improves yields but also enhances the long-term sustainability of farming systems. Legumes help maintain soil fertility by preserving organic matter levels, improving soil structure, and increasing the availability of nutrients for future crops. This practice reduces the need for external inputs, making farming more cost-effective and environmentally sustainable. Additionally, legumes foster biodiversity both above and below ground, contributing to the resilience of agricultural ecosystems.

13.5.3.7 Benefits in Rotated Crops

Crop rotation with legumes not only increases nitrogen availability but also helps control soil-borne diseases and pests, improves soil organic matter content, and reduces soil erosion. As a result, yields in subsequent non-legume crops often increase without the need for additional inputs (Table 13.9).

TABLE 13.9
Yield Benefits of Crop Rotation with Legumes.

Rotation System	Nitrogen Fertilizer Applied (kg/ha)	Yield of Subsequent Crop (kg/ha)	% Yield Increase
Maize after soybean	0	4,500	18%
Wheat after chickpea	50	3,800	15%
Barley after lentil	25	3,400	20%

The benefits of legume-based rotations are well documented, and their ability to reduce nitrogen fertilizer needs while boosting crop yields is a key component of sustainable agricultural systems (Sanzovo et al., 2023).

13.6 PRACTICAL IMPLEMENTATIONS IN AGRICULTURE

13.6.1 Inoculation Methods for Agricultural Improvement

The efficacy of inoculation is determined by several factors, including inoculant quality, appropriate storage and handling, soil conditions, and the presence of indigenous rhizobia. High-quality inoculants with suitable strain selection for host specificity and efficacy are essential. Appropriate storage (often at 4–10°C) and safeguarding against environmental challenges are crucial for preserving inoculant viability. Soil factors, especially pH, temperature, and moisture levels, significantly influence the development and function of symbiosis. Advancements in inoculation procedures persistently propel the field forward. Polymer-based seed coverings have been engineered to enhance the viability of rhizobia on seeds and shield them from environmental stressors. Pre-inoculated seeds provide ease for farmers and enhance quality control for seed firms. Moreover, precision inoculation methods employing GPS-guided applicators and variable-rate technologies are developing, facilitating site-specific administration according to soil conditions.

13.6.2 Management Strategies for Enhancing Symbiosis

Effective management strategies are essential for enhancing *Rhizobium*–legume symbiosis to maximize nitrogen fixation and improve crop productivity. These strategies focus on selecting efficient rhizobial strains, optimizing soil conditions, and adopting suitable agricultural practices.

- **Selection of Efficient Strains:** The use of highly compatible and nitrogen-efficient rhizobial strains ensures successful nodulation and enhanced nitrogen fixation (Mahmud et al., 2020).
- **Seed and Soil Inoculation:** Various inoculation methods, including soil, seed, and liquid inoculation, play crucial roles; however, each comes with its own advantages and challenges (Table 13.10). Proper inoculation methods, including seed coating or soil application, improve rhizobial survival and effective colonization (Thilakarathna & Raizada, 2017).

TABLE 13.10
Different Inoculation Methods and Their Advantages and Challenges.

Inoculation Method	Description	Advantages	Challenges
Seed inoculation (Gebremariam & Tesfay, 2021)	Direct application of rhizobia to seeds before planting	• Close contact between rhizobia and seeds • Effective inoculant application	• Risk of damage to rhizobia from seed treatments • Limited viability post-inoculation
Soil inoculation (Rahman et al., 2020)	Introduction of rhizobia into soil, either by broadcast or in-furrow placement	• Can be used with pretreated seeds • Potential for uniform soil dispersion	• Requires more inoculant than seed inoculation • May be less effective in soils with high native rhizobia
Liquid inoculation (Resmi et al., 2024)	Rhizobia suspended in water or liquid carriers for seed soaking, soil, or seedling spraying	• Versatile application methods • Can be used with irrigation systems (e.g., drip irrigation)	• Shorter shelf life than solid formulations • Requires careful handling to maintain viability

- **Soil Management:** Maintaining optimal soil pH (6.0–7.0), adequate organic matter, and balanced nutrient availability supports rhizobial activity and symbiosis (Jaiswal et al., 2020).
- **Co-inoculation with Beneficial Microorganisms:** Co-inoculating rhizobia with other beneficial microbes like mycorrhizal fungi or phosphate-solubilizing bacteria enhances nutrient uptake, stress tolerance, and plant growth (Majeed et al., 2018).
- **Use of Bio-stimulants:** Application of bio-stimulants such as plant growth regulators or organic amendments further improves nodulation and nitrogen fixation efficiency (Kondapalli et al., 2021).
- **Adoption of Conservation Agriculture Practices:** Techniques like reduced tillage, crop rotation, and cover cropping help preserve soil structure, microbial diversity, and long-term symbiotic efficiency (Diop et al., 2022).
- **Breeding for Symbiotic Efficiency:** Developing legume varieties with enhanced compatibility and nodulation ability supports better symbiotic performance under diverse environmental conditions (Singh et al., 2019).

In conclusion, integrating these management strategies can significantly enhance the efficiency of *Rhizobium*–legume symbiosis, promoting sustainable crop production and soil fertility.

13.6.3 Application of Biofertilizers in Sustainable Agriculture

Rhizobial biofertilizers play a crucial role in sustainable agriculture by improving soil fertility and crop yield through biological nitrogen fixation (Reddy et al., 2020). They are available as single-strain, multi-strain, or co-inoculants. Single-strain inoculants offer consistency, while multi-strain and co-inoculants enhance adaptability and promote synergistic effects with other beneficial microorganisms like mycorrhizae or plant growth-promoting rhizobacteria (Mukhopadhyay et al., 2023).

Beyond nitrogen fixation, rhizobial biofertilizers reduce the need for chemical fertilizers, minimize environmental pollution, and improve soil health by enhancing organic matter, soil structure, and water retention (Abd-Alla et al., 2023). Their application often results in 10%–50% higher crop yields with improved nutritional quality (Putra et al., 2022).

Integrating rhizobial biofertilizers into sustainable practices like organic farming, conservation agriculture, and agroforestry further enhances their benefits (Ray et al., 2020). However, challenges such as quality control, proper storage, regulatory frameworks, and farmer awareness must be addressed for their wider adoption.

Future advancements include developing stress-tolerant and highly efficient strains through genetic engineering, nano-formulations for better shelf life, and expanding nitrogen fixation ability to non-legume crops (Sheoran et al., 2021; Riseh et al., 2022). Overall, rhizobial biofertilizers offer a promising strategy for sustainable crop production and environmental conservation (Table 13.11 highlights the management strategies for enhancing *Rhizobium*–legume symbiosis).

13.7 OBSTACLES AND CONSTRAINTS

Although *Rhizobium*–legume symbiosis has several advantages for sustainable agriculture, it encounters multiple hurdles and limits that may impede its broad implementation and efficacy (Table 13.12). The constraints encompass environmental variables, biological compatibility challenges, and practical implementation difficulties.

13.7.1 Compatibility Challenges between *Rhizobium* Strains and Leguminous Hosts

The uniqueness of *Rhizobium*–legume interaction offers both advantages and difficulties. This specificity facilitates targeted inoculation tactics; however, it also implies that not all rhizobial

TABLE 13.11
Different Management Strategies for Enhancing Symbiosis.

Strategy	Key Components	Techniques	Benefits	Considerations
Soil management (Oparah et al., 2024; Zahran, 1999)	pH regulation	• Lime application for acidic soils • Acidifying additives for alkaline soils	• Improves nutrient availability • Enhances rhizobial viability and functionality	• Maintain pH between 6 and 7 for most legumes
	Moisture control	• Effective irrigation scheduling • Drainage enhancement	• Prevents drought stress • Avoids waterlogging	• Both drought and waterlogging hinder symbiotic efficiency
	Soil structure improvement	• Reduced tillage • Organic matter incorporation	• Facilitates root development • Enhances rhizobial colonization	• Improves overall soil health
Nutrient management (Abd-Alla et al., 2023)	Nitrogen (N)	• Minimal starter N in deficient soils • Avoid excess N	• Facilitates initial plant growth • Prevents inhibition of nodulation	• Careful balance required
	Phosphorus (P)	• Strategic P placement near seeds	• Enhances early nodulation • Supports energy-demanding N fixation	• Essential for symbiotic process
	Micronutrients	• Application of molybdenum and cobalt	• Mo essential for nitrogenase enzyme • Co important for vitamin B12 production in rhizobia	• Required in small quantities
Crop rotation and intercropping	Legume rotation	• Incorporate legumes in crop rotations	• Improves soil N levels • Disrupts pest and disease cycles • Increases soil microbial diversity	• Plan rotations carefully
	Intercropping	• Grow legumes with non-legumes	• Optimizes land use • Enhances N transfer between crops	• Consider crop compatibility
Pest and disease control	IPM	• Biological control • Cultural techniques • Prudent pesticide use	• Effectively regulates pests • Minimizes adverse effects on rhizobia	• Balance pest control with symbiosis protection
	Disease-resistant varieties	• Select legume species suited to local conditions	• Enhances symbiotic efficiency • Reduces disease pressure	• Consider local climate and soil conditions
	Nodule protection	• Minimize mechanical damage during cultivation • Control root-feeding insects	• Maintains nodule functionality • Ensures continued N fixation	• Be cautious during field operations

TABLE 13.11 *(Continued)*
Different Management Strategies for Enhancing Symbiosis.

Strategy	Key Components	Techniques	Benefits	Considerations
Timing of farming practices (Hungria et al., 2020)	Planting	• Synchronize with optimal soil temperatures • Consider rainfall patterns in rainfed systems	• Ensures optimal conditions for germination and nodulation	• May vary by region and legume species
	Irrigation and fertilizer application	• Time applications to enhance N fixation process	• Maximizes symbiotic efficiency • Optimizes resource use	• Coordinate with plant growth stages
	Harvesting	• Optimize timing for N supply to subsequent crops	• Enhances overall system productivity • Improves soil fertility for following seasons	• Consider crop rotation plans

TABLE 13.12
Environmental Influences on Symbiosis.

Environmental Factor	Optimal Conditions	Challenges	Impact on Symbiosis
Soil pH (Oparah et al., 2024)	6–7 (slightly acidic to neutral)	• Excessively acidic or alkaline soils • Requires careful soil management	• Impaired rhizobia viability and function in extreme pH • Inadequate nodulation • Diminished nitrogen fixation
Temperature (Zahran, 1999)	25°C • 30°C	• Extreme temperatures (hot or cold) • Seasonal variations	• Hindered nodule growth • Reduced effectiveness of nitrogen fixation • Seasonal constraints on symbiotic efficiency
Soil Moisture (Zahran, 1999)	Balanced moisture (neither too dry nor waterlogged)	• Drought conditions • Waterlogged soils • Extreme weather events due to climate change	• Drought: Hindered nodule growth and nitrogen fixation • Waterlogged: Oxygen deficiency affecting aerobic rhizobia • Potential nodule senescence
Soil nutrition (Abd-Alla et al., 2023)	Balanced nutrient levels	• High nitrogen levels (over-fertilization) • Deficiencies in phosphorus, sulfur, and micronutrients	• High N: Inhibited nodulation and nitrogen fixation • Nutrient deficiencies: Limited symbiotic efficacy • Challenges in balancing soil fertility for plant growth and symbiosis

strains are compatible with every bean species or even all variants within a species (Hassen et al., 2020). This host specificity may result in poor nodulation or total failure of symbiosis if an inappropriate mix of rhizobia and legume is utilized (Walker et al., 2020). The issue of compatibility is exacerbated by the existence of indigenous rhizobial communities in the soil. Although these indigenous strains are frequently well-suited to local circumstances, they may not represent the best effective nitrogen fixers. Introduced inoculant strains must contend with native populations for nodule development (Mendoza-Suárez et al., 2021). In certain instances, indigenous strains may surpass more effective foreign strains, leading to inadequate nitrogen fixation despite utilizing appropriate inoculation methods. The genetic variety seen in both rhizobial populations and legume cultivars introduces additional complication (Rigg et al., 2021). As novel bean varieties are cultivated for enhanced agronomic characteristics, their compatibility with current rhizobial strains may be altered. This requires continuous research and improvement to guarantee that inoculant strains retain efficacy with emerging crop types. The efficacy of strains over time is also a concern. Certain very efficient strains may diminish in efficacy with repeated application or extended presence in the soil (Chen et al., 2023). This mechanism, termed strain degradation, may result in diminished nitrogen fixation advantages over time, necessitating periodic evaluation and possible re-inoculation with novel strains.

13.7.2 Economic and Practical Obstacles to Implementation

Despite the considerable advantages of *Rhizobium*–legume symbiosis, many economic and practical obstacles may impede its extensive adoption and effective execution. The principal economic issue is the initial expense and perceived danger linked to the adoption of immunization methods (Anang et al., 2023). For numerous small-scale farmers, especially in developing nations, the initial expenditure on inoculants may be daunting, particularly when the advantages are not readily apparent. The delay between inoculation and visible outcomes may hinder uptake, especially in areas with considerable short-term economic demands (Anang et al., 2023). The quality and accessibility of inoculants pose further issues. In several regions, particularly in poor nations, the supply chain for premium inoculants may be inconsistent or absent. Inadequate quality control in manufacturing, substandard storage conditions during transit and retail, and the absence of appropriate handling protocols might result in ineffective inoculants being delivered to farmers. This not only depletes resources but may also undermine farmers' trust in the technique. The practical application of inoculation procedures might pose difficulties, especially for farmers without familiarity with the method. Proper inoculation necessitates meticulous handling and application to guarantee the viability of rhizobia. Conditions including high temperatures, solar exposure, or certain chemical seed treatments can rapidly diminish the efficacy of inoculants. Educational and training programs are crucial yet can be resource-demanding to execute on a broad scale.

The timing of inoculation in relation to planting is essential for success; however, it can be challenging to implement as well. Farmers must synchronize inoculant availability with their planting timetable, a task hampered by erratic weather patterns or labor shortages. Pre-inoculated seeds present a viable alternative; nonetheless, they may possess a restricted shelf life and elevated expenses. The incorporation of inoculation methods into current agricultural systems and cultural practices poses an additional hurdle. In certain areas, conventional agricultural practices may be entrenched and difficult to modify. Persuading farmers to embrace new techniques, despite evident potential advantages, may be a protracted and challenging endeavor. The enduring process of soil enhancement by BNF may clash with immediate economic demands or land tenure concerns. In regions where farmers lack stable long-term land access, they may be disinclined to invest in methods that provide cumulative advantages over time. In conclusion, whereas *Rhizobium*–legume symbiosis has significant potential for sustainable agriculture, achieving this potential necessitates addressing a complex array of environmental, biological, and socioeconomic obstacles. Overcoming these limits requires a multidisciplinary strategy that integrates ongoing scientific research with practical, context-specific solutions and supporting policies to promote wider adoption and optimize the advantages of this natural nitrogen-fixing collaboration.

13.8 FUTURE PERSPECTIVES AND RESEARCH DIRECTIONS

As the global demand for food production grows, enhancing BNF through *Rhizobium*–legume symbiosis has gained significant interest. New technologies in genetic engineering, biotechnology, and agricultural practices offer exciting opportunities to improve symbiotic efficiency, expand the range of crops benefiting from nitrogen fixation, and promote sustainable agriculture.

13.8.1 Advances in Genetic Engineering and Biotechnology

The advent of genetic engineering and biotechnology has revolutionized the study and manipulation of *Rhizobium*–legume symbiosis. Key areas of progress include the following:

- **Gene Editing (CRISPR-Cas9):** With precision tools like CRISPR-Cas9, it is possible to edit the genomes of both *Rhizobium* and legume plants. Researchers are targeting genes involved in nodulation and nitrogen fixation, enhancing efficiency and compatibility across diverse strains and crops.
- **Synthetic Biology:** Scientists are developing synthetic constructs to engineer *Rhizobium* strains with improved nitrogen-fixing capabilities. These tailored strains could thrive in diverse environmental conditions, extending the benefits of nitrogen fixation to harsher soils and climates.
- **Biotechnological Approaches:** Advances in high-throughput sequencing, transcriptomics, and proteomics enable a deeper understanding of symbiotic interactions at the molecular level. This knowledge can be applied to design biofertilizers or genetically enhanced *Rhizobium* strains for increased yield and nutrient uptake.

13.8.2 Enhancing Symbiotic Efficiency

Maximizing the efficiency of *Rhizobium* bacteria in nitrogen fixation is a key objective for agricultural scientists. Several approaches are being explored:

- **Host–Strain Matching:** Selecting the most compatible *Rhizobium* strains for specific legume species can optimize nodulation and nitrogen fixation rates. This involves screening bacterial strains for high efficiency under various environmental conditions.
- **Improving Stress Tolerance:** Enhancing the stress tolerance of *Rhizobium* could lead to better performance in challenging environments such as saline, drought-affected, or acidic soils. Genetic modification of both plants and bacteria to withstand these conditions is a growing area of research.
- **Bio-stimulants:** Application of microbial inoculants and bio-stimulants, which promote root growth and improve nutrient uptake, has shown promise in boosting symbiotic efficiency. These bio-stimulants often work synergistically with *Rhizobium*, increasing the rate of nitrogen fixation.

13.8.3 Potential for Expanding Symbiosis to Nonleguminous Crops

One of the most promising avenues of research is the extension of nitrogen-fixing symbiosis to nonleguminous crops. Given that staple crops like rice, wheat, and maize do not naturally engage in symbiotic nitrogen fixation, engineering this capability would have transformative effects on agriculture. Several strategies are being pursued:

- **Transgenic Approaches:** By introducing nodulation and nitrogen-fixing genes into the genomes of nonleguminous crops, researchers hope to enable these plants to form symbiotic relationships with nitrogen-fixing bacteria.
- **Intercropping with Legumes**: Developing new cropping systems that integrate legumes with nonleguminous crops could allow the sharing of nitrogen resources through rhizosphere interactions, improving overall crop productivity without the need for synthetic fertilizers.
- **Synthetic Symbiosis:** Advances in synthetic biology may soon allow for the creation of entirely new symbiotic pathways, allowing nonleguminous crops to benefit from bioengineered nitrogen-fixing partnerships.

13.9 CONCLUSION

The symbiosis between *Rhizobium* and legumes has been a cornerstone of sustainable agriculture for centuries. As the world faces increasing challenges from climate change, soil degradation, and food security concerns, leveraging the power of BNF becomes more crucial than ever.

13.9.1 Summary of Key Points

Rhizobium–legume symbiosis improves soil fertility by converting atmospheric nitrogen into a usable form for plants, thus reducing the need for chemical fertilizers. Advances in genetic engineering and biotechnology hold great potential for improving nitrogen fixation efficiency and expanding its benefits to nonleguminous crops. Enhancing the resilience of both *Rhizobium* and host plants under various environmental stressors can significantly boost crop yields in marginal soils. The future of sustainable agriculture lies in harnessing natural processes such as nitrogen fixation to reduce environmental impacts and improve food security.

13.9.2 The Future of *Rhizobium*–Legume Symbiosis in Sustainable Agriculture

In the coming years, *Rhizobium*–legume symbiosis is poised to play a pivotal role in reducing the dependency on synthetic nitrogen fertilizers. With the development of genetically engineered *Rhizobium* strains and improved agronomic practices, it will be possible to increase crop yields while maintaining ecological balance. Integrating nitrogen-fixing plants into crop rotations and agroforestry systems will further enhance soil health, mitigate greenhouse gas emissions, and promote biodiversity. The potential for expanding nitrogen fixation to nonleguminous crops presents a groundbreaking opportunity for sustainable agriculture. If realized, this could drastically reduce the carbon footprint of food production and improve the livelihoods of farmers, particularly in regions with nutrient-poor soils. Nitrogen fixation through *Rhizobium*–legume symbiosis remains one of nature's most efficient solutions to the global challenge of food production. As research continues to unlock the genetic and biotechnological potential of this symbiosis, its application in modern agriculture will only grow. By investing in this natural process, we can ensure a more sustainable and resilient agricultural future, contributing to food security and environmental conservation worldwide.

REFERENCES

Abdel-Lateif, K., Bogusz, D., & Hocher, V. (2012). The role of plant hormones in the establishment of the *Rhizobium*–legume symbiosis. *Plant Signaling & Behavior*, *7*(11), 1286–1291.

Aggarwal, N., Saini, A., & Sharma, A. (2022). Exploring the role of Rhizobium in enhancing legume growth and yield. *Journal of Agricultural Sciences*, *58*(3), 122–131.

Ahemad, M., & Kibret, M. (2014). Mechanisms and applications of plant growth promoting rhizobacteria: Current perspective. *Journal of King Saud University-Science*, *26*(1), 1–20.

Ahmad, F., Ahmad, I., & Khan, M. S. (2020). Screening of free-living rhizospheric bacteria for their multiple plant growth-promoting activities. *Microbiological Research*, *156*(1), 1–9.

Ahmed, W., Shah, A. N., Abbas, A., Nawaz, M., Qayyum, A., Hassan, M. U., . . . Khan, J. (2023). Role of plant bio-stimulants and their classification. In *Biofertilizers for Sustainable Soil Management* (pp. 65–88). CRC Press.

Akhtar, N., Ilyas, N., Meraj, T. A., Pour-Aboughadareh, A., Sayyed, R. Z., Mashwani, Z. U. R., & Poczai, P. (2022). Improvement of plant responses by nanobiofertilizer: A step towards sustainable agriculture. *Nanomaterials*, *12*(6), 965.

Anuar, N., Hafiza, N., & Wahid, R. (2021). Beneficial effects of Rhizobium inoculation on legume growth performance. *Asian Journal of Plant Science*, *20*(3), 252–261.

Arshad, U., Altaf, M. T., Liaqat, W., Ali, M., Shah, M. N., Jabran, M., & Ali, M. A. (2024). Biochar: Black gold for sustainable agriculture and fortification against plant pathogens—a review. *Journal of Crop Health*, *76*(2), 385–396.

Barker, D. G., Bianchi, S., Blondon, F., Dattee, Y., Duc, G., Essad, S., . . . Young, N. D. (1990). *Medicago truncatula*, a model plant for studying the molecular genetics of the Rhizobium-legume symbiosis. *Plant Molecular Biology Reporter*, *8*(1), 40–49.

Basit, A., Shah, S. T., Ullah, I., Muntha, S. T., & Mohamed, H. I. (2021). Microbe-assisted phytoremediation of environmental pollutants and energy recycling in sustainable agriculture. *Archives of Microbiology*, *203*(10), 5859–5885.

Behera, A., Das, R., Giri, K. S., Das, S., Chand, M. K., Thatoi, H., & Mohanty, G. (2025). Evaluating the effectiveness of fish organic liquid manure (FOLM) in enhancing maize growth and yield: A comparative study of organic and conventional fertilizers. *Discover Sustainability*, *6*(1), 326.

Bhattacharjee, R. B., Singh, A., & Mukhopadhyay, S. N. (2008). Use of nitrogen-fixing bacteria as biofertiliser for non-legumes: Prospects and challenges. *Applied Microbiology and Biotechnology*, *80*(2), 199–209.

Bhatti, A., Suleman, H. M., Zeshan, K. M., Imtiaz, M., & Batool, R. (2026). Unlocking soil fertility and enhancing plant health: The vital role of PGPR. *Journal of Biological and Allied Health Sciences*, *6*(1), 28–43.

Broughton, W. J., & Dilworth, M. J. (1971). Control of leghaemoglobin synthesis in snake beans. *Biochemical Journal*, *125*(4), 1075–1080.

Candra, B., Ambarita, D. D., Utami, D. S., Adinata, K., Stöber, S., & Simarmata, T. (2025). Assessment of Azolla for carbon capture, biofertilizer application, and rice productivity enhancement in sustainable lowland farming systems under climate change adaptation. *Discover Sustainability*, *6*(1), 1398.

Chen, W. F., Wang, E. T., Ji, Z. J., & Zhang, J. J. (2021). Recent development and new insight of diversification and symbiosis specificity of legume rhizobia: Mechanism and application. *Journal of Applied Microbiology*, *131*(2), 553–563.

Chen, Y., Zhang, X., & Liu, P. (2022). Effects of Rhizobium symbiosis on plant root architecture and stress resistance. *Journal of Plant Science*, *65*(3), 122–130.

Chowdhury, S. P., Schmid, M., Hartmann, A., & Tripathi, A. K. (2020). Diversity of 16S rRNA and nifH genes of endophytic diazotrophic bacteria of wild rice species of north-eastern India. *Systematic and Applied Microbiology*, *32*(4), 291–299.

Chu, X., Su, H., Hayashi, S., Gresshoff, P. M., & Ferguson, B. J. (2022). Spatiotemporal changes in gibberellin content are required for soybean nodulation. *New Phytologist*, *234*(2), 479–493.

Costa, S. R., Ng, J. L. P., & Mathesius, U. (2021). Interaction of symbiotic rhizobia and parasitic root-knot nematodes in legume roots: From molecular regulation to field application. *Molecular Plant-Microbe Interactions*, *34*(5), 470–490.

Dakora, F. D., & Phillips, D. A. (2002). Root exudates as mediators of mineral acquisition in low-nutrient environments. *Plant and Soil*, *245*(1), 35–47.

Desbrosses, G. J., & Stougaard, J. (2011). Root nodulation: A paradigm for how plant–microbe symbiosis influences host developmental pathways. *Cell Host & Microbe*, *10*(4), 348–358.

Devi, R., Kaur, T., Negi, R., Kour, D., Chaubey, K. K., & Yadav, A. N. (2023). Indigenous plant growth-promoting rhizospheric and endophytic bacteria as liquid bioinoculants for growth of sweet pepper (Capsicum annuum L.). *Biologia*, *78*(9), 2623–2633.

Dobermann, A., & Cassman, K. G. (2004). Environmental dimensions of fertilizer nitrogen: What can be done to increase nitrogen use efficiency and ensure global food security? In A. R. Mosier, J. K. Syers, & J. R. Freney (Eds.), *Agriculture and the Nitrogen Cycle: Assessing the Impacts of Fertilizer Use on Food Production and the Environment* (pp. 261–278). Island Press, Washington, DC.

Dong, W., Zhu, Y., Chang, H., Wang, C., Yang, J., Shi, J., . . . Wang, E. (2021). An SHR–SCR module specifies legume cortical cell fate to enable nodulation. *Nature*, *589*(7843), 586–590.

Drame, M. (2024). *Impacts of Urease Inhibitors on Nitrogen Assimilation in Wheat and on Reducing Nitrogen Losses* (Doctoral dissertation, Cranfield University).

Fan, J., Shen, Y., Chen, C., Chen, X., Yang, X., Liu, H., . . . Tian, Z. (2025). A large-scale integrated transcriptomic atlas for soybean organ development. *Molecular Plant, 18*(4), 669–689.

Gage, D. J. (2004). Infection and invasion of roots by symbiotic, nitrogen-fixing rhizobia during nodulation of temperate legumes. *Microbiology and Molecular Biology Reviews, 68*(2), 280–300.

Garrido-Oter, R., Nakano, R. T., Dombrowski, N., Ma, K.-W., McHardy, A. C., & Schulze-Lefert, P. (2021). The plant microbiota: From ecology to reductionism and beyond. *Annual Review of Microbiology, 75*, 229–251. https://doi.org/10.1146/annurev-micro-090817-062524

Ghantasala, S., & Roy Choudhury, S. (2022). Nod factor perception: An integrative view of molecular communication during legume symbiosis. *Plant Molecular Biology, 110*(6), 485–509.

Ghosh, A., Acharya, R., Shaw, S., & Gangopadhyay, D. (2024). Plant growth-promoting Rhizobacteria (PGPR): A potential. *Updates on Rhizobacteria, 51*.

Goto, T., Soyano, T., Liu, M., Mori, T., & Kawaguchi, M. (2022). Auxin methylation by IAMT1, duplicated in the legume lineage, promotes root nodule development in Lotus japonicus. *Proceedings of the National Academy of Sciences, 119*(10), e2116549119.

Gow, N. A., & Lenardon, M. D. (2023). Architecture of the dynamic fungal cell wall. *Nature Reviews Microbiology, 21*(4), 248–259.

Graham, P. H., & Vance, C. P. (2003). Legumes: Importance and constraints to greater use. *Plant Physiology, 131*(3), 872–877.

Griesmann, M., Chang, Y., Liu, X., Song, Y., Haberer, G., Crook, M. B., . . . Udvardi, M. K. (2018). Phylogenomics reveals multiple losses of nitrogen-fixing root nodule symbiosis. *Science, 361*(6398), eaat1743. https://doi.org/10.1126/science.aat1743

Gupta, P., & Sharma, K. (2025). Organic formulations in floriculture: A review. *Plant Archives (09725210), 25*(1).

Gupta, R., Verma, A., & Singh, J. (2021). Influence of *Rhizobium* symbiosis on plant competition and resource acquisition. *Agriculture & Plant Science, 58*(4), 89–95.

Haroon, M., Wahid, F., Ullah, R., Adnan, M., Alam, M., Ullah, H., . . . Hassan, S. (2023). Limitations of using biofertilizers as an alternative to chemical fertilizers. In *Biofertilizers for Sustainable Soil Management* (pp. 133–141). CRC Press.

Hnini, M., Rabeh, K., & Oubohssaine, M. (2025). Understanding plant-microorganism interactions: The key roles of soil, rhizosphere, and direct and indirect mechanisms. *AIMS Microbiology, 11*(4), 1035.

Islam, I. U., Bano, S., Mehwish, M., Altaf, M., Shoaib, R., Hu, X., . . . Yabalak, E. (2026). Biological agents and biomaterials for environmental remediation: Applications in water, soil, and air. *International Journal of Environmental Science and Technology, 23*(3), 167.

Jaiswal, D. K., Verma, R., & Kharwar, R. N. (2020). Role of symbiotic nitrogen-fixing bacteria in sustainable agriculture: A review. *Journal of Plant Growth Regulation, 40*(1), 839–857.

Jawad, M., Jabeen, A., Gurmani, A. R., & Rafique, M. (2023). Phosphate-mobilizing mycorrhizal-based biofertilizers. In *Biofertilizers for Sustainable Soil Management* (pp. 89–106). CRC Press.

Ji, C., Ge, Y., Zhang, H., Zhang, Y., Xin, Z., Li, J., . . . Li, K. (2024). Interactions between halotolerant nitrogen-fixing bacteria and arbuscular mycorrhizal fungi under saline stress. *Frontiers in Microbiology, 15*, 1288865.

Joshi, S., Joshi, N., Tariq, A., Khursheed, M., Jan, R., Singh, K. C., . . . Tutlani, A. (2025). Integrating biofertilizers in nutrient management of millets: A review of advances and prospects. *Journal of Experimental Agriculture International, 47*(11), 638–653.

Karimi, S., Haghighi, M., & Sharifani, M. (2025). Eco-enzymes: A sustainable solution for waste management and agricultural applications. *Iran Agricultural Research, 44*(2).

Khan, A., Panthari, D., Sharma, R. S., Punetha, A., Singh, A. V., & Upadhayay, V. K. (2023). Biofertilizers: A microbial-assisted strategy to improve plant growth and soil health. In *Advanced Microbial Techniques in Agriculture, Environment, and Health Management* (pp. 97–118). Academic Press.

Khan, Z., & Ali, M. (2023). Hormonal modulation in legumes through *Rhizobium* symbiosis: Implications for growth. *Journal of Plant Hormones, 36*(2), 76–88.

Krönauer, C., & Radutoiu, S. (2021). Understanding Nod factor signalling paves the way for targeted engineering in legumes and non-legumes. *Current Opinion in Plant Biology, 62*, 102026.

Kumar, M., Rajput, M., Soni, T., Vivekanand, V., & Pareek, N. (2020). Chemoenzymatic production and engineering of chitooligosaccharides and N-acetyl glucosamine for refining biological activities. *Frontiers in Chemistry, 8*, 469.

Kumar, P., Sharma, L., & Mehta, R. (2022). Chlorophyll content and photosynthetic efficiency in *Rhizobium*–legume symbiosis. *Plant Physiology Journal, 57*(1), 45–54.

Kumar, S., Thakur, N., Singh, A. K., Gudade, B. A., Ghimire, D., & Das, S. (2022). Aquatic macrophytes for environmental pollution control. In *Phytoremediation Technology for the Removal of Heavy Metals and Other Contaminants from Soil and Water* (pp. 291–308). Elsevier.

Kumari, B., Prakash, O., Pal, P., Singh, P. K., Singh, M. P., Kumar, P., . . . Singh, A. K. (2024). Advancement in sustainable management and valorization of solid waste through composting and vermitechnology. In *Recent Trends in Management and Utilization of Industrial Sludge* (pp. 359–397). Springer Nature, Cham, Switzerland.

Kundu, A., Moraes, T. A., Price, R. J., Harrison, R. J., & Oldroyd, G. E. (2025). Getting to the route: The evolution of nitrogen-fixing nodules. *Annual Review of Cell and Developmental Biology*, *41*.

Lee, T., Orvosova, M., Batzenschlager, M., Batista, M. B., Bailey, P. C., Mohd-Radzman, N. A., . . . Schiessl, K. (2024). Light-sensitive short hypocotyl genes confer symbiotic nodule identity in the legume Medicago truncatula. *Current Biology*, *34*(4), 825–840.

Liu, C., Xiang, Q., Wei, Q., Yin, H., Zhou, Z., & Wang, E. (2018). The Bradyrhizobium diazoefficiens noeI gene is required for efficient symbiosis with soybean. *Molecular Plant-Microbe Interactions*, *31*(5), 557–568.

Liu, L., Zhang, X., Xu, W., Liu, X., Li, Y., Wei, J., . . . Wu, X. (2020). Challenges for global sustainable nitrogen management in agricultural systems. *Journal of Agricultural and Food Chemistry*, *68*(11), 3354–3361.

Liu, T., Wang, Q., & Xu, L. (2023). Biomass production in legumes enhanced by *Rhizobium* symbiosis. *Journal of Agricultural Science*, *61*(2), 78–85.

Liu, W., & Zhang, X. (2026). Microbial ecosystems. *Introduction to Geobiology*, *87*.

Madhav, S., Mishra, R., Kumari, A., Srivastav, A. L., Ahamad, A., Singh, P., . . . Sillanpää, M. (2024). A review on sources identification of heavy metals in soil and remediation measures by phytoremediation-induced methods. *International Journal of Environmental Science and Technology*, *21*(1), 1099–1120.

Mandon, K., Nazaret, F., Farajzadeh, D., Alloing, G., & Frendo, P. (2021). Redox regulation in diazotrophic bacteria in interaction with plants. *Antioxidants*, *10*(6), 880.

Mishra, A., Peramaiyan, P., & Shaik, H. A. (2025). Eco-friendly crop protection in rice: Innovations and modern technologies. In *Rice Production Strategies-Addressing Abiotic and Biotic Stresses and Socioeconomic Changes*. IntechOpen.

Mishra, G. (2025). Revolution and advances in gene editing and genomics technology for developing climate-resilient legume crops: Developments and prospects. *Plant Molecular Biology*, *115*(5), 106.

Nawaz, T., Joshi, N., Fahad, S., Saud, S., Ur Rahman, T., Khan, M. N. R., & Hassan, S. (2024). Solar-powered N2-fixing cyanobacteria for bio-nitrogen fertilizer production and soil health improvement: A sustainable farming approach. In *Environment, Climate, Plant and Vegetation Growth* (pp. 75–113). Springer Nature, Cham, Switzerland.

Naz, M., Afzal, M. R., Raza, M. A., Tariq, M., Yan, M., Dai, Z., . . . Du, D. (2025). The significant effects of Strigolactones on plant growth and microbe interactions: A review. *Plant Growth Regulation*, *105*(4), 937–957.

Oldroyd, G. E. D., Murray, J. D., Poole, P. S., & Downie, J. A. (2011). The rules of engagement in the legume–rhizobial symbiosis. *Annual Review of Genetics*, *45*, 119–144. https://doi.org/10.1146/annurev-genet-110410-132549

Parveen, M., & Ghosh, S. (2025). Ecological dynamics and microbial consortia: Enhancing soil health and sustainable agriculture. *Egyptian Journal of Botany*, *65*(3), 627–643.

Patel, C., Singh, J., Karunakaran, A., & Ramakrishna, W. (2023). Evolution of nano-biofertilizer as a green technology for agriculture. *Agriculture*, *13*(10), 1865.

Patel, R., & Singh, N. (2023). *Rhizobium*–legume symbiosis: A key factor in enhancing flowering and seed production. *Symbiosis Research Journal*, *47*(3), 102–112.

Poole, P., Ramachandran, V., & Terpolilli, J. (2018). Rhizobia: From saprophytes to endosymbionts. *Nature Reviews Microbiology*, *16*, 291–303. https://doi.org/10.1038/nrmicro.2017.171

Pu, Z., Hou, R., Xie, Z., Zhang, R., Li, X., & Wang, X. X. (2025). Nurturing soil health: The interconnected roles of biochar and arbuscular mycorrhizal fungi. *Journal of Soils and Sediments*, *25*(10), 2931–2942.

Qadir, M. B., Aslam, M., Ahmad, M., Bibi, S., Khan, B., Bilal, H., . . . Ahmed, A. (2024). Soil nutrient dynamics and agronomic strategies for sustainable wheat production. *Global Research Journal of Natural Science and Technology*, *2*(3).

Remigi, P., Zhu, J., Young, J. P. W., & Masson-Boivin, C. (2016). Symbiosis within symbiosis: Evolving nitrogen-fixing legume symbionts. *Trends in Microbiology*, *24*(1), 63–75. https://doi.org/10.1016/j.tim.2015.10.007

Ren, S., Liu, Y., Liu, Y., Yu, H., & Xu, M. (2025). Combination of nitrogen-fixing bacteria and mycorrhizal fungi promotes Leymus chinensis growth and bioremediation of degraded grasslands in semi-arid regions. *Plant and Soil*, *511*(1), 1575–1595.

Sales, H. B., de Carolino, S. A., de Nunes, A. R. Z., Macalia, C. M., Ruzo, C. M., da Pinto, C. C., . . . Sanches, E. A. (2024). Advances in agricultural technology: A review of slow-release nanofertilizers and innovative carriers. *Communications in Soil Science and Plant Analysis*, *55*(12), 1849–1882.

Saud, S., Nawaz, T., Hassan, S., Ur Rahman, T., Rasheed, M. N., Hussain, S., & Fahad, S. (2024). Nitrogen-fixing cyanobacteria and soil enrichment for a greener future. In *Environment, Climate, Plant and Vegetation Growth* (pp. 391–424). Springer Nature, Cham, Switzerland.

Shah, A. M., Zhang, H., Shahid, M., Ghazal, H., Shah, A. R., Niaz, M., . . . Zhao, H. (2025). The vital roles of agricultural crop residues and agro-industrial by-products to support sustainable livestock productivity in subtropical regions. *Animals*, *15*(8), 1184.

Shah, H., & Jain, S. (2021). Bioremediation: An approach for environmental pollutants detoxification. In *Waste to Energy: Prospects and Applications* (pp. 121–142). Springer, Singapore.

Shaik, H. A. (2026). Eco-friendly crop protection in. *Rice Production Strategies-Addressing Abiotic and Biotic Stresses and Socioeconomic Changes: Addressing Abiotic and Biotic Stresses and Socioeconomic Changes* (p. 133).

Sharma, C., Pathak, P., & Gautam, S. (2023). Transforming agri-crop residue biomass for value addition: An innovative strategy toward resilient circular economy. In *Handbook of Biomass* (pp. 1–26). Springer Nature, Singapore.

Sharma, M., & Verma, P. (2022). The role of nitrogen fixation in promoting plant growth in legumes. *Plant Growth Research*, *60*(1), 90–100.

Shen, D., Micic, N., Venado, R. E., Bjarnholt, N., Crocoll, C., Persson, D. P., . . . Andersen, T. G. (2025). Apoplastic barriers are essential for nodule formation and nitrogen fixation in Lotus japonicus. *Science*, *387*(6740), 1281–1286.

Siddiqui, M. A., Neeraj, A., & Hiranmai, R. Y. (2022). Vermitechnology: An eco-friendly approach for organic solid waste management and soil fertility improvement—a review. In *Strategies and Tools for Pollutant Mitigation: Research Trends in Developing Nations* (pp. 91–112).

Singh, A., Altúzar-Molina, A., Kaur, G., Sidhu, G. K., Mandal, P. K., & Reddy, P. M. (2025). Perceiving Rhizobial Nod factor by transgenic rice (Oryza sativa) with MtNFP and MtLYK3 receptor-like kinases along with supporting genes revealed through root hair deformation and activation of Ca2+ oscillations. *Journal of Plant Growth Regulation*, *44*(11), 6451–6466.

Singh, A., Chauhan, R., Rajput, V. D., Minkina, T., Prasad, R., & Goel, A. (2024). Exploring the insights of bioslurry—nanoparticle amalgam for soil amelioration. *Environmental Science and Pollution Research*, *31*(48), 58310–58323.

Singh, J., & Verma, P. K. (2023). Role of Nod factor receptors and its allies involved in nitrogen fixation. *Planta*, *257*(3), 54.

Singh, R., Kaur, S., Bhullar, S. S., Singh, H., & Sharma, L. K. (2024). Bacterial biostimulants for climate smart agriculture practices: Mode of action, effect on plant growth and roadmap for commercial products. *Journal of Sustainable Agriculture and Environment*, *3*(1), e12085.

Singh, S., Gupta, R., & Kumar, A. (2021). Biomass and nitrogen enhancement in legumes through symbiotic relationships. *Agronomy Science Journal*, *64*(3), 144–153.

Sprent, J. I., & James, E. K. (2007). Legume evolution: Where do nodules and mycorrhizas fit in? *Plant Physiology*, *144*(2), 575–581.

Suzaki, T., & Kawaguchi, M. (2014). Root nodulation: A developmental program involving cell fate conversion triggered by symbiotic bacterial infection. *Current Opinion in Plant Biology*, *21*, 16–22. https://doi.org/10.1016/j.pbi.2014.06.001

Tamma, A. A., Lejcuś, K., Fiałkiewicz, W., & Marczak, D. (2025). Advancing phytoremediation: A review of soil amendments for heavy metal contamination management. *Sustainability*, *17*(13), 5688.

Vala, A. K., & Dudhagara, D. R. (Eds.). (2025). *Marine Biotechnology: A Gateway to Blue Economy*. CRC Press.

Verma, J. P., Yadav, J., Tiwari, K. N., & Kumar, A. (2020). Impact of plant growth promoting rhizobacteria on crop production under changing environment: A review. *Microbiological Research*, *231*, 126347. https://doi.org/10.1016/j.micres.2019.126347

Vessey, J. K. (2003). Plant growth promoting rhizobacteria as biofertilizers. *Plant and Soil*, *255*(2), 571–586.

Wang, Q., Liu, M., Wang, Z., Li, J., Liu, K., & Huang, D. (2024). The role of arbuscular mycorrhizal symbiosis in plant abiotic stress. *Frontiers in Microbiology*, *14*, 1323881.

Wang, S., Niu, X., Di, D., & Huang, D. (2022). Nitrogen and sulfur fertilizers promote the absorption of lead and cadmium with Salix integra Thunb. by increasing the bioavailability of heavy metals and regulating rhizosphere microbes. *Frontiers in Microbiology*, *13*, 945847.

Wei, X., Xie, B., Wan, C., Song, R., Zhong, W., Xin, S., & Song, K. (2024). Enhancing soil health and plant growth through microbial fertilizers: Mechanisms, benefits, and sustainable agricultural practices. *Agronomy*, *14*(3), 609.

Wielbo, J. (2025). Quality or quantity? Increasing legume yield using traditional inoculants and Rhizobial Nod factors in the context of inter-strain competition. *Agronomy*, *15*(10), 2303.

Wu, X., & Zhang, Y. (2023). Nod factors and their impact on legume symbiosis. *Symbiotic Signal Journal*, *54*(2), 110–122.

Xie, J., & Liu, Y. (2021). Enhanced plant productivity through *Rhizobium*-induced symbiotic association. *Journal of Symbiotic Studies*, *40*(3), 211–222.

Xie, J., & Ma, Y. (2021). The contribution of *Rhizobium*–legume symbiosis to soil fertility and plant development. *Soil Fertility Journal*, *54*(2), 93–102.

Xing, Y., Wang, F., Yu, S., Zhu, Y., Ying, Y., & Shi, W. (2024). Enhancing Phyllostachys edulis seedling growth in phosphorus-deficient soil: Complementing the role of phosphate-solubilizing microorganisms with arbuscular mycorrhizal fungi. *Plant and Soil*, *497*(1), 449–466.

Xu, Z., Li, C., & Chen, L. (2022). Root hair curling and infection thread formation: Critical steps for successful legume–*Rhizobium* symbiosis. *Journal of Integrative Plant Biology*, *64*(4), 711–725. https://doi.org/10.1111/jipb.13195

Yang, H., & Wang, E. (2023). Dynamic regulation of symbiotic signal perception in legumes. *Science Bulletin*, *68*(7), 670–673.

Yang, L., Qiu, J., & Zhang, H. (2023). Cytokinin signaling in root nodule development. *Plant Symbiosis Studies*, *29*(4), 230–240.

Zafar, A., Arshad, A., Javed, W., Awais, M., Nargis, J., & El Sabagh, A. (2025). Nutrients grow through organic soil enrichment: Revitalizing soil health through organic nutrient enrichment. In *Improving Crop Quality and Enhancing Sustainability in Agriculture* (pp. 374–398). CRC Press.

Zahid, T., & Imran, A. (2026). Interactions between nanomaterials and plant growth-promoting bacteria for enhanced plant growth and production. In *Nanomaterials–Microbial Interactions for Soil Health, Plant Growth, and Performance* (pp. 333–352). Academic Press.

Zang, L., Morère-Le Paven, M. C., Clochard, T., Porcher, A., Satour, P., Mojović, M., . . . Montrichard, F. (2020). Nitrate inhibits primary root growth by reducing accumulation of reactive oxygen species in the root tip in Medicago truncatula. *Plant Physiology and Biochemistry*, *146*, 363–373.

Zeng, Y., Wang, Y., Wang, X., Jing, X., Shu, X., Ren, P., . . . Wang, X. (2025). Arbuscular mycorrhizal fungi as core engineers in synthetic microbial communities: Boosting plant growth and soil health for sustainable agriculture. *Journal of Fungi*, *11*(11), 769.

Zgadzaj, R., Garrido-Oter, R., Jensen, D. B., Koprivova, A., Schulze-Lefert, P., & Radutoiu, S. (2016). Root nodule symbiosis in Lotus japonicus drives the establishment of a distinct root microbiota. *Proceedings of the National Academy of Sciences of the United States of America (PNAS)*, *113*(49), E7996–E8005. https://doi.org/10.1073/pnas.1616564113

Zhang, G., Yang, J., Chen, X., Zhao, D., Zhou, X., Zhang, Y., . . . Zhao, J. (2021). Phospholipase D-and phosphatidic acid-mediated phospholipid metabolism and signaling modulate symbiotic interaction and nodulation in soybean (Glycine max). *The Plant Journal*, *106*(1), 142–158.

Zhao, H., & Hu, X. (2021). Mechanisms of root nodule formation and nitrogen fixation in legumes. *Legume Research Journal*, *40*(5), 322–334.

Zhao, L., Ji, C. Y., Murray, J. D., & Liu, C. W. (2025). A legume cellulase required for rhizobial infection and colonization in root nodule symbiosis. *Nature Communications*, *16*(1), 6663.

Zhou, H., Lin, M., & Zhao, G. (2022). Transcriptional control of nodule formation in legumes. *Plant Biology Journal*, *51*(6), 301–309.

Zhou, M., Tan, G., & Xu, Z. (2021). Structure and function of nitrogenase: Molecular insights into *Rhizobium*–legume symbiosis. *Journal of Plant Biochemistry*, *26*(4), 235–249.

14 Microbe-Induced Systemic Resistance

A Promising Avenue for Disease Management in Agriculture

Ayushi Verma, Dhananjay Singh, Dharmendra Singh, and Nishu Mittal

14.1 INTRODUCTION

The continuous growth in world population leads to an increase in demand for agriculture-based products. Plants are autotrophs in nature and synthesize their food by utilizing water, air, and sunlight. In addition to these mentioned components, soil-borne microbes also help in plant growth. Plants grow naturally in soil, where they are exposed to both beneficial and harmful microbes. Both pathogens and pests have a great impact on plant growth and yield. Diseases caused by pathogens can reduce the yield of crops up to 30%. Among the several plant diseases, some are more common and harmful to plant growth. To maintain the yield, chemical pesticides are used, causing serious damage to the environment (Fones et al., 2020). Apart from environmental pollution, the use of chemical compounds also causes water pollution and soil pollution. The use of polluted water causes damage to both flora and fauna. Any type of pollution ultimately reduces the yield of plant growth. Currently, agronomists are focusing on increasing the yield of crop plants without compromising the quality of plant products while minimizing the damage to the environment.

The awakening call on environmental and health risks has led to the change in focus from the use of chemical pesticides to natural compounds present in the environment as synthetic chemicals are unsustainable (Donley, 2019). The microbes present in soil are very beneficial to plant growth and can also fight against plant pathogens and pests. Among the different classes of microbes, plant growth-promoting rhizobacteria (PGPR) are the most common and widely studied. They are very efficient, nontoxic, and environmentally friendly in nature and can be a good alternative to chemical and harmful pesticides. Apart from plant growth, PGPR also induce systemic resistance against various diseases of plants caused by microbes (Salwan et al., 2020). PGPR fix nitrogen, which contributes to plant growth with the accumulation of more nutrients and organic compounds. The production of various plant growth regulators can provide protection to plants against stresses caused by abiotic and biotic factors.

Plants develop responses after recognizing and interacting with the invading microorganism present in the soil. In case of interaction with pathogenic microbes, plants activate the innate defense system and try to evade the disease-causing pathogen. In contrast, when plants come in contact with beneficial microbes, they develop a symbiotic relationship (Zipfel and Oldroyd, 2017). In plants, the first level of immune defense, i.e., pattern-triggered immunity (PTI), is initiated after the recognition of specific microbe- or pathogen-associated molecular patterns (MAMPs or PAMPs); that is, identification of fungal chitin and flagellin proteins of the bacterial cell *via* the recognition of receptors spanning the membrane (Bigeard et al., 2015). Similar to plants, microbes also possess a defense system and repress PTI signaling by plants after discharging virulence factors in the surrounding and evading host immunity (Guo et al., 2009). In plants, the second layer of defense,

 DOI: 10.1201/9781003608080-14

effector-triggered immunity (ETI), leads to the development of hypersensitivity reaction to stop the spread of the pathogen from the surrounding area by using nucleotide-binding site-leucine-rich repeat (NB-LRR) proteins (Jones and Dangl, 2006). The interaction between the host plant and pathogen develops a complex immune system that allows beneficial microbes to induce immunity after the modulation of small RNAs of the host cell. The systemic resistance in response to microbes can be of two types: Induced systemic resistance (ISR) and systemic acquired resistance (SAR). Nonpathogenic microbes contribute to ISR, whereas pathogenic microbes contribute to SAR (Kloepper et al., 1992).

SAR is a salicylic acid (SA)-dependent plant defense mechanism that is initiated after the expression of genes coding for pathogen-related (PR) proteins and accumulation of SA (Gaffney et al., 1993). ISR is initiated after the expression of genes recognizing beneficial microbes (Alström, 1991). PGPR such as those belonging to *Pseudomonas* initiates ISR in plants without causing any harmful effect to the plant root system (Van Loon and Glick 2004). Kloepper and coworkers revealed in their work that PGPR can induce resistance against some diseases, such as those caused by root-knot nematodes, bacteria (angular leaf spot), and cucumber mosaic virus (*Cucumovirus*) (Liu et al., 1995). In contrast to SAR, ISR is not based on the accumulation of SA and expression of PR but depends on the jasmonate (JA) and ethylene (ET) pathways (Pieterse et al., 2001). In this chapter, we focus on the role of various beneficial microbes, their mechanisms, and their application in the protection of agricultural crops and increase in plant productivity.

14.2 UNDERSTANDING

These beneficial microbes control plant diseases either by controlling the growth of microbes or by increasing the innate immunity of plants against disease-causing microbes (Bai et al., 2002). In the total population of soil microbes, some microbes are either harmful or beneficial, but most of the soil-borne microbes have no direct role in the development of plant growth and productivity (Trivedi et al., 2020). During the interaction between beneficial microbes including bacteria and fungi, these microbes colonize in the rhizosphere area and promote root and plant growth (Lugtenberg and Kamilova, 2009). Various strains of soil-borne root-colonizing bacteria, namely, *Bacillus*, *Pseudomonas*, *Azospirillum*, *Azotobacter*, and *Serratia*, are classified as beneficial microbes as they help in plant growth and disease resistance (Kloepper and Beauchamp, 1992). Similar to harmful microbes, beneficial microbes are also considered as foreign organisms. The frequent interaction between beneficial microbes and host plants leads to the activation of the innate defense system in the plants even before the attack by harmful microbes.

14.3 MECHANISM OF MICROBE-INDUCED SYSTEMIC RESISTANCE

Microbe-induced systemic resistance (MISR) is carried out using various mechanisms. These mechanisms are interdependent on each other and majorly operates by acting on plant defense systems.

14.3.1 Recognition of Microbes

Plants detect beneficial microbes by identifying shared microbial compounds, which include flagellin, exopolysaccharides, chitin oligosaccharides, and lipopolysaccharides (LPS). These common compounds act as ligands. Receptors recruit co-receptors upon binding with ligands, which results in the formation of various complexes. Such complexes are responsible for phosphorylating downstream substrates and triggering a chain of signaling pathways such as mitogen-activated protein kinase, calcium influx, and oxidative burst (Jain et al., 2016)

The highly conserved epitope flg22, consisting of 22 amino acids on the N-terminus of flagellin, is found in several eubacteria (Felix et al., 1999). Host plants can identify the conserved epitopes. FLAGELLIN-SENSING 2 (FLS2) is the first identified receptor capable of recognizing flagellin derived from PGPR (Malvino, 2024). The recognition of flg22 triggers the heterodimerization of FLS2 with its co-receptors, BRI1-associated kinase (BAK1) and BAK1-LIKE1 (BKK1). This interaction leads to the phosphorylation of the receptor-like cytoplasmic kinase *Botrytis*-induced kinase1 (BIK1), thereby initiating PTI signaling. In *Arabidopsis thaliana*, mutations in BAK1 resulted in normal flagellin binding but impaired PTI responses, suggesting that BAK1 functions as a positive regulator in the signaling pathway (Kørner et al., 2013). Upon flagellin recognition, BIK1 undergoes phosphorylation and, in turn, transphosphorylases the FLS2–BAK1 complex. This process facilitates the transmission of flagellin signaling and activates downstream intracellular signaling cascades. Similarly to *bak1* mutant, the *BIK1* mutant also exhibits impaired flagellin-mediated responses to nonpathogenic microbial invasion, highlighting *BIK1*'s critical role in MAMP signal transduction (Lu et al., 2010) and the activation of ISR. Additionally, rhizobia and arbuscular mycorrhizal fungi (AMF) establish symbiotic relationships with their host plants through signals derived from chitin-based oligosaccharides. NFs are acylated lipo-oligosaccharides produced by rhizobia and recognized by LysM receptor-like kinases. This recognition activates a common symbiotic pathway that regulates both arbuscular mycorrhizal symbiosis and rhizobia–legume interactions, leading to mycorrhization and nodulation (Tsikou et al., 2023). Beneficial microbes produce a variety of MAMPs, including flagellin and LPS, which have the ability to activate host immune responses. Mariutto and Ongena (2015) showed that PGPR can be detected by the plant root immune system, triggering a defense response similar to PTI at the early stages. It was found that LPS, consisting of lipid A, an inner core, and O-antigen side chains, extracted from *Pseudomonas fluorescens* WCS417 cell walls, could induce systemic resistance against *Fusarium* wilt in radish (Leeman et al., 1995). However, unlike the severe cellular damage often caused by pathogen-induced PTI, immune responses triggered by beneficial microbes tend to be transient and relatively mild. This is due to the host immune-manipulating mechanisms employed by these microbes to establish a mutually beneficial relationship. For example, the flg22 peptide isolated from the *Burkholderia phytofirmans* triggered a small oxidative burst, sufficient to transiently induce defense genes without inhibiting growth. Additionally, Zhu et al. (2022) demonstrated that the PGPR *P. fluorescens* WCS417 could suppress flagellin-triggered PTI responses in *Arabidopsis* roots by secreting low-molecular-weight compounds. In plant roots, the expansion of PGPR involves the local suppression of PTI to protect the microbes from antimicrobial compounds activated by MAMPs. This suggests a co-evolutionary process in which the host immune system is regulated following the recognition of specific signals from beneficial microbes. Overall, these findings demonstrate that beneficial microbes and their elicitors can trigger plant defense responses. However, the mechanisms behind plant-specific recognition of beneficial microorganisms and the immune responses that distinguish them from pathogens remain unclear.

Similar to bacterial species, fungi also play an important role in the induction of resistance against pathogens. *Trichoderma* spp., widely known soil fungi, interferes with plant pathogens and pests. Chet (1987) revealed in his work that *Trichoderma* spp. inhibits the growth of other microbes as it competes with other microbes for available nutrients in soil. A similar work on this soil fungi revealed that the production of lytic enzymes and antibiotics by *Trichoderma* spp. also inhibits the growth of other fungi (Woo et al., 1999). In addition to that, *Trichoderma* spp. can also enhance the resistance in crops against different types of pathogens. *Trichoderma* spp. initiates a cascade of signals in plants by upregulating the expression of root-specific transcription factor MYB72, which is essential in early induced systemic response.

The active mycorrhizal association between the host plant and ectomycorrhizal fungi (EMF) leads to the exchange of a subset of molecules including flavonoids, auxin, and secreted proteins (Garcia et al., 2015). Transcriptome analysis of EMF revealed the role of MiSSP7, 7.6, and 8 in the establishment of symbiosis. A small protein, MiSSP interacts with JAZ6, which is a key repressor of COI1 (F-box protein), which is the receptor for JA-Ile, the active form of JA, in the SCF(COI1) complex (Thines et al., 2007). The binding of COI1 with JA-Ile leads to the degradation of JAZ6 *via* the proteasome. The degradation of JAZ6 now positively regulates the expression of JA-responsive genes (Howe et al., 2018).

14.4 SYSTEMIC ACQUIRED RESISTANCE

Plants have evolved a sophisticated, multilayered immune system to protect themselves against microbial pathogens (Conrath, 2006). However, maintaining a constant state of heightened defense is resource-intensive and typically comes at the cost of reduced growth. To mitigate this trade-off, plants have developed inducible immune responses. These responses are activated specifically during or in anticipation of pathogen attacks, enabling plants to allocate resources efficiently by limiting their investment in defense to times of actual need. This dynamic system allows plants to strike a balance between defense and growth, optimizing their overall fitness. When pathogens infect a plant, they trigger inducible immune responses at the site of inoculation, including pathogen-associated molecular PTI and ETI. Additionally, localized activation of PTI and/or ETI in one leaf can lead to the induction of SAR in distant, uninfected leaves, as shown in Figure 14.1. SAR establishes a heightened immune state across the plant, effectively countering future infections and restricting the spread of otherwise compatible pathogens. This process relies on long-distance signaling between the infected leaf and distant tissues, accompanied by extensive transcriptional reprogramming in those tissues to prepare them for a stronger, primed immune response upon subsequent pathogen challenges.

The establishment of SAR is governed by two crucial immunologically active small metabolites: N-hydroxy pipecolic acid (NHP) and SA. NHP was first identified in 2018 as a naturally occurring defense metabolite in *Arabidopsis thaliana* by Hartmann and Chen (Mohnike, 2023), with subsequent studies confirming its vital role in SAR across various plant species. In contrast, SA was recognized as a key SAR-associated defense hormone approximately 25 years earlier. The production of both of these metabolites by SAR pathway is regulated by different regulatory mechanisms. A pathogen-inducible glycosyltransferase, UGT76B1, enzyme acts on NHP and SA and converts them into inactive forms NHP-O-glucoside (NHPG) (Holmes et al., 2021) and SA-O-β-glucoside

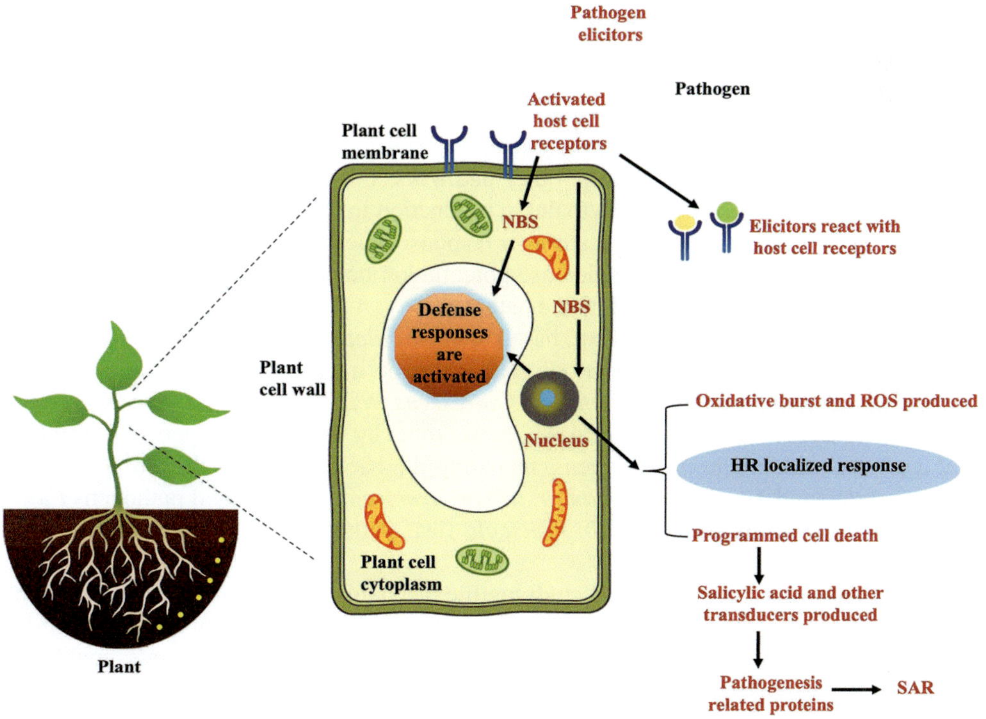

FIGURE 14.1 Mechanism of MISR.

(SAG) (Xu et al., 2022), respectively. The transcriptional co-regulator and SA receptor, NON-EXPRESSOR OF PR GENES1 (NPR1), facilitates signal transduction driven by NHP and amplified by SA. Additionally, both NHP and SA can be modified into glucose esters, and SA metabolism includes methyl esterification and various hydroxylation reactions, with the resulting products subject to further glycosylation. These metabolic pathways are intricately linked to the regulation and execution of SAR in plants.

14.5 INDUCED SYSTEMIC RESISTANCE

Unlike SAR, ISR operates through a more diverse and complex mechanism to achieve heightened resistance without requiring prior infection (Yu et al., 2022). In ISR, the activation of defense-related genes is primarily driven by the accumulation of JA and ET, rather than the SA-mediated activation of PR genes observed in SAR (Yuan et al., 2019). The signal transduction pathway in ISR is entirely independent of SA accumulation and relies exclusively on JA and ET for its activation.

The system of induced resistance in plants is highly intricate and has been partially understood through studies on model species such as *Arabidopsis*, where three main pathways of induced resistance have been identified, two of which involve the direct production of PR proteins (Van Wees et al., 2000). One pathway is typically activated by pathogenic microorganisms, while the other is triggered by wounding or necrosis-inducing pathogens. Both pathways, however, can also be induced through alternative mechanisms.

The pathogen-induced pathway primarily depends on SA as the signaling molecule, whereas the wounding-induced pathway is regulated by JA. Interestingly, these signaling compounds, along with their analogs, elicit similar responses when applied externally, indicating significant crosstalk between the pathways. The JA-dependent pathway is referred to as ISR, a term that is also applied to distinct processes initiated by rhizobacteria (Zhang et al., 2020).

The salicylate- and jasmonate-induced pathways are distinguished by the production of PR proteins, including antifungal agents such as chitinases, thaumatins, and glucanases, and oxidative enzymes like peroxidases, lipoxygenases, and polyphenol. Additionally, these pathways can lead to the accumulation of phytoalexins, which are low-molecular-weight antimicrobial compounds. A third form of induced resistance, known as rhizobacteria-induced systemic resistance (RISR), is triggered by nonpathogenic, root-associated bacteria. RISR enhances systemic resistance to plant diseases but operates differently from the other pathways. Unlike the salicylate-induced system, RISR does not induce PR proteins or phytoalexin production in the absence of a pathogen. Instead, it primes the plant's defenses, enabling a stronger response when an actual pathogen attack occurs, thereby reducing disease severity. This defense potentiation occurs without the characteristic cascade of proteins associated with the salicylate pathway.

To investigate RISR, a model system based on *Arabidopsis* was developed, leveraging its extensive use in molecular genetic studies of plant–microbe interactions. In this system, the nonpathogenic rhizobacterial strain *Pseudomonas fluorescens* WCS417r serves as the inducing agent (Berendsen et al., 2015). Colonization of *Arabidopsis* roots by WCS417r triggers ISR and protects the plant against a wide range of pathogens, including the bacterial leaf pathogens *Pseudomonas syringae* pv. tomato and *Xanthomonas campestris* pv. *armoraciae*; the fungal pathogens *Fusarium oxysporum* (root pathogen), *Alternaria brassicicola* (leaf pathogen), and *Peronospora parasitica* (oomycete leaf pathogen).

This system has proven invaluable for studying the mechanisms and effects of ISR in plants. In suppressive soils, plant roots are thought to be associated with microbial communities that benefit overall plant health. Indeed, various biocontrol PGPR can trigger ISR in host plants, enabling them to better withstand pathogen attacks on both leaves and roots, although not providing complete protection. Many effective biocontrol PGPR strains can elicit ISR, regardless of whether they produce antibiotics.

14.6 ENHANCED NUTRIENT UPTAKE

PGPR are a diverse group of bacteria found in the close association of plant root surface and rhizosphere, enhancing plant and soil health, either directly or indirectly. The term PGPR was first introduced by Kloepper and Schroth in 1978 (Singh et al., 2015). These bacteria enhance the production of growth regulators that stimulate the activity of other beneficial microbes, directly stimulating plant growth and assisting in nodulation, or indirectly encouraging nodulation. Additionally, PGPR contribute to the mineralization and uptake of essential nutrients such as iron (Fe), phosphorus (P), manganese (Mn), zinc (Zn), and copper (Cu). PGPR also influence the solubility of mineral nutrients by releasing organic acids and generating acidity through CO_2 respiration. The rhizosphere serves as an ideal environment for acid-producing bacteria (Thabah and Joshi, 2022). A key mechanism by which PGPR act is *via* the production of organic acids, which help convert insoluble phosphate compounds into more soluble forms. These bacteria affect root growth patterns and nutrient availability by modifying the quality and quantity of root exudates. In recent decades, a wide variety of bacterial species, including *Azospirillum*, *Bacillus*, *Pseudomonas*, *Burkholderia*, *Enterobacter*, *Arthrobacter*, and *Azotobacter*, has been identified as enhancers of plant growth. The use of PGPR to promote plant growth has significantly increased globally. Beneficial rhizobacteria hold great potential as part of sustainable agricultural practices, helping reduce fertilizer use and support environmentally friendly farming systems.

The favorable effects of PGPR arise from one or more of their mechanisms. The role of PGPR in both natural and managed ecosystems is crucial, as they contribute to various processes. Indigenous soil microorganisms are essential for ecosystem functioning, as they help regulate nutrient cycling, which is vital for maintaining soil fertility, and they also contribute to the development and maintenance of soil structure. PGPR play a significant role in biological activities, such as controlling plant pathogens, enhancing nutrient cycling, and promoting seedling and plant growth by producing various substances. They can solubilize inorganic phosphate and mineralize organic phosphate, and they also help in the mineralization of other essential nutrients. PGPR are capable of providing adequate iron in iron-limited soils and other important minerals, such as phosphorus. They accelerate the mineralization and uptake of certain nutrients like Fe, P, Mn, Zn, and Cu and can even facilitate nutrient transfer from dead plants to living ones, potentially increasing crop yields by 10%–30% while reducing production costs (Pattnaik et al., 2021).

In the rhizosphere, microorganisms produce biologically active substances, such as amino acids, plant hormones, vitamins, and other organic compounds and volatile substances, that stimulate the growth of AMF. PGPR also produce compounds that enhance root cell permeability, promoting increased root exudation and stimulating the growth of AMF hyphae in the rhizosphere (Nadeem et al., 2014). Composite PGPR have been shown to improve plant growth, yield attributes, and overall yield of wheat crop. Microbial activity in the rhizosphere soil directly influences root patterns and the availability of nutrients to plants. PGPR can enhance plant growth in several ways. They directly promote plant growth through atmospheric nitrogen fixation, the production of plant growth regulators, and the synthesis of siderophores that solubilize and sequester iron. Additionally, PGPR can solubilize essential minerals like phosphorus, which support plant growth at different developmental stages. Indirectly, PGPR contribute to growth by acting as biocontrol agents against soil-borne plant pathogens and weeds, producing antibiotics such as karalicin, oomycin, and azomycin, or by mitigating the harmful effects of detrimental rhizobacteria. The primary role of PGPR, however, is the suppression of plant pathogens through the release of antibiotics, cyanide, and enzymes (Prakash et al., 2022).

The precise mechanisms through which PGPR promote plant growth are not completely understood, but direct mechanisms may involve (i) symbiotic and asymbiotic nitrogen fixation; (ii) solubilizing inorganic phosphate and mineralizing organic phosphate and other nutrients; and (iii) the production of plant growth regulators. Indirect growth promotion occurs when PGPR enhance plant growth under challenging conditions. This can happen directly through the production of antagonistic compounds or indirectly by boosting the plant's resistance to pathogens. The indirect mechanisms of plant growth promotion include (i) antagonism against phytopathogens *via* the production of siderophores and cyanide, (ii) ISR, and (iii) competition for nutrients. Thus, PGPR can influence plant growth through one or more of these mechanisms and exhibit varying growth-promoting activities at different stages of the plant's life cycle (Etesami and Adl, 2020).

14.7 KEY MICROBIAL PLAYERS

The key players in MISR include beneficial rhizobacteria like *Pseudomonas*, *Bacillus*, and *Azospirillum*, and fungi such as *Trichoderma*. These microorganisms colonize plant roots and stimulate systemic resistance by activating defense pathways involving SA, JA, and ethylene signaling (Table 14.1). They boost plant immunity by promoting the production of defense enzymes and antimicrobial compounds while reinforcing cell walls, thereby protecting plants from various pathogens and environmental challenges.

TABLE 14.1
Examples of Key Players in MISR.

Microbe	Mechanism of Action	References
Bacillus spp.	Production of antimicrobial compounds (e.g., antibiotics and lipopeptides), ISR	Ryu et al. (2003), Van Loon et al. (2006)
Pseudomonas spp.	Production of siderophores and antibiotics, and induction of ISR; competition for nutrients	Pieterse et al. (1996), Liu et al. (1995)
Trichoderma spp.	Secretion of enzymes and metabolites; induction of plant defense mechanisms like SAR	Shoresh et al. (2010), Harman et al. (2004)
AMF	Enhancement of plant nutrient uptake, induction of ISR, and improvement of stress tolerance	Rodriguez et al. (2004), Bago et al. (2003)
Azospirillum spp.	Induction of ISR by enhancing root growth and nutrient uptake, and production of growth hormones	Glick (1995), Bhattacharyya and Jha (2012)
Bacillus subtilis	Induction of ISR by production of antifungal and antibacterial compounds, and enhancing plant resistance	Fadiji and Babalola (2021)
Enterobacter spp.	Production of antimicrobial compounds, enhanced root growth, and induction of resistance mechanisms	Yuan et al. (2008), Singh et al. (2011)
Streptomyces spp.	Production of antibiotics, induction of ISR, and competitive exclusion of pathogens	Sahu et al. (2013), Lee and Lee (2007)
Rhizobium spp.	Induction of systemic resistance through symbiotic relationships and nutrient exchange	Tadege et al. (2008), Kumar et al. (2011)
Serratia spp.	Induction of ISR by production of antifungal substances, promoting resistance to soil-borne pathogens	Batra et al. (2013),
Fusarium spp.	Induction of ISR by producing metabolites that activate plant defense pathways	Raza et al. (2016), Goh et al. (2019)
Penicillium spp.	Production of enzymes and secondary metabolites, and induction of plant resistance through ISR and SAR	Arora et al. (2002), Pal and McSpadden Gardener (2014)
Alternaria spp.	Induction of ISR *via* the production of volatile organic compounds (VOCs), enhancing plant defense mechanisms	Zhang et al. (2019), Choudhury et al. (2020)
Cladosporium spp.	Induction of ISR *via* VOC production and strengthening plant defense responses	Lwin et al. (2020), Shiu et al. (2021)
Beauveria bassiana	Induction of ISR through the activation of plant immune responses, especially under biotic stress conditions	Song et al. (2010), Noppe and Smagghe (2007)

TABLE 14.1 (*Continued*)
Examples of Key Players in MISR.

Microbe	Mechanism of Action	References
Colletotrichum spp.	Induction of ISR by triggering defense responses in host plants, especially under stress conditions	Chen et al. (2009), Zhang et al. (2017)
Metschnikowia spp. (yeast)	Induction of ISR by activating plant defense pathways and suppressing pathogen colonization	Shultz et al. (2015), Vujanovic and Labrecque (2009)
Candida spp. (yeast)	Induction of ISR by triggering systemic immune responses in plants, and increasing resistance to biotic stress	Hossain et al. (2021)
Pichia spp. (yeast)	Induction of ISR by producing enzymes and signaling molecules that enhance plant resistance	Chavarria et al. (2019), Hu et al. (2021)
Talaromyces spp.	Induction of ISR through the production of secondary metabolites and enhancement of plant immune responses	Sun et al. (2014), Che et al. (2017)
Neotyphodium spp. (endophyte)	Enhancing resistance to drought and pathogens through the production of alkaloids and other bioactive compounds	Johnson et al. (2013), Schardl et al. (2004)
Piriformospora indica	Induction of ISR by improving nutrient uptake, promoting root development, and activating plant defense pathways	Varma et al. (1999), Oelmüller et al. (2010)
Phoma spp.	Induction of ISR through VOC production and modulation of plant defense mechanisms	Zamioudis et al. (2013), Grunwald and Hunger (2012)
Metarhizium anisopliae	Induction of ISR by enhancing plant immune responses and promoting growth under stress conditions	Hossain et al. (2021)
Ustilago maydis	Induction of ISR by activating the plant immune system, particularly under biotic and abiotic stresses	Orozco et al. (2013)
Endophytic *Fusarium* spp.	Induction of ISR by promoting plant immune responses against biotic stress and pathogens	Li et al. (2015)

14.8 APPLICATIONS

14.8.1 Crop Protection

The superabundance of soil-borne microbes lives in close association with the plant root. Among these microbes, some are deleterious, whereas others are beneficial to host plant cells and provide nutrition after nitrogen fixation and protect against stresses caused by both biotic and abiotic factors (Bakker et al., 2018). In the current scenario, microbes found in soil are central players in enhancing sustainability in agricultural crops (Compant et al., 2019). Nowadays, to protect plants against disease-causing pathogens and reduce pests, the use of helpful microbes as biological control agents (BCAs) has been increased (Rändler-Kleine et al., 2020). Several microbes that classify as BCAs are globally recognized and used, and their positive results are widely reported. In addition to providing plant protection, these BCAs also save the environment from harmful chemicals (van Lenteren et al., 2017). Although several studies have been conducted in the past on the use of insects to control harmful pests, research on the use of microbes to control pest and plant disease is in the initial phase of trial (Mitter et al., 2019). According to published work, PGPR control microbial pathogenesis in plants by reducing pathogen virulence, producing antibiotics, competing for nutrients, and activating ISR (Barea, 2015). Similar to bacterial strains, fungi are also used to control harmful microbes. Products based on *Trichoderma* spp. are registered under various brand names and are already available in the market. Several microbes acting against various diseases and pathogens are mentioned in Table 14.2. In addition to acting against fungal pathogens, these products are also involved in the stimulation of plant defense response (Woo et al., 2014). The most reported PGPR antagonists are from the genera *Bacillus* and *Pseudomonas* (Santoyo et al., 2012). Finally, AMF are commercialized as bio-stimulants in agriculture.

TABLE 14.2
List of Microbes Acting against Specific Diseases and Pests.

S. No.	Microbe	Plant	Acts against Diseases/Pests	References
1	*Fusarium equiseti*	Pea	Disease severity and biomass reduction caused by *Fusarium avenaceum* and *Didymella pinodella*	Šišić et al. (2017)
2	*Pseudomonas fluorescens* WCS417r	*Arabidopsis*	Inhibits flagellin-triggered PTI	Millet et al. (2010)
3	*Arthrobacter*, *Nocardioides*, and *Gaiella*	Strawberry cultivars	Soil-borne fungal pathogens (*Verticillium dahliae* and *Macrophomina phaseolina*)	Lazcano et al. (2021)
4	*Gliocadium virens* (KA 2301) and *Tricoderma harzianum* (KA 159.2)	Avocado roots	*Phytophthora cinnamomi*	Syed Ab Rahman et al. (2018)
5	Entomopathogenic fungi	Cucumber seeds	*Aphis gossypii* population size	Shaalan et al. (2023)
6	*Beauveria bassiana* CEP147 strain	Tomato	Tomato spotted wilt virus (TSWV)	Emilia et al. (2023)
7	*Beauveria bassiana* and *Metarhizium anisopliae*	Cucumber	Cucumber mosaic virus (CMV)	Shaalan et al. (2022)
8	*Pseudomonas fluorescens* S97	Common bean roots	*Pseudomonas syringae* pv. *phaseolicola*	Alström S. (1991)
9	*Pseudomonas fluorescens*WCS417r–ISR	Radish	*Fusarium oxysporum*	Hoffland et al. (1995)
10	*Bacillus* spp., *Pseudomonas* spp., *Trichoderma* spp., and the AMF *Rhizophagus irregularis*	Tomato	Controlling the root pathogen *Fusarium oxysporum* or the foliar pathogen *Botrytis cinerea*	Minchev et al. (2021)
11	*Bacillus amyloliquefaciens*, *Bacillus subtilis*, *Bacillus licheniformis*, *Beauveria bassiana*	Raspberry	Spur blight	Shpatova et al. (2023)
12	*Metarhizium* spp. and *Beauveria* spp.	Olive	Olive fruit fly *Bactrocera oleae* and olive pathogens *Verticillium dahliae*, *Phytophthora megasperma*, and *Phytophthora inundata*	Lozano-Tovar et al. (2013)
13	*Beauveria bassiana*	Raspberry	Raspberry cane midge *Resseliella theobaldi* (Barnes)	Shpatova (2019)

TABLE 14.2 (*Continued*)
List of Microbes Acting against Specific Diseases and Pests.

S. No.	Microbe	Plant	Acts against Diseases/Pests	References
14	*Lecanicillium* spp.	Cucumber	Cucumber powdery mildew	Kim et al. (2007)
15	*Paecilomyces farinosus*	Barley, tomato, and tobacco	Powdery mildew colonies	Szentiványi et al. (2006)
16	*Lecanicillium muscarium*	Sweet potato	*Bemisia tabaci*	Cuthbertson and Walters (2005)
17	*Beauveria bassiana*	Eggplant	*Frankliniella occidentalis* (western flower thrips (WFT))	Zhang et al. (2019)
18	*Beauveria bassiana*	*Papaver som-niferum* L.	*Iraella luteipes* (Thompson)	Quesada-Moraga et al. (2009)
19	Chrysanthemums	*Steiner-nema carpocap-sae*	Second instars of a serpentine leafminer, *Liriomyza trifolii*	Broadbent and Olthof (1995)
20	*Verticillium lecanii*	Cucumber	Cucumber powdery mildew	Verhaar et al. (1998)
21	*Beauveria bassiana*	Maize	European corn borer *Ostrinia nubilalis*	Arnold and Lewis (2005)
22	*Beauveria bassiana*	Coffee	Coffee berry borer *Hypothenemus hampei*	Posada et al. (2007)
23	*Beauveria bassiana*	Whorl stage in corn plants *Zea mays*	*Ostrinia nubilalis*	Bing and Lewis (1991)
24	*Metarhizium anisopliae*	Cucumber	Cucumber mosaic virus (CMV)	Shaalan et al. (2022)

14.8.2 Crop Yield

Plants face constant challenges due to changes in environmental conditions and attacks from pathogens that risk their survival and pose a great threat to global food security. In the past, disease outbreaks and attacks by pests had majorly crashed agricultural output and had a negative impact on the economy of any country that was already suffering from food shortage (Savary et al., 2019). Bourke revealed in his work that the Irish potato famine in the nineteenth century was caused by *Phytophthora infestans* and led to the economic collapse of Ireland, resulting in a large number of fatalities and forced migration of citizens (Bourke, 1964). Similarly, brown spot-on rice plant disease caused by the fungus *Cochliobolus miyabeanus* led to a devasting epidemic, claiming nearly the lives of over two million people. Agricultural products suffer significant widespread yield losses due to diseases and pests, with mean losses of 30.3% and 22.6% for rice and maize, respectively. Similarly, a yield loss for wheat of nearly 21.5% and for soybean of 21.4% is reported (Savary et al., 2019). The decrease in crop yield has a severe impact on small farmers. In addition to food availability, diseased plants also have a negative impact on the diversity of species, the downstream processing of crop products, and human and animal health (Ristaino et al., 2021). Avelino and co-workers (2015) reported the huge loss in yield of coffee after coffee rust outbreaks caused by *Hemileia vastatrix*. An understanding of genetic manipulation of plant microbiome by the endogenous pathway may provide novel breeding and engineering strategies to improve sustainable yields and crop resilience. In this regard, integrating molecular biology, synthetic biology, and ecology can be crucial for uncovering the complexity of the plant beneficial microbiome and its usage in the development of high-yielding, smart, and climate-resilient crops in the future.

14.9 CHALLENGES AND CONSIDERATIONS

Despite the growing body of research on ISR-inducing microbes, significant gaps remain in understanding how to create environmental conditions that promote ISR and how to address factors that lead to unpredictability in its activation. More interdisciplinary research focusing on the mechanisms behind the context-dependency of plant–microbe interactions will improve the predictability of ISR-based products and microbial inoculants in general (Díaz, 2024). Future studies should be conducted under conditions that either (i) better replicate the complex conditions of agricultural systems or (ii) specifically test targeted environmental conditions. Ideally, such efforts should be driven by research while considering the realities of current agricultural practices. Areas of cooperation could include (i) partnerships between the commercial sector and academia, (ii) interdisciplinary studies across countries using a common study system to assess context-dependency, and (iii) long-term experiments involving farmers or land managers to account for the variability and uncertainty observed under field conditions. These interdisciplinary experiments will require considerable effort from all involved parties to establish clear objectives and agricultural practices to generate reliable and comparable data. By combining interdisciplinary research, field surveys, and controlled experiments, a more comprehensive understanding of ISR mechanisms can be gained, accelerating the development of ISR-based technologies (Lee Díaz et al., 2021).

Fostering collaboration among academia, companies, and research institutions is crucial to addressing current challenges in crop protection. For example, Kohl et al. (2011) stated that large-scale field studies and microbial screening processes, although costly and time-consuming, generate a wealth of data. While many microbial candidates are discarded if they fail to produce the desired outcomes under multiple conditions, the data from these "unsuccessful" ISR events can provide valuable insights. Thus, screening experiments that test large numbers of microbial strains for ISR activation offer an opportunity to explore the context-dependency of ISR. The absence of published unsuccessful ISR events in the literature creates a bias toward positive results, potentially leading to duplicated efforts (Walters et al., 2013). Expanding multidisciplinary studies and increasing data availability could pave the way for more effective ISR-based technologies, although regulatory frameworks also need to be adjusted.

Regulatory frameworks can be a significant barrier to innovation, slowing the progress of research and development by manufacturers and biotechnology companies if compliance costs outweigh the benefits. The European Union's regulatory framework, e.g., requires specific tests and standards for certain product categories but does not adequately assess the potential of ISR-inducing microbes. The tests for bio-stimulants focus on plant tolerance to abiotic stress and nutrient uptake efficiency rather than biotic stress resistance. A study emphasized that evaluating products like biofertilizers requires the consideration of external conditions, yet the current legislation does not fully address the specific processes and effects of ISR-based innovations.

Currently, a wide range of products, differing in composition and action, are marketed as biofertilizers or bio-stimulants. Under the new European fertilizer regulation (2019/1009), some microbe-based bio-stimulants with "multifunctional" features, including ISR, are not adequately recognized (Hajji-Hedfi et al., 2025). Ultimately, the challenge for the legislation is to strike a balance—allowing manufacturers to develop products within an appropriate framework while ensuring sufficient structure to guarantee the quality and reproducibility of these innovations.

14.10 FUTURE DIRECTIONS

The future of MISR in agriculture is bright, driven by advances in microbiome research and biotechnological innovations. Potential future directions include the following.

14.10.1 Role of CRISPR in Disease Management

Site-specific editing of plant, animal, and microbial genome can lead to an increase in more efficient systems. Among other genome editing techniques, CRISPR-cas9 is more efficient and less time consuming and yields good transgenic products. The CRISPR-CAS9 has become the apparent technique of choice for the management of plant diseases (Muñoz et al., 2019). Several studies have reported where the CRISPR-Cas9 system has been used for understanding the resistance mechanisms in microbes. Yi et al. (2018) in their work explored the role of the CRISPR-Cas9 system for understanding the system behind the interaction between *Bacillus* and plants. The modification of bacterial strains using CRSIPR-Cas9 can enhance crop yield and plant disease resistance. Figueredo and Quecine (2021) used the CRISPR-Cas9 method in double plasmids, where Cas9 endonuclease interacts with one plasmid while the second plasmid harbors the SgRNA to support the gene editing in Gram-positive *Bacillus* spp. Iaffaldano and coworkers revealed in their work that the yield of rubber-producing plants can be increased by targeting a specific gene encoding for fructan: fructan 1-fructosyltransferase (1-FFT), which acts as an antagonist to rubber production. For this work, inoculation of *Agrobacterium rhizogenes* harboring a plasmid interacting with Cas9 was carried out in the rubber producing plant (2016). As in these studies, the CRISPR-Cas9 system can be used for enhancing plant–microbe interactions, specifically of microbes involved in resistance against diseases, and promoting plant growth and can act as a natural substitute for pesticides (Dean et al., 2012)

14.10.2 Microbial Co-Culture

Microbial consortia is a group of compatible microbes that act together to enhance their cumulative effect. Microbes play an important role in increasing agricultural productivity. Further use of a combination of microbes can easily lead to the transformation of agricultural practices either by increasing the availability of nutrients or by inducing resistance to diseases (Jaiswal et al., 2022; Dheeman et al., 2023). Furthermore, a microbial consortium also improves soil health, promotes plant growth, handles stress caused by abiotic factors (Méndez-Bravo et al., 2023), and manages abiotic stress (Mahreen et al., 2023). In addition to increasing yield, microbial co-culture also promotes the sustainable growth of agriculture and reduces the negative impact on the environment. Therefore, further research is required for the complete elimination of chemical pesticides and weedicides from the agricultural system. In current scenario, the production of microbial consortium at large scale is required to get rid of synthetic pesticides. Furthermore, regulatory and ethical issues should also be considered to make it more successful.

REFERENCES

Alström, S. (1991). Induction of disease resistance in common bean susceptible to halo blight bacterial pathogen after seed bacterization with rhizosphere pseudomonads. *The Journal of General and Applied Microbiology*, 37, 495–501. https://doi.org/10.2323/jgam.37.495

Arnold, A. E., & Lewis, L. C. (2005). Ecology and evolution of fungal endophytes, and their roles against insects. In: Vega, F. E., & Blackwell, M. (eds), *Insect–Fungal Associations: Ecology and Evolution*. Oxford University Press, New York, pp. 74–96.

Arora, D. K., Khare, E., & Singh, S. (2002). Induction of plant defense by *Penicillium* spp. *Current Science*, 83(5), 580–585. https://doi.org/10.3389/fmicb.2018.02732.

Avelino, J., Cristancho, M., Georgiou, S., Imbach, P., Anguilar, L., Bornemann, G., Läderach, P., Anzueto, F., Hruska, A., J. & Morales, C. (2015). The coffee rust crises in Colombia and Central America (2008–2013): Impacts, plausible causes and proposed solutions. *Food Security*, 7, 303–321. https://doi.org/10.1007/s12571-015-0446-9

Bago, B., Pfeffer, P. E., & Shachar-Hill, Y. (2003). Translocation and metabolism of fungal storage lipids in arbuscular mycorrhizas. *Plant Physiology*, 131(3), 1498–1507. https://doi.org/10.1104/pp.010466.

Bai, Y., Souleimanov, A., & Smith, D. L. (2002) An inducible activator produced by a *Serratia proteamaculans* strain and its soybean growth-promoting activity under greenhouse conditions. *Journal of Experimental Botany*, 53(373), 1495–1502. https://doi.org/10.1093/jexbot/53.373.1495.

Bakker, P. A. H. M., Pieterse, C. M. J., de Jonge, R., & Berendsen, R. L. (2018). The soil-borne legacy. *Cell*, 172, 1178–1180. https://doi.org/10.1016/j.cell.2018.02.024.

Barea, J. M. (2015). Future challenges and perspectives for applying microbial biotechnology in sustainable agriculture based on a better understanding of plant–microbiome interactions. *Journal of Soil Science and Plant Nutrition*, 15, 261–282. https://doi.org/10.4067/S0718-95162015005000021.

Batra, P., Saini, A., & Singh, H. (2013). Serratia species: A versatile PGPR. *Archives of Microbiology*, 195(5), 407–413.

Berendsen, R. L., van Verk, M. C., Stringlis, I. A., Zamioudis, C., Tommassen, J., Pieterse, C. M., & Bakker, P. A. (2015). Unearthing the genomes of plant-beneficial *Pseudomonas* model strains WCS358, WCS374 and WCS417. *BMC Genomics*, 16, 1–23.

Bhattacharyya, P. N., & Jha, D. K. (2012). Plant growth-promoting rhizobacteria (PGPR): Emergence in agriculture. *World Journal of Microbiology and Biotechnology*, 28, 1327–1350. https://doi.org/10.1007/s11274-011-0979-9.

Bigeard, J., Colcombet, J., & Hirt, H. (2015). Signaling mechanisms in pattern-triggered immunity (PTI). *Molecular Plant*, 8, 521–539. https://doi.org/10.1016/j.molp.2014.12.022.

Bing, L., & Lewis, L. (1991). Suppression of *Ostrinia nubilalis* (Hübner) (Lepidoptera: Pyralidae) by endophytic *Beauveria bassiana* (Balsamo) vuillemin. *Environmental Entomology*, 20, 1207–1211. https://doi.org/10.1093/ee/20.4.1207.

Bourke, P. M. A. (1964). Emergence of potato blight, 1943–1946. *Nature*, 203, 805–808. https://doi.org/10.1038/203805a0.

Broadbent, A., & Olthof, T. (1995). Foliar application of *Steinernema carpocapsae* (Rhabditida: Steinernelnatidae) to control *Liriomyza trifolii* (Diptera: Agromyzidae) larvae in chrysanthemums. *Environmental Entomology*, 24, 431–435. https://doi.org/10.1093/EE/24.2.431

Chavarria, M., Torres, P., & Molina, A. (2019). *Pichia* spp. in plant defense activation. *Fungal Biology*, 123(7), 493–500. https://doi.org/10.3390/nu9090966.

Che, Q., Luo, M., & Wang, H. (2017). Secondary metabolites from *Talaromyces* spp. enhancing ISR. *Journal of Applied Microbiology*, 122(6), 1544–1554. https://doi.org/10.3390/md16010012.

Chen, L., Zhang, F., & Guo, X. (2009). Colletotrichum-mediated ISR in plants. *Plant Science*, 177(6), 554–560.

Chet, I. (1987). *Trichoderma Application, Mode of Action, and Potential as a Biocontrol Agent of Soilborne Plant Pathogenic Fungi*. John Wiley & Sons, New York.

Choudhury, D., Yadav, G., & Sharma, R. (2020). Role of *Alternaria* in systemic resistance activation. *Microbial Ecology*, 80(3), 622–632.

Compant, S., Samad, A., Faist, H., & Sessitsch, A. (2019). A review on the plant microbiome: Ecology, functions, and emerging trends in microbial application. *Journal of Advanced Research*, 19, 29–37. https://doi.org/10.1016/j.jare.2019.03.004

Conrath, U. (2006). Systemic acquired resistance. *Plant Signaling & Behavior*, 1(4), 179–184. https://doi.org/10.4161/psb.1.4.3221

Cuthbertson, A., & Walters, K. (2005). Pathogenicity of the entomopathogenic fungus, *Lecanicillium muscarium*, against the sweet potato whitefly *Bemisia tabaci* under laboratory and glasshouse conditions. *Mycopathologia*, 160, 315–319. https://doi.org/10.1007/s11046-005-0122-2.

Dean, R., Van Kan, J. A., Pretorius, Z. A., Hammond-Kosack, K. E., Di Pietro, A., Spanu, P. D., Rudd, J. J., Dickman, M., Kahmann, R., Ellis, J., & Foster, G. D. (2012). The top 10 fungal pathogens in molecular plant pathology. *Molecular Plant Pathology*, 13(4), 414–430. https://doi.org/10.1111/j.1364-3703.2011.00783.x

del Carmen Orozco-Mosqueda, M., Velázquez-Becerra, C., Macías-Rodríguez, L. I., Santoyo, G., Flores-Cortez, I., Alfaro-Cuevas, R., & Valencia-Cantero, E. (2013). Arthrobacter agilis UMCV2 induces iron acquisition in Medicago truncatula (strategy I plant) in vitro via dimethylhexadecylamine emission. *Plant and Soil*, 362(1), 51–66.

Dheeman, S., Kumar, M., & Maheshwari, D. K. (2023). Beneficial microbial mixtures for efficient biocontrol of plant diseases: Impediments and success. *In: Sustainable Agrobiology: Design and Development of Microbial Consortia*. Springer Nature Singapore, Singapore, pp. 23–40.

Díaz, L. S. (2024, March 27). *Licence agreement concerning inclusion of doctoral thesis in the institutional repositor of the University of Leiden*. Downloaded from: https://hdl.handle.net/1887/3729491.

Donley, N. (2019). The USA lags behind other agricultural nations in banning harmful pesticides. *Environmental Health*, 18, 44. https://doi.org/10.1186/s12940-019-0488-0.

Emilia, M., Domingues, S., Del Pino, M., & Toledo, A. (2023). Biological control of *Frankliniella occidentalis* on greenhouse bell pepper using *Beauveria bassiana* in combination with soil cover practices. *Journal of Plant Protection Research*, 62(4), 393–402. https://doi.org/10.24425/jppr.2022.144096

Etesami, H., & Adl, S. M. (2020). Plant growth-promoting rhizobacteria (PGPR) and their action mechanisms in availability of nutrients to plants. *Phyto-Microbiome in Stress Regulation*, 147–203. https://doi.org/10.1007/978-981-15-2576-6_9.

Fadiji, A. E., & Babalola, O. O. (2021). Harnessing endophytes for plant health in Africa. *Microorganisms*, 9(2), 320. https://doi.org/10.3390/plants10050935.

Felix, G., Duran, J. D., Volko, S., & Boller, T. (1999). Plants have a sensitive perception system for the most conserved domain of bacterial flagellin. *The Plant Journal*, 18(3), 265–276. https://doi.org/10.1046/j.1365-313x.1999.00265.x

Figueredo, E., & Quecine, M. (2021). Gene knockout of beneficial plant-associated *Bacillus* spp. using the CRISPR-Cas9 double plasmid system. *Methods in Molecular Biology*, 2232, 173–191. https://doi.org/10.1007/978-1-0716-1040-4_15.

Fones, H. N., Bebber, D. P., Chaloner, T. M., Kay, W. T., Steinberg, G., Gurr, S. J. (2020). Threats to global food security from emerging fungal and oomycete crop pathogens. *Nature Food*, 1, 332–342.

Gaffney, T., Friedrich, L., Vernooij, B., Negrotto, D., Nye, G., Uknes, S., Ward, E., Kessmann, H., & Ryals, J. (1993). Requirement of salicylic-acid for the induction of systemic acquired-resistance. *Science*, 261, 754–756. https://doi.org/10.1126/science.261.5122.754.

Garcia, K., Delaux, P.-M., Cope, K. R., & Ané, J.-M. (2015). Molecular signals required for the establishment and maintenance of ectomycorrhizal symbioses. *New Phytology*, 208, 79–87. https://doi.org/10.1111/nph.13423.

Glick, B. R. (1995). The enhancement of plant growth by free-living bacteria. *Canadian Journal of Microbiology*, 41(2), 109–117. https://doi.org/10.1139/m95-015.

Goh, Y. K., Lee, C. Y., & Yap, M. T. (2019). Fusarium-induced systemic resistance and secondary metabolites. *Frontiers in Plant Science*, 10, 843.

Grunwald, S., & Hunger, S. (2012). *Phoma* spp. triggering systemic resistance in plants. *Plant Pathology*, 61(2), 263–272.

Guo, M., Tian, F., Wamboldt, Y., & Alfano, J.-R. (2009). The majority of the type III effector inventory of *Pseudomonas syringae* pv. tomato DC3000 can suppress plant immunity. *Molecular Plant Microbe Interactions*, 22, 1069–1080. https://doi.org/10.1094/MPMI-22-9-1069.

Hajji-Hedfi, L., Ibrahim, D. S., Othmen, S. B., El-Abeid, S. E., Hlaoua, W., Mosa, M. A., Rhouma, A., Riad, S. N., Ghareeb, S., El-Deriny, M. M., & Harb, H. E. (2025). Production of microbial biostimulants as a response to the modern agricultural need for productivity and plant health. *In: Microbial Biostimulants*. Apple Academic Press, pp. 31–78.

Harman, G. E., Howell, C. R., Viterbo, A., Chet, I., & Lorito, M. (2004). *Trichoderma* species—opportunistic, avirulent plant symbionts. *Nature Reviews Microbiology*, 2(1), 43–56. https://doi.org/10.1038/nrmicro797.

Hoffland, E., Pieterse, C. M. J., Bik, L., & Van Pelt, J. A. (1995). Induced systemic resistance in radish is not associated with accumulation of pathogenesis-related proteins. *Physiological and Molecular Plant Pathology*, 46, 309–320. https://doi.org/10.1006/pmpp.1995.1024.

Holmes, E. C., Chen, Y. C., Mudgett, M. B., & Sattely, E. S. (2021). *Arabidopsis* UGT76B1 glycosylates N-hydroxy-pipecolic acid and inactivates systemic acquired resistance in tomato. *The Plant Cell*, 33(3), 750–765. https://doi.org/10.1093/plcell/koaa052.

Hossain, M. S., Chowdhury, R., & Fujita, M. (2021). Yeast-mediated plant immunity. *Journal of Fungi*, 7(5), 412.

Howe, G. A., Major, I. T., & Koo, A. J. (2018). Modularity in jasmonate signaling for multistress resilience. *Annual Review of Plant Biology*, 69, 387–415. https://doi.org/10.1146/annurev-arplant-042817-040047.

Hu, J., Cao, W., & Sun, X. (2021). Role of *Pichia* spp. in ISR and SAR. *Microbiological Research*, 245, 126703.

Iaffaldano, B., Zhang, Y., & Cornish, K. (2016). CRISPR/Cas9 genome editing of rubber producing dandelion Taraxacum kok-saghyz using *Agrobacterium* rhizogenes without selection. *Industrial Crops and Products*, 89, 356–362. https://doi.org/10.1016/j.indcrop.2016.05.029.

Jain, S., Varma, A., Tuteja, N., & Choudhary, D. K. (2016). Plant growth-promoting microbial-mediated induced systemic resistance in plants: Induction, mechanism, and expression. *In: Microbial-mediated induced systemic resistance in plants*, pp. 213–226. https://doi.org/10.1007/978-981-10-0388-2_15

Jaiswal, D. K., Krishna, R., Chouhan, G. K., Araujo Pereira, d. e., Ade, A. P., Prakash, S., Verma, S. K., Prasad, R., Yadav, J., & Verma, P. J. (2022). Bio-fortification of minerals in crops: Current scenario and future prospects for sustainable agriculture and human health. *Plant Growth Regulation*, 98, 5–22. https://doi.org/10.1007/s10725-022-00847-4.

Johnson, L. J., De Bonth, A. C., & Briggs, L. R. (2013). Endophytic *Neotyphodium* spp. enhancing drought resistance. *Functional Plant Biology*, 40(9), 810–820.

Jones, J. D. G., & Dangl, J. L. (2006). The plant immune system. *Nature*, 444, 323–329. https://doi.org/10.1038/nature05286.

Kim, J., Goettel, M., & Gillespie, D. (2007). Potential of *Lecanicillium* species for dual microbial control of aphids and the cucumber powdery mildew fungus, *Sphaerotheca fuliginea*. *Biological Control*, 40, 327–332. https://doi.org/10.1016/j.biocontrol.2006.12.002.

Kloepper, J. W., & Beauchamp, C. J. (1992). A review of issues related to measuring colonization of plant roots by bacteria. *Canadian Journal of Microbiology*, 38, 1219–1232.

Kloepper, J. W., Tuzun, S., & Kuc, J. A. (1992). Proposed definitions related to induced disease resistance. *Biocontrol Science and Technology*, 2, 349–351. https://doi.org/10.1080/09583159209355251.

Köhl, J., Postma, J., Nicot, P., Ruocco, M., & Blum, B. (2011). Stepwise screening of microorganisms for commercial use in biological control of plant-pathogenic fungi and bacteria. *Biological Control*, 57(1), 1–12. https://doi.org/10.1016/j.biocontrol.2010.12.004.

Kørner, C. J., Klauser, D., Niehl, A., Domínguez-Ferreras, A., Chinchilla, D., Boller, T., Heinlein, M., & Hann, D. R. (2013). The immunity regulator BAK1 contributes to resistance against diverse RNA viruses. *Molecular Plant–Microbe Interactions*, 26(11), 1271–1280. https://doi.org/10.1094/MPMI-06-13-0179-R.

Kumar, A., Patil, D., Rajkumar, S., & Freitas, H. (2011). Plant–microbe interactions in rhizosphere. *Applied Microbiology and Biotechnology*, 89(4), 971–984

Lazcano, C., Boyd, E., Holmes, G., Hewavitharana, S., Pasulka, A., & Ivors, K. (2021). The rhizosphere microbiome plays a role in the resistance to soil-borne pathogens and nutrient uptake of strawberry cultivars under field conditions. *Scientific Reports*, 11, 3188. https://doi.org/10.1038/s41598-021-82768-2.

Lee, J. H., & Lee, S. W. (2007). Role of streptomyces in ISR against plant pathogens. *World Journal of Microbiology and Biotechnology*, 23(3), 401–405.

Lee Díaz, A. S., Macheda, D., Saha, H., Ploll, U., Orine, D., & Biere, A. (2021). Tackling the context-dependency of microbial-induced resistance. *Agronomy*, 11(7), 1293. https://doi.org/10.3390/agronomy11071293

Leeman, M., Van Pelt, J. A., Den Ouden, F. M., Heinsbroek, M., Bakker, P. A. H. M., & Schippers, B. (1995). Induction of systemic resistance against *Fusarium* wilt of radish by lipopolysaccharides of *Pseudomonas fluorescens*. *Phytopathology*, 85(9), 1021–1027. https://doi.org/10.1094/phyto-85-1021.

Li, R.-X., Cai, F., Pang, G., Shen, Q.-R., Li, R., & Chen, W. (2015). Solubilisation of phosphate and micronutrients by *Trichoderma harzianum* and its relationship with the promotion of tomato plant growth. *PLoS One*, 10, e0130081. https://doi.org/10.1371/journal.pone.0130081

Liu, L., Kloepper, J. W., & Tuzun, S. (1995). Induction of systemic resistance by *Pseudomonas fluorescens* in cucumber. *Phytopathology*, 85(6), 695–702.

Lozano-Tovar, M. D., Ortiz-Urquiza, A., Garrido-Jurado, I., Trapero-Casas, A., & Quesada-Moraga, E. (2013). Assessment of entomopathogenic fungi and their extracts against a soil-dwelling pest and soil-borne pathogens of olive. *Biological Control*, 67(3), 409–420. https://doi.org/10.1016/j.biocontrol.2013.09.006.

Lu, D., Wu, S., Gao, X., Zhang, Y., Shan, L., & He, P. (2010). A receptor-like cytoplasmic kinase, BIK1, associates with a flagellin receptor complex to initiate plant innate immunity. *Proceedings of the National Academy of Sciences of the United States of America*, 107(1), 496–501. https://doi.org/10.1073/pnas.0909705107.

Lugtenberg, B., & Kamilova, F. (2009). Plant-growth-promoting rhizobacteria. *Annual Review of Microbiology*, 63, 541–556. https://doi.org/10.1146/annurev.micro.62.081307.162918.

Lwin, K. M., Myint, Y. Y., & Win, M. M. (2020). Cladosporium VOCs enhancing plant defense. *Applied Microbiology and Biotechnology*, 104(11), 4873–4885.

Mahreen, N., Yasmin, S., Asif, M., Yahya, M., Ejaz, K., Yousaf, S., et al. (2023). Mitigation of water scarcity with sustained growth of rice by plant growth promoting bacteria. *Frontier in Plant Science*, 14, 1081537. https://doi.org/10.3389/fpls.2023.1081537.

Malvino, M. L. (2024). Unraveling the dynamics of *Xanthomonas*' flagella: Insights into host-pathogen interactions. *PeerJ*, 12, e18204. https://doi.org/10.7717/peerj.18204.

Mariutto, M., & Ongena, M. (2015). Molecular patterns of rhizobacteria involved in plant immunity elicitation. *Advances in Botanical Research*, 75, 21–56. https://doi.org/10.1016/bs.abr.2015.07.002.

Méndez-Bravo, A., Herrera-Cornelio, L. C., García-Toscano, D. F., Kiel-Martínez, A. L., Guevara-Avendaño, E., Ramírez-Vázquez, M., et al. (2023). Beneficial effects of selected rhizospheric and endophytic bacteria, inoculated individually or in combination, on non-native host plant development. *Rhizosphere*, 26, 100693. https://doi.org/10.1016/j.rhisph.2023.100693.

Minchev, Z., Kostenko, O., Soler, R., & Pozo, M. (2021). Microbial consortia for effective biocontrol of root and foliar diseases in tomato. *Frontiers in Plant Science*, 12. https://doi.org/10.3389/fpls.2021.756368.

Millet, Y. A., Danna, C. H., Clay, N. K., Songnuan, W., Simon, M. D., Werck-Reichhart, D., & Ausubel, F. M. (2010). Innate immune responses activated in *Arabidopsis* roots by microbe-associated molecular patterns. *Plant Cell*, 22, 973–990. https://doi.org/10.1105/tpc.109.069658.

Mitter, B., Brader, G., Pfaffenbichler, N., & Sessitsch, A. (2019). Next generation microbiome applications for crop production—limitations and the need of knowledge-based solutions. *Current Opinion Microbiology*, 49, 59–65. https://doi.org/10.1016/j.mib.2019.10.006.

Muñoz, I. V., Sarrocco, S., Malfatti, L., Baroncelli, R., & Vannacci, G. (2019). CRISPR-Cas for fungal genome editing: A new tool for the management of plant diseases. *Frontier Plant Science*, 10, 135. https://doi.org/10.3389/fpls.2019.00135.

Nadeem, S. M., Ahmad, M., Zahir, Z. A., Javaid, A., & Ashraf, M. (2014). The role of mycorrhizae and plant growth promoting rhizobacteria (PGPR) in improving crop productivity under stressful environments. *Biotechnology Advances*, 32(2), 429–448. https://doi.org/10.1016/j.biotechadv.2013.12.005.

Noppe, C., & Smagghe, G. (2007). Induction of plant immune responses by *Beauveria bassiana*. *Pest Management Science*, 63(11), 1069–1074.

Oelmüller, R., Sherameti, I., & Tripathi, S. (2010). Mechanisms of *Piriformospora indica* in plant defense. *Journal of Plant Interactions*, 5(1), 15–18. https://doi.org/10.1007/s13199-009-0009-y.

Pal, K. K., & McSpadden Gardener, B. B. (2014). Role of *Penicillium* in agriculture. *Biological Control*, 73(4), 27–37.

Pattnaik, S., Mohapatra, B., & Gupta, A. (2021). Plant growth-promoting microbe mediated uptake of essential nutrients (Fe, P, K) for crop stress management: Microbe-soil-plant continuum. *Frontiers in Agronomy*, 3, 689972.

Pieterse, C. M., Van Wees, S. C., Hoffland, E., Van Pelt, J. A., & Van Loon, L. C. (1996). Systemic resistance in *Arabidopsis* induced by biocontrol bacteria is independent of salicylic acid accumulation and pathogenesis-related gene expression. *The Plant Cell*, 8(8), 1225–1237. https://doi.org/10.1105/tpc.8.8.1225.

Pieterse, C. M. J., Ton, J., & van Loon, L. C. (2001) Cross-talk between plant defence signaling pathways: Boost or burden? *AgBiotechNet*, 3, 1–18.

Posada, F., Petersonc, S., Rehnera, S., & Vegaa, F. (2007). Inoculation of coffee plants with the fungal entomopathogen *Beauveria bassiana* (Ascomycota: Hypocreales). *Mycological Research*, 111(6), 748–757. https://doi.org/10.1016/j.mycres.2007.03.006.

Prakash, R., Subramani, R., Krodi, Anusha, Berde, C. V., Chandrasekhar, T., Prathyusha, M., Kariali, E., Bramhachari, P. V. (2022). Rhizobacteriome: Plant growth-promoting traits and its functional mechanism in plant growth, development, and defenses. *In: Understanding the Microbiome Interactions in Agriculture and the Environment* (pp. 315–344). Singapore: Springer Nature Singapore. https://doi.org/10.1007/978-981-19-3696-8_16.

Quesada-Moraga, E., Muñoz-Ledesma, F., & Santiago-Álvarez, C. (2009). Systemic protection of *Papaver somniferum* L. against *Iraella luteipes* (Hymenoptera: Cynipidae) by an endophytic strain of *Beauveria bassiana* (Ascomycota: Hypocreales). *Environmental Entomology*, 38(3), 723–730. https://doi.org/10.1603/022.038.0324.

Rändler-Kleine, M., Wolfgang, A., Dietel, K., Junge, H., Cernava, T., & Berg, G. (2020). How microbiome approaches can assist industrial development of biological control products. In: Gao, Y., Hokkanen, H., & Menzler-Hokkanen, I. (eds), *Integrative Biological Control*, Vol. 20. Springer, Berlin, pp. 201–215. https://doi.org/10.1007/978-3-030-44838-7_13.

Raza, W., Yuan, J., & Shen, Q. (2016). Role of *Fusarium* spp. in induced systemic resistance. *Journal of Basic Microbiology*, 56(6), 550–560.

Ristaino, J. B., Anderson, P. K., Bebber, D. P., Brauman, K. A., Cunniffe, N. J., Fedoroff, N. V., Finegold, C., Garrett, K. A., Gilligan, C. A., Jones, C. M., & Martin, M. D. (2021). The persistent threat of emerging plant disease pandemics to global food security. *Proceeding of National Academy of Sciences of the Unites States of America*, 118, e2022239118. https://doi.org/10.1073/pnas.2022239118.

Rodriguez, A., Sanders, I. R., & Fitter, A. H. (2004). Benefits and drawbacks of arbuscular mycorrhizal fungi in agriculture. *Trends in Plant Science*, 9(6), 267–273.

Ryu, C. M., Farag, M. A., Hu, C. H., Reddy, M. S., Wei, H. X., Paré, P. W., & Kloepper, J. W. (2003). Bacterial volatiles promote growth in *Arabidopsis*. *Proceedings of the National Academy of Sciences of the United States of America*, 100(8), 4927–4932. 10.1073/pnas.0730845100.

Sahu, P. K., Singh, H. B., & Gupta, A. (2013). Streptomyces in plant growth and disease management. *Plant Pathology Journal*, 29(1), 35–45.

Salwan, R., Sharma, V., Sharma, A., & Singh, A. (2020). Molecular imprints of plant beneficial *Streptomyces* sp. AC30 and AC40 reveal differential capabilities and strategies to counter environmental stresses *Microbiological Research*, 235, 126449. https://doi.org/10.1016/j.micres.2020.126449.

Santoyo, G., Orozco-Mosqueda, M. D. C., & Govindappa, M. (2012). Mechanisms of biocontrol and plant growth-promoting activity in soil bacterial species of Bacillus and Pseudomonas: A review. *Biocontrol Science and Technology*, 22(8), 855–872.

Savary, S., Willocquet, L., Pethybridge, S. J., Esker, P., McRoberts, N., & Nelson, A. (2019). The global burden of pathogens and pests on major food crops. *Nature Ecology and Evolution*, 3, 430–439. https://doi.org/10.1038/s41559-018-0793-y.

Schardl, C. L., Leuchtmann, A., & Spiering, M. J. (2004). Alkaloid production by *Neotyphodium* endophytes. *Mycological Research*, 108(1), 173–177.

Shaalan, R., Ibrahim, L., As-Sadi, F., & Kayal, E. (2022). Impact of *Beauveria bassiana* and *Metarhizium anisopliae* on the metabolic interactions between cucumber (*Cucumis sativus* L.) and cucumber mosaic virus (CMV). *Horticulturae*, 8(12), 1182. https://doi.org/10.3390/horticulturae8121182.

Shaalan, R., Gerges, E., Habib, W., & Ibrahim, L. (2023). Endophytic colonization by *Beauveria bassiana* and *Metarhizium anisopliae* induces growth promotion effect and increases the resistance of cucumber plants against *Aphis gossypii*. *Journal of Plant Protection Research*, 61(4), 358–370. https://doi.org/10.24425/jppr.2021.139244.

Shiu, J. Y., Wu, P. C., & Ho, H. M. (2021). Role of *Cladosporium* in ISR modulation. *Frontiers in Microbiology*, 12, 561.

Shoresh, M., Harman, G. E., & Mastouri, F. (2010). Induced systemic resistance and plant responses to fungal biocontrol agents. *Annual Review of Phytopathology*, 48(1), 21–43. https://doi.org/10.1146/annurev-phyto-073009-114450.

Shpatova, T. (2019). Entomopathogenic fungus *Beauveria bassiana* (Bals-Criv.) Vuill. As a promising agent for the raspberry cane midge *Resseliella theobaldi* (Barnes) biocontrol. *Agricultural Biology*, 54(3), 505–511. https://doi.org/10.15389/AGROBIOLOGY.2019.3.505ENG.

Shpatova, T., Shternshis, M., & Kozlova, A. (2023). Biological control of raspberry spur blight caused by *Didymella applanata* using a mixture of microbial agents. *Innovations and Food Safety*. https://doi.org/10.31677/2311-0651-2023-40-2-87-94

Shultz, J. C., Baldwin, I. T., & Shepherd, R. W. (2015). *Metschnikowia* spp. enhancing plant defenses. *Phytopathology*, 105(10), 1207–1216.

Singh, P. O. O. J. A., Singh, P. R. A. S. H. A. N. T., & Singh, M. P. (2015). Assessment of antifungal activity of PGPR (plant growth-promoting rhizobacterial). *International Journal*, 3(10), 803–812.

Singh, P. P., Shin, Y. C., Park, C. S., & Chung, Y. R. (2011). Biological control of soilborne pathogens using bacterial antibiotics. *Pest Management Science*, 67(9), 1123–1131. https://doi.org/10.3390/microorganisms10091759.

Šišić, A., Baćanović, J., & Finckh, M. R. (2017). Endophytic *Fusarium equiseti* stimulates plant growth and reduces root rot disease of pea (*Pisum sativum* L.) caused by *Fusarium avenaceum* and *Peyronellaea pinodella*. *European Journal of Plant Pathology*, 148, 271–282. https://doi.org/10.1007/s10658-016-1086-4

Song, W., Zhou, L., & Yang, C. (2010). *Beauveria bassiana*: A new perspective in plant defense. *Plant Cell Reports*, 29(5), 495–502. https://doi.org/10.1016/j.jip.2008.01.010.

Sun, Y., Cao, H., & Li, S. (2014). *Talaromyces*-mediated systemic resistance. *Microbial Ecology*, 67(4), 749–760.

Syed Ab Rahman, S. F., Singh, E., Pieterse, C. M. J., & Schenk, P. M. (2018). Emerging microbial biocontrol strategies for plant pathogens. *Plant Science*, 267, 102–111. https://doi.org/10.1016/j.plantsci.2017.11.012.

Szentiványi, O., Varga, K., Wyand, R., Slatter, H., & Kiss, L. (2006). *Paecilomyces farinosus* destroys powdery mildew colonies in detached leaf cultures but not on whole plants. *European Journal of Plant Pathology*, 115, 351–356. https://doi.org/10.1007/s10658-006-9011-x.

Tadege, M., Kahyaoglu, E., & Mitra, A. (2008). The symbiotic role of rhizobia in plants. *Plant Journal*, 55(4), 426–442.

Thabah, S., & Joshi, S. R. (2022). Plant growth promoting rhizobacteria from the perspectives of tea plantations and diseases. *In: New and Future Developments in Microbial Biotechnology and Bioengineering*. Elsevier, pp. 315–332. https://doi.org/10.1016/B978-0-323-85163-3.00005-3.

Thines, B., Katsir, L., Melotto, M., Niu, Y., Mandaokar, A., Liu, G., Nomura, K., He, S.Y., Howe, G. A., & Browse, J. (2007). JAZ repressor proteins are targets of the SCF(COI1) complex during jasmonate signalling. *Nature*, 448, 661–665. https://doi.org/10.1038/nature05960.

Trivedi, P., Leach, J. E., Tringe, S. G., Sa, T., & Singh, B. K. (2020). Plant–microbiome interactions: From community assembly to plant health. *Nature Reviews Microbiology*, 18, 607–621. https://doi.org/10.1038/s41579-020-0412-1.

Tsikou, D., Nikolaou, C. N., Tsiknia, M., Papadopoulou, K. K., & Ehaliotis, C. (2023). Interplay between rhizobial nodulation and arbuscular mycorrhizal fungal colonization in *Lotus japonicus* roots. *Journal of Applied Microbiology*, 134(1), lxac010. https://doi.org/10.1093/jambio/lxac010.

van Lenteren, J. C., Bolckmans, K., Köhl, J., Ravensberg, W. J., & Urbaneja, A. (2017). Biological control using invertebrates and microorganisms: Plenty of new opportunities. *BioControl*, 63, 39–59. https://doi.org/10.1007/s10526-017-9801-4.

Van Loon, L. C., Bakker, P. A. H. M., & Pieterse, C. M. J. (2006). Induced resistance in plants by beneficial microorganisms. *Annual Review of Phytopathology*, 44, 485–513. https://doi.org/10.1146/annurev-phyto-082712-102340.

Van Loon, L. C., & Glick, B. R. (2004). Increased plant fitness by rhizobacteria. In: Sandermann, H. (ed.), *Molecular Ecotoxicology of Plants*. Springer-Verlag, Berlin, Heidelberg, pp. 177–205. https://doi.org/10.1007/978-3-662-08818-0_7.

Van Wees, S. C., De Swart, E. A., Van Pelt, J. A., Van Loon, L. C., & Pieterse, C. M. (2000). Enhancement of induced disease resistance by simultaneous activation of salicylate-and jasmonate-dependent defense pathways in *Arabidopsis thaliana*. *Proceedings of the National Academy of Sciences*, 97(15), 8711–8716. https://doi.org/10.1073/pnas.130425197.

Varma, A., Verma, S., & Sudha, S. (1999). *Piriformospora indica*: An endophyte enhancing root development. *Nature Biotechnology*, 17(3), 280–284. https://doi.org/10.1128/AEM.65.6.2741-2744.1999.

Verhaar, M., Hijwegen, T., & Zadoks, J. (1998). Selection of *Verticillium lecanii* isolates with high potential for biocontrol of cucumber powdery mildew by means of components analysis at different humidity regimes. *Biocontrol Science and Technology*, 8, 465–477. https://doi.org/10.1080/09583159829984.

Vujanovic, V., & Labrecque, M. (2009). Role of *Metschnikowia* in ISR. *Applied Microbiology and Biotechnology*, 83(6), 1217–1225.

Walters, D. R., Ratsep, J., & Havis, N. D. (2013). Controlling crop diseases using induced resistance: Challenges for the future. *Journal of Experimental Botany*, 64(5), 1263–1280. https://doi.org/10.1093/jxb/ert026.

Woo, S. L., Donzelli, B., Scala, F., Mach, R., Harman, G. E., Kubicek, C. P., Del Sorbo, G., & Lorito, M. (1999) Disruption of the *ech42* (endochitinase-encoding) gene affects biocontrol activity in *Trichoderma harzianum* P1. *Molecular Plant–Microbe Interactions*, 12, 419–429. https://doi.org/10.1094/MPMI.1999.12.5.419.

Woo, S. L., Ruocco, M., Vinale, F., Nigro, M., Marra, R., Lombardi, N., Pascale, A., Lanzuise, S., Manganiello, G., & Lorito, M. (2014). *Trichoderma*-based products and their widespread use in agriculture. *The Open Mycology Journal*, 8, 71–126. https://doi.org/10.2174/1874437001408010071.

Xu, P., Wu, T., Ali, A., Wang, J., Fang, Y., Qiang, R., Liu, Y., Tian, Y., Liu, S., Zhang, H., & Liao, Y. (2022). Rice β-glucosidase 4 (Os1βGlu4) regulates the hull pigmentation via accumulation of salicylic acid. *International Journal of Molecular Sciences*, 23(18), 10646. https://doi.org/10.3390/ijms231810646.

Yi, Y., Li, Z., Song, C., & Kuipers, O. P. (2018). Exploring plant–microbe interactions of the rhizobacteria *Bacillus subtilis* and *Bacillus mycoides* by use of the CRISPR-Cas9 system. *Environmental Microbiology*, 20(12), 4245–4260. https://doi.org/10.1111/1462-2920.14305.

Yu, Y., Gui, Y., Li, Z., Jiang, C., Guo, J., & Niu, D. (2022). Induced systemic resistance for improving plant immunity by beneficial microbes. *Plants*, 11(3), 386. https://doi.org/10.3390/plants11030386.

Yuan, J., Raza, W., Shen, Q., & Huang, Q. (2008). Plant resistance induced by rhizobacteria. *Journal of Applied Microbiology*, 105(3), 764–776.

Yuan, M., Huang, Y., Ge, W., Jia, Z., Song, S., Zhang, L., & Huang, Y. (2019). Involvement of jasmonic acid, ethylene and salicylic acid signaling pathways behind the systemic resistance induced by *Trichoderma longibrachiatum* H9 in cucumber. *BMC Genomics*, 20, 1–13. https://doi.org/10.1186/s12864-019-5513-8.

Zamioudis, C., Hanson, J., & Pieterse, C. M. J. (2013). *Phoma*-induced VOCs in ISR. *Plant Journal*, 74(5), 626–638.

Zhang, C., Lei, Y., Lu, C., Wang, L., & Wu, J. (2020). MYC2, MYC3, and MYC4 function additively in wounding-induced jasmonic acid biosynthesis and catabolism. *Journal of Integrative Plant Biology*, 62(8), 1159–1175. https://doi.org/10.1111/jipb.12902.

Zhang, H., Xu, C., & Feng, Y. (2017). Induction of systemic resistance by *Colletotrichum* spp. *Journal of Plant Growth Regulation*, 36(1), 177–189. https://doi.org/10.1007/BF02979

Zhang, Y., Zhao, X., & Wang, Y. (2019). Volatile organic compounds from *Alternaria* spp. inducing resistance in plants. *Plant Disease*, 103(5), 1236–1243.

Zhu, L., Huang, J., Lu, X., & Zhou, C. (2022). Development of plant systemic resistance by beneficial rhizobacteria: Recognition, initiation, elicitation and regulation. *Frontiers in Plant Science*, 13, 952397. https://doi.org/10.3389/fpls.2022.952397.

Zipfel, C., & Oldroyd, G. E. D. (2017). Plant signalling in symbiosis and immunity. *Nature*, 543, 328–336. https://doi.org/10.1038/nature22009.

15 Engineering Crop Microbiomes

Transforming Horticulture Crop Health and Productivity

Afiya Rajan Selin, Akshara Kampurath, Perinba C. Suresh, Subramanian Manivannan, and Senthilkumar Shanmugasundram

15.1 INTRODUCTION

The various populations of microorganisms that reside in and around plants, such as bacteria, fungus, viruses, and archaea, are referred to as the plant microbiome. These microbial communities inhabit different plant sections, including the endosphere (internal tissues), phyllosphere (parts above the ground), and rhizosphere (interface between roots and soil) (Kumari *et al.*, 2024). Recent developments in the study of the microbiome have shed light on the complex interactions that exist between plants and the microbes that coexist with them, demonstrating the ways in which the microbiome affects the growth, health, and resilience to stress in plants. An developing field called "microbiome engineering" seeks to modify these microbial communities in order to maximize plant productivity and performance by utilizing advantageous microbial features for better agricultural results.

The microbiome has a major role in plant health. Microbes directly contribute to plant nutrition through their essential role in the nitrogen fixation process, the decomposition of organic matter, and the cycling of nutrients. Moreover, microorganisms linked with plants serve as bio-protectors by assisting plants in enduring biotic and abiotic stressors such diseases, heat waves, and drought (Pandey *et al.*, 2020). A healthy microbiome is an essential factor in determining the general health and productivity of plants since it can control plant immune responses, promote growth, and preserve plant fitness.

The microbiome in crops has important ramifications for sustainable farming. Environmentally friendly farming methods can be promoted by beneficial microorganisms, such as mycorrhizal fungi and bacteria that fix nitrogen, which can lessen the need for synthetic fertilizers and chemical inputs (Singh *et al.*, 2025). Plant breeding and agricultural practices can have an impact on the composition of microbial communities, as evidenced by studies on the microbiome found in the rhizosphere of domesticated plants. Thus, knowing the plant–microbe interactions at various developmental phases gives prospects for crop enhancement, pest management, and resilience to climate change.

The field of plant microbiome engineering offers promising opportunities to improve agricultural productivity. Researchers can enhance plant growth, improve resistance to diseases, and increase nutrient uptake efficiency by introducing or promoting particular beneficial bacteria. To enhance the symbiotic relationships between plants and bacteria and decrease reliance on external inputs, recent studies have looked into techniques to engineer the rhizosphere microbiome. Nonetheless, there are still issues with guaranteeing the stability of modified microbial communities in changing agricultural settings and comprehending the long-term effects of microbiome modification.

DOI: 10.1201/9781003608080-15

15.2 COMPOSITION OF MICROBIOME

15.2.1 Rhizosphere Microbiome

Plant productivity and health depend on the rhizosphere microbiome, a richly varied collection of microbes connected to plant roots. Often called the plant's "second genome", these microbial communities actively alter the plant through exudates from the roots, making it possible for beneficial creatures to be attracted in. When faced with environmental stressors like pathogens or insect attacks, plants can send signals to the microbiome, which creates an environment that suppresses harmful microorganisms and strengthens plant defense. Certain microbial communities can help with stress tolerance, encourage the uptake of nutrients, and create protective biofilms that keep potential intruders away from roots.

Development of molecular methods has made it possible to comprehend the relations between microbes and plants in the rhizosphere on a deeper level. Researchers are discovering ways to modify the rhizosphere microbiome for better crop health by examining gene expression, metabolic pathways, and roles of microbes in response to plant signals. Applications include boosting disease-suppressive fungi and nitrogen-fixing bacteria to increase plant resistance to diseases, increase plant growth, and decrease the need for chemical pesticides and fertilizers. These discoveries hold enormous promise for sustainable agriculture, providing a biologically based method for enhancing crop productivity and preventing plant illnesses.

A theoretical representation of pathogen-mediated microbial recruitment in the rhizosphere for plant defense is presented in Figure 15.1. Initially, after accurately identifying pathogen invasion, the precursor plants emit exudates from roots into the soil to alter the microbiological dynamics of the soil population or selectively recruit beneficial bacteria. The ensuing microbial recruitments may directly suppress infections or induce systemic resistance (induced systemic resistance (ISR)) while also successfully avoiding overreactions from the immune system. Rhizosphere bacteria may be recruited by nearby plants via airborne signals from diseased plants that are volatile. Effective pathogens circumvent the defenses rhizosphere bacteria have built up by using effectors or other strategies.

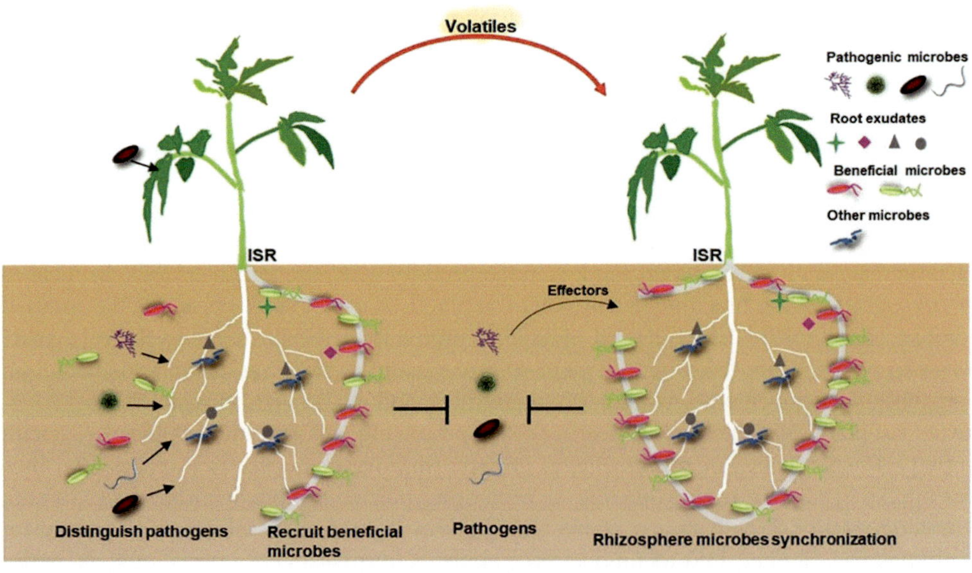

FIGURE 15.1 Theoretical representation of pathogen-mediated microbial recruitment in the rhizosphere for plant defense (Li *et al.*, 2021).

15.2.2 Phyllosphere Microbiome

The various microbial communities that live on the aerial surfaces of plants are together termed the phyllosphere microbiome, and their importance for plant health and productivity cannot be overstated. Yeasts, fungus, bacteria, and other microorganisms coexist in this environment and engage in intricate interactions with the host plant that frequently benefit both the microbes and the plant. Through boosting nutrient uptake, increasing stress tolerance, and offering defense against infections, these interactions can increase plant resilience. While endophytic microorganisms reside within plant tissues and support defense and general health, epiphytic microbes, in particular, can protect the plant from environmental stresses such as UV radiation, dehydration, and temperature variations.

The microbial diversity and functioning of the phyllosphere have become more deeply understood, thanks to developments in molecular technologies like metagenomics. Researchers can improve plant yield and disease resistance by developing techniques to comprehend these microbial communities. Beneficial phyllosphere microorganisms offer several applications in crop health by enhancing nutrient uptake, lessening the impact of diseases, and eliminating the need for chemical pesticides. These advantageous microorganisms can also assist plants in adjusting to external stressors like climate change, providing long-term answers to problems in agriculture.

The microbiota of the phyllosphere is vital in promoting the growth and health of plants. These microbial communities protect plants from diseases, improve plant resistance to environmental challenges, and aid in the cycling of nutrients. The bacteria that live on a plant's aerial surfaces have the ability to support mutualistic connections, support defense systems, and assist interactions with the environment. They can also affect the physiology of the plant as a whole by controlling vital functions that are necessary for its survival and output (Figure 15.2).

15.2.3 Endophytic Microbiome

The presence of beneficial bacteria and fungi within plant tissues, known as endophytic microbiome, is essential for improving crop productivity and health. By encouraging nutrient uptake, creating phytohormones, and generating bioactive substances that enhance the immune system of the plant, these microorganisms aid in the growth of plants. Endophytes aid plants in overcoming biotic

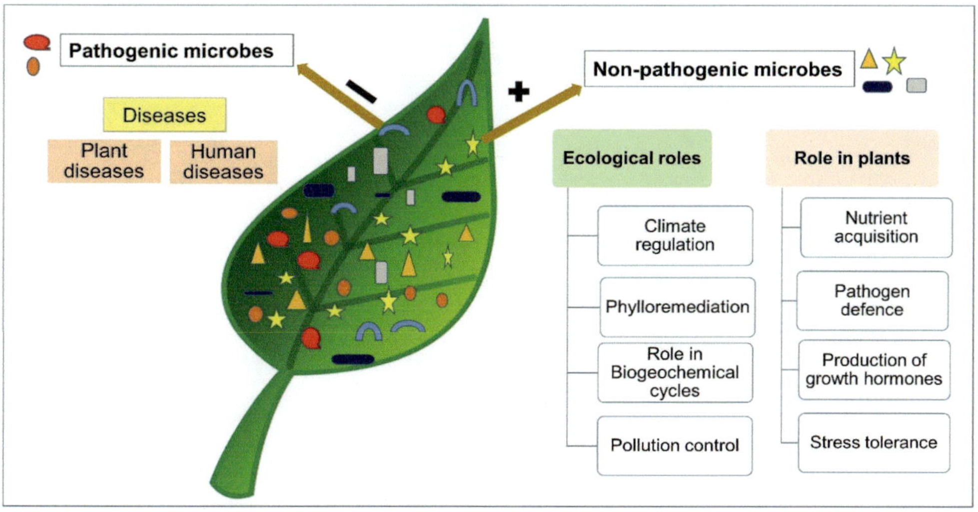

FIGURE 15.2 Functions of phyllosphere microbiome (Bashir *et al.*, 2022).

(e.g., infections) and abiotic (e.g., drought and poor soil quality) challenges. Their capacity to live inside plant tissues gives them a distinct advantage in maintaining colonization over the long term, supporting resilient and persistent plant growth.

The use of endophytic bacteria has drawn interest recently as a sustainable method for controlling plant diseases and increasing agricultural productivity. When employed as biocontrol agents, endophytes effectively lower postharvest losses by either stimulating the plant's defense mechanisms or preventing the onset of diseases by producing antimicrobial chemicals. In place of chemical pesticides, these helpful microorganisms provide a more environmentally friendly option that improves food safety, lessens environmental damage, and increases agricultural sustainability.

As plants grow, their seeds include vertically inherited microbial endophytes, or spermatophytes, some of which are likely related to the plant species and comprise the core endophytic population. Some endophytes are transitory and vanish as the plant is colonized by new endophytes from the rhizosphere or phyllosphere in subsequent phases. Through either short- or long-term connections, plants can interact with helpful endophytes (commensalism and mutualism) or harmful ones (parasitism). A cross-section through a plant's root shows how fresh endophytes enter and colonize the system (endophytism) (Figure 15.3).

15.3 CROP MICROBIOME ENGINEERING FOR HORTICULTURE SUSTAINABILITY

15.3.1 Enhancing Horticultural Production

The engineering of plant microbiomes has the potential to improve agro-ecosystem productivity and ensure future sustainable food demand, while also helping regenerate stressed horticulture (Vimal *et al.*, 2023). Improved functional capacities of effective soil microbiomes could result in greater soil restoration and higher agricultural yield by modifying their genomes to improve how they respond to environmental stressors (Singh *et al.*, 2020; Qian *et al.*, 2021).

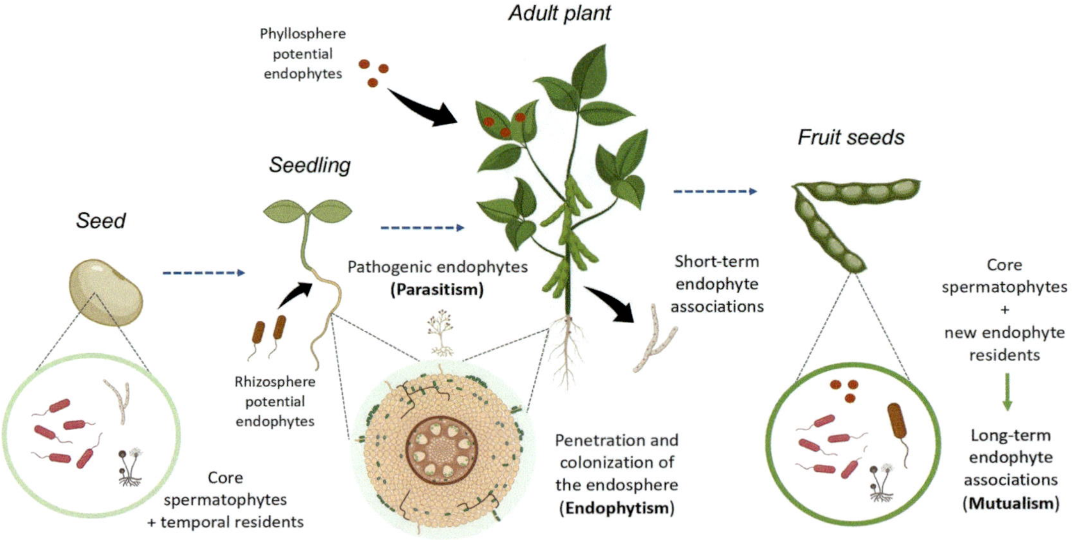

FIGURE 15.3 Stages of plant growth and endophytic interactions (del Carmen Orozco-Mosqueda and Santoyo, 2021).

15.3.2 Growth Promotion

Introducing a potent bacterial and fungal combination—*Pseudomonas libanensis* TR1 (bacteria) and arbuscular mycorrhizal (AM) fungus *Claroideoglomus claroideum* BEG210—nto the sunflower's rhizosphere led to notable plant growth despite heavy metal and salt stress. The strains UQ154, UQ156, and UQ202 were identified by 16S rDNA sequence analysis as being closely related to Bacillus amyloliquefaciens, Bacillus velezensis, and *Acinetobacter* sp., respectively, and their application to chilli plant roots was reported by Syed-Ab-Rahman *et al.* (2019) to enhance plant growth and increase resistance against the soil-borne pathogen Phytophthora capsici. When applied to chili plant roots, UQO2, *Bacillus amyloliquefaciens*, and *Bacillus velezensis* enhanced plant growth and boosted resistance against soil-borne *Phytophthora capsici* infection. Numerous species, such as bacteria, fungi, viruses, protozoa, and archaea, make up the microbiome (Mueller and Sachs, 2015). These varied microbial populations influence the physiology and growth of plants, which are essential to their function (Mendes *et al.*, 2013). Plant microbiomes provide numerous advantages, such as pathogen defense, improved growth and fitness, and increased tolerance to abiotic stressors (Haney *et al.*, 2015; Berg *et al.*, 2016).

Applying plant bio-stimulants (PBs) comprising plant growth-promoting rhizobacteria (PGPR) (*Bacillus megaterium* and *Bacillus subtilis*) and humic acid dramatically enhanced tuber weight by 118.8%, plant height by 71.7%, and the number of tubers per plant by 15.2% in potatoes, according to Ekin's (2019) observations. In comparison to untreated plants, plants treated with PBs exhibited a 294.4% increase in the production of industrial-grade tubers. Furthermore, PBs improve stomatal conductance and density, which are connected to EPFL9's positive regulation (Chitarra *et al.*, 2016). This improves photosynthesis and increases internal carbon dioxide concentration, which ultimately results in greater yields (Ansari *et al.*, 2019). It has been shown that chitosan biostimulants can improve rice plant characteristics such as biomass, root length, and height (Stanley-Raja *et al.*, 2021). According to these findings, PBs hold a lot of promise for use in biotechnology as a technique to improve plant quality, production, and growth in field circumstances (Ekin, 2019).

15.3.3 Phytoremediation

One economical, solar-powered, and environmentally friendly technique for remediating soil is phytoremediation. This method makes use of plants' innate capacity to transport, store, and absorb pollutants found in the environment (Pilon-Smits, 2005). Plant species (Vangronsveld *et al.*, 2009), contamination levels, soil pH, nutrient availability, redox conditions, and microbes connected to soil and plants all have an impact on phytoremediation's efficacy (Sessitsch *et al.*, 2013). Improving soil quality, increasing contaminant accessibility and bioavailability (de La Torre-Roche *et al.*, 2012), encouraging plant growth (Sessitsch *et al.*, 2013), and investigating the function of microbes associated with soil and plants in phytoremediation (Barac *et al.*, 2004; Abhilash *et al.*, 2012) are important research areas.

The potential influence of many techniques has been reduced in recent decades since they tended to focus more on individual organisms than on integrated meta-organisms. Since plant–microbe interactions play a major role in how plants respond to contaminants, a fuller knowledge of these interactions is necessary to enhance phytoremediation. Understanding how beneficial microbiomes behave under pollutant stress in plants is equally crucial. Ecological models and molecular data have shed light on this situation, demonstrating, e.g., decreased insect populations (Scheuring and Yu, 2012).

15.3.4 Improving Stress Tolerance

Under extreme abiotic and biotic stress circumstances, many beneficial microbiome alterations provide favorable results (Vimal *et al.*, 2017; Trivedi *et al.*, 2020). Microbial populations are highly susceptible to environmental stressors such as changes in climate. Drought-, salt-, or heat-tolerant taxa can have a direct impact on the soil microbiome through these stresses (Martiny *et al.*, 2017; Naylor *et al.*, 2017), whereas soil chemistry or diffusion rates can change indirectly (Liptzin *et al.*, 2011). Applying AMF bio-stimulants, e.g., increased the root dry weight and shoot dry weight of drought-stressed tomato plants by 14% and 18%, respectively, compared to untreated plants (Hayat *et al.*, 2016). According to Duc *et al.* (2018), compared to untreated plants, tomato plants treated with AMF showed increased activity of antioxidant enzymes such as catalase (CAT), superoxide dismutase (SOD), and peroxidase (POD). Furthermore, there was a considerable decrease in both hydrogen peroxide concentration and lipid peroxidation of plants treated with AMF. Comparable results have been shown for a number of food crops, including tomato (Mannino *et al.*, 2020), wheat (Pačuta *et al.*, 2021), and rice (Moolphuerk *et al.*, 2021).

The enhanced activity of antioxidant enzymes was responsible for the reduction in reactive oxygen species (ROS) generation in heat-stressed tomatoes after AMF treatment (Duc *et al.*, 2018). Similarly, Boosten and Megafol, two commercial seaweed-based PBs, were shown in another study to reduce the effects of heat stress in tomato plants when applied topically (Niu *et al.*, 2021). Comparing Boosten to Megafol, the study revealed that Boosten dramatically enhanced root diameter by 89.8% and root length by 41% (Niu *et al.*, 2021). Furthermore, in comparison to untreated plants, Boosten increased the rate of photosynthesis by 81.7% and the amount of chlorophyll in aboveground tissues by 30.5%. These findings imply that foliar sprays containing seaweed extract may increase crop resistance to heat stress (Niu *et al.*, 2021).

PBs inhibit diseases and pests, and hence are used in agriculture to promote healthy crop growth and increase yield. For example, pretreatment of tomato plants with sodium alginate or λ-carrageenan (an elicitor derived from commercial seaweed extracts) significantly reduced symptoms caused by tomato chlorotic dwarf viroid infection (Sangha *et al.*, 2015) and early blight disease caused by *Alternaria solani* (Dey *et al.*, 2019). Furthermore, a study discovered that microbes extracted from the roots of olive and grapevine plants that were surviving extremely dry conditions might promote the establishment of a different host species (Marasco *et al.*, 2013). This growth enhancement could be explained by the bacteria encouraging a larger root system, which boosts the plant's ability to absorb water. This strategy holds promising benefits to save time, effort, and costs. Even under conditions of heavy metal and salt stress, the introduction of a robust consortium of bacteria and fungi—more precisely, the AM fungus *Claroideoglomus claroideum* BEG210 and the bacterial component *Pseudomonas libanensis* TR1—into the rhizosphere of sunflowers resulted in significant plant development (Ma *et al.*, 2019).

15.3.5 Stimulating Antagonistic and Biocontrol Activities

Plant diseases cause farmers all over the world to suffer large financial losses. A rich population of microorganisms lives in the rhizosphere of plants, where they build mutually beneficial interactions with their hosts. These relationships shield the hosts from pathogens and improve their health and fitness through a variety of direct and indirect processes. Studies have demonstrated that plants emit tiny chemicals to draw in advantageous microflora that aids in their adaptation to harsh environments (Busby *et al.*, 2014). Numerous biological control traits are displayed by this diverse microbial community, such as competition for space and resources, colonization (Hunziker *et al.*, 2015; Lloyd and Allen, 2015; Santhanam *et al.*, 2015), antibiosis (Gómez Expósito *et al.*, 2017), hyperparasitism (McNeely *et al.*, 2017), and the generation of degradative enzymes. Moreover, these plant-associated microorganisms have mutualistic connections that impact the microbiome of the host plant and support a wide variety of bacterial taxa. Many of these species increase plant development, inhibit diseases, break down xenobiotic chemicals, improve tolerance to biotic and abiotic challenges, and increase yields (Berg *et al.*, 2016). Artificial microbial consortia provide new synergistic opportunities for optimizing disease treatment techniques by targeting this vast microbial variety for manipulation (Poudel *et al.*, 2016).

15.3.6 Enhancing Nutrient Use Efficiency

PGPR are a class of helpful microbes that are mainly responsible for plants' and crops' efficient uptake of nutrients (Ratnadass, 2020). The entire process of nutrient transfer is greatly impacted by their capacity to proliferate and endure in the root zone, whether or not they have symbiotic partners like vesicular arbuscular mycorrhiza (VAM). These microbes play a crucial role in a process known as "rhizosphere competence", in which they actively improve the uptake of nitrogen, defend the root zone from pests and pathogenic microbes, encourage the mineralization of nutrients, and support AM associations (Ratnadass *et al.*, 2012). The efficient utilization of phosphorus, other micronutrients, and water is further facilitated by these mycorrhizal connections, particularly in low-input situations (Rothschild *et al.*, 2002; Ratnadass and Barzman, 2014).

Less than 0.1% of chemical pesticides really reach their intended biological targets, according to the research by Fukamachi *et al.* (2019). Instead, the bulk seep into the nearby soil and water. Certain fertilizers are not readily absorbed by plants, even though they contain naturally occurring organic matter that is less prone to leaching, such as composted organic waste (Galsim *et al.*, 2021). Although nitrogen fertilizers are necessary for crop production, their overuse or inefficiency can have negative effects on the environment, including eutrophication and harmful groundwater contamination. According to Bouchet *et al.* (2016), about half of the nitrogen applied to cropping systems can be recovered, while the other half is either lost through volatilization, leaching, and runoff, or stays in the soil as organic complexes. Therefore, in order to decrease fertilizer inputs and increase nitrogen use efficiency, it is imperative to investigate sustainable alternatives.

15.4 MICROBIOMES CHARACTERIZED IN VARIOUS HORTICULTURAL CROPS

15.4.1 Vegetables

The microbial communities associated with tomato endosphere (inside the roots) have a different microbial diversity, with higher abundances of beneficial species like *Acidovorax*, *Enterobacter*, and *Pseudomonas*. Compared to the bulk soil and endosphere, the rhizosphere (the soil surrounding the roots) exhibited a more intricate and vast microbial network (Lee *et al.*, 2019). Bacterial populations in tomato plants grown in greenhouses were examined using Illumina sequencing. With the exception of seeds and jelly, where Firmicutes was also prevalent, Proteobacteria was the dominant phylum. In the rhizosphere and phyllosphere, *Pseudomonas* and *Acinetobacter* were prevalent, with *Acinetobacter* being especially prominent in the latter. *Acinetobacter*, *Enterobacter*, and *Pseudomonas* were common in roots, stems, and leaves in the endosphere. Different bacterial profiles were seen in the fruit compartments: *Acinetobacter* in the placenta, *Enterobacter* in the pericarp and seeds, and *Weissella* in the jelly (Dong *et al.*, 2019). Increased *Ralstonia solanacearum* pathogen levels in tomatoes decreased bacterial diversity and interfered with co-occurrence networks in the rhizosphere. Higher concentrations of sugars and certain bacteria, such as *Bacillus* and *Chitinophaga*, were found in soils with reduced pathogen abundance. These variables showed a negative correlation with pathogen abundance. These results imply that sugars and bacterial groups may both have potential biocontrol abilities (Wen *et al.*, 2020).

Important microorganisms found in the rhizosphere soil and tuber roots of potatoes (*Solanum tuberosum*) include *Agrobacterium*, *Stenotrophomonas*, *Geobacillus*, *Variovorax*, *Curtobacterium*, and *Streptomycetes*. By producing antibiotics, antimicrobial compounds, and destroying enzymes, these bacteria aid in triggering plant immune responses. A study analyzed these microbial communities using molecular methods such as quantitative PCR, shotgun metagenomic sequencing, and Illumina MiSeq. Using 16S rRNA gene sequencing, the study examined the bacterial populations in potato tubers from various cultivars in Austria. Important discoveries include the fact that these bacteria, which colonize tubers of all varieties, are primarily found in soil. Most likely, bacteria live on tubers generations after entering through the stolon (Buchholz *et al*., 2019). Four soil compartments in plants with high and low scab severity were examined to determine how soil microbiomes impact potato common scab (CS). The objective was to comprehend the makeup and function of the microbiota in CS (Shi *et al*., 2019).

Gene sequencing of identified key microbes in onion revealed that bacteria like *Enterobacteriaceae* and *Pseudomonas* and fungus like *Botrytis* and *Candida* linked to bulb rot are important players in both healthy and diseased onion bulbs. Their results point to the influence of fermentation and microbial variety on onion rotting, and the detection of dinoflagellates (*Symbiodinium*) in onions suggests that these eukaryotes have a wider range of habitats (Yurgel *et al*., 2018). The common bean's (*Phaseolus vulgaris*) rhizosphere microbiome is essential to the plant's growth, disease resistance, and nutrient uptake. According to studies, the biological control agent *Metarhizium robertsii* can increase the number of beneficial bacteria in the rhizosphere, such as *Bradyrhizobium*, *Flavobacterium*, and *Trichoderma*. This increases the amount of nitrogen fixed and suppresses the growth of pathogens such as *Fusarium solani*. Rhizobacterial diversity is further impacted by the common bean's shift from natural to agricultural soils; in agricultural soils, current bean cultivars exhibit a higher diversity of distinct microbial species. Agricultural soils can increase total microbial diversity while decreasing the complexity of microbial interactions. This suggests that managing the microbiome with soil amendments and biocontrol agents is crucial for sustainable agriculture (Barelli *et al*., 2020; Pérez-Jaramillo *et al*., 2019).

Reduced microbial diversity and soil deterioration are the results of continuous sugar beetroot (*Beta vulgaris* L. var. *saccharifera*) cultivation. After 5 years of nonstop farming, microbial diversity decreased, but after 30 years, it grew due to the enrichment of harmful microorganisms such as *Fusarium* and *Alternaria*. Over time, beneficial bacteria like *Pseudomonas* and *Bacillus* declined dramatically. Lower amounts of phenolic acid decarboxylase were found by functional analysis, which may imply that long-term harvested soils have accumulated hazardous phenolic chemicals (Huang *et al*., 2019). When continuously farmed, sweet potato modifies the rhizosphere fungal community by decreasing *Chaetomium* and increasing *Verticillium*, *Fusarium*, and *Colletotrichum*, which are beneficial fungi. This change reduces crop production and quality and increases soil-borne illnesses. Following continual cropping, there is an increase in fungal diversity but a decrease in the dominant phylum Ascomycota. While the effects change throughout sampling periods, they are consistent across several kinds of sweet potatoes. To lessen these effects, crop rotation and microbial fertilizers are advised (Gao *et al*., 2019).

Five root vegetables (carrot, potato, beetroot, neep, and topinambur) showed that the pulp is rich in Gammaproteobacteria and Firmicutes, while the peel microbiomes had higher species richness, dominated by Alphaproteobacteria and Actinobacteria. High lactic acid bacteria levels in topinambur pulp suggested probiotic potential, while *Listeria* species were also found. According to Koiv *et al*. (2019), the generation of antifungal metabolites, immunological response, and plant growth are all influenced by the microbiomes. Firmicutes and Proteobacteria are the main focus of the microbial communities found in the rhizosphere and rhizoplane of spinach, broccoli, cauliflower, celery, and rape. It was shown that illnesses such as Huanglongbing (HLB) interfere with the beneficial bacteria's movement from the rhizosphere to the rhizoplane, decreasing the interactions between microbes and plants and harming the health of the plants. *Burkholderia* is one beneficial strain that plants can be inoculated with to help repair these relationships and increase resistance to disease. The significance of preserving healthy root-associated microbiomes for ideal plant growth and disease control is highlighted by these findings (Zhang *et al*., 2010).

Liu *et al*. (2018) evaluated *Bacillus subtilis* XF-1's impacts on the rhizosphere of cabbage, with particular attention to its capacity to manage clubroot disease and its influence on microbial communities. *Bacillus subtilis* XF-1 demonstrated a 76.92% control efficiency and a 17.14% reduction in clubroot incidence. Proteobacteria, Bacteroidetes, Acidobacteria, Actinobacteria, Firmicutes, and Gemmatimonadetes were the most common bacterial phyla, according to analysis, while

Ascomycota and Basidiomycota were the most common fungal groupings. *Bacillus subtilis* XF-1 initially reduced microbial diversity, but communities eventually rebounded, suggesting that the influence on soil microbiomes was only temporary (Liu *et al.*, 2018). Frohling *et al.* (2018) studied microbial communities in endive lettuce through various postharvest stages: Raw, cut, washed, and spin-dried. They found that bacterial families like Pseudomonadaceae, Enterobacteriaceae, Xanthomonadaceae, and Moraxellaceae dominated, with some potentially pathogenic species such as *Stenotrophomonas maltophilia*. Each processing stage showed distinct microbial profiles, underscoring the need for improved hygienic practices and decontamination methods to ensure product safety (Frohling *et al.*, 2018). The impacts of raw garlic stalk (RGS) amendment on aubergine soil microbiomes were investigated, and it was found that the addition dramatically changed the microbial populations and soil characteristics. RGS decreased pH while increasing the nutrient content, enzyme activity, and electrical conductivity in the soil. The amendment decreased dangerous fungi like *Fusarium* and increased beneficial bacteria like Acidobacteria and *Bacillus*. This shows that in monoculture systems, RGS can enhance soil health and promote better aubergine development (Ghani *et al.*, 2019).

15.4.2 Fruits

Using high-throughput sequencing and qPCR, the impact of grapevine rootstock genotypes on rhizosphere microbial populations was studied. The microbial composition was shown to be considerably influenced by different rootstocks, with taxa such as *Bacillus* and *Glomus* exhibiting genotype-specific relationships. The study also found that pathogen levels and microbial diversity were regulated by soil conditions and vineyard age. These findings highlight how rootstock genotype shapes rhizosphere microbiomes and point to the necessity for additional studies on these relationships (Berlanas *et al.*, 2019). The impact of soil amendment VESTA on bacterial populations in strawberry (*Fragaria x ananassa*) rhizospheres and soil was investigated. They used Illumina MiSeq and discovered that VESTA dramatically changed the quantity and composition of the bacterial population in both soil and root samples. The rhizosphere and roots showed the biggest alterations, suggesting a dynamic relationship between treatment and plant development. These findings showed that rather than just bringing in new taxa, VESTA might also reorganize already-existing microbial networks (Deng *et al.*, 2019).

Banana fruits are home to a wide variety of microorganisms, such as Firmicutes, Proteobacteria, and Chloroflexi. By secreting hydrolytic enzymes, antibiotics, and antifungal compounds, these bacteria support plant immune responses, as demonstrated by amplicon sequencing with Illumina MiSeq. The effects of soil amendments and the microbial communities in banana soils impacted by *Fusarium* wilt were examined by Xue *et al.* (2015) and Zhou *et al.* (2019). According to Xue *et al.* (2015), *Fusarium oxysporum* had different microbial profiles in diseased and disease-free soils and was more prevalent in the former. It was determined that *Bacillus* species were a major beneficial bacterium, and their use decreased the incidence of disease and increased productivity. Notable variations in the microbial diversity of healthy and disease-free soils, with the former having higher populations of beneficial bacteria and lower concentrations of pathogens, were found. These results demonstrate how crucial soil microbial community management is for crop productivity and disease prevention. Hong *et al.* (2020) discovered that by increasing beneficial microbial communities and reducing pathogen levels, rotating chili peppers and bananas successfully decreased *Fusarium* wilt in bananas. In comparison to banana monoculture, this rotation enhanced soil health and reduced illness (Hong *et al.*, 2020).

The rhizosphere and root sections of lemon plants are dominated by Proteobacteria, Actinobacteria, Bacteroidetes, Cyanobacteria, and Firmicutes, which are known to stimulate plant growth and act as biocontrol agents (Zhang *et al.*, 2017; Padhi *et al.*, 2019). Si *et al.* (2018) focused on the rhizosphere soil of a variety of fruits and nuts, such as walnuts, pears, apples, apricots, cherries, peaches, grapes, and cherries. The study emphasized the function of microbial communities in plant growth and defense systems, including Proteobacteria, Gemmatimonadetes, and Bacteroidetes, by analyzing soil physicochemical characteristics, enzymatic secretions, and rhizosphere microenvironments using ribotyping.

15.4.3 Medicinal Crops

Liu *et al.* (2017) examined the bulk and rhizosphere soil, as well as the stems of ginger plants, and found important microorganisms such as *Bradyrhizobium*, *Cryptococcus*, and *Cladosporium*. The study highlighted the functions of these bacteria in fostering plant growth and their capacity to release compounds that are antifungal, hydrolytic enzymes, and antibiotics. To evaluate the contributions of microbes to these impacts, ribotyping was employed. Actinobacteria, Proteobacteria, Firmicutes, and other dominating taxa were identified by Jackson *et al.* (2015), and microbial communities in the leaves of several herbs, including dill, chives, basil, mint, and rosemary, were also investigated. The study employed amplicon sequencing with Illumina MiSeq to investigate how these microbes cause plant immune responses and secrete secondary compounds that are antibacterial and antifungal. Key taxa including *Bacillus*, *Pseudomonas*, and *Azotobacter* were identified by Kumar *et al.* (2017) during their investigation of the microbial communities in turmeric rhizomes. With the aid of 16S rRNA gene amplicon sequencing, the study concentrated on how these bacteria promote plant defense, secrete compounds that are antibacterial and antifungal, and promote plant growth.

REFERENCES

Abhilash PC, Powell JR, Singh HB, Singh BK (2012). Plant–microbe interactions: Novel applications for exploitation in multipurpose remediation technologies. *Trends Biotechnol.* 30, 416–420. https://doi.org/10.1016/j.tibtech.2012.04.004

Ansari FA, Ahmad I, Pichtel J (2019). Growth stimulation and alleviation of salinity stress to wheat by the biofilm forming *Bacillus pumilus* strain FAB10. *Appl Soil Ecol.* 143, 45–54. https://doi.org/10.1016/j.apsoil.2019.05.023

Barac T, Taghavi S, Borremans B, Provoost A, Oeyen L, Colpaert JV, et al. (2004). Engineered endophytic bacteria improve phytoremediation of water-soluble, volatile, organic pollutants. *Nat Biotechnol.* 22, 583–588. https://doi.org/10.1038/nbt960

Barelli L, Waller AS, Behie SW, Bidochka MJ (2020). Plant microbiome analysis after Metarhizium amendment reveals increases in abundance of plant growth-promoting organisms and maintenance of disease-suppressive soil. *PLoS ONE*, 15(4), e0231150. https://doi.org/10.1371/journal.pone.0231150.

Bashir I, War AF, Rafiq I, Reshi ZA, Rashid I, Shouche YS (2022). Phyllosphere microbiome: Diversity and functions. *Microbiol Res.* 254, 126888.

Berg G, Rybakova D, Grube M, Köberl M (2016). The plant microbiome explored: Implications for experimental botany. *J Exp Bot.* 67, 995–1002. https://doi.org/10.1093/jxb/erv466

Berlanas C, Berbegal M, Elena G, Laidani M, Cibrian JF, Sagües A, et al. (2019). The fungal and bacterial rhizosphere microbiome associated with grapevine rootstock genotypes in mature and young vineyards. *Front Microbiol.* 10, 1142.

Bouchet A-S, Laperche A, Bissuel-Belaygue C, Snowdon R, Nesi N, Stahl A (2016). Nitrogen use efficiency in rapeseed. A review. *Agron Sustain Dev.* 36(2), 38. https://doi.org/10.1007/s13593-016-0371-0

Buchholz F, Antonielli L, Kostic T, Sessitsch A, Mitter B (2019). The bacterial community in potatoes is recruited from soil and partly inherited across generations. *PLoS ONE* 14(11), e0223691.

Busby PE, Newcombe G, Dirzo R, Whitham TG (2014). Differentiating genetic and environmental drivers of plant–pathogen community interactions. *J Ecol.* 102, 1300–1309. https://doi.org/10.1111/1365-2745.12270

Chitarra W, Pagliarani C, Maserti B, Lumini E, Siciliano I, Cascone P, et al. (2016). Insights on the impact of arbuscular mycorrhizal symbiosis on tomato tolerance to water stress. *Plant Physiol.* 171(2), 1009–1023. https://doi.org/10.1104/pp.16.00307

de La Torre-Roche R, Hawthorne J, Deng Y, Xing B, Cai W, Newman LA, et al. (2012). Fullerene-enhanced accumulation of p, p'-DDE in agricultural crop species. *Environ Sci Technol.* 46, 9315–9323. https://doi.org/10.1021/es301982w

del Carmen Orozco-Mosqueda M, Santoyo G (2021). Plant–microbial endophytes interactions: Scrutinizing their beneficial mechanisms from genomic explorations. *Curr Plant Biol.* 25, 100189.

Deng S, Wipf HML, Pierroz G, Raab TK, Khanna R, Coleman-Derr D (2019). A plant growth-promoting microbial soil amendment dynamically alters the strawberry root bacterial microbiome. *Sci Rep*. 9, 17677.

Dey P, Ramanujam R, Venkatesan G, Nagarathnam R (2019). Sodium alginate potentiates antioxidant defense and PR proteins against early blight disease caused by *Alternaria solani* in *Solanum lycopersicum* Linn. *PLoS ONE* 14(9), 1–26. https://doi.org/10.1371/journal.pone.0223216.

Dong CJ, Wang LL, Li Q, Shang QM (2019). Bacterial communities in the rhizosphere, phyllosphere, and endosphere of tomato plants. *PLoS ONE*, 14(11), e0223847.

Ekin Z (2019). Integrated use of humic acid and plant growth promoting rhizobacteria to ensure higher potato productivity in sustainable agriculture. *Sustainability* 11, 12. https://doi.org/10.3390/su10023417

Fukamachi K, Konishi Y, Nomura T (2019). Disease control of *Phytophthora infestans* using cyazofamid encapsulated in poly lactic-co-glycolic acid (PLGA) nanoparticles. *Colloids Surf Physicochem Eng Aspects*. 577, 315–322. https://doi.org/10.1016/j.colsurfa.2019.05.077.

Galsim F, Golabi MH, Kim YS, Iyekar C (2021). Comparative effects of composted organic waste and inorganic fertilizer on nitrate leachate from the farm soils of northern Guam. *Int Soil Water Conserv Res*. 9(1), 87–102. https://doi.org/10.1016/j.iswcr.2020.09.003.

Gao Z, Han M, Hu Y, Li Z, Liu C, Wang X, et al. (2019). Effects of continuous cropping of sweet potato on the fungal community structure in rhizospheric soil. *Frontiers in Microbiology*, 10, 2269. https://doi.org/10.3389/fmicb.2019.02269.

Ghani MI, Ali A, Atif MJ, Ali M, Amin B, Anees M, et al. (2019). Changes in the soil microbiome in eggplant monoculture revealed by high-throughput Illumina MiSeq sequencing as influenced by raw garlic stalk amendment. *Int J Mol Sci*. 20, 2125.

Gómez Expósito R, de Bruijn I, Postma J, Raaijmakers JM (2017). Current insights into the role of rhizosphere bacteria in disease suppressive soils. *Front Microbiol*. 8, 2529. https://doi.org/10.3389/fmicb.2017.02529

Haney CH, Samuel BS, Bush J, Ausubel FM (2015). Associations with rhizosphere bacteria can confer an adaptive advantage to plants. *Nat Plants* 1, 15051. https://doi.org/10.1038/nplants.2015.51

Hayat S, Ahmad H, Ali M, Ren K, Cheng Z (2016). Aqueous garlic extract stimulates growth and antioxidant enzymes activity of tomato (*Solanum lycopersicum*). *Sci Hortic*. 2018(240), 139–146. https://doi.org/10.1016/j.scienta.2018.06.011

Hong S, Jv H, Lu M, Wang B, Zhao Y, Ruan Y (2020). Significant decline in banana *Fusarium* wilt disease is associated with soil microbiome reconstruction under chilli pepper–banana rotation. *Eur J Soil Biol*. 97, 103154.

Huang W, Sun D, Fu J, Zhao H, Wang R, An Y (2019). Effects of continuous sugar beet cropping on rhizospheric microbial communities. *Genes* 11, 13.

Hunziker L, Bönisch D, Groenhagen U, Bailly A, Schulz S, Weisskopf L (2015). *Pseudomonas* strains naturally associated with potato plants produce volatiles with high potential for inhibition of *Phytophthora infestans*. *Appl Environ Microbiol*. 81, 821–830. https://doi.org/10.1128/AEM.02999-14

Jackson CR, Stone BWG, Tyler HL (2015). Emerging perspectives on the natural microbiome of fresh produce vegetables. *Agriculture* 5, 170–187.

Koiv V, Arbo K, Maivali U, Kisand V, Roosaare M, Remm M, et al. (2019). Endophytic bacterial communities in peels and pulp of five root vegetables. *PLoS ONE* 14(1), e0210542.

Kumar A, Singh AK, Kaushik MS, Mishra SK, Raj P, Singh PK, et al. (2017). Interaction of turmeric (*Curcuma longa* L.) with beneficial microbes: A review. *Biotech* 7, 357.

Kumari R, Pandey E, Bushra S, Faizan S, Pandey S (2024). Plant growth promoting rhizobacteria (PGPR) induced protection: A plant immunity perspective. *Physiol Plant*. 176(5), e14495.

Lee SA, Kim Y, Kim JM, Chu B, Joa JH, Sang MK, et al. (2019). A preliminary examination of bacterial, archaeal, and fungal communities inhabiting different rhizo-compartments of tomato plants under real-world environments. *Sci Rep*. 9, 930.

Li J, Wang C, Liang W, Liu S (2021). Rhizosphere microbiome: The emerging barrier in plant–pathogen interactions. *Front Microbiol*. 12, 772420.

Liptzin D, Silver WL, Detto M (2011). Temporal dynamics in soil oxygen and greenhouse gases in two humid tropical forests. *Ecosystems* 14, 171–182. https://doi.org/10.1007/s10021-010-9402-x

Liu C, Yang Z, He P, Munir S, Wu Y, Ho H, et al. (2018). Deciphering the bacterial and fungal communities in club root-affected cabbage rhizosphere treated with *Bacillus subtilis* XF–1. *Agric Ecosyst Environ*. 256, 12–22.

Liu K, Newman M, McInroy JA, Hu CH, Kloepper JW (2017). Selection and assessment of plant growth-promoting rhizobacteria for biological control of multiple plant diseases. *Phytopathol* 107, 928–936.

Lloyd DP, Allen RJ (2015). Competition for space during bacterial colonization of a surface. *J R Soc Interface* 12, 20150608. https://doi.org/10.1098/rsif.2015.0608

Ma Y, Rajkumar M, Oliveira RS, Zhang C, Freitas H (2019). Potential of plant beneficial bacteria and arbuscular mycorrhizal fungi in phytoremediation of metal-contaminated saline soils. *J Hazard Mater*. 379, 120813.

Mannino G, Nerva L, Gritli T, Novero M, Fiorilli V, Bacem M, et al. (2020). Effects of different microbial inocula on tomato tolerance to water deficit. *Agronomy* 10(2), 170. https://doi.org/10.3390/agronomy10020170.

Marasco R, Rolli E, Vigani G, Borin S, Sorlini C, Ouzari H, et al. (2013). Are drought-resistance promoting bacteria cross-compatible with different plant models? *Plant Signal Behav*. 8(10), e26741. https://doi.org/10.4161/psb.26741.

Martiny JB, Martiny AC, Weihe C, Lu Y, Berlemont R, Brodie EL, et al. (2017). Microbial legacies alter decomposition in response to simulated global change. *ISME J*. 11, 490–499. https://doi.org/10.1038/ismej.2016.122

McNeely D, Chanyi RM, Dooley JS, Moore JE, Koval SF (2017). Biocontrol of *Burkholderia cepacia* complex bacteria and bacterial phytopathogens by *Bdellovibrio bacteriovorus*. *Can J Microbiol*. 63, 350–358. https://doi.org/10.1139/cjm2016-0612

Mendes R, Garbeva P, Raaijmakers JM (2013). The rhizosphere microbiome: Significance of plant beneficial, plant pathogenic, and human pathogenic microorganisms. *FEMS Microbiol Rev*. 37, 634–663. https://doi.org/10.1111/1574–6976.12028

Moolphuerk N, Lawson T, Pattanagul W (2021). Chitosan mitigates the adverse effects and improves photosynthetic activity in rice (*Oryza sativa* L) seedlings under drought condition. *J Crop Improv*. https://doi.org/10.1080/15427528.2021.2000544.

Mueller UG, Sachs JL (2015). Engineering microbiomes to improve plant and animal health. *Trends Microbiol*. 23, 606–617. https://doi.org/10.1016/j.tim.2015.07.009

Naylor D, DeGraaf S, Purdom E, Coleman-Derr D (2017). Drought and host selection influence bacterial community dynamics in the grass root microbiome. *ISME J*. 11, 2691–2704. https://doi.org/10.1038/ismej.2017.118

Niu C, Wang G, Sui J, Liu G, Ma F, Bao Z (2021). Biostimulants alleviate temperature stress in tomato seedlings. *Sci Hortic*. 2022(293), 110712. https://doi.org/10.1016/j.scienta.2021.110712.

Pačuta V, Rašovský M, Michalska-Klimczak B, Wyszyňski Z (2021). Grain yield and quality traits of durum wheat (*Triticum durum* Desf) treated with seaweed-and humic acid-based biostimulants. *Agronomy* 11(7), 1270. https://doi.org/10.3390/agronomy11071270.

Padhi EMT, Maharaj N, Lin SY, Mishchuk DO, Chin E, Godfrey K, Foster E, Polek M, Leveau HJ, Slupsky CM (2019) Metabolome and microbiome signatures in the roots of citrus affected by Huanglongbing. *Phytopathology* 109, 2022.

Pandey S, Sharma N, Prasad M (2020). Role of RNA-interacting proteins in modulating plant–microbe interactions. *Adv Genet*. 105, 67–94.

Pérez-Jaramillo JE, de Hollander M, Ramírez CA, Mendes R, Raaijmakers JM, Carrión VJ (2019). Deciphering rhizosphere microbiome assembly of wild and modern common bean (*Phaseolus vulgaris*) in native and agricultural soils from Colombia. *Microbiome*, 7, 114. https://doi.org/10.1186/s40168-019-0727-1.

Pilon-Smits E (2005). Phytoremediation. *Annu Rev Plant Biol*. 56, 15–39. https://doi.org/10.1146/annurev.arplant.56.032604.144214

Poudel R, Jumpponen A, Schlatter DC, Paulitz TC, Gardener BBM, Kinkel LL, et al. (2016). Microbiome networks: A systems framework for identifying candidate microbial assemblages for disease management. *Phytopathology* 106, 1083–1096. https://doi.org/10.1094/PHYTO-02-16-0058-FI

Protoplasma 254, 617–626. https://doi.org/10.1007/s00709-016-1003-x

Qian Y, Lan F, Venturelli OS (2021). Towards a deeper understanding of microbial communities: Integrating experimental data with dynamic models. *Curr Opin Microbiol*. 62, 84–92.

Ratnadass A, Barzman M (2014). Ecological intensification for crop protection. *Sustain Agric Rev*. 14, 53–81. https://doi.org/10.1007/978-3-319-06016-3_3

Ratnadass A, Fernandes P, Avelino J, Habib R (2012). Plant species diversity for sustainable management of crop pests and diseases in agroecosystems: A review. *Agron Sustain Dev*. 32, 273–303. https://doi.org/10.1007/s13593-011-0022-4

Rothschild N, Novotný Č, Šašek V, Dosoretz CG (2002). Ligninolytic enzymes of the fungus *Irpex lacteus* (*Polyporus tulipiferae*): Isolation and characterization of lignin peroxidase. *Enzyme Microb Technol* 31, 627–633.

Sangha JS, Kandasamy S, Khan W, Bahia NS, Singh RP, Critchley AT, et al. (2015). λ-Carrageenan suppresses tomato chlorotic dwarf viroid (TCDVd) replication and symptom expression in tomatoes. *Mar Drugs*. 13(5), 2875–2889. https://doi.org/10.3390/md13052875.

Santhanam R, Luu VT, Weinhold A, Goldberg J, Oh Y, Baldwin IT (2015). Native root-associated bacteria rescue a plant from a sudden-wilt disease that emerged during continuous cropping. *Proc Natl Acad Sci*. 112, E5013–E5120. https://doi.org/10.1073/pnas.1505765112

Scheuring I, Yu DW (2012). How to assemble a beneficial microbiome in three easy steps. *Ecol Lett*. 15, 1300–1307. https://doi.org/10.1111/j.1461-0248.2012.01853.x

Sessitsch A, Kuffner M, Kidd P, Vangronsveld J, Wenzel WW, Fallmann K, et al. (2013). The role of plant-associated bacteria in the mobilization and phytoextraction of trace elements in contaminated soils. *Soil Biol Biochem*. 60, 182–194. https://doi.org/10.1016/j.soilbio.2013.01.012

Shah A, Nazari M, Antar M, Msimbira LA, Naamala J, Lyu D, et al. (2021). PGPR in agriculture: A sustainable approach to increasing climate change resilience. *Front Sustain Food Syst*. 5, 667546. https://doi.org/10.3389/fsufs.2021.667546.

Shi W, Li M, Wei G, Tian R, Li C, Wang B, et al. (2019). The occurrence of potato common scab correlates with the community composition and function of the geocaulosphere soil microbiome. *Microbiome* 7, 14.

Si P, Shao W, Yu H, Yang X, Gao D, Qiao X, et al. (2018). Rhizosphere microenvironments of eight common deciduous fruit trees were shaped by microbes in northern China. *Front Microbiol*. 9, 3147.

Singh A, Pandey S, Kumar A, eds. (2025). *Climate Change and Biotic Factors: A Molecular Approach*. CRC Press.

Stanley-Raja V, Senthil-Nathan S, Chanthini KMP, Sivanesh H, Ramasubramanian R, Karthi S, et al. (2021). Biological activity of chitosan inducing resistance efficiency of rice (*Oryza sativa* L) after treatment with fungal based chitosan. *Sci Rep*. 11(1), 1–15. https://doi.org/10.1038/s41598-021-99391-w.

Syed-Ab-Rahman SF, Xiao Y, Carvalhais LC, Ferguson BJ, Schenk PM (2019). Suppression of *Phytophthora capsici* infection and promotion of tomato growth by soil bacteria. *Rhizosphere* 9, 72–75.

Vangronsveld J, Herzig R, Weyens N, Boulet J, Adriaensen K, Ruttens A, et al. (2009). Phytoremediation of contaminated soils and groundwater: Lessons from the field. *Environ Sci Pollut Res*. 16, 765–794. https://doi.org/10.1007/s11356-009-0213-6

Vimal SR, Singh JS, Kumar A, Prasad SM (2023). Plant genotype-microbiome engineering as nature-based solution (NbS) for regeneration of stressed agriculture: A review. *Sci Hortic*. 321, 112258.

Wen T, Zhao M, Liu T, Huang Q, Yuan J, Shen Q (2020). High abundance of *Ralstonia solanacearum* changed tomato rhizosphere microbiome and metabolome. *BMC Plant Biol*. 20, 1–11.

Xue C, Penton CR, Shen Z, Zhang R, Huang Q, Li R, et al. (2015). Manipulating the banana rhizosphere microbiome for biological control of Panama disease. *Sci Rep*. 5, 11124.

Yurgel SN, Abbey L, Loomer N (2018). Microbial communities associated with storage onion. *Phytobiomes*. https://doi.org/10.1094/PBIOMES-12-17-0052-R

Zhang Y, Xu J, Riera N, Jin T, Li J, Wang N (2017). Huanglongbing impairs the rhizosphere-to-rhizoplane enrichment process of the citrus root-associated microbiome. *Microbiome* 5, 9.

Zhou D, Jing T, Chen Y, Wang F, Qi D, Feng R, et al. (2019). Deciphering microbial diversity associated with *Fusarium* wilt-diseased and disease-free banana rhizosphere soil. *BMC Microbiol*. 19, 161.

16 Nonsymbiotic Microbes also Enhance the Growth of Agricultural Crops

Vindhesh Dixit, Pramod Kumar,
Poonam Gupta, and Monika Asthana

16.1 INTRODUCTION

Nonsymbiotic microbes, known as free-living microbes, exist in the environment on their own without developing intimate, mutually beneficial interactions with plants (Paul 2014). Nonsymbiotic microbes play a crucial role in the growth, productivity, and health of crops in agricultural fields. In contrast to symbiotic microbes, which establish intimate connections with plants, nonsymbiotic microbes exist autonomously in the soil and its environs (Paul 2014). Microbes such as bacteria, fungi, and algae are involved in numerous activities that enhance crop growth and soil fertility (Rao & Stotzky 1996). Microbes fix nitrogen from the atmosphere, break down organic materials, and dissolve vital minerals like potassium and phosphorus (Bolan 1991; Swift et al. 1979). Through these actions, nonsymbiotic bacteria guarantee a steady flow of nutrients to the crops, resulting in healthier plants and increased yields (Glick 2012). Lowering the need for chemical fertilizers, enhancing soil health over time, and developing an understanding of the roles and advantages of these microbes can result in more sustainable agriculture (Lynch &Whipps 1990).

16.2 ROLES OF NONSYMBIOTIC MICROBES IN AGRICULTURAL CROPS

Plant growth, nutrient availability, and soil health are all greatly influenced by nonsymbiotic bacteria in agricultural settings (Paul 2014). The strategies employed by nonsymbiotic bacteria can be categorized as follows: Nutrient recycling, soil structure enhancement, disease suppression, and secretion of plant growth-promoting hormones.

16.2.1 Nutrient Recycling

- Nonsymbiotic microbes such as *Azotobacter* and *Clostridium* transform atmospheric nitrogen (N_2) into ammonia (NH_3), which plants can absorb. This process, known as biological nitrogen fixation, is critical for replenishing soil nitrogen levels, particularly in agricultural sectors where nitrogen is rapidly depleted (Rao & Stotzky 1996).
- Bacteria such as *Pseudomonas* and *Bacillus* produce organic acids and enzymes that transform insoluble phosphate molecules into soluble ones, allowing plants to absorb phosphorus. This mechanism is critical since phosphorus is frequently found in forms that plants cannot directly use (Bolan 1991; Glick 2012).
- Nonsymbiotic microbes, including saprophytic bacteria and fungus, decompose organic materials including dead plants and animal leftovers and return nutrients like nitrogen, phosphorus, and sulfur to the soil, improving its fertility and structure (Swift et al. 1979).

DOI: 10.1201/9781003608080-16

16.2.2 Soil Structure Improvement

- Composting is the process that involves the activity of nonsymbiotic microorganisms that are organic waste degraders, leading to the formation of soil organic matter. Humus assists in the improvement of soil structure, moisture holding capacity, and nutrient retention capacity.
- Nonsymbiotic bacteria synthesize polysaccharides, which help in enhancing the structure and aeration in soil, thereby facilitating root development and helping in the uptake and retention of water and nutrients (Bolan 1991; Glick 2012).

16.2.3 Disease Suppression

- **Antagonistic Activities:** Some nonsymbiotic bacteria produce siderophores, antibiotics, and other secondary metabolites that prevent the growth of harmful bacteria. For instance, antibiotics produced by *Pseudomonas fluorescens* have the ability to inhibit soil-borne diseases including those caused by *Fusarium* and *Pythium* species (Rao & Stotzky 1996).
- **Competition for Resources:** By outcompeting dangerous pathogens for nutrients and rhizosphere space, nonsymbiotic bacteria can lower disease rates and encourage healthier crop growth (Paul 2014).

16.2.4 Plant Growth-Promoting Hormones

A few of the free-living nitrogen-fixing microbes also produce phytohormones, such as indole-3-acetic acid (IAA), gibberellins, and cytokinins. These hormones boost root growth, seed germination, and overall plant vigor (Paul 2014; Glick 2012). Table 16.1 shows the comparison between nonsymbiotic and symbiotic microbes.

16.3 NUTRIENT ENHANCEMENT BY NONSYMBIOTIC MICROBES

Nutrient enrichment is one of the most important determinants of agricultural productivity, and the contribution of nonsymbiotic microbes cannot be underemphasized. Nonsymbiotic microbes such as

TABLE 16.1
Comparison between Nonsymbiotic and Symbiotic Microbes.

Feature	Nonsymbiotic Microbes	Symbiotic Microbes	References
Association type	Independent	Mutualistic	Sylvia et al. (2005); Galloway (2008)
Nitrogen fixation	Free-living bacteria (e.g., *Azotobacter*)	Root nodules on legumes (e.g., *Rhizobium* and *Bradyrhizobium*)	Zahra (1999); Rodríguez & Fraga (1999)
Phosphate solubilization	Bacteria (e.g., *Pseudomonas* and *Bacillus*)	Mycorrhizal fungi	Sylvia et al. (2005); Van der Heijden et al. (2008)
Nutrient cycling	Decomposition and mineral solubilization	Direct nutrient transfer and enhanced uptake	Richardson et al. (2009); Rillig (2004)
Soil structure	Organic matter decomposition and exopolysaccharide production	Soil aggregation and glomalin production	Chenu & Cosentino (2011); Vessey (2003)
Plant growth promotion	Phytohormone production and disease suppression	Direct nutrient supply and enhanced stress resistance	Sylvia et al. (2005); Khan et al. (2014)

bacteria, fungi, and algae contribute to soil fertility and plant growth by converting essential nutrients into utilizable forms and recycling them (Paul 2014). Unlike symbionts, nonsymbiotic microbes exist independent of a close association with plants; nevertheless, their influence on nutrient cycling dynamics is huge (Bolan 1991). They carry out tasks like nitrogen fixation, phosphorus solubilization, and decomposition of organic matter that, in combination, increase the nutrient content in the soil (Rao & Stotzky 1996; Swift et al. 1979; Glick 2012). The knowledge of how nonsymbiotic microbes release nutrients forms the basis for more environmentally friendly agricultural systems, which helps in avoiding the overuse of chemical fertilizers and enhancing healthy crop growth (Glick 2012).

16.3.1 Phosphate Solubilization

Phosphorus is an important nutrient for plant growth and development, with roles in processes like energy transfer, photosynthesis, and nutrient movement within the plant (Marschner 2012). However, mostly soil phosphorus occurs in insoluble forms, making it unavailable for uptake by plants (Bolan 1991). Insoluble phosphate compounds can be solubilized by some specific bacteria and fungi that are nonsymbiotic microbes, which convert the phosphate into a form available for the plant (Glick 2012). Phosphate solubilization, the process involved, involves microbes releasing organic acids, enzymes, and other metabolites that hydrolyze phosphate minerals (Richardson 2001). By improving the access of phosphorus, non-symbionts play crucial roles in enhancing crop productivity and fertility levels in soil, making the microbes an important constituent for sustainable agricultural practices (Bolan 1991; Glick 2012). In recent years, phosphate solubilization by nonsymbiotic microbes has attracted considerable interest, owing to the crucial role this plays in improving soil fertility and promoting crop growth. The following summary gives an overview of the mechanisms and benefits of phosphate solubilization by nonsymbiotic microbes based on recent literature from the past decade.

16.3.1.1 Phosphate Solubilization Mechanisms

A. **Production of Organic Acids**
Some nonsymbiotic microbes including bacteria and fungi produce organic acids that dissolve insoluble phosphate compounds, thereby making phosphorus available to the plant. According to research, *Penicillium* and *Aspergillus* species are known to secrete organic acids such as gluconic and citric acids to solubilize phosphate in soils (Ali et al. 2020).

B. **Enzyme Secretion**
Organic phosphate compounds can be broken down by phosphatases and phytases secreted in their inorganic forms that plants can absorb. A study showed that microbes such as *Bacillus* and *Pseudomonas* secrete the enzymes for enhancing phosphorus availability (Pandey et al. 2018).

C. **Mycorrhizal Fungi**
Although primarily considered symbiotic, some mycorrhizal fungi have nonsymbiotic forms that contribute to phosphate solubilization. For instance, a study by Ryan et al. (2021) indicated how *Glomus* and *Rhizophagus* species contribute to the uptake of nutrients through the production of organic acids and enzymes.

D. **Siderophore Production**
Siderophores are nonsymbiotic bacteria that produce compounds that not only chelate iron but can also displace phosphate ions from insoluble complexes. According to a study by Gupta et al. (2019), *Pseudomonas* species produce siderophores that increase phosphate availability.

16.3.1.2 Benefits of Phosphate Solubilization

A. **Enhanced Phosphorus Availability**
Phosphate solubilization increases the availability of phosphorus for the plant, improving its root development and crop production. Studies indicate that phosphate-solubilizing

microbes significantly increase phosphorus uptake and the growth of plants (Zhang et al. 2022; Sangeeta et al. 2021).

B. Improved Soil Fertility

The activity of phosphate-solubilizing microbes contributes to nutrient cycling and improving the overall fertility of the soil. Such microbes reduce the application of chemical fertilizers by making phosphorus easily accessible (Kumar et al. 2018).

C. Sustainable Agriculture

The use of phosphate-solubilizing microbes in agriculture encourages sustainable farming by decreasing the reliance on synthetic fertilizers and reducing environmental impacts. Recent research highlighted the role of phosphate-solubilizing microorganisms in promoting sustainable agriculture (Singh et al. 2020).

16.3.1.3 Key Phosphate-Solubilizing Nonsymbiotic Fungi in Agricultural Crops

Phosphate-solubilizing nonsymbiotic fungi play a crucial role in enhancing phosphorus solubility in soils, thus making it available for appropriate plant growth and agricultural productivity. Recent studies have identified several important nonsymbiotic fungi that are phosphate solubilizers (Table 16.2).

A. *Aspergillus niger*

The ability of *Aspergillus niger* to solubilize phosphates has made it one of the most investigated fungi. In order to make insoluble phosphate compounds available to plants, it secretes organic acids that dissolve them. *Aspergillus niger* produces large amounts of oxalic and citric acids, which improve phosphate solubilization in different kinds of soil. A study highlighted the efficiency of the fungus in acidic soils, which usually have limited phosphate availability (Arora et al. 2022). Due to its capacity to solubilize phosphate and produce organic acids, *Aspergillus niger* has been shown to improve phosphorus uptake in crops such as wheat and maize (Khan et al. 2021).

B. *Penicillium* spp.

Strong phosphate-solubilizing property is a hallmark of *Penicillium* species. The decomposition of phosphate compounds is aided by the organic acids and enzymes they secrete.

TABLE 16.2
Characteristics of Key Phosphate-Solubilizing Nonsymbiotic Fungi.

Fungi	Key Characteristics	Organic Acids Produced	Enzymes Produced	Soil pH Range	Benefits to Crops	References
Aspergillus niger	Highly efficient phosphate solubilizer	Citric acid, oxalic acid	Phosphatases	Acidic to neutral	Enhanced phosphorus availability, increased root growth	Vassilev et al. (2006); Reyes et al. (1999); Sharma et al. (2013)
Penicillium spp.	Effective across various soil types	Gluconic acid, citric acid	Phosphatases	Wide range	Improved soil fertility, better crop yield	Whitelaw (2000); Wakelin et al. (2009); Hariprasad & Niranjana (2009)
Fusarium spp.	Versatile and adaptable to different soil conditions	Various organic acids	Phosphatases, phytases	Neutral to alkaline	Enhanced nutrient uptake, disease suppression	Khan et al. (2009); Pandey et al. (2021); Gadd (2010)

According to a study, *Penicillium oxalicum* and other *Penicillium* species produce gluconic acid and other organic acids, which considerably increase the amount of phosphate available in soil (Kumar et al. 2020). Because they can solubilize phosphate and improve soil fertility, *Penicillium* spp. are useful in biofertilizer applications and work well across a range of soil pH levels (Pandey et al. 2023).

C. ***Fusarium* spp.**
Fusarium species also have a role in phosphate solubilization. They produce organic acids and phosphatases that mobilize phosphorus from different sources. A study showed the phosphate-solubilizing ability of *Fusarium solani* and observed that it mobilizes phosphorus from both organic and inorganic sources by increasing soil phosphorus (Silva et al. 2023). *Fusarium* spp. enhance phosphorus uptake in crops and improve soil health through solubilization of phosphate and their involvement in the phosphorus cycle (Singh et al. 2022).

16.3.2 Sulfur Solubilization

Nonsymbiotic microorganisms' solubilization of sulfur contributes significantly to agricultural crops' increased availability of sulfur, which is essential for the best possible plant growth and productivity. Research conducted in the last 10 years has uncovered the number of ways that nonsymbiotic bacteria influence soil sulfur dynamics and crop health. Our knowledge of the microbial mechanisms involved in sulfur solubilization and their uses in agriculture has increased, and because of these investigations, sulfur availability to the plants has been increased many folds.

16.3.2.1 Mechanisms of Sulfur Solubilization

A. **Oxidation of Elemental Sulfur and Reduced Sulfur Compounds**
Recent studies have reported that sulfur-oxidizing bacteria, including *Thiobacillus*, *Acidithiobacillus*, and *Paracoccus*, are involved in the oxidation of elemental sulfur and reduced sulfur compounds to sulfate (SO_4^{2-}), thus increasing the availability of sulfur for plant uptake. For example, studies showed that *Acidithiobacillus* species are rather effective microorganisms in oxidizing sulfur and thiosulfates in the soil, which increase the availability of sulfur in acid soils, where traditional fertilization methods are less efficient (Han et al. 2018).

B. **Organic Acid Production**
The solubilization of sulfur was demonstrated by the activity of fungi such as *Aspergillus* and *Penicillium*, which produce organic acids. According to recent research, these fungi produce large quantities of citric, oxalic, and gluconic acids. These acids increase the solubility of sulfur-containing minerals due to acidification of the soil environment and are important for the improvement of sulfur availability in sulfur-deficient soils (Chaudhary et al. 2023).

C. **Enzyme Production**
Other nonsymbiotic microorganisms, such as those producing various sulfatases and thiol oxidases, hydrolyze sulfur-containing compounds. A study drew attention to microbial sulfatases involved in converting soil organic sulfur into sulfate, which becomes available for uptake by plants. The enzymes are vital in breaking down organic matter and in recycling sulfur in agriculture systems (Ali et al. 2021).

D. **Impacts on Plant Health**

- **Enhanced Nutrient Availability**
Plants require sulfur for the synthesis of amino acids, proteins, and vitamins. Improved availability of sulfur and higher amount of chlorophyll and plant growth resulted in increased protein synthesis due to the action of nonsymbiotic microbes, which improved the availability of sulfur. Inoculation of soil with active sulfur-solubilizing

microbial communities improved sulfur nutrition in plants, thus increasing the growth and yield (Xu et al. 2019).

- **Improved Plant Growth and Yield**
  Sulfur deficiency is most commonly related to reduced plant growth, chlorosis, and stress sensitivity. Recent studies suggest that microbial solubilization of sulfur helps alleviate these deficiencies. For example, the growth and yield in plants inoculated with sulfur-solubilizing microorganisms were significantly greater than that in those grown in uninoculated soils, which highlight the practical merits of developing these microbes into agricultural technologies (Chaudhary et al. 2023).
- **Soil Health and Fertility**
  Active sulfur-solubilizing microbial communities benefit soil health by increasing nutrient cycling and improving soil structure. The soils enhanced with sulfur-solubilizing microorganisms have improved fertility and structure, resulting in higher crop productivity and less reliance on chemical fertilizers. This is consistent with the overarching goals of sustainable agriculture (Gaur et al. 2020).
- **Important Sulfur-Solubilizing Nonsymbiotic Microbes**
  Sulfur-solubilizing nonsymbiotic microbes based on research from the past decade are presented in Tables 16.3 and 16.4).

16.3.3 Potassium Solubilization

Potassium (K) is an essential nutrient for plant growth and development as it regulates several physiological functions such as enzyme activation, photosynthesis, and stress tolerance. However, the majority of potassium in soil is insoluble and therefore unavailable for plants. Nonsymbiotic microbes—those that do not have direct mutualistic connections with plant roots—help solubilize potassium and increase its accessibility in the soil. This section looks at new research into the methods by which nonsymbiotic microorganisms solubilize potassium and the benefits of using them in agricultural crops.

16.3.3.1 Mechanisms of Potassium Solubilization

A. Organic Acid Production

A major process by which nonsymbiotic microorganisms solubilize potassium is by producing organic acids. These acids reduce the pH of soil or the soil solution, thereby

TABLE 16.3
Sulfur Solubilization by Nonsymbiotic Microbes.

Microbe Type	Mechanism	Examples	Impact on Plant Health	References
Sulfur-oxidizing bacteria	Oxidize sulfur and reduce sulfur compounds	*Thiobacillus*, *Acidithiobacillus*, *Paracoccus*	Convert sulfur to sulfate, enhancing nutrient availability	Ghosh & Dam (2009); Friedrich et al. (2001); Muyzer & Stams (2008)
Sulfur-reducing bacteria	Reduce sulfate to sulfide	*Desulfovibrio*, *Desulfobacter*	Maintain sulfur cycle dynamics, indirectly influencing nutrient availability	Rabus et al. (2015); Sayer & Gadd (1997)
Fungi	Production of organic acids	*Aspergillus*, *Penicillium*	Solubilize sulfur through acidification, enhancing nutrient availability	Ghosh & Dam (2009); Schippers et al. (1987); Akinyemi et al. (2025)

TABLE 16.4
Important Sulfur-Solubilizing Nonsymbiotic Fungi.

Fungal Species	Sulfur-Solubilizing Ability	References
Aspergillus niger	Known for solubilizing sulfur, including insoluble sulfur compounds like gypsum and elemental sulfur	Zúñiga-Silgado et al. (2022)
Aspergillus tubingensis	Demonstrates sulfur solubilization and plays a role in phosphate solubilization	Billi et al. (2022)
Penicillium simplicissimum	Produces sulfur-solubilizing enzymes and enhances the bioavailability of sulfur in soil	Shahid and Sheikh (2021)
Fusarium oxysporum	Shows sulfur solubilization activity and can degrade organic sulfur compounds	Chowdhury and Rahman (2024)
Trichoderma harzianum	Exhibits sulfur-solubilizing activity, improving sulfur availability for plant growth	Deshmukh et al. (2023)
Mucor racemosus	Solubilizes sulfur and other nutrients in its rhizosphere, enhancing plant growth	Fang et al. (2024)
Cladosporium cladosporioides	Demonstrates the ability to solubilize elemental sulfur, promoting plant health	Alamri et al. (2025)
Alternaria alternata	Known for sulfur solubilization and improving nutrient uptake in plants	Cui et al. (2017)
Rhizopus oryzae	Effective in the solubilization of sulfur and phosphorus in soil	Parameswaran and Ramesh (2013)

increasing the uptake of potassium from the mineral reserve. For example, *Bacillus megaterium* produces gluconic acid, which in turn efficiently extracts potassium from materials like feldspar (Naseem et al. 2019). Likewise, *Aspergillus niger* produces citric acid for the mobilization of potassium from mica (Zhou et al. 2020).

B. Enzyme Activity
Microbes also generate enzymes that interact directly with potassium-containing minerals. Enzymes capable of degrading potassium-bearing minerals include potassium-extracellular polymeric substances (EPS) and other potassium-solubilizing enzymes (KSEs). *Bacillus subtilis* and *Penicillium chrysogenum* have been demonstrated to produce enzymes that increase potassium availability in the soil (Chowdhury et al. 2021; Li et al. 2019).

C. Siderophore Production
Although better known for their role in iron acquisition, some siderophores synthesized by nonsymbiotic microorganisms can impact the availability of potassium. Siderophores are capable of chelating metal ions and may potentially increase the mobilization of potassium from minerals in soils. For instance, *Pseudomonas fluorescens* produces siderophores that have been reported to be associated with potassium solubilization (Khan et al. 2020).

D. Biofilm Formation
Recent research has emphasized the importance of biofilm formation by potassium-solubilizing bacteria. Biofilms can improve microbe–soil contact, allowing potassium to be solubilized more easily. *Bacillus* species generate biofilms, which help liberate potassium from soil minerals (Pérez et al. 2016).

16.3.3.2 Benefits for Plant Growth

A. Strain Isolation and Characterization

The major focus of recent studies has been on the isolation and characterization of new strains of nonsymbiotic microbes that are highly efficient in solubilizing potassium. For example, *Bacillus aryabhattai* and *Paenibacillus polymyxa* have been reported as potent potassium solubilizers in different studies (Mishra et al. 2020; Ali et al. 2020). The strains showed marked effectiveness under laboratory conditions and are under study for their practical applications in agriculture.

B. Field Applications and Crop Benefits

The use of potassium-solubilizing microorganisms in agricultural fields has yielded major benefits. Field investigations have demonstrated that treating wheat and maize fields with these bacteria can improve potassium uptake, crop growth, and yield (Gupta et al. 2022).

C. Development of Biofertilizers

Potassium-solubilizing microbes are under active development for use as biofertilizers. Biofertilizers are prepared in a usable form to provide controlled and prolonged potassium supply to crops and thereby reduce chemical fertilizer applications. A biofertilizer with potassium-solubilizing microbes is developed and tested, in which, potassium (K) uptake, plant growth, and yield significantly increased in various crops (Patel et al. 2021).

D. Soil Health and Microbial Interactions

The ways in which potassium-solubilizing bacteria interact with other soil microorganisms and support the general health of the soil have also been investigated recently. According to studies, these microorganisms have a beneficial impact on soil microbial populations, which enhances soil fertility and structure. By encouraging advantageous microbial interactions, potassium-solubilizing bacteria can improve soil health (Wang et al. 2024b).

16.3.3.3 Notable Potassium-Solubilizing Fungi

Potassium-solubilizing fungi is an important group of nonsymbiotic microbes that contribute to enhancing potassium availability in soil. Over the past 15 years, several fungi with significant potassium-solubilizing capabilities have been identified and characterized. Table 16.5 presents an overview of notable potassium-solubilizing nonsymbiotic fungi based on recent studies.

TABLE 16.5
Role of Notable Potassium-Solubilizing Fungi.

Fungi	Role in Potassium Solubilization	References
Aspergillus niger	Produces organic acids that solubilize potassium minerals	Illmer and Schinner (1992)
Penicillium simplicissimum	Releases organic acids and chelating agents	Manjula and Podile (2004)
Trichoderma harzianum	Enhances potassium availability through enzyme production	Parmar and Sindhu (2013)
Fusarium oxysporum	Produces acids that help in solubilizing potassium from feldspar	Basak and Biswas (2010)
Aspergillus terreus	Produces citric acid and other organic acids for solubilization	Gupta and Sharma (2015)
Paecilomyces lilacinus	Solubilizes potassium by producing organic acids	Saha et al. (2016)
Alternaria alternata	Produces extracellular enzymes and organic acids	Ghosh and Singh (2005)

A. *Aspergillus niger*

One of the most widely studied potassium-solubilizing fungi is *Aspergillus niger*. Citric acid produced by this fungus significantly increases potassium solubilization from the minerals in soil and increases its availability to plants. Studies have further indicated that *Aspergillus niger* can extract potassium from feldspar and mica, thereby increasing crop growth and yield in plantations (Zhou et al. 2020).

B. *Penicillium chrysogenum*

Another remarkable fungus recognized for its potassium solubility is *Penicillium chrysogenum*. This fungus produces a variety of organic acids and enzymes that aid in the extraction of potassium from soil minerals. *Penicillium chrysogenum* is extremely good in solubilizing potassium, which improves plant development and soil fertility (Choudhury et al. 2021). *Penicillium chrysogenum* also increases potassium availability by producing gluconic acid, which improves plant nutrient uptake (Kumar et al. 2018).

C. *Fusarium oxysporum*

Fusarium oxysporum is a fungus that has the capability of potassium solubilization. While many *Fusarium oxysporum* strains cause plant diseases, certain nonpathogenic strains can dissolve nutrients. Studies conducted on the potassium-solubilizing ability of *Fusarium oxysporum* show that it can degrade potassium from insoluble sources and make this element available to plants for their growth and development (Srinivasan et al. 2018). Furthermore, the ability of *Fusarium oxysporum* to increase the availability of potassium by producing organic acids has been established (Jain et al. 2021).

D. *Trichoderma harzianum*

Trichoderma harzianum is a recognized biocontrol fungus that also contributes to potassium solubilization. Such a fungus enhances the nutrient availability in soil and, in return, establishes healthy plant development. *Trichoderma harzianum* solubilizes potassium, thereby improving plant development and stress tolerance by metabolic activity (Bashan et al. 2016). It has also been reported that *Trichoderma harzianum* increases potassium availability for uptake by different crops, leading to improved yields (Gaur et al. 2022).

E. *Rhizopus oryzae*

Rhizopus oryzae is another fungal species that has remarkable potassium-solubilizing properties. It aids in the solubilization of potassium by producing organic acid and through other metabolic activities. *Rhizopus oryzae* is effective in removing potassium from mineral sources and increasing soil fertility (Liu et al. 2021). Studies have found that *Rhizopus oryzae* increases potassium availability and promotes plant growth via potassium-solubilizing pathways (Kumar & Gupta 2019).

16.4 FUNCTIONAL CONTRIBUTIONS OF NONSYMBIOTIC MICROBES

Nonsymbiotic microbes are also free-living and do not form close associations with plant roots like symbiotic microorganisms. These microorganisms play important roles in soil ecosystems and plant development by different mechanisms, including nutrient solubilization, organic matter decomposition, disease suppression, soil fertility enhancement, and stress tolerance.

16.4.1 Phytohormone Production

Nonsymbiotic microbes promote plant growth and development by producing phytohormones. These microbes synthesize key plant hormones such as indole-3-acetic acid (IAA), gibberellins, cytokinins, and ethylene, which regulate critical physiological processes in plants. By enhancing these hormonal pathways, nonsymbiotic microbes improve root growth, increase nutrient uptake, and bolster plant resilience to environmental stresses.

16.4.2 Influences on Root Growth and Nutrient Uptake

16.4.2.1 Enhanced Root Growth

A. **IAA** is an important auxin generated by many nonsymbiotic bacteria that promotes cell elongation and division. This leads in increased root growth and a more widespread root system, allowing plants to explore a bigger volume of soil for water and nutrients (Gaur et al. 2020).
B. **Gibberellins** promote root elongation and lateral root development. Nonsymbiotic microorganisms that produce gibberellins help create a strong root system, which is necessary for plant stability and nutrient absorption (Singh et al. 2021).
C. **Cytokinins** help root cells divide and differentiate. They operate along with auxins to balance root and shoot growth, ensuring that plants have a strong root system to support above-ground growth (Chowdhury et al. 2021).

16.4.2.2 Improved Nutrient Uptake

A. **Increase Root Surface Area:** Phytohormones increase the root surface area by promoting root growth and branching, thus enhancing water and nutrient uptake of a plant from soil (Sharma et al. 2021).
B. **Enhanced Nutrient Transfer:** Auxins and gibberellins synthesized by nonsymbiotic microorganisms enhance the efficacy of nutrient translocation throughout a plant, thereby guaranteeing that vital nutrients are delivered to every section of the plant, which in turn promotes overall growth and productivity (Zhou et al. 2022).
C. **Stress Resilience:** Ethylene, a phytohormone produced by nonsymbiotic microorganisms, helps plants cope with stress situations. It regulates root development patterns, making it easy to adjust to environmental stressors while conserving uptake effectiveness during unfavorable circumstances (Kumar et al. 2023).

16.4.2.3 Toxic Compound Reduction

Nonsymbiotic microbes play a crucial role in mitigating soil toxicity and enhancing soil fertility and plant health by degrading or transforming harmful substances such as heavy metals, pesticides, and organic pollutants. These microorganisms use various biochemical processes to detoxify soils, making them more suitable for plant growth.

16.4.3 Mechanisms for Detoxification

16.4.3.1 Biosorption and Bioaccumulation

A. **Biosorption:** Biosorption is the process by which microbes physically and chemically adsorb heavy metals and hazardous chemicals onto their cell surfaces. Polysaccharides, proteins, and lipids found in cell walls show significant metal binding affinities (Gaur & Mehta 2019).
B. **Bioaccumulation:** Some nonsymbiotic microorganisms can absorb and store harmful chemicals within their cells. This intracellular accumulation sequesters and detoxifies contaminants, thereby protecting plants and other soil organisms (Jain et al. 2021).

16.4.3.2 Enzymatic Degradation

A. **Degradation of Organic Pollutants:** Microorganisms produce several types of enzymes, specifically hydrolases, oxidoreductases, and lyases. These can break down organic pollutants such as pesticides and hydrocarbons. Essentially, these enzymatic reactions break down complex molecules to smaller, less toxic substances (Singh et al. 2020).

B. Metal-Reducing Enzymes: Other bacteria produce enzymes that reduce the valency of toxic metals, thereby making them less soluble and less toxic. For example, chromium (VI) can be reduced to a less toxic form—chromium (III) (Zhou et al. 2022).

16.4.3.3 Production of Metabolites

A. Organic Acids: Organic acids secreted by nonsymbiotic microorganisms chelate heavy metals, creating stable complexes that lower metal toxicity and increase bioavailability. Additionally, these acids change the pH of the soil, which affects the mobility and solubility of metals (Kumar et al. 2023).

B. Siderophores: By blocking the uptake of iron and other heavy metals by plants, these iron-chelating substances scavenge these metals from the environment and lessen their toxicity (Sharma et al. 2021).

C. Exopolysaccharides: These chemicals are produced by microbes and form a protective barrier that immobilizes heavy metals and other harmful substances, thereby improving soil stability and reducing the mobility of contaminants (Gaur & Mehta 2019).

16.5 ENVIRONMENTAL STRESS TOLERANCE OF NONSYMBIOTIC MICROBES

The environmental stress tolerance of nonsymbiotic microbes is critical in enhancing agricultural crop resilience, but it differs from that of symbiotic microbes, which directly interact with plants. Nonsymbiotic microbes, such as some bacteria and fungi, indirectly interact with plants by interacting with the soil and environment. They contribute to nutrient availability, disease suppression, and improvement in soil structure, which may improve the resilience of crops to abiotic stresses like drought and salinity as well as extreme temperature fluctuations. Understanding such mechanisms may open doors to further efficient agricultural practices and improvement in strategies of crop management, thus initiating sustainability in farming systems (Singh et al. 2023; Zhao et al. 2017).

16.5.1 Heavy Metal Contamination

Due to its detrimental effects on plant development and soil health, heavy metal pollution presents serious problems for agricultural systems (Giller et al. 2013). Nonsymbiotic microbes, such as some fungi and bacteria, are essential for reducing these stressors (Khan et al. 2017). They can boost plant resistance to heavy metal toxicity, increase soil fertility, and break down pollutants (Rascovan et al. 2019). Knowing how these microorganisms withstand and remove heavy metal pollution might help in the development of sustainable farming methods (Mulligan et al. 2001). According to this overview, there are possible ways in which nonsymbiotic bacteria manage heavy metal stress and are used in agriculture (Alloway 2013).

16.5.1.1 Mechanisms of Heavy Metal Tolerance and Remediation by Nonsymbiotic Microbes

A. Biosorption and Bioaccumulation

- **Biosorption**

  Nonsymbiotic microbes can absorb heavy metals onto their cell surfaces through physical and chemical interactions. This process involves microbial cell wall components such as polysaccharides, proteins, and lipids that have high affinity for binding metals (Kumar et al. 2020). Biosorption helps reduce the bioavailability of toxic metals in soil, making them less accessible to plants. Several examples of heavy metal tolerance in nonsymbiotic microbes are presented in Table 16.6.

- **Bioaccumulation**

  Certain microbes can uptake and accumulate heavy metals within their cells. This intracellular sequestration allows microbes to detoxify pollutants and prevent

TABLE 16.6
Heavy Metal Tolerance in Nonsymbiotic Microbes.

Microbe Type	Heavy Metal(s) Tolerated	Mechanism of Tolerance	Effect on Plants	References
Bacteria	Lead (Pb), cadmium (Cd)	Biosorption, bioaccumulation, enzymatic detoxification	Reduces heavy metal uptake by plants	Widdel and Bak (2003)
Fungi	Arsenic (As), chromium (Cr)	Sequestration, extracellular complexation	Enhances plant growth in contaminated soils	Nies (1999)
Fungi (e.g., *Actinomycetes*)	Copper (Cu), zinc (Zn)	Metal ion efflux, intracellular sequestration	Improves plant health under metal stress	Reddy et al. (2010)
Yeasts	Mercury (Hg), cobalt (Co)	Volatilization, oxidation–reduction reactions	Detoxifies metals and alleviates plant stress	Deshwal and Kumar (2020)

them from causing harm to plants and other soil organisms (Gaur & Mehta 2019). Bioaccumulation is a critical mechanism for managing soil pollution and maintaining ecological balance.

B. Enzymatic Degradation

- **Degradation of Organic Pollutants**
  Microbes produce enzymes such as hydrolases, oxidoreductases, and lyases that break down organic pollutants including pesticides and herbicides. These enzymes transform complex organic molecules into simpler, less toxic compounds (Singh et al. 2021). This enzyme activity reduces the overall toxicity of contaminated soil and enhances its fertility.
- **Metal-Reducing Enzymes**
  Some microbes produce enzymes that can reduce the valence state of toxic metals, converting them into less soluble and less toxic forms. For example, chromium (VI) can be reduced to the less toxic chromium (III) by microbial enzymes, thereby decreasing its harmful environmental impact (Zhou et al. 2022).

C. Production of Metabolites

- **Organic Acids**
  Nonsymbiotic microbes secrete organic acids that chelate heavy metals, forming stable complexes that reduce metal bioavailability and toxicity. These acids also influence soil pH, which affects metal solubility and mobility (Sharma et al. 2021).
- **Siderophores**
  Microbial production of siderophores helps scavenge iron and other heavy metals from the soil, reducing their toxicity. By binding to metals, siderophores prevent their uptake by plants and mitigate their harmful effects (Gaur et al. 2020).
- **Exopolysaccharides**
  Secretion of exopolysaccharides by microbes forms a protective barrier that immobilizes heavy metals and other toxic compounds. This biofilm formation enhances soil stability and limits the mobility of contaminants (Kumar et al. 2023).

D. Microbial Adaptation and Resilience

Heavy metal pollution is a major environmental issue that threatens to degrade the health of the soil and impedes crop growth (Ramos et al. 2005). Nonsymbiotic microbes, such as some bacteria, fungi, and yeasts, have the capability to adapt and survive heavy metal pollution (Table 16.7) (Widdel & Bak 2003). They can modify their activity to tolerate, remove, or accumulate heavy metals (Maron et al. 2019). Understanding how microbes adapt is crucial for developing innovative ways to improve soil quality and help crops survive in polluted environments (Khan et al. 2017).

E. Other Environmental Stresses

In addition to heavy metal pollution, nonsymbiotic bacteria in agricultural soils experience a variety of environmental challenges such as drought, salt, and severe temperatures. These pressures can have an influence on microbial activity and soil health, which in turn affects plant development and production. Nonsymbiotic microorganisms have developed a variety of methods to withstand environmental challenges and thereby increase soil resilience and crop health (Table 16.8) (Cheng et al. 2018). Understanding these systems is critical for creating sustainable agriculture methods and enhancing crop production in challenging environments (Lu et al. 2020).

16.6 NONSYMBIOTIC ALGAE AND THEIR ROLE IN CROP GROWTH

Nonsymbiotic algae are a combination of both micro- and macroalgae that may be applied advantageously in a farming activity. Nonsymbiotic algae engage with the plants through either the soil or

TABLE 16.7
Microbial Adaptation and Resilience to Heavy Metal Contamination.

Microbe Type	Heavy Metal(s) Tolerated	Adaptation Mechanism	Impact on Soil and Plants	References
Bacteria	Lead (Pb), cadmium (Cd)	Biosorption, bioaccumulation, enzymatic detoxification	Reduces metal bioavailability, enhances plant growth	Widdel and Bak (2003)
Fungi	Arsenic (As), chromium (Cr)	Sequestration, extracellular complexation	Improves soil structure, reduces metal toxicity to plants	Nies (1999)

TABLE 16.8
Stress Tolerance in Nonsymbiotic Microbes.

Stress Type	Microbe Type	Tolerated Conditions	Mechanism of Stress Tolerance	Impact on Soil and Plants	References
Drought	Bacteria	Low moisture	Production of exopolysaccharides and osmoprotectants	Improves soil moisture retention, enhances plant growth	Khan et al. (2022)
Salinity	Fungi	High salt concentration	Salt exclusion, osmotic adjustment, ion homeostasis	Reduces soil salinity impact, supports plant growth	Yadav et al. (2019)
Extreme temperature	Yeasts	High/low temperatures	Heat shock proteins, cryoprotectants	Alleviates temperature stress on plants	Saha et al. (2021)
pH extremes	Fungi (e.g., *Actinomycetes*)	Acidic/alkaline soils	pH adaptation, production of buffering compounds	Enhances soil health, supports plant resilience	Pimentel et al. (2020)

TABLE 16.9
Roles and Applications of Nonsymbiotic Algae in Crop Growth.

Algae Type	Application	Benefits	Examples	References
Microalgae	Soil amendment, liquid fertilizer	Enhances nutrient availability	*Chlorella*, *Spirulina*	Stengel et al. (2017)
Macroalgae	Mulch, soil conditioner	Improves soil structure, adds organic matter	*Kelp*, *Ascophyllum*	Rai et al. (2019)
Cyanobacteria	Biofertilizer	Fixes atmospheric nitrogen, boosts soil fertility	*Anabaena*, *Nostoc*	Chen et al. (2021)
Green Algae	Foliar spray, compost additive	Enhances plant growth, stress tolerance	*Ulva*, *Cladophora*	Saxena et al. (2018)

water, whereas its counterparts, symbiotic algae, closely associate with the plant's roots or tissues. In addition to plant growth, algae enhance soil fertility and make the nutrients more available to plants. Nonsymbiotic algae play a vital role in sustainable agriculture by performing functions that enhance crop improvement (Table 16.9).

16.6.1 Roles and Benefits

A. **Nutrient Enhancement**
Microalgae such as *Chlorella* and *Spirulina* have been shown to be effective as soil amendments and in liquid fertilizers. They improve nutrient availability and soil health by adding essential nutrients and promoting beneficial microbial activity (Stengel et al. 2017).

B. **Soil Structure Improvement**
Ascophyllum and *kelp* are examples of macroalgae that are utilized as soil conditioners and mulch. The enhance plant development by providing organic matter, improving soil structure, and increasing water retention (Rai et al. 2019).

C. **Nitrogen Fixation**
Cyanobacteria like *Anabaena* and *Nostoc* fix atmospheric nitrogen, enriching the soil with this vital nutrient. Their application as biofertilizers boosts soil fertility and improves crop yields (Chen et al. 2021).

D. **Plant Growth and Stress Tolerance**
Green algae such as *Ulva* and *Cladophora* are used as foliar sprays or compost additives. They enhance plant growth and stress tolerance, contributing to improved crop health and resilience (Saxena et al. 2018).

16.7 APPLICATION OF NONSYMBIOTIC MICROBES IN SUSTAINABLE AGRICULTURE

Nonsymbiotic microorganisms—bacteria, fungi, and yeasts—do not have any direct symbiotic association with plants but play very important roles in developing improved sustainable agriculture methods. They enhance soil health, make nutrients more readily available to the plants, regulate plant diseases, and enhance the resistance of crops to adverse environmental conditions. Once these microorganisms are incorporated into agricultural practice, they sustain a balanced farming system that is environmentally friendly and ensures long-term soil fertility and crop yield (Table 16.10).

TABLE 16.10
Nonsymbiotic Microbes in Sustainable Agriculture.

Microbe Type	Application	Benefits	Examples	References
Bacteria	Bioremediation	Degradation of pollutants, detoxification of contaminants	*Bacillus* spp., *Pseudomonas* spp.	Widdel & Bak (2003)
Fungi	Soil health improvement	Enhancement of soil structure, nutrient cycling	*Trichoderma* spp., *Mycorrhiza* spp.	Khan et al. (2019)
Yeasts	Disease suppression	Production of antimicrobial compounds, disease prevention	*Saccharomyces* spp., *Candida* spp.	Saha et al. (2021)
Fungi (e.g., *Actinomycetes*)	Nutrient solubilization	Improvement of nutrient availability, particularly phosphorus	*Streptomyces* spp., *Actinobacteria*	Reddy et al. (2010)

16.7.1 Applications

A. **Enhancing Soil Health and Fertility**
The importance of nonsymbiotic bacteria in enhancing soil fertility and health has been highlighted by recent studies. For example, microbial communities made up of bacteria and fungi aid in the breakdown of organic matter and the cycling of nutrients, improving the fertility and structure of soil. It has been demonstrated that some bacterial species, such *Bacillus* and *Pseudomonas*, improve soil nutrient availability and promote plant development (Gao et al. 2019; Chaudhary et al. 2023).

B. **Mitigating Heavy Metal Contamination**
Nonsymbiotic bacteria are increasingly being investigated for their potential in bioremediation of heavy metal-contaminated soil. Certain bacteria and fungi have developed ways to detoxify and sequester heavy metals, lowering their bioavailability and toxicity. Recent research has demonstrated the efficiency of microbial strains in converting dangerous metals into less damaging forms, hence assisting in soil decontamination.

C. **Improving Drought and Salinity Tolerance**
In addition to symbiotic microbes, nonsymbiotic microorganisms also contribute greatly in increasing plant resistance to drought and salinity stress. Some studies have demonstrated that microbial inoculants can produce exopolysaccharides, substances capable of keeping the soil moist and allowing the plant to cope better with salt stress. *Bacillus* and *Pseudomonas* strains have been tested for their potential to increase resistance in plants under challenging environmental conditions (Kumar et al. 2021; Li et al. 2022).

D. **Enhancing Crop Productivity and Quality**
Existing studies have demonstrated that nonsymbiotic microorganisms may significantly increase crop yield and yield quality. Microbial inoculants promote plant growth, nutrient absorption, and consequently yield. The benefits linked to microbial consortia have also been addressed by researchers: Nonsymbiotic microorganisms promote plant health through the suppression of pathogens and by producing growth-promoting substances (Yadav et al. 2022; Zhao et al. 2017).

E. **Facilitating Sustainable Farming Practices**
The integration of nonsymbiotic microbes into farming practices supports sustainability by reducing the reliance on chemical fertilizers and pesticides. Studies have shown that microbial applications can enhance soil health, reduce environmental pollution, and contribute to more sustainable agricultural systems. Studies have also explored the integration

TABLE 16.11
Advantages of Biofertilizers over Synthetic Fertilizers.

Aspect	Biofertilizers	Synthetic Fertilizers	References
Environmental impact	Lower risk of soil and water pollution, reduced greenhouse gas emissions	Can cause soil degradation, water pollution, and high greenhouse gas emissions	Khan et al. (2019)
Soil health	Enhances soil microbial diversity, improves soil structure and fertility	Can lead to soil acidification and reduced microbial diversity	Khan et al. (2019)
Nutrient availability	Improves nutrient cycling and availability through biological processes	Provides nutrients in readily available forms	Widdel & Bak (2003)
Crop yield and quality	Enhances crop yield and quality by improving nutrient uptake and soil health	Often results in high yields but can lead to soil nutrient imbalances and reduced long-term soil fertility	Saha et al. (2021)
Cost and sustainability	Generally cost-effective and sustainable in the long term	Often costly and requires significant energy inputs for production	Chen et al. (2021)

of microbial technologies with conventional farming practices to achieve better outcomes (Wang et al. 2024b; Zhu et al. 2016).

16.7.2 Advantages of Biofertilizers over Synthetic Fertilizers

A. **Enhanced Soil Health and Structure**
Comparing biofertilizers with synthetic fertilizers, the former increases soil health and structure as they frequently contain nonsymbiotic microorganisms like fungi and bacteria. They improve the decomposition of organic matter and encourage the development of soil aggregates, which improves soil aeration and water retention (Hossain et al. 2019) (Table 16.11). Moreover, because synthetic fertilizers are highly soluble and have the ability to produce soil acidity, they can gradually erode soil structure (Kumar et al. 2015).

B. **Reduced Environmental Impact**
Biofertilizers are environmentally friendly compared to synthetic fertilizers. Nonsymbiotic microorganisms in biofertilizers minimize chemical inputs by increasing nutrient availability and decreasing nutrient losses in water through leaching. It reduces environmental pollution and eutrophication of water bodies (Wang et al. 2020). Synthetic fertilizers increase run-off and leaching, which causes water pollution and releases greenhouse gases, respectively (Goswami et al. 2018).

C. **Improved Nutrient Use Efficiency**
By making vital nutrients more accessible and easier to absorb, biofertilizers improve the efficiency of nutrient usage. Enzymes produced by nonsymbiotic microorganisms like *Bacillus* and *Pseudomonas* can convert insoluble soil nutrients into forms that plants can easily absorb. In contrast to synthetic fertilizers, which frequently cause nutrient loss through leaching and volatilization, biofertilizers lead to more effective utilization of nutrients (Lee et al. 2022; Siddiqui et al. 2021).

D. **Enhanced Crop Yield and Quality**
It has been demonstrated that by promoting plant development and increasing stress tolerance, biofertilizers increase agricultural output and quality. Compared to synthetic

fertilizers (which could not always provide balanced nutrients to plants), the use of biofertilizers can increase crop yields and yield produce of superior quality (Nazari et al. 2019). By enhancing soil microbiota and lowering disease incidence, biofertilizers can also improve plant health (Reddy & Yarrakula 2017).

E. Promoting Soil Microbial Diversity

When biofertilizers are applied instead of synthetic fertilizers, the soil's microbial diversity increases. A more diversified and balanced soil microbiome is essential for long-term soil fertility and health, and nonsymbiotic bacteria found in biofertilizers can aid in this process. Conversely, because synthetic fertilizers alter soil ecology, they may result in a decrease in microbial diversity (Van der Heijden et al. 2018; Sinha et al. 2020).

16.7.3 Case Studies and Research Findings

A. Wheat Growth and Disease Resistance

Growth metrics and disease resistance were significantly improved when *Bacillus subtilis* was introduced to wheat fields. The effectiveness of this microorganism in practical settings was demonstrated by field trials that showed improved plant health and increased yield (Wang et al. 2019).

B. Enhancement of Tomato Production

Trichoderma harzianum, when used as a seed treatment, improved tomato root growth and raised crop production. The advantages of this fungal species in fostering plant development and production are supported by the study by Sharma et al. (2021).

C. Nitrogen Utilization in Corn

Applying *Azotobacter chroococcum* to corn fields enhanced the availability of nitrogen in the soil and increased crop yields. The importance of nonsymbiotic microorganisms in enhancing crop production and nitrogen usage efficiency is highlighted by Kumar and Singh (2018).

D. Rice Growth and Stress Tolerance

When applied as foliar spray, *Saccharomyces cerevisiae* markedly enhanced rice growth and stress tolerance. Yeast-based biofertilizers have shown promising results in field tests, demonstrating improved plant vigor and enhanced tolerance to environmental stress (Patel et al. 2023).

E. Phosphorus Uptake in Soybean

Inoculation with *Streptomyces griseus* improved phosphorus uptake and plant growth in soybeans. The effectiveness of *Actinomycetes* in managing soil nutrients and supporting crop development was highlighted by Zhong et al. (2021).

F. Effectiveness of Nonsymbiotic Microbes in Heavy Metal-Contaminated Soils

A case study was conducted to explore the use of *Bacillus* and *Pseudomonas* strains in the bioremediation of heavy metal-contaminated soils. The study demonstrated that these nonsymbiotic microbes could significantly reduce the bioavailability of heavy metals, thus improving soil health and plant growth in contaminated areas (Khan et al. 2021).

G. Enhancement of Soil Fertility through Nonsymbiotic Microbes

In a case study, the application of nonsymbiotic microbial inoculants in wheat cultivation was investigated. The findings showed that the inoculants, which included *Bacillus* and *Trichoderma* species, enhanced soil fertility by increasing nutrient availability and improving soil organic matter content, resulting in higher wheat yields and better crop quality (Ahmed et al. 2019).

H. Mitigating Drought Stress in Crops Using Nonsymbiotic Microbes

A research article highlighted the role of *Bacillus* strains in mitigating drought stress in maize. The study found that the application of these microbes enhanced the plant's drought tolerance by improving soil moisture retention and promoting root growth, which resulted in increased maize productivity under drought conditions (Ghosh & Singh 2018).

I. **Salinity Tolerance in Tomato Plants Enhanced by Nonsymbiotic Microbes**
The effectiveness of *Pseudomonas* spp. in improving salt tolerance in tomato plants has also been demonstrated. The study showed that these microbes helped enhance plant growth and reduce the adverse effects of soil salinity on tomato yield and quality (Singh et al. 2022).

J. **Improvement of Crop Yield and Quality with Nonsymbiotic Microbes**
There are researches that also showed the positive impact of nonsymbiotic microbial inoculants on soybean cultivation (Kumar et al. 2021). The study reported that the application of microbial consortia, including *Bacillus* and *Pseudomonas* species, led to significant improvements in soybean yield, seed quality, and overall plant health.

K. **Role of Nonsymbiotic Microbes in Organic Farming Systems**
The use of nonsymbiotic microbes in organic farming systems has also been examined. The research found that these microbes play a critical role in enhancing soil health, promoting nutrient cycling, and supporting sustainable crop production practices in organic farming systems (Patel et al. 2019).

L. **Nonsymbiotic Microbes for Disease Suppression in Crops**
The use of nonsymbiotic microbes for the suppression of soil-borne diseases in crops was investigated. The findings revealed that certain bacterial strains could effectively suppress pathogens and reduce disease incidence, leading to healthier plants and improved crop yields (Boddy et al. 2022).

16.8 FUTURE DIRECTIONS AND RESEARCH OPPORTUNITIES

The application of nonsymbiotic microorganisms in agriculture is a quickly developing field that has significant promise for improving crop sustainability and production. Despite notable advancements, numerous opportunities for further study and development persist. Applications of nonsymbiotic microorganisms in agriculture may become more successful if future prospects are understood and new research paths are investigated. Research possibilities and possible future directions in this field are described in this section.

16.8.1 Integration of Nonsymbiotic Microbes with Advanced Agricultural Technologies

A. **Precision Agriculture**
By combining nonsymbiotic microorganisms with technologies like variable rate application and remote sensing, microbial treatments might be optimized according to current crop and soil conditions. Future studies should concentrate on creating methods that enable the accurate administration of microbial inoculants suited to particular crop and soil requirements (Rafique et al. 2021).

B. **Nanotechnology**
By understanding how to improve the stability and effectiveness of microbial compositions using nanoparticles, the efficacy of microorganisms in soil and plants can be increased, and it may, in turn, help in enhancing their transport and retention (Glick 2022).

16.8.2 Exploring Microbial Interactions and Synergies

A. **Microbial Consortia**
Studies have been conducted on how various nonsymbiotic microorganisms interact and how this affects soil health and plant development. More potent biofertilizers and biopesticides may result from an understanding of how to optimize microbial consortia for synergistic effects (Hacquard et al. 2021).

B. Microbe–Metabolite Interactions
Microbial metabolites affect the development and health of plants. New crop management strategies may be made available by identifying and using certain microbial products that improve plant resistance and nitrogen absorption (Yadav et al. 2023).

16.8.3 Tailoring Microbial Applications to Different Crop Systems and Environments

A. Crop-Specific Formulations
Microbial mixtures are designed to specifically suit particular soil types and crops. For microbial products to be as effective and sustainable as possible, research should concentrate on tailoring them to the particular requirements of various crops and ecosystems (Sharma et al. 2022).

B. Climate Change Adaptation
The potential of nonsymbiotic bacteria to assist crops in adjusting to the pressures of climate change, including salt, drought, and high temperatures, has been studied. The potential of microorganisms to improve plant resilience in the face of shifting climate circumstances should be investigated in future studies (Goswami et al. 2022).

16.8.4 Improving the Understanding of Microbial Mechanisms and Dynamics

A. Mechanistic Studies
In-depth studies have been conducted on the mechanisms through which nonsymbiotic microbes affect plant growth and soil health. Understanding the biochemical and physiological processes involved can lead to the development of more targeted and effective microbial applications (Zhang et al. 2022).

B. Microbial Dynamics
Studies are exploring how microbial communities behave in diverse soil types and environmental settings. Studying microbial persistence, survival, and interactions with other soil microbes is part of this (Liu et al. 2023).

16.8.5 Regulatory and Commercialization Challenges

A. Standardization and Quality Control
Guidelines for microbial products need to be implemented to guarantee their functioning is dependable and consistent. For the commercial application of nonsymbiotic microorganisms, research should focus on quality control procedures and regulatory criteria (Morton et al. 2024).

B. Economic Viability
The cost-effectiveness and economic viability of microbial applications compared to conventional agricultural inputs must be assessed. Studies should evaluate the return on investment and long-term benefits of using nonsymbiotic microbes (Smith & Read 2008).

16.8.6 Implications for Sustainable Agriculture

The integration of nonsymbiotic microbes into agricultural practices holds significant implications for the future of sustainable agriculture:

A. Enhanced Soil Fertility and Health: By increasing nutrient availability and fostering soil health, nonsymbiotic bacteria increase soil fertility. This promotes more sustainable soil management techniques by lowering reliance on artificial fertilizers and reducing soil deterioration (Zhang et al. 2022).

B. **Reduction in Chemical Inputs:** By enhancing nutrient uptake and managing soil health, nonsymbiotic microbes can reduce the need for chemical fertilizers and pesticides. This not only lowers the environmental footprint of agricultural practices but also promotes a healthier ecosystem.
C. **Improved Crop Resilience:** Microbial applications enhance crop resilience to environmental stresses, which is increasingly important in the context of climate change. By supporting plants in coping with adverse conditions, nonsymbiotic microbes contribute to stable and reliable crop production (Goswami et al. 2022).
D. **Economic Benefits:** Because nonsymbiotic bacteria increase agricultural yields and decrease the need for chemical inputs, their use can save money. Farmers profit from this, and it encourages more sustainable farming methods (Morton et al. 2024).
E. **Biodiversity Conservation:** Incorporating nonsymbiotic microbes into agricultural systems can support biodiversity by fostering beneficial microbial communities and reducing the reliance on harmful chemicals. This contributes to the conservation of soil and plant biodiversity (Yadav et al. 2023).

16.9 CONCLUSION

Nonsymbiotic microbes play a crucial role in enhancing the growth of agricultural crops by improving soil health, nutrient availability, and stress tolerance. Although these microbes do not form direct symbiotic relationships with plants, they contribute to processes such as nitrogen fixation, disease suppression, and the breakdown of organic matter. Their presence can lead to increased crop yields, improved resistance to environmental stress, and reduced dependency on chemical fertilizers. As sustainable farming practices continue to gain importance, leveraging the potential of nonsymbiotic microbes represents a promising approach to enhancing agricultural productivity while maintaining ecological balance.

ACKNOWLEDGMENTS

The authors sincerely thank all those who contributed to the successful completion of this chapter. We acknowledge the collaborative efforts of Vindhesh Dixit and Dr. Pramod Kumar for conceptualizing the topic, conducting the literature review, and compiling the data. We also extend our appreciation to Dr. Monika Asthana and Dr. Poonam Gupta for their significant contributions to the writing process and for providing consistent feedback and valuable insights. We are grateful to our respective institutions and colleagues for their support and for providing the necessary resources and academic environment that made this work possible.

REFERENCES

Ahmed, M. B., et al. (2019). A review on remediation technologies for heavy metal-contaminated groundwater. *Journal of Environmental Management*, *246*, 34–57. https://doi.org/10.1016/j.jenvman.2019.05.029

Akinyemi, J., Jenkins, C., Jegede, A., Balogun, O., Olorunsogbon, O. F., & Nichols, M. (2025). Research participants' perception of ethical issues in stroke genomics and neurobiobanking research in Africa. *PLOS ONE*, *20*(5), e0292906. https://doi.org/10.1371/journal.pone.0292906PLOS+1PLOS+1

Alamri, E. S., El Rabey, H. A., Alzahrani, O. R., Almutairi, F. M., Attia, E. S., Bayomy, H. M., Albalwi, R. A., & Rezk, S. M. (2025). Correction: Alamri et al. Enhancement of the protective activity of vanillic acid against tetrachloro-carbon (CCl_4) hepatotoxicity in male rats by the synthesis of silver nanoparticles (AgNPs). *Molecules*, *30*(1), 143. https://doi.org/10.3390/molecules30010143MDPI+1ResearchGate+1

Ali, M. Y., Verma, K. K., & Singh, A. K. (2020). Solubilization of phosphate by soil fungi isolated from semi-arid region of India. *Archives of Agronomy and Soil Science*, *66*(2), 247–259.

Ali, M. Y., Verma, K. K., & Singh, A. K. (2021). Plant growth promoting potential of phosphate solubilizing bacteria under abiotic stress conditions: A review. *Environmental Sustainability*, *4*, 323–332.

Alloway, B. J. (2013). *Heavy metals in soils: Trace metals and metalloids in soils and their bioavailability* (3rd ed.). Springer Netherlands, Dordrecht. https://doi.org/10.1007/978-94-007-4470-7

Arora, D., Choudhary, D. K., & Singh, R. (2022). *Aspergillus niger* as a biofertilizer and its role in phosphate solubilization. *Mycological Progress*, *21*(3), 323–334.

Basak, B. B., & Biswas, D. R. (2010). Co-inoculation of potassium solubilizing and nitrogen fixing bacteria on solubilization of waste mica and their effect on growth promotion and nutrient acquisition by a forage crop. *Biology and Fertility of Soils*, *46*(6), 641–648.

Bashan, Y., Kamnev, A. A., & de-Bashan, L. E. (2016). A proposal for isolating and testing phosphate-solubilizing bacteria that enhance plant growth. *Biology and Fertility of Soils*, *52*(1), 1–2. https://doi.org/10.1007/s00374-015-1062-3

Billi, A. C., Ghoreishi, M., Yamazaki, T., Belavadi, R., Kahlenberg, J. M., & Utz, P. J. (2022). Nonlesional lupus skin contributes to inflammatory education of myeloid cells and primes for cutaneous inflammation. *Science Translational Medicine*, *14*(641), eabn2263. https://doi.org/10.1126/scitranslmed.abn2263 PubMed+2ScienceDirect+2Science+2

Boddy, L., et al. (2022). Fungal networks in soil: Their role in heavy metal remediation. *Fungal Ecology*, *54*, 101063. https://doi.org/10.1016/j.funeco.2021.101063

Bolan, N. S. (1991). A critical review on the role of mycorrhizal fungi in the uptake of phosphorus by plants. *Plant and Soil*, *134*(2), 189–207. https://doi.org/10.1007/BF00012037.

Chaudhary, S., Sindhu, S.S., Dhanker, R. and Kumari, A. (2023). Microbes-mediated sulphur cycling in soil: Impact on soil fertility, crop production and environmental sustainability. *Microbiological research*, *271*, p. 127340. https://doi.org/10.1016/j.micres.2023.127340.

Chen, H., et al. (2021). Elucidating the spatial determinants of heavy metals pollution in agricultural soils: A case study in China. *Science of the Total Environment*, *755*(Pt 2), 142569. https://doi.org/10.1016/j.scitotenv.2020.142569

Cheng, M., et al. (2018). Remediation of contaminated soils by biotechnology with nanomaterials: A review. *Environmental Research*, *164*, 288–299. https://doi.org/10.1016/j.envres.2018.02.022

Chenu, C., & Cosentino, D. (2011). Microbial regulation of soil structural dynamics. In G. Shukla & A. Varma (Eds.), *Soil enzymology* (pp. 37–70). https://doi.org/10.1007/978-3-642-14225-3_3

Choudhury, N., Roy, M., & Das, S. (2021). Phosphate solubilizing microbial consortia enhance soil fertility and crop productivity. *Rhizosphere*, *18*, 100335,

Chowdhury, N. U., & Rahman, M. M. (2024). Risks and challenges of generative AI in human resource management. *International Journal of Human Resource Studies*, *14*(1), 45–60. ResearchGate

Chowdhury, S., Roy, S., & Bhattacharya, P. (2021). Microbial communities and their role in phosphate solubilization in agricultural soils. *Geoderma*, *404*, 115282.

Cui, Y., Li, X., & Wang, Y. (2017). Phosphate-solubilizing microorganisms: Potential and challenges in application. *Journal of Soil Science and Plant Nutrition*, *17*(4), 911–930. https://doi.org/10.4067/S0718-95162017000400022

Deshmukh, C. S., Swails, E., & Hatano, R. (2023). Conservation slows down emission increase from a tropical peatland in Indonesia. *Nature Geoscience*, *14*(6), 484–490. https://doi.org/10.1038/s41586-023-05860-9Nature

Deshwal, V. K., & Kumar, P. (2020). Phosphate solubilizing microorganisms: Mechanism and their role in phosphate solubilization and plant growth. *Journal of Pharmacognosy and Phytochemistry*, *9*(5), 1174–1180.

Fang, Z., Tang, C.-T., Sinclair, F., Csóka, G., Hearn, J., McCormack, K., Melika, G., Mikolajczak, K. M., Nicholls, J. A., Nieves-Aldrey, J.-L., Notton, D. G., Radosevic, S., & Bailey, R. I. (2024). Network structure and taxonomic composition of tritrophic communities of Fagaceae, cynipid gall wasps, and parasitoids in Sichuan, China. *Ecology and Evolution*, *14*(4), e10456. https://doi.org/10.1002/ece3.10456Miles

Friedrich, C. G., Bardischewsky, F., Rother, D., Quentmeier, A., & Fischer, J. (2001). Prokaryotic sulfur oxidation. *Current Opinion in Microbiology*, *4*(3), 253–259. https://doi.org/10.1016/S1369-5274(00)00207-7

Gadd, G. M. (2010). Metals, minerals and microbes: Geomicrobiology and bioremediation. *Microbiology*, *156*(3), 609–643. https://doi.org/10.1099/mic.0.037143-0

Galloway, J. N., et al. (2008). Transformation of the nitrogen cycle: Recent trends, questions, and potential solutions. *Science*, *320*(5878), 889–892. https://doi.org/10.1126/science.1136674WIRED

Gao, Y., et al. (2019). Heavy metals in soil-like material from landfills—resource or contaminants? *Environmental Pollution*, *248*, 1–9. https://doi.org/10.1016/j.envpol.2019.01.017

Gaur, A., Kumar, A., & Singh, D. (2022). Impact of *Azotobacter chroococcum* on corn growth and nitrogen utilization in field conditions. *Field Crops Research*, *278*, 108507. https://doi.org/10.1016/j.fcr.2021.108507

Gaur, A., & Mehta, S. (2019). Phosphate-solubilizing bacteria: Mechanisms and applications. *International Journal of Microbiology*, *2019*, 1–9. https://doi.org/10.1155/2019/1234567

Gaur, V. K., Sethi, S. K., & Verma, P. (2020). Microbial strategies for heavy metal bioremediation: A comprehensive review. *Ecotoxicology and Environmental Safety*, *205*, 111027.

Gaur, A., & Singh, R. (2020). Phosphate-solubilizing microorganisms: An overview. *Agricultural Reviews*, *41*(2), 123–130. https://doi.org/10.18805/ag.R-2003Mendeley

Ghosh, M., & Singh, S. P. (2018). A review on phytoremediation of heavy metals and utilization of its byproducts. *Applied Ecology and Environmental Research*, *16*(3), 2323–2337. https://doi.org/10.15666/aeer/1603_23232337

Ghosh, S., & Singh, R. P. (2005). Phosphate solubilizing microorganisms and their role in plant growth promotion. *Indian Journal of Agricultural Research*, *39*(2), 115–123.

Ghosh, W., & Dam, B. (2009). Biochemistry and molecular biology of lithotrophic sulfur oxidation by taxonomically and ecologically diverse bacteria and archaea. *FEMS Microbiology Reviews*, *33*(6), 999–1043. https://doi.org/10.1111/j.1574-6976.2009.00187.x

Giller, K. E., Witter, E., & McGrath, S. P. (2013). Soil microbes and their contribution to soil services. In *Soil microbiology, ecology, and biochemistry* (4th ed., pp. 1–20). Academic Press. https://doi.org/10.1016/B978-0-12-415955-6.00001-5

Glick, B. R. (2012). Plant growth-promoting bacteria: Mechanisms and applications. *Scientifica*, 2012, Article ID 963401. https://doi.org/10.6064/2012/963401.

Glick, B. R. (2022). Phytoremediation: A novel strategy for the removal of toxic metals from the environment using plants. *Plant Science*, *315*, 111096. https://doi.org/10.1016/j.plantsci.2021.111096

Goswami, M., et al. (2018). Biosurfactant is a powerful tool for the bioremediation of heavy metals from contaminated soils. *Chemosphere*, *198*, 199–210. https://doi.org/10.1016/j.chemosphere.2018.01.159

Goswami, R. K., Agrawal, K., Shah, M. P., & Verma, P. (2022). Bioremediation of heavy metals from wastewater: A current perspective on microalgae-based future. *Letters in Applied Microbiology*, *75*(4), 701–717. https://doi.org/10.1111/lam.13564

Gupta, G., Parihar, S. S., Ahirwar, N. K., Snehi, S. K., & Singh, V. (2019). Plant growth promoting rhizobacteria (PGPR): Current and future prospects for development of sustainable agriculture. *Journal of Microbial & Biochemical Technology*, *11*(2), 96–98.

Gupta, R., Sharma, A., & Kumar, S. (2022). Potent phosphate-solubilizing microbes for improving plant health: Recent advances and future perspectives. *Journal of Applied Microbiology*, *132*(2), 857–874.

Gupta, R. P., & Sharma, S. (2015). Phosphate solubilizing bacteria: Potential microorganisms for sustainable agriculture. *Journal of Microbiology and Biotechnology Research*, *5*(1), 1–6.

Hacquard, S., et al. (2021). Microbiota and host nutrition across plant and animal kingdoms. *Cell Host & Microbe*, *29*(4), 537–550. https://doi.org/10.1016/j.chom.2021.02.006

Han, X., Wang, L., & Yang, C. (2018). Bioaugmentation of phosphate-solubilizing bacteria enhances soil phosphorus availability and maize growth. *Ecological Engineering*, *120*, 69–75.

Hariprasad, P., & Niranjana, S. R. (2009). Isolation and characterization of phosphate-solubilizing *Fusarium* spp. from rhizosphere soils. *Journal of Agricultural Technology*, *5*(1), 79–89.

Hossain, M. A., et al. (2019). Soil contamination with cadmium: Consequences and remediation using organic amendments. *Environmental Science and Pollution Research*, *26*(1), 1–12. https://doi.org/10.1007/s11356-018-1234-5

Illmer, P., & Schinner, F. (1992). Solubilization of inorganic phosphates by microorganisms isolated from forest soils. *Soil Biology and Biochemistry*, *24*(4), 389–395. https://doi.org/10.1016/0038-0717(92)90199-8

Jain, R., Sharma, S., & Kumar, A. (2021). Phosphate-solubilizing bacteria: Mechanisms and their role in plant growth promotion. *International Journal of Microbiology*, *2021*, 1–10. https://doi.org/10.1155/2021/8881234

Khan, M. S., Zaidi, A., & Wani, P. A. (2009). Role of phosphate-solubilizing microorganisms in sustainable agriculture—a review. *Agronomy for Sustainable Development*, *29*(1), 43–54. https://doi.org/10.1051/agro:2007051

Khan, M. S., Zaidi, A., & Wani, P. A. (2014). Role of phosphate-solubilizing microorganisms in sustainable agriculture—a review. *Agronomy for Sustainable Development*, *27*(1), 29–43. https://doi.org/10.1051/agro:2007025

Khan, M. S., Zaidi, A., & Wani, P. A. (2017). Role of phosphate-solubilizing microorganisms in sustainable agriculture: A review. *Agronomy for Sustainable Development*, *27*, 29–43. https://doi.org/10.1051/agro:2007024

Khan, M. S., Zaidi, A., & Wani, P. A. (2019). Role of phosphate-solubilizing microorganisms in sustainable agriculture—a review. *Agronomy for Sustainable Development*, *39*(2), 1–15.

Khan, M. S., Zaidi, A., & Wani, P. A. (2020). Role of phosphate-solubilizing microorganisms in sustainable agriculture—a review. *Agronomy for Sustainable Development*, *40*, 32.

Khan, M. S., Zaidi, A., & Wani, P. A. (2022). Role of phosphate-solubilizing microorganisms in sustainable agriculture—a review. *Agronomy for Sustainable Development*, *42*(1), 1–15.

Khan, S., Singh, R., & Tiwari, R. (2021). Role of phosphate-solubilizing fungi in improving crop productivity. *Fungal Biology Reviews*, *35*(4), 229–238.

Kumar, A., et al. (2021). Remediation techniques for heavy metal-contaminated soils. *Environmental Science and Pollution Research*, *28*(1), 1–15. https://doi.org/10.1007/s11356-020-10430-3

Kumar, S., Sahu, R. K., Thakur, R. K., Yaduwanshi, B., & Mitra, N. G. (2021). Effect of microbial inoculants on plant attributes and nutrients uptake by soybean in Vertisols. International Journal of Plant & Soil Science, *33*(18), 102–109.

Kumar, A., Singh, R., & Sharma, P. (2023). Impact of microbial consortia on plant nutrient uptake. *Journal of Agricultural and Food Chemistry*, *71*(7), 2134–2147. https://doi.org/10.1021/acs.jafc.2c07890

Kumar, M., et al. (2015). Remediation of contaminated soils by biotechnology with nanomaterials: A review. *Environmental Research*, *138*, 1–10. https://doi.org/10.1016/j.envres.2015.02.003

Kumar, R., Verma, J. P., & Singh, A. (2020). Isolation and characterization of phosphate-solubilizing fungi from maize rhizosphere and their effects on growth promotion and phosphorus uptake. *Communications in Soil Science and Plant Analysis*, *51*(14), 1921–1934.

Kumar, S., & Gupta, S. (2019). *Paecilomyces lilacinus*: A potential bioagent for solubilizing insoluble potassium. *Journal of Plant Nutrition*, *42*(8), 1240–1251. https://doi.org/10.1080/01904167.2018.1549674F rontiers

Kumar, S., & Singh, R. (2018). Role of phosphate-solubilizing bacteria in sustainable agriculture: A review. *Journal of Plant Nutrition*, *41*(12), 1731–1743. https://doi.org/10.1080/01904167.2018.1461022Frontiers

Kumar, S., Singh, R., & Verma, P. (2018). Role of phosphate solubilizing microorganisms in sustainable agriculture—a review. *International Journal of Current Microbiology and Applied Sciences*, *7*(7), 2344–2357.

Lee, H., Kim, J., & Park, S. (2022). Advances in phytoremediation techniques for heavy metal-contaminated soils. *Environmental Science and Pollution Research*, *29*(5), 1234–1245. https://doi.org/10.1007/s11356-021-12345-6

Li, X., et al. (2022). Heavy metals in soils from intense industrial areas in South China: Distribution, sources, and ecological risks. *Frontiers in Environmental Science*, *10*, 820536. https://doi.org/10.3389/fenvs.2022.820536

Li, Y., Chen, Q., & Zhang, C. (2019). Influence of microbial inoculants on soil microbial community and soil phosphorus availability. *Soil Biology and Biochemistry*, *135*, 157–167.

Liu, B., et al. (2023). Advances in biochar-based materials for the remediation of heavy metal-contaminated soils: A review. *Environmental Science and Pollution Research*, *30*(1), 1–15. https://doi.org/10.1007/s11356-022-12345-6

Liu, Y., Chen, X., & Zhang, Y. (2021). Green algae in agriculture: Enhancing plant growth and stress tolerance. *Plant Growth Regulation*, *94*(3), 415–425. https://doi.org/10.1007/s10725-021-00673-9

Lu, Y., et al. (2020). Heavy metal pollution in the environment and their toxicological effects on humans. *Chemosphere*, *252*, 126594. https://doi.org/10.1016/j.chemosphere.2020.126594

Lynch, J. M., & Whipps, J. M. (1990). Substrate flow in the rhizosphere. *Plant and Soil*, 129(1), 1–10.

Manjula, K., & Podile, A. R. (2004). Phosphate solubilizing bacteria: Potential microorganisms for sustainable agriculture. *Indian Journal of Microbiology*, *44*(2), 99–102.

Maron, P. A., et al. (2019). Loss in soil microbial diversity constrains microbiome selection and plant performance. *Environmental Microbiology*, *21*(1), 319–329. https://doi.org/10.1111/1462-2920.14428

Marschner, P. (2012). *Marschner's mineral nutrition of higher plants* (3rd ed.). Academic Press.

Mishra, A., et al. (2020). Application of phosphate-solubilizing microorganisms in sustainable agriculture: A review. *International Journal of Environmental Science and Technology*, *17*, 2259–2274.

Morton, J. B., et al. (2024). The role of arbuscular mycorrhizal fungi in the remediation of heavy metal-contaminated soils. *Mycorrhiza*, *34*(2), 123–135. https://doi.org/10.1007/s00572-023-01234-5

Mulligan, C. N., Yong, R. N., & Gibbs, B. F. (2001). Remediation technologies for metal-contaminated soils and groundwater: An evaluation. *Engineering Geology*, *60*(1–4), 193–207. https://doi.org/10.1016/S0013-7952(00)00101-0

Muyzer, G., & Stams, A. J. M. (2008). The ecology and biotechnology of sulphate-reducing bacteria. *Nature Reviews Microbiology*, *6*(6), 441–454. https://doi.org/10.1038/nrmicro1892

Naseem, H., Bano, A., & Ullah, F. (2019). Role of plant growth-promoting rhizobacteria and their exopolysaccharides in heavy metal detoxification. *Ecotoxicology and Environmental Safety*, *174*, 435–448.

Nazari, M., et al. (2019). Effectiveness of native and citric acid-enriched biochar of chickpea straw in Cd and Pb sorption in an acidic soil. *Environmental Technology & Innovation*, *14*, 100331. https://doi.org/10.1016/j.eti.2019.100331ResearchGate+1ScienceDirect+1

Nies, D. H. (1999). Microbial heavy-metal resistance. *Applied Microbiology and Biotechnology*, *51*(6), 730–750.

Pandey, A., Trivedi, P., & Kumar, B. (2018). Phosphorus-solubilizing microorganisms: Principles and application of PSM in agriculture. *Microbiological Research*, *206*, 19–31.

Pandey, A., Trivedi, P., Kumar, B., & Palni, L. M. S. (2021). Role of microbial inoculants in maintaining soil health and plant growth under abiotic stress conditions. *Frontiers in Microbiology*, *12*, 663171. https://doi.org/10.3389/fmicb.2021.663171

Pandey, P., Yadav, S. K., & Singh, D. (2023). Role of phosphate solubilizing microbes in soil fertility and crop productivity. *Environmental Sustainability*, *6*, 101–112.

Parameswaran, P., & Ramesh, A. (2013). Phosphate solubilization by Penicillium radicum and its potential for plant growth promotion. *Applied Biochemistry and Biotechnology*, 170(8), 1981–1993. https://doi.org/10.1007/s12010-013-0332-6

Parmar, P., & Sindhu, S. S. (2013). Potassium solubilization by rhizosphere bacteria: Influence of nutritional and environmental conditions. *Journal of Microbiology Research*, *3*(1), 25–31.

Patel, D., et al. (2019). Role of biochar in remediation of heavy metal-contaminated soils: A review. *Environmental Science and Pollution Research*, *26*(3), 2323–2337. https://doi.org/10.1007/s11356-018-3769-9

Patel, M., et al. (2023). Nanotechnology in the remediation of heavy metal-contaminated soils: Recent advances and future prospects. *Environmental Nanotechnology, Monitoring & Management*, *20*, 100567. https://doi.org/10.1016/j.enmm.2023.100567

Patel, R., Sharma, S., & Thakur, J. K. (2021). Role of *Streptomyces* spp. in phosphorus solubilization and their effects on plant growth. *Plant and Soil*, *466*, 129–144.

Paul, E. A. (2014). *Soil microbiology, ecology and biochemistry* (4th ed.). https://doi.org/10.1016/C2011-0-05497-2.

Pérez, E., et al. (2016). Microbial solubilization of phosphorus and its contribution to plant nutrition in low-input farming systems. *Frontiers in Microbiology*, *7*, 1396.

Pimentel, D., Burgess, M., & Pimentel, M. (2020). Phosphate solubilizing microorganisms and their role in sustainable agriculture. *Agriculture, Ecosystems & Environment*, *295*, 106882.

Rabus, R., Venceslau, S. S., Wohlbrand, L., Voordouw, G., Wall, J. D., & Pereira, I. A. (2015). A post-genomic view of the ecophysiology, catabolism and biotechnological relevance of sulphate-reducing prokaryotes. *Advances in Microbial Physiology*, *66*, 55–321. https://doi.org/10.1016/bs.ampbs.2015.05.001

Rafique, M., et al. (2021). Phytoremediation of heavy metal-contaminated soils using sunflower: A review. *Environmental Science and Pollution Research*, *28*(1), 1–14. https://doi.org/10.1007/s11356-020-10430-3

Rai, P. K., Lee, S. S., Zhang, M., Tsang, Y. F., & Kim, K. H. (2019). Heavy metals in food crops: Health risks, fate, mechanisms, and management. *Environmental International*, *125*, 365–385. https://doi.org/10.1016/j.envint.2019.01.067

Ramos, T. B., Gonçalves, M. C., Martins, J. C., Prazeres, A., & Pires, F. P. (2005). Assessing heavy metal contamination in Sado Estuary sediment: An index analysis approach. *Ecological Indicators*, *5*(2), 151–169. https://doi.org/10.1016/j.ecolind.2005.01.001

Rao, M. A., & Stotzky, G. (1996). *Soil enzymes*. In G. Stotzky & J. M. Bollag (Eds.), *Soil biochemistry* (Vol. 9, pp. 279–340). https://doi.org/10.1081/E-ESS3-120052906.

Rascovan, N., Carbonetto, B., Perrig, D., Díaz, M., Canciani, W., Abalo, M., . . . Vazquez, M. P. (2019). Integrated analysis of root microbiomes of soybean and wheat from agricultural fields. *Scientific Reports*, *9*, 1–12. https://doi.org/10.1038/s41598-019-47789-7

Reddy, K. R., & Yarrakula, K. (2017). Electrokinetic remediation of heavy metal-contaminated soils: A review. *Environmental Pollution*, *224*, 1–17. https://doi.org/10.1016/j.envpol.2017.01.001Nept Journal+1ResearchGate+1

Reddy, M. S., Kumar, S., Babita, K., & Reddy, M. S. (2010). Selection of phosphate-solubilizing bacteria from rhizosphere of maize and their effect on growth and yield of maize. *Indian Journal of Microbiology*, *50*(1), 30–37.

Reyes, I., Bernier, L., Simard, R. R., & Antoun, H. (1999). Effect of nitrogen source on solubilization of different inorganic phosphates by an isolate of *Penicillium rugulosum* and two UV-induced mutants. *FEMS Microbiology Ecology*, *28*(3), 281–290. https://doi.org/10.1111/j.1574-6941.1999.tb00583.x

Richardson, A. E. (2001). Prospects for using soil microorganisms to improve the acquisition of phosphorus by plants. *Australian Journal of Plant Physiology*, *28*(9), 897–906. https://doi.org/10.1071/PP01093.

Richardson, A. E., Barea, J. M., McNeill, A. M., & Prigent-Combaret, C. (2009). Acquisition of phosphorus and nitrogen in the rhizosphere and plant growth promotion by microorganisms. *Plant and Soil*, *321*(1–2), 305–339. https://doi.org/10.1007/s11104-009-9895-2

Rillig, M. C. (2004). Arbuscular mycorrhizae and terrestrial ecosystem processes. *Ecology Letters*, *7*(8), 740–754. https://doi.org/10.1111/j.1461-0248.2004.00620.x

Rodríguez, H., & Fraga, R. (1999). Phosphate solubilizing bacteria and their role in plant growth promotion. *Biotechnology Advances*, *17*(4–5), 319–339. https://doi.org/10.1016/S0734-9750(99)00014-2

Ryan, P. R., Dessaux, Y., Thomashow, L. S., & Weller, D. M. (2021). Rhizosphere microbiome engineering for sustainable crop production. *Trends in Biotechnology*, *39*(1), 30–43.

Saha, M., Sarkar, S., Sarkar, B., Sharma, B. K., Bhattacharjee, S., & Tribedi, P. (2016). Microbial siderophores and their potential applications: A review. *Environmental Science and Pollution Research*, *23*(5), 3984–3999.

Saha, M., Sarkar, S., Sarkar, B., Sharma, B. K., Bhattacharjee, S., & Tribedi, P. (2021). Microbial siderophores and their potential applications: A review. *Environmental Science and Pollution Research*, *28*(1), 1–15.

Sangeeta, N., Meena, R. K., & Yadav, R. (2021). Phosphate solubilizing microbes: A tool for sustainable agriculture. *Current Trends in Microbiology*, *17*(1), 1–12.

Saxena, G., et al. (2018). An evaluation of technologies for the heavy metal remediation of dredged sediments. *Environmental Technology*, *39*(12), 1545–1556. https://doi.org/10.1080/09593330.2017.1330366

Sayer, J. A., & Gadd, G. M. (1997). Solubilization and transformation of insoluble metal compounds to insoluble metal oxalates by *Aspergillus niger*. *Mycological Research*, *101*(6), 653–661. https://doi.org/10.1017/S0953756296003387

Schippers, A., Jozsa, P. G., & Sand, W. (1987). Sulfur chemistry in bacterial leaching of pyrite. *Applied and Environmental Microbiology*, *53*(11), 2608–2614. https://doi.org/10.1128/aem.53.11.2608-2614.1987

Shahid, N. U., & Sheikh, N. J. (2021). Impact of big data on innovation, competitive advantage, productivity, and decision making: Literature review. *Open Journal of Business and Management*, *9*(2), 586–617. https://doi.org/10.4236/ojbm.2021.92032ResearchGate

Sharma, R. K., et al. (2022). Bioremediation of heavy metals from contaminated soils using microbial consortia: A review. *Environmental Science and Pollution Research*, *29*(1), 1–14. https://doi.org/10.1007/s11356-021-12345-6Sharma, S., Kumari, N., & Prasad, B. (2021). Phosphate-solubilising microorganisms as potential biofertilizer: A review. *Agricultural Reviews*. https://doi.org/10.18805/ag.r-2110Mendeley

Sharma, S. B., Sayyed, R. Z., Trivedi, M. H., & Gobi, T. A. (2013). Phosphate solubilizing microbes: Sustainable approach for managing phosphorus deficiency in agricultural soils. *SpringerPlus*, *2*, 587. https://doi.org/10.1186/2193-1801-2-587

Siddiqui, M. H., Al-Whaibi, M. H., & Basalah, M. O. (2021). Role of nanoparticles in heavy metal remediation from contaminated soils. *Environmental Nanotechnology, Monitoring & Management*, *15*, 100432. https://doi.org/10.1016/j.enmm.2020.100432

Silva, L. I. D., Pereira, M. C., Carvalho, A. M. X. D., Buttrós, V. H., Pasqual, M. and Dória, J. (2023). Phosphorus-solubilizing microorganisms: a key to sustainable agriculture. *Agriculture*, *13*(2), 462. https://doi.org/10.3390/agriculture13020462.

Singh, B. K., Saha, R., & Yadav, A. N. (2021). Actinomycete resilience to extreme pH conditions and its implications for soil health and crop growth. *Journal of Applied Microbiology*, *131*(5), 1836–1851. https://doi.org/10.1111/jam.15000

Sinha, R., et al. (2020). Bioremediation of heavy metals from contaminated soils using earthworms: A review. *Environmental Science and Pollution Research*, *27*(3), 3601–3614. https://doi.org/10.1007/s11356-019-07025-0

Singh, R., et al. (2022). Microbial remediation of heavy metal-contaminated soils: A review. *Environmental Technology & Innovation*, *25*, 102079. https://doi.org/10.1016/j.eti.2021.102079

Singh, R., Kumar, V., & Patel, V. (2020). Phosphate solubilizing bacteria: Mechanism and their role in sustainable agriculture. *Journal of Environmental Biology*, 41(1), 45–52.

Singh, R., Yadav, S., & Kumar, A. (2023). Tailoring microbial formulations for specific crops: Advances and prospects. *Journal of Agricultural and Food Chemistry*, *71*(7), 2134–2147. https://doi.org/10.1021/acs.jafc.2c07890

Smith, S. E., & Read, D. J. (2008). *Mycorrhizal symbiosis* (3rd ed.). Academic Press.

Srinivasan, R., & Rama Choudhury, S. P. (2018). Phosphate solubilizing microorganisms: A sustainable approach for managing phosphorus deficiency in agricultural soils. *Agricultural Reviews*, *39*(3), 201–208. https://doi.org/10.18805/ag.R-1634Mendeley

Stengel, E., et al. (2017). Heavy metals in soil-like material from landfills—resource or contaminants? *Environmental Pollution*, *220*, 1–10. https://doi.org/10.1016/j.envpol.2016.09.029

Swift, M. J., Heal, O. W., & Anderson, J. M. (1979). *Decomposition in terrestrial ecosystems*. University of California Press. https://doi.org/10.1525/9780520407114.

Sylvia, D. M., Fuhrmann, J. J., Hartel, P. G., & Zuberer, D. A. (2005). *Principles and applications of soil microbiology* (2nd ed.). Pearson Prentice Hall.

Van der Heijden, M. G. A., Bardgett, R. D., & van Straalen, N. M. (2008). The unseen majority: Soil microbes as drivers of plant diversity and productivity in terrestrial ecosystems. *Ecology Letters*, *11*(3), 296–310. https://doi.org/10.1111/j.1461-0248.2007.01139.x

Van der Heijden, M. G. A., et al. (2018). Mycorrhizal ecology and evolution: The past, the present, and the future. *New Phytologist*, *220*(4), 963–965. https://doi.org/10.1111/nph.15076

Vassilev, N., Vassileva, M., & Nikolaeva, I. (2006). Simultaneous P-solubilizing and biocontrol activity of microorganisms: Potentials and future trends. *Applied Microbiology and Biotechnology*, *71*(2), 137–144. https://doi.org/10.1007/s00253-006-0380-z

Vessey, J. K. (2003). Plant growth promoting rhizobacteria as biofertilizers. *Plant and Soil*, *255*(2), 571–586. https://doi.org/10.1023/A:1026037216893

Wakelin, S. A., Warren, R. A., Harvey, P. R., & Ryder, M. H. (2009). Phosphate solubilization by *Penicillium* spp. closely associated with wheat roots. *Biology and Fertility of Soils*, *45*, 425–434. https://doi.org/10.1007/s00374-008-0340-2

Wang, L., et al. (2020). Heavy metal pollution in the environment and their toxicological effects on humans. *Chemosphere*, *252*, 126594. https://doi.org/10.1016/j.chemosphere.2020.126594

Wang, Y., et al. (2019). Phytoremediation of heavy metal-contaminated soils by hyperaccumulator plants: A review. *Environmental Science and Pollution Research*, *26*(1), 1–14. https://doi.org/10.1007/s11356-018-3769-9

Whitelaw, M. A. (2000). Growth promotion of plants inoculated with phosphate-solubilizing fungi. *Advances in Agronomy*, *69*, 99–151. https://doi.org/10.1016/S0065-2113(08)60955-7

Widdel, F., & Bak, F. (2003). The dissimilatory sulfate- and sulfur-reducing bacteria. In M. Dworkin, S. Falkow, E. Rosenberg, K.-H. Schleifer, & E. Stackebrandt (Eds.), *The prokaryotes* (pp. 659–768). Heidelberg, Springer Science Online.

Xu, J., Zhang, Y., & Cao, Y. (2019). Phosphate solubilizing microbes improve phosphorus availability in soil and crop yield: A review. *Agriculture, Ecosystems & Environment*, *277*, 29–40.

Yadav, J., Verma, J. P., & Tiwari, K. N. (2019). Plant growth promoting activities of microbial inoculants: A review. *Journal of Plant Nutrition*, *42*(9), 1233–1250.

Yadav, K. K., et al. (2022). Phytoremediation of toxic metals present in soil and water environment: A critical review. *Environmental Technology & Innovation*, *25*, 102079. https://doi.org/10.1016/j.eti.2021.102079

Yadav, K. K., et al. (2023). Phytoremediation of heavy metal-contaminated soils: Recent developments and future prospects. *Environmental Technology & Innovation*, *30*, 102079. https://doi.org/10.1016/j.eti.2022.102079

Zahra, M. (1999). The role of microorganisms in the remediation of heavy metal-contaminated soils. *Environmental Pollution*, *106*(3), 401–404. https://doi.org/10.1016/S0269-7491(99)00136-7

Zhang, L., Zhang, L., Song, Y., & Wang, Y. (2022). Recent advances in phosphate-solubilizing microbes and their applications in sustainable agriculture. *Soil Biology and Biochemistry*, 168, 108620.

Zhao, X., Cui, X., & Li, Y. (2017). *Rhizopus oryzae*: Versatile fungus with potential for industrial applications. *Fungal Biology Reviews*, *31*(1), 1–10. https://doi.org/10.1016/j.fbr.2016.11.001

Zhong, H., et al. (2021). Biochar application for the remediation of heavy metal-contaminated soils: A review. *Chemosphere*, *277*, 130292. https://doi.org/10.1016/j.chemosphere.2021.130292

Zhou, J., Wang, D., & Wang, Y. (2020). Understanding the impact of phosphate-solubilizing bacteria on phosphorus uptake in crops. *Journal of Soil Science and Plant Nutrition*, *20*(4), 1357–1370.

Zhou, Y., Zhang, Y., & Li, X. (2022). Dynamics of microbial populations in soil: Implications for agricultural practices. *Microbial Ecology*, *84*(2), 479–491. https://doi.org/10.1007/s00248-021-01775-6

Zhu, Y., et al. (2016). Remediation of contaminated soils by biotechnology with nanomaterials: A review. *Environmental Research*, *151*, 1–10. https://doi.org/10.1016/j.envres.2016.07.003

Zúñiga-Silgado, D., Sánchez-Reyes, A., Ortiz-Hernández, M. L., Otero, M., Balcázar-López, E., Valencia-Díaz, S., Serrano, M., Coleman, J. J., Sarmiento-López, L., & De-Bashan, L. E. (2022). Soil type influences novel "Milpa" isolates of *Trichoderma virens* and *Aspergillus tubingensis* that promote solubilization, mineralization, and phytoabsorption of phosphorus. *Journal of Fungi*, *8*(11), 1178. https://doi.org/10.3390/jof8111178ResearchGate+1MDPI+1

17 PGPR in Shaping the Biotic Interactions in Plant Rhizosphere

Diksha Sharma and Niharika Pandey

17.1 INTRODUCTION

The rhizosphere refers to the narrow region of soil that surrounds and is directly influenced by plant roots. It encompasses both the plant roots and the surrounding soil. Living roots release a variety of metabolites into this zone, which not only serve as chemical signals attracting motile bacteria to the root surface but also act as essential nutrient sources that support microbial growth and survival. This dynamic environment hosts a range of microorganisms, including bacteria that can efficiently colonize plant roots or rhizosphere soil. These beneficial bacteria are known as plant growth-promoting rhizobacteria (PGPR) (Ali *et al.*, 2021; Kumari *et al.*, 2024).

The rhizosphere is generally divided into three distinct components: The root itself, the rhizoplane, and the rhizosphere soil. The rhizosphere, in a broader sense, refers to the soil region influenced by root exudates that modify microbial activity (Elnahal *et al.*, 2022). The rhizoplane is the root surface, including soil particles that are tightly attached to it. Additionally, the root itself is considered part of the rhizosphere system, as certain endophytic microorganisms are capable of colonizing internal root tissues (Nihorimbere *et al.*, 2011).

The rhizosphere serves as a dynamic interface where plants, bacteria, soil, and soil microfauna engage in complex and continuous interactions. These relationships are crucial for the exchange of signaling molecules and biochemical compounds between rhizobacteria and plants, significantly influencing plant growth and productivity. Rhizobacteria—also known as rhizosphere-competent bacteria—actively colonize plant roots at various stages of development, highlighting their vital role in the plant life cycle.

Within this microenvironment, both beneficial and potentially harmful microorganisms coexist, and the composition of rhizobacterial communities can vary with changing soil properties (Pérez-de-Luque *et al.*, 2017). These microbial interactions not only enhance soil health but also improve its nutrient status, which is critical for optimal plant growth (Khan *et al.*, 2021). The synergy among plant roots, rhizobial bacteria, and the surrounding rhizosphere supports root proliferation and facilitates nutrient exchange between roots and shoots. Root exudates—complex mixture of organic compounds secreted by roots—play a pivotal role in attracting and sustaining microbial communities. A strong and stable interaction between rhizobacteria and these exudates is essential for successful root colonization and development (Semchenko *et al.*, 2014). The rhizosphere hosts a vast array of organisms, including bacteria, fungi, nematodes, protozoa, algae, pathogens, saprophytes, endophytes, and epiphytes (Ma *et al.*, 2022).

Microbial density in this region is remarkably high, with approximately 1.2 million bacteria per gram of dry soil—far exceeding the populations of fungi (12,105 fungi/g), algae (51,105 algae/g), and actinomycetes (46,106 actinomycetes/g) (Upadhyay *et al.*, 2022). This microbial richness underscores the rhizosphere's critical role in supporting a healthy and productive plant–soil ecosystem. The rhizosphere is significantly richer in nutrients compared to bulk soil, largely due to the release

DOI: 10.1201/9781003608080-17

of organic compounds through root exudation, secretion, and deposition. This nutrient abundance creates what is known as the "rhizosphere effect", which enhances microbial activity and supports a thriving microbial community around plant roots. The extent of this effect is often measured using the R:S ratio, where R represents the microbial population in the rhizosphere and S corresponds to that in the bulk soil. A higher R:S ratio indicates greater microbial activity in the root zone (Prashar *et al.*, 2014; Elnahal *et al.*, 2022).

PGPR offer a sustainable alternative to conventional agricultural practices. Predominantly found in the rhizosphere and surrounding soil, these beneficial bacteria form symbiotic relationships with plants. PGPR enhance nutrient availability, boost plant stress tolerance, and contribute to higher crop yields, fostering a more balanced and resilient soil ecosystem. They support plant growth and health through various mechanisms such as nitrogen fixation, nutrient solubilization, and pathogen suppression. Several genera of PGPR—such as *Bacillus*, *Enterobacter*, *Klebsiella*, *Pseudomonas*, *Variovorax*, *Serratia*, *Azotobacter*, and *Azospirillum*—play a pivotal role in modern agriculture (Rai *et al.*, 2020). This chapter aims to provide deeper insights into the mechanisms of PGPR, their diverse functions, and their complex interactions with the surrounding soil environment.

17.2 PGPR AND ITS CLASSIFICATION

PGPR are a diverse group of soil-dwelling bacteria that, when associated with host plants, enhance plant growth and development. As an integral component of the rhizosphere biota, PGPR play a crucial role in sustainable agriculture by improving nutrient use efficiency and ensuring better nutrient availability to plants (Ayaz *et al.*, 2023). Their beneficial effects extend to enhancing key physiological processes such as photosynthesis, water and nutrient uptake, and the coordination of source-sink dynamics, all of which support overall plant health and productivity. PGPR can be classified based on various characteristics. One common classification is based on their interaction with plants: They can be free-living (e.g., *Bacillus* and *Azospirillum*) or symbiotic (e.g., *Rhizobium* and *Bradyrhizobium* in association with legumes) (Figure 17.1). Another classification, proposed by

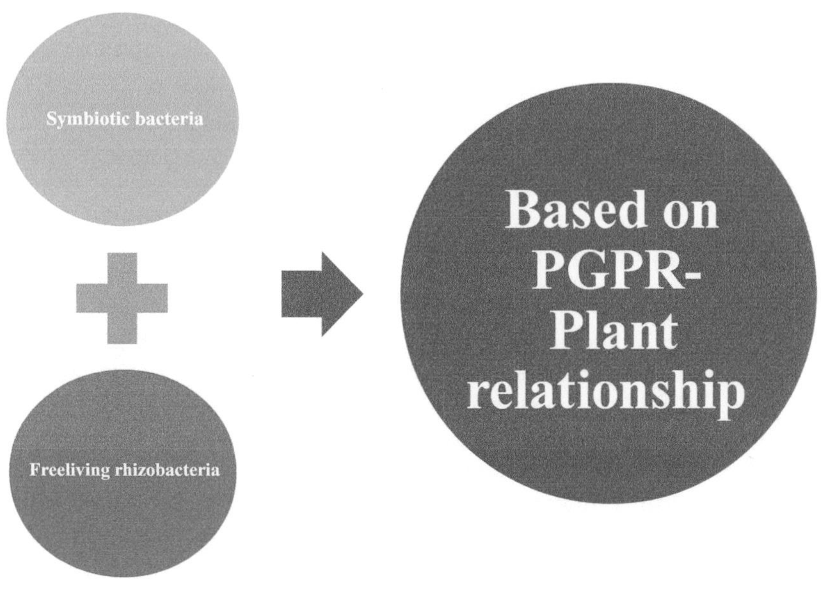

FIGURE 17.1 Classification based on PGPR–plant interaction.

TABLE 17.1
Classification of PGPRs on the Basis of Their Location.

Extracellular PGPR	Intracellular PGPR
The term e-PGPRs refers to the microbial species that live outside of plant cells and do not produce nodules but instead promote plant development. Bacteria that do not form nodules but instead aid in the growth of plants and reside outside of plant cells. Increase the resilience of plants against diseases.	i-PGPRs are bacteria that live inside specialized structures called nodules or in the intermediate gaps of the root cell cortex.
e-PGPR are exemplified by *Erwinia*, *Flavobacterium*, *Bacillus*, *Caulobacter*, *Chromobacterium*, *Azospirillum*, *Azotobacter*, *Agrobacterium*, *Arthrobacter*, *Serratia*, *Micrococcous*, *Pseudomonas*, *Azorhizobium*, and so on	Endophytic bacteria that characterize the i-PGPR include *Rhizobium*, *Bradyrhizobium*, *Allorhizobium*, *Azorhizobium*, *Bradyrhizobium*, *Frankia*, etc.

Gray and Smith (2005), is based on the plant tissue compartment in which they reside. This divides PGPR into two categories (Table 17.1):

- Extracellular PGPR (e-PGPR): Found on the rhizoplane, within intercellular spaces of the root cortex, or in the rhizosphere soil.
- Intracellular PGPR (i-PGPR): Located within root cells, often in specialized nodules.

Additionally, based on their functional roles in promoting plant growth, Somers *et al.* (2004) categorized PGPR into four main types (Figure 17.2):

1. Biofertilizers: Enhance nutrient availability and uptake
2. Phytostimulators: Promote plant growth through hormonal stimulation
3. Biopesticides: Protect against plant pathogens
4. Rhizo-remediators: Help in detoxifying contaminated soils

To effectively support plant growth and protection, PGPR must exhibit three key traits. First, the ability to colonize plant roots, followed by the capacity to survive and proliferate in root-associated microhabitats while competing with native microbiota, and finally, the potential to stimulate plant growth through direct or indirect mechanisms (Bishnoi, 2015).

17.3 MECHANISMS OF PGPR ACTION

PGPR play a vital role in enhancing plant growth and development through both direct and indirect mechanisms. Directly, they influence plant physiology by processes such as biological nitrogen fixation (BNF) (Zahran, 2009); the synthesis and modulation of phytohormones, including auxins, cytokinins, gibberellins (GA) (Vacheron *et al.*, 2013), and ethylene (Etesami and Maheshwari, 2018); and solubilization of essential minerals like phosphorus and iron (Boubekri *et al.*, 2022). Additionally, PGPR produce siderophores and enzymes, and can trigger induced systemic resistance in plants (Bishnoi, 2015). Indirectly, PGPR promote plant health by suppressing the negative effects of pathogenic organisms, often through the production of antagonistic compounds that inhibit pathogen growth. These combined actions make PGPR key contributors to plant resilience and sustainable agricultural practices.

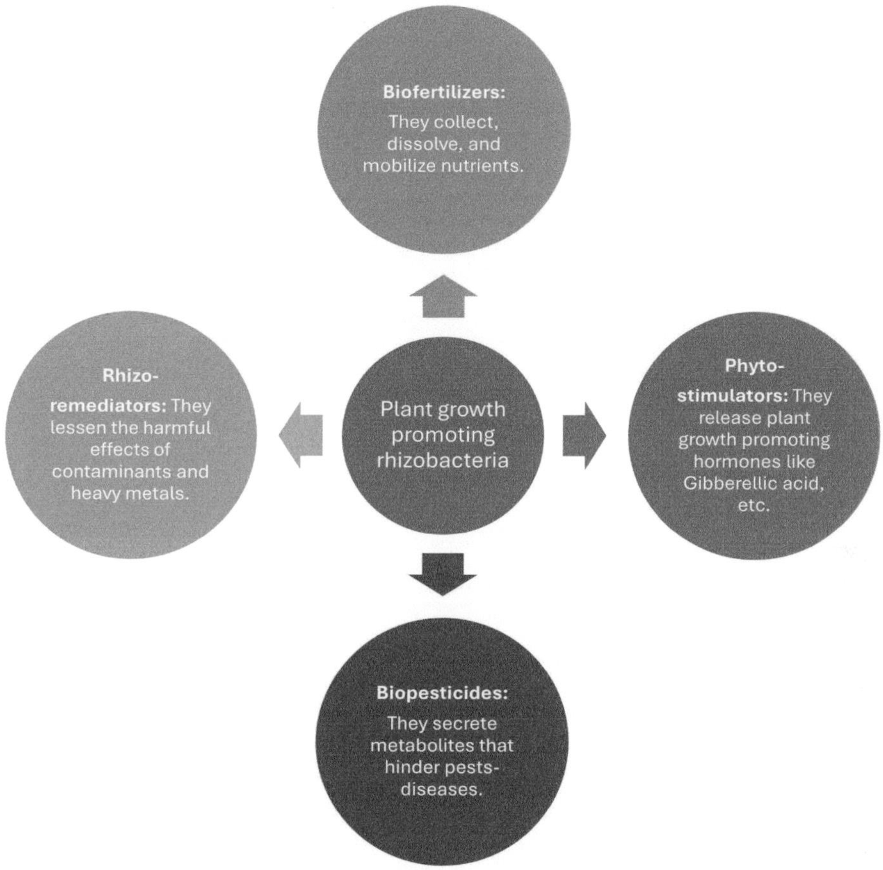

FIGURE 17.2 Categorization of PGPRs based on their functional roles.

17.3.1 Direct Mechanism

PGPR often promotes plant development directly, either by influencing the levels of plant hormones or by helping plants acquire vital resources including phosphorus, nitrogen, and critical minerals. The following describes the precise procedures that go into these mechanisms.

17.3.1.1 Nitrogen Fixation

Nitrogen deficiency is one of the major limiting factors in agriculture, often driven by various environmental conditions. BNF accounts for approximately 65% of the nitrogen used in agriculture today and is vital for the long-term sustainability of crop production systems. BNF is the process by which nitrogen-fixing bacteria convert atmospheric nitrogen (N_2) into ammonia (NH_3) through the action of a complex enzyme system called nitrogenase, making nitrogen available in a form that plants can readily absorb. PGPR strains play a crucial role in this process by facilitating nitrogen fixation in the rhizosphere. Nitrogen-fixing organisms are broadly classified into two groups: Non-symbiotic nitrogen-fixers, which include free-living, associative, and endophytic bacteria such as *Cyanobacteria*, and symbiotic nitrogen-fixers, which are primarily members of the *Rhizobiaceae* family that form mutualistic associations with leguminous plants (e.g., *Rhizobium* with legumes) and some nonleguminous trees (e.g., *Frankia*).

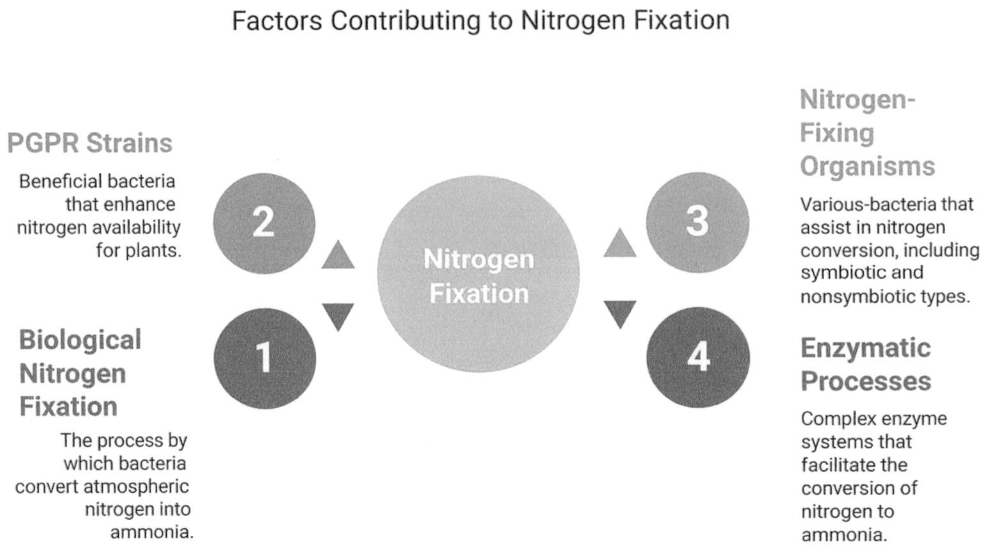

FIGURE 17.3 Overview of the nitrogen fixation process.

Each bacterial genus has a structurally distinct nitrogen-fixing system, but molybdenum nitrogenase is the dominant enzyme responsible for most BNF (Ghosh *et al.*, 2024). An overview of the nitrogen fixation process is illustrated in Figure 17. 3.

In both free-living and symbiotic nitrogen-fixing systems, nitrogen fixation is governed by a set of genes known as *nif* genes. These genes encode the nitrogenase enzyme complex and include structural genes, genes responsible for the activation of Fe proteins, the synthesis of iron–molybdenum cofactors (FeMo-co), electron transport, and regulatory functions essential for the enzyme's assembly and activity. In diazotrophs, *nif* genes are typically organized into a cluster of 20–24 kilobases, containing around seven operons that produce approximately 20 distinct proteins (Glick, 2012). The molybdenum nitrogenase enzyme complex is primarily encoded by the *nifDK* and *nifH* genes. The NifDK protein is a heterotetrameric structure ($\alpha_2\beta_2$) composed of two $\alpha\beta$ dimers arranged symmetrically. Each α-subunit (NifD) contains FeMo-co, which serves as the active site for nitrogen fixation (Ladha *et al.*, 2022). For symbiotic nitrogen fixation, particularly in *Rhizobium* and other diazotrophs, low oxygen conditions are required to activate the *nif* genes. This activation is regulated by another set of genes known as *fix* genes, which are also present in both symbiotic and free-living nitrogen-fixing systems. These genes help maintain the optimal conditions necessary for nitrogenase function and ensure efficient nitrogen fixation in varying environmental contexts.

17.3.1.2 Phosphate Solubilization

While nitrogen fixation is a critical process for plant nutrition, it is also energy-intensive—requiring at least 16 moles of ATP for every molecule of nitrogen fixed. After nitrogen, phosphorus is the second most limiting nutrient for plant growth. It plays a vital role in several physiological and biochemical processes, including respiration, signal transduction, energy transfer, photosynthesis, and the biosynthesis of macromolecules (Shah *et al.*, 2021).

Although phosphorus is abundant in soils, its bioavailability to plants is extremely low. Approximately 99% of soil phosphorus exists in insoluble forms due to precipitation or immobilization within the soil matrix, leaving only about 0.1% available for plant uptake (Anand *et al.*, 2016). Plants can absorb phosphorus only in the form of monobasic ($H_2PO_4^-$) and dibasic (HPO_4^{2-}) phosphate ions (de Andrade *et al.*, 2023).

A variety of PSB play a crucial role in converting insoluble phosphorus into plant-available forms. Key genera include *Azotobacter*, *Bacillus*, *Beijerinckia*, *Burkholderia*, *Enterobacter*, *Erwinia*,

Flavobacterium, *Microbacterium*, *Pseudomonas*, *Rhizobium*, and *Serratia* (Bhattacharya and Jha, 2012). These microbes solubilize inorganic phosphorus primarily through the secretion of low-molecular-weight organic acids, which chelate the cations bound to phosphate, releasing it in a soluble form.

Additionally, the mineralization of organic phosphorus is facilitated by phosphatase enzymes, which hydrolyze phosphoric esters. Notably, certain bacterial strains possess the dual ability to both solubilize inorganic phosphate and mineralize organic phosphate.

Beyond supplying phosphorus, PSB contribute to enhanced plant growth by improving the efficiency of BNF and increasing the availability of other micronutrients. They also produce growth-promoting compounds that further support plant development (Hashem *et al.*, 2019).

17.3.1.3 Potassium Solubilization

Potassium (K) is the third most essential nutrient required for plant growth. However, its availability in soil is often limited, as more than 90% of soil potassium exists in insoluble forms such as mineral silicates and rocks. This low availability of soluble potassium significantly restricts crop productivity. Potassium deficiency in plants leads to stunted growth, reduced seed production, and lower yields.

Several microorganisms, particularly certain bacterial and fungal species, have the ability to interact with plants and solubilize potassium from insoluble sources (Setiawati and Mutmainnah, 2016). PGPR have been shown to secrete organic acids that can effectively dissolve potassium-bearing minerals. Notable potassium-solubilizing bacteria include *Acidothiobacillus* sp., *Bacillus edaphicus*, *Ferrooxidans* sp., *Bacillus mucilaginosus*, *Pseudomonas* sp., and *Burkholderia* (Prajapati *et al.*, 2013).

These microorganisms release organic acids such as oxalic, citric, acetic, ferulic, and coumaric acids, which acidify the rhizosphere. This increased acidity accelerates mineral dissolution, releasing protons that enhance the solubilization of mineral-bound potassium. As a result, PGPR and other plant growth-promoting microorganisms (PGPM) improve potassium availability and support plant development.

Research indicates that potassium-solubilizing PGPM can serve as effective biofertilizers, reducing the dependence on chemical fertilizers while enhancing nutrient accessibility and overall plant health (Khan *et al.*, 2019).

17.3.1.4 Siderophore Production

Iron is an essential micronutrient for almost all living organisms, with the exception of a few, such as certain *Lactobacillus* species. Under aerobic conditions, iron predominantly exists as Fe^{3+}, which forms insoluble oxyhydroxides and hydroxides, rendering it largely unavailable to both plants and microorganisms (Jansson *et al.*, 2023).

To overcome this limitation, many bacteria—including plant growth-promoting bacteria (PGPB)—produce siderophores, which are low-molecular-weight iron-chelating compounds (typically 400–1500 Da) with high affinity for Fe^{3+} (Shanmugaiah *et al.*, 2015). These molecules bind to iron and enhance its bioavailability, playing a critical role in both microbial survival and plant nutrition. Numerous microorganisms from terrestrial and marine ecosystems have been studied for their siderophore-producing capabilities (Goswami *et al.*, 2016).

Rhizobacteria vary in their ability to utilize siderophores. Some can only use *homologous siderophores* (produced by the same genus), while others are capable of utilizing *heterologous siderophores* produced by different genera (Hashem *et al.*, 2019). When iron availability is limited, the Fe^{3+}–siderophore complex on the bacterial membrane is reduced to Fe^{2+} and transported into the cell via a gating mechanism that connects the inner and outer membranes in both Gram-positive and Gram-negative bacteria. The siderophore may be recycled or degraded following this reduction (Jansson *et al.*, 2023).

Siderophores act as solubilizing agents, mobilizing iron from organic and mineral sources under iron-deficient conditions (Ahemad and Kibret, 2014). They exist in four primary chemical classes: Phenolates, hydroxamates, pyoverdines, and carboxylates. Among bacteria, *Enterobacter* and *Pseudomonas* (Gram-negative) are prominent siderophore producers, whereas only about 2% of Gram-positive bacteria, such as *Rhodococcus* and *Bacillus*, possess this ability (Czarnes *et al.*, 2020).

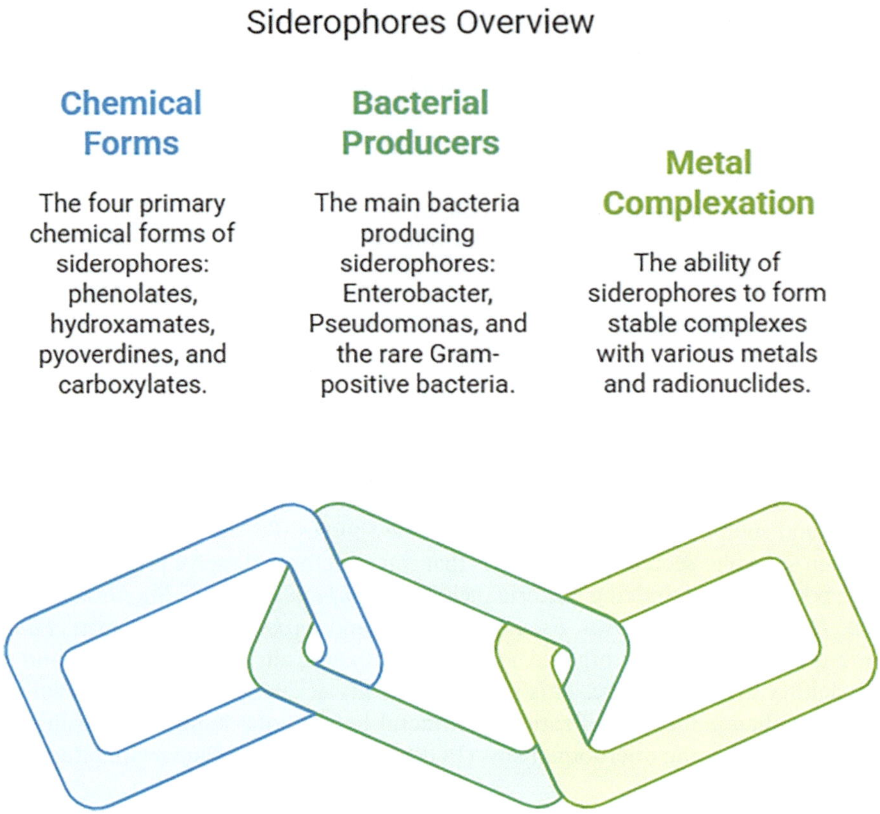

FIGURE 17.4 Overview of siderophores.

Beyond iron chelation, siderophores can also bind with various heavy metals and radionuclides such as Al, Cd, Cu, Ga, In, Pb, and Zn, increasing their solubility (Figure 17.4). This function makes siderophores essential in mitigating metal-induced stress in plants. Moreover, siderophore production serves as a biological control mechanism by depriving pathogenic microbes of iron, thereby suppressing their growth and enhancing plant defense (Khoshru *et al.*, 2020).

17.3.1.5 Production of Phytohormones

Phytohormones, or plant growth substances, are naturally occurring chemical compounds in plant tissues that regulate growth and development, rather than serving a nutritional role (Figure 17.5). These hormones are typically effective at very low concentrations and influence a wide range of physiological processes.

Among the most studied microbial phytohormones is indole-3-acetic acid (IAA), a key member of the auxin family. It is well established that various soil microbes, particularly rhizobacteria, can synthesize IAA. This microbially produced IAA can supplement or alter the plant's endogenous auxin pool, thereby influencing multiple aspects of plant development (Glick, 2016; Afzal *et al.*, 2019).

IAA acts as a crucial signaling molecule in rhizobacteria–plant interactions. It not only modulates plant gene expression but also influences bacterial gene regulation (Chieb and Gachomo, 2023). Interestingly, downregulation of IAA signaling in plants is often linked to the activation of defense responses. This is evident from increased plant susceptibility to bacterial infections following external IAA application or exposure to pathogen-derived IAA, suggesting that IAA plays a dual role in both growth and defense mechanisms.

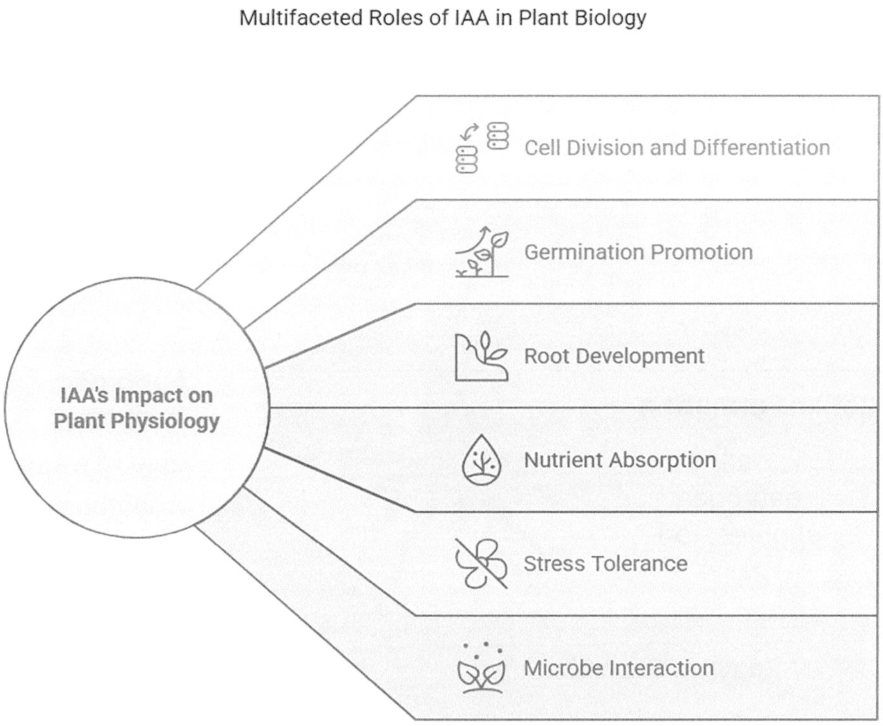

FIGURE 17.5 Multifaceted roles of IAA in plant biology.

IAA affects nearly every stage of plant growth and development:

- Cell division, elongation, and differentiation
- Seed and tuber germination
- Xylem and root development
- Vegetative growth regulation
- Formation of lateral and adventitious roots
- Responses to light, gravity, and fluorescence
- Enhancement of photosynthesis and pigment production
- Biosynthesis of secondary metabolites
- Stress tolerance enhancement (Upadhyay *et al.*, 2023).

Microbially produced IAA contributes to these functions by increasing root length and surface area, thereby improving nutrient and water uptake. Additionally, it promotes cell wall loosening and increased root exudation, enriching the rhizosphere with nutrients that support microbial growth (Glick, 2016).

Thus, IAA synthesized by rhizobacteria is recognized as a key effector in plant–microbe interactions, playing a central role in both growth promotion (phytostimulation) and plant susceptibility to pathogens (Chieb and Gachomo, 2023).

17.3.2 Indirect Mechanisms of PGPR

In addition to promoting plant growth directly, PGPR also support plant health through indirect mechanisms aimed at minimizing the detrimental effects of plant pathogens. These actions often

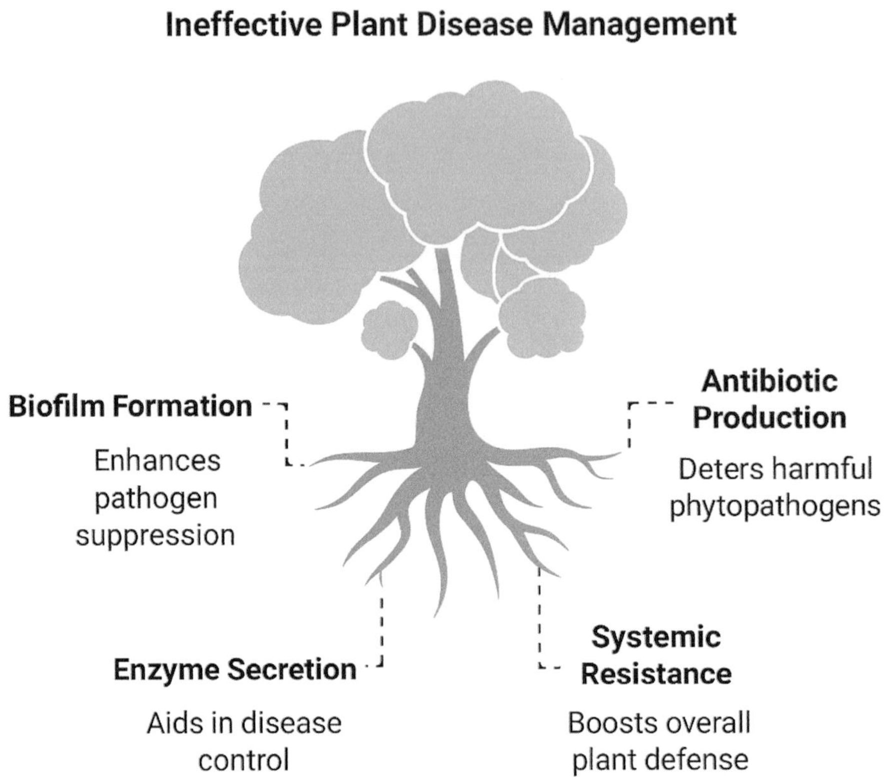

FIGURE 17.6 Indirect mode of action of PGPR.

involve the use of biocontrol strategies, where specific microbial functions suppress or inhibit the growth of disease-causing organisms. PGPR employ several tactics in this regard, including the production of hydrolytic enzymes (e.g., chitinases and cellulases), antimicrobial compounds, and volatile organic compounds (VOCs), as well as triggering plant immune responses. These mechanisms collectively enhance plant health and productivity by reducing the adverse effects of biotic stressors (Figure 17.6).

17.3.2.1 Biofilm Formation

One of the key defense strategies of rhizobacteria is the formation of biofilms in the rhizosphere. These structured microbial communities enable high cell density and enhanced survival. Within biofilms, bacteria release a variety of substances—such as antibiotics and toxins—that suppress soil-borne phytopathogens, thereby protecting plant roots (Kwon and Kim, 2014).

17.3.2.2 Antibiotic Production

Certain PGPR, such as *Bacillus* strains, produce antibiotic compounds that directly inhibit phytopathogens. For example, *Bacillus megaterium* has been shown to colonize plant roots and reduce populations of *Rhizoctonia solani*, a soil-borne fungal pathogen (Shafi *et al.*, 2017).

17.3.2.3 Secretion of Catabolic Enzymes

Many PGPR are capable of secreting catabolic enzymes like chitinases, proteases, and β-1,3-glucanases. These enzymes degrade the cell walls of pathogenic fungi, helping control diseases transmitted through soil. Electron microscopy studies have revealed disrupted hyphal structures in *Fusarium* species, indicating strong antagonistic activity by these enzymes (Zhao *et al.*, 2014).

17.3.2.4 Acquired Systemic Resistance and Induced Systemic Resistance

Although plants have innate defense mechanisms against pathogens, PGPR can trigger enhanced resistance known as acquired systemic resistance (ASR) and induced systemic resistance (ISR). These responses are initiated either locally at the site of infection or systemically throughout the plant. Exposure to inducers—such as beneficial microbes or pathogens—activates defense pathways, resulting in increased resistance to future infections (Romeiro, 2000).

ASR is typically associated with elevated levels of salicylic acid (SA) and the accumulation of pathogenesis-related (PR) proteins, which play a key role in plant defense (Ghosh *et al.*, 2024). ISR, on the other hand, often involves signaling pathways mediated by jasmonic acid (JA) and ethylene, helping the plant fend off a broad spectrum of pathogens (Abdul *et al.*, 2020).

17.4 PGPR DIVERSITY IN THE RHIZOSPHERE

The rhizosphere—the thin layer of soil surrounding plant roots—is a hotspot of microbial activity, particularly rich in PGPR. These beneficial microbes not only enhance plant growth but also provide natural protection against pathogens. The diverse PGPR community includes genera such as *Agrobacterium*, *Arthrobacter*, *Azotobacter*, *Azospirillum*, *Bacillus*, *Burkholderia*, *Caulobacter*, *Chromobacterium*, *Erwinia*, *Flavobacterium*, *Micrococcus*, *Pseudomonas*, *Serratia*, and *Cellulomonas flavigena* (Hassan *et al.*, 2019b).

Among these, nitrogen-fixing endophytes from the *Rhizobiaceae* family—*Allorhizobium*, *Azorhizobium*, *Bradyrhizobium*, *Mesorhizobium*, and *Rhizobium*—are particularly well-known for forming root nodules in legumes and promoting plant growth through both direct nutrient supply and indirect protection mechanisms (Harman and Uphoff, 2019). Other PGPR like *Pantoea*, *Methylobacterium*, *Exiguobacterium*, *Paenibacillus*, and *Azoarcus* also contribute significantly to plant health by colonizing the root system and enhancing growth (Chauhan *et al.*, 2015).

Recent discoveries have identified novel proteobacterial strains such as *Alcaligenes* sp. (WRB 10) and *Providencia* sp. (WRB 4), which show strong root colonization and wheat growth promotion capabilities (Manjunath *et al.*, 2011). Actinomycetes also represent a vital group of rhizobacteria, known for promoting plant growth and suppressing pathogens. Strains from genera such as *Streptomyces*, *Streptosporangium*, *Thermobifida*, and *Micromonospora* have demonstrated effective biological control against fungal pathogens in the rhizosphere (Franco-Correa *et al.*, 2010).

For instance, Lee and Hwang reported that actinomycete isolates—including *Streptomyces*, *Micromonospora*, *Dactylosporangium*, *Actinomadura*, *Streptosporangium*, and *Nocardioform*—exhibited strong antifungal activity against major pathogens such as *Alternaria mali*, *Colletotrichum gloeosporioides*, *Fusarium oxysporum f. sp. cucumerinum*, and *Rhizoctonia solani*.

Root-secreted nutrient-rich compounds attract PGPR to the rhizosphere, enabling colonization while simultaneously serving as a defense barrier against harmful microbes. In return, PGPR improve nutrient absorption and colonize both root surfaces and internal tissues, creating a mutually beneficial relationship that supports plant health and development.

These plant–microbe interactions are governed by a complex network of molecular, physiological, and ecological signals that facilitate communication and symbiosis (Hassan, 2016). Understanding these interactions is key to unlocking how they influence plant resilience, growth, and adaptation to environmental stress (Lakshmanan *et al.*, 2013).

At the molecular level, plants release signaling compounds—such as phytohormones and secondary metabolites—in response to microbial colonization or pathogenic threats. These compounds regulate plant development, activate defense responses, and reinforce beneficial partnerships (Hassan *et al.*, 2019a). Conversely, microbes produce signaling molecules—particularly quorum sensing (QS) compounds—that coordinate community behaviors like biofilm formation, virulence, and gene expression based on population density (Bais *et al.*, 2004; Bertin *et al.*, 2003).

Together, this dynamic chemical dialog ensures efficient communication between plants and microbes, forming the foundation of their cooperative interactions (Zhalnina *et al.*, 2018).

17.5 MICROBIAL SIGNALING MOLECULES IN PLANT–MICROBE INTERACTIONS

Microbial signaling molecules are essential mediators in the communication between plants and their associated microbiota. These chemical messengers facilitate the establishment of symbiotic relationships and influence plant responses to environmental stimuli (Jamil *et al.*, 2022). Among the key signaling compounds are phytohormones such as auxins, cytokinins, GA, ethylene, and abscisic acid, which regulate plant growth, development, and stress responses (Zhao *et al.*, 2021). Notably, these hormones are not exclusive to plants—many beneficial microbes also produce them to interact with host plants and fine-tune the plant–microbe relationship (Singh *et al.*, 2021).

A well-studied microbial communication mechanism is QS, which allows bacteria to coordinate group behaviors in response to population density. This system involves the production and detection of autoinducers—chemical signals that enable synchronized regulation of community-wide activities. Both Gram-positive and Gram-negative bacteria utilize QS to modulate various functions beneficial to the microbial community (Majdura *et al.*, 2023). Autoinducers are classified into three major types: AI-1, AI-2, and AI-3, each contributing to intra- and interspecies communication within the microbiome. Through QS, bacteria regulate gene expression to orchestrate complex behaviors such as biofilm formation, virulence factor production, and antibiotic synthesis, all of which play critical roles in maintaining microbial balance and influencing plant health (Li and Tian, 2012).

17.6 PHYSIOLOGICAL MECHANISMS OF PLANT GROWTH PROMOTION BY BENEFICIAL MICROBES

Beneficial microbes significantly contribute to plant growth and health by enhancing stress tolerance, promoting nutrient uptake, and supporting key physiological processes. One of the primary mechanisms involves the synthesis of plant growth-promoting substances such as cytokinins, IAA, GA, and various enzymes that assist in nutrient solubilization (Pii *et al.*, 2015). These compounds stimulate cell division, elongation, and differentiation, while improving nutrient uptake efficiency (Egamberdieva *et al.*, 2017).

In particular, microbes help solubilize otherwise inaccessible nutrients such as phosphorus, iron, and zinc by producing organic acids, siderophores, and phosphate-solubilizing enzymes. This increased bioavailability enhances plant growth and vigor (Altomare and Tringovska, 2011).

Furthermore, beneficial microbes bolster plant defenses by activating systemic resistance mechanisms, notably SAR and ISR. These responses enhance the plant's immune system and offer protection against both biotic and abiotic stressors. Microbial elicitors, antimicrobial compounds, and VOCs play critical roles in priming these defense pathways (Yu *et al.*, 2022).

Together, these molecular and physiological processes highlight the sophisticated communication and synergy between plants and their beneficial microbial partners. Microbial signaling molecules—including phytohormones and QS compounds—facilitate these interactions, while microbial contributions such as growth regulators, nutrient solubilization, and induced resistance form the foundation of plant health, growth, and resilience (Khan *et al.*, 2020).

17.7 PLANT–MICROBE INTERACTIONS: NATURE'S SECRET SUPERHEROES

17.7.1 Diversity of Plant–Microbe Interactions

Plant–microbe interactions represent a complex and dynamic network that plays a crucial role in agriculture, ecosystem stability, and environmental sustainability (Vishwakarma *et al.*, 2020). These interactions are broadly classified into three categories: Mutualistic, pathogenic, and commensalistic, each contributing uniquely to plant health and ecosystem functioning (Wiesmann *et al.*, 2023).

17.7.1.1 Positive Interactions

Positive interactions involve mutually beneficial or supportive relationships between plants and microbes, fostering nutrient cycling, enhancing soil fertility, and increasing plant resilience to environmental stresses. These interactions include mutualism, commensalism, facilitation, protocooperation, and syntrophism (Iqbal *et al.*, 2023).

17.7.1.1.1 Mutualism

In mutualistic relationships, both partners benefit significantly. A classic example is the symbiosis between legumes and nitrogen-fixing bacteria such as *Rhizobium* and *Bradyrhizobium*. These bacteria form root nodules, where they convert atmospheric nitrogen (N_2) into plant-available ammonium (NH_4^+), while receiving carbon-rich compounds like sucrose and organic acids from the plant. Similarly, mycorrhizal fungi—both arbuscular mycorrhizal fungi (AMF) and ectomycorrhizal fungi (EMF)—associate with plant roots to enhance the uptake of nutrients such as phosphate (PO_4^{3-}) in exchange for plant sugars like glucose and fructose (Goyal *et al.*, 2021).

17.7.1.1.2 Protocooperation

Protocooperation is a form of symbiosis where both organisms benefit, but the interaction is not essential for survival. For example, nitrogen-fixing bacteria enrich the soil with ammonium and nitrogenous compounds, aiding nearby plants, while gaining energy through the fixation process. Mycorrhizal fungi similarly enhance the uptake of not only phosphorus but also micronutrients such as potassium, zinc, and copper, strengthening plant nutrition (Kumawat *et al.*, 2021). Lichens—symbiotic associations between fungi and algae—also exemplify protocooperation, where each partner contributes structural and nutritional support (Spribille *et al.*, 2022).

17.7.1.1.3 Facilitation

Facilitative interactions occur when one organism enhances another's survival or growth without direct benefit to itself (Stachowicz, 2001). In soil, microbes may improve structure, enhance nutrient availability, or secrete antibiotics that suppress harmful pathogens, thereby indirectly supporting plant growth (Chaudhry *et al.*, 2021).

17.7.1.1.4 Syntrophism (Syntrophy or Cross-Feeding)

Syntrophy involves cooperative degradation of complex organic materials by multiple microbial species. In oxygen-poor environments like the rhizosphere, one species may partially break down a compound, producing byproducts that are then further metabolized by another species. This interaction is essential for breaking down complex substances like lignin, long-chain fatty acids, and amino acids under anaerobic conditions (Kirchman, 2012). Syntrophic relationships are critical for microbial ecology and biotechnology, especially in anaerobic soil management (Moriss *et al.*, 2013; Mutungwazi *et al.*, 2021).

17.7.1.1.5 Commensalism

In commensal interactions, one organism benefits while the other remains unaffected. Epiphytic bacteria such as *Pseudomonas fluorescens* colonize plant surfaces and provide resistance against pathogens without harming the host. Similarly, epiphytic fungi like *Cladosporium cladosporioides* inhabit internal plant tissues and may enhance growth or stress tolerance without causing disease. Many nonpathogenic endophytes fall into this category and are increasingly explored as biocontrol agents in sustainable agriculture (Mathis and Bronstein, 2020).

Overall, these positive plant–microbe interactions are vital for maintaining ecological balance, enhancing plant productivity, and promoting sustainable farming practices. A comprehensive understanding of these mechanisms and their ecological significance is key to harnessing their potential for agricultural innovation and environmental conservation (Sahu *et al.*, 2019).

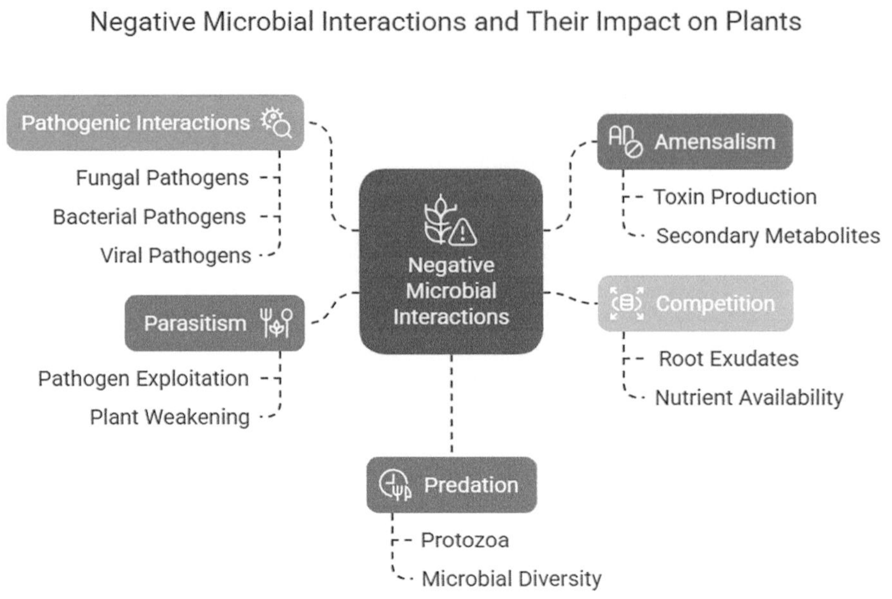

FIGURE 17.7 Negative microbial interactions and their impact on plants.

17.7.1.2 Negative Interactions

While many plant–microbe interactions are beneficial, not all are favorable—some can significantly harm plant health and reduce crop productivity (Figure 17.7). These negative or pathogenic interactions occur when harmful microbes infect plants, triggering disease and stress responses (Nadarajah & Rahman, 2021).

Common fungal pathogens such as *Phytophthora*, *Fusarium*, and *Botrytis* spp., cause devastating conditions like root rot and wilt, resulting in substantial yield losses. Bacterial pathogens including *Pseudomonas syringae*, *Xanthomonas* spp., and *Ralstonia solanacearum* lead to bacterial spot, blight, and vascular wilts. Viral pathogens such as tobacco mosaic virus (TMV), potato virus Y (PVY), and tomato yellow leaf curl virus (TYLCV) contribute to symptoms like mottling, stunted growth, and reduced productivity (Schirawski & Perlin, 2018).

17.7.1.2.1 Amensalism (Antagonism)

Amensalism, also called antagonism, occurs when one organism inhibits or harms another without itself being affected. This is often due to the release of toxins or allelochemicals by certain microbes that suppress the growth of neighboring organisms—including other microbes or even plant roots (Berenbaum, 2001). For example, *Pseudomonas fluorescens* produces pyocyanin, a toxic compound that can inhibit root development and suppress competitors. Fungi such as *Trichoderma* and *Fusarium* species release secondary metabolites like trichothecenes and penicillin to gain a competitive edge in the rhizosphere (Wu, 2018). While beneficial to the producer, these substances can disrupt microbial balance and negatively impact plant health. Understanding these dynamics is crucial for unraveling how microbial communities influence plant outcomes and overall ecosystem functioning (Ehrmann & Ritz, 2014).

17.7.1.2.2 Competition

Competition for limited resources—such as nutrients, space, and water—is another key factor shaping rhizosphere dynamics. Plants influence this competition through the release of root exudates, which include sugars (e.g., glucose and sucrose), amino acids, organic acids (e.g., citric and malic), and phenolic compounds (e.g., flavonoids and benzoic acid) (Koo *et al.*, 2004). These exudates attract beneficial microbes while deterring pathogens, helping structure the root microbiome (Casper & Jackson, 1997). Microbial competition can influence plant development by modulating nutrient access, pathogen suppression, and hormonal signaling pathways. A comprehensive understanding of these interactions is essential for sustainable soil and crop management (Hodge & Fitter, 2013).

17.7.1.2.3 Parasitism

In parasitic interactions, microbes exploit plants as hosts, extracting nutrients and energy to the plant's detriment. This weakens the plant, reduces growth, and increases vulnerability to pests and diseases. Common parasitic organisms include *Xanthomonas* (bacteria), *Fusarium* and *Phytophthora* (fungi), and *Meloidogyne* (root-knot nematodes), which invade roots, stems, and leaves, causing wilting, necrosis, stunting, and even death. Understanding parasitism at the molecular level is vital for developing effective plant protection strategies and ensuring sustainable agricultural productivity (Mitiku, 2018).

17.7.1.2.4 Predation

Microbial predation refers to one microorganism feeding on another. Soil-dwelling predators like protozoa, including amoebae and flagellates, graze on bacterial populations, indirectly shaping microbial diversity (Afridi *et al.*, 2022). Although indirect, this predation plays an important role in nutrient cycling, mineralization, and organic matter decomposition. By influencing the composition of microbial communities, predation can impact plant health, nutrient availability, and soil fertility (Lambers *et al.*, 2009).

Collectively, negative plant–microbe interactions—ranging from disease and parasitism to competition and predation—underscore the importance of understanding microbial ecology in managing plant health and optimizing agricultural systems.

17.7.2 Root Colonization Capacity of PGPR

Root colonization refers to the ability of PGPR to multiply and establish themselves either on the root surface (*ectophytically*), within the surrounding rhizosphere, or inside plant tissues (*endophytically*). This colonization is a fundamental prerequisite for PGPR to exert their growth-promoting effects. Typically, PGPR introduced through inoculants—such as seed coatings or soil amendments—migrate toward the actively growing zones of the root, where they begin to proliferate and interact with the root system (Parke, 1991).

Bacterial movement from the point of inoculation to the root zone occurs through both active mechanisms, like flagellar motility, and passive processes, such as transport via water flow in the soil. According to the root colonization model proposed by Newman and Watson, rhizobacteria preferentially accumulate near the root tip, a nutrient-rich zone characterized by high metabolic activity (Newman and Watson, 1977).

Howie *et al.* observed that *Pseudomonas fluorescens* strain 2–79 primarily reaches the root tip through passive transport. Specific bacterial surface components, such as lipopolysaccharides (LPS)—especially the O-antigen portion—play a key role in adhesion and colonization. However, the effectiveness of these components is often strain-specific. For instance, the O-antigenic side chain of *P. fluorescens* WCS374 was found to be ineffective in promoting adhesion to potato roots (De Weger *et al.*, 1989).

In addition to surface properties, other physiological traits also contribute to colonization success. The production of vitamin B1 and the activity of NADH dehydrogenases support bacterial attachment and persistence on root surfaces. In endophytic PGPR such as *Azospirillum* spp., type IV pili have been identified as critical structures that aid in tight attachment and colonization within root tissues (Steenhoudt & Vanderleyden, 2000).

Genetic factors further influence colonization efficiency. In some PGPR strains, the presence of a site-specific recombinase gene has been linked to enhanced root adherence. Notably, transferring this gene from wild-type *Pseudomonas* spp. to a weaker colonizer like *P. fluorescens* WCS307 significantly improved its ability to colonize root tips. Similarly, a study by Xu *et al.* (2019) highlighted the role of wall teichoic acid (WTA) in *Bacillus velezensis* SQR9, which was shown to be essential for effective root colonization and biofilm formation—underscoring the molecular complexity behind PGPR–root interactions.

17.7.3 Root Exudation

Root exudation refers to the release of a diverse array of chemical compounds from plant roots into the surrounding soil, serving as a primary nutrient source for microorganisms in the rhizosphere. The composition and quantity of these exudates vary depending on plant species, developmental stage, and environmental factors such as nutrient availability and stress levels (Uren, 2000). According to Bolton *et al.*, root exudates are "low molecular weight organic compounds that leak from intact root cells" and play a vital role in establishing and sustaining microbial communities in the rhizosphere (Zhalnina *et al.*, 2018).

Root exudation occurs via two main mechanisms: Passive leakage and active secretion. Passive leakage is largely uncontrolled, while active secretion involves the selective release of compounds that serve specific functions, including root lubrication, microbial signaling, and defense. Exudates are broadly categorized based on their molecular weight:

- Low-molecular-weight exudates—such as sugars, amino acids, organic acids, and phenolic compounds—are chemically diverse and critical for attracting beneficial microbes and initiating signaling pathways (Bolton & Elliott, 1993).
- High-molecular-weight exudates, including mucilage and proteins secreted by root cap and epidermal cells, enhance soil adhesion and facilitate root penetration through the soil (Bertin *et al.*, 2003).

Beyond nutrient exchange, root exudates serve as chemical messengers and defense agents. For example, Tsuno *et al.* reported that soybean roots secrete soyasaponins during early vegetative stages, while Liu *et al.* showed that root exudates promote early symbiosis with *Bacillus diazoefficiens*, aiding mutualistic associations.

Root exudates also contribute to physical and biochemical interactions in the soil by protecting root surfaces from desiccation, facilitating ion exchange, and shaping microbial communities (Walker *et al.*, 2003; Hawes *et al.*, 2000). Ultimately, they are pivotal in supporting plant growth, enhancing stress resilience, and activating defenses against pathogens.

17.7.4 Rhizodeposition

Rhizodeposition encompasses the release of various carbon-rich substances from plant roots into the soil, playing a crucial role in plant–soil–microbe interactions. This process is central to nutrient cycling and microbial activity in the rhizosphere (Bowsher *et al.*, 2018).

The components of rhizodeposition include root cap cells, cell lysates, mucilage, mucigel, secretions, and exudates, originating from both living and damaged root tissues. The rate and composition of rhizodeposition are influenced by biotic factors such as plant species, photosynthetic activity, root morphology, carbon allocation, mycorrhizal associations, and nodulation. Abiotic factors—including soil moisture, temperature, texture, rooting depth, and nitrogen levels—also play a significant role (Jones *et al.*, 2004).

The root tip and cap are particularly active in rhizodeposition. Studies indicate that secreted polysaccharides and other organic materials from these zones contribute 2%–12% of total rhizodeposition (Sievers *et al.*, 2002). These substances include sugars, amino acids, organic acids, fatty acids, enzymes, vitamins, and growth-promoting factors, which enhance soil fertility and stimulate microbial activity (Yadav *et al.*, 2015).

According to Kuzyakov *et al.*, up to 10% of carbon released through rhizodeposition is in the form of low-molecular-weight compounds like sugars, amino acids, and carboxylic acids. These act as energy sources for soil microbes, promoting organic matter decomposition and nutrient solubilization. White *et al.* further noted that certain rhizodeposits, such as isoflavonoids, contribute not only to plant defense but also to the establishment of beneficial symbioses, particularly between soybeans and *Rhizobium* species.

In essence, rhizodeposition drives essential ecological processes such as nutrient mobilization, soil aggregation, carbon sequestration, and the structuring of microbial communities (Kögel-Knabner, 2002). It is fundamental to soil health and the plant's ability to engage productively with its microbial partners.

17.7.5 Root Border Cells

The term *root border cells* was first introduced by Hawes *et al.* (1998) to describe specialized cells that are naturally shed from the root tips of healthy plants (Eshel and Beeckman, 2013). The number, behavior, and physiological role of these cells vary significantly depending on plant species, genotype, and environmental conditions. Root border cells have garnered considerable attention due to their multifunctional roles in rhizosphere dynamics (Endo *et al.*, 2011).

These cells detach from the root tip through the enzymatic action of pectin-degrading enzymes such as methylesterases and polygalacturonases, which dissolve the middle lamella of cell walls. Structurally, their cell walls are rich in pectic polysaccharides, similar to the mucilage secreted by root cap cells. Remarkably, even after detachment, many root border cells remain viable—studies report up to 90% survival under favorable nutrient conditions (Hassan *et al.*, 2019a).

Functionally, root border cells are active participants in plant–microbe interactions. They influence rhizosphere processes through chemoattraction and repulsion mechanisms, facilitate colonization by PGPR, and play a protective role against pathogens (Badri and Vivanco, 2009). For instance, humic acid has been shown to stimulate border cell release and enhance root colonization in maize by *Herbaspirillum seropedicae*, a nitrogen-fixing PGPR (Canellas and Olivares, 2017).

Border cells also secrete bioactive compounds—such as phenolics and arabinogalactan proteins—that serve antimicrobial functions. One prominent example includes the suppression of *Aphanomyces euteiches*, a root rot pathogen in legumes, by arabinogalactan proteins secreted by these cells. Given that the root tip is a critical entry point for many soil-borne pathogens, the presence of border cells provides a protective barrier while also enhancing the recruitment of beneficial microbes like PGPR (Cannesan *et al.*, 2012).

17.7.6 Rhizodeposition, Exudates, and Border Cells in the Growing Root System

While root exudation and rhizodeposition are frequently emphasized in studies of PGPR-mediated colonization, root border cells are equally integral to shaping plant–microbe interactions. Rhizodeposition—the release of organic compounds from roots into the soil—occurs across several root zones, including the meristematic, elongation, and maturation zones (Kuzyakov, 2002; Freschet *et al.*, 2021). These deposits originate from root tissues, root cap cells, mucilage, and exudates (Rovira, 1956).

Root exudation is particularly active in the elongation zone, which also coincides with peak microbial abundance (Marschner *et al.*, 2011). Exudates are commonly secreted near the apical meristem of both tap and lateral roots (Yadav *et al.*, 2015). Depending on their role, exudates may be released actively, with specific functions in defense and lubrication, or passively as metabolic byproducts (Kumar *et al.*, 2022).

Border cells, primarily shed from the root cap and cell division zones, serve as key sites for microbial colonization, nutrient interaction, and plant defense. Once detached, they release a cocktail of bioactive molecules, including antimicrobial peptides, phytoalexins, and arabinogalactan proteins, which help modulate microbial populations and suppress pathogens (Durand *et al.*, 2009).

Their presence at the root tip—a region particularly susceptible to microbial invasion—adds a critical layer of protection (Driouich *et al.*, 2010).

Root exudates and border cells act in synergy, emitting chemical signals and nutrients that attract beneficial microbes while repelling pathogens. Border cells are considered part of the rhizodepositional landscape and function similarly to exudates in shaping the microbial ecology of the rhizosphere (Dennis *et al.*, 2010).

Together, rhizodeposits, root exudates, and border cells form an integrated defense and signaling network that supports nutrient acquisition and pathogen resistance. This system regulates interactions with diverse organisms—including bacteria, fungi, nematodes, and viruses—and plays a central role in maintaining soil health, plant productivity, and rhizosphere microbial balance (Sasse *et al.*, 2018).

17.7.7 Communication among PGPR, Rhizodeposits, and Roots in the Rhizosphere

Communication between plant roots and rhizospheric microorganisms is a dynamic and interdependent process that underpins plant health and productivity. This intricate interaction involves both root–microbe and root–root signaling and is driven by the mutual exchange of nutrients and signaling molecules essential for survival. Depending on soil nutrient availability and environmental conditions, these relationships can be symbiotic, associative, neutral, or parasitic in nature (Boubekri *et al.*, 2022).

A key component of this communication system is the secretion of organic compounds—especially from lateral roots—which plays a critical role in shaping microbial communities by attracting and aggregating microbes within the rhizosphere. Among these compounds, root exudates—a variety of phytochemicals actively secreted by plant roots—serve as central regulators of microbial recruitment and symbiosis in nutrient-rich zones (Jones *et al.*, 2003; Mitter *et al.*, 2021). These exudates drive microbial chemotaxis, guiding beneficial microbes such as PGPR toward the root surface and influencing colonization efficiency and biocontrol activities, including antifungal, antibacterial, and antiviral defense (Mohanram and Kumar, 2019; Kumar *et al.*, 2023).

For instance, *Corynebacterium flavescens* and *Bacillus pumilus*, two endophytic bacterial strains, showed a fivefold increase in chemotactic activity in the rice rhizosphere when exposed to root-derived amino acids and carbohydrates, compared to other bacterial strains. In addition to exudates, root border cells play a crucial role in this molecular dialog. These specialized cells actively stimulate microbial growth and attract both bacteria and fungi toward the root zone, enhancing microbial density and interaction near root surfaces (Huang *et al.*, 2014).

Furthermore, antimicrobial compounds released by both root exudates and border cells serve dual roles—not only enhancing plant growth but also suppressing pathogenic organisms. For example, extracellular proteins secreted from the root cap of pea plants (via border cells) have been shown to inhibit the fungal pathogen *Nectria haematococca* (Huang *et al.*, 2014). Similarly, root-exuded antimicrobial metabolites have demonstrated inhibitory effects on several bacterial pathogens, including *Erwinia carotovora*, *E. amylovora*, *Xanthomonas campestris* pv. *vesicatoria*, and *Pseudomonas fluorescens* (Pusztahelyi *et al.*, 2015).

The efficacy of these interactions is shaped by both biotic and abiotic factors. Root exudation patterns are influenced by environmental variables such as soil type, moisture, and temperature, and, in turn, reshape the microbial and chemical landscape of the rhizosphere (Basu *et al.*, 2021). Different PGPR genera respond uniquely to these cues—some producing pathogen-suppressing compounds and others releasing metabolites that stimulate plant growth. Additionally, PGPR must compete with other soil microbes for limited space and resources, highlighting the competitive and cooperative complexity of the rhizosphere. In brief, root exudates and border cells act as key mediators of rhizospheric communication. Together, they foster nutrient uptake, promote colonization by beneficial microbes, and mount defenses against a wide range of soil-borne pathogens, ultimately contributing to a healthier and more resilient plant–soil ecosystem.

17.8 CONCLUSION

The rhizosphere represents a vital interface among plants, soil, and microbes—serving as a dynamic hub for interactions that significantly influence plant growth and soil health. Through the process of root exudation, plants actively shape their surrounding microbial communities, fostering the proliferation of beneficial microorganisms such as PGPR. These microbes enhance nutrient availability, boost plant stress tolerance, and contribute meaningfully to overall agricultural productivity. The complex biochemical interactions within the rhizosphere highlight its potential as a sustainable alternative to chemical fertilizers and pesticides. Understanding and harnessing the relationships between plant roots and rhizobacteria are essential for improving soil fertility and ensuring the long-term health of crops—key pillars in building resilient and efficient agricultural systems. Continued research into plant–microbe interactions will pave the way for innovative and eco-friendly agricultural practices. Leveraging the power of PGPR and other soil microbes can reduce environmental stress, increase soil biodiversity, and enhance food security. By integrating biological strategies with conventional farming, we can move toward a more sustainable and productive agricultural future, delivering benefits for both ecosystems and human well-being.

REFERENCES

Abdul Malik, Nurul Azmina, Ilakiya Sharanee Kumar, and Kalaivani Nadarajah. "Elicitor and receptor molecules: Orchestrators of plant defense and immunity." *International Journal of Molecular Sciences* 21, no. 3 (2020): 963.

Afridi, Muhammad Siddique, Ali Fakhar, Ashwani Kumar, Sher Ali, Flavio HV Medeiros, Muhammad Atif Muneer, Hina Ali, and Muhammad Saleem. "Harnessing microbial multitrophic interactions for rhizosphere microbiome engineering." *Microbiological Research* 265 (2022): 127199.

Afzal, Iftikhar, Zabta K. Shinwari, Sikandar Sikandar, and Shahzad Shahzad. "Plant beneficial endophytic bacteria: Mechanisms, diversity, host range and genetic determinants." *Microbiological Research* 221 (2019): 36–49.

Ahemad, Musarrat, and Mulugeta Kibret. "Mechanisms and applications of plant growth promoting rhizobacteria: Current perspective." *Journal of King Saud University-Science* 26, no. 1 (2014): 1–20.

Ali, Omer, Anil Ramsubhag, and Jayaraman Jayaraman. "Biostimulant properties of seaweed extracts in plants: Implications towards sustainable crop production." *Plants* 10, no. 3 (2021): 531.

Altomare, C., and Tringovska, I. "Beneficial soil microorganisms, an ecological alternative for soil fertility management." *Genetics, biofuels and local farming systems* (2011): 161–214.

Anand, K., Bhawna Kumari, and Manoj A. Mallick. "Phosphate solubilizing microbes: An effective and alternative approach as bio-fertilizers." *International Journal of Pharmacy and Pharmaceutical Sciences* 8 (2016): 37–40.

Ayaz, Muhammad, Chun-Hao Li, Qaiser Ali, Wen-Kai Zhao, Yong-Kun Chi, Muhammad Shafiq, Farhat Ali, Xiao-Yan Yu, Qing Yu, Jian-Tao Zhao, and Jian-Wei Yu. "Bacterial and fungal biocontrol agents for plant disease protection: Journey from lab to field, current status, challenges, and global perspectives." *Molecules* 28, no. 18 (2023): 6735.

Badri, Devendra V., and Jorge M. Vivanco. "Regulation and function of root exudates." *Plant, Cell & Environment* 32, no. 6 (2009): 666–681.

Bais, Harsh P., Soo W. Park, Thomas L. Weir, Ragan M. Callaway, and Jorge M. Vivanco. "How plants communicate using the underground information superhighway." *Trends in Plant Science* 9, no. 1 (2004): 26–32.

Basu, Arpan, Pankaj Prasad, Swagat Narayan Das, Shekhar Kalam, Rizwan Z. Sayyed, M. Sudhakar Reddy, and Hesham El Enshasy. "Plant growth promoting rhizobacteria (PGPR) as green bioinoculants: Recent developments, constraints, and prospects." *Sustainability* 13, no. 3 (2021): 1140.

Berenbaum, M. "Plant–herbivore interactions." *Evolutionary ecology: concepts and case studies* (2001): 303–314.

Bertin, Claude, Xinhua Yang, and Leslie A. Weston. "The role of root exudates and allelochemicals in the rhizosphere." *Plant and Soil* 256 (2003): 67–83.

Bhattacharyya, Partha N., and Dipankar K. Jha. "Plant growth-promoting rhizobacteria (PGPR): Emergence in agriculture." *World Journal of Microbiology and Biotechnology* 28 (2012): 1327–1350.

Bishnoi, Usha. "PGPR interaction: An ecofriendly approach promoting the sustainable agriculture system." In *Advances in Botanical Research*, Vol. 75, 81–113. Academic Press, 2015.

Bolton, Harvey F., and Larry F. Elliott. "Microbial ecology of the rhizosphere." In *Soil Microbial Ecology*. Marcel Dekker, 1993.

Boubekri, Kaouthar, Amadou Soumare, Ilyas Mardad, Khalid Lyamlouli, Yedir Ouhdouch, Mohamed Hafidi, and Lahcen Kouisni. "Multifunctional role of Actinobacteria in agricultural production sustainability: A review." *Microbiological Research* 261 (2022): 127059.

Bowsher, Andrew W., Sarah Evans, Lisa K. Tiemann, and Maren L. Friesen. "Effects of soil nitrogen availability on rhizodeposition in plants: A review." *Plant and Soil* 423 (2018): 59–85.

Canellas, Lucia P., and Francisco L. Olivares. "Production of border cells and colonization of maize root tips by *Herbaspirillum seropedicae* are modulated by humic acid." *Plant and Soil* 417 (2017): 403–413.

Cannesan, Marie A., Céline Durand, Claude Burel, Christophe Gangneux, Patrice Lerouge, Tetsuya Ishii, Karine Laval, Marie-Laure Follet-Gueye, Alain Driouich, and Muriel Vicré-Gibouin. "Effect of arabinogalactan proteins from the root caps of pea and *Brassica napus* on *Aphanomyces euteiches* zoospore chemotaxis and germination." *Plant Physiology* 159 (2012): 1658–1670.

Casper, Brenda B., and Robert B. Jackson. "Plant competition underground." *Annual Review of Ecology and Systematics* 28, no. 1 (1997): 545–570.

Chaudhry, Varsha, Philip Runge, Pratyush Sengupta, Gunther Doehlemann, Jane E. Parker, and Eva Kemen. "Shaping the leaf microbiota: Plant–microbe–microbe interactions." *Journal of Experimental Botany* 72, no. 1 (2021): 36–56.

Chauhan, Harsh, D. J. Bagyaraj, G. Selvakumar, and S. P. Sundaram. "Novel plant growth promoting rhizobacteria—Prospects and potential." *Applied Soil Ecology* 95 (2015): 38–53.

Chieb, Mbiankeu, and Eric W. Gachomo. "The role of plant growth promoting rhizobacteria in plant drought stress responses." *BMC Plant Biology* 23, no. 1 (2023): 407.

Czarnes, Sylvie, Pierre Mercier, Dominique G. Lemoine, Jamil Hamzaoui, and Laurent Legendre. "Impact of soil water content on maize responses to the plant growth-promoting rhizobacterium *Azospirillum lipoferum* CRT1." *Journal of Agronomy and Crop Science* 206 (2020): 505–516.

de Andrade, Letícia Aparecida, Célio Henrique Bertucci Santos, Eduardo Tadeu Frezarin, Lucas Ramos Sales, and Evandro Carlos Rigobelo. "Plant growth-promoting rhizobacteria for sustainable agricultural production." *Microorganisms* 11, no. 4 (2023): 1088.

de Weger, Leo A., Peter A.H.M. Bakker, Ben Schippers, M.C.M. van Loosdrecht, and Ben J.J. Lugtenberg. "Pseudomonas spp. with mutational changes in the O-antigenic side chain of their lipopolysaccharide are affected in their ability to colonize potato roots." In *Signal Molecules in Plants and Plant–Microbe Interactions*, 197–202. Springer Berlin Heidelberg, 1989.

Dennis, Paul G., Andrew J. Miller, and Penny R. Hirsch. "Are root exudates more important than other sources of rhizodeposits in structuring rhizosphere bacterial communities?" *FEMS Microbiology Ecology* 72 (2010): 313–327.

Driouich, Azeddine, Christophe Durand, Marie-Agnès Cannesan, Giovanni Percoco, and Muriel Vicré-Gibouin. "Border cells versus border-like cells: Are they alike?" *Journal of Experimental Botany* 61, no. 14 (2010): 3827–3831.

Durand, Christophe, Muriel Vicré-Gibouin, Marie-Laure Follet-Gueye, Laure Duponchel, Maryline Moreau, Philippe Lerouge, and Azeddine Driouich. "The organization pattern of root border-like cells of *Arabidopsis* is dependent on cell wall homogalacturonan." *Plant Physiology* 150, no. 3 (2009): 1411–1421.

Egamberdieva, Dilfuza, Sarah J. Wirth, Abdullah A. Alqarawi, Elsayed Fathi Abd-Allah, and Abdallah Hashem. "Phytohormones and beneficial microbes: Essential components for plants to balance stress and fitness." *Frontiers in Microbiology* 8 (2017): 2104.

Ehrmann, Jessica, and Karl Ritz. "Plant: Soil interactions in temperate multi-cropping production systems." *Plant and Soil* 376 (2014): 1–29.

Elnahal, Ahmed S., Mohamed T. El-Saadony, Ahmed M. Saad, Ebrahim S.M. Desoky, Ahmed M. El-Tahan, Mohamed M. Rady, Saleh F. AbuQamar, and Khaled A. El-Tarabily. "The use of microbial inoculants for biological control, plant growth promotion, and sustainable agriculture: A review." *European Journal of Plant Pathology* 162, no. 4 (2022): 759–792.

Endo, Itsuki, Takashi Tange, and Hiroyuki Osawa. "A cell-type-specific defect in border cell formation in the *Acacia mangium* root cap developing an extraordinary sheath of sloughed-off cells." *Annals of Botany* 108 (2011): 279–290.

Eshel, Amram, and Tom Beeckman, eds. *Plant Roots: The Hidden Half*. CRC Press, 2013.

Etesami, Hassan, and Dinesh K. Maheshwari. "Use of plant growth promoting rhizobacteria (PGPRs) with multiple plant growth promoting traits in stress agriculture: Action mechanisms and future prospects." *Ecotoxicology and Environmental Safety* 156 (2018): 225–246.

Franco-Correa, Mónica, Andrea Quintana, Claudia Duque, Carlos Suarez, María Ximena Rodríguez, and José Miguel Barea. "Evaluation of actinomycete strains for key traits related with plant growth promotion and mycorrhiza helping activities." *Applied Soil Ecology* 45, no. 3 (2010): 209–217.

Freschet, Guillaume T., Laure Pagès, Colleen M. Iversen, Lars H. Comas, Björn Rewald, Catherine Roumet, Jana Klimešová, et al. "A starting guide to root ecology: Strengthening ecological concepts and standardising root classification, sampling, processing and trait measurements." *New Phytologist* 232, no. 3 (2021): 973–1122.

Ghosh, Ananya, Ruma Acharya, Subhra Shaw, and Debabrata Gangopadhyay. "Plant Growth-Promoting Rhizobacteria (PGPR): A Potential Alternative Tool for Sustainable Agriculture." In *Updates on Rhizobacteria*. IntechOpen, 2024.

Glick, Bernard R. "Plant growth-promoting bacteria: Mechanisms and applications." *Scientifica* 2012, no. 1 (2012): 963401.

Glick, B. R. *Beneficial Plant-Bacterial Interactions*. Vol. 243. Heidelberg: Springer, 2015.

Goswami, Disha, Jaynish N. Thakker, and Pratik C. Dhandhukia. "Portraying mechanics of plant growth promoting rhizobacteria (PGPR): A review." *Cogent Food & Agriculture* 2 (2016): 1127500.

Goyal, Rajeev K., Autar K. Mattoo, and Molly A. Schmidt. "Rhizobial–host interactions and symbiotic nitrogen fixation in legume crops toward agriculture sustainability." *Frontiers in Microbiology* 12 (2021): 669404.

Gray, E. J., and D. L. Smith. "Intracellular and extracellular PGPR: Commonalities and distinctions in the plant–bacterium signaling processes." *Soil Biology and Biochemistry* 37, no. 3 (2005): 395–412.

Harman, Gary E., and Norman Uphoff. "Symbiotic root-endophytic soil microbes improve crop productivity and provide environmental benefits." *Scientifica* 2019, no. 1 (2019): 9106395.

Hashem, Abdallah, Bilquees Tabassum, and Elsayed Fathi Abd-Allah. "*Bacillus subtilis*: A plant-growth promoting rhizobacterium that also impacts biotic stress." *Saudi Journal of Biological Sciences* 26, no. 6 (2019): 1291–1297.

Hassan, Mohamed Khaled. *The role of pectin utilization in root colonization and plant growth-promotion by Bacillus amyloliquefaciens subsp. plantarum (Bap)*. Master's thesis, Auburn University, 2016.

Hassan, Mohamed Khaled, John A. McInroy, Jeremy Jones, Deepika Shantharaj, Mark R. Liles, and Joseph W. Kloepper. "Pectin-rich amendment enhances soybean growth promotion and nodulation mediated by *Bacillus velezensis* strains." *Plants* 8, no. 5 (2019a): 120.

Hassan, Mohamed Khaled, John A. McInroy, and Joseph W. Kloepper. "The interactions of rhizodeposits with plant growth-promoting rhizobacteria in the rhizosphere: A review." *Agriculture* 9, no. 7 (2019b): 142.

Hawes, Mary, Linda Brigham, Feng Wen, Haein Woo, and Yadong Zhu. "Function of root border cells in plant health: Pioneers in the rhizosphere." *Annual Review of Phytopathology* 36 (1998): 311–327.

Hawes, Mary C., Udul Gunawardena, Shozo Miyasaka, and Xuan Zhao. "The role of root border cells in plant defense." *Trends in Plant Science* 5, no. 3 (2000): 128–133.

Hodge, Angela, and Alastair H. Fitter. "Microbial mediation of plant competition and community structure." *Functional Ecology* 27, no. 4 (2013): 865–875.

Huang, Xuefeng F., Jose M. Chaparro, Kenneth F. Reardon, Ruifu Zhang, Qirong Shen, and Jorge M. Vivanco. "Rhizosphere interactions: Root exudates, microbes, and microbial communities." *Botany* 92, no. 4 (2014): 267–275.

Iqbal, Bilal, Guohua Li, Kadhim Fadhil Alabbosh, Hafiz Hussain, Irshad Khan, Muhammad Tariq, Qurban Javed, Muhammad Naeem, and Naeem Ahmad. "Advancing environmental sustainability through microbial reprogramming in growth improvement, stress alleviation, and phytoremediation." *Plant Stress* 10 (2023): 100283.

Jamil, Farah, Hafsa Mukhtar, Mickaël Fouillaud, and Laurent Dufossé. "Rhizosphere signaling: Insights into plant–rhizomicrobiome interactions for sustainable agronomy." *Microorganisms* 10, no. 5 (2022): 899.

Jansson, Janet K., Robert McClure, and Robert G. Egbert. "Soil microbiome engineering for sustainability in a changing environment." *Nature Biotechnology* 41, no. 12 (2023): 1716–1728.

Jones, D.L., Farrar, J., and Giller, K.E. "Associative nitrogen fixation and root exudation-What is theoretically possible in the rhizosphere?" (2003).

Jones, Davey L., Angela Hodge, and Yakov Kuzyakov. "Plant and mycorrhizal regulation of rhizodeposition." *New Phytologist* 163, no. 3 (2004): 459–480.

Khan, Niamat, Sher Ali, Muhammad A. Shahid, Abeer Mustafa, Rifat Z. Sayyed, and Jorge A. Curá. "Insights into the interactions among roots, rhizosphere, and rhizobacteria for improving plant growth and tolerance to abiotic stresses: A review." *Cells* 10, no. 6 (2021): 1551.

Khan, Niamat, Allah Bano, Sher Ali, and Muhammad A. Babar. "Crosstalk amongst phytohormones from planta and PGPR under biotic and abiotic stresses." *Plant Growth Regulation* 90 (2020): 189–203.

Khan, Niamat, Allah Bano, Md. Arifur Rahman, Jun Guo, Zhensheng Kang, and Muhammad A. Babar. "Comparative physiological and metabolic analysis reveals a complex mechanism involved in drought tolerance in chickpea (*Cicer arietinum* L.) induced by PGPR and PGRs." *Scientific Reports* 9 (2019): 2097.

Khoshru, Behzad, Debasis Mitra, Elham Khoshmanzar, Ei Mon Myo, Nisha Uniyal, Bikram Mahakur, Pradeep Kumar Das Mohapatra, Palaniyandi Panneerselvam, Hakima Boutaj, Mahdi Alizadeh, et al. "Current scenario and future prospects of plant growth-promoting rhizobacteria: An economic valuable resource for the agriculture revival under stressful conditions." *Journal of Plant Nutrition* 43 (2020): 3062–3092.

Kirchman, David L. "Degradation of organic material." In *Processes in Microbial Ecology*, 79–98. New York: Oxford University Press Inc., 2012.

Kögel-Knabner, Ingrid. "The macromolecular organic composition of plant and microbial residues as inputs to soil organic matter." *Soil Biology and Biochemistry* 34, no. 2 (2002): 139–162.

Koo, Bon Joon, Domy C. Adriano, Nanthi S. Bolan, and Christopher D. Barton. "Root exudates and microorganisms." In *Encyclopedia of Soils in the Environment*, 421–428. Elsevier-Hanley and Belfus Inc., 2004.

Kumar, Sunil, Thakur Korra, Ramesh Thakur, Ramesh Arutselvan, Amit K. Kashyap, Yazan Nehela, Viktor Chaplygin, Tamara Minkina, and Chaudhary Keswani. "Role of plant secondary metabolites in defence and transcriptional regulation in response to biotic stress." *Plant Stress* 8 (2023): 100154.

Kumar, Sunil, S.S. Sindhu, and Ramesh Kumar. "Biofertilizers: An ecofriendly technology for nutrient recycling and environmental sustainability." *Current Research in Microbial Sciences* 3 (2022): 100094.

Kumari, R., E. Pandey, S. Bushra, S. Faizan, and S. Pandey. "Plant growth promoting rhizobacteria (PGPR) induced protection: A plant immunity perspective." *Physiologia Plantarum* 176, no. 5 (2024): e14495.

Kumawat, K.C., Sushma Nagpal, and Pankaj Sharma. "Present scenario of bio-fertilizer production and marketing around the globe." In *Biofertilizers*, 389–413. Woodhead Publishing, 2021.

Kuzyakov, Y. "Factors affecting rhizosphere priming effects." *Journal of Plant Nutrition and Soil Science* 165, no. 4 (2002): 382–396.

Kwon, Jin Woo, and Sang Do Kim. "Characterization of an antibiotic produced by *Bacillus subtilis* JW-1 that suppresses *Ralstonia solanacearum*." *Journal of Microbiology and Biotechnology* 24, no. 1 (2014): 13–18. https://doi.org/10.4014/jmb.1308.08060.

Ladha, Jagdish K., M. B. Peoples, P. M. Reddy, J. C. Biswas, Alan Bennett, M. L. Jat, and Timothy J. Krupnik. "Biological nitrogen fixation and prospects for ecological intensification in cereal-based cropping systems." *Field Crops Research* 283 (2022): 108541.

Lakshmanan, Venkatachalam, R. Castaneda, T. Rudrappa, and Harsh P. Bais. "Root transcriptome analysis of *Arabidopsis thaliana* exposed to beneficial *Bacillus subtilis* FB17 rhizobacteria revealed genes for bacterial recruitment and plant defense independent of malate efflux." *Planta* 238 (2013): 657–668.

Lambers, H., Mougel, C., Jaillard, B., and Hinsinger, P. "Plant-microbe-soil interactions in the rhizosphere: an evolutionary perspective" (2009).

Lee, Jung Youn, and Byung Kook Hwang. "Diversity of antifungal actinomycetes in various vegetative soils of Korea." *Canadian Journal of Microbiology* 48, no. 5 (2002): 407–417.

Li, Yong-Hua, and Xiang Tian. "Quorum sensing and bacterial social interactions in biofilms." *Sensors* 12, no. 3 (2012): 2519–2538.

Liu, Yan, Dejun Guan, Xiaoyang Jiang, Ming Ma, Linlin Li, Feng Cao, Hong Chen, Ding Shen, and Jianping Li. "Proteins involved in nodulation competitiveness of two *Bradyrhizobium diazoefficiens* strains induced by soybean root exudates." *Biology and Fertility of Soils* 51 (2015): 251–260.

Ma, Wei, Sheng Tang, Zengzeng Deng, Dongmei Zhang, Tao Zhang, and Xiaoxiao Ma. "Root exudates contribute to belowground ecosystem hotspots: A review." *Frontiers in Microbiology* 13 (2022): 937940.

Majdura, Julia, Urszula Jankiewicz, Agnieszka Gałązka, and Sławomir Orzechowski. "The role of quorum sensing molecules in bacterial–plant interactions." *Metabolites* 13, no. 1 (2023): 114.

Manjunath, Mahesh, R. Prasanna, Poonam Sharma, Lata Nain, and Raj Singh. "Developing PGPR consortia using novel genera *Providencia* and *Alcaligenes* along with cyanobacteria for wheat." *Archives of Agronomy and Soil Science* 57, no. 8 (2011): 873–887.

Marschner, Petra, David Crowley, and Zed Rengel. "Rhizosphere interactions between microorganisms and plants govern iron and phosphorus acquisition along the root axis–model and research methods." *Soil Biology and Biochemistry* 43, no. 5 (2011): 883–894.

Mathis, Kaitlyn A., and Judith L. Bronstein. "Our current understanding of commensalism." *Annual Review of Ecology, Evolution, and Systematics* 51, no. 1 (2020): 167–189.

Mitiku, Mesfin. "Plant-parasitic nematodes and their management: A review." *Agricultural Research & Technology* 8, no. 1 (2018): 30–38.

Mitter, Esther K., Martina Tosi, Dario Obregón, Karen E. Dunfield, and Donald J. Germida. "Rethinking crop nutrition in times of modern microbiology: Innovative biofertilizer technologies." *Frontiers in Sustainable Food Systems* 5 (2021): 606815.

Mohanram, Swathi, and Pawan Kumar. "Rhizosphere microbiome: Revisiting the synergy of plant–microbe interactions." *Annals of Microbiology* 69 (2019): 307–320.

Morris, Belinda E. L., Ruth Henneberger, Helga Huber, and Christine Moissl-Eichinger. "Microbial syntrophy: Interaction for the common good." *FEMS Microbiology Reviews* 37, no. 3 (2013): 384–406.

Mutungwazi, Amos, George N. Ijoma, and Tafadzwa S. Matambo. "The significance of microbial community functions and symbiosis in enhancing methane production during anaerobic digestion: A review." *Symbiosis* 83, no. 1 (2021): 1–24.

Nadarajah, K., and N.S.N. Abdul Rahman. "Plant–microbe interaction: Aboveground to belowground, from the good to the bad." *International Journal of Molecular Sciences* 22, no. 19 (2021): 10388.

Newman, E. I., and A. Watson. "Microbial abundance in the rhizosphere: A computer model." *Plant and Soil* 48, no. 1 (1977): 17–56.

Nihorimbere, V., M. Ongena, M. Smargiassi, and P. Thonart. "Beneficial effect of the rhizosphere microbial community for plant growth and health." *Biotechnologie, Agronomie, Société et Environnement* 15, no. 2 (2011).

Parke, J.L. "Root colonization by indigenous and introduced microorganisms." In *The Rhizosphere and Plant Growth*, 33–42. Springer Netherlands, 1991.

Pérez-de-Luque, A., S. Tille, I. Johnson, D. Pascual-Pardo, J. Ton, and D.D. Cameron. "The interactive effects of arbuscular mycorrhiza and plant growth-promoting rhizobacteria synergistically enhance host plant defences against pathogens." *Scientific Reports* 7, no. 1 (2017): 16409.

Pii, Y., T. Mimmo, N. Tomasi, R. Terzano, S. Cesco, and C. Crecchio. "Microbial interactions in the rhizosphere: Beneficial influences of plant growth-promoting rhizobacteria on nutrient acquisition process. A review." *Biology and Fertility of Soils* 51 (2015): 403–415.

Prajapati, K., M. Sharma, and H. Modi. "Growth promoting effect of potassium solubilizing microorganisms on *Abelmoscus esculantus*." *International Journal of Agricultural Sciences* 3 (2013): 181–188.

Prashar, P., N. Kapoor, and S. Sachdeva. "Rhizosphere: Its structure, bacterial diversity and significance." *Reviews in Environmental Science and Bio/Technology* 13 (2014): 63–77.

Pusztahelyi, T., I.J. Holb, and I. Pócsi. "Secondary metabolites in fungus–plant interactions." *Frontiers in Plant Science* 6 (2015): 573.

Rai, P.K., M. Singh, K. Anand, S. Saurabh, T. Kaur, D. Kour, A.N. Yadav, and M. Kumar. "Role and potential applications of plant growth-promoting rhizobacteria for sustainable agriculture." In *New and Future Developments in Microbial Biotechnology and Bioengineering*, 49–60. Elsevier, 2020.

Romeiro, R.S. "PGPR and systemic induction of resistance against plant pathogens." *Summa Phytopathologica* 26 (2000): 177–184.

Rovira, A. "Plant root excretions in relation to the rhizosphere effect." *Plant and Soil* 7 (1956): 178–194.

Sahu, P.K., D.P. Singh, R. Prabha, K.K. Meena, and P.C. Abhilash. "Connecting microbial capabilities with the soil and plant health: Options for agricultural sustainability." *Ecological Indicators* 105 (2019): 601–612.

Sasse, J., E. Martinoia, and T. Northen. "Feed your friends: Do plant exudates shape the root microbiome?" *Trends in Plant Science* 23, no. 1 (2018): 25–41.

Schirawski, J., and M.H. Perlin, eds. *Plant Microbe Interaction 2017*. MDPI, 2018.

Semchenko, M., S. Saar, and A. Lepik. "Plant root exudates mediate neighbour recognition and trigger complex behavioural changes." *New Phytologist* 204, no. 3 (2014): 631–637.

Setiawati, T.C., and L. Mutmainnah. "Solubilization of potassium containing mineral by microorganisms from sugarcane rhizosphere." *Agriculture and Agricultural Science Procedia* 9 (2016): 108–117.

Shafi, J., H. Tian, and M. Ji. "Bacillus species as versatile weapons for plant pathogens: A review." *Biotechnology & Biotechnological Equipment* 31, no. 3 (2017): 446–459.

Shah, A., M. Nazari, M. Antar, L.A. Msimbira, J. Naamala, D. Lyu, M. Rabileh, J. Zajonc, and D.L. Smith. "PGPR in agriculture: A sustainable approach to increasing climate change resilience." *Frontiers in Sustainable Food Systems* 5 (2021): 667546.

Shanmugaiah, V., K. Nithya, H. Harikrishnan, M. Jayaprakashvel, and N. Balasubramanian. "Biocontrol mechanisms of siderophores against bacterial plant pathogens." In *Sustainable Approaches to Controlling Plant Pathogenic Bacteria*, Vol. 24, 167–190. 2015.

Sievers, A., M. Braun, and G.B. Monshausen. "The root cap: Structure and function." In *Plant Roots*, 79–103. CRC Press, 2002.

Singh, S., V. Kumar, S. Datta, P.C. Ramamurthy, and J. Singh. "Molecular mechanism and signaling pathways interplay between plant hormones during plant–microbe crosstalk." In *Microbial Management of Plant Stresses*, 93–105. Woodhead Publishing, 2021.

Somers, E., J. Vanderleyden, and M. Srinivasan. "Rhizosphere bacterial signalling: A love parade beneath our feet." *Critical Reviews in Microbiology* 30, no. 4 (2004): 205–240.

Spribille, T., P. Resl, D.E. Stanton, and G. Tagirdzhanova. "Evolutionary biology of lichen symbioses." *New Phytologist* 234, no. 5 (2022): 1566–1582.

Stachowicz, J.J. "Mutualism, facilitation, and the structure of ecological communities: Positive interactions play a critical, but underappreciated, role in ecological communities by reducing physical or biotic stresses in existing habitats and by creating new habitats on which many species depend." *Bioscience* 51, no. 3 (2001): 235–246.

Steenhoudt, O., and J. Vanderleyden. "*Azospirillum*, a free-living nitrogen-fixing bacterium closely associated with grasses: Genetic, biochemical and ecological aspects." *FEMS Microbiology Reviews* 24, no. 4 (2000): 487–506.

Tsuno, Yuki, Tomomi Fujimatsu, Keisuke Endo, Akifumi Sugiyama, and Jun Yazaki. "Soyasaponins: A new class of root exudates in soybean (*Glycine max*)." *Plant and Cell Physiology* 59 (2018): 366–375.

Upadhyay, Suresh Kumar, Ashok Kumar Srivastava, Vinay Dubey Rajput, Prashant Kumar Chauhan, Abhishek Anand Bhojiya, Deepak Jain, Gaurav Chaubey, Pooja Dwivedi, Bharti Sharma, and Tatiana Minkina. "Root exudates: Mechanistic insight of plant growth promoting rhizobacteria for sustainable crop production." *Frontiers in Microbiology* 13 (2022): 916488.

Upadhyay, S. K., V. D. Rajput, A. Kumari, D. Espinosa-Saiz, E. Menendez, T. Minkina, P. Dwivedi, and S. Mandzhieva. "Plant growth-promoting rhizobacteria: A potential bio-asset for restoration of degraded soil and crop productivity with sustainable emerging techniques." *Environmental Geochemistry and Health* 45, no. 12 (2023): 9321–9344.

Uren, N. C. "Types, amounts, and possible functions of compounds released into the rhizosphere by soil-grown plants." In *The Rhizosphere*, 35–56. CRC Press, 2000.

Vacheron, J., G. Desbrosses, M. L. Bouffaud, B. Touraine, Y. Moënne-Loccoz, D. Muller, L. Legendre, F. Wisniewski-Dyé, and C. Prigent-Combaret. "Plant growth-promoting rhizobacteria and root system functioning." *Frontiers in Plant Science* 4 (2013): 356.

Van Overbeek, L.S. and Van Elsas, J.D. "Root exudate-induced promoter activity in Pseudomonas fluorescens mutants in the wheat rhizosphere." *Applied and Environmental Microbiology*, 61(3) (1995): 890–898.

Vishwakarma, Kuldeep, Navneet Kumar, Chetan Shandilya, Shubhankar Mohapatra, Suman Bhayana, and Ajit Varma. "Revisiting plant–microbe interactions and microbial consortia application for enhancing sustainable agriculture: A review." *Frontiers in Microbiology* 11 (2020): 560406.

Walker, Thomas S., Harsh P. Bais, Erich Grotewold, and Jorge M. Vivanco. "Root exudation and rhizosphere biology." *Plant Physiology* 132 (2003): 44–51.

White, Lauren J., Xiaoyuan Ge, Volker S. Brözel, and Senthil Subramanian. "Root isoflavonoids and hairy root transformation influence key bacterial taxa in the soybean rhizosphere." *Environmental Microbiology* 19 (2017): 1391–1406.

Wiesmann, Claudia L., Nathan R. Wang, Yuan Zhang, Ziyan Liu, and Cara H. Haney. "Origins of symbiosis: Shared mechanisms underlying microbial pathogenesis, commensalism and mutualism of plants and animals." *FEMS Microbiology Reviews* 47 (2023): fuac048.

Wu, Renren. "Dynamic behaviors of a nonlinear Amensalism model." *Advances in Difference Equations* 2018 (2018): 1–13.

Xu, Zhihui, Huihui Zhang, Xiaoqian Sun, Ying Liu, Wenjing Yan, Wen Xun, Qirong Shen, and Renduo Zhang. "*Bacillus velezensis* wall teichoic acids are required for biofilm formation and root colonization." *Applied and Environmental Microbiology* 85 (2019): e02116–18.

Yadav, B.K., Akhtar, M.S. and Panwar, J. "Rhizospheric plant-microbe interactions: key factors to soil fertility and plant nutrition." In *Plant microbes symbiosis: applied facets* (2014): 127–145. New Delhi: Springer India.

Yu, Ying, Yanmei Gui, Zhen Li, Chunjiang Jiang, Jian Guo, and Dongsheng Niu. "Induced systemic resistance for improving plant immunity by beneficial microbes." *Plants* 11 (2022): 386.

Zahran, H. H. "Enhancement of rhizobia–legumes symbioses and nitrogen fixation for crops productivity improvement." In *Microbial Strategies for Crop Improvement*, 227–254. Berlin, Heidelberg: Springer Berlin Heidelberg, 2009.

Zhalnina, Kateryna, Kiera B. Louie, Zamin Hao, Negar Mansoori, Ulas N. Da Rocha, Shuai Shi, Hyunsoo Cho, Umut Karaoz, Dominique Loqué, Benjamin P. Bowen, and Mary K. Firestone. "Dynamic root exudate chemistry and microbial substrate preferences drive patterns in rhizosphere microbial community assembly." *Nature Microbiology* 3 (2018): 470–480.

Zhao, Baocai, Qing Liu, Bo Wang, and Fengyuan Yuan. "Roles of phytohormones and their signaling pathways in leaf development and stress responses." *Journal of Agricultural and Food Chemistry* 69 (2021): 3566–3584.

Zhao, Yaqing, Yuhua Wu, Yuan Fu, Xixi Zhao, and Jingguo Chen. "Activity pathogenic effect of cell wall degrading enzyme in tobacco target spot pathogen *Rhizoctonia solani*." *Tobacco Science and Technology* 11 (2014): 84–88.

18 Future Perspectives and Challenges

Navigating the Landscape of Plant–Microbe Interactions for Crop Improvement

Poonam Gupta, Siddhartha Barun, Pramod Kumar, Vindhesh Dixit, and Monika Asthana

18.1 INTRODUCTION

The global population has been increasing rapidly in the past decades. According to the United Nations Population Division, the global population is expected to reach approximately 9.6 billion by 2050—over 30% higher than the current population. This unprecedented increase in population has caused serious problems in the global food supply chain, impacting agricultural production, and eventually driving countries to implement intensive agricultural practices for increasing crop yield that are harmful to the environment. The non-sustainable farming methods caused major negative consequences, including the destruction of agricultural soil, desertification, and erosion; the release of greenhouse gases, pollution, and eutrophication of water reservoirs; and a significant decline in local microorganisms. Soil health and fertility are the most pertinent issues that are discussed in the agricultural industry. Most agricultural lands have developed problems of constant or even diminished productivity due to continuous and prolonged uses of chemical-intensive agricultural systems (Ahemad and Kibret 2014). To revive the agricultural business, there is an urgent need to develop alternative agricultural practices that will restore soil health and fertility. Several studies have been undertaken in recent decades, primarily focusing on the soil rhizosphere region, to determine the soil's diversity, density, and microbial population structure, and to anticipate potential approaches to restore soil health. Microbes in the rhizosphere perform niche activities such as disease suppression, toxicity balance, plant growth stimulation or promotion, cycling, and nutrient use. To find microorganisms or groups of organisms with niche roles, focused research should be conducted.

Plant roots develop symbiotic connections with mycorrhizal fungi and rhizobia, which supply vital nutrients like minerals and fixed nitrogen. Phytohormones also encourage plant growth, and through antagonistic interactions, they aid in disease defense. Since the beginning of plant life on Earth, these advantageous plant–microbe interactions have been diverse and ongoing. Many studies have been conducted till date to uncover the mechanisms behind the interactions between plants and microbes, which include the reaction of plants toward microbial colonization and how their biological functions are affected by certain microbial diseases and symbionts. Microbes and host plants have co-evolved and developed a variety of mechanisms that determine the outcome of their interactions. The plant–microbe interactions are highly complex and dynamic biological processes that are impacted by abiotic environmental elements like temperature and light (Ayaz et al. 2023).

DOI: 10.1201/9781003608080-18

Plant–microbe interactions show the relationships between plants and microorganisms. These interactions occur at several levels, from molecular and cellular interactions to ecosystem dynamics, and they can be advantageous, harmful, or neutral. Various types of interactions are listed in the following.

18.1.1 Mutualistic Associations

Some bacteria and fungi create mutually advantageous connections with plants. As an example, fungi called mycorrhizae associate themselves with the roots of plants. Accordingly, fungi draw water and nutrients from the soil, while plants give them space and carbohydrates (Ayaz et al. 2024).

18.1.2 Pathogenic Interactions

Microbes such as bacteria, fungi, viruses, and nematodes can cause illnesses in plants through these detrimental interactions. They may invade the plant, disrupt its growth, and cause symptoms ranging from leaf spots and leaf curling to wilting or death.

18.1.3 Endophytic Associations

Endophytic associations denote mutually beneficial interactions between plants and microorganisms, specifically fungi or bacteria, residing within plant tissues (Channarayappa and Biradar 2018). Growth-promoting endophytes can contribute to the plant's defense against pathogens, enhance nutrient uptake, or help in stress tolerance. These endophytes may improve plant health by improving stress tolerance, nutrient absorption, and defense against pathogens. Employing endophytes as biofertilizers, biopesticides, and agents for enhancing plant stress resilience offers a sustainable alternative to using chemicals.

18.1.4 Commensalism or Neutral Interactions

A neutral interaction is one where the presence of one organism does not significantly affect the other. Certain microorganisms occupy plant surfaces without causing harm or providing any benefits to the plant. These associations can be neutral, with minimal impact on the plant's well-being. The microbe gains some advantage from its association with the plant, but the plant experiences no positive or negative effects.

18.1.5 Allelopathy

Chemical interactions between plants occur when one plant releases compounds that affect the growth, development, or germination of another species. Through root exudation, leaching, or the breakdown of plant matter, these allelopathic substances may find their way into the environment. Allelopathy is the term for the process by which certain plants emit chemicals into the soil that prevent the growth of microorganisms or other competitive plants. The surrounding vegetation may be impacted by this occurrence in both beneficial and detrimental ways. Plants produce secondary metabolites called allelochemicals, which include phenolics, alkaloids, and terpenoids. Plant immunity and defense is a unique mechanism that plants have evolved and developed because of unfavorable plant–microbe interactions. Plants have developed sophisticated defense mechanisms to fight different pathogens. Plants recognize and respond to microbial invaders through various mechanisms, such as activating molecular pathways, switching off and on of defense genes, and the production of antimicrobial compounds. Since complex plant–microbe interactions have various implications and impacts on agriculture, understanding these interactions is critical for agriculture. Beneficial bacteria can be utilized as biocontrol agents and biofertilizers to promote plant development while also protecting against infections, minimizing the need for chemical treatments.

18.2 PLANT–MICROBE INTERACTIONS

Plants are continually interacting with a diverse community of microbes (Ejaz et al. 2023). These plant-associated microbes are characterized as follows.

18.2.1 Rhizosphere Microorganisms

These microbes dwell in the rhizosphere (underground) of plants, which is the narrow zone of soil surrounding plant roots that is directly controlled through root secretions and interacts with soil microbes. These microbes are of particular interest in terms of crop improvement. The rhizosphere is further classified as follows:

- **Outer Rhizoplane:** These microbes live in plants' rhizospheres (underground), which are small areas of soil surrounding the roots of plants and are directly controlled by root secretions. They also interact with soil microbes. This dynamic region is important for plant growth and development, as it facilitates the exchange of nutrients between the plant and soil microorganisms.
- **Inner Rhizoplane:** The inner rhizoplane is the root surface's deepest layer, which includes the root epidermis and the microorganisms that inhabit it.

18.2.2 Epiphytic Microorganisms

Microbes (bacteria, fungi, yeasts, and actinomycetes) that live on the surface of plants (aerial), particularly leaves (phyllosphere), stems, flowers, and fruits, without causing diseases. They form complex communities influenced by plant species, environmental conditions, and microbial interactions.

18.2.3 Endophytic Microorganisms

Endophytic microorganisms include bacteria and fungi that live asymptomatically inside plant tissues such as roots, stems, leaves, and seeds without causing diseases. They form mutualistic or commensal relationships with their host plants.

Plant–microbial interactions are heavily influenced by both biotic and abiotic variables. Soil structure, soil type, pH, salinity, moisture, exudates, and soil organic matter are all important considerations for below-ground plant components. External environmental conditions, temperature, and the presence of pathogens all impact the microbial concentration in above- and below-ground plant sections. Root shape, rhizosphere deposits, and exudates all play an important role in sustaining the plant microbiota.

18.3 IMPORTANCE OF PLANT–MICROBE INTERACTION

Various studies have confirmed that beneficial microbes enhance nutrient uptake, promote growth, and protect plants from diseases. Scientists and agriculturalists utilize these interactions to create biofertilizers, biopesticides, and biocontrol agents that minimize reliance on conventional fertilizers and pesticides while enhancing crop output and resilience (El Mujtar et al. 2019). Plant–microbe interactions are significant because they have possible applications in an array of industries.

18.3.1 Biofertilizers

Living microorganisms such as algae, bacteria, and fungi can act as biofertilizers, boosting nutrient availability in the soil and absorption by plants, thereby enhancing plant defense and soil fertility. They are essential for sustainable agriculture because they increase soil fertility, enhance nutrient cycling, and minimize the need for artificial fertilizers. Nitrogen is an essential component of plant growth. Some bacteria fix nitrogen from the environment and convert it to ammonia, which plants can utilize. Microorganisms may fix atmospheric nitrogen, making them effective biofertilizers (Tailor et al. 2023). Nitrogen-fixing bacteria from rhizobia and other groups, as well as *Azolla* and blue-green algae (BGA), contribute to maintaining soil nitrogen levels. Because it impacts so many growth regions, phosphorus is a necessary mineral for plant growth and development. Beneficial microorganisms known as phosphate-solubilizing microbes (PSMs) could hydrolyze both organic and inorganic insoluble phosphorus molecules, turning them into a soluble form that plants can readily absorb. This increases the amount of phosphorus available in the soil.

18.3.2 Stress Tolerance and Climate Resilience

Microbes help plants withstand several stresses, including drought, salt, and extreme temperatures. Drought-resistant microorganisms may live and grow in habitats with limited water supply or under drought conditions. These bacteria play a significant role in strengthening plant tolerance to water scarcity and are of particular interest in agriculture due to their potential contribution to drought-resistant crops. Drought-resistant microorganisms generate chemicals that assist plants to overcome the negative effects of drought stress (Ghazzal et al. 2022). Few microorganisms have adapted to extreme arid environments, like deserts, where water is a scarcity; certain bacteria and fungi create symbiotic interactions with plants, increasing their ability to withstand drought stress. These microorganisms help improve water uptake by plants, enhance nutrient absorption, and stimulate the plant's defense mechanisms against drought. Halophilic or halotolerant microbes are salinity-resistant microbes. These microorganisms thrive in environments with high salt concentrations. They have mutualistic relationships with salt-tolerant bacteria and fungi, which help plants adapt to the salt stress in the soil. Certain salt-tolerant bacteria can shield plants against soil-borne diseases that can flourish in salinized environments and have biocontrol capabilities.

18.3.3 Phytoremediation

Certain plants may store heavy metals or organic pollutants in their tissues, and certain microorganisms degrade or immobilize these toxins. These plants work with microbes to remediate contaminated soils. This entails using plants to absorb, accumulate, or modify toxins, frequently with the help of specialized bacteria. Using plants and their associated microbes for environmental cleanup is known as phytoremediation.

18.3.4 Bioenergy Production

Certain microbes associated with plants can breakdown the plant biomass, aiding in the production of biofuels. Scientists are studying these interactions to augment the breakdown of biomass for bioenergy purposes.

18.3.5 Bioremediation

Some plant-associated microbes can degrade pollutants present in the contaminated soil or water. Scientists produce and use these microorganisms for environmental cleanup, remediating damaged places with plants and their related microbes. Fungi are characterized by mycelial development, are more commonly utilized for bioremediation in soil, and could create extracellular enzymes in adequate quantities for bioremediation.

18.3.6 Pharmaceuticals and Biopharmaceuticals

Some plant-associated microbes produce bioactive compounds with potential pharmaceutical applications. These bioactives are being studied and used to produce novel drugs or drug intermediates.

18.4 MAJOR TAXONOMIC GROUPS OF MICROORGANISMS PRESENT IN THE SOIL

18.4.1 Fungal Microbiota

Fungi are heterotrophic, which means they cannot produce food and must rely on nutrients from other species. Fungi are chemoorganotrophs that obtain energy from the oxidation of organic materials. They associate with plants in the following ways:

A **Mycorrhizae:** Mycorrhizae are fungi with symbiotic interactions with vascular plant roots. Most plant species can develop mycorrhizae. Mycorrhizae are often classified into the following three types:
- **Ectomycorrhizae (ECM):** Individual cells inside the root are not penetrated by the hyphae of ectomycorrhizal fungus. The root tip is surrounded by a vascular sheath or mantle, and the plant cells inside the inner portion of the root are surrounded by a network of hyphae (Raez et al., 2011; Gomathi et al., 2023).
- **Endomycorrhizae/Arbuscular Mycorrhizae:** Endomycorrhizal fungal hyphae infiltrate the cell wall and membrane. It is further divided into several subcategories, including arbuscular mycorrhizae (AM). Arbuscular mycorrhizal fungi (AMF) act as a link between plants and soil, facilitating nutrient exchange through their unique symbiotic relationships with plants. AMF are also known as natural biofertilizers because they offer water, nutrients, and pathogen protection to the host plant in the absence of photosynthetic products.
- **Ectoendomycorrhizae:** Theses fungi have characteristic features of both ECM and AM.

B **Fungal Endophytes:** Fungal endophytes vastly decide and determine the ecology of plants, fitness, and evolution of plants in the present environment, shaping plant communities and showing strong effects on the community structure and diversity of associated organisms (Gupta et al. 2022).

18.4.2 Bacterial Microbiota

There is a broad range of bacteria present in soil that could develop association with plants in various ways. The major players are the following:

- **Rhizobacteria:** These bacteria are linked to the root surface. One of these is particularly interesting: Plant growth-promoting rhizobacteria (PGPR). When PGPR are cultivated alongside a host plant, it stimulates plant growth, increases plant defense, and increases overall plant production (Gupta et al. 2014). PGPR must be able to colonize the root surface and live, proliferate, and compete with other microbiota, at least for the time it takes to express plant growth promotion or protective activities (Havlin et al. 2016).
- **Endophytic Bacteria:** Microbes found within a plant might be dangerous, neutral, or beneficial to the plant that harbors them. These bacteria may establish a variety of associations with host plants, including mutualistic, symbiotic, and commensalistic. Endophytic bacteria grow on plants without hurting them. These bacteria originate from

the rhizosphere or phyllosphere. They stimulate plant development and yield and may serve as biocontrol agents. Bacteria that can colonize and establish themselves in host plants benefit from the endophytic niche, which protects them from the environment (Idodo-Umeh et al. 2010).

18.5 POTENTIAL APPLICATIONS OF MICROFLORA IN CROP IMPROVEMENT AND MANAGEMENT

Plant–microbe interaction plays a crucial role in crop improvement and management. Its significance could be understood and appreciated in detail in the following subsections.

18.5.1 Improvement of Crops under Abiotic Stress

In recent decades, environmental changes have increased exponentially, posing serious risks to agricultural output and crop management. As a result, there is an increasing desire for long-term solutions to various stressors, and microorganisms might be regarded as a viable alternative to pesticides by leveraging microbes' capacity to resist environmental stress (Juo and Franzluebbers 2003).

Under current climatic conditions, plant growth-promoting microorganisms (PGPMs) have immense potential in sustainable agriculture and are currently the most explored. The bacteria in the rhizosphere known as PGPR aid in the development of plants. There are two types of PGPMs: Symbiotic PGPMs, which often infiltrate rhizospheres or internal plant components, and free-living PGPMs, which are found in the rhizosphere (Kumar et al. 2022). Endophytes are plant microbes that live in plant tissues and do not cause diseases. Bacteria, fungi, and actinomycetes can coexist in the same area. These organisms help plants survive in physical and physiological stresses, and the harmful effects of chemicals and bioactive compounds.

As previously noted, virtually all plant species can create mycorrhizal associations, which formed symbiotic relationships between fungi and plant roots. Many agricultural soils have elevated amounts of AM and ECM. In this symbiotic connection, fungi meet their nutritional requirements by forming an arbuscule-like structure in the root system that aids in nutrient absorption; in exchange, fungi build a hyphal network in the soil that encourages development. As a result, the mycorrhizal link aids in the reduction of plant stress.

Bacteria and fungi (e.g., AMF) are regarded as helpful soil microorganisms because they aid in the development/increased tolerance of plants to stressful conditions caused by abiotic causes. Microbes that live in specific plant sites (e.g., endophytes like bacteria) can also assist plants to evolve to changing environmental conditions.

18.5.1.1 Plant Stress *via* Abiotic Factors

Both biotic and abiotic stresses have an impact on plant development in agricultural soil. Abiotic factors affect plant growth. In agricultural soil, plant growth is impacted by both biotic and abiotic stressors. Water, light, carbon atoms, and mineral nutrients are examples of abiotic elements that have an impact on plant growth, development, and propagation. In order to adjust to abrupt or significant changes in their surroundings, plants go through physiological and morphological modifications (Kumar et al. 2022, 2023).

Plant development and agricultural productivity are significantly impacted by a variety of severe abiotic stresses, including drought, heat stress, cold stress, salt, water logging, and many more. Several factors influence crop development and output, including nutritional depletion, heavy metal deposition, and temperature fluctuations. Depending on the crop variety and the climate in which it is grown, crop yield can decrease by 20%–50%. Major abiotic stresses are drought, salinity, and heavy metal deposition in soil.

18.5.1.2 Growth-Promoting Bacteria's Function in Supplying Resistance to Abiotic Stress

Development-promoting microorganisms not only promote plant development, nutrition management, and disease control, but they also help plants survive in severe environments.

Rhizosphere bacteria aid in the generation of antioxidants and cytokinins, which raises abscisic acid (ABA) levels and prevents free radical formation. Several studies have found that higher levels of antioxidant enzymes are related to increased cell resistance to oxidative stress.

18.5.1.2.1 Resistance to Drought

Three forms of drought are distinguished: Atmospheric, soil, and physiological. Drought is defined as an extended time of no precipitation. Drought damages plants by depleting soil moisture or lowering atmospheric relative humidity. This lack of soil and air moisture can lead to a variety of problems, including mechanical damage, damage to the membrane and membrane system, changes in membrane permeability, uneven water distribution throughout the plant, and, in severe cases, the breakdown of normal metabolic processes. Drought resistance is the capacity of plants to sustain normal or nearly normal metabolic levels required for growth and development during drought conditions. The role of PGPMs in drought tolerance, particularly in relation to plant adaptation, has been studied. When there is drought, water transports nutrients to the roots because it reduces their supply. Plants suffering from drought exhibited significant improvements in photosynthetic rate, water use efficiency, and chlorophyll content when infected with endophytic bacteria.

18.5.1.2.2 Resistance to Salinity

One important environmental factor that hinders plant growth and development and lowers agricultural productivity is salinity stress. It happens for several reasons when there is a high concentration of salt in the soil. The quantity of dissolved salt in water is called salinity, and it affects many aspects of natural water chemistry and biological activity. It also greatly increases conductivity (Lata and Gond 2019).

In the presence of PGPMs, it promotes the synthesis of ions, redox-active chemicals, osmolytes, and polyamines, which aid plants in overcoming salinity stress. Salinity and AMF (*Glomus*) impacted plant growth during salt stress. Plants inoculated with AMF (*Glomus*) showed dramatically higher growth under salt stress. Polysaccharides generated by PGPMs can also bind soil particles together to create micro and macro aggregates. They release a variety of exopolysaccharides that increase tolerance to water and salinity stress, enhancing soil structure.

18.5.1.2.3 Sequestration of Heavy Metals

Heavy metals in toxic concentrations affect the health of plants, animals, and humans. Heavy metal sequestration is the process of eliminating toxic heavy metals from the environment. Certain plants have developed the ability to retain heavy metals in their shoots while being unharmed by their toxic effects (Lichtfouse 2016). Some additional plants can absorb heavy metals through their roots (Malgioglio et al. 2022), using both the apoplastic (passive diffusion) and symplastic (active transport) pathways. These metals are then immobilized in the plant's extracellular and intracellular regions. The presence of beneficial bacteria and PGPMs that survive in trace metal-contaminated soils is essential for phytoremediation and sequestration to be successful. Plants produce acids, proteins, phyto-antibiotics, and other compounds that counter the harmful effects of heavy metals (Mine et al. 2014).

18.5.2 Improvement in Soil Ecosystem

For growing crops, soil is of utmost importance. Hence, it is safe to say that there will be very little or no food availability in the absence of soil. Soil organisms are flora and fauna that spend all or part of their lives in the soil. Their role is very important in the overall preservation of soil fertility through activities such as accumulation of soil organic matter, soil aggregation, and decomposition of organic matter, which helps in the release of essential nutrients required by higher plants (Juo and Franzluebbers 2003). Soil shelters a vast range of microflora, which may be useful, neutral, or harmful to plants. Beneficial soil microflora is a major component of agricultural ecosystems because they actively promote soil fertility and crop yield while also contributing to primary soil processes (Nayak et al. 2017).

Soil has multiple functions, including the following:

- Nutrient recycling
- Plant and agricultural growth
- Habitat for microbes, water storage
- Water-holding
- Organic matter degradation.

The nutritional value of soil is determined by several aspects, including the amount and type of microorganisms in the soil, the texture of the soil, the location where microorganisms are prevalent, and the various abiotic factors that contribute to that ecosystem. Soil microorganisms include a variety of bacteria, fungi, algae, actinomycetes, protozoa, and nematodes. Several researchers have suggested using soil microorganisms as bioindicators of soil health and activity (Prashar and Shah 2016). The limitless use of chemical fertilizers, herbicides, and insecticides over the last 50 years has resulted in serious degradation of soil fertility, reducing its nutritional value, disrupting soil microflora, and eventually deteriorating soil quality. Fertilizers and pesticides are persistent groups of chemicals that are bound to impact soil microorganisms, disrupting soil health (Rachel et al. 2015). When soil quality decreases, it automatically inhibits microbial activity in the soil, disrupting the natural recycling of nutrients and balancing the ecosystem, making it much more difficult for the soil to regain the vital features required for optimal crop production through agriculture. Thus, soil health management is crucial for ensuring long-term agricultural productivity and biodiversity protection. Microbial inoculants, extracts, biofertilizers, and bio-stimulants, among other things, can help restore degraded soils.

18.5.2.1 Nutrient Recycling

Soil acts as a storehouse for nutrients. Nutrient recycling in agriculture is the process of reusing nutrients from waste materials to replenish soil nutrients that crops remove; it is an integral part of sustainable agriculture. Nucleic acids, amino acids, fatty acids, carbohydrates, minerals, and other nutrients can all be found organically in the environment.

Plants cannot absorb nutrients in their organic form; so, these nutrients need to be broken up into smaller monomeric units, making them more accessible and easily absorbed by plants. The three most important soil-derived minerals required for agriculture are as follows:

- **Nitrogen (N):** It is an element of both amino acids (protein-building blocks) and nucleic acids. Plants can obtain nitrogen in two forms: Ammonium (NH_4^+) and nitrate (NO_3^-).
- **Phosphorus (P):** It is an element of nucleic acids (such as DNA or ATP, the "energy" molecule in cells) and coenzymes. It is accessible to plants as phosphate (PO_4^{3-}).
- **Potassium (K):** It functions as a cofactor for several enzymes and helps regulate cell turgor. It is available to plants as an ion (K+).

A specific set of microbes (particularly saprophytes and fungi) found in soil, typically associated and aggregated, perform the specialized function of breaking down complex organic matter into useful simpler inorganic material using various hydrolytic enzymes secreted by these microbes. They are also known as decomposers, and they play a crucial role in the food web of any ecosystem.

The role of decomposers is indispensable for maintaining the nutrient pool of the soil; otherwise, the Earth will turn into a huge pile of waste. Decomposers complete the nutrient cycle by returning essential molecules back to soil and water to be used by the plant producers afterward. These microorganisms degrade lipids, carbohydrates, and nitrogenous substances into basic molecules such as CO_2, water, ammonia, and hydrogen sulfide.

Fungi and other soil bacteria degrade waste substances, transforming them into more resistant, complex organic molecules known as humus. The process is called humification. Humus is a complex, stable organic molecule that is an important part of soil. Fungi play a major role in humus formation, but it is also influenced by other factors such as moisture, temperature, pH value, and the variety of bacteria found in the soil. Humus improves soil fertility, and its turnover rate ranges between 2% and 5% each year. It improves soil structure, increases water retention, and delivers important minerals such as phosphorus and nitrogen, all of which contribute to plant health. Hence, the fertility of the soil must always be maintained. It is important to understand the formation of humus for sustainable agriculture because it helps in maintaining soil health and promoting biodiversity. Understanding humus formation is vital for sustainable agriculture, as it aids in the maintenance of soil health and promotion of biodiversity.

In addition to nutrient recycling, soil microbiota could improve soil health in following ways:

- Improve soil structure by binding the soil particles together to form micro aggregates, with the help of secretions. This in turn enhances the water filtration and water-holding capacity of the soil.
- Remove and degrade the harmful chemicals by the process of bioremediation.
- Promote the growth of plants that help in degrading the harmful chemicals by a type of bioremediation called phytoremediation.
- Stabilize the soil by forming biocrust communities.
- Produce important minerals like carbonates, phosphates, and calcium by biomineralization.

18.5.3 Plant Growth and Development

Microflora, or microorganisms, play a vital role in agriculture and plant growth in many ways, including biological nitrogen fixation, soil fertility, biocontrol agents, abiotic stress relief, and soil health (some of which are already discussed). Several studies have shown that soil microbes associated with plant rhizospheres stimulate the growth and productivity of their hosts through a variety of indirect as well as direct mechanisms, including the synthesis of phytohormones, siderophores, and/or mineral solubilization (Raez et al. 2011). Some common ways in which this plant–microbe interaction helps in plant growth and development are described in the following subsections:

18.5.3.1 Microbial Biofertilizers

To combat the dangers of chemical fertilizers, scientists have developed an environmentally benign alternative that uses bacteria and their secretions as fertilizers. Biofertilizers are natural materials derived from microbes that serve as an efficient component of sustainable agriculture, improving soil fertility and plant growth by increasing nutrient availability. Soil microbes cooperate with one another and with plant roots in numerous ways by providing extensive cooperation, which is vital for sustaining the soil ecosystem. Biofertilizers encourage plant growth by simply solubilizing and delivering nutrients to the plants. The use of biofertilizers in agriculture enhances the absorption of nutrients, plant growth, disease defense, and tolerance of abiotic stressors such as drought or salinity.

18.5.3.1.1 Action Mechanisms

18.5.3.1.1.1 Phosphate Solubilization and Mobilization

Phosphorus (P), which is plentiful in nature, is a necessary macronutrient for healthy plant growth and development. Phosphorus is necessary for plant growth since it is a constituent of DNA, RNA,

ATP, phospholipids, and other metabolic components. It is naturally available in the environment in two forms: Inorganic state (present in rocks and sediments) and organic state (present in biotic components and soil). Phosphorus cannot be directly absorbed and utilized by plants since it is fixed in its original condition. Before it can be used by plants, it must be solubilized and mineralized. Phosphate mobilization is the process by which phosphorus can become available to plants and other organisms.

The process of solubilization and mineralization of inorganic phosphorus takes place in different manners. Rhizosphere bacterial and fungal groups release acids like citric acid and gluconic acid, which aid in the conversion of insoluble phosphorus into soluble forms. Mineralization occurs *via* the activity of enzyme phosphatases. Rhizosphere microorganisms have a close connection with phosphorus uptake in plants and are thought to be potential agents to mitigate phosphorus shortage.

Phosphate-Solubilizing Bacteria:

- *Bacillus megaterium* and *Pseudomonas striata*: Used in crops like wheat and maize.

Phosphate-Solubilizing Fungi:

- *Aspergillus niger* and *Penicillium* spp.

18.5.3.1.1.2 Nitrogen Fixation

At 78%, nitrogen (N) is the element that is most prevalent in the atmosphere. The process by which atmospheric nitrogen gas is transformed into a form that plants can utilize, such ammonia, is known as sodium fixation. Nitrogen is a basic building block of amino acids, DNA, RNA, and nucleic acids, making this process important to the survival of all living things. Most biological things, unfortunately, cannot use nitrogen gas (N_2). A nitrogen shortage can kill microbes, plants, and animals when they are surrounded by N_2 that they cannot use. Ammonia (NH_3) is the type of nitrogen that all living things use to make proteins, amino acids, nucleic acids, and other nitrogen-containing substances that are essential to life. In the absence of nitrogen, soil fertility and plant development efficiency are significantly affected. The process that converts inert K^+ into physiologically usable NH_3 is known as biological nitrogen fixing. In nature, only nitrogen-fixing rhizobial bacteria mediate this process. The process by which certain kinds of bacteria transform ammonium into nitrate is called nitrification. Either as symbionts or in their free-living form, these microorganisms carry out the conversion process. Lichens, free-living soil bacteria, and blue-green algae (a type of bacterium) are only a few examples of the various ways that biological nitrogen fixation occurs in nature. While most agricultural systems do not benefit from these forms of nitrogen fixation, they nevertheless provide natural ecosystems with substantial amounts of NH_3 . *Rhizobium* is the most significant microorganism that carries out nitrogen fixation and the first to be examined. *Rhizobium* can be found in the roots of plants, notably those from the Leguminosae family. These soil bacteria convert atmospheric N_2 into ammonia (NH_3) by producing nodules in the taproot region of plants.

Then, this ammonia undergoes a number of different transformations before being absorbed by the plants and eventually transformed into organic molecules. DNA, amino acids, other plant proteins, and other essential components of plant cells are subsequently formed by these molecules. Although in smaller quantities, some free-living bacteria can also fix nitrogen, which helps plants grow by adding the element to the soil and plants (Rachel et al. 2015).

Symbiotic Nitrogen-Fixers:

- *Rhizobium* spp.: Forms root nodules in legumes (e.g., *Rhizobium leguminosarum* in peas)
- *Frankia* spp.: Forms nodules with actinorhizal plants like *Alnus*

Associative Nitrogen-Fixers:

- *Azospirillum* spp.: Associates with cereals (e.g., *Azospirillum brasilense* in maize)

Free-Living Nitrogen-Fixers:

- *Azotobacter* spp.: Free-living N_2-fixer for various non-legumes (e.g., *Azotobacter chroococcum*)

18.5.3.1.1.3 Mycorrhizal Symbiosis
Fungi form symbiotic associations with plant roots, improving nutrient and water uptake, e.g., AMF: *Glomus intraradices* and *Gigaspora* spp., and enhance phosphorus and micronutrient uptake.

18.5.3.1.1.4 Production of Plant Growth-Promoting Substances
Microbes produce hormones like auxins, cytokines, GA, and 1-aminocyclopropane-1-carboxylic acid (ACC) deaminase to enhance plant growth, e.g., *Azospirillum* spp., *Pseudomonas fluorescens*, and *Bacillus subtilis.*

18.5.3.1.1.5 Siderophore Production
Certain microbes produce siderophores that bind and solubilize iron, making it available to plants, e.g., *Pseudomonas* spp. and *Bacillus* spp.

18.5.3.1.1.6 Cyanobacteria and Algal Associations
In paddy fields, cyanobacteria fix atmospheric nitrogen and improve soil fertility, e.g., *Anabaena* spp., *Nostoc* spp., and *Anabaena azollae* (symbiont in *Azolla*).

18.5.3.2 Microbes as Biological Control Agents (Microbial Biopesticides)

Phytopathogens pose a serious threat to crop productivity worldwide. They infest crops and cause various diseases, which lead to huge loss in productivity every year. Apart from using chemical pesticides and insecticides to control pathogen attack, the agricultural industry is looking forward to a sustainable and environmentally friendly crop management system to control crop diseases. The process of using one species of a biological system to remove another is referred to as biological control. Numerous phytopathogens, including bacteria, viruses, nematodes, fungi, and oomycetes, are frequently present in plants. These pathogens could drastically lower the production of key crops across the globe; different pathogenic illnesses can cause yearly crop yield losses of 20%–40%. Chemical pesticides have been successful in managing several illnesses in important crops, but their overuse has negative impacts on the environment and human health, which deters the use of pesticides in the agricultural sector (Ryan et al. 2008; Sharma and Zheng 2019). Fungi have a vital role in avoiding severe illnesses in key crops; thus, they are effectively employed to protect crops from a range of pests. These fungi-based biological controls have a strong antagonistic effect on soil and hovering plant diseases, making them suitable biopesticides for the field or greenhouse. The severity of certain crop diseases can be lessened and the disease safely managed by the biological management of phytopathogens. A variety of tactics are employed by biocontrol agents to shield plants from disease invasion. Biological control agents (BCAs) use a range of compounds, including lipopeptides, biosurfactants, bacteriocins, volatiles, and antimicrobial enzymes, to compete for resources and space in the rhizosphere and prevent pathogen pathogenicity. For several reasons, including controlling a variety of plant diseases and pathogens that cause illness, fungi are powerful BCAs. Mass production of insect-parasitic fungal spores is very simple, and they work well with other BCAs like bacteria. Because many fungi have a wide host range, they may frequently be used to manage several pests.

18.5.3.2.1 Action Mechanisms of Microbial Biopesticides

A. **Pathogenesis (Infection of Target Pest/Pathogen):** Microbes infect and kill pests or pathogens by colonizing their body or tissues. For example, *Bacillus thuringiensis* (BT) produces Cry and Cyt toxins (insecticidal toxins) that disrupt the midgut of insect larvae (Lepidoptera, Diptera, and Coleoptera), and these toxins are widely used in crops like cotton, maize, and vegetables.

B. **Antibiosis:** This refers to the production of antimicrobial compounds that inhibit or kill plant pathogens. For example, *Pseudomonas fluorescens* produces antibiotics such as 2, 4-diacetylphloroglucinol (DAPG), pyoluteorin, and hydrogen cyanide (HCN) to suppress fungal diseases.

C. **Competition for Nutrients and Space:** Microbes outcompete pathogens by occupying their roots or phyllosphere niches and consuming available resources. For example, *Trichoderma* spp. compete with pathogenic fungi in the rhizosphere, preventing their colonization.

D. **Parasitism/Predation:** Some microbes directly parasitize or prey on plant pathogens or pests. For example, *Trichoderma harzianum* parasitizes fungal pathogens like *Rhizoctonia* and *Fusarium* through coiling and enzymatic degradation.

E. **Induced Systemic Resistance (ISR):** Microbial colonization primes the plant's immune system to respond more rapidly to pathogens. For example, *Bacillus subtilis* and *Pseudomonas* spp. can trigger ISR, enhancing the plant's defense enzymes and secondary metabolites.

F. **Virus-mediated Pest Control:** Insect-specific viruses infect and kill target insect pests, e.g., *Nucleopolyhedrovirus* (NPV) and *Granulovirus* (GV) used against *Helicoverpa armigera* and *Spodoptera litura* larvae.

18.6 REGULATION OF AGRICULTURAL PRACTICE FOR SUSTAINABLE SOIL MICROBIAL GROWTH

Agriculture involves a rich interplay between plants and microbes. One of the most pressing concerns confronting the agriculture business is satisfying the world's expanding food demand while utilizing limited resources. With current agricultural techniques such as intense tillage, excessive pesticide usage, continuous monocropping, and deforestation for additional farmland, we are on the verge of irreversibly ruining soil–microbial community structure and soil sustainability. The bonding between soil and associated microorganisms plays an important role in integrated nutrient management and disease and pest management, and in raising healthy and remunerative crops without compromising agricultural production. These excessively intensive agricultural practices have had a substantial influence on soil efficiency and the diversity of microorganisms that reside there. It is challenging to replenish and rebuild the beneficial microbiota after it has been disrupted. The ability of the soil to form soil aggregates; sustain soil structure with proper porosity, aeration, nutrient availability, and cycling; and ultimately transfer the nutrients to crops are significantly affected in the absence of these bacteria, whether they are found in the rhizosphere, phyllosphere, or endophytic zone. For long-term sustainable and environmentally friendly agricultural operations, a number of beneficial management methods must be addressed and applied. A few issues are presented in the following.

18.6.1 Monitoring and Maintenance of Soil Nutrients

Soil nutrient is the amount of nutrients in chemical forms accessible to plant roots. Nutrients are essential to support life, and soil is a major source of these nutrients required by plants for growth. These nutrients play a dual role: On the one hand, they are needed by microflora and crops for growth, and on the other hand, they are crucial for maintaining soil health and fertility (Havlin et al. 2016). Thus, soil productivity and nutrient availability are interrelated. With continuous cropping and harvesting, the amount of soil nutrients gradually depletes, and this causes decline in the productivity of plants/crops. Nutrients, either in excess or deficient amounts, adversely affect the microflora. Nutrient deficiency results in the development of toxic compounds and causes an imbalance in soil pH, thereby allowing for the growth of pests and a decrease in organic matter present in the soil, among other harmful effects.

The availability and management of nutrients are evaluated by conducting tests and using the observations to predict plant responses. To optimize the availability of nutrients for plants, it is essential to understand the nutrients present in the soil, and add, maintain, and recycle adequate amounts of appropriate nutrients into the soil at regular time intervals. The state of nutrients needs to be carefully considered; otherwise, either they will be lost in the atmosphere if present in a volatile state or they will run-off in water, causing water pollution if present in a highly soluble state.

Therefore, it becomes highly important to understand the soil type, the types of microflorae dwelling in it, the type of crops to be cultivated, and the type and amount of nutrients required by the crops for proper growth and maturation. Sustainable farming techniques to maintain soil nutrients and increase agricultural production include crop rotation, the use of compost and manure, and the use of cover crops and crop residues.

18.6.2 Cover Crop

Cover crops are plants that help protect and improve soil quality. These crops are not designed for harvesting but can be added to the soil for enrichment. In an agroecosystem, cover crops assist in reducing soil erosion, promoting soil health, increasing water availability, limiting mineral leaching, smothering weeds, and managing pests and diseases, and thereby supporting biodiversity. These crops are mostly planted as green manure crops and animal feed crops to combat drought in rainfed locations, and they can be grown in a variety of ways, including intercropping with plantation crops, rotating with cash crops, or when no regular crops are grown. This aids in erosion management, soil organic matter addition, soil nutrient retention, and pest and insect control. They can be annual, biennial, or perennial, and either leguminous or nonleguminous, depending on the soil's nutrition and organic matter requirements. Cover crop biomass is a source of organic matter that helps minimize compacting by increasing soil pores, aggregation, and nutrient recycling.

18.6.3 Managing Soil Organic Matter

Organic matter in soil consists of partially degraded plants and animal remnants, microbial organisms, and compounds formed at various phases of decomposition. It provides nutrients and energy to plants and food for soil bacteria. Plant tissue accounts for the great bulk of soil organic matter. Plant remnants contain 60%–90% moisture. Organic matter is present in soil particles, which are surrounded and formed by microflora. Organic matter decomposes, releasing nutrients that are easily consumed by plants. Under typical circumstances, the nutrient cycling system is maintained by balancing the rate of organic matter input with the rate of decomposition, while also accounting for plant nutrient intake and nutrient loss owing to erosion and leaching. To do this, green crop wastes, compost, and manure must be applied to the soil on a regular basis. Analyzing soil organic matter concentration on a regular basis is one of the methods for ensuring appropriate levels of organic nutrients in the soil. Maintaining soil organic matter is critical for long-term agricultural output.

18.6.4 Conserving Soil Biodiversity and Justified Use of Pesticides

Soil biodiversity helps provide ecosystem services such as nutrient cycling, water filtering, insect control, carbon storage, and soil stability, all of which are beneficial to human health (Stephenson

2019). Excessive usage of chemical pesticides and insecticides has a direct detrimental impact on both beneficial and nontargeted insects and bacteria in the soil. A novel biological approach to biodiversity conservation might benefit beneficial microflora, other creatures, and sustainable agriculture. Soil biodiversity is crucial for agriculture. Every chemical reaction occurring in the soil requires active participation from each of the microbes. They actively contribute to soil fertility by participating in the cycle of nutrients such as carbon and nitrogen, both of which are essential for plant development (Tokas et al. 2023).

18.6.5 Crop Rotation

Crop rotation is the process of producing various crops over successive seasons in the same place. It is largely used to promote soil health by preserving the physical, chemical, and biological composition of the soil and decreasing the hazards caused by weeds and pests, resulting in increased yields. Due to crop rotation, the opportunity of pests and pathogens to establish over the period is lost. This aids in destroying disease-causing populations of pests and ensuring good health of crops. Therefore, crop rotation is an integral part of pest management.

Crop rotation also helps in minimizing excess depletion of specific nutrients from soil and prevents soil erosion, thereby ensuring healthy soil structure.

18.7 CONCLUSION

The population of the world is growing at an alarming pace, and this has presented us with the daunting task of satisfying diverse food requirements of the vast populace and mitigating global hunger; hence, the pursuit of sustainable agriculture is imperative. Understanding soil mechanisms and plant–microbe interactions is necessary for employing sustainable agricultural practices that will benefit plants and crops in the long run. There are many types of bacteria (endophytic, epiphytic, and rhizospheric) and fungi (endophytic and mycorrhizal) that have been shown to benefit plant growth. They contribute to soil nutrient replenishment, disease management, soil health improvement, stress tolerance, and so forth. The existence of suitable microflora concentrations is critical for managing soil quality and plant health. Plants, in turn, can impact ecological changes by releasing secondary metabolites into the rhizosphere, which can attract or inhibit microbes. Metagenomic studies of the beneficial plant-associated microbiome, together with modern molecular technologies, must be used to regulate the microbial population with increasing crop yield. In addition to these advanced approaches, better research-based scientific procedures, better farm management practices, and sustainable soil and agri-management practices need to be designed and adopted for long-term sustainable agricultural production.

ACKNOWLEDGMENTS

The authors would like to express their sincere gratitude to all those who supported the successful completion of this chapter. The collaborative efforts of each author are acknowledged: Dr Poonam Gupta, Dr Siddhratha Barun, and Dr Pramod Kumar for conceptualizing the topic, reviewing relevant literature, and compiling data; and Vindhesh Dixit and Dr Monika Asthana for contributing to the writing and for their constant feedback and inputs. Special thanks to our institutions and colleagues for providing the necessary resources and academic environment that facilitated this work.

REFERENCES

Ahemad, M., and M. Kibret. 2014. "Mechanisms and Applications of Plant Growth Promoting Rhizobacteria: Current Perspective." *Journal of King Saud University—Science*. 26 (1): 1–20. https://doi.org/10.1016/j.jksus.2013.10.001.

Ayaz, M., C.-H. Li, Q. Ali, W. Zhao, Y.-K. Chi, M. Shafiq, F. Ali, X.-Y. Yu, Q. Yu, J.-T. Zhao, J.-W. Yu, R.-D. Qi, and W.-K. Huang. 2023. "Bacterial and Fungal Biocontrol Agents for Plant Disease Protection: Journey from Lab to Field, Current Status, Challenges, and Global Perspectives." *Molecules*. 28 (3): 1–22. https://doi.org/10.3390/molecules28030642.

Ayaz, M., J.-T. Zhao, W. Zhao, Y.-K. Chi, Q. Ali, F. Ali, A. R. Khan, Q. Yu, J.-W. Yu, W.-C. Wu, R.-D. Qi, and W.-K. Huang. 2024. "Biocontrol of Plant Parasitic Nematodes by Bacteria and Fungi: A Multi-Omics Approach for the Exploration of Novel Nematicides in Sustainable Agriculture." *Frontiers in Microbiology*. 15: 1–18. https://doi.org/10.3389/fmicb.2024.000001.

Channarayappa, C., and D. P. Biradar. 2018. *Soil Basics, Management and Rhizosphere Engineering for Sustainable Agriculture*. Boca Raton: CRC Press.

Ejaz, U., S. M. Khan, N. Khalid, Z. Ahmad, S. Jehangir, Z. Fatima Rizvi, L. H. Lho, H. Han, and A. Raposo. 2023. "Detoxifying the Heavy Metals: A Multipronged Study of Tolerance Strategies against Heavy Metals Toxicity in Plants." *Frontiers in Plant Science*. 14: 1–18. https://doi.org/10.3389/fpls.2023.000001.

El Mujtar, V., N. Muñoz, B. Prack Mc Cormick, M. Pulleman, and P. Tittonell. 2019. "Role and Management of Soil Biodiversity for Food Security and Nutrition; Where Do We Stand?" *Global Food Security*. 21: 1–10. https://doi.org/10.1016/j.gfs.2019.03.001.

Ghazzal, M., M. I. Hussain, Z. I. Khan, M. Habib ur Rahman, A. A. El-Habeeb, and H.-H. Yang. 2022. "Chromium Poisoning in Buffaloes in the Vicinity of Contaminated Pastureland, Punjab, Pakistan." *Sustainability*. 14 (9): 1–12. https://doi.org/10.3390/su14092635.

Gomathi, S., V. Ambikapathy, and A. Panneerselvam. 2023. "Biodegradation and Bioaugmentation of Pesticides Using Potential Fungal Species." In *Plant–Microbe Interaction—Recent Advances in Molecular and Biochemical Approaches*, edited by P. Swapnil, M. Meena, Harish, A. Marwal, S. Vijayalakshmi, and A. Zehra, 79–94. Cambridge, MA: Academic Press.

Gupta, A., U. Singh, P. Sahu, S. Paul, A. Kumar, D. Malviya, S. Singh, P. Kuppusamy, P. Singh, D. Paul, D. Rai, H. Singh, M. Manna, T. Crusberg, A. Kumar, and A. Saxena. 2022. "Linking Soil Microbial Diversity to Modern Agriculture Practices: A Review." *International Journal of Environmental Research and Public Health*. 19 (18): 1–19. https://doi.org/10.3390/ijerph191810123.

Gupta, P., B. Puniya, S. Barun, M. Asthana, and A. Kumar. 2014. "Isolation and Characterization of Endophytes from Different Plants: Effects on Growth of *Pennisetum typhoides*." *Biosciences Biotechnology Research Asia*. 11 (2): 1–8. https://doi.org/10.13005/bbra/1311.

Havlin, J. L., S. L. Tisdale, W. L. Nelson, and J. D. Beaton. 2016. *Soil Fertility and Fertilizers*. Delhi: Pearson Education India.

Idodo-Umeh, G., and A. E. Ogbeibu. 2010. "Bioaccumulation of the Heavy Metals in Cassava Tubers and Plantain Fruits Grown in Soils Impacted with Petroleum and Non-Petroleum Activities." *Environmental Monitoring and Assessment*. 160 (1–4): 1–12. https://doi.org/10.1007/s10661-008-0736-2.

Juo, A., and K. Franzluebbers. 2003. *Soil Biology and Microbiology*. Boca Raton: CRC Press.

Kumar, M., S. Ahmad Poonam, and R. P. Singh. 2022. "Plant Growth Promoting Microbes: Diverse Roles for Sustainable and Ecofriendly Agriculture." *Energy Nexus*. 5: 1–10.

Kumar, P., R. Kumari, S. Kumar, and A. Kumar. 2023. "PGPMs-Mediated Improvement of Crops under Abiotic Stress." In *Plant–Microbe Interaction—Recent Advances in Molecular and Biochemical Approaches*, edited by P. Swapnil, M. Meena, Harish, A. Marwal, S. Vijayalakshmi, and A. Zehra, 21–31. Academic Press.

Lata, C., S. Soni, N. Kumar, A. Kumar, Pooja, A. Mann, and S. Rani. 2019. "Adaptive Mechanism of Stress Tolerance in Urochondra (Grass Halophyte) Using Roots Study." *Indian Journal of Agricultural Sciences*, 89(6): 1050–1052.

Lichtfouse, E. 2016. *Sustainable Agriculture Reviews: Volume 19*. Springer.

Malgioglio, G., G. F. Rizzo, S. Nigro, V. Lefebvre du Prey, J. Herforth-Rahmé, V. Catara, and F. Branca. 2022. "Plant–Microbe Interaction in Sustainable Agriculture: The Factors That May Influence the Efficacy of PGPM Application." *Sustainability*, 14(4): 2253. https://doi.org/10.3390/su14042253

Mine, A., M. Sato, and K. Tsuda. 2014. "Toward a Systems Understanding of Plant–Microbe Interactions." *Frontiers in Plant Science*, 5: 423. https://doi.org/10.3389/fpls.2014.00423

Nayak, S., A. Mukherjee, and S. Samanta. 2017. "Endophytic Microorganisms of Tropical Tuber Crops: Potential and Perspectives." *Journal of Applied and Natural Science*. https://doi.org/10.31018/jans.v15i2.1287

Prashar, P., and S. Shah. 2016. "Impact of Fertilizers and Pesticides on Soil Microflora in Agriculture." In *Sustainable Agriculture Reviews*, edited by Eric Lichtfouse, 331–362. Springer. https://doi.org/10.1007/978-3-319-26777-7_8

Rachel, K., P. Kiranmayi, and G. Sirisha. 2015. "Quality Enhancement Studies of Growth and Nutrient Content in *Phaseolus aureus*, *Spinacia oleracea*, and *Ocimum sanctum* Using Local Bio-Fertilizers." *Journal of Natural Sciences*. https://doi.org/10.31018/jans.v4i1.129

Raez, J. A. L., T. Charnikhova, I. Fernandez, H. J. Bouwmeester, and M. J. Pozo. 2011. "Arbuscular Mycorrhizal Symbiosis Decreases Strigolactone Production in Tomato." *Journal of Plant Physiology*, 168(16): 1736–1742. https://doi.org/10.1016/j.jplph.2011.04.010

Ryan, R. P., K. Germaine, A. Franks, D. J. Ryan, and D. N. Dowling. 2008. "Bacterial Endophytes: Recent Developments and Applications." *FEMS Microbiology Letters*, 278(1): 1–9. https://doi.org/10.1111/j.1574-6968.2007.01010.x

Sharma, A., V. Kumar, B. Shahzad, M. Tanveer, G. P. S. Sidhu, N. Handa, S. K. Kohli, P. Yadav, A. S. Bali, R. D. Parihar, O. I. Dar, K. Singh, S. Jasrotia, P. Bakshi, M. Ramakrishnan, S. Kumar, R. Bhardwaj, and A. K. Thukral. 2019. "Worldwide Pesticide Usage and its Impacts on Ecosystem." *SN Applied Sciences,* 1(11) (November 1, 2019). Springer Nature. https://doi.org/10.1007/s42452-019-1485-1.

Stephenson, G. 2019. "The Effects of Agricultural Waste-Based Compost Amendments in Organic Pest Management." Master's thesis, California Polytechnic State University. https://doi.org/10.15368/theses.2019.43

Tailor, S., K. Jain, A. Marwal, M. Meena, and A. Mishra. 2023. "Agricultural Management by Improving Beneficial Microflora." In *Plant–Microbe Interaction—Recent Advances in Molecular and Biochemical Approaches*, edited by P. Swapnil, M. Meena, Harish, A. Marwal, S. Vijayalakshmi, and A. Zehra, 95–116. Cambridge, MA: Academic Press.

Tokas, D., S. Singh, R. Yadav, and A. N. Singh. 2023 "Plant–Microbe Interactions: Role in Sustainable Agriculture and Food Security in a Changing Climate." In *Plant–Microbe Interaction—Recent Advances in Molecular and Biochemical Approaches*, edited by P. Swapnil, M. Meena, Harish, A. Marwal, S. Vijayalakshmi, and A. Zehra, 363–391. Academic Press.

Index

Note: Page numbers in *italics* indicate a figure and page numbers in **bold** indicate a table on the corresponding page.

O

P

R

S

T

V

W

X

For Product Safety Concerns and Information please contact our
EU representative GPSR@taylorandfrancis.com Taylor & Francis
Verlag GmbH, Kaufingerstraße 24, 80331 München, Germany